THE
MATHEMATICAL PAPERS OF
ISAAC NEWTON
VOLUME V
1683–1684

Arithmetica Universalis
Liber primus.

Arithmetica vulgaris et Algebra ijsdem computandi fundamentis innituntur et ad eandem metam collimant. At Arithmetica definite et particulariter, Algebra autem indefinite et universaliter, sicut enunciata feri omnia quæ in ejus computo habentur, et præsertim conclusiones, Theoremata dici possint. Verùm Algebra maximè præcellit quod cum in Arithmetica Quæstiones tantùm resolvantur, progrediendo a datis ad quæsitas quantitates, hæc a quæsitis tanquam datis ad datas tanquam quæsitas plerumq̃ regreditur: ut ad conclusionem aliquam, seu æqualionem, quocunq̃ demum modo perveniatur, ex qua quantitatem quæsitam elicere liceat. Eoq̃ pacto conficiuntur difficillima problemata quorum resolutiones ex Arithmetica sola frustra peterentur. Arithmetica tamen Algebræ in omnibus ejus operationibus ita inservit, ut non nisi unicam perfectam computandi scientiam constituere videantur, et utramq̃ propterea tanquam unam universalem conjunctim explicabo.

Quisquis hanc Scientiam aggreditur imprimis vocum et notarum significationes intelligat et fundamentales addiscat operationes, Additionem, Subductionem, Multiplicationem, Divisionem, Extractionem radicum, Reductiones fractionum & radicalium quantitatum, et modos ordinandi terminos Æquationum

ac

The 'Arithmeticæ Universalis Liber primus' (2, [1]).

THE
MATHEMATICAL PAPERS OF
ISAAC NEWTON

VOLUME V
1683-1684

EDITED BY

D. T. WHITESIDE

WITH THE ASSISTANCE IN PUBLICATION OF
M. A. HOSKIN AND A. PRAG

CAMBRIDGE
AT THE UNIVERSITY PRESS
1972

CAMBRIDGE UNIVERSITY PRESS
Cambridge, New York, Melbourne, Madrid, Cape Town, Singapore, São Paulo

Cambridge University Press
The Edinburgh Building, Cambridge CB2 8RU, UK

Published in the United States of America by Cambridge University Press, New York

www.cambridge.org
Information on this title: www.cambridge.org/9780521082624

Notes, commentary and transcriptions
© Cambridge University Press 1972

First published 1972
This digitally printed version 2008

A catalogue record for this publication is available from the British Library

Library of Congress Catalogue Card Number: 65–11203

ISBN 978-0-521-08262-4 hardback
ISBN 978-0-521-04584-1 paperback
ISBN 978-0-521-72054-0 paperback set (8 volumes)

TO ADOLPH PAVLOVICH YUSHKEVICH
WHO FIRST INTRODUCED THIS TO RUSSIAN EYES

PREFACE

This fifth volume of Newton's mathematical papers is largely devoted to re-producing the unfinished algebraic treatise which its author, soon after its composition in the winter of 1683–4, deposited in Cambridge University Library as the polished script of professorial lectures given out during the previous decade, and which is traditionally (since William Whiston gave it this title in 1707) known as his 'Arithmetica Universalis'. Its autograph is now presented for the first time in public in its original sequence, freed from a number of accretions and transpositions incautiously introduced by Whiston into his *princeps* edition, preceded by certain preliminary fragments, backed by several unprinted ancillary documents elucidating and exploring points of textual detail, and supplemented by an incomplete (and unpublished) 'Arithmeticæ Universalis Liber primus' in which Newton began its radical revision. When set against the wide variety of his contemporary interests as evinced in the fourth volume, the narrowness of its theme is clearly but relative. In the grand scheme of Newton's mathematical pursuits during his middle years at Cambridge his algebraic studies play no dominating rôle, and it is merely the contingent dictate of a format unable to tolerate a single book of some 1300 large pages which now prevents their joint edition with his remaining papers of the period 1674–84.

For allowing reproduction of the manuscripts here reproduced I am once more heavily indebted to the Librarian and Syndics of the University Library, Cambridge, and also—in the case of three small items—to their equals in the Bodleian and Christ Church, Oxford, and in the Royal Greenwich Observatory, Herstmonceux. Let me again, also, express my appreciation of the courtesy and efficiency of the staff of the Anderson Room during the period in which the content of this volume has been assembled. The very necessary financial support for its production continues to be generously provided by the Sloan Foundation, the Leverhulme Trust, and the Master and Fellows of Trinity College, Cambridge. Subsidy of a less tangible sort has come from Churchill College, Cambridge, which has recently admitted me to its fellowship. To Sir Harold Hartley, now at last retired from vigorous activity, my most grateful thanks for all that he has done on my behalf over the years: I need not add that the warmth of his encouragement remains undimmed.

My colleagues in this edition have again helped unstintingly in a variety of ways to perfect the fine structure and technical detail of this work. Without all their effort it would have been much the poorer. As before, I may add, Dr M. A. Hoskin has concerned himself more particularly with ensuring the accuracy of the figures reproduced, while the index has been compiled by Mr A. Prag.

To the Syndics of Cambridge University Press, finally, and to the University Printer my continuing gratitude on behalf of the many who, editorially and in design, printing and production, have joined their various talents and skills to create yet one more book of consummate beauty.

D.T.W.

24 September 1971

EDITORIAL NOTE

This volume narrowly follows the style of its predecessors, and for a more elaborate discussion of the main conventions used we may refer the reader to pages x–xiv of the first. It remains, as ever, our ideal to be as completely faithful to the lay-out of the manuscripts reproduced as the limits of page-size and typographical format permit; in the present instance the prevailing good quality of the autographs reset has, with one or two somewhat messy exceptions, made that task a particularly simple one and we have taken minimal liberties with the words and symbols written down by Newton. The text-figures reproduced have been slightly stylized and, in trivial cases, silently corrected, but significant irregularities and misdrawings are nowhere eliminated in this way without an accompanying explanation (in footnote) of the pertinent defect in the original. As usual, all editorial insertion in the source text is made within square brackets and introduced merely to smooth out trivial illogicalities and slips of Newton's pen which it would be out of balance to stress. We have, within the confines of modern fluency and idiom, kept the English translations (mostly on right-hand pages facing their Latin originals, but sometimes accompanying the latter in footnote) deliberately literal, conceiving our primary purpose to be the clarification and illumination of the Latin text rather than the contriving and polishing of an elegant paraphrase. For both backwards and forwards reference within the present volume we use the formulas '**1**, 2, §1' (by which understand '[Part] **1**, [Section] 2, [Subsection] 1') and '**2**, [2]' (understand '[Part] **2**, [Division] 2'). Citations of earlier volumes are throughout made by the code 'IV: 274–82', meaning '[Volume] IV: [pages] 274–82'. In both Latin and English texts two thick vertical bars in the left margin denote that the lines alongside are cancelled in the original: this symbol must not be confused with the two thin strokes ' ‖ ' (invariably accompanied in the margin alongside by a twin sign, followed by the new page number in square brackets) here used in our reproduction of the main text of the 'Arithmetica' (**1**, 2, §1) to mark a division in the manuscript's paging at the point so marked. It should be clear throughout that the frequent citations of 'Newton's pages' are keyed to those of the manuscript 'Arithmetica' thus distinguished, and in no case to the pagination of the present volume. A few non-standard mathematical symbols used by Newton are explained in appropriate footnotes.

GENERAL INTRODUCTION

In this fifth volume we reproduce, in complement to the miscellany of researches comprehended in the fourth, the extant record of Newton's algebraic studies during the ten years from 1673, culminating in the winter and spring of 1683–4 (or so we believe) in the much reworked but unfinished compilation which he afterwards deposited in the Cambridge archives as his principal Lucasian *lectiones* of the preceding decade. The text of these ninety-seven self-styled 'lectures'—the accuracy of this description and, still more, of the chronology which Newton sought to impose upon them in the margin of their manuscript is very much open to query[1]—fills the main bulk of the present tome. To it, apart from a number of preliminary fragments, we have adjoined several short related papers which amplify and illustrate certain passages and problems within it, and have also set in sequel an incomplete 'Arithmeticæ Universalis Liber primus' in which Newton began extensively to remodel it. As ever, into the bare text of these pieces we have larded a considerable quantity of footnotes which we hope will adequately explain pertinent points of technical and historical detail, and have also added on facing pages a line-by-line rendering of Newton's Latin into modern English. To these two species of running commentary we optimistically refer requests for particular enlightenment. The little we have been able to find out regarding the general background to these papers—and in the case of the parent *Arithmetica* its subsequent printing history and contemporary impact—is presented in preceding introductions, padded out with certain conjectures we have dared tentatively to formulate in lieu of any pertinent documentary evidence on matters at issue. The reader will make his own mind up about the cogency and relevance of the latter.

Regretfully, the historical circumstance which led Newton to compile the gist of his professorial lectures on 'universal arithmetic' is to us as shadowy as the impulse which caused him peremptorily to leave off their revision a few months later with so much still to do. In regard to the first we may reasonably conjecture that pressure was in some manner put upon him in late 1683 to fulfil, however tardily, his statutory obligation annually to deposit a fair copy of ten of his lecture scripts,[2] and be all but sure that the arrival in Cambridge the next spring of his young amanuensis Humphrey Newton, able to take over from Isaac the dreary, time-consuming chore of rewriting his much cancelled and corrected worksheets in legible and coherent form, gave him new heart to

---

(1) This point is discussed on pages 4–6 below.
(2) See III: xviii.

codify and expand his previous mathematical investigations.[3] But these surmises remain as unproved (and essentially undemonstrable?) as the plausible suggestion that it was Edmond Halley's famous first visit to Cambridge in the late summer of 1684 to talk about the unsolved problem of elliptical planetary motion[4] which provoked him abruptly to relinquish his restructuring of the *Arithmetica*. We may guess still more tenuously that the appearance in mid-1685 of John Wallis' voluminous *Treatise of Algebra, Both Historical and Practical*[5] would have long deterred Newton from making any efforts to have his own rival studies made publicly available. In later years, certainly, he grew increasingly soured with the often cumbersome computations and techniques of Cartesian algebra—at one point, indeed (if we may believe David Gregory), he qualified it as 'the Analysis of the Bunglers in Mathematicks'[6]—and we may be certain that his reluctance during 1705–6 to have Whiston edit the deposited text of his algebraic lectures was not merely the manifestation of a growing personal antagonism to his successor in the Lucasian chair.[7] While, however, there are already distinct traces in his *Arithmetica* of Newton's incipient preference for the elegance and logical deftness of the demonstrative techniques of classical geometrical synthesis[8] this could surely have played no major rôle in his decision in 1684 to abandon his intended introduction to algebra 'for the use of academic youth'.[9]

(3) Compare 2, [1]: note (1) below. We have already observed (IV: 169, note (2)) that Humphrey Newton was set to work almost straightaway on transcribing portions of Newton's intended edition of St John Hare's 'Trigonometria', no doubt so that Newton himself might press on with the revision of his Lucasian *lectiones* as well as be able to devote more time to the chemical experiments he was then busy making in his Trinity elaboratory (on which see note (12)).

(4) See IV: xx and 657, note (4).

(5) It seems unlikely that Newton had any detailed knowledge of the content of Wallis' book before its publication (compare IV: 416, note (30)), though he would probably already have been sent a copy of the *Proposal about Printing a Treatise of Algebra* which Wallis circulated in 1683 (see IV: 413, note (20)).

(6) A stray note in Gregory's diary, dated 'May 1708' (Christ Church College, Oxford. MS 346: 96, first published by W. G. Hiscock in his *David Gregory, Isaac Newton and their Circle. Extracts from David Gregory's Memoranda 1697–1708* (Oxford, 1937): 42).

(7) Whiston's rôle in preparing the *editio princeps* of Newton's lectures for publication during 1705–6 (as the anonymously authored *Arithmetica Universalis; sive De Compositione et Resolutione Arithmetica Liber*) is discussed on pages 8–12 below.

(8) See 1, 2, §1: Newton's pages 130–4 and (more forcefully) 214–15, 242.

(9) 'In Usum Juventutis Academicæ' (Whiston's subtitle to his 1707 *princeps* edition of Newton's *Arithmetica*).

(10) See IV: xi–xiii, xvii–xix.

(11) John Conduitt in the 'Memoirs of Sir Isaac Newton' which he drafted for Fontenelle in late 1727 (King's College, Cambridge. Keynes MS 129 (A): 15–16, first published by Edmund Turnor in his *Collections for the History of the Town and Soke of Grantham* (London, 1806): 163; on Conduitt's 'Memoirs' see I: 3, note (2)). In a partially cancelled preliminary passage (*ibid.* MS 129 (B): 11–12) Conduitt earlier wrote that 'He used to write down any thoughts

We see no good reason to modify the pen-picture which we have already[10] sketched of Newton's character and interests at this period of his life, one approaching near enough the traditional view that

At the University he spent the greatest part of his time in his closet & when he was tired with his severer studies of Philosophy his only releif & amusement was going to some other study as History Chronology Divinity & Chymistry[,] all wch he examined & searched thoroughly as appears by the many papers he has left on those subjects,... & he was hardly ever alone without a pen in his hand & a book before him,[11]

though we should never stress his stern bookishness and personal isolation to the total exclusion of good humour and social amicability, rarely manifested in him though they might have been. Whether or not Conduitt's present reference to the intensity of Newton's studies at this time in biblical chronology and divinity are wholly valid, the numerous dated entries in his contemporary papers and notebooks[12] indelibly underline the profusion and depth of his chemical (and alchemical) researches in the early 1680's. On the variegated quality and abundance of his companion explorations of mathematical topics we need not here insist, though in hindsight we may, with regard to the unexciting and far from novel nature of so many of Newton's professorial lectures on algebra, agree with Rigaud that, whereas

Newton lecturing on his discoveries in Physics and Optics[13] only conveys to us the idea of the instructor of mankind opening the paths of knowledge to those, who were most immediately connected with him;... there is something melancholy in the thought of such a man's time being taken up in teaching the elements of Mathematics.[14]

---

wch occurred upon the books he was reading, & made large abstracts of them, & has left behind him many rheams of loose papers being foul draughts of his Mathematical treatises & Chronology & abstracts of the history. The volumes of foul papers he has left behind besides many he burnt...not long before he died [compare ɪ: x], some of wch are the same thing writt over five or six times,...shew his patience & perseverance in the several studies he was pursuing"

(12) Notably ULC. Add. 3973 and 3975. These are discussed at some length by A. R. and M. B. Hall in 'Newton's Chemical Experiments', *Archives Internationales d'Histoire des Sciences*, **11**, 1958: 113–52.

(13) Rigaud refers obliquely to the manuscripts deposited by Newton in Cambridge University Library as the polished record of his optical and dynamical lectures during 1670–3 and 1684–5 (now ULC. Dd. 9. 67 and Dd. 9. 46 respectively). Mathematical extracts from the former (published posthumously in 1728 as his *Lectiones Opticæ, Annis MDCLXIX, MDCLXX & MDCLXXI In Scholis publicis habitæ*) have been printed in ɪɪɪ: 474–512 and 544–9, while much of the latter (comprehending portions of two preliminary states of what was later to appear as Liber I of Newton's *Philosophiæ Naturalis Principia Mathematica* in 1687) will be reproduced in the sixth volume.

(14) Stephen Peter Rigaud, *Historical Essay on the First Publication of Sir Isaac Newton's Principia* (Oxford, 1838): 97, note e. He went on: 'The beginning of the *Arithmetica Universalis* is of this kind; but so much of the work is taken up with higher speculations, that we are reconciled to what seems only introductory to them....Flamsteed ([Baily] p. 166) could taunt

But, of course, we must remember that such introductory instruction—and not the creative research for which Newton was so supremely fitted—was the prime purpose for which the Lucasian chair had been founded[15] and that he had a legal and moral duty to undertake it from time to time. If only he had been a good teacher with a realistic understanding of the near-total mathematical incapacity of the few undergraduates who even bothered to attend his lectures, this abiding impression that he here all but wasted the valuable hours it took him to pen them would not have been so strong. It is one of the ironies of his posthumous appreciation that the printed version of his *Arithmetica* was to become his most often read and republished mathematical work, while his more advanced papers on calculus and analytical geometry were relatively little studied or even ignored by the world at large till the present day. Such is fame.

The even tenor and obscurity of these uneventful middle Cambridge years were very soon to change. We have already suggested[16] that the 'Matheseos Universalis Specimina Mathematica' and its sequel 'De Computo Serierum' on which Newton was working during July and early August of 1684 were summarily abandoned after Halley paid his celebrated visit to Newton, seeking —successfully—to divert his attention to the dynamics of planetary motion. Significantly, his chemical researches at this period were equally suddenly broken off for almost two years.[17] With those few crowded intervening months, during which he composed the various preliminary states of his mighty *Principia*, we shall, in a mathematical way, be extensively concerned in the next volume. Let us now, meanwhile, pass without further ado to the reproduced record of his algebraic studies during the previous decade.

---

him with reading "mathematics for a salary at Cambridge"; as if any thing but a deep sense of public duty could have induced him to undergo such unworthy drudgery.' (Compare Francis Baily, *An Account of the Rev^d. John Flamsteed* (London, 1835): 166, where quotation is made of Newton's curt remark to Flamsteed on 6 January 1698/9 [= *Correspondence of Sir Isaac Newton*, **4**, 1967: 296–7] that 'I do not love to be printed upon every occasion much less to be dunned & teezed by forreigners about Mathematical things or thought by our own people to be trifling away my time about them when I should be about y^e Kings business', and of Flamsteed's note upon the letter at this point: 'Was Mr Newton a *trifler* when he read Mathematicks for a sallery at Cambridge[?]'.)

(15) The pertinent phrase in the professorial statutes (reproduced on III: xx–xxv) is '...promovendis...studijs mathematicis [in dicta Academia]' (*ibid.*: xxi), an 'officium' further defined (*ibid.*: xxii) as that 'horum studiorum quà publice, quà privatim excolendorum'.

(16) See IV: xx, 418–19.

(17) The last entry in his chemical notebook till 26 April 1686 is a short one, terminating a crowded series of jottings over the previous winter and spring: 'Friday May 23 [1684]. Jovem super aquilā volare feci....' (ULC. Add. 3975: 149–50). Understand, in conventional alchemical parlance of the period, that Newton added 'Jupiter' [= *stannum* or tin] to 'eagle' [= *sal armoniacum*, in the present instance perhaps a sublimate of mercury] to produce an appropriately volatile compound.

ANALYTICAL TABLE
OF CONTENTS

PREFACE   *page* vii

EDITORIAL NOTE   ix

GENERAL INTRODUCTION   xi

LIST OF PLATES   xxiii

PART 1
THE DEPOSITED LUCASIAN LECTURES
ON ALGEBRA
(Winter 1683–1684)

INTRODUCTION   3

Newton's previous researches into algebra: his concept of a 'universal arithmetick' mirroring the structure of common (particularized) arithmetic and geometry, 4. The marginal chronology imposed by him on the deposited autograph 'copy' of his professorial *lectiones*, though broadly acceptable, is suspect in detail, 5. No contemporary reaction to the lectures as given is recorded, but must have been minimal, 6. Renewed interest in the manuscript from 1700: Cotes' undergraduate transcription, 7. During 1705–6 Whiston edits the deposited text for publication, 9. Newton is much concerned that an unfinished, perhaps already obsolescent work of his should appear in public without his being allowed to revise it, 11. The 1707 *editio princeps* under (Newton's?) title of *Arithmetica Universalis; sive De Compositione et Resolutione Arithmetica Liber*: a last-minute 'Lectori S.' points to certain deficiencies in its typography, evidently at Newton's instruction, 12. The revisions and excisions to be made in its re-edition, as listed by Newton in his library copy, 13. The edition has little initial impact in Britain, 15. On the continent Leibniz is much intrigued and, finding difficulty with the section on factorization, seeks help from Johann Bernoulli and Jakob Hermann in his recension of the book, 16. Raphson's English translation of the *Arithmetica* (1720) and Newton's (unattributed) second Latin edition (1722) create a wave of interest in Newtonian algebra, 17. Posthumous editions of the *Arithmetica*, 18. Extant fragments pertaining to Newton's lectures on algebra: the 'paper' given by him to Flamsteed in 'Midsummer 1674' and other minor drafts, 19. The 'synthetic resolution of problems in my lectures': later extensions of particular problems (to the general theory of the loaded string and the computation of cometary orbits in Wren's rectilinear hypothesis), 20.

APPENDIX 1. 'W[histon]'s' preface to the 1707 *editio princeps*. Certain exaggerations (stemming from Newton?) are identified, 21.

APPENDIX 2. Leibniz' review of the *Arithmetica* (in *Acta Eruditorum*, November 1708). Prior emphasis is placed on Newton's rules for factorizing given equations, 24. On his modified Cartesian technique for determining roots of surd binomials, 26. On his algorithm for delimiting the number of complex roots possessed by an equation, 28. And on his technique for

resolving a quartic algebraic polynomial into 'surd' quadratic factors, 29. His concluding remarks support Newton's criterion of geometrical (in contrast to algebraic) 'simplicity' in constructing the real roots of cubics by intersecting conics or equivalent conchoidal neusis, 30.

1. PRELIMINARY NOTES AND DRAFTS FOR THE 'ARITHMETICA'   32

§1 (Royal Greenwich Observatory. Flamsteed MS 42 [front insert]). Newton's 'paper' of notes for a lecture in 'Midsummer 1674' (later allotted in the deposited copy to 'October' and there split between two consecutive *lectiones*). The text on 'refining' simple algebraic equations into standard form then given to Flamsteed, 32.

§2 (ULC. Add. 3964. 2: 5^r–6^v). Delimitation of the roots of equations and their factorization. Representing the real roots of an equation 'on a chart' as the meet(s) of a base line with the Cartesian curve defined by it: 'impossible roots are usually expressed by folds in the curve', 34. All (rational) roots of an equation are divisors of its lowest term: testing for integral roots on this basis, 36. Numbers which, when substituted in an equation, yield partial sums which are all positive or all negative are limits to its roots, 38. The familiar (equivalent) Newtonian 'derivative' test, 40. Finding linear factors of a given 'equation' with numerical coefficients by substituting numbers in algebraic progression and then seeking to determine divisors of the resulting values which have a constant difference, 42. Simple cubic, quartic and quintic examples, 44. Extension of the technique to locating quadratic factors by a similar finite-difference algorithm: instanced in a quartic and a quintic, 46.

§3 (ULC. Add. 3959. 3: 23^r). Newton's derivation of the 'Heronian' formula for a triangle's area in terms of its sides. The geometrical construction demonstrated, 50.

2. THE COPY DEPOSITED IN THE CAMBRIDGE ARCHIVES   54

§1 (ULC. Dd. 9. 68: 1–251). The main text. Algebra is arithmetic universalized by the introduction of 'specious' (free) variables in place of constants: in the solution of problems it excels in the analysis which precedes their (synthetic) composition, 54. Newton defines the notations he will employ: comma, point and 'angle' decimal separatrixes, 56. Algebraic positives and negatives: the signs $+$, $-$ and $\pm$ ($\mp$), 58. Multiplicative combination by a comma or the symbols $\times$ or 'in', 60. The fraction-bar and bracket used to denote division: radical signs, and those for equality and proportion, 62. The operation of algebraic addition (modelled on its arithmetical equivalent), 64. More complicated examples, 68. Algebraic subtraction (likewise styled on the arithmetical): examples, 70. Arithmetical multiplication of decimals, 72. The analogous algebraic operation exemplified, 74. Arithmetical division of decimals, 76. Simple algebraic equivalents, 80. The procedure exemplified in more complicated instances, 82. 'There are other methods, but this is easiest', 86. Extraction of a (non-terminating) arithmetical square root, 88. And of cube and fifth roots, 94. The equivalent algebraic technique exemplified, 96. The reduction of algebraic fractions to least terms, 100. And to a common denominator, 104. The reduction of algebraic radicals to their least terms, 106. And to a common denomination, 108.

The algebraic equation: listing its canonical forms, 110. Seven rules for 'refining' single equations to standard form (with all terms on one side equal to zero on the other): first, by elimination of fractions, 112. Next, by squaring (or cubing) to eliminate radicals, 114. And by removal of a common factor, or (rarely) by extracting roots: the general solution of a quadratic equation, 116. Elimination of unknowns between two or more equations given simultaneously, 118. Simple cases (reduced by equation or direct substitution) where one unknown is of first power only, 120. A (complicated) general technique of elimination, 122. Typical short cuts sometimes possible, 124. Four formulas for eliminating an unknown

between given quadratic, cubic and quartic equations, two by two, 126. Elimination of surd quantities from equations (by appropriate reordering and then squaring or cubing), 128.

'How any question is to be reduced to an equation.' As many (independent) equations are to be deduced from the conditions of the problem as there are unknowns: two arithmetical examples, 130. In such arithmetical cases translation into algebraic 'language' is straightforward, 132. Seventeen worked arithmetical 'questions' (partly taken from Kinckhuysen and Schooten): Problems 1–3 relate to quantities satisfying simple numerical conditions, 134. Problems 4 and 5 invoke simple money-distribution and time-of-journey conditions, 136. The latter generalized, 138. Problems 6 and 7: work done jointly by 'agents', 140. Problem 8: compounding substances to derive a mix in given proportion, 142. Problems 9 and 10: the mix to have a given cost or specific gravity (Hiero's crown), 144. Problem 11: pasturing cattle on grass which grows at a given rate till eaten, 146. Problem 12: motions of colliding spheres after impact, 148. Problem 13 (cancelled): Kinckhuysen's horse-sale puzzle, 150. Problems 14 and 15: two numerical questions (from Kinckhuysen), 152. Problem 16: the general problem of four continued proportionals whose sum and square sum are given, 154. Problem 17: the purchase price of a 5-year annuity (solution of the resulting quintic is referred to subsequent *regulæ* not in the extant manuscript), 156.

'How given problems are to be reduced to an equation.' The 'translation' of geometrical relationships into algebraic equivalents, 158. The same derived algebraic equation can be used to achieve solutions to a variety of related problems, 160. The prime necessity is to find an algebraic rendering which isomorphically models the structure of a problem, not to worry about deducing one particular quantity from another: Schooten's problem of a quadrilateral inscribed in a semicircle is introduced to serve as a running illustration, 162. Euclidean propositions useful to this end, 164. Their application to Schooten's problem, 168. No distinction to be made between known and unknown quantities in the algebraic translation, 170. Reducing the algebraic equivalent to a single equation, 172. Different avenues of approach lead to the same resolvent equation, 174. Algebraic reduction will be much simplified if the problem itself is neatly defined, 178. The use of Cartesian techniques of analytical geometry: illustrated by finding the intersection of a straight line with a kappa curve, 180. And with an ellipse, 182.

In further illustration Newton works through sixty-one geometrical 'questions' (reordered into sequence of difficulty in his 1722 Latin revise), 184. Problems 1–6: algebraic determination of altitudes, base-segments, sides and areas in given triangles, 186. Problems 7–8: similar computation of the diagonal of a parallelogram (given its sides and the second diagonal), and of the sides of a quadrilateral (given its angles, perimeter and area), 192. Problem 9: Schooten's fish-pond promenade, 194. Problem 10: to cut off a given area in an angle by a transversal through a given point, 196. Problem 11: intersections of a bipolar locus (Apollonius' circle, hyperbola,...) with a line, 198. Problem 12: a determinate variant on Descartes' construction of the Greek 4-line locus, 200. Problem 13: the Apollonian 'inclination' of a line-segment through the corner of a square and to lie in the angle of its opposite sides, 202. Pertinent remarks on choosing as algebraic variable line-lengths which are 'similarly' related to pairs of points or lines in the geometrical problem: Lagrange's criticism that this implies foreknowledge of the form of the solution, 204. Problem 14: to set off a parabolic chord of given length through a given point, 206. Problem 15: the Newtonian geometrical model for angle-section, and the algebraic equations deriving therefrom, 208. Problem 16: from three observations, to fix the position of a comet moving uniformly along a straight line of given inclination, 210. Problem 17: to find the principal focus of rays issuing from a point and refracted at a spherical interface, 212. Problem 18: the Cartesian defining equation of a plane section of a right circular cone, 214. Problem 19: the equation of a plane section of a hyperboloid of revolution (Newton

anticipates Sluse and Towneley), 216. Problem 20: the equation of the locus (a cissoid) of the mid-point of one 'leg' of a 'sector' moving along a line, where the other leg is through a given pole, 218. Problem 21: the locus (an ellipse) of a point fixed in a line-segment moving in a given angle, 220. Problem 22: the locus (also an ellipse) of a point fixed in the segment's (moving) plane, 222. Problem 23: the 'Heronian' rule relating a triangle's area to its sides, and derived trigonometrical formulas, 224. Problem 24: the Cartesian equation of a conic derived from its generalized focus-directrix definition, 226. Problem 25: that of the (hyperbolic) locus which divides an angle's general transversal, inclined through a given point, in a given ratio, 228. Problem 26: that of an Apollonius' circle, 230. Problems 27/8: fixing a point whose distances from three lines/points are in given ratio, 230–2. Problems 29/ 30: determining a triangle whose sides and altitude are in arithmetical/geometrical progression, 234–6. Problem 31: determining a triangle, given its altitude and that its base is the arithmetic mean of its (inclined) sides, 238. Problem 32: to construct a transversal to three given lines, given that its intercepts shall be of given length, 240. Problem 33: to incline a line-segment in an angle so as to touch a given circle, 240. Problem 34: determination in a plane interface of the point at which a light-ray passing between given points is refracted, 242. Problem 35: the (hyperbolic) locus of the vertex of a triangle whose base angles have a given difference, 244. 'The same more shortly' (by choosing a better base-line in retrospect), 248. Problem 36: the (cubic) locus of the vertex of a triangle, one of whose base angles differs from twice the other by a given amount, 250. The degenerate Pappus locus (a hyperbola) when the difference is nil, 252. Problems 37–41: cases of Apollonius' problem of circle tangencies. To draw a circle through two points to touch a given line, 252. . . . through a point to touch two lines, 254. . . . through two points to touch a second circle, 256. . . . through a point to touch a line and a second circle, 258. And hence (Viète's reduction) to touch a line and two given circles, 260. . . . through a point to touch two other circles (Newton's solution is wholly geometrical), 262. And hence (Viète) to touch three other circles, 264. Problem 42: the Descartes–Schooten problem of determining the latitude and solar declination for which three 'staves' of given height shall (in a polar region) cast an (elliptical) shadow of given proportion, 266. The 'first part of the analysis' (despite minor differences, structurally the same as Descartes'), 270. The 'second part': Descartes' solution is obtained in improved form, 274. Problem 43: weights in (oblique) equilibrium round a (frictionless) pin, 278. Problem 44: a weighted string obliquely balances two weights, each round a pin, 282. The shape assumed by a (weightless) string rigidly attached at its end-points and supporting weights at two points: Newton will later (see Appendix 2) derive from this the differential equation of the generally loaded string, 284. Problem 45: to compute the depth of a well by determining the time between a stone's release and the returning sound of its striking the bottom, 286. Problems 46–9: determining a triangle given various combinations of its sides and altitude, 286–92. Problem 50: construction of a rectilinear configuration determined by the equality of two of its angles, 294. Problem 51 (an *early* version of Problem 23): to ascertain the angles and area of a triangle, given its sides, 296. Problem 52: given four timed angular observations of it (from the orbiting earth), to fix a comet's (solar) orbit in Wren's hypothesis that it moves uniformly in a straight line, 298. The possibility (raised independently by Huygens and Boscovich) that it is effectively indeterminate, 302. Problem 53: the Cartesian equation of the (conic) locus described 'organically' from a rectilinear describend, 304. Problem 54: construction of a parabola through four points by determining its Cartesian equation, 306. Problems 55–7: the geometrical construction of a conic through five points, 308. . . . through four points, touching a given line at one of them, 310. . . . through three points, touching given lines at two of them, 312. Their construction by Cartesian algebraical techniques (in the latter two problems the condition of tangency is rendered by determining the 'last' ratios of vanishing quantities *ab initio*), 314–20. Problem 58: determining the motion of a spherical globe, bounding back from a wall after colliding with a second globe so as to hit

this again, 322. Problem 59: where two globes, simultaneously released, fall under simple gravity in the same vertical, to find the point at which one, after colliding with the base, meets the second in its rebound, 324. Problem 60: computing the distance apart of these two globes when they are released at differing times, 328. Problem 61: computation of the distance of collision of one of these globes perpetually rebounding from the base to strike the other (as it descends) and recoiling back again to the base, 332.

'How equations [reduced to numerical form] are to be resolved.' The roots of an equation: possibility of multiple (real) roots, 338. 'An equation can have as many roots as it has dimensions, and no more', 340. Positive, negative and 'impossible' roots, 342. Real roots of the derived algebraic equation do not necessarily (because of some limitation) yield a corresponding solution to a problem, 344. Bounding the number of positive/negative roots by the Cartesian rule of signs, 346. Newton's (incomplete) rule for bounding the number of complex roots, 348. His complete rule (distinguishing 'positive' from 'negative' impossibles), 350. Transmuting an equation by changing the sign of its roots, and by increasing or diminishing them by a given quantity, 352. Eliminating first or second terms from an equation by this means, 354. Multiplying and dividing the roots: removing radical roots in this way, 356. The 'generation' of equations by multiplying linear factors together: their coefficients as homogeneous functions of the roots, 360. Delimiting the bounds of the roots: the generalized root-squaring method and a related Newtonian technique, 362. Further ingenious (but highly impractical) corollaries, 364. The Newtonian 'derivative' test, 366. Determining rational roots of a numerical equation by finding its factors, 368. Isolating linear factors by substituting a sequence of numbers in arithmetical progression and (by inspection) discerning a related sequence of divisors of the resulting quantities whose (first) differences are constant: instanced in a cubic, 370. Quartic and quintic examples, 372. Extension of this finite-difference technique to determining quadratic factors, 374. Exemplified in equations of fourth, fifth and seventh degree, 376. The technique can be further extended to finding higher-order factors (but this is not for the novice), 378. The technique applied to literal equations, 380. Examples, 382. The (occasionally) simpler technique of determining highest common factors of components of an equation, 384. Exemplified, 386. Newton's procedure for deriving 'surd' factors (when they exist) of equations of even dimension: the quartic case, 388. An example is worked, 390. 'Literal equations also are reduced by the same method': an instance, 392. The technique is valid in determining rational factors also: a quartic example is worked, 394. The analogous procedure for determining 'surd' cubic factors of a sextic: an instance, 396. The extension to equations of eight, ten and twelve dimensions is outlined, 400. And the case of the octic elaborated, 402. An example, 404. Analogous procedures for 'cubic-surd' factorization of polynomials is here neglected as being rarely of use, 408. 'Well-known' special resolutions: the Cardan solution of the reduced cubic, 408. The 'impossible' case (all three roots real), involving cube roots of complex binomials, is intractable by algebraic methods not admitting these, 410. The Cartesian solution of the reduced quartic, splitting it into a pair of quadratics by means of an auxiliary cubic, 412. A worked example, 414. The (Euclidean) extraction of the square root of a surd binomial: a shorter way which may sometimes be used, 416. Newton's refinement (still not universally valid, as Euler later showed) of the Descartes–Waessenaer rule for extracting its cube root (when possible): examples, 418–20.

'The linear construction of equations': since, once it is known roughly, exact numerical computation of an equation's root is easy, geometrical constructions for effecting its approximate determination are appended, 420–90. Types of curve (straight line, circle and the conics) admitted by the ancients into ('plane' and 'solid') geometry, 422. The recent acceptance of all (algebraic) curves which can be defined by a Cartesian equation, and their classification by the latter's dimension: this is, however, no true index of their geometrical 'simplicity' (where ease of construction, not definability, is paramount), 424. Hence the (transcendental) cycloid ought to be employed in angle-section in preference to

complicated algebraic curves, 426. On this basis Newton urges the superiority of the Archimedean trisection by a conchoidal neusis: 'equations belong to arithmetical computation and have no place in geometry,' 428. A rectilinear conchoidal neusis is constructable by the meets of a circle and hyperbola, 430. But the neusis is itself easier (approximately) to effect: whence Newton prefers to construct cubic and quartic problems by its aid, rather than by intersecting conics, 432. The equation $x^3 + qx + r = 0$ constructed by such a neusis, 434. Analogous construction of $x^3 + px^2 + r = 0$, 436. A conchoidal neusis used to construct the full cubic, 438. Proof of the method, 440. An equivalent circular neusis used to construct $x^3 + qx + r = 0$, 442. Its demonstration, 444–6. Analogous construction of $x^3 + px^2 + r = 0$, 448. A variant circular neusis constructs the full cubic, 450. Its proof, 452–4. All these neusis constructions are effectable by explicitly drawing a conchoid, 454. The rectilinear neusis applied to insert two mean proportionals (essentially mirrors Nicomedes' neusis), 456. Equivalent employment of a circular neusis (unconsciously(?) following Viète), 458. The circular neusis used to trisect an angle, 458–60. The equivalent rectilinear neusis (Pappus'), 460. The minimally variant Archimedean trisection by a circular neusis, 462. A variant circular neusis used to insert two mean proportionals, 464. The cissoid (traced in Newtonian style by the mid-point of one 'leg' of a 'sector' moving on a line, while the other leg is perpetually inclined through a pole) used to the same purpose, 464–6. Construction (bungled) of the cubic $x^3 \pm px^2 - qx + r = 0$ by the general cissoidal locus of an arbitrary point in the sector's leg, 466. The importance of 'mechanical ease' in geometrical construction, 468–70. The admission into geometry of curves classified according to the dimension of their Cartesian equation is arbitrary: the contrary (Greek) distinction of 'plane' loci (straight lines/circles) from 'solid' ones (conics) is geometrically more realistic than, by reason of the identity of their algebraic degree, to list circles and conics together, 470–2. Constructions employing conics are not intrinsically superior to ones using (algebraically) higher curves, 474. The parabola is geometrically less simple than the ellipse, and still less so than the circle, even though its Cartesian equation is simplest, 476. The full cubic equation solved by determining the meets of a circle and ellipse (constructed by a trammel), 476–8. Demonstration of this, 478–82. The ellipse may be replaced by a hyperbola or parabola, 484. An equivalent construction by a circle and fixed ellipse (by homothety), 486–8. Again, the ellipse may be replaced by a hyperbola or parabola, 488. The method applied to insert two mean proportionals, 490.

§2 (ULC. Dd. 9. 68: 253–68). Newton's 'corrections' to the main text. [1] An augmented section 'on notation', 492–500. A geometrical model of multiplication (by similarity of triangles), 494. The analogous model for division, 496. Construction of the geometrical mean (by a semicircle), 498. [2] The 'finding of divisors' (a little variant preliminary draft): 'equations' are here named 'quantities', 500–4. [3] 'The reduction of radicals by the extraction of roots' (again a preliminary draft): Newton's refined Euclidean/Cartesian methods for extracting the square/cube roots of a surd binomial, 506–8.

§3 (ULC. Add. 3963. 1ʳ/1ᵛ). The synthetic resolution of certain (geometrical) problems 'in my lectures'. Those of Problems 1, 3 and 4 are immediate: that of Problem 6 is (by oversight) circular, 508–10. That of Problem 9 is a straightforward geometrical recasting of the previous algebraic solution, 510–12. The resolution of Problem 10 has a lacuna (and to fill it requires a circuitous argument): that of Problem 13 is taken over from Newton's Waste Book, 512. Problem 32 is given an elegant Euclidean solution: Problem [52?] is merely enunciated, 514. A neat (but, in context, irrelevant) equant model of planetary motion in an ellipse round a solar focus, 514–16.

APPENDIX 1 (Bodleian. MS New College 361. 3: 2ᵛ). Two draft outlines of Newton's technique for eliminating an unknown between two given cubic equations, 518/519.

APPENDIX 2 (ULC. Add. 3965. 13: 375ᵛ). The theory of the generally loaded string hanging under simple gravity (derived *c.* 1720 from the addendum to Problem 44). [1] The form of a (weightless) string loaded at one or more of its points, 520–1. Extension (independently of Huygens) to the continuous case: the suspension bridge (Huygens)/catenary (the Bernoullis, Leibniz) as particular instances of uniform (horizontal/longitudinal) loading, 521. [2] A slightly augmented revise.

APPENDIX 3. Attempted computations of a cometary path in Wren's hypothesis (Problem 52) that this is a uniformly traversed straight line. [1] (ULC. Add. 3965. 14: 589ʳ). A first essay at fixing the orbit of the 1680/1 comet: the four sightings (taken too far apart) yield an impossible site, 524–7. [2] (ULC. Add. 3965. 14: 587ʳ). Revised computation of the 1680/1 comet: narrower sightings yield a more realistic path, 528–9. [3] (ULC. Add. 3965. 11: 155ʳ). Outline of an algebraic solution of the general problem, 530–1.

APPENDIX 4 (ULC. Add. 3970. 3: 570ᵛ). Draft computations for an unidentified geometrical problem: a related figure is (uniquely?) restored, 532.

PART 2
THE 'ARITHMETICÆ UNIVERSALIS LIBER PRIMUS'
(1684)

INTRODUCTION 535

The extent of Humphrey Newton's secretarial help in preparing the manuscript: identification of the principal passages in the parent *Arithmetica* subsumed into it, 535. Conjectures regarding Newton's intentions in the unwritten portion of the planned text, 536. The much amplified Newtonian 'Rule of brothers', 537.

THE 'FIRST BOOK OF UNIVERSAL ARITHMETIC' 538

[1] (ULC. Add. 3993: 1–93). The main text. The lightly revised opening, 538. 'On notation' (all but identical with its parent), 540–6. Following sections on addition, subtraction, multiplication and division (only trivially changed) are here omitted: the few variants are given in footnotes, 546. 'The finding of divisors' (largely unaltered, but a few changes are made in its central portion), 546–54. 'Extraction of roots' (again mostly here omitted as trivially variant), 554. 'The reduction of fractions and radicals' (condenses and lightly revises equivalent passages in the parent *Arithmetica*), 554–60. 'An equation and its roots' (much reworked from its preceding 'De forma Æquationis'), 560–2. 'The finding of equations': its opening refashions a corresponding passage in the parent text on reducing arithmetical 'questions' to algebraic equations, 564. In illustration Newton works at length seven arithmetical problems (all borrowed from the earlier *Arithmetica*), 566–72. Useful points to remember in 'translating' a geometrical configuration into algebraic equations, 572–4. Five lengthy illustrative problems (all but one taken unchanged from the parent text), 574–84. Problem 11: the classical 3-line locus makes its (disguised) Newtonian bow, 578. Problem 12: seven algebraic 'modes' of rendering Schooten's problem of a quadrilateral inscribed in a semicircle into equations, 580–4. 'The reduction of equations': thirteen *regulæ* (four new) for effecting this by standard operations, 584–94. Refinements in elimination between simultaneous equations, 590–2. Particular techniques of reduction sometimes useful, 594. The *regulæ* applied to reducing the algebraic 'translations' of the preceding problems to a single equation (the cubic which results in Problem 12 is not solved), 596–602. The answers are listed, 602–4. Developing a 'method' of reducing to algebraic equivalent which is skilful, elegant and concise, 604. An example (the merchant's

doubled profit once more) in illustration, 606. Twenty-one further (more sketchily worked) examples: Problems 1–20, transcribed by Humphrey Newton from the parent text without addition by Isaac, are here omitted, 608. Problem 21, the square case of Apollonius' verging problem, is somewhat recast, but again illustrates Newton's rule of 'twins', 608–10.

[2] (ULC. Add. 3963. 5: 40^r–41^r). The much amplified 'Rule of brothers'. The refinement which ensues in algebraic reduction from choosing as base variable one related 'in the same way' to 'twin' magnitudes in a problem, 612. A simple instance in a class of triangle problems, 612–14. A second example: the Apollonian verging problem yet again, 614. A third: the (reduced) Apollonian circle-tangency problem, 616–18. Variant ways of constructing its solution, 620.

APPENDIX (ULC. Add. 3964. 3: 22^r/22^v). The grading of problems by the degree of the algebraic equations into which they 'translate' (in conflict with Newton's remarks in his *Arithmetica*), 622. Simple instances, 623. The grading equation must be in its simplest possible form, 624.

INDEX OF NAMES   625

LIST OF PLATES

The 'Arithmeticæ Universalis Liber primus' (**2**, [1])   *frontispiece*

I   Plane sections of the right circular cone and one-sheet
hyperboloid of revolution (**1**, **2**, §1)   *facing p.* 218

II   A rectilinear cometary orbit constructed from four angular
observations (**1**, **2**, §1)   298

III   Faulty construction of the cissoid of Diocles by a moving
angle (**1**, **2**, §1)   466

IV   Opening page of the 'Regula Fratrum' (**2**, [2])   612

PART 1

THE DEPOSITED LUCASIAN LECTURES ON ALGEBRA
(Winter 1683-1684)

INTRODUCTION

At some undetermined time following[1] the autumn of 1683 (when their final 'Octob.' series was delivered) Isaac Newton deposited in the University archives, as the enduring record of his professorial teaching at Cambridge, the corrected copy of ninety-seven purported 'lectures' on the foundations of algebra and its application to the resolution of problems in arithmetic and geometry by solving the equations which are their equivalent algebraic re-statement. This deposited text we now reproduce[2] liberated from a number of errors and accretions introduced by Whiston in his subsequent *editio princeps*, lightly underpinning it with the few preliminary fragments known[3] and annexing certain later amplifications.[4]

For a detailed listing, section by section, of the content of this central manu-script and its present appendages we may refer the reader to the analytical index which precedes, while for our explanation of particular difficulties we direct him to pertinent footnotes at appropriate points in sequel. In broad outline it derives partly (in its discussion of the elemental algebraic operations and of the reduction and exact solution of equations) from Newton's earlier, unpublished 'Observations'[5] on the introduction to Cartesian algebra presented by Gerard Kinckhuysen in his 1661 *Stelkonst*, partly (in its techniques for delimiting the number and nature, real or complex, of the roots of equations and for reducing these by factorization) from his own independent researches as a young post-graduate student into the theory of equations,[6] and partly (in its approximate geometrical construction of cubics) from his previously elaborated 'Problems for construing æquations'.[7] Apart from novelties in detail and the fabrication of new illustrative 'questions', what is most notable is Newton's developing awareness—still far from completely expressed[8]—of the fundamental structural equivalence which exists between the elements (constants and free variables, and their functional relationships) of algebra and those (given lines and un-determined line-lengths, and their coordinate interconnections) of geometry,

---

(1) Probably not long after, though not straightaway since the deposited manuscript was used as parent text, about the spring or early summer of 1684, to father an unfinished revise, the 'Arithmeticæ Universalis Liber primus' (reproduced as **2**, [1]).

(2) ULC. Dd. 9.68, reproduced as **2**, §§1/2.

(3) See **1**, §§1–3 and Appendix 1 following.

(4) **2**, §3 and Appendixes 2/3.

(5) See II: 364–444. Nicolaus Mercator's Latin translation of the *Algebra ofte Stelkonst* is reproduced on II: 295–362 preceding.

(6) See I: 489–539.

(7) Reproduced on II: 450–516.

(8) Compare **2**, §1: note (206).

and his deepening grasp of the still more general isomorphism which permits a two-way 'translation' between mathematical 'speech' and the 'language' of exact science in all its manifestations.[9] His guiding doctrine that algebra is 'universal arithmetick' embroiders a theme stated briefly in an opening phrase[10] of his 1671 treatise on infinite series and fluxions, and expounded (in a geometrical context) earlier still in preface to James Gregory's study of universal mathematical principles.[11] Now also, however, he reaches tentatively forward to Barrow's notion that algebra is in its essence the abstract logic of relationships between quantities in divorce from their particular setting, and hence to be developed as an independent, metamathematical system.[12] But we should beware of any anachronistic interpretation in assessing these vague insights on Newton's part into the nature of mathematics: for him the only viable concepts and admissible operations in algebra remain uniquely the universalized images of those in common arithmetic, while such concepts as the non-commutativity of a relationship or a non-standard model of an algebraic structure are wholly foreign to his viewpoint. Nor, given the cautious, imprecise manner in which he expounds them, is it perhaps profitable to attempt too fine an analysis of the thoughts Newton here expresses.

On the historical background to the manuscript and its composition we may be somewhat more expansive, though at certain points reasoned conjecture must inevitably fill the place of documented knowledge. Just as we earlier doubted the extant record of Newton's preceding lectures on optics,[13] it will be clear that the textual sequence of their deposited algebraic continuation and the

(9) See 2, §1: Newton's pages 39–41, and note (128). The analogy is further elaborated in the revised 'Arithmeticæ Universalis Liber primus'; compare 2, [1]: note (59).

(10) 'Cùm in numeris et speciebus operationes computandi persimiles sint, neqʒ differre videantur nisi in characteribus quibus quantitates in istis definitè, in his indefinitè designantur: ...operationes Additio, Subtractio, Multiplicatio, Divisio et extractio Radicum exinde addisci possunt modo lector...et Arithmeticæ et Algebræ vulgaris peritus fuerit et noverit correspondentiam...' (III: 32–4).

(11) *Geometriæ Pars Universalis, Inserviens Quantitatum Curvarum transmutationi & mensuræ* (Padua, 1668): Procemium: [†2ʳ] 'Observatum est à nostri seculi geometris mathematicam ab antiquis malè fuisse divisam in geometriam, arithmeticam, &c, sed potius in universalem & particularem: Matheseos pars universalis tractat de proportione in communi abstrahendo ab omni quantitatis specie [*sc.* particulari], cui affinis est recentiorum analytica: matheseos pars particularis dividitur; in geometriam propriè sic dictam, quæ nihil aliud est nisi matheseos pars universalis figuræ restricta; in arithmeticam, quæ eadem est mathesis universalis numero,... &c.' (Newton's library copy of Gregory's *Geometria* is now at Trinity College, Cambridge. NQ.9.48².)

(12) Most clearly, Newton asserted in his revised 'Arithmeticæ Universalis Liber primus' (see note (9)) that 'Est...Algebra nihil aliud quàm sermo mathematicus designandis quantitatum relationibus accommodatus'. At the end of the second lecture (given in spring 1664 in the Cambridge schools, no doubt with Newton in the audience) of his inaugural series of

chronology imposed in its margins are to be regarded with no small suspicion. Statute required of the Cambridge Lucasian professor only that the 'lectures' so registered in the archives of the University be polished versions of the (ten) best of those annually delivered, and Newton interpreted the provision no less tolerantly than his predecessor.[14] Our examination of the present autograph record leads us firmly to conclude that it was compiled over a period of but a few months (during the winter of 1683–4?) and hence is not a strictly contemporary report of the lectures delivered by him—and repeated at intervals with various improvements and restylings?—during the period 1673–83. Independent support can be given to the related inference that Newton's marginal datings of the manuscript 'lectures', and indeed their very division as such paragraph by paragraph, do not correspond exactly to the original times of delivery and the texts then read out. Thus when, during a visit paid to Cambridge in June and early July of 1674, the astronomer John Flamsteed renewed an earlier brief contact with Newton, he was given a 'paper'[15] containing notes for a lecture in 'Midsummer' of that year, whose content, except for an omitted paragraph, is reproduced without essential change in the section 'De concinnanda Æquatione solitaria' of the deposited version[16] but is there split between *Lectiones 6* and *7* of a series of ten which, according to the marginal dating, he began to deliver only in 'Octob. 1674'. Five years later he there claimed to have begun his six recorded lectures for 1679 in 'Octob[er]',[17] whereas the list of his exits and redits at Trinity College reveals that he was away from Cambridge between 28 July and 27 November (in Lincolnshire in fact, busy settling the estate of his

---

Lucasian *Lectiones Mathematicæ XXIII*; *In quibus Principia Matheseôs generalia exponuntur* (London, ₁1683: 55), Isaac Barrow had earlier affirmed that 'Algebra, quam vocant, vel...Analysis (eatenus intellecta, quatenus a Geometriæ vel Arithmeticæ pronunciatis et regulis distincti quid innuit) non magis ad Mathematicam, quam ad Physicam...aut aliam quamvis scientiam videtur spectare. Est enim duntaxat pars quædam aut species Logicæ, seu modus quidam utendi ratione circa quæstionum solutionem, inventionemque vel probationem conclusionum, qualis in aliis omnibus scientiis exercetur haud rarò. Quare non est pars, aut species, sed instrumentum potius Mathematicæ subministrans'.

(13) Compare III: 435–7.

(14) See III: xviii, xxii. The pertinent statutory clause decreed 'ut Professor semel quotannis... non pauciorum quàm decem ex illis, quos præcedente anno publicè habuerit, lectionum exemplaria nitidè descripta Procancellario exhibeat, in publicis Academiæ archivis asservanda'. The first Lucasian professor Isaac Barrow had, in the case of his initial sequence of lectures (subsequently published as *Lectiones Mathematicæ XXIII...Habitæ Cantabrigiæ Annis 1664, 1665, 1666* (see note (12)) but delivered, it would seem, in three batches between March 1664 and early summer of 1665), already introduced the tolerance of making only periodic public deposit of 'lectures' drastically reordered in sequence, extensively rewritten, and subdivided into a frequency of delivery of some 7–8 *per annum* only.

(15) Reproduced as 1, §1.

(16) See 2, §1: Newton's pages 32–3.

(17) See 2, §1: Newton's page 111.

recently deceased mother).[18] We have likewise good reason to suspect that
Newton's exposition of the Wrennian construction of a comet's path,[19] purportedly introduced to his Cambridge audience in 'Lect. 2' in 'Oct. 1680', was
provoked by the appearance of the brilliant comet of 1680–1 in the heavens over
England in the following month. The only reasonable inference from these and
other[20] inconsistencies in the content and chronology of the deposited copy is
that Newton there summarized the substance of lectures on algebra given out
by him over more than a decade and imposed on this compendium a time-scale
which is markedly inaccurate in detail. Whether the manuscript was even
nominally addressed to a public academic audience is unclear,[21] and it may
well be that in composing it Newton had an eye to its future circulation in
printed form. But, in default of any documented information on Newton's
intentions in preparing his text, we must continue to assume that its content
agrees in main outline with the elements of algebra and their application to
solving problems which he had earlier inculcated from his professorial rostrum.

No contemporary reaction to the lectures as delivered has seemingly been
preserved, and we may assume that they engendered little excitement among
Cambridge students of the day, undergraduate or senior. (On the 'paper' of
notes which Flamsteed took away from Cambridge in July 1674 Flamsteed
could only bleakly record that it was 'given at one of [Mr Newton's] lectures'
without commenting on the size or quality of the assembled audience or their

---

(18) See J. Edleston, *Correspondence of Sir Isaac Newton and Professor Cotes* (London, 1850):
lxxxv. On the day following his return to Cambridge Newton wrote to Robert Hooke that 'I
have been this last half year in Lincolnshire cumbred wth concerns amongst my relations till
yesterday when I returned hither; so y^t I have had no time to entertein Philosophical medita-
tions, or so much as to study or mind any thing els but Countrey affairs' (*Correspondence of Isaac
Newton*, **2**, 1960: 300).

(19) See 2, §1: Newton's pages 121–4, and compare Appendix 3: note (1) following.

(20) For instance, Newton had some doubt where exactly in the deposited copy the
division 'Lect 8' of the final 'Octob 1683' series should go; see 2, §1: note (698).

(21) In an unguarded moment on page 203 of the deposited text Newton addresses himself
to a single 'Lector'; compare 2, §1: note (580). The pretence of speaking to an assembled
audience is wholly abandoned in the revised 'Arithmeticæ Universalis Liber primus'; see
2, [1]: note (120).

(22) 'I have lately tryed to looke into Cartes 3^d booke [of the *Geometrie*]..., thinking to
understand what goes before by ye helpe of his rules delivered there: & I begin to hope I may
by my owne strength, & I judge it is better to find one conclusion out than have 20 shewed me,
wch made me deferr moving questions to you so long, & partly because I cannot move my
many doubts in proper termes. But I know you are to good & wise to deride me' (*Correspondence
of Isaac Newton*, **2**: 86). Since Horne was writing from 'Hadly Suffolk. Aug. 22^d', he was evidently using a summer vacation away from Cambridge to try and master Descartes' *Geometrie*
on his own and we should not necessarily presume that a prior attendance at Newton's lectures
had stimulated him to do so. As for the level of Horne's understanding of algebra, this is
sufficiently revealed by his 'factorisation' of $x^2 = 6$ as the product of $x = 2$ and $x = 3$, and by
his inability to 'find out y^e order' of Descartes' division by a quadratic of the sextic in y

understanding of what was communicated.) With the possible exception of a young King's don, Thomas Horne, who in the middle 1670's addressed to Newton a plea for help in understanding 'Cartes 3^d booke of ye nature of æquations',[22] Cambridge seemed to wish to remain indifferent to its Lucasian professor's efforts to arouse interest in the fundaments of algebra, and Newton might just as well have kept his silence. Away in London the indefatigable John Collins heard (through Isaac Barrow?) only the vaguest rumour of Newton's 'having read Lectures [of ye doctrine of infinite Series] and of Algebra, and put them into the public Library at Cambridge',[23] while in Oxford the Savilian professor John Wallis—then busy preparing his own treatise of algebra for publication (and ever watchful for material to incorporate in it)—appears to have remained wholly ignorant of the highly pertinent lectures of his Cambridge equivalent.[24] Likewise, it would seem, Newton's public deposit of his algebraic lectures in the middle 1680's went unremarked by the English mathematical world (Collins died in November 1683 just as the last series was coming to its close in the Cambridge schools) and ignorance of their availability for study and inspection was complete for the next decade and a half. Not till the winter of 1701/2, when Roger Cotes, then still an undergraduate at Trinity, completed for his private purposes a still extant transcript[25] of the deposited copy, have we

subsequently repeated by Newton on page 18 of his deposited 'lectures' (see 2, §1: note (52)). From the draft of Newton's reply (*Correspondence*, 2: 187) a later letter perhaps also from Horne— or was it Collins?—similarly appears to have sought information regarding Descartes' presentation of the Cardan solution of the reduced cubic.

(23) Collins to Oldenburg for Leibniz, 14 June 1676 (*Correspondence of Isaac Newton*, 2: 51). He was soon disabused on the former by Newton himself, who wrote on 5 September following that 'though about 5 years agoe I wrote a discourse in wch I explained ye doctrine of infinite æquations, yet I have not hitherto read it but keep it by me' (*ibid.*: 95; compare III: xix, note (36)). The next year Collins was to suggest to Wallis that 'Mr Newton...intends a full treatise of Algebra' (*ibid.*: 242), but from the listing of its content according to Collins' best 'app[rehension]s' it is clear that he knew only of Newton's earlier researches up to 1672; see II: xiv–xv.

(24) Though Wallis was quick to include Newton's researches in infinite series in his book (see IV: 672, note (54)), neither the English original (London, 1685) of his *Treatise of Algebra, Both Historical and Practical* nor its augmented Latin revise (in his *Opera Mathematica*, 2 (Oxford, 1693): 1–482) contain any reference to Newton's current algebraical investigations. Collins for the most part kept Wallis in the dark about the activities of his Cambridge equal 'in regard you lye under a censure from diverse for printing discourses that come to you in private Letters without permission or consent as is said of the parties concerned' (*Correspondence of Isaac Newton*, 2: 242).

(25) Trinity College, Cambridge. MS R.16.39 [first tract]. At its end Cotes has noted: 'Descripsi ex Autographo anno 1701/2. R.C.' The volume also contains his transcripts of Newton's other deposited Lucasian 'lectures': namely, as he detailed them for William Jones on 30 September 1711 (Edleston, *Correspondence* (note (18) above): 209), 'the first draught of his *Principia* [ULC. Dd.4.18] as he read it in his Lectures...& his Optick Lectures [ULC. Dd.9.67; see III: 435–42], the substance of which is for ye most part contained in his printed [*Opticks*], but with further Improvements'.

any evidence that it was ever consulted by anyone. Whatever ameliorations Cotes may then have wished to introduce into its text are unrecorded, but his continuing interest in its improvement is evidenced in an incidental remark of his to Newton, several years after Whiston's *editio princeps* was published, that 'I intended to have importun'd you, to review your *Algebra* for a better Edition of it'.[26]

Of the back-cloth to that first printed edition much more is known. When Newton resigned his Lucasian professorship to his deputy William Whiston in December 1701, it was natural that the latter should wish to familiarize himself with the deposited lectures of his predecessor.[27] We have indeed seen[28] that Whiston did not hesitate to introduce portions of Newton's earlier optical lectures concerning the mathematical theory of the rainbow into his own Lucasian *prælectiones physico-mathematicæ* in the spring of 1706, and about that time also he turned his attention to the succeeding ones on algebra and began to consider their publication. In the meantime—whether or not these were started solely to counter Whiston's intention is not known—rumours began to spread in both Oxford and London that 'Mr Newton's friends solicit him to publish a

---

(26) Cotes to Newton, 26 April 1712 (ULC. Add. 3963.28). Cotes' little variant draft (now in Trinity College. MS R.16.38) is reproduced by Edleston in his *Correspondence* (note (18) above): [100–2 +]119.

(27) Newton's signed and sealed autograph letter of resignation (Cambridge University Archives. Register Book 39.8.4) reads: 'I Isaac Newton Fellow of Trinity College in the University of Cambridge being in possession of the place of Mathematick Lecturer founded by Henry Lucas Esq in the said University: do hereby resigne quit & make voyd the said place of Mathematick Lecturer together with all my right & title to the same & to all Lands Houses Tenements rents profits & perquisites thereunto belonging. And by this Resignation I do enable the R^t Worshippful the Vicechancellour & other Electors of the said University as fully as if I were dead to elect ordein & constitute a new Mathematick Lecturer in my room. In witness whereof I have hereunto set my hand & seale this tenth day of December...1701'. Flamsteed wrote maliciously to Abraham Sharp the following 31 March that 'He [Newton] has lost his professorship at Cambridge & put Mr Whiston into it$_{[,]}$ tis sayd by some...that Augustus left a Tiberius to succeed him purposely to render his own fame the more illustrious' (compare *The Correspondence of Isaac Newton*, 4, 1967: 472, note (8)), but this is decidedly unfair. As Newton's deputy in receipt of 'the full profits of the place' (see Edleston's *Correspondence* (note (18) above): xxxvi) Whiston had by then delivered almost a year of his own professorial lectures and proved his academic competence. (These lectures, subsequently published as his *Prælectiones Astronomicæ Cantabrigiæ in Scholis Publicis Habitæ* (Cambridge, 1707): 8–111, are introduced by an '*Anteloquium*' (ibid.: 1–8) which opens with a revealing comment on his predecessor's esoteric lectures on the topic: '... dum inter nubila, inter sydera Professores vestri secundo numine tranarunt, plebem Mathematicam immane quantum supervolarunt. Dum illi non usitata nec tenui ferebantur penna, juventutem Academicam ab aliis artibus, utpote inferioribus, occupandum reliquerunt. Testor ego vosmet ipsos, Auditores, hosce parietes alias testaturus, Quoties Dignissimi Professores vestri vacua spatiati sunt in Aula, & hasce scholas obambularunt solitarii? Quam rari enim ex turba togata qui tam alta audiendi, aut adeo sublimia Matheseως capita sibi intelligenda proposuerunt? Quam exigui numero qui scientias hasce, nisi in transcursu, sibi salutandas duxerunt?')

Treatise of Algebra which he wrote long since'.[29] If such ill-founded whispers penetrated to Cambridge, Whiston ignored them and went quietly ahead, arranging with the London stationer to underwrite the expense of printing the deposited manuscript and then subsequently, between September 1705 and the following June, correcting both specimen and proof sheets[30] as they emerged from the compositor's bench at the University Press. In February 1706 David Gregory accurately noted in his memoranda that 'Its talked that there is now printing at Cambridge, Elements or Principles of Algebra, written long since by Sr Isaac Newton',[31] but withdrew his added remark that it was 'lately revised by him' when he saw Newton in London in July and was given a first-hand account of the history of the 'Algebra, that is a printing and near printed at Cambridge'. Though, as Whiston was later to announce publicly,[32] Newton had given his reluctant approval for the edition,

He was forced seemingly to allow of it, about 14 months agoe, when he stood for Parliament-man at that [Cambridge] University.[33] He has not seen a sheet of it, nor

---

(28) See III: 476, note (2).

(29) According to David Gregory, who was told it in these words 'at Oxon the 14 March 1703/4 by Mr Smalbrook' (Christ Church, Oxford. MS 346:123, reproduced by W. G. Hiscock in *David Gregory, Isaac Newton and their Circle. Extracts from David Gregory's Memoranda 1677–1708* (Oxford, 1937): 16). Ten days later at London Gregory noted correspondingly that 'M^r [Nicolas] Fatio [de Duillier] told me he believed that Mr Newton...would publish his Algebra' (*ibid.*: 124). At this period Newton's attention was directed to revising his *Principia* for publication (compare *ibid.*: 124, 173 [=Hiscock, *Gregory*: 16, 36–7]) and to seeing the first English (1704) and Latin (1706) editions of his *Opticks* through the press. It will be clear from the sequel that in 1706 he retained only vague memories of the content of the copy of his algebraic lectures which he had deposited twenty years before.

(30) Those (namely A–V) at least which were composed by 29 June 1706. D. F. McKenzie has given a careful analysis of the existing printing records of the *editio princeps* in his account of *The Cambridge University Press 1696–1712*, **1** (Cambridge, 1966): 278–9; compare also **2**: 270–83.

(31) Christ Church, Oxford. MS 346: 80 [=Hiscock, *Gregory* (note (29)): 32].

(32) *Memoirs of the Life and Writings of Mr. William Whiston...Written by himself in the 79th, 80th, 81st and 82nd Years of his Age* (London, $_1$1749): 135: 'In the...Year 1707, I published, by the Author's Permission, Sir *Isaac Newton*'s *Arithmetica Universalis*, or *Algebra*, from that Copy which was laid up in the Archives of the University, as all Mr. *Lucas*'s Professor's Lectures are obliged to be....: Which *Algebra* had been nine [*sic*] Years Lectures of Sir *Isaac Newton*'s...'. An unpublished earlier note by John Conduitt, doubtless reporting Newton's own version of the incident, insists that 'His Arithmetick was first printed by Whiston ag^t Sr I. inclination' (King's College, Cambridge. Keynes MS 130.5), adding that '[it] being full of errors, he afterwards [in 1722] printed it himself [&] corrected the faults'.

(33) In May 1705, that is. Having been knighted by Queen Anne in a lavish ceremony at Trinity only a month before, Newton evidently thought he stood an especially good chance of election. In the poll on 17 May, however, he came a very poor fourth after three other un-distinguished (but 'locally' more active) candidates; see Edleston's *Correspondence* (note (18) above): lxxiv, note (153). The sourness of political defeat no doubt to some extent hardened his heart against Whiston's edition of a manuscript, his approval of which had been extracted to so little effect on his aspiration to take up his old member's seat in the Commons once more.

knows he what volume it is in, nor how many sheets it will make, nor does he well remember the contents of it. He intends to go down to Cambridge this summer and see it, & if it doe not please him, to buy up the Coppyes. It was read by way of Lectures many years agoe, and put up in the publick Library according to the Statute.[34]

Gregory's following précis[35] of 'What he remembers of the Contents of it' would seem to indicate that, in all but its main outlines at least, Newton's impression of the lectures he had deposited twenty years before was very blurred, if indeed it was not conflated in his mind with the unfinished 'Arithmeticæ Universalis Liber primus'[36] which he soon afterwards began to draft in its revise and whose incomplete text he still kept by him:

There is a Chapter concerning every one of the operations, Addition, Subtraction, Multiplication, Division, & Extraction of Roots. In each he teaches that operation in all sorts of quantitys, integers, fractions, &c. In the Chapter of the Extraction of Roots, after the Extraction of the Square root, he has one universal Rule of extracting all roots.[37] The next Chapter is concerning the Divisors of a quantity, which he does by an universal new methode. The next is about the Invention of Equations,[38] or the reducing of a Problem to one or more Equations. Which he shews to be nothing more than writing down the Problem plainly in the Algebraical Language. This notion he exemplifys in 20[!] problemes. The next is about reduction of more Equations to one fewer by turning out one unknown quantity. This also he elucidates by the forsaid Examples, rehandled.[39] These Problems are some Arithmetical, some Geometrical, but all remarkably usefull, as well as fitt Examples to these purposes. One of them[40] is that *in Additamento ad Appendicem* [*ad Cartesii Geometriam*] *de Cubicarum Equationum Resolutione* of Francis Schooten, concerning the Shadows of three perpendicular stakes, but much more simply solved.

---

(34) Christ Church MS 346: 175–6 [=Hiscock, *Gregory* (note (29) above): 36].

(35) Christ Church MS 346: 176–7.

(36) See **2**, Introduction below. The text of this manuscript (now ULC. Add. 3993) is reproduced in [1] following.

(37) Namely, the Cartesian *regula* elaborated on **2**, §1: Newton's pages 207–10.

(38) Gregory's following description agrees hardly at all with the relevant section 'De forma Æquationis' in the deposited copy (**2**, §1: Newton's pages 30 ff) but accords loosely with the revised 'Inventio Æquationum' in the 'Arithmeticæ Universalis Liber primus' (compare **2**, [1]: note (55)).

(39) This accords well with the section on 'Reductio Æquationum' in the revised *Arithmetica* (compare **2**, [1]: note (88)): there is no equivalent in the deposited text (**2**, §1).

(40) Namely, Problem 42 in the deposited copy (**2**, §1: Newton's pages 106–11). The example does not occur in the text of the (unfinished) revised *Arithmetica* (**2**, [1]).

(41) Corresponding to pages 211–51 of the deposited copy (**2**, §1). Newton's following summary, however, departs widely from its content and (with Gregory's present distortions eliminated) seems to be based on his earlier 'Problems for construing æquations' (see II: 450–516), then still in his possession.

(42) This 'next...easy *Postulatum*' is manifestly the product of Gregory's misunderstanding of what Newton told him. If, in the 'simplest' case, a straight line is to be drawn between three given straight lines—defined, say, by $y = 0$, $y = ax + b$ and $y = cx + d$ in some standard

Next he comes to the Construction of Equations.[41] Here he shews the Description of a right Line and a Circle, or Euclids *Postulata*[3] not to have been received because of the Simplicity of the Lines, but for the mechanical Easiness of them: so that toward the Construction of Problems above the reach of Euclids *Postulata*, the Problem next most mechanically easy must be received as a *Postulatum*. This he shews to be the following. *Ducere rectam datæ longitudinis inter duas datas lineas interceptam, quæ (producta saltem) per datum transeat punctum.* If the *datæ lineæ* be right lines, (which is the simplest case) by this *Postulatum* he constructs Cubick Equations. The next mechanically easy Problem, or *Postulatum*, he takes to be this. *Rectam ducere, cujus lineæ segmenta, inter tres datas lineas intercepta, sint datæ longitudinis.* This when simplest, (that is when the *datæ lineæ* are *rectæ*,) constructs Biquadratick[!] Equations.[42] But this he does not shew. Here he breaks off, there being further only a Title of a Chapter, *De Seriebus Infinitis*.[43]

Despite its minor inconsistencies and confusions Gregory's report vividly conveys Newton's concern that an unfinished text composed so long before should now be presented to the world as though it represented his latest researches into the structure and applications of algebra. Whether or not he paid his intended visit to the Cambridge Press is not known, but he contrived to hold up the setting of Whiston's edition for several months while he determined what to do. On 1 September Gregory was shown the existing proofs of the new 'Algebra' and remarked:

There are just 20 sheets[44] or 320 pages in 8ᵛᵒ printed. The title is *Arithmeticæ Universalis sive Algebræ Elementa*. The running title is, *Algebræ Elementa*. He is not pleased with the Titles, as not agreeing with the Introduction. The whole will not be above 2 or 3 sheets more.[45] He intends a Preface himself, in which he is to tell all that he would have changed (which chiefly consists in putting out) and that against the next edition. I hope we shall have it published shortly;[46]

subsequently he added that 'The title is at least agreed to be this. *Arithmetica Universalis, sive Tractatus de Resolutione et Compositione Arithmetica....* There is to

---

(perpendicular) Cartesian system of coordinates—such that its intercepted segments are of lengths α and β, then a pair of quadratic equations in x (or X) results on eliminating X (or x), y and Y between $y = ax+b$, $Y = cX+d$, $y/Y = \pm\alpha/\beta$ and $\alpha \mp \beta = \sqrt{[(x-X)^2 + (y-Y)^2]}$. Newton himself would scarcely have made such a clumsy mistake as to think that a general quartic equation could thereby be constructed.

(43) The main deposited text has no such appended title, but the preliminary addition 'Ad pag 29 l. 37' (reproduced in 2, §2) incorporates the unimplemented title of a further section 'De Reductione fractionum et Radicalium ad series convergentes'; see 2, §1: note (614) and §2: note (36).

(44) Signatures A–V, that is. Compare note (30).

(45) The published 'Algebra' contains in fact only six further pages (321–6) of Newtonian text, but to this Whiston appended (pages 327–43) a reprint of Edmond Halley's 'Methodus Nova Accurata & facilis inveniendi Radices Æquationum....' (see note (50)), requiring altogether not quite two octavo sheets (X and signatures 1ʳ–4ʳ of Y) more.

(46) Christ Church MS 346: 178 [=Hiscock, *Gregory*: 37].

be subjoyned to the end... the Construction of Solid Problemes by an Ellipsis & a Circle: but no other alterations'.[47]

In the event the completed *editio princeps* finally appeared in the bookshops in May 1707[48] with Whiston's running head changed only on the first sheet[49] and an appended article by Edmond Halley on 'A new, accurate and easy method for finding the roots of any equations generally, without prior reduction'[50] but lacking any preface signed by its author. In lieu of the last a not uninformative *Ad Lectorem*[51] (initialled by Whiston but bearing traces of Newton's guiding hand) briefly sets the background to the definitively titled *Arithmetica Universalis; sive De Compositione et Resolutione Arithmetica Liber*,[52] stressing that it reproduces the unaltered manuscript record of lectures delivered

---

(47) Christ Church MS 346: 176, 184 [=Hiscock, *Gregory*: 36, 40]. It is natural to suppose that the 'Construction... by an Ellipsis & a Circle' is that, by trammel and homothety, still outstanding on pages 247–51 of the deposited text (2, §1) and subsequently added on pages 321–6 of the printed *Arithmetica* (see note (45)), though Newton may have had it in mind to insert allied constructions of the general quartic equation from the parent 'Problems for construing æquations' (see II: 490–8). Whiston's (last-minute?) decision to append a reprint of Halley's 'Methodus Nova...' evidently conflicts with Newton's present determination in autumn 1706 to include 'no other alterations' from the text of the deposited manuscript. In his own 1722 revised edition of the *Arithmetica* he was to omit the alien appendix.

(48) Compare McKenzie's *Cambridge Press* (note (30) above), **1**: 278, where references in the contemporary *Term Catalogues* and the *London Gazette*, No. 4335 [for 26–9 May 1707] are quoted. The final sheets X and Y, the prelims (sheet a) and the revised sheet A were printed off by late March. The run was a thousand copies, each retailing at 4*s*. 6*d*. A collation, 'lectio' by 'lectio', of the deposited manuscript—used as the press copy (see 2, §1: note (2))—with the corresponding pages of the *editio princeps* is given in Edleston's *Correspondence* (note (18) above): xcii–xcv.

(49) To read (over the appropriate pages) 'Notatio', 'Additio' and 'Subductio': this leads to an illogical transition from the new head 'Subductio' (replacing 'Algebræ') on page 16 to the old head 'Elementa' retained on page 17 opposite. In the preliminary pages ([a 4ᵛ]) an editorial 'Lectori S[cholium]' (reproduced fully in Appendix 1: note (1)) informs the reader—surely at Newton's pressing insistence?—that more accurate 'tituli perpetui', summarizing the argument on each page, need to be inserted in future editions 'prout in prioribus aliquot paginis alia de causa recusis jamjam fecimus'.

(50) 'Methodus Nova Accurata & facilis inveniendi Radices Æquationum quarumcunque generaliter, sine prævia Reductione. Per *Edm. Halley*', first published in *Philosophical Transactions* **18** (1694), No. 210 [for May 1694]: §IV: 136–48. (In his 1707 reprint Whiston delicately added Halley's newly acquired title of 'Geom. Prof. Savil'.) In generalization of the variant on Newton's method for solving the algebraic equation $f(z) = 0$ published by Joseph Raphson in his *Analysis Æquationum Universalis* in 1690 (see IV: 665, note (24)), Halley, making use of a *Speculum Analyticum Generale* in which the binomial expansions of $(a+e)^k$, $k = 1, 2, 3, ..., 7$, are listed, presupposes the expansion $f(a+e) = b + es + e^2t + e^3v + e^4w$ ... (without explicitly identifying $b = f(a)$, $s = f'(a)$, $t = \frac{1}{2}f''(a)$, $v = \frac{1}{6}f'''(a)$, $w = \frac{1}{24}f^{iv}(a)$, ...) and then refines the approximation $z \approx a$ by iterating the 'errors' $e \approx (-b/(s+et) \approx) -bs/(s^2-bt)$ and, equivalently to $O(e^3)$, $e \approx (-\frac{1}{2}s + \sqrt{(\frac{1}{4}s^2 - bt)})/t[\approx -b/s - b^2t/s^3 ...]$. (In illustration he approaches the root $z = \sqrt[k]{(a^k+b)} \approx a$, $k = 3, 5$ and then a general positive integer, by iterating the successive errors $e = z - a \approx (-ka^{k-1} + \sqrt{(k^2a^{2(k-1)} + 2bk(k-1)a^{k-2})})/k(k-1)a^{k-2}$, further applying the methods to several particular cubic and quartic equations $f(z) = 0$.) Whiston, as he affirmed

'almost thirty years before', composed 'with a hasty pen' and never intended for the printer, and pointing to its consequent incompleteness, particularly in regard to its promised construction of quartics and higher equations.[53] A not inconsiderable number of small misprints spatter the following text, but Newton's only real complaint regarding Whiston's editorial capacity could have been that he too faithfully and impercipiently followed the parent manuscript, incorporating in his *princeps* edition its several inconsistencies and lapses into error[54] without, in the main, even bringing them to the reader's notice. As its author knew better than anyone, the *Arithmetica*'s larger faults of poorly con-

---

in his preface *Ad Lectorem* (see Appendix 1: note (9)), seized on Halley's article to fill an evident lacuna in Newton's discussion of the solution of equations *in numeris* (compare 2, §1: note (170)). We should not assume that Newton hesitated to concur solely because he resented an alien intrusion in the printed *Arithmetica*, one which seemed—none too justly—to point to his own deficiency. It may well be that he saw Halley's method as affording no effective computational improvement over his own simpler procedure by iterating the error

$$e(= z - a) \approx -b/s \ [= -f(a)/f'(a)]:$$

if so, he would have full support in the modern world of the fast electronic computer. (Thus, two iterations of $e \approx b/ka^{k-1}$ yield the error $b/k(a + b/ka^{k-1})^{k-1} \approx b/ka^{k-1} - (k-1) b^2/k^2 a^{2k-1} \dots$, scarcely more than that, $b/ka^{k-1} - \frac{1}{2}(k-1) b^2/k^2 a^{2k-1} \dots$, deriving from a single application of Halley's much more cumbrous formula above.) We suggest (compare 2, §1: note (746)) that Newton himself originally intended to conclude his deposited 'lectures' with an outline of the technique for obtaining the root of an equation *in proximis numeris* which he had expounded earlier in his *De Analysi* (II: 218–20) and, in improved form, in his 1671 tract (III: 42–6), and which had been published by John Wallis in his *Treatise of Algebra* (London, 1685): 338–40. Of the last Halley remarked somewhat dismissively in his article that 'Qualia...in hoc negotio præstiterit sagacissima ingenii *Newtoniani* vis, ex contractiore Specimine à Clarissimo *Wallisio*, Cap. xcɪv *Algebræ* suæ, edito, potius conjecturâ assequi quam pro certo comperiri licet. Ac dum obstinata Authoris modestia amicorum precibus devicta cedat, inventaque hæc sua pulcherrima in lucem promere dignetur, expectare cogimur' (*Philosophical Transactions*, **18**: 136–7 = *Arithmetica Universalis*, ₁1707: 328).

(51) Reproduced in Appendix 1 following.

(52) The subtitle continues: *Cui accessit Halleiana Æquationum Radices Arithmetice inveniendi methodus. In Usum Juventutis Academicæ.* Newton's name appears neither in the title nor the body of the work; even in Whiston's *Ad Lectorem* he is referred to by the circumlocution 'nostr[æ] Academi[æ] Professor Mathematicus tunc temporis [*sc.* triginta fere abhinc annis] Celeberrimus' and yet more evasively as 'Cl. Author'. As we have seen, however, the *Arithmetica*'s authorship was an open secret, even before publication. If the omission of Newton's name was contrived on purpose, Leibniz surely guessed the reason when he wrote on 15 March 1708 (N.S.) to Johann Bernoulli that '*Newtoni* nomen nuspiam [in libro] memoratur. Credo quod ipse ex sua dignitate non putavit profiteri se Authorem libri ante triginta annos conscripti, cum Prælectiones publicas *Cantabrigiæ* haberet, ex quibus est concinnatus; quanquam imperfectum ipse Editor profiteatur' (*Got. Gul. Leibnitii et Johan. Bernoullii Commercium Philosophicum et Mathematicum*, **2** (Lausanne/Geneva, 1745): 182 [= C. I. Gerhardt, *Leibnizens Mathematische Schriften*, **3** (Halle, 1855): 822]).

(53) In the 'regulæ post docendæ' referred to on page 52 of the deposited text, but never subsequently incorporated; see 2, §1: notes (170) and (726).

(54) On the most important of these see 2, §1: notes (71)/(517), (85)/(96)/(596) and (681)/(686).

ceived presentation and muddled notions of what a novice would find both comprehensible and useful[55] were irreparably built into its fabric. In his private library copy of the edition[56] Newton corrected many minor misprints, inserted more appropriate running heads ('Multiplicatio', 'Divisio', 'Extractio Radicum', 'De Forma Æquationis', 'Reductio Æquationum', 'Resolutio Quæstionum Arithmeticarum (Geometricarum)' and the like) and on the *Arithmetica*'s page 279 deleted an unwarranted half-title 'Æquationum Constructio linearis';[57] more radically, he mapped out a large-scale reordering of the sixty-one geometrical problems comprising its central portion,[58] seeking to grade them into a more logical sequence and in increasing levels of difficulty, while in the concluding section on the 'curvilinear' construction of equations he pared away all not directly needed flesh, reducing it to two skeletal conchoidal neuses, now denuded of their proof.[59] That last savage act of butchery apart, all these improvements were incorporated in the Latin revise—future parent (and rightfully so) of all subsequent editions—which he himself brought to publication in 1722, careful as ever not to let knowledge of his energetic participation circulate beyond a small privy group.[60]

For a book that afterwards came to be Newton's most often republished mathematical work, the *Arithmetica* had initially an exceedingly small impact.

(55) The beginning student is required, for instance, to be able to comprehend the *ad hoc* derivation of simple 'ultimate' ratios (see 2, §1: Newton's pages 133–5) and to manipulate Newton's complicated algorithm for delimiting the number of complex roots possessed by an equation (*ibid.*: pages 155–9) without being afforded any insight into its possible justification; subsequently, following an excessively long digression (*ibid.*: pages 187–203) on the splitting of polynomials of even degree into surd factors, he is warned off from practising such a technique 'cum tantarum reductionum perexiguus sit usus et rei possibilitatem potius quam praxin commodissimam voluerim exponere' (*ibid.*: page 203).

(56) See 2, §1: note (1). This copy was sold at the Thame Park auction of Newton's library in the 1920's (see R. de Villamil, *Newton the Man* (London, 1931): 6) and passed later into the possession of W. J. Greenstreet. It is accurately entered in the Musgrave list as 'Newtoni Arithmetica Universalis, 8vo. 1707' (*ibid.*: 88).

(57) Compare 2, §1: note (615).

(58) Pages 67–146 of the deposited copy (2, §1). A check-list of the renumberings is given in 2, §1: note (216).

(59) See 2, §1: note (615) for details of Newton's excisions.

(60) *Arithmetica Universalis; sive De Compositione et Resolutione Arithmetica Liber. Editio Secunda. In qua multa immutantur & emendantur, nonnulla adduntur* (London, 1722). Neither Whiston's previous *Ad Lectorem* nor Halley's appended 'Methodus Nova...' (see notes (47) and (50) above) are included in this standard text of the *Arithmetica*, published (much as before) 'Impensis Benj. & Sam. Tooke'. It appeared to Newton's first editor that 'that acute Mathematician Mr. [*John*] *Machin*, Professor of Astronomy at *Gresham* College...and one of the Secretaries of the Royal Society,...published this Work again, by the Author's later Desire or Permission; I lay no claim to it' (Whiston, *Memoirs* (note (32) above): 135); John Conduitt's memorandum (King's College, Cambridge. Keynes MS 130.5) again adds the clarification that 'Machin overlooked the press[,] for wᶜʰ [Sʳ I.] intended to have given him 100 Guineas

In England, where the cult of Newton's scientific *persona* was still just beginning, it made no mark at all and was not even graced by a review in the *Philosophical Transactions* (then admittedly at a low ebb and desperately short of contributors of quality). At Cambridge, as far as we can gather, Whiston made no attempt to introduce any algebraic topics into his Lucasian lectures for the 'mathematical plebs' but kept rigidly to physical and theoretical astronomy as his theme, while Roger Cotes (since January 1706 first occupant of the newly created Plumian Professorship of Astronomy) was likewise for the moment absorbed in problems of contemporary astronomy and physical science and soon, from mid-1709, to be equally heavily engaged in editing the second edition of Newton's *Principia* for publication. In Oxford their companion Savilian professors, Edmond Halley and (till 1708) David Gregory were engrossed in preparing the equally magnificent *princeps* edition of Apollonius' *Conics* finally brought out under the former's name in 1710. In London the widely capable Huguenot mathematician Abraham de Moivre was, in between heavy bouts of migraine, hard put to it to earn a living as an itinerant teacher,[61] and his fellow algebraist John Harris was similarly occupied in cramming elementary 'Mathematick & Mechanick Exercises' into anyone who would pay

but he made him wait 3 years for a preface & then did not write one...'. (Elsewhere (Keynes MS 130.6 (2)) Conduitt noted that 'Sʳ I. told me that Machin understood his *Principia* better than anybody[,] that Halley was the best Astronomer [but] Machin the best Geometer'.) Newton's active stage-managing of the 1722 revise can be documented in several ways: most notably, a stray autograph sentence on an otherwise clean folio page (now ULC. Add. 3960.7: 95) differs by only a single, trivial adverb from an inserted footnote clarifying the reference to the unimplemented 'Regulæ post docendæ' on page 52 of the deposited copy; see 2, §1: note (170). The ubiquitous presence of its author's editorial hand was not missed by W. J. 's Gravesande when he came to reissue the *Arithmetica* ten years later: 'Secunda vice [liber] in lucem prodiit Londini 1722; sed in statu perfectiore, ut quis facile percipiat non omnino fœtum abdicasse virum Celeberrimum; ordo propositionum non tantum mutatus est, sed in ipsis solutionibus & demonstrationibus correctiones multæ reperiuntur, non nisi ipsi Auctori tribuendæ' (*Arithmetica Universalis*;...*Auctore Is. Newton, Eq. Aur.* (Leyden, 1732): L[ectori] S[alutem]: *2ʳ). The less perceptive review of the revised edition which appeared in the *Acta Eruditorum* (February 1723): 75 is quoted in Appendix 2: note (29) following.

(61) On 2 December 1707 he wrote to Johann Bernoulli that 'Un mal de tête qui m'a tenu pendant 6 mois m'a empeché de répondre à [votre] derniere lettre...: J'attribuois mon mal de tête à l'application que j'avois apportée à quelques calculs assez longs; j'ai donc cessé de calculer pendant quelque temps, excepté que j'étois obligé de la faire légerement pour l'usage de mes écoliers; mais cela ne remédioit à rien.... Je suis obligé de travailler presque du matin au soir, c'est à dire d'enseigner mes écoliers et de marcher; mais comme la ville est fort grande, une partie considérable de mon temps est employée uniquement à marcher; c'est ce qui...m'ôte le loisir d'étudier' (K. Wollenschläger, 'Der mathematische Briefwechsel zwischen Johann I Bernoulli und Abraham de Moivre' [*Verhandlungen der Naturforschenden Gesellschaft in Basel*, **43**, 1933: 151–317]: 239–40). No mention of the *Arithmetica* occurs in de Moivre's correspondence with Bernoulli: the latter probably heard of its publication from Halley (who had sent him a presentation copy of his edition of Apollonius' *De Sectione Rationis* (Oxford, 1706) the previous spring, so beginning a brief interchange of letters).

him: in any case, a work 'In usum juventutis Academicæ' was too theoretical a beginning text for the young trade apprentices who would form its London audience, and the thirteen years' delay before it appeared in Raphson's more widely acceptable vernacular version[62] is further indication that the *Arithmetica* achieved little immediate popularity in England or indeed Scotland.

On the Continent a more interested, if not wholly favourable, reception awaited it. Already in November 1707 Johann Bernoulli could write to Leibniz that he looked forward 'avidly' to reading a book which, or so an English informant had told him, was 'small in bulk' but 'immensely superior' to French works of greater volume.[63] In turn, when Leibniz a few weeks later received a review copy of the *Arithmetica* from the editors of the *Acta Eruditorum*, he lost little time in conveying his own first impressions of the book both to Bernoulli and to Jakob Hermann, including an excerpt from its section 'On finding [rational] divisors' accompanied by a request that each should try to extend Newton's rule to higher dimensions.[64] Bernoulli in a typical *volte-face* chose abruptly to dismiss the technique as 'immediately' comprehensible, returning a 'script' by his young nephew Niklaus to show how readily it could be generalized;[65] but Hermann reacted in a more modest and helpful manner, smoothing out for Leibniz between April and August 1708 his several difficulties in comprehending Newton's involved finite-difference algorithm and his following technique, yet

---

(62) *Universal Arithmetick: or, a Treatise of Arithmetical Composition and Resolution. To which is added, Dr. Halley's Method of finding the Roots of Æquations Arithmetically. Translated from the Latin by the late Mr. Raphson, and revised and corrected by Mr. Cunn* (London, 1720). An extract from Cunn's preface is reproduced in Appendix 1: note (10) following.

(63) 'Ex Anglia [probably from Edmond Halley; see note (61)] interim intellexi, Newtonum recentissime in lucem dedisse Algebram aliquam sub titulo *Arithmeticæ Universalis*; libellum esse parvi voluminis, sed tanto majoris momenti, ut in illo a Gallis quidem superetur, in hoc vero eosdem immane quantum superet. Ejus libri copiam mihi promissam ut nanciscar, avide expecto' (C. I. Gerhardt, *Leibnizens Mathematische Schriften*, 3 (Halle, 1855): 821).

(64) Leibniz to Bernoulli, 15 March 1708 (N.S.): 'Liber Newtonianus sub titulo *Arithmeticæ Universalis* Cantabrigiæ hoc ipso anno editus, ex Anglia ad me missus est....Percurri festina lectione, & quædam deprehendi non spernenda, præsertim in Exemplis. Unum placuit, quod pro Te exscribi jussi, ut si vacat, consideres & examines: etsi enim non magnopere ad praxin facere videatur, videtur tamen speculationis causa dignum consideratu, præsertim si processus in infinitum haberetur, pro divisore gradus cujuscunque' (Gerhardt, *Mathematische Schriften*, 3 (1855): 822); and to Hermann, 16 December 1707/29 February 1708: 'Pervenit ad me ex Anglia nova Algebra ex veteribus Newtoni prælectionibus concinnata; sunt in ea non tantum utilia exempla, sed et præcepta quædam peringeniosa ad investigandos divisores, etsi enim praxi nonnihil sunt perplexa, ingenium tamen indicant.' / 'Mitto excerptum...de modo investigandi divisores unius aut duarum dimensionum. Optem res produci ad plures dimensiones' (*Mathematische Schriften*, 4 (1858): 322/325).

(65) The relevant passages from Bernoulli's response on 16 May 1708 (Gerhardt, *Mathematische Schriften*, 3: 824–5) are quoted in Appendix 2: note (5) following; see also 2, §1: note (530).

(66) See Appendix 2: notes (5) and (16) below.

more intricate, for deriving the 'surd' divisors of polynomials of even degree.[66] The substance of these explanatory remarks was subsequently incorporated by Leibniz in a long, painstaking recension of the *Arithmetica* which appeared, as usual without attribution of author, in the *Acta Eruditorum* the following November,[67] effectively establishing its excellence in foreign eyes. Let us accord due recognition to this disproof of the tradition that Leibniz showed an unrelenting antipathy to all things Newtonian in his old age.

We have already spoken of the Latin revise which Newton himself prepared for publication in 1722. It is difficult to assess whether this was published to meet a genuine demand or, as would seem more likely, merely to print a corrected text to oust Whiston's *editio princeps*. Newton's death five years later in March 1727 led more obviously to a new wave of interest in his mathematical works, published and unpublished, and not least in his algebraic lectures, now half a century old. A widely popular revision of Raphson's 1720 English translation, incorporating the improvements made by Newton in the Latin revise, was published at London in the summer of 1728.[68] In October of that year, in a further change of front, Johann Bernoulli began at his University of Basel a lengthy series of lectures on 'Elements of Algebra'[69] which were based narrowly

(67) *Acta Eruditorum* (November 1708): 519–26. Because of its exceptional interest we reproduce the text in Appendix 2.

(68) *Universal Arithmetick: or, A Treatise of Arithmetical Composition and Resolution. To which is added Dr. Halley's Method of finding the Roots of Equations Arithmetically.... The Second Edition, very much Corrected* (London, 1728); reprinted in photo-facsimile in *The Mathematical Works of Isaac Newton*, **2** (New York, 1967): 3–134. As Samuel Cunn observes in his 'Advertisement' on page [iv], 'This New Edition, in *English*, of Sir *Isaac Newton*'s *Algebra*, has been very carefully compared with the correct Edition of the Original, that was published in 1722; and suitable Emendations have been every where made accordingly. What was there wanting, is a general Method of finding, in Numbers, the Roots of Equations; and...is supplied by Dr. *Halley*'s Method, which is here annexed....' An inferior 'third' edition of Raphson's translation—essentially a typographically faulty repeat of the second, accompanied somewhat irrelevantly by *A Treatise upon the Measures of Ratios* by a certain James Maguire, A.M., and very inadequately by a commentary wherein *The whole [is] illustrated and explained, in a Series of Notes* by the equally obscure Rev. Theaker Wilder, 'Senior Fellow of Trinity College, Dublin', but now once more omitting Halley's 'New, Exact and Easy Method, of finding the Roots of any Equations Generally...'—appeared at London in 1769, primarily intended as a university teaching text.

(69) These lectures are preserved in a transcript (now in the Biblioteca Nazionale Braidense at Milan) made at the time by his student Albrecht von Haller; see H. M. Nobis, 'Über einige Haller-Handschriften, welche verlorene Vorlesungen des Johann (I) Bernoulli betreffen', *Gesnerus* **26**, 1969: 53–62, especially 56, 59–61 and 65–9. (Nobis fails to notice, however, that the main portion of the third manuscript volume is in large part an exact copy—made by Haller in lieu of purchasing the printed original?—of the text of Newton's *Arithmetica* and erroneously attributes its content to Bernoulli.) The 'In Isaaci Newton *Arithmeticam universalem* Prælecta' were delivered by Bernoulli from 1 December following: their expository rather than critical character may be seen in the facsimile of their opening page reproduced by Nobis (*ibid.*: 61).

on the book to which Whiston had once given that short title. Four years later, building on a short 'Specimen of a commentary' which he had circulated in 1727,[70] W. J. 's Gravesande published the standard Continental edition of the *Arithmetica*, again appending Halley's 'New Method' for deriving the roots of an equation and further adding 'all writings in the *Transactions* which seem usefully to illustrate Newton's book': namely, seven further articles on the algebraic theory and geometrical construction of equations by Halley, John Colson, de Moivre, Colin Maclaurin and George Campbell.[71] With an excess of overblown pedestrian commentary its text was republished by G. A. Lecchi at Milan in 1752[72] and nine years later by G. Castiglione at Amsterdam[73] and G. A. Décoré at Leyden,[74] and further reprinted by Samuel Horsley in the first volume of his *Newtoni Opera Omnia* in 1779.[75] From the middle of the eighteenth century onwards, however, the creative impact of the *Arithmetica* is more to be traced in the standard texts on algebra by Maclaurin,[76] Leonhard Euler[77] and Edward Waring,[78] which both absorbed its insights, adding their own polishes, and repaired its several deficiencies of content and presentation. Already by the

(70) *Specimen Commentarii in Arithmeticam Universalem Newtoni* (appended to his *Matheseos Universalis Elementa* (Leyden, 1727): an English translation appeared the next year and was reprinted in 1752).

(71) *Arithmetica Universalis*; *sive de Compositione et Resolutione Arithmetica Liber. Auctore Is. Newton, Eq. Aur.* (Leyden, 1732). In his 'L[ectori] S[alutem]' 's Gravesande observed ([∗2ᵛ]) that 'Cum Newtoni *Arithmetica* primum in lucem prodiret, adjecerat Editor, propter convenientiam materiæ, Hallei Methodum extrahendi radices Æquationum, desumtam ex *Transactionibus Philosophicis Societatis Regiæ Londinensis*; Nos hoc exemplum secuti, non tantum hoc[c]e Hallei scriptum, sed omnia adjecimus, quæ in dictis *Transactionibus* reperiuntur, & quæ nobis ad Newtoni librum illustrandum utilia visa sunt. Quæ Anglico sermone conscripta erant latinè reddidit vir Rev. Joh. Petr. Bernard...'. In sequel to the text of the *Arithmetica* there follow thereafter (on pages 245–344) reprints of Halley's 'Methodus Nova Accurata & facilis inveniendi Radices Æquationum...' (see note (50)) [245–57], Colson's 'Æquationum Cubicarum & Biquadraticarum, tum Analytica, tum Geometrica & Mechanica, Resolutio Universalis' (*Philosophical Transactions* **25** (1706–7), No. 309 [for January–March 1707]: 2253–68) [258–69], de Moivre's 'Æquationum quarundam Potestatis tertiæ, quintæ, septimæ, nonæ, & superiorum, ad infinitum usque pergendo, in terminis finitis, ad instar Regularum pro Cubicis quæ vocantur *Cardani*, Resolutio Analytica' (*ibid.*: 2368–71) [270–73], and Halley's Cartesian 'De Constructione Problematum Solidorum, sive Æquationum tertiæ vel quartæ Potestatis, unica data Parabola ac Circulo efficienda; dissertatiuncula' (*Philosophical Transactions* **16** (1686–92), No. 188 [for July/August 1687]: 335–43) [274–81] and his related tract 'De Numero Radicum in Æquationibus Solidis ac Biquadraticis, sive tertiæ ac quartæ potestatis, earumɋ limitibus, tractatulus' (*ibid.*, No. 190 [November 1687]: 387–402) [282–97]; followed by Bernard's Latin renderings of Maclaurin's 'Letter...to *Martin Folkes*... concerning Æquations with impossible Roots' (*Philosophical Transactions* **34** (1726–7), No. 394 [for May–July 1726]: 104–12) [298–305] and 'A second Letter...concerning the Roots of Equations, with the Demonstration of other Rules in Algebra' (*ibid.* **36** (1729–30), No. 408 [for March/April 1729]: 59–96) [305–32], and of Campbell's 'A Method for determining the Number of impossible Roots in adfected Æquations' (*ibid.*: **35** (1727–8), No. 404 [October 1728]: 515–31; compare ɪ: 525, note (40)) [333–44].

time Beaudeux' first French translation appeared in 1802[79] the *Arithmetica* was acquiring the veneration accorded to an increasingly outmoded *œuvre maîtresse*, and when Delambre came to analyse it in his *Histoire de l'astronomie* a quarter of a century afterwards[80] its working life had effectively come to an end. The recent commented Russian translation by A. P. Yuschkevich[81] serves (like the present volume) a purely historical purpose.

With regard to the remaining papers now reproduced we may be brief since, with one exception, the circumstances of their composition are undocumented and little illumination is to be derived from the internal evidence of their texts. What we may further conjecture is rapidly summarized.

In the first section is printed the little which now exists of the various, loosely connected notes and drafts which Newton subsumed into the polished version subsequently deposited. We have already discussed the opening autograph 'paper'[82] given by Newton to Flamsteed 'at one of his lectures in Midsummer 1674' and since preserved in the latter's papers: it is, as we observed, invaluable as the only independent document now known which casts light on the content of Newton's Lucasian lectures as they were originally delivered. The two

(72) *Arithmetica Universalis. Perpetuis Commentariis illustrata et aucta a J. A. Lecchi* (3 volumes).

(73) *Arithmetica Universalis.... Cum Commentario Johannis Castillionei* (2 volumes: 1761).

(74) *Arithmetica Universalis summi Newtoni, contracta, illustrata, et locupletata: Præeunte Logica Analitica, a Godef. Ant. Décoré* (Leyden, 1761).

(75) *Isaaci Newtoni Opera quæ exstant Omnia*, 1 (London, 1779): 1–229. On pages viii–xi of the preceding *Præfatio* Horsley lists the heads of nineteen chapters of an 'Appendix in *Arithmeticam Universalem*' which he had prepared in commentary but which—fortunately for his reader overtired with Newtonian excrescences—proved too bulky to annex. In the first 'Traditur Methodus inveniendi numeros primos per cribrum Eratosthenis', in the last 'Quæritur causa, unde Cardani Regulæ unquam deficiant', and the intervening chapters are scarcely less conservative in their summarised themes.

(76) *A Treatise of Algebra in Three Parts* (London, ₁1748). Whole chapters of this posthumous compilation by Patrick Murdoch are virtually word-for-word repeats of portions of Newton's earlier text, others narrow commentaries upon it. Four more editions appeared during the next forty years.

(77) *Vollständige Anleitung zur Algebra* (St Petersburg, ₁1770): the Russian original appeared two years earlier. It was quickly translated into Dutch (1773), French (1774, with additions by Lagrange), Latin (1790) and English (1797).

(78) *Meditationes Algebraicæ* (Cambridge, 1770). Waring summarizes Newton's algebraical innovations in his historical *Præfatio* (ibid.: i–viii, especially iv).

(79) N. Beaudeux, *Arithmétique de Newton, traduite avec des notes* (2 volumes: Paris, An X).

(80) J.-B. Delambre, *Histoire de l'astronomie au dix-huitième siècle* (Paris, 1827): 34–42. Delambre was essentially interested only in those of the central geometrical worked examples, notably Problems 16, 42 and 52 in the 1707 *editio princeps* [= 30, 55 and 56 in the 1722 renumbering; compare 2, §1: note (216)], which had an astronomical setting.

(81) *Isaak N'juton: Vseobščaja arifmetica ili kniga ob arifmetičeskikh sinteze i analize. Perevod, stat'ja i kommentarii A. P. Juškeviča* (Moscow/Leningrad, 1948).

(82) See page 5. The text itself is reproduced in 1, §1 below.

following drafts,[83] roughly datable by their handwriting to the late 1670's, are more finished pieces—respectively expounding techniques for delimiting the roots of an equation and deriving them accurately by factorization, and geometrically deducing the classical 'Heronian' formula which expresses the area of a plane triangle in terms of its sides—which yet differ significantly in important details from their revises in the deposited manuscript, and were no doubt retained by Newton for that reason.

Next, in sequel to the main text of the *Arithmetica* and its preliminary 'corrections', we reproduce a fragmentary 'synthetic resolution' by Newton 'of problems in my lectures'.[84] Here there are no new insights—indeed its demonstrations are at times both circular and incomplete[85]—but it is pleasantly concise and businesslike in its approach and includes at its end a problem, not foreshadowed in the main text of the Lucasian *lectiones*, comprehending an exceedingly elegant equant hypothesis regulating the 'mean' motion of a solar planet as it is carried round in a Keplerian elliptical orbit.[86]

The first of four short concluding appendixes is an unfinished exposition by Newton of his method for eliminating a given variable between two simultaneous equations, sketched out, it would appear, at about the time (1721–2) he was preparing his revision of Whiston's *editio princeps* for publication. Its schematic clarity is its only virtue. The second, dating from the same period, represents an extension of Problem 44 in the deposited text of the *Arithmetica* to comprehend, in Huygenian style,[87] the theory of a generally loaded string hanging loose under simple gravity. There is, however, nothing to indicate that Newton foresaw this application till some time after 1690–1 when Jakob and Johann Bernoulli, Leibniz and Huygens had each achieved their independent constructions of the catenary:[88] it may well be that David Gregory's unwise attempt during 1697–9 to support, in the face of Leibniz' hostile but accurate criticism, an irremediably defective deduction of its basic differential equation[89] was the stimulus which led Newton to augment his earlier deduction of the slope of a link in a hanging chain. The third appendix reproduces worksheets from the middle 1680's, in which Newton sought to apply the Wrennian construction of cometary orbits (supposed to be straight lines traversed at a uniform rate) to

---

(83) 1, §§2/3.

(84) 2, §3.

(85) Compare 2, §3: notes (10) and (17).

(86) See 2, §3: note (29).

(87) But wholly independently: Huygens' equivalent approach was not published in Newton's lifetime nor indeed till the present century (see Appendix 2: note (3) following).

(88) Published (without proofs) in the *Acta Eruditorum* (June 1691): 274–82; compare Appendix 2: notes (1) and (7).

(89) In his 'Catenaria' (*Philosophical Transactions*, **19**, No. 231 [August 1697]: 637–52): 637–8; compare Appendix 2: note (7).

suitable quadruplets of observations of the comet of 1680/1 both made by
himself at Cambridge and also passed on to him later by Flamsteed. Such trials
quickly convinced him of the method's inherent inability to compute accurately
the location even of small sections of orbit, and he thereafter employed the
more viable simplified hypothesis that a comet travels in a parabola round the
sun set at its focus.[90] In the present context it remains nevertheless interesting to
see that a problem obscurely incorporated in his algebraic *lectiones* held, in fact,
much more than a passing theoretical importance for him. The final appendix
repeats Newton's rough computations for an unidentified geometrical problem
in the late 1670's, one perhaps intended to be set as an example in his Lucasian
lectures of the period.

APPENDIX 1. WHISTON'S PREFACE TO HIS '*EDITIO PRINCEPS*' OF NEWTON'S '*ARITHMETICA*'.[1]

[29 April 1707]

AD LECTOREM.

Cum post haud paucos Doctorum Virorum[2] in Arte Analytica tradenda
labores Liber aliquis materia plenus, mole parvus, in regulis necessariis brevis,
in exemplis certo consilio electis longus, & primis Tyronum conatibus accom-

---

(90) See Appendix 3: notes (1) and (3). The many subtleties and ambiguities which
complicate our modern understanding of Newton's changing views on comets during 1680–6
are discussed at length by J. A. Ruffner, taking account of much unpublished manuscript
evidence, in 'The Background and Early Development of Newton's Theory of Comets'
(unprinted Ph.D. thesis, University of Indiana, 1966).

(1) *Arithmetica Universalis; sive De Compositione et Resolutione Arithmetica Liber. Cui accessit
Halleiana Æquationum Radices Arithmetice inveniendi methodus. In Usum Juventutis Academicæ*
(Cambridge, 1707): the following editorial 'Ad Lectorem' is found on signatures a3r–a4r. We
have argued in the preceding introduction that Newton intervened in the printing of the
Arithmetica when, in the summer of 1706, all but the last two sheets had been composed (and
printed off?), delaying further work on it for several months before finally authorizing its
publication with a few minor changes. A codicil to the reader regarding the inadequate
running heads in the main text ([a4v]: 'Lectori S[cholium?]. Notes velim Titulum Perpetuum
paginæ cuique, pro Typographorum consuetudine, hic appositum, in Distinctos, cujusque
paginæ argumentum indicaturos sequentibus editionibus esse mutandum: prout in prioribus
aliquot paginis alia de causa recusis jamjam fecimus') was, as we have remarked above
(Introduction: note (49)) almost certainly inserted at his insistence. Here, likewise, several of
the sentences under which Whiston has signed his initials have a distinctly Newtonian ring to
them, and their detail has a sharpness which originates in the memory of the 'Cl. Autor' of
the text they introduce rather than in vague editorial conjecture and interpretation. We
should accordingly treat them as having the stamp of Newton's approval.

(2) In particular John Kersey and John Wallis, who published their several thick volumes
on algebra at London in 1673 and 1685 respectively.

modatus etiamnum desiderari videretur; interque κειμήλια[3] nostra Academica hujusmodi Tractatus M.S. publicas Professoris Mathematici tunc temporis Celeberrimi[4] Prælectiones, triginta fere abhinc annis in Scholis habitas continens, mihi statim occurreret; Dedi Operam ut Libellus iste, imperfectus licet, & currente calamo pro officii urgentis[5] ratione compositus, nec prælo ullatenus destinatus,[6] tamen in usum studiosæ juventutis nunc in publicum prodiret. In quo quidem Quæstiones haud paucæ è variis Scientiis adductæ multiplicem Arithmeticæ Usum satis ostendunt. Animadvertendum tamen Constructiones illas sive Geometricas sive Mechanicas prope finem adpositas inveniendis solum duabus tribusve Radicum figuris prioribus, uti suo loco[7] dicitur, inservire: Opus enim Cl. Autor ad umbilicum nunquam perduxit; Cubicarum Æquationum Constructionem hic loci tradidisse contentus; dum interea in animo habuerit Biquadraticarum aliarumque superiorum potestatum Constructionem methodo generali exponendam adjicere, & qua ratione reliquæ Radicum Figuræ essent extrahendæ sigillatim docere.[8] Cum autem summo Viro hisce minutiis postmodo vacare minime placuerit, defectum hunc aliunde supplere volui; atque eum in finem generalem planeque egregiam Cl. Hallei Æqua-

---

(3) 'treasures', perhaps 'heirlooms'.

(4) A manifest exaggeration: till his *Principia* appeared in 1687 Newton was a relatively obscure Cambridge don, widely known only for the papers on optics published by him in the *Philosophical Transactions* during 1672–6 and respected as a mathematician only by a tight circle of London *cognoscenti* (notably Collins and Oldenburg) in the 1670's.

(5) Again a decided exaggeration, if one which Newton himself would wish to promote thirty years afterwards: in practice (compare III: xviii–xix) the Cambridge Lucàsian professor was only very lightly encumbered with formal duties of his office. Perhaps Whiston was thinking of Newton's remark on page 146 of the deposited manuscript (2, §1) that, in his set of worked geometrical 'Problemata,...et aliqua quæ inter scribendum occurrebant immiscui sine Algebra soluta...'.

(6) Even this may not be wholly true, however much Newton would have wished in retrospect to urge it in excuse; see note (21) of the preceding introduction.

(7) Page 211 of the deposited manuscript [= page 279 of the printed *editio princeps*].

(8) Compare 2, §1: notes (170), (614) and (726).

(9) *Philosophical Transactions*, **18**, No. 210 [for May 1694]: 136–48. Even if Newton in 1707 accepted Whiston's stop-gap, he subsequently changed his mind and eliminated Halley's 'Methodus Nova Accurata & facilis...' (reproduced on pages 327–43 of the *editio princeps*) in his own revise of the *Arithmetica* in 1722. See note (47) of the preceding introduction.

(10) That is, 'G[ulielmus] W[histon]'. In Samuel Cunn's corresponding preface 'To the Reader' which introduces Raphson's English translation of the *Arithmetica* (published as *Universal Arithmetick: Or, A Treatise of Arithmetical Composition and Resolution. To which is added, Dr. Halley's Method of finding the Roots of Æquations Arithmetically* at London in 1720, and reissued in 1728...*very much Corrected*) Whiston's 'Ad Lectorem' appears in slightly paraphrased form:

'The Book was originally writ for the private Use of the Gentlemen of *Cambridge*, and was deliver'd in Lectures, at the publick Schools, by the Author, then *Lucasian* Professor in that University. Thus, not being immediately intended for the Press, the Author had not prosecuted his Subject so far as might otherwise have been expected; nor indeed did he ever find Leisure to bring his Work to a Conclusion: So that it must be observ'd, that all the Constructions, both

tionum Radices extrahendi methodum ex Actis nostris Philosophicis,[9] exorata prius utrobique venia, huc transferendam judicavi. Vale Lector, & conatibus nostris fave.

Dabam Cantabrigiæ   G. W.[10]
   III. Kal. Mai. A.D. MDCCVII.

APPENDIX 2. LEIBNIZ' REVIEW OF THE '*ARITHMETICA*'.[1]

[*c.* October 1708]

Extracted from the *Acta Eruditorum* for November 1708[2]

Arithmetica Universalis, sive De Compositione & Resolutione Arithmeticæ[!] *Liber.* Cantabrigiæ & Londini, impensis Benj. Tooke, 1707, in 8. Constat Alph. 1.[3]

   Hæc Algebræ Elementa in gratiam Prælectionum publicarum triginta fere abhinc annis conscripsit celeberrimus Geometra, *Isaacus Newtonus*, cum adhuc

---

Geometrical and Mechanical, which occur towards the End of the Book, do only serve for finding the first two or three Figures of Roots; the Author having here only given us the Construction of Cubick Equations, though he had a Design to have added, a general Method of constructing Biquadratick, and other higher Powers, and to have particularly shewn in what Manner the other Figures of Roots were to be extracted. In this unfinish'd State it continued till the Year 1707, when Mr. *Whiston*, the Author's Successor in the *Lucasian* Chair, considering that it was but small in Bulk, and yet ample in Matter, not too much crowded with Rules and Precepts, and yet well furnished with choice Examples, (serving not only as *Praxes* on the Rules, but as Instances of the great Usefulness of the Art it self; and, in short, every Way qualified to conduct the young Student from his first setting out on this Study) thought it Pity so noble and useful a Work should be doom'd to a College-Confinement, and obtain'd Leave to make it Publick. And in order to supply what the Author had left undone, subjoyn'd the General and truly Noble Method of extracting the Roots of Æquations, publish'd by Dr. *Halley* in the *Philosophical Transactions*, having first procur'd both those Gentlemen's Leave for his so doing.'

   (1) Like other reviews in the *Acta Eruditorum*, the present recension of Whiston's *editio princeps* of the deposited text of Newton's algebraic lectures at Cambridge is unsigned, but its authorship—as following footnotes emphasize—is amply confirmed both by the style of its writing and by the repetition of the main emphases of its content in Leibniz' letters of the period to Johann I Bernoulli and Jakob Hermann. Not only is this the only serious public assessment of the *Arithmetica* to appear in Newton's lifetime, but it will serve to reveal how highly Leibniz estimated Newton's mathematical expertise in the classical fields of algebra and synthetic geometry, however much he was soon to decry it in infinitesimal calculus. The date of its composition is effectively bounded by that (29 August 1708) of Hermann's letter to him explaining Newton's 'surd' factorization of a quartic (see note (16)) and that of its publication in the *Acta* for the following November (which very probably appeared in that same month or perhaps early in December).

   (2) *Acta Eruditorum Anno MDCCVIII publicata...Lipsiæ* (*Mensis Novembris*): 519–26. Except for correcting a few misprints and dividing the original, continuously written text into paragraphs, we have kept narrowly to its orthography and typographical conventions.

   (3) 'Alph[abeto] l[atino].'

Professione Mathematica in Academia Cantabrigiensi fungeretur. Utut autem libellum, pro officii urgentis ratione conscriptum, prelo nullatenus destinaverit Autor; cum tamen in Arte Analytica adhuc desiderari animadverteret Cl. *Whistonus* libellum materia plenum, mole parvum, in regulis necessariis brevem, in exemplis certo consilio electis longum & primis tyronum conatibus accommodatum, illum cum publico communicari e re duxit. Nec immerito: reperies enim in hoc libello quædam singularia, quæ in vastis de Analysi voluminibus frustra quæsiveris. Non jam commemorabimus exempla selecta,[4] quibus ex Arithmetica, Geometria, Statica & Astronomia transumtis ipsiusque ingenio dignis Analyticam Artem illustrat, ad sublimiora tendentibus apprime profutura; sed quædam saltem proferre libet præcepta, quibus Analysin locupletavit.

Referimus huc regulam de inventione divisorum rationalium æquationum.[5] Sit e. gr. $x^3 - x^2 + 10x + 6 = 0$.[6] Substitui jubet sigillatim pro x terminos hujus progressionis Arithmeticæ 1. 0. -1. Numeros hinc ortos -4. 6. -14 cum omnibus ipsorum divisoribus e regione terminorum progressionis ita collocat.

1	4	1. 2. 4	$+4$
0	6	1. 2. 3. 6	$+3$
-1	14	1. 2. 7. 14	$+2$

Jam quoniam terminus primus x^3 nonnisi per unitatem divisibilis, ex divisoribus excerpit progressionem terminorum unitate differentium 4. 3. 2, eosque collocat e regione terminorum progressionis primum assumtæ. Quo facto, terminum ex ista progressione respondentem termino 0 in progressione prima pro valore ipsius x assumit, tentansque divisionem per $x + 3$ reperit $x^2 - 4x + 2$. Quodsi

---

(4) Pages 41–52/67–146 in Newton's manuscript (reproduced as 2, §1 below) = *Arithmetica*, ₁1707: 80–96/116–234.

(5) Leibniz' contemporary correspondence reveals that he had some difficulty in comprehending this Newtonian rule. Both to Johann Bernoulli on 15 March 1708 (*Got. Gul. Leibnitii et Johan. Bernoulli Commercium Philosophicum et Mathematicum*, **2** (Lausanne/Geneva, 1745): 182 [= ed. C. I. Gerhardt, *Leibnizens Mathematische Schriften*, **3** (Halle, 1855): 822]) and to Jakob Hermann on the previous 29 February (Gerhardt, *Leibnizens Mathematische Schriften*, **4** (Halle, 1858): 325) he sent an excerpt—perhaps little different from his summary in the present review—'de modo investigandi divisores unius aut duarum dimensionum', accompanied by a request that each should try to extend it 'ad plures dimensiones' and 'pro divisore gradus cujuscumque'. In a typical piece of attempted one-up-manship, Bernoulli responded on 16 May that 'non tantùm statim fundamentum detexi inventi Newtoniani pro dividendis æquationibus per alias simpliciores; sed etiam regulam quam habet pro invenienda æquatione duarum dimensionum, quæ propositam dividat, reddidi simpliciorem, intelligibiliorem, & ad praxim magis accommodatam', adding with regard to its generalization that 'facile vidi posse haberi processum in infinitum pro divisore gradus cujuscumque, si quis vellet incæptum attentè persequi. Nec mea me fefellit opinio; Agnatus enim meus [Niklaus I], juvenis vix 20 annorum, cui id curæ commiseram, mox omnia felicissime executus est...; Imo non tantum processus ostenditur in infinitum, sed ipsa traditur Regula generalis pro invenienda æquatione dividente

æquatio istiusmodi divisorem simplicem non habeat, sitque plurium quam trium dimensionum, e. gr. $x^4 - x^3 - 5x^2 + 12x - 6 = 0$,[7] hanc regulam pro inveniendo divisore duarum dimensionum, si talem habeat, commendat. Substituit in æquatione proposita ut ante pro x quatuor vel plures terminos progressionis Arithmeticæ 3. 2. 1. 0. −1. −2. −3, & resultantes numeros 39. 6. 1. −6. −21. −26 cum eorum divisoribus e regione disponit, hisque quadrata terminorum progressionis subjungit.

3	39	1. 3. 13. 39	9	−30. −4. 6. 8. 10. 12. 22. 48	−4. 6.
2	6	1. 2. 3. 6	4	−2. 1. 2. 3. 5. 6. 7. 10	−2. 3.
1	1	1.	1	0. 2.	0. 0.
0	6	1. 2. 3. 6	0	−6. −3. −2. −1. 1. 2. 3. 6	2. −3.
−1	21	1. 3. 7. 21	1	−20. −6. −2. 0. 2. 4. 8. 22	4. −6.
−2	26	1. 2. 13. 26	4	−22. −9. 2. 3. 5. 6. 17. 30	6. −9.

Ab his quadratis primum subducit, postea iisdem addit singulos divisores, & tam differentias, quam aggregata e regione ipsorum collocat. Quodsi terminus primus fuerit divisibilis per aliquem numerum, e.gr. per 3, si habeatur $3x^4$, pro quadratis assumi jubet facta ex quadratis in divisorem numeralem termini primi 3. Ex summis atque differentiis inventis excerpit progressiones Arithmeticas, quotquot excerpere licet, atque e regione terminorum progressionis assumtæ terminos ejus singulos disponit. Si his factis fiat terminus primus[8] istiusmodi progressionis $\mp C$, differentia, quæ habetur per subtractionem ipsius $\mp C$ ex termino proxime superiori, qui stat e regione termini 1 progressionis primæ[,] $\mp B$, termini primi æquationis divisor numeralis A, & l litera incognitam quantitatem in eadem designans, divisorem tentandum fore ait

---

cujuscunque gradus. Forte non indignum judicabis hoc Schediasma, quod *Actis* vestris *Beroliniensibus* inseratur...' (*Commercium*, **2**: 185–6 [= *Mathematische Schriften*, **3**: 824]). (Leibniz evidently did not share this high opinion of a fairly straightforward extension of Newton's procedure, and Niklaus' *Regula Generalis*—of which two widely variant versions are known (see 2, §1: note (530))—appeared in neither the *Acta Eruditorum* nor the *Miscellanea Berolinensia* of the period.) Hermann was more appreciative, replying first on 19 April that 'Newtoni Regulam...nondum quantum satis est examinare vacavit, interim tamen licet prolixa nonnihil sit elegans admodum mihi videtur, adeo ut omnino operæ pretium mihi videatur in ejusdem demonstrationem inquirendi' (*Mathematische Schriften*, **4**: 327) and then, after Leibniz' repeated insistence on 11 May that he should examine it, communicating a detailed exposition of Newton's method of divisors on 12 July following (see *ibid.*: 329–32); in generalization of the technique, he then added, 'ex hisce principiis jam liquere arbitror, quo pacto divisores altiorum graduum indagari debeant, atque inquisitionis laborem longum admodum futurum esse....Cl. Abbas Fardelli...mihi innotuit, Dn. Nicolaum Bernoullium Newtonianas regulas pro inventione divisorum etiam demonstrasse. Cum igitur ejus ratiocinia nondum viderim, gratum mihi esset sciendi num similibus cum meis fundamentis superstructa sit ejus disquisitio.'

(6) (2, §1: Newton's page 174 =) *Arithmetica*: 44.
(7) (2, §1: Newton's pages 177–8 =) *Arithmetica*: 45–6.
(8) Understand that in line with 0 in the first column.

$All \pm Bl \pm C$. E.gr. cum in proposito exemplo termini progressionum secundarum termino 0 in progressione prima respondentes sint 2 & -3, erit $+C = 2$ & $-C = -3$, reperietur porro $-B = -2$ & $+B = 2$. Cumque sit $A = 1$ & $l = x$, divisores tentandi habentur $xx + 2x - 2 = 0$ & $xx - 3x + 3 = 0$, ambo quæsito satisfacientes. Hæ regulæ non sunt spernendæ, & licet praxis ob multitudinem combinationum tentandarum sit sæpe difficilis, præsertim in altioribus, utile tamen foret, similes regulas pro altioribus divisoribus dari, vel potius unam pro omnibus, quod etiam præstari posse videtur.

Tradit quoque Autor regulam extrahendi radices quantitatum, ex integris ac surdis compositarum, quanquam de ea re etiam regula apud *Schotenium* in additis ad Geometriam *Cartesii* extet.[9] Sit nempe A pars quantitatis major, B minor; fore ait[10] $\frac{1}{2}A + \frac{1}{2}\sqrt{(AA - BB)}$ quadratum majoris partis radicis & $\frac{1}{2}A - \frac{1}{2}\sqrt{(AA - BB)}$ quadratum partis minoris. Ita reperietur $\sqrt{(3 + \sqrt{8})} = 1 + \sqrt{2}$. Addit[11] regulam aliam altiores radices ex quantitatibus numeralibus duarum potentia commensurabilium partium extrahendi, sequentis tenoris: Sit quantitas $A \pm B$. Ejus pars major A. Index radicis extrahendæ c. Quære minimum numerum N, cujus potestas N^c dividitur per $AA - BB$ sine residuo, & quotus sit Q. Computa $\sqrt[c]{(A + B, \sqrt{Q})}$ in numeris integris proximis. Sit illud r. Divide $A\sqrt{Q}$ per maximum divisorem rationalem. Sit quotus s, sitque $(r + N:r):2s = t$ in numeris integris proximus. Et erit $(ts \pm \sqrt{(ttss - N)}):2\sqrt[c]{Q}$ radix quæsita, si modo radix extrahi potest. E.gr. sit radix cubica extrahenda ex $\sqrt{968} + 25$, erit

---

(9) See 2, §1: note (602). Newton's rule in the *Arithmetica* is, of course, a refinement of the Cartesian one expounded by Schooten in his *Additamentum* to Descartes' Latin *Geometria* ($_1$1649: 329–36 = $_2$1659: 394–400).

(10) (2, §1: Newton's page 207 =) *Arithmetica*: 58. On this classical Euclidean rule (*Elements*, x, 90–6) see II: 313, note (37).

(11) (2, §1: Newton's page 209 =) *Arithmetica*: 59–60.

(12) (2, §1: Newton's pages 147–9 =) *Arithmetica*: 235–7.

(13) The Cartesian rule of signs stated by Newton on pages 149–50 of his deposited manuscript (2, §1) [= *Arithmetica*: 237]. Leibniz' misattribution of it to Harriot derives seemingly from his misreading of John Wallis' extended description of Harriot's researches in algebra (as edited by Warner and Aylesbury in the posthumously assembled *Artis Analyticæ Praxis*, London, 1631) in his *Treatise of Algebra, Both Historical and Practical* (London, 1685): Chapters XXX–LIV: 125–207. Wallis, correctly, does not assert that the Cartesian rule is to be found stated in Harriot's *Praxis*, but does not scruple to add his impression—of very dubious validity— that '... (upon a survey of the several forms [of equations for delimiting roots expounded in the *Praxis*],) it will be found, that... as many times as in the order of Signs + −, you pass from + to −, and contrariwise; so many are the Affirmative Roots: But as many times as + follows +, or − follows −; so many are the Negative Roots:...' (*Algebra*: 158). In his review of Wallis' book (*Acta Eruditorum* (June 1686): 283–9) Leibniz converted this to a firm statement that Harriot himself had enunciated the rule of signs, observing its truth 'ex inductione, ut videtur' (*ibid*.: 285)—a belief in which Baillet's following strictures (in his *La Vie de Monsieur Des-Cartes*, Horthemels, 1691) only served to confirm him; in his subsequent review of the augmented Latin edition of Wallis' *Algebra* [= *Opera Mathematica*, 2 (Oxford, 1693): 1–482] he reiterated his support of the Robervallian charge (Wallis, *Algebra*, $_1$1685: 198 = $_2$1693:

$AA - BB = 343$, $N = 7$, $Q = 1$, $A + B$, $\sqrt{Q} = \sqrt{968} + 25$, $r = 4$, $A\sqrt{Q} = \sqrt{968} = 22\sqrt{2}$, $s = \sqrt{2}$, $(r + N : r) : 2s = 5 : 2\sqrt{2} = 2$ in numeris integris proximis, $t = 2$, $ts = 2\sqrt{2}$, $\sqrt{(ttss - N)} = 1$ & $\sqrt[2c]{Q} = 1$. Tentetur itaque per multiplicationem, num ipsius $2\sqrt{2} + 1$ cubus sit $\sqrt{968} + 25$, remq ita se habere deprehendes.

Cum de numero radicum quas quælibet æquatio admittit, disserit,[12] eleganter post alios rationem reddidit, cur una æquatio plures radices habere queat, quia scilicet plures ejusdem problematis dantur solutiones, quas omnes eadem æquatio æque repræsentat. Monet vero æquationum radices sæpius impossibiles esse debere, ne casus problematum, qui impossibiles sunt, tanquam possibiles exhibeant. Regulæ Hariotti[13] de numero radicum verarum & falsarum in qualibet æquatione a nemine hactenus demonstratæ (& cujus demonstrationem ut nobis dedisset Autor, inprimis optandum fuisset) annectit[14] aliam licet paulo restrictiorem, sed demonstratione itidem destitutam, pro cognoscendo numero radicum impossibilium. Constitue, inquit, seriem fractionum, quarum denominatores sunt numeri in hac progressione 1. 2. 3. 4. 5 &c. pergendo ad numerum usque, qui est dimensionum æquationis; numeratores vero eadem series numerorum in ordine contrario. Divide unamquamque

---

204–5 [+205–20]) that in his *Geometrie* Descartes had heavily plagiarized Harriot (see *Acta Eruditorum* (June 1696): 256–9, especially 256–7). Just a few months before the appearance of Whiston's *editio princeps* of the *Arithmetica* he wrote in the same vein to Jakob Hermann on 18 January 1707: 'Nescio an Tibi aliquando significaverim, quantopere optarem ab aliquo demonstrari Regulam ab Harrioto olim inventam (unde videtur descripsisse Cartesius), quod signorum mutationes in æquationibus non nisi radices reales habentibus sint tot, quot radices veræ, et signorum consecutiones tot, quot radices falsæ. Harriotus eam inductione veram comperit. Cartesius rationem ejus nullam assignavit, nec quisquam post ipsum. Is non mediocris est analyticæ scientiæ defectus. Si hæc demonstrari posset propositio: *æquatione multiplicata per veram (falsam) radicem, unitate augeri numerum mutationum (consecutionum) in signis qui prius erat,* etiam propositum theorema demonstratum foret' (C. I. Gerhardt, *Leibnizens Mathematische Schriften,* 4: 308). Hermann answered two months later on 19 March, tardily but with some hope: 'Quantum ad theorema Harrioti circa radices veras ex mutatione signorum in æquationibus dignoscendas, quodque Cartesius in sua Geometria quoque usurparat nulla addita demonstratione, miror utique a nullo adhuc demonstratum esse. Verumque est theorema demonstratum fore si probari potest, quod *multiplicata æquatione per veram aut falsam radicem, unitate augeatur numerus mutationum aut consecutionum signorum qui prius erat, si nullæ adsint radices imaginariæ.* . . . Cogitavi nonnihil de hoc theoremate et aliquid observavi, quo ut spero probari posset veritas propositionis, sed tempus mihi nondum fuit' (*ibid.*: 310). Writing again on 24 June, however, he could only weakly affirm that 'Præstabit. . . aliquam ejus demonstrationem haberi, quam nullam, qua non sine magno Scientiæ defectu hactenus caremus' (*ibid.*: 317). In hindsight, we will readily see that Hermann's attempted simplification of Leibniz' rider by restricting it to equations without complex roots is all but irrelevant, but it is surprising to find that Leibniz himself could not obtain its entirely straightforward proof. (The inductive proof which results therefrom had already been suggested by Jean Prestet in his *Elemens de Mathematiques* (Paris, ₁1675): 368 and was subsequently improved by De Gua and Kästner in the 1740's; see G. Eneström's notes in *Bibliotheca Mathematica* (3) **7**, 1906–7: 300, 307.)

(14) (2, §1: Newton's pages 155–6 =) *Arithmetica*: 242.

fractionem posteriorem per priorem. Fractiones prodeuntes colloca super terminis mediis æquationis. Et sub quolibet mediorum terminorum, si quadratum ejus ductum in fractionem capiti imminentem sit majus quam rectangulum terminorum utrinꝗ consistentium, colloca signum $+$; sin minus, signum $-$. Sub primo vero & ultimo termino colloca signum $+$. Et tot erunt radices impossibiles, quot sunt in subscriptorum signorum serie mutationes de $+$ in $-$ & $-$ in $+$. E.gr. si fuerit $x^3 + pxx + 3ppx - q = 0$, habebitur ex præscripto regulæ

$$\overset{\frac{1}{3}}{x^3} + px x + \overset{\frac{1}{3}}{3pp}x - q = 0.$$
$$\;\;\; + \quad - \quad\; + \quad\; +$$

Quare cum signorum subscriptorum $+\; -\; +\; +$ duæ deprehendantur mutationes, altera de $+$ in $-$, altera de $-$ in $+$, concludendum erit, duas esse radices impossibiles. Sed fatetur Autor,[15] non omnes radices impossibiles hac methodo dignosci. Itaque optandum fuisset, ut limites, ad quas porrigitur hæc methodus, nobis assignasset: quod ipsi difficile non fuisset.

Cæterum ubi methodum communem reducendi æquationes per divisores rationales tradidit; ipsi annectit[16] aliam easdem (si fieri potest) per surdos reducendi. Sit æquatio biquadratica $x^4 + px^3 + qxx + rx + s = 0$. Fiat ex ipsius mente $q - \frac{1}{4}pp = a, r - \frac{1}{2}a = b, s - \frac{1}{4}aa = c$. Ponatur n communis aliquis terminorum b & $2c$ divisor integer, non quadratus, isque impar, & qui per 4 divisus unitatem relinquit, si terminorum q & r alteruter impar fuerit. Ponatur porro pro R divisor aliquis quantitatis $b:n$, si p sit par; vel imparis divisoris dimidium, si p sit impar; vel nihil, si dividuum b sit nihil. Auferatur quotus de $\frac{1}{2}pk$, & reliqui dimidium dicatur l. Fiat $Q = (a + nk^2):2$, & tentetur, num n dividat $Q^2 - s$, & quoti radix sit rationalis atque æqualis l. Quod ubi contigerit, ad utramque æquationis partem addi debere $nkkxx + 2nklx + nll$, extractam utrobique radicem fore $xx + \frac{1}{2}px + Q = n^{\frac{1}{2}}(kx + l)$. Hanc methodum etiam ad æquationes literales

(15) See (2, §1: Newton's page 159 $=$) *Arithmetica*: 245: 'Possunt plures [radices impossibiles] esse [quam per regulam allatam deteguntur], licet id perrarò eveniat.' In sequel (compare 2, §1: note (483)) Newton gives the counter-instance $x^3 - 3a^2x - 3a^3 = 0$.

(16) (2, §1: Newton's pages 187–203 $=$) *Arithmetica*: 257–72. Leibniz was greatly troubled by this following rule of Newton's for finding surd divisors and once more (compare note (13)) had recourse to Jakob Hermann for enlightenment, sending him, about mid-July 1708, a (now lost?) 'Schediasma Newtonianæ methodi divisores irrationales inveniendi' for perusal and comment. With an apology for 'aliquot jam elapsæ septimanæ ex quo...Tuæ litteræ cum adjuncto Schediasmate...mihi redditæ sunt', Hermann replied on 29 August, enclosing a careful explanation of Newton's 'methodus...satis egregia' for determining surd factors of given quartic and sextic equations (when this is possible) and adding the observation that 'Simili modo procedendum est in æquationibus altiorum graduum sed parium, ubi tamen conditionum numerus crescente in immensum calculo augetur, ut fere hujusmodi artificia in praxi vix adhiberi possint' (Gerhardt, *Leibnizens Mathematische Schriften*, 4: 332–4, especially 333–4).

extendi, exemplo ostendit: nec minus alio monstrat, quod, si pro n sumatur 1, eadem ad extractionem quoque radicum rationalium applicari valeat. Similiter si æquatio fuerit sex dimensionum, $x^6 + px^5 + qx^4 + rx^3 + sx^2 + tx + v = 0$, fieri jubet $q - \frac{1}{4}pp = a$, $r - \frac{1}{2}pa = b$, $s - \frac{1}{2}pb = c$, $c - \frac{1}{4}aa = d$, $t - \frac{1}{2}ab = e$, $v - \frac{1}{4}bb = f$, $fd - \frac{1}{4}ee = g$. Hinc sumi debere pro n communem aliquem terminorum $2d$, e & $2f$ divisorem integrum, non quadratum, nec per numerum quadratum divisibilem, qui per 4 divisus unitatem relinquat, si modo terminorum p, r, t aliquid sit impar; pro k vero divisorem integrum quantitatis $g : 2nn$,[17] si p sit par; vel divisoris imparis dimidium, si p sit impar; vel nihil si $g = 0$;[17] pro Q quantitatem $\frac{1}{2}a + \frac{1}{2}nkk$; pro l divisorem aliquem quantitatis $(Qr - QQp - t) : n$, si Q sit integer[,] vel divisoris imparis dimidium, si Q sit fractus denominatorem habens numerum 2; vel nihil, si dividuum istud $(Qr - QQp - t) : n = 0$; & denique pro R quantitatem $\frac{1}{2}r - \frac{1}{2}Qp + \frac{1}{2}nkl$. Hinc tentandum esse, num $R^2 - v$ dividi possit per n, & quoti radix extrahi, & præterea, num radix ista æqualis fuerit tam quantitati $(QR - \frac{1}{2}t) : nl$, quam quantitati $(QQ + pR - nll - s) : 2nk$. His enim positis, si radix illa dicatur m, pro æquatione proposita scribi posse hanc

$$x^3 + \tfrac{1}{2}pxx + Qx + r = \pm \sqrt{(n, \ kxx + \sqrt{lx} + m)}. \text{[18]}$$

Simili ratione procedit in æquationibus octo, decem, duodecim &c. dimensionum. Et licet non exponat regulam generalem aut processum in infinitum, fortasse tamen ex dictis erui posset. Addit alia ad hoc argumentum spectantia.[19] Quoniam tamen ipse p. 272[20] fatetur, tantarum reductionum perexiguum esse usum, & se impossibilitatem[!] potius quam praxin commodissimam exponere voluisse; pluribus quoque specialius enarrandis supersedemus. Cur vero demonstrationes plerumque omiserit, rationem hanc reddit,[21] quod vel satis faciles ipsi fuerint visæ, vel nonnunquam absque nimiis ambagibus tradi non potuissent.

---

(17) The printed text reads in error '9: 2*nn*' and '9 = 0' at these respective places. (Leibniz' handwritten 'g' is virtually indistinguishable from his '9'.)

(18) Read '$x^3 + \frac{1}{2}pxx + Qx + R = \pm \sqrt{n}, \ (kxx + lx + m)$'! The manuscript (Newton's page 195) —copied blindly by Whiston in his 1707 *editio princeps* (*Arithmetica*: 265)—has, by a slip of its author's pen, '$x^3 + \frac{1}{2}pxx + Qx + r = \pm \sqrt{n} \times \overline{kxx + lx + m}$' at the corresponding point; compare 2, §1: note (555).

(19) (2, §1: Newton's page 203 =) *Arithmetica*: 272: 'Adjungere jam liceret reductiones æquationum per extractionem surdæ radicis cubicæ, sed...has, ut quæ perrarò utiles sint, brevitatis gratia prætereo.' In his preceding letter of explanation to Leibniz on 29 August 1708 (see note (16)) Jakob Hermann had hazarded the unconsidered opinion that 'Similiter [cum reductione per surdas quadraticas] tentari posset reductio æquationum per extractiones Radicis Cubicæ, sed hos casus nondum examinare vacavit' (C. I. Gerhardt, *Leibnizens Mathematische Schriften*, 4: 334); see, however, 2, §1: note (578).

(20) (2, §1: Newton's page 203 =) *Arithmetica*: 272, lines 8–10.

(21) (2, §1: Newton's page 211 =) *Arithmetica*: 279.

Ubi ad æquationum constructionem linearem progreditur,[22] Curvarum in genera juxta dimensiones æquationum in lineis contemplandis & earum proprietatibus eruendis laudat; sed carpit legem *Cartesii* & eorum, qui eum sequuntur,[23] qua sanxerunt, non licere problema per lineam superioris gradus construere, quod construi potest per lineam inferioris. Neque enim æquationem, sed descriptionem esse, quæ Curvam aptam efficiat, nec æquationis simplicitatem, sed descriptionis facilitatem esse, ob quam in constructione problematum Curva una alteri præferatur. Æquationes esse expressiones computi Arithmetici, nec in Geometria locum proprie habere. Multiplicationes, divisiones & ejusmodi computationes in Geometriam recens introductas esse, idque inconsulto, & contra primum institutum hujus scientiæ. Geometriam enim excogitatam esse, ut expedito linearum ductu computandi tædium effugeremus. Laudat veterum in Arithmetica & Geometria distinguenda industriam: taxat recentiorum in utraque confundenda inadvertentiam, hosque amisisse ait simplicitatem, in qua Geometriæ elegantia omnis consistat. Atque hæc causa est, quod ipse cum *Archimede* aliisque veteribus problematum solidorum constructionem non per sectiones Conicas, sed per Conchoidem edoceat,[24] subjungens tamen nonnulla[25] de constructione problematum solidorum per Ellipsin, cum ea ob facilitatem descriptionis eodem jure, quo circulus Parabolæ præferri debeat, utut simpliciorem æquationem amanti. Monet[26] hac occasione, compositionis leges non dandas esse ex Analysi, quæ quidem ad compositionem conducat, a qua tamen omnis compositio liberanda sit, antequam vera evadat, utpote a mixtura speculationum Analyticarum abhorrens. *Leibnitianam* a

(22) (2, §1: Newton's pages 211–15 =) *Arithmetica*: 279–83. Some such word as 'distinctionem' needs to be understood immediately in sequel.

(23) Newton himself speaks only of 'Recentiores', though it is, of course, obvious whom he has principally in mind; compare 2, §1: note (620).

(24) (2, §1: Newton's pages 215–38 =) *Arithmetica*: 283–311. In a wholly unprecedented burst of enthusiasm Newton there refers to 'magnus ille Archimedes', dubbing him 'princ[eps] Mathematicorum': none of his own contemporaries singly ever rated more than an occasional 'acutissimus', though at one point in his *Philosophiæ Naturalis Principia Mathematica* (London, ₁1687: 20: *Axiomata sive Leges Motus*, Scholium) he did link Christopher Wren, John Wallis and Christiaan Huygens as 'hujus ætatis Geometrarum facile Principes'.

(25) (2, §1: Newton's pages 238–51 =) *Arithmetica*: 311–26.

(26) (2, §1: Newton's page 242 =) *Arithmetica*: 315.

(27) In his previously published papers in the *Acta Eruditorum* (conveniently listed by Émile Ravier in his *Bibliographie des Œuvres de Leibniz* (Paris, 1937): 'Les articles dans les journaux savants, 1675–1716': 47–87, especially 49–80) Leibniz made only passing reference to the merits of classical geometrical analysis as against those of its more modern algebraic equivalent, but he had already clearly stated his viewpoint—one, as he now says, closely akin to that of Newton in the *Arithmetica*—in a letter early in 1676 to Antonio Magliabecci 'de usu analyseos Veterum linearis et imperfectione analyseos per algebram hodiernæ' (C. I. Gerhardt, *Leibnizens Mathematische Schriften*, 7 (Halle, 1863): 301–16, especially 312): 'Velim autem in his problematibus…animadverti usum analyseos linearis Veterum, qui sane tantus est, ut si quis ea neglecta sine discrimine solam recentiorum algebram adhibeat, in calculos ingentes se

Newtoniana hac in re non discrepantem sententiam aliquoties attigimus,[27] quæ methodum veterum conservandam judicat, & primaria theoremata problemataque a recentioribus per novas artes inventa secundum eam a viris doctis *Archimedeo* exemplo demonstrari vellet, etsi ipsi pariter ac *Newtono* hoc agere non vacaverit. In constructionibus etiam elegantiam & simplicitatem inprimis spectandam censet, quoniam in ea re magis exercitium ingenii & promotio artis inveniendi, quam praxis quæritur. Nam fatetur, cum ad praxin perveniendum est, Geometriaque non in charta, sed campo majoribusque operibus exercenda, ad computum utilius deveniri, & quæsitum innumeris vero proximis, exhiberi debere: quod novis artibus plerumque jam præstari potest; ut adeo Geometria practica hodie parum a perfectione absit, etsi scientia ipsa (ipsius judicio) infinitis adhuc modis locupletari queat. Quod etsi fortasse non adeo sit necessarium ad Mathesin practicam, mirifice tamen proderit Geometriæ ad Physicam magis magisque applicandæ, quod ipsum aliter in praxes vitæ humanæ utiles refundi censet: eaque in re campus immensus industriæ & felicitati posterorum patet.

Sub finem tandem subjungitur ex Transactionibus Anglicanis methodus *Hallejana* generalis radices æquationum per approximationem sine ulla prævia reductione inveniendi,[28] cui specimen Cl. *Lagny*, Mathematici Parisini, ab ipso Dn. H*allejo* laudatam, occasionem dederat.[29]

---

induere possit...; plerumque etiam in constructiones contortas et minime naturales, quas evitabit si cum Veteribus subinde usum Datorum adhibebit et cum analysi recentiorum opportune miscebit. Sciendum enim est...duplicem esse analysin, unam qua problema unumquodque resolvitur per se, et incognitæ habitudo ad cognitas investigatur, alteram qua problema propositum reducitur ad aliud problema facilius, quod fit usu Datorum, quando ostenditur uno dato haberi et aliud. Et prior quidem Methodus Algebraica est, quam a Vieta et Cartesio maxime celebratam, recentiores hodie solam analysin esse putant, cum tamen altera methodus et Veteribus usitata fuerit...et suas quoque certas et constantes regulas habeat, et difficultatem magis dividat in partes, atque ideo soleat feliciores exhibere solutiones magisque naturales, et intellectum non per symbola sed ipsas rerum ideas ducat. Unde æstimandum... quam longe adhuc absit analysis, quæ hodie passim in usu est, a perfectione vulgo jactata.'

(28) *Arithmetica*: 327–43: 'Methodus Nova Accurata & facilis inveniendi Radices Æquationum quarumcunque generaliter, sine prævia Reductione. Per *Edm. Halley*, Geom. Prof. Savil. (*Edita in Actis Philosoph. Nº· 210. A.D. 1694.*).' See note (50) of the preceding introduction.

(29) Of Newton's own revised edition of the *Arithmetica* in 1722 the *Acta* carried the briefest of notices, commenting only that 'Posteriorem [editionem] cum [priore] conferentes non aliam immutationem factam esse deprehendimus, nisi quod ordo problematum fuerit in tyronum gratiam immutatus, ut difficiliora ultimo loco reserventur & regulæ diverso charactere exprimantur. Methodus vero extrahendi radicem ex æquationibus algebraicis *Hallejana*, quæ priori editioni subjungebatur, & præfatio, quam præmiserat *Whistonus*, omittuntur' (*Acta Eruditorum* (February 1723): 75). No mention is there made of pages 315/16 being a cancel (see 2, §1: note (686))—doubtless through an editorial oversight, though it is just possible that the review copy sent to the *Acta* had only the original, faulty leaf bound into it.

1

LECTURE NOTES AND PRELIMINARY DRAFTS FOR THE 'ARITHMETICA'

[Mid/late 1670's]

§1. NOTES FOR A LECTURE ON ALGEBRA GIVEN BY NEWTON IN 'MIDSUMMER 1674'.[1]

From the autograph[2] in the Royal Greenwich Observatory, Herstmonceux Castle

I. $\textcircled{a}$ $\dfrac{ax}{a-x}+b=x$ per reductionem fit $ax+ab-bx=ax-xx$ seu $xx=bx-ab$.

$\textcircled{\beta}$ $\dfrac{a^3-abb}{2cy-cc}=y-c$ fit $\dfrac{a^3-abb}{c}=[2]yy^{(3)}-3cy+cc$ seu $\dfrac{a^3-abb-c^3}{c}+3cy=2yy$.

$\textcircled{\gamma}$ $\dfrac{aa}{x}-a=x$ fit $aa-ax=xx$. $\textcircled{\delta}$ $\dfrac{aabb}{cxx}=\dfrac{xx}{a+b-x}$ fit $\dfrac{a^3bb+aab^3-aabbx}{c}=x^4$.

---

(1) As attested by John Flamsteed, who, having no doubt personally attended Newton's lecture in the Schools at Cambridge, acquired the slip of paper bearing these notes and subsequently recorded at its bottom that it was 'Mr Newtons paper given at one of his lectures Midsummer 1674'. Flamsteed had earlier 'visited Dr Barrow and Mr Newton' at Cambridge in December 1670 during a short stay, *en route* from London to Derby, whose main purpose was to be enrolled as a pensioner at Jesus College (see Francis Baily, *An Account of the Revd John Flamsteed* (London, 1835): 29); this was clearly only a formal social call. Four years later he was resident there for six weeks, between 29 May and 13 July 1674, and on 5 June received the degree of M.A. conferred upon him by royal mandate (Joseph Edleston, *Correspondence of Sir Isaac Newton and Professor Cotes* (London, 1850): 253, note *). Unfortunately, his private diaries and autobiographical memoranda (compare Baily, *Flamsteed*: 35–6) have nothing to say of his interests, contacts and activities during this second visit. We may assume that Flamsteed's second meeting with Newton was as uneventful as the first, for no (known) correspondence between the two immediately ensued.

The content of the present lecture on 'refining' given algebraic equations into standard form is, it will be evident, in large part repeated in the section 'DE CONCINNANDA ÆQUATIONE SOLITARIA' of the revised, polished copy of Newton's contemporary Lucasian lectures which he deposited ten years afterwards in Cambridge University Library, but the correspondence is far from perfect. The present paragraphs I, III and IV (except for (γ)) furnish the examples which illustrate the transformations employed by Newton in Rules 3, 4 and 5 on pages 32–3 of the deposited copy (ULC. Dd. 9. 68; reproduced as 2, §1), while paragraph II is there wholly discarded. More confusingly, to fit the pre-ordained chronology of the deposited copy (whose marginal datings break its content cleanly, but manifestly none too accurately, into eleven sets of *lectiones* purportedly delivered in 'Octob.' of each of the years 1673–83) its delivery is there

II. Ⓐ $\dfrac{aa-xx}{a+b}+a=[-]x$ fit $xx=\dfrac{a}{+b}x\dfrac{+2aa}{+ab}$. Ⓑ $\dfrac{y^3-aby}{aa+a\sqrt{aa-bb}}+a=\sqrt{aa-bb}$

fit $y^3-aby+abb=0$.[4]

III. Ⓐ $\sqrt{aa-xx}+a=x$ fit $aa-xx=xx-2ax+aa$ seu $x=a$.

$$\text{Ⓑ} \sqrt{3}:\overline{aax+2axx-x^3}-a+x=0$$

fit $aax+2axx-x^3=a^3-3aax+3axx-x^3$ seu $xx=4ax-aa$.

$$\text{Ⓖ} \; y=\sqrt{ay+yy-a\sqrt{ay-yy}}$$

primo fit $y=\sqrt{ay-yy}$ d[e]in $2y=a$.

IV. Ⓐ $2y=a$ fit $y=\frac{1}{2}a$. Ⓑ $\dfrac{bx}{a}=a$ fit $x=\dfrac{aa}{b}$. Ⓖ $ax-cx=ac$ fit $x=\dfrac{ac}{a-c}$.

Ⓓ $\dfrac{2ac}{-cc}x^3\dfrac{+a^3}{+aac}xx\dfrac{-2a^3c}{+aacc}x-a^3cc=0$, fit $x^3\dfrac{+a^3+aac}{2ac-cc}xx-aax-\dfrac{a^3c}{2a-c}=0$.

---

postponed in time to 'October 1674' and straddles 'Lect. 6' and 'Lect. 7' of that series. Flamsteed's dating of 'Midsummer', entirely consistent with known historical event and unfettered by any need to satisfy a chronology imposed by the format of the deposited copy, would seem to be unchallengeable.

(2) Flamsteed MSS, Volume 42 (insert pasted on the front flyleaf).

(3) By a momentary slip of his pen (the mistake is not continued into the sequel) Newton wrote ' $=yy$ ' simply.

(4) Since, of course, $a(a+\sqrt{[a^2-b^2]})\,(a-\sqrt{[a^2-b^2]})=ab^2$. This paragraph is omitted in the deposited copy (see note (1)).

§2. DELIMITATION OF THE ROOTS OF EQUATIONS AND THEIR FACTORIZATION.[1]

[Late 1670's?]

From the original[2] in the University Library, Cambridge

QUOMODO RADICES ÆQUATIONIS IN CHARTA
EXHIBERI POSSINT.

Sed æquationis alicujus constitutio optimè dignoscetur exponendo radices in charta aliqua hoc modo.[3] Pro radice ejus substitue in ea successive numeros quosvis[4] ut 0, 1, 2, 5, 10, 20 et his intermedios. Numeros istos expone per lineas

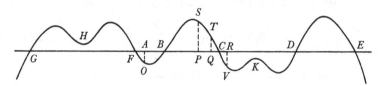

AP, *AQ*, *AR* &c in recta aliqua ab eodem puncto *A* ductos et numeros qui horum substitutione resultant, divisos per datū aliquem numerū ne chartæ magnitudinē excedant expone per perpendicula *AO*, *PS*, *QT*, *RV* &c ad priorum terminos erecta, ducendo affirmativa ut *PS*, [*Q*]*T* ad unam plagam, negativa ut *AO*, [*R*]*V* ad alteram. Dein per terminos perpendiculorum *O*, *S*, *T*, *V* manu duc lineam æquabilem *OSTV* & lineæ *AB*, *AC*, *AD*, *AE* &c quæ ad intersectiones hujus cum recta *AP* terminantur erunt radices æquationis. Similiter substituendo pro radice seriem numerorum negativorum possis lineam ad alteram partem puncti *A* producere et ejus intersectionibus cum eadem recta *AP* radices negativas ut *AF*, *AG* exprimere.[5] Radices impossibiles exprimi solent[6] per flexuras concavo-convexas quibus curva appropinquat rectam sed recedit iterū antequam attingit, quales sunt eæ ad *K* et *H*.[7]

---

(1) These observations on the 'constitution' of the roots of an algebraic equation, their delimitation and—in the case of rational roots—their identification were manifestly part of a preliminary version, now otherwise lost, of the central, algebraic section of the deposited revise (2, §1) of Newton's contemporary Lucasian lectures. To be sure, the first section of the present manuscript and the opening paragraphs of the second were later discarded in the revise—and Newton may well have preserved the present isolated sheet because its content was not then wholly absorbed—but the latter half of its second section 'De limitibus æquationum' and its final part 'De inventione divisorum' reappear with little modification in the text of the deposited copy (on Newton's pages 170–2 and 172–80 respectively), whose imposed marginal chronology asserts that their content was given out to his Cambridge audience in the second half of 'Lect. 10' of the 'Oct. 1681' series of Newton's professorial lectures and in 'Lect. 1' and 'Lect. 2' of that of 'Octob. 1682'. However, in assigning the time of composition

Translation

HOW THE ROOTS OF AN EQUATION MIGHT BE DISPLAYED ON A CHART

But the character of some equation will best be distinguished by representing its roots on a chart in this manner.[3] In it in place of its root substitute in succession any numbers you will,[4] say 0, 1, 2, 5, 10, 20 and intermediate ones. Represent those numbers by the lines AP, AQ, AR and so on drawn in some straight line from the same point A, and the numbers resulting from their substitution (divided by some given number so as not to exceed the size of the paper) represent by the perpendiculars AO, PS, QT, RV and so on raised at the end-points of the preceding lines, drawing positive lengths such as PS and QT in one direction, negative ones such as AO and RV in the other. Then through the end-points O, S, T, V of the perpendiculars draw free-hand the smooth curve $OSTV$, and the lines AB, AC, AD, AE, ... terminating at its intersections with the straight line AP will be the equation's roots. Similarly, by substituting a series of negative numbers for the root you might extend the curve on the other side of the point A and by its meets with the same straight line AP express[5] the negative roots such as AF and AG. 'Impossible' roots are usually[6] expressed by concavo-convex folds at which the curve approaches the base-line but retreats again before reaching it: here, in instance, those at K and H.[7]

---

of this paper (which, together with its lost companion sheets, may well have been the very text on which he initially lectured) we have been swayed by our assessment of Newton's autograph hand slightly to antedate the provably inexact chronology of the deposited revise.

(2) Add. 3964.2: 5^r–6^v, a folded folio sheet crowded on all its four sides with the autograph text now reproduced. Its initial state is carefully written out, but Newton later—doubtless at the time (early 1684?) he incorporated the bulk of its content in his deposited version (see note (1))—revised it in a much rougher fashion, with multiple cancellations and interlineations not easy to read.

(3) Namely, the (real) roots of $f(x) = 0$ are represented, in the model of the plane curve defined in Cartesian coordinates by the equation $f(x) = y$, by its intersections with the base-line $y = 0$.

(4) Newton first wrote 'scribe successive quantitates quasvis' (write in succession any quantities you will).

(5) 'cog[noscere]' (determine) is cancelled. Newton first began the next sentence 'Et flexuræ multiplices' (And multiple folds...).

(6) Not by any means always, as the counter-instance of the general parabola $y = x^n + a$ illustrates.

(7) Compare I: 522, note (32).

[De limitibus æquationis.][8]

Diximus[9] radices omnes qui integri sunt numeri divisores esse infimi termini æquationis. Unde si in æquatione substituas sigillatim divisores ōes[10] istius infimi termini, et nullus eorum efficiat ejus terminos evanescere, concludas nullam ejus radicem esse numerum integrum. Omnes autem ultimi termini inveniuntur dividendo eum per minimum ejus divisorem et quotum perpetim per minimum divisorem ejus donec quotus nullum amplius admittat divisorem præt^r seipsum; deinde multiplicando in se invicem singulos binos ternos quaternos & quinos divisores[11] &c. Ut si infimus terminus fuerit 60. divide hunc per 2 et quotum 30 per 2 & quotum 15 per 3 & restabit Quotus 5 per seipsum solummodo divisibilis[,] Ergo divisores ōes[10] primi sunt 1, 2, 3, 5,[12] ex binis compositi, 4, 6, 10, 15, ex ternis 12, 20, 30, ex quaternis 60. Atcq ita si infimus terminus fuerit $21aa - 63ab$. Divide eum per 3, & Quotum $7aa - 21ab$ per 7, & Quotum $aa - 3ab$ per a et restabit $a - 3b$ Quotus per se solum divisibilis, Ergo divisores primi erunt 1, 3, 7, a, $a - 3b$, ex binis compositi 21, $3a$, $7a$, $3a - 9b$, $7a - 21b$, $aa - 3ab$, ex ternis $21a$, $21a - 63b$, $3aa - 9ab$, $7aa - 21ab$, ex quaternis $21aa - 63ab$.

Numerus autem pro radice hoc modo commode substitui potest.[13] Sit æquatio $x^4 - 3x^3 - 28xx + 132x - 144 = 0$ et 4[14] numerus pro x substituendus. et erit $x^4 = x \times x^3 = 4x^3$, & $4x^3 - 3x^3 = x^3 = x \times xx = 4xx$, &

$$4xx - 28xx = -24xx = -96x$$

& $-96x + 132x = 36x = 144$. & $144 - 144 = 0$. Cum igitur nihil restet concludo 4 esse radicem æquationis. Eodem modo si 7[15] substituendum esset dicerem $x^4 = 7x^3$. $7x^3 - 3x^3 = 4x^3 = 28xx$. $28xx - 28xx = 0$. $0 + 132x = 132x = 924$.

$$924 - 144 = 780.$$

Cùm igitur 780 restet concludo 7 non esse radicem.[16] Si negativus numerus substituendus est, mutentur signa terminorum imparium dimensionum et substituatur affirm[ativus]. Ut si -6 substituendus esset substituo $+6$ in hanc $x^4 + 3x^3 - 28xx - 132x - 144 = 0$.

---

(8) This heading is inserted by us in line with that prefacing the corresponding section in Newton's revise (see page 362 below).

(9) Evidently in the lost preceding sheets of the present manuscript (see note (1)): at the equivalent place in his revise (2, §1: Newton's page 165) he was to write that 'omnes æquationis cujuscuncq terminos nec fractos nec surdos habentis, radices non surdas...esse aliquos ex divisoribus integris ultimi termini'.

(10) Read 'omnes'.

(11) Newton understands 'præter unum' (except 1).

(12) This should be '1, 2, 2, 3, 5' strictly.

(13) Compare I: 490 and II: 222.

[THE LIMITS OF AN EQUATION][8]

We have said[9] that all roots which are integers are divisors of the lowest term of an equation. Hence if you separately substitute in an equation all the divisors of its lowest term, but none of them makes its terms vanish, you are to conclude that it possesses no root which is an integer. All those, however, of the lowest term are found by dividing it by its least divisor, and the quotient repeatedly by its least divisor, till a quotient allows no divisor beyond itself; and then multiplying these divisors[11] one into another singly and two, three, four, five at a time, and so on. For instance, if the lowest term were 60, divide this by 2, and the quotient 30 by 2, and the quotient 15 by 3, and there will remain the quotient 5 divisible only by itself; consequently, the prime divisors are in total 1, 2, 3, 5,[12] those compounded two at a time are 4, 6, 10, 15, those three at a time 12, 20, 30, that of all four 60. And thus if the lowest term were $21a^2 - 63ab$, divide it by 3, and the quotient $7a^2 - 21ab$ by 7, and the quotient $a^2 - 3ab$ by a, and there will remain the quotient $a - 3b$ divisible by itself alone; in consequence, the prime divisors will be 1, 3, 7, a, $a - 3b$, those compounded two at a time 21, $3a$, $7a$, $3a - 9b$, $7a - 21b$, $a^2 - 3ab$, those three at a time $21a$, $21a - 63b$, $3a^2 - 9ab$, $7a^2 - 21ab$, that of all four $21a^2 - 63ab$.

A numerical value can, however, conveniently be substituted for a root in this manner.[13] Let the equation be $x^4 - 3x^3 - 28x^2 + 132x - 144 = 0$ and 4[14] the number to be substituted for x. Then will there be $x^4 = x \times x^3 = 4x^3$, and $4x^3 - 3x^3 = x^3 = x \times x^2 = 4x^2$, and $4x^2 - 28x^2 = -24x^2 = -96x$, and

$$-96x + 132x = 36x = 144,$$

and $144 - 144 = 0$. Since therefore nothing is left, I conclude that 4 is a root of the equation. In the same way if 7[15] had had to be substituted, I would say:

$$x^4 = 7x^3, \quad 7x^3 - 3x^3 = 4x^3 = 28x^2, \quad 28x^2 - 28x^2 = 0,$$

$$0 + 132x = 132x = 924, \quad 924 - 144 = 780.$$

Since therefore 780 remains, I conclude that 7 is not a root.[16] If a negative number has to be substituted, let the signs of terms of odd dimension be changed and then substitute its positive [equal]. So if -6 needed to be substituted, I substitute $+6$ in this equation $x^4 + 3x^3 - 28x^2 - 132x - 144 = 0$.

---

(14) Newton first substituted $x = 6$ in the equation, computing '$x^4 = x \times x^3 = 6x^3$, & $6x^3 - 3x^3 = 3x \times xx = 18xx$, & $18xx - 28xx = -10xx = -60x$ & $-60x + 132x = 72x$', but abandoned the calculation when he saw that $72x - 144 = 288 \neq 0$.

(15) '5' was first written, producing '$x^4 = 5x^3$. $5x^3 - 3x^3 = 2x^3 = 2x \times xx$.

$10xx - 28xx = -18xx = -90x$. $-90x + 132x = 42x = 210$. $210 - 144 = 66$'.

(16) In fact, $x^4 - 3x^3 - 28x^2 + 132x - 144 = (x - 2)(x - 3)(x - 4)(x + 6)$ and so the roots of the equation are 2, 3, 4 and -6.

Interim dum numerus aliquis substituitur notandum[17] est quod si quant[it]-atum quæ per additionē et subductionem generantur nulla sit diversi signi ab altissimo[18] termino æquationis, numerus iste si affirmativus est major erit maxima æquationis radice affirmativa; si negativus, negativâ.[19] Ut in ultimo exemplo quantitatum $4x^3$, 0, $132x$ & 780 nulla erat diversi signi a termino x^4, proinde maxima radix[20] æquationis minor est quam 7. Atꝗ ita substituendo[21] −6 patebit nullam radicem negativam hunc numerum transcendere. Quare omnes radices inter hos duos limites 7 et −6 consistunt dempta una quæ est ipsa limes −6. Quamvis igitur infimi termini 144 multi sint divisores, sufficiet[22] tamen eos paucos tentare qui inter hos limites consistunt.

Possunt autem limites strictius definiri per hanc regulam. Si ultimus numerus qui substituendo numerum aliquem resultat non sit diversi signi ab altissimo termino æquationis,[23] duc unumquemꝗ æquationis terminum in numerum dimensionum ejus ut et in alium numerum qui sit isto minor unitate.[24] Dein eundem numerum prædictum ut prius substitue & si omnes jam resultant cum signo altissimi termini, numerus iste limes erit.[25] Ut in exemplo superiori[26] ubi altissimus terminus erat $+x^4$ quoniam substituendo numerum 4 pro x, in ultimo loco resultabat 0, duco terminos æquationis in numeros dimensionum inꝗ alios numeros unitate minor et prodit 4, $3x^4−3$, 2, $3x^3−2$, 1, $28xx+1$, 0, $132x−0$, $−1 \times 144$ sive dividendo omnia per $2xx$, $6xx$, $6xx−9x−28$. Substituo jam 4 pro x dicendo $6xx=6x \times x=24x$. $24x−9x=15x=60$. $60−28=32$. Cum igitur termini omnes per subductionem resultantes nempe $15x$ & 32 sint ejusdem signi cum altissimo termino æquationis, concludo nullam radicem esse majorem quam 4. Rursus si æquatio esset $x^5−5x^4−2x^3−xx+50x+30=0$. Quoniam substituendo 5 pro x prodit ultimo loco $+5$ numerus ejusdem signi cum altissimo termino x^5, multiplico terminos æquationis ut supra et fit $20x^3−60xx−12x−2$.

---

(17) Newton first wrote 'cognoscere'.

(18) This replaces 'primo' (first).

(19) A subtle observation. In proof, if the given equation is $\sum_{0 \leqslant i \leqslant n} (a_i x^{n-i}) = 0$ and the partial sums are therefore $\sum_{0 \leqslant i \leqslant j} (a_i x^{n-i}) \equiv f(x)_j$, say, $j = 0, 1, 2, ..., n$, then the p-th derivative, $f^{[p]}(x)_n$, of $f(x)_n$ is

$$\sum_{0 \leqslant i \leqslant n-p} ((n-i)_p a_i x^{n-i-p}) = \sum_{0 \leqslant i \leqslant n-p} (n-i)_p x^{-p}(f(x)_i - f(x)_{i-1})$$

$$= \sum_{0 \leqslant i \leqslant n-p} x^{-p}[(n-i)_p - (n-i-1)_p]f(x)_i$$

$$= p(n-i-1)_{p-1} x^{-p} \cdot \sum_{0 \leqslant i \leqslant n-p} f(x)_i.$$

Hence if $x > 0$ and all $f(x)_i \geqslant 0$, $i = 0, 1, 2, ..., n$, then *a fortiori* each of the derivatives $f^{[p]}(x)_n \geqslant 0$, $p = 1, 2, 3, ..., n$, and Newton's observation reduces to the derivative test expounded below. Similarly, if $x < 0$ and all $f(x)_i \leqslant 0$, then all $f^{[p]}(x)_i \leqslant 0$.

(20) Understand 'affirmativa' (positive): the largest negative root is, of course, −6.

You should notice[17] during the process of substituting some number that, if none of the quantities generated by addition and subtraction be opposite in sign to the highest[18] term in the equation, then that number will, if it is positive, be greater than the greatest positive root of the equation; but, if negative, greater than the greatest negative.[19] Thus in the last example none of the quantities $4x^3$, 0, $132x$ and 780 was opposite in sign to x^4, and accordingly the greatest[20] root of the equation is less than 7. And likewise by substituting[21] -6, it will appear that no negative root surpasses this number. Consequently all roots are situated between these two limits 7 and -6, except for one which is this very limit -6. Therefore, although the lowest term 144 has many divisors, it will yet suffice[22] to test those few which lie between these limits.

But the limits can be more narrowly defined by this rule. If the last number which results on substituting some number be not opposite in sign to the highest term of the equation,[23] multiply each term of the equation into the number of its dimensions and also into a second number which is one unit less.[24] Then substitute the same above-mentioned number as before and if all now prove to have the sign of the highest term, that number will be a limit.[25] As in the above example,[26] where the highest term was $+x^4$, since 0 resulted in the final place on substituting the number 4 in place of x, I multiply the terms of the equation into the numbers of their dimensions and then into other numbers which are a unit less, and there arises

$$4 \times 3x^4 - 3 \times 2 \times 3x^3 - 2 \times 1 \times 28x^2 + 1 \times 0 \times 132x - 0 \times -1 \times 144,$$

that is, on dividing through by $2x^2$, $6x^2 - 9x - 28$. I now substitute 4 in place of x, saying $6x^2 = 6x \times x = 24x$, $24x - 9x = 15x = 60$, and $60 - 28 = 32$. Since therefore all the terms, namely $15x$ and 32, resulting from the subtraction are of the same sign as the highest term of the equation, I conclude there is no root greater than 4. Again, had the equation been $x^5 - 5x^4 - 2x^3 - x^2 + 50x + 30 = 0$, since on substituting 5 in place of x there results $+5$ in the final place, a number with the same sign as the highest term x^5, I multiply the terms of the equation as

---

(21) '-7 vel etiam' (-7 or even) is deleted.

(22) In finding integral roots, that is.

(23) A following cancelled phrase reads 'et tamen numerus unus vel plures ante ultimum istum cum signo diverso resultant' (and yet one or more numbers before that last one come out with a differing sign).

(24) This, of course, yields the second derivative of the equation, multiplied by x^2.

(25) The rule is not exact, as Newton saw almost at once (for he immediately cancelled the present paragraph in favour of the one which follows). For instance, substitution of $x = 3\frac{1}{2}$ in his present example yields, for the part-sums of (half) its second derivative $6x^2 - 9x = 42 > 0$ and $6x^2 - 9x - 28 = 14 > 0$, even though $3\frac{1}{2}$ is not greater than the greatest root (4) of the equation.

(26) Namely, $x^4 - 3x^3 - 28x^2 + 132x - 144 = 0$.

Ubi substituo $x=5$ hoc modo $20x^3=100xx$. $100xx-60xx=40xx=200x$. $200x-12x=188x=940$. $940-2=938$. Et cum termini omnes $40xx$, $188x$ & 938 prodeunt cum signis $+$ concludo 5 esse majorem quavis radice affirmativa. Rursus substituendo -1 in eandem æquat^m vel quod perinde est $+1$ in hanc $-x^5-5x^4+2x^3-xx-50x+30$ prodeunt omnes numeri negativi. Quare[27] erit -1 limes quem nulla radix negativa transcendit. Idem evenit substituendo $-\frac{2}{3}$. Quare cum altissimus terminus $-x^5$ sit negativus erit -1 vel strictius $-\frac{2}{3}$ limes alter[,] adeo ut radices omnes inter 5 et $-\frac{2}{3}$ consistant.[28]

Possunt autem limites arctius[29] definiri hoc modo. Si numerus aliquis substitutus producit ultimo loco numerum ejusdem signi cum altissimo æquationis termino, sed non in omnibus locis intermedijs[,] multiplica terminos æquationis per numerum dimensionum ac divide per radicem ejus. Dein substitue in hac nova æquatione præfatum numerum & si rursus ultimò prodit numerus ejusdem signi cum altissimo termino, multiplica terminos hujus per numerū dimensionum ac divide per duplum radicis & eundem numerum in hac substitue, et si iterum prodit numerus ejusdem signi cum altissimo signo, rursus multiplica terminos[30] per numerum dimensionū divideꝗ per triplum radicis & eundem numerum denuò substitue. Et si hoc opus iterando semper prodeat numerus ejusdem signi cum altissimo termino æquationis, numerus iste limes erit radicum.[31] Ut in exemplo allato,[32] ubi altissimus terminus est $+x^4$, substituendo 4 prodit ultimo loco 0, cui signum $+$ æque ac signum $-$ præfigi potest. Multiplico igitur terminos æquationis per numerum dimensionum cujusꝗ divideꝗ per x et prodit $4x^3-9xx-56x+132$ et substituendo 4 pro x oritur $+20$. Rursus igitur multiplico terminos per numerū dimensionum divideꝗ per $2x$ et prodit $6xx-9x-56$ et scribendo 4 pro x oritur $+4$. Et multò magis si termini iterum multiplicentur per dimensionū suarum numerum, orietur numerus affirmativus eo quod $6xx$ majorem obtinebit rationem ad

---

(27) Newton has cancelled 'cum altissimus terminus $-x^5$ sit negativus' (since the highest term $-x^5$ is negative).

(28) In fact, the three real roots of the quintic lie in the intervals $[-\frac{2}{3}, 1]$, $[2, 3]$ and $[4, 5]$.

(29) 'exactiùs' (more exactly) was first written. Neither adverb is accurate since (see note (19)) the present derivative test is equivalent to the preceding rule.

(30) 'æquationem' (the equation) is justly cancelled.

(31) In modern terms, if each of the p-th derivatives $f^{[p]}(x) \equiv \sum_{0 \leqslant i \leqslant n-p} ((n-i)_p a_i x^{n-i-p})$ of $f(x) \equiv f^{[0]}(x) \equiv \sum_{0 \leqslant i \leqslant n} (a_i x^{n-i})$ are positive, $p = 0, 1, 2, ..., n$, for some $x = X$, then

$$f(X+h) = f(X) + hf^{[1]}(X) + \frac{1}{2!}h^2 f^{[2]}(X) + ... + h^n a_0$$

ispositive for all $h \geqslant 0$ and so $X+h$ cannot be a root of $f(x) = 0$. Although, as we shall see in the next volume, the general 'Taylor' expansion of a function first appears in Newton's papers in Proposition XII (ULC. Add. 3960.10: 173–7) of the unpublished first version, written during the winter of 1691–2, of his 'Tractatus de Quadratura Curvarum', we would insist that

above directed and there comes $20x^3 - 60x^2 - 12x - 2$. In it I substitute $x = 5$ in this manner:

$$20x^3 = 100x^2,\ 100x^2 - 60x^2 = 40x^2 = 200x,\ 200x - 12x = 188x = 940,$$

and $940 - 2 = 938$. Then, since all the terms $40x^2$, $188x$ and 938 turn out to have their signs $+$, I conclude that 5 is greater than any positive root. Further, on substituting -1 in the same equation, or, what is equivalent, $+1$ in this $-x^5 - 5x^4 + 2x^3 - x^2 - 50x + 30$, all the resulting numbers prove to be negative. Consequently[27] -1 will be a limit which no negative root transcends. The same happens on substituting $-\frac{2}{3}$. Hence, since the highest term $-x^5$ is negative, -1 or more strictly $-\frac{2}{3}$ will be a second limit, with the result that all roots are situated between 5 and $-\frac{2}{3}$.[28]

But the limits can be more tightly[29] defined in this way. If some number, when substituted, produces in the final place a number having the same sign as the highest term in the equation, but not so in all intermediate places, multiply the equation's terms by the number of their dimensions and divide by its root. Then substitute the previous number in this fresh equation, and if again there finally ensues a number of the same sign as the highest term's, multiply this one's terms by the number of their dimensions and divide through by twice the root, and in it substitute the self-same number; if once more there ensues a number of the same sign as the highest term's, again multiply its terms[30] by the number of their dimensions and divide by three times the root, and substitute the same number afresh. And if on repeating this procedure there should always result a number identical in sign with the highest term of the equation, that number will be a limit to the roots.[31] So in the example instanced,[32] where the highest term is $+x^4$, on substituting 4 there results 0 in the final place: to this the sign $+$ or the sign $-$ may equally be attached. I therefore multiply the terms of the equation by the number of their dimensions and divide each by x; there results $4x^3 - 9x^2 - 56x + 132$, yielding $+20$ on substituting 4 for x. I therefore again multiply the terms by the number of their dimensions and divide by $2x$; there results $6x^2 - 9x - 56$, yielding $+4$ on substituting 4 for x. And much more so, if the terms be once again multiplied by the number of their dimensions, will a positive number ensue since $6x^2$ will acquire a yet greater ratio to the negative

---

here (as indeed in Book 2, Proposition 10 of his *Principia*) only the knowledge of the finite algebraic expansion

$$\sum_{0 \leqslant i \leqslant n} (a_i(x+h)^{n-i}) = \sum_{0 \leqslant i \leqslant n} (a_i x^{n-i}) + h . \sum_{0 \leqslant i \leqslant n-1} ((n-i)\, a_i x^{n-i-1})$$

$$+ \frac{1}{2!} h^2 . \sum_{0 \leqslant i \leqslant n-2} ((n-i)\,(n-i-1)\, a_i x^{n-i-2}) + \ldots + h^n a_0$$

is presumed by Newton.

(32) $x^4 - 3x^3 - 28x^2 + 132x - 144 = 0$ (with roots 2, 3, 4 and -6).

negativos terminos $-9x-56$. Cum igitur numeri semper prodeant affirmativi concludo [nullam] radicem majorem esse quam 4, adeoꝗ radices omnes inter limites 4 et -6 consistere præter limites ipsos qui in hoc casu sunt radices. Rursus si æquatio esset $x^5-[2]x^4-10x^3+30xx+6[5]x-[120]=0$[33] et scire vellem an radix aliqua sit major quam 2

In $x^5-2x^4-10x^3+30x^2+65x-120$ substituendo 2 oritur $+50$.
In $5x^4-8x^3-30xx+60x+65$ substituendo 2 oritur $+81$.
In $10x^3-12xx-30x+30$ sub[s]ti[tu]endo 2 oritur $+2$.
In $10x^2-8x-10$ substituendo 2 oritur $+14$.

Concludo itaꝗ 2 majorem esse quavis radice affirmativâ. Similiter si limitem negativarum scire vellem

De $-x^5-2x^4+10x^3+30xx-65x-120$ subst[it]uendo 2 oritur -114.
At de $-5x^4-8x^3+30xx+60x-65$ substituendo 2 oritur $+31$ numerus diversi signi ab alt[issimo] term[ino] $-5x^4$.

Unde limes paulo major quam 2 statuendus est. Et quidem substi[tu]endo 3 prodibunt semper numeri negativi.[34] Adeoꝗ radices omnes int^r limites 2 et -3[35] consistunt, nec opus fuerit aliquos ultimi termini 120 divisores pro radice tentare præter eos paucos qui limites hosce non transcendunt.

[De inventione divisorum.][36]

At ubi limites hosce non animus est investigare vel limitum tanta est distantia ut multi divisores possint intercedere, vel altissimus terminus per numerum aliquem multiplicatur it[a] ut radix possit esse numerus fractus: pro radice substitue tres [vel] plures terminos hujus progressionis Arithmeticæ 3. 2. 1. 0. -1. -2 ac terminos resultantes una cum omnibus eorum divisorib^s statue e regione respondentium terminorum progressionis, positis divisorum signis tam affirmativis quam negativis. Dein e regione etiam statue progressiones arithmeticas quæ per omnium numerorum divisores percurrunt, pergentes a majoribus terminis ad minores eodem ordine quo termini progressionis 3, 2, 1, 0, -1 &c[37] pergunt, et quarum termini differunt vel unitate vel numero aliquo qui dividit coefficientem altissimi termini æquationis. Siqua occurrit ejusmodi progressio iste terminus ejus qui stat e regione termini 0 progressionis primæ,

---

(33) The manuscript has the uncorrected original choice of equation

$$`x^5-3x^4-10x^3+30xx+60x-7=0',$$

replaced in the sequel by the variant reproduced; correspondingly, Newton first went on to substitute $x=2$ in his original equation and its first and (halved) second derivatives, '$5x^4-12x^3-30xx+60x+60$' and '$10x^3-18xx-30x+30$', correctly deriving the values '$+137$', '$+44$' and '-22'.

terms $-9x-56$. Since therefore the numbers turn out always to be positive, I conclude that no root is greater than 4 and accordingly all roots (apart from the limits themselves which in this case are roots) are situated between the limits 4 and -6. Again, had the equation been

$$x^5 - 2x^4 - 10x^3 + 30x^2 + 65x - 120 = 0^{(33)}$$

and I wished to know whether some root is greater than 2:

on substituting 2 in $x^5 - 2x^4 - 10x^3 + 30x^2 + 65x - 120$ there ensues $+50$;

- - - - - - - $5x^4 - 8x^3 - 30x^2 + 60x + 65$ - - - - $+81$;

- - - - - - - $10x^3 - 12x^2 - 30x + 30$ - - - - $+2$;

- - - - - - - $10x^2 - 8x - 10$ - - - - $+14$.

I conclude consequently that 2 is greater than any positive root. Similarly, should I have wished to know a limit to the negative ones:

from $-x^5 - 2x^4 + 10x^3 + 30x^2 - 65x - 120$ on substituting 2 there ensues -114;

but from $-5x^4 - 8x^3 + 30x^2 + 60x - 65$ - - - - - - - - - - - $+31$,

a number opposite in sign to the highest term $-5x^4$.

Hence the limit has to be set a little greater than 2. In fact, on substituting 3 negative numbers result in all cases.[34] Consequently all the roots are situated between the limits 2 and -3,[35] and there will be no need to test any divisors of the final term 120 as a root except those few which do not surpass these limits.

[THE FINDING OF DIVISORS][36]

But when you have no mind to investigate these limits, or the interval between the limits is so large that numerous divisors can lie between them, or the highest term is multiplied by some number and the root may accordingly be a [rational] fraction, then in place of the root substitute three or more terms of this arithmetical progression [...] 3, 2, 1, 0, -1, -2, [...] and put the resulting terms together with all their divisors in line with corresponding terms of the progression, setting both positive and negative signs on the divisors. Next, also in line, place arithmetical progressions which run through the divisors of all the numbers, proceeding from the greater terms to the smaller ones in the same sequence as the terms of the progression 3, 2, 1, 0, -1, ...[37] go, and whose terms differ by unity or some number dividing the coefficient of the highest term of the equation. If any progression of this kind does occur, that term in it which is stationed in line with the term 0 of the first progression will, when divided by the difference

---

(34) In the adjusted derivatives $\frac{1}{2}f^{[2]}(x)$, $\frac{1}{6}f^{[3]}(x)$ and $\frac{1}{24}f^{[4]}(x)$ respectively.

(35) In fact, the equation has only one (non-rational) real root $x \approx 1\frac{1}{2}$.

(36) The heading is here introduced in line with that on page 259 of the deposited revise (see 2, §2: note (17)).

(37) The superfluous phrase 'e regione positi' (set in line) is cancelled.

divisus per differentiam terminorum, et cum signo suo annexus literæ per quam radix designatur, componet divisorem per quem reductio æquationis tentanda est. Et si *p* nullum hac methodo inventum divisorem res succedit concludere potes æquationem reduci non posse per divisorem unius dimensionis. Ut si æquatio fuerit $x^3 - xx - 10[x] + 6 = 0$, pro x substituendo sigillatim terminos progressionis 2. 1. 0. −1, orientur numeri 10, −4, 6, −14[38] quos cum omnibus eorum divisoribus colloco e regione terminorum progressionis 2. 1. 0. −1 hoc modo.

2	10	1, 2, 5, 10	5
1	4	1, 2, 4	4
0	6	1, 2, 3, 6	3
−1	14	1, 2, 7, 14	2

Deinde quoniam coefficiens altissimi termini æquationis est unitas, quæro in divisoribus progressionem cujus termini differunt unitate, et a superioribus ad inferiora pergendo decrescunt perinde ac termini progressionis lateralis 2, 1, 0, −1. Et hujusmodi progressionem unicam tantum invenio nempe 5, 4, 3, 2. Cujus itaǧ terminum +3 seligo[39] qui stat e regione termini 0 progressionis primæ 2, 1, 0, −1, tentoǧ divisionem per $x+3$. Et res succedit prodeunte $xx - 4x + 2 = 0$.

Rursus si æquatio fuerit $6y^4 - y^3 - 21yy + 3y + 20 = 0$. Pro y substituo sigillatim 2, 1, 0, −1 & numeros resultantes 30, 7, 20, 1[40] cum omnibus eorum divisoribus e regione colloco ut sequitur.

2	30	1, 2, 3, 5, 6, 10, 15, 30	10
1	7	1, 7	7
0	20	1, 2, 4, 5, 10, 20	4
−1	1[40]	1	1

Et in divisoribus hanc solam animadverto progressionem arithm. 10, 7, 4, 1. Hujus terminorum differentia 3 dividit altissimi æquationis termini coefficientem 6. Quare sumendo terminum ejus 4 qui stat e regione termini 0, divisum per differentiā terminorum 3 adjungens cum signo suo + speciei y per quam radix significatur, tento divisionem per $[y] + \frac{4}{3}$ vel quod perinde est per $3[y] + 4$ et res succedit prodeunte $2y^3 - 3yy - 3y + 5$.

---

(38) A trivial slip for the correct '+14'.

(39) Newton first wrote 'annecto literæ x qua radix denotatur' (I attach to the letter x by which the root is denoted).

(40) Read '3'! The extra factor 3 in the bottom line permits a second final column '3. 1. −1. −3' which has to be examined: in fact, the substitution of $y = -2$ yields the value 34, which has no factor −5, and so the latter column can safely be discarded. The oversight was

of the terms and attached with its sign to the letter by which the root is denoted, make up the divisor by which reduction of the equation is to be attempted. And if no test by a divisor found by this method succeeds, you may conclude that the equation cannot be reduced by a divisor of one dimension. If the equation were, for instance, $x^3 - x^2 - 10x + 6 = 0$, on separately substituting for x the terms of the progression 2, 1, 0, -1 the numbers 10, -4, 6, [+] 14 will ensue; these together with all their divisors I locate in line with the terms of the progression 2, 1, 0, -1 in this way:

2	10	1, 2, 5, 10	5
1	4	1, 2, 4	4
0	6	1, 2, 3, 6	3
-1	14	1, 2, 7, 14	2.

Next, because the coefficient of the highest term in the equation is unity, I look among the divisors for a progression whose terms differ by unity and decrease in their sequence from upper to lower levels in the same manner as the terms of the side-progression 2, 1, 0, -1. Of this sort I find but a single sequence, namely 5, 4, 3, 2. Accordingly I select[39] its term $+3$ standing in line with the term 0 of the first progression 2, 1, 0, -1 and attempt division by $x + 3$. The process succeeds, yielding $x^2 - 4x + 2 = 0$.

Again, if the equation were $6y^4 - y^3 - 21y^2 + 3y + 20 = 0$, in y's place I separately substitute 2, 1, 0, -1 and the resulting numbers 30, 7, 20, 1[40] along with all their divisors I station in line, as follows:

2	30	1, 2, 3, 5, 6, 10, 15, 30	10
1	7	1, 7	7
0	20	1, 2, 4, 5, 10, 20	4
-1	1[40]	1	1.

Among the divisors I notice this arithmetical progression alone: 10, 7, 4, 1. The difference 3 of its terms divides the coefficient 6 of the highest term in the equation. Hence, taking the term 4 in it which stands in line with the term 0, dividing by the difference 3 of the terms and adjoining to the quotient the sign $+$ of the variable y by which the root is denoted, I attempt division by $y + \frac{4}{3}$ or, what is exactly the same, by $3y + 4$ and the process succeeds, yielding

$$2y^3 - 3y^2 - 3y + 5.$$

continued into page 175 of the deposited revise (see 2, §1: note (523)), and thence by Whiston into print in Newton's *Arithmetica Universalis*; *sive De Compositione et Resolutione Arithmetica Liber* (Cambridge, ₁1707): 44, but was at length caught by Newton himself and entered as a correction in his library copy of the book, subsequently incorporated in the second (1722) edition of the *Arithmetica*.

Atɋ ita si æquatio esset $24x^5 - 50x^4 + 49x^3 - 140xx + 64x + 30 = 0$, Opus se haberet ut sequitur.[41]

2	42	1, 2, 3, 6, 7, 14, 21, 42.	3	3	7
1	23	1, 23.	1	−1	1
0	30	1, 2, 3, 5, 6, 10, 15, 30.	−1	−5	−5
−1	297	1, 3, 9, 11, 33, 99, 297.	−3	−9	−11

Tres occurrunt hic progressiones quorum termini −1, −5, −5 divisi per differentias terminorum 2, 4, 6, dant tres divisores tentandos $x - \frac{1}{2}$, $x - \frac{5}{4}$ & $x - \frac{5}{6}$ [seu $2x - 1$, $4x - 5$ & $6x - 5$],[42] quorum ultimus æquationem dividit prodeunte $4x^4 - 5x^3 + 4xx - 20x - 6 = 0$.

Si nullus occurrit hac methodo divisor vel nullus qui dividit æquationē[,] concludendum est æquationem non posse reduci per divisorem unius dimens[ionis]. Potest tamen forte, si plurium sit quam trium dimensionum, reduci per divisorem duarum, & si ita, divisor investigabitur hac methodo. Pro radice substitutis ut ante quatuor vel pluribus terminis hujus progressionis 3, 2, 1, 0, −1, −2. divisores omnes numerorum resultantium adde et subduc quadratis terminorū progressionis istius ductis in divisorem aliquem coefficientis altissimi æquatˢ termini et e regione colloca. Dein terminos omnes nota qui sunt in progressionibus Arithmeticis. Sit C terminus istiusmodi progressionis qui stat e regione termini 0, B differentia quæ oritur subducendo C de termino[43] proxime inferiori qui stat e regione −1, & A prædictus coefficientis termini altissimi divisor et z litera radicem significans et erit $Azz. Bz. C$.[44] divisor tentandus, ubi C signum habebit diversum ab eo quod habuit in progressione et B signum differentiæ quam significat. Ut si æquatio reducenda sit $x^4 - x^3 - 5xx + 12x - 6 = 0$ pro x scribo successive 3, 2, 1, 0, −1, −2 & prodeuntium terminorum 39, 6, 1, −6, −21, −26, divisores e regione dispono addoɋ et subduco terminis ijsdem quadratis viz 9, 4, 1, 0, 1, 4, et progressiones quæ in ijsdem obveniunt e latere etiam scribo ad hunc modum.

3	39	1, 3, 13, 39.	9	−30, −4, 6, 8, 10, 12, 22, 48.	−4.	6.
2	6	1, 2, 3, 6.	4	−2, 1, 2, 3. 5. 6. 7. 10.	−2.	3.
1	1	1.	1	0. 2.	0.	0.
0	6	1, 2, 3, 6.	0	−6. −3. −2. −1. 1. 2. 3. 6.	2.	−3.
−1	21	1, 3, 7, 21.	1	−20. −6. −2. 0. 2. 4. 8. 22.	4.	−6.
−2	26	1, 2, 13, 26.	4	−22. −9. 2. 3. 5. 6. 17. 30.	6.	−9.

(41) Strictly, the second column should read '−42. −23. +30. −297', but Newton is here interested only in their absolute values. Note also that he has omitted the possible factor 27 (of 297) in the bottom line, but this does not affect the completeness of his final columns. This trivial oversight was carried over into page 176 of the deposited revise (see 2, §1: note (524)) and into print in 1707, being repaired by Newton in a late correction in his library copy of the *Arithmetica* (note (40) above): 44, which was duly incorporated into the 1722 edition.

(42) We insert these integral multiples for convenience's sake.

And thus, had the equation been $24x^5 - 50x^4 + 49x^3 - 140x^2 + 64x + 30 = 0$, the operation would have gone as follows:[41]

2	42	1, 2, 3, 6, 7, 14, 21, 42	3	3	7
1	23	1, 23	1	−1	1
0	30	1, 2, 3, 5, 6, 10, 15, 30	−1	−5	−5
−1	297	1, 3, 9, 11, 33, 99, 297	−3	−9	−11 |.

Here three progressions occur, and their terms $-1, -5, -5$ divided by the differences 2, 4, 6 of the terms give three divisors to be tested: $x - \frac{1}{2}$, $x - \frac{5}{4}$ and $x - \frac{5}{6}$[, that is, $2x - 1$, $4x - 5$ and $6x - 5$].[42] Of these the last divides the equation, producing $4x^4 - 5x^3 + 4x^2 - 20x - 6 = 0$.

If no divisor—or at least none which divides the equation—is suggested by this method, the conclusion must be that the equation cannot be reduced by a divisor of one dimension. But it can perhaps, if it is of more than three dimensions, be reduced by a divisor of two, and, if so, the divisor will be discovered by this method. Having, as before, substituted in place of the root four or more terms of this progression [...] 3, 2, 1, 0, −1, −2 [...], add and subtract all divisors of the resulting numbers to and from the squares of the terms of that progression multiplied by some divisor of the coefficient of the highest term of the equation, and set the results out in line. Then note all terms which are in arithmetical progression. Let C be the term in a progression of that type which stands in line with the term 0, B the difference which arises on subtracting C from the next lower term[43] standing in line with -1, and A the above-mentioned divisor of the coefficient of the highest term, and then (with the letter z denoting the root) the divisor to be attempted will be $Az^2 \pm Bz \pm C$,[44] in which C will have its sign opposite to that which it had in the progression, B the sign of the difference which it denotes. So if the equation to be reduced is $x^4 - x^3 - 5x^2 + 12x - 6 = 0$, in x's place I write successively 3, 2, 1, 0, −1, −2 and the divisors of resulting terms 39, 6, 1, −6, −21, −26 I arrange in line, then add them to and subtract them from the squares, namely 9, 4, 1, 0, 1, 4, of those same terms, and progressions which turn up in these I write also on the side in this manner:

3	39	1, 3, 13, 39	9	−30, −4, 6, 8, 10, 12, 22, 48	−4	6
2	6	1, 2, 3, 6	4	−2, 1, 2, 3, 5, 6, 7, 10	−2	3
1	1	1	1	0, 2	0	0
0	6	1, 2, 3, 6	0	−6, −3, −2, −1, 1, 2, 3, 6	2	−3
−1	21	1, 3, 7, 21	1	−20, −6, −2, 0, 2, 4, 8, 22	4	−6
−2	26	1, 2, 13, 26	4	−22, −9, 2, 3, 5, 6, 17, 30	6	−9 |.

(43) Namely, $-B + C$ (the value when $z = -1$ of $Bz + C$).

(44) Cartesian notation for '$Azz + Bz + C$' where the coefficients A, B, C are allowed to take on all positive and negative values; compare IV: 652, note (35). In practice it will be convenient to divide through the given equation by the coefficient of its highest term and then set $A = 1$, of course.

Dein progressionum inventarum terminos 2 et − 3 qui stant e regione termini 0, signis mutatis usurpo pro C, eosque subducens de terminis inferioribus, 4, & − 6 qui stant e regione termini − 1 residua 2 & − 3 usurpo pro B et unitatem pro A. Et sic habeo divisores duos tentandos $xx + 2x − 2$ & $xx − 3x + 3$, per quorum utrumque res succedit.

Rursus si æquatio reducenda sit $3y^5 − 6y^4 + y^3 − 8yy − 14y + 14 = 0$. Opus erit ut sequitur.

3	170		27		−7.	17
2	38	1. 2. 9. 38	12	−26. −7. 10. 11. 13. 14. 31. 50	−7.	11
1	10	1. 2. 5. 10	3	−7. −2. 1. 2. 4. 5. 8. 13	−7.	5
0	14	1. 2. 7. 14	0	−14. −7. −2. −1. 1. 2. 7. 14	−7.	−1
−1	10	1. 2. 5. 10	3	−7. −2. 1. 2. 4. 5. 8. 13	−7.	−7
−2	190		12		−7.	−13

Primo rem tento addendo et subducendo divisores quadratis terminorum progressionis 4. 1. 0. 1.[45] usurpato 1 pro [A] sed res non succedit. Quare pro [A] usurpo 3 alterum nempe coefficientis[46] divisorem & progressiones in terminis resultantibus hasce duas invenio −7, −7, −7, −7 & 11. 5. −1. −7. Expeditionis causa neglexeram divisores extimorum numerorum 170 et 190. Quare continuatis progressionibus sumo proximos earum hinc inde terminos vizt −7 & 17 superius & −7. −13 inferius ac tento si differentiæ horum & numerorum 27 ac 12 qui stant e regione in quarta columna, dividunt istos extimos numeros 170 & 190 qui stant e regione in columna secunda. Et quidem differentia inter 27 & −7 id est 34 dividit 170 et differentia inter −7 & 12 id est 19 dividit 190 sed differentia inter 17 & 27 id est 44 non dividit 170.[47] Posteriorem progressionem itaque rejicio. Juxta priorem C est +7 & B nihil, terminis progressionis nullam habentibus differentiam. Quare Divisor tentandus erit $3yy + 7$. Et divisio succedit prodeunte $y^3 − 2yy − 2y + 2$.

---

(45) A final '4.' is cancelled.
(46) Understand 'altissimi termini' (of the highest term).

Then in the progressions found I use in place of C, with signs changed, the terms 2 and -3 standing in line with the term 0; and, taking them from the lower terms 4 and -6 standing in line with the term -1, I use the residues 2 and -3 for B; and put unity in place of A. I thus have two divisors, x^2+2x-2 and x^2-3x+3, to be attempted and the reduction succeeds through either of them.

Again, if the equation to be reduced be $3y^5-6y^4+y^3-8y^2-14y+14=0$, the procedure will be as follows:

3	170		27		-7	17
2	38	1, 2, 19, 38	12	$-26, -7, 10, 11, 13, 14, 31, 50$	-7	11
1	10	1, 2, 5, 10	3	$-7, -2, 1, 2, 4, 5, 8, 13$	-7	5
0	14	1, 2, 7, 14	0	$-14, -7, -2, -1, 1, 2, 7, 14$	-7	-1
-1	10	1, 2, 5, 10	3	$-7, -2, 1, 2, 4, 5, 8, 13$	-7	-7
-2	190		12		-7	-13

I first attempt a reduction by adding and subtracting the divisors to and from the squares of the terms of the progression 4, 1, 0, 1,[45] employing 1 in A's place, but to no avail. Consequently, in A's place I employ 3 (namely the second divisor of the coefficient)[46] and I find two progressions present in the resulting terms: $-7, -7, -7, -7$ and $11, 5, -1, -7$. To expedite matters I had hitherto neglected the divisors of the outermost numbers 170 and 190. Consequently, continuing the progressions, I take their next terms in either direction (viz: -7 and 17 above, -7 and -13 below) and test if the differences between these and the numbers 27 and 12 standing in line in the fourth column divide the outermost numbers 170 and 190 standing in line in the second column. And, to be sure, the difference between 27 and -7 (that is, 34) divides 170 and the difference between -7 and 12 (that is, 19) divides 190, but the difference between 17 and 27 (that is, 44) does not divide 170.[47] I accordingly reject the latter progression. By the first one C is $+7$ and B zero, the terms of the progression having no difference. Hence the divisor to be attempted will be $3y^2+7$. And the division succeeds, yielding y^3-2y^2-2y+2.

---

(47) Read '...et differentia inter 17 & 27 id est 10 dividit 170' (and the difference between 17 and 27, that is, 10, divides 170)! The correct counter-instance is 'sed differentia inter 12 et -13 id est 25 non dividit 190', and this was duly inserted on page 179 of the revise (2, §1) below.

§3. DERIVATION OF THE 'HERONIAN' FORMULA FOR THE AREA OF A TRIANGLE.[1]

[c. 1677?]

From the original[2] in the University Library, Cambridge

PROPOSITIO.

Si de dimidio collectorum laterum dati trianguli latera sigillatim subducantur, latus continuè facti e dimidio et reliquis erit area trianguli.

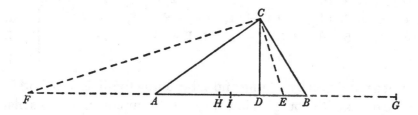

CONSTRUCTIO.

Trianguli cujuscunꝗ propositi *ABC* bisecetur latus quodvis *AB* in *I*, et in eo utrinꝗ producto capiantur tum *AE* et *AF* æquales *AC*, tum *BG* et *BH* æquales *BC*. Jungantur *CE* et *CF*, et in latus illud *AB* ab angulo opposito *C* demittatur[3] perpendiculum *CD*. Dico jam quod si de dimidio collectorum laterum $\frac{1}{2}GF$ latera *BC*, *AC*, *AB* sigillatim subducantur, latus continuè facti e dimidio $\frac{1}{2}GF$ et reliquis $\frac{1}{2}FH$, $\frac{1}{2}GE$, $\frac{1}{2}EH$ erit area trianguli $\frac{1}{2}AB \times DC$.

DEMONSTRATIO.

Sunt enim (per Prop 12 vel 13 lib. II Elem.) $2AB \times AD$ & $AB^q + AC^q - BC^q$ æqualia. Hæc aufer de æqualibus $2AB \times AE$ & $2AB \times AC$ et restabunt æqualia $2AB \times DE$ & $BC^q - AB^q + 2BAC - AC^q = BC^q - BE^q$

$$= \overline{BC + BE} \text{ in } \overline{BC - BE} = GEH.$$

---

(1) This Newtonian proof of a theorem probably first discovered by Archimedes but long identified in European eyes with the name of the Alexandrian mathematician Heron, in whose *Dioptra* and *Metrica* it is to be found (compare J. Tropfke, *Geschichte der Elementar-Mathematik in systematischer Darstellung*, **5** (Berlin/Leipzig, ₂1923): 86–8) is manifestly the original version of the much amplified (but structurally identical) 'Prob. 23' which Newton set down— almost at the last moment—about spring 1684 on page 257 of the 'polished' revise of his contemporary Lucasian lectures on algebra (see 2, §1: note (272)). We should, in consequence, not place too firm a trust in the dating of its composition implied by Newton's setting that revise between 'Lect. 2' and 'Lect. 3' of the 'Octob. 1677' series of his deposited lectures;

Translation

PROPOSITION

If from half the collected sum of the sides of a given triangle the sides are individually subtracted, the square root of the continued product of that half and the remainders will be the area of the triangle.

CONSTRUCTION

Of an arbitrary triangle ABC proposed let any side AB be bisected at I, and in its extension either way take AE and AF equal to AC and also BG and BH equal to BC. Join CE and CF, and onto the side AB from the opposite corner C let fall[3] the perpendicular CD. I now assert that, if from the half-sum of the sides $\frac{1}{2}GF$ there be separately taken the sides BC, AC, AB, then the square root of the continued product of that half $\frac{1}{2}GF$ and the remainders $\frac{1}{2}FH$, $\frac{1}{2}GE$, $\frac{1}{2}EH$ will be the triangle's area $\frac{1}{2}AB \times DC$.

DEMONSTRATION

For (by *Elements*, II, 12/13) $2AB \times AD$ and $AB^2 + AC^2 - BC^2$ are equal. Take these from the equals $2AB \times AE$ and $2AB \times AC$, and there will remain the equals $2AB \times DE$ and

$$BC^2 - AB^2 + 2AB \times AC - AC^2 = BC^2 - BE^2 = (BC + BE)(BC - BE)$$

$$= GE \times EH.$$

---

nevertheless, it is reasonable to suppose that the present text is close in style and content to the theorem given out—if indeed it ever was—to Newton's Cambridge audience in the late 1670's. We may equally readily conjecture that Newton's knowledge of the theorem was derived in the first instance from reading in his last undergraduate year (compare 1: 6, 16) Oughtred's modified presentation of its Heronian proof, making use of the triangle's in-centre, which was inserted as Theorema 20 in Caput XVIII of the third edition of his *Clavis Mathematicæ denuo Limata* (Oxford, ₃1652: 71–2). In Problema V of the following Caput XIX, equally new in the third edition (*Clavis*: 76–8), Oughtred also proved a variant form of this 'theorema de areâ trianguli plani' (to be found in still unprinted papers of Thomas Harriot penned about 1600, but first published by Woldegk Weland in Problema II of his *Strena Mathematica* (Leyden, 1640): 8–9) which an 'amicus quidam, vir doctus' had brought to his attention many years previously, but this *idem aliter* is not pursued by Newton.

(2) Add. 3959.3: 23ʳ, a stray autograph half-sheet whose original location in the corpus of Newton's papers can no longer, since its nineteenth-century cataloguer set it without annotation in a folder of miscellaneous mathematical items, be determined with accuracy.

(3) Newton first wrote less definitely '...et demittatur a vertice trianguli in basem' (let fall from the triangle's vertex onto its base).

Restantia aufer de æqualibus $2AB \times FE$ & $4BAC$, et restabunt æqualia $2AB \times DF$ &

$$AB^q + 2BAC + AC^q - BC^q = BF^q - BC^q = \overline{BF + BC} \times \overline{BF - BC} = GFH.$$

Unde contentum sub GEH et GFH est $4AB^q \times DE \times DF$ seu $4AB^q \times DC^q$. Nam[4] DC medium est proportionale inter DE ac DF. Est igitur $\dfrac{GF}{2} \times \dfrac{FH}{2} \times \dfrac{GE}{2} \times \dfrac{EH}{2}$ æquale $\frac{1}{4}AB^q \times DC^q$ et ejus latus quadratum æquale $\frac{1}{2}AB \times DC$.[5] **Q**.E.D.

---

(4) Since $AE = AF = AC$ and so EF is the diameter of the semi-circle through E, C and F.

Take the remainders from the equals $2AB \times FE$ and $4AB \times AC$, and there will be left the equals $2AB \times DF$ and

$$AB^2 + 2AB \times AC + AC^2 - BC^2 = BF^2 - BC^2 = (BF + BC)(BF - BC)$$
$$= GF \times FH.$$

Hence the 'solid' product of $GE \times EH$ and $GF \times FH$ is $4AB^2 \times DE \times DF$, that is, $4AB^2 \times DC^2$. For[4] DC is the mean proportional between DE and DF. Therefore $\frac{1}{2}GF \times \frac{1}{2}FH \times \frac{1}{2}GE \times \frac{1}{2}EH$ is equal to $\frac{1}{4}AB^2 \times DC^2$, and its square root to $\frac{1}{2}AB \times DC$.[5] As was to be demonstrated.

---

(5) That is, 'area trianguli ABC' (the area of the triangle ABC).

2

NEWTON'S LECTURES ON ALGEBRA DURING 1673-1683: THE COPY LATER DEPOSITED IN THE CAMBRIDGE ARCHIVES[1]

[Winter 1683–4?]

From the autograph[2] in the University Library, Cambridge

§1. THE MAIN TEXT.

[ARITHMETICA UNIVERSALIS.][3]

‖[1]
Octob. 1673.
Lect 1

‖ Computatio vel fit per numeros ut in vulgari Arithmetica vel per species ut Analystis mos est. Utraꝗ ijsdem innititur fundamentis et ad eandem metam collimat, Arithmetica quidem definitè et particulariter, Algebraica[4] autem

---

(1) Though this revise, polished and partly augmented, of Newton's professorial lectures on algebra during the decade from 1673 came, in the version edited by his successor William Whiston at Cambridge in 1707 and subsequently reissued at London in 1722 in corrected form by Newton himself, ultimately to be the most widely read and influential of his writings, it had no immediate impact upon his contemporaries. With the possible, transient exception of John Flamsteed (see 1, §1: note (1)), we know the name of no student who attended the original lectures; nor do we know the circumstances in which, after a ten-year gap, Newton felt called upon suddenly to obey the lapsed provision in the Lucasian statutes (see III: xviii) which ordained that he make deposit in the University Library of 'polished copies' of (no fewer than) ten such professorial lectures *per annum*. No reference to these lectures on algebra occurs anywhere in his extant correspondence of the period, nor is the deposited copy, here reproduced for the first time in its original state, recorded by any contemporary till Roger Cotes made a full transcript of its text as a Trinity undergraduate in the winter of 1701/2. (At the end of his copy, now in Trinity College, Cambridge. R.16.39, Cotes added the note: 'Descripsi ex Autographo anno 1701/2. R.C.') Even the date of its deposit is unrecorded, and the few preliminary drafts of its content which have survived (see 1, §§1–3 preceding) throw no light on the external reasons for its composition. In default of exact documentation, we have to trust the internal evidence of the text that it represents the uneventful codification of the essence of what he had delivered on algebra and analytical geometry in his professorial lectures over the decade 1673–83. It will be evident that whole sections of it are in fact taken over virtually word for word from earlier researches, notably from his 'Observationes'

Translation

[Universal Arithmetic][3]

Computation is conducted either by means of numbers, as in common arithmetic, or through general variables, as is the habit of analytical mathematicians. Both types rest on the same foundations and are directed to the same goal, the arithmetical proceeding indeed by an approach which is definite and particular, but the algebraic[4] advancing in an indefinite, universal way so that

(II: 364–444) on Gerard Kinckhuysen's *Algebra* and the following 'Problems for construing æquations' (II: 450–516) and also (in Latin rephrase) from his earliest studies (I: 489–539) of the theory of equations and the delimitation of their roots. Detailed reference to these earlier passages, and also to variants introduced into the two primary published editions (of 1707 and 1722) and in Newton's annotated library copy of the former (now in Leicester) will be made at appropriate places in following footnotes. His strong criticism of the deficiencies of Whiston's unauthorized 1707 edition, partially implemented in his 1722 revise, is discussed at length in the preceding introduction.

(2) Dd. 9.68: pages 1–251 (of the main text here given) are so numbered in Newton's hand; pages 253–68 (of the 'Correctiones' reproduced in §2 following) are left unmarked by him. The manuscript was evidently used without intermediary transcript as printer's copy for the 1707 *editio princeps*, for paginating marks (corresponding to the *paginæ primæ* [pages 145, 177 and 305 of the published version] of sheets K, M and V) are inserted at the appropriate divisions on Newton's pages 82, 103 and 233 respectively: this will explain certain untutored misreadings of Newton's hand which appear in the printed version, such as '&' (='et') for 'ut' (note (5)) and 'cum' for 'eum' (note (57)). (Compare D. F. McKenzie, *The Cambridge University Press, 1696–1712*, **1** (Cambridge, 1966): 278–9.) Our assessment of Newton's handwriting suggests that the text was penned by him, with care apart from some lapses, over a narrow span of time in the early 1680's. Narrower bounds to its date of composition are yielded by external evidence: since Newton's immediate revise of its text, the unfinished 'Arithmeticæ Universalis Liber Primus' (ULC. Add. 3993, reproduced in 2, 1 below), is in the hand of his amanuensis Humphrey Newton, we may be reasonably sure that the present manuscript was written shortly before the latter's arrival in Cambridge in early 1684 (see IV: 169 note (2)). From its margins we faithfully reproduce Newton's imposed division of its text into eleven annual 'Octob.' series between 1673 and 1683, and their further breakdown into $(7+10+10+10+10+6+6+8+10+10+10=)97$ 'lectiones', but we would again caution the reader that—as with his other deposited sets of Lucasian lectures—this preferred chronology is not narrowly to be trusted, being merely a framework created to satisfy a statutory requirement.

(3) As distinct from 'Arithmetica particularis' (numerical arithmetic) in which particular, invariant numbers are combined by the same fundamental operations of addition, subtraction, multiplication, division and root-extraction. There is no heading in the manuscript itself, but in introducing this familiar short form of Whiston's 1707 appellation, *Arithmetica Universalis; seu De Compositione et Resolutione Arithmetica Liber* (which, if not then sanctioned by Newton, was certainly retained by him in his 1722 revised edition of the work) we both anticipate the title of Newton's unfinished redraft (2, [1]: note (3)) and adhere to a tradition now more than two centuries old.

(4) Understand 'computatio' (computation) once more; the equivalent 'Algebra' is cancelled.

indefinitè et universaliter ita ut[5] enuntiata ferè omnia quæ in hac computa-tione[6] habentur, et præsertim conclusiones, Theoremata dici possint. Verùm Algebra maximè præcellit quòd cùm in Arithmetica Quæstiones tantùm resolvantur progrediendo a datis ad quæsitas quantitates, hæc a quæsitis tanquam datis ad datas tanquam quæsitas quantitates plerumcȝ regreditur ut ad conclusionem aliquam, seu æquationem, quocuncȝ demum modo perveniatur, ex quâ quantitatem quæsitam elicere liceat. Eocȝ pacto conficiuntur[7] difficillima Problemata quorum resolutiones ex Arithmetica sola frustra peterentur. Arithmetica tamen Algebræ in omnibus ejus operationibus ita subservit, ut non nisi unicam perfectam computandi scientiam constituere videantur; et utramcȝ propterea conjunctim explicabo.

Quisquis hanc scientiam aggreditur, imprimis notas[8] intelligat, et funda-mentales addiscat operationes, Additionem nempe, Subductionem, Multiplica-tionem, Divisionem, Extractionem Radicum, Reductiones fractionum et radicalium quantitatum, et modos ordinandi terminos Æquationum ac in-cognitas quantitates (ubi plures sunt) exterminandi. Deinde has operationes reducendo Problemata ad æquationes exerceat, et ultimò naturam et resolu-tionem æquationum contempletur.

‖[2]　　　　　　　　　　‖De Notatione.[9]

Integrorum numerorum notas (0, 1, 2, 3, 4 &c) et notarum, ubi plures inter se nectuntur, valores nemo non intelligit. Quemadmodum verò numeri in primo loco ante unitatem, sive ad sinistram, scripti denotant denas unitates, in secundo centenas, in tertio millenas &c: sic numeri in primo loco post unitatem scripti denotant decimas partes unitatis, in secundo centesimas, in tertio millesimas, &c. Priores dici solent integri numeri, posteriores vero fracti decimales. Et ad distinguendum integros a decimalibus[10] interjici solet comma vel punctum vel etiam lineola. Sic numerus 732'569 denotat septingentas triginta duas unitates, una cum quincȝ decimis, sex centesimis, et novem millesimis partibus unitatis. Qui et sic 732,569, vel sic 732˙569[11] vel etiam sic 732|569 nonnunquam scribitur. Atcȝ ita numerus 57104'2083 denotat quin-

---

(5) The 1707 *editio princeps* here (page 1) has '&'; Whiston's printer evidently misread Newton's 'ut' as 'et' (compare note (2)).

(6) The equivalent 'in hoc computo' was first written.

(7) Newton has cancelled 'transiguntur' (are transacted).

(8) When Newton subsequently determined to replace the following section by one 'De vocum quarundam et notarum significatione' (see next note), he changed this correspondingly to read 'vocum et notarum significationes' (the meanings of appellations and symbols).

(9) In later correction (see §2: note (2) following) Newton was to replace this passage by a lengthier equivalent section bearing the title 'De vocum quarundam et notarum significa-tione' (The meaning of certain appellations and symbols).

(10) 'a numeris post unitatem' (from the numbers after the unit) was first written.

almost all pronouncements which are made in this style of computation—and especially its conclusions—may be called theorems. However, algebra most excels, in contrast with arithmetic where questions are solved merely by progressing from given quantities to those sought, in that for the most part it regresses from the sought quantities, treated as given, to those given, as though these were the ones sought, so as ultimately and in any manner to attain some conclusion—that is, an equation—from which it is permissible to derive the quantity sought. By this technique the most difficult problems are accomplished,[7] ones whose solution it would be useless to seek of arithmetic alone. Yet arithmetic is so instrumental to algebra in all its operations that they seem jointly to constitute but a single, complete computing science, and for that reason I shall explain both together.

Whoever enters upon this science should in the first place understand the symbols[8] and also learn the fundamental operations: namely, addition, subtraction, multiplication, division, the extraction of roots, the reductions of fractions and radical quantities, and the ways of ordering the terms of equations and of eliminating unknown quantities in them (when they have several). Then let him practise these operations by reducing problems to equations, and, finally, consider the nature and solution of equations.

NOTATION[9]

The symbols (0, 1, 2, 3, 4, ...) for integers, and the values of those symbols when several are joined together, everyone understands. To be sure, numbers written, for instance, in the first place immediately before a unit (on its left, that is) denote tens of units, those in the second place hundreds, in the third thousands, and so on; likewise, numbers written in the first place after a unit denote tenths of a unit, those in the second place hundredths, in the third thousandths, and so forth. The former are usually called integers, but the latter, however, decimal fractions. And to distinguish integers from decimals[10] a comma, a point or even a short line is usually interposed. Thus the number 732'569 denotes seven hundred and thirty-two units, together with five tenths, six hundredths and nine thousandths of a unit. This is also sometimes written 732,569 or thus 732˙569[11] or even in this way 732|569. And similarly the

---

(11) '732: 569' is cancelled. The comma on the line is François Viète's in the tabulations in his *Canon Mathematicus, Seu Ad Triangula cum Adpendicibus* (Paris, 1579) and is, following his lead, common in seventeenth-century trigonometrical tables. The point or (more usually) comma above the line—replacing an original prime?—is found in several contemporary arithmetics; Newton's cancelled variant of the colon as decimal divider occurs in Richard Balam's *Algebra* (London, 1653) and Richard Rawlyns' *Practical Arithmetick* (London, 1656). The 'angle' separatrix which follows (Newton's favoured form as an undergraduate—see I: 489 note (3)—and in many of his worksheet computations) was introduced by Oughtred

quaginta septem mille, centum et quatuor unitates una cum duabus decimis, octo millesimis, et tribus decimis millesimis partibus unitatis. Et numerus 0'064 denotat sex centesimas et quatuor millesimas partes.

Cùm rei alicujus quantitas ignota est vel indeterminatè spectatur, ita ut per numeros non liceat exprimere, solemus per speciem aliquam seu literam designare. Et siquando cognitas quantitates tanquam indeterminatas spectemus, discriminis causa designamus initialibus Alphabetæ literis *a*, *b*, *c*, *d*, et incognitas finalibus *z*, *y*, *x*, &c. Aliqui[12] pro cognitis substituunt consonantes vel majusculas literas, et vocales vel minusculas pro incognitis.

Numerus speciei alicui immediatè præfixus denotat speciem illam toties sumendam esse. Sic $2a$ denotat duo *a*, $3b$ tria *b*, $15x$ quindecim *x*.

Quantitates vel affirmativæ sunt seu majores nihilo, vel negativæ seu nihilo minores. Sic in rebus humanis possessiones dici possunt bona affirmativa, debita verò bona negativa.[13] Incꝗ motu locali[14] progressus dici potest motus affirma-

‖[3] tivus, et regressus motus negativus, quia prior auget et posterior diminuit ‖ iter confectum. Et ad eundem modum in Geometria, si linea versus plagam quamvis ducta pro affirmativa habeatur, negativa erit quæ versus plagam oppositam ducitur. Veluti si *AB* dextrorsum ducatur et *BC* sinistrorsum ac *AB* statuatur affirmativa tunc *BC* pro negativa habebitur, eò quòd inter ducendum diminuit

AB redigitꝗ vel ad breviorem *AC*, vel ad nullam si forte *C* inciderit in ipsum *A*, vel ad minorem nulla si *BC* longior fuerit quam *AB* de qua aufertur. Negativæ quantitati designandæ nota $-$[5] affirmativæ nota $+$ præfigi solet. Et ubi neutra notarum præfigitur nota $+$ subintelligenda est. Adhoc $\mp$ signum incertum est, & $\pm$ signum etiam incertum sed priori contrarium.[15]

In aggregato quantitatum nota $+$ significat quantitatem suffixam esse cæteris addendam, & nota $-$ esse subducendam. Et has notas vocabulis *plus* et *minus* exprimere solemus. Sic $2+3$, sive 2 plus 3, valet summam numerorum 2 et 3, hoc est 5. Et $5-3$, sive 5 minus 3, valet differentiam quæ oritur subducendo 3 a 5, hoc est 2. Et $-5+3$ valet differentiam quæ oritur subducendo 5 a 3, hoc est -2. Et $6-1+3$ valet 8. Item $2a+3a$ valet $5a$. Et $3b-2a-b+3a$ valet

---

in his *Clavis Mathematicæ* (London, ₁1631). In Whiston's 1707 *editio princeps* (page 2) the variants with a comma (on and above the line) as separatrix were garbled into the forms 732, |569 and 732'|569 with short intruding vertical rules, while the 'angle' notation was printed 732L569, the separatrix now appearing as a distinct symbol; these bastards were accepted by Newton in 1722 and repeated—with further distortions and the widespread omission of the comma where it is set as a division above the line!—in all subsequent editions. See F. Cajori, *A History of Mathematical Notations*, **1** (La Salle, Illinois, 1928): 314–27, 329. In our English version we will hereafter use the conventional Napierian point as decimal separatrix.

(12) Notably Albert Girard and Oughtred, following the lead set in 1591 by Viète; see Cajori's *History of Mathematical Notations*, (note (11) above), **1**: 181–3, 380. The preceding notation is, of course, that introduced by Descartes in 1637 in his *Geometrie*.

number 57104'2083 denotes fifty-seven thousand, one hundred and four units, together with two tenths, eight thousandths and three ten-thousandths of a unit. And the number 0'064 denotes six hundredths and four thousandths.

When the quantity of some object is unknown or indeterminately regarded (so that it is impermissible to express it by means of numbers), we usually designate it by some species or letter. And should we ever regard known quantities as indeterminate, for distinction's sake we designate these by initial letters of the alphabet a, b, c, d, [...] and the (true) unknowns by terminal ones z, y, x, ... Some[12] in place of known quantities substitute consonants or capital letters, using vowels or small letters for unknowns.

A number set immediately in front of some variable denotes that the variable is to be taken an equivalent number of times. Thus $2a$ denotes two a, $3b$ three b, $15x$ fifteen x.

Quantities are either positive, that is, greater than zero, or negative, that is, less than zero. So in human affairs possessions can be called positive goods, but debts negative ones.[13] And in local motion[14] an advance can be called a positive motion and a retreat a negative one, since the former increases the course completed and the latter diminishes it. And in much the same manner in geometry, if a line drawn in some direction be considered as positive, then its negative is one drawn in the opposite direction. For instance, if AB be drawn rightwards and BC leftwards, and AB is decreed to be positive, then BC will be considered as negative because, as it is being drawn, it diminishes AB, reducing it to the shorter length AC, or to nothing should C by chance coincide with A, or to less than nothing were BC longer than the length AB from which it is taken. To denote a negative quantity the symbol $-$ is usually prefixed, to a positive one the symbol $+$. And when neither symbol is prefixed, the sign $+$ must be understood. In addition, the sign $\mp$ is uncertain, and also the sign $\pm$ but converse to the first.[15]

In a collection of quantities the symbol $+$ signifies that the quantity appended to it must be added to the rest, the symbol $-$ that it must be taken away from them. And these symbols we usually express in words as 'plus' and 'minus'. So $2+3$, that is, 2 plus 3, has for its value the sum of the numbers 2 and 3, namely 5. And $5-3$, that is, 5 minus 3, is in value the difference which arises on taking 3 from 5, namely 2. And $-5+3$ is in value the difference which arises on taking 5 from 3, namely -2, while the value of $6-1+3$ is 8. That of $2a+3a$, likewise, is $5a$. And the value of $3b-2a-b+3a$ is $2b+a$; for that of $3b-b$ is $2b$ and that of

---

(13) The same phrasing occurs in Newton's contemporary 'Geometria Curvilinea' in the scholium to Definition 4 of its 'Lib. 1' (see IV: 424).

(14) Compare III: 70, note (80).

(15) This last sentence is a late insertion in the manuscript.

$2b + a$; nam $3b - b$ valet $2b$ et $-2a + 3a$ valet a, quorum aggregatum[16] est $2b + a$. Et sic in alijs. Hæ autem notæ $+$ et $-$ dicuntur signa. Et ubi neutrum initiali quantitati præfigitur signum $+$ subintelligi debet.

Duæ vel plures species immediatè connexæ designant factum, seu quantitatem quæ fit per multiplicationem omnium in se invicem. Sic ab denotat quantitatem quæ fit multiplicando a per b. Et abx denotat quantitatem quæ fit multiplicando a per b, et factum illud per x. Puta si a sit 2, et b sit 3, et x sit 5, tum ab erit 6 et abx 30.

Inter multiplicantes quantitates comma vel nota $\times$ vel vocabulum *in*, ad factum designandum nonnunquam interscribitur. Sic 3, 5 3$\times$5 vel 3 in 5 denotat 15. Sed usus harum notarum præcipuus est, ubi compositæ quantitates sese multiplicant. Veluti si $y - 2b$ multiplicet $y + b$, terminos utriusꝗ multiplicatoris lineolâ superimposita connectimus et scribimus $\overline{y - 2b}$ in $\overline{y + b}$, vel $\overline{y - 2b}$, $\overline{y + b}$.[17]

|[4] ||[18] Si quantitas seipsam multiplicet, numerus factorum, compendij gratia, suffigi solet. Sic pro aaa scribimus a^3, pro $aaaa$ scribimus a^4, pro $aaaaa$ scribimus a^5, et pro $aaabb$ scribimus a^3bb vel a^3b^2.[19] Puta si a sit 5 et b sit 2, tum a^3 erit $5 \times 5 \times 5$ sive 125, et a^4 erit $5 \times 5 \times 5 \times 5$ sive 625, atꝗ a^3b^2 erit $5 \times 5 \times 5 \times 2 \times 2$ sive 500. Ubi nota quod numerus inter duas species immediatè scriptus, ad priorem semper pertinet. Sic 3 in quantitate a^3bb non denotat bb ter capiendum esse sed a in se bis[20] ducendum. Nota etiam quod hæ quantitates tot dimensionum vel potestatum vel dignitatum esse dicuntur quot factoribus seu quantitatibus se multiplicantibus constant, et numerus suffixus vocatur index potestatum vel dimensionum. Sic aa est duarum dimensionum vel potestatum, & a^3 trium, ut indicat suffixus numerus 3. Dicitur etiam aa quadratum, a^3 cubus, a^4 quadrato-quadratum, a^5 quadrato-cubus, a^6 cubo-cubus, a^7 quadrato-quadrato-cubus, et sic porro. Et quantitas a ex cujus in se multiplicatione hæ potestates generantur dicitur earum radix, nempe radix quadratica quadrati aa, cubica cubi a^3, &c.

Quantitas infra quantitatem cum lineola interjecta[21] denotat quotum,[22] seu quantitatem quæ oritur ex divisione superioris quantitatis per inferiorem. Sic $\frac{6}{2}$ denotat quantitatem quæ oritur dividendo 6 per 2, hoc est 3. et $\frac{5}{8}$ quanti-

---

(16) This replaces 'summa' (sum).

(17) In sequel, in our English version we will employ modern brackets to render Newton's superscript braces and multiplying comma (on which see Cajori's *History*, 1: 267).

(18) This and the three paragraphs next following are interchanged in Newton's revise (see §2: note (14) following).

(19) We will use the latter form with 'b^2' (rather than the equivalent 'bb') in the English version, in line with modern practice. While only rarely preferring 'bbb' to 'b^3'—and '$bbbb$' to 'b^4' almost never—, like many of his contemporaries Newton writes 'bb' interchangeably with 'b^2' in his manuscript papers.

$-2a+3a$ is a, and their aggregate[16] is $2b+a$. And so in other cases. These symbols $+$ and $-$, however, are called 'signs'. And when neither is prefixed to an initial quantity, the sign $+$ should be understood.

Two or more variables immediately connected denote their product, or the quantity which results by multiplying all together in turn. Thus ab denotes the quantity which comes from multiplying a by b. And abx the quantity which comes from multiplying a by b, and then that product by x. If, say, a be 2, b be 3 and x 5, then ab will be 6 and abx 30.

Between multiplying quantities a comma or the symbol '$\times$' or the word 'into' is not infrequently written to designate the product. Thus '3,5' or '3×5' or '3 into 5' denotes 15. But the most important use of these symbols is when compound quantities multiply one another. If, for instance, $y-2b$ should multiply $y+b$, we connect the terms in each multiplier by a short, overhead line and write '$\overline{y-2b}$ into $\overline{y+b}$' or '$\overline{y-2b}, \overline{y+b}$'.[17]

[18]If a quantity should multiply itself, for shortness the number of factors is usually appended. Thus in place of aaa we write a^3, in place of $aaaa$ we write a^4, in place of $aaaaa$ we write a^5, and in place of $aaabb$ we write a^3bb or a^3b^2.[19] If, say, a be 5 and b 2, then a^3 will be $5 \times 5 \times 5$ or 125, a^4 will be $5 \times 5 \times 5 \times 5$ or 625, and a^3b^2 $5 \times 5 \times 5 \times 2 \times 2$ or 500. Here note that a number written immediately between two variables relates always to the preceding one. Thus '3' in the quantity a^3b^2 does not signify that b^2 is to be taken three times, but denotes that a has twice[20] to be multiplied into itself. Note also that these quantities are said to be of as many dimensions, powers or 'dignities' as they comprise factors or quantities multiplying one another, and that the number appended is called the index of their power or dimension. Thus a^2 is of two dimensions or powers, while a^3, as its appended number '3' indicates, is of three. Also, a^2 is said to be the square, a^3 the cube, a^4 the square-square (fourth power), a^5 the square-cube (fifth power), a^6 the cube-cube (sixth power), a^7 the square-square-cube (seventh power), and so forth. And the quantity a from whose self-multiplication these powers are generated is said to be their root—precisely, the square root of the square a^2, the cube one of the cube a^3, and so on.

A quantity set below a quantity with a short intervening rule[21] denotes their quotient, that is, the quantity which arises from the division of the upper quantity by the lower one. Thus $\frac{6}{2}$ denotes the quantity which arises on dividing 6 by 2, namely 3; and $\frac{5}{8}$ the quantity which arises on dividing 5 by 8, that is, an

---

(20) Since $a^3 = (a \times a) \times a$.

(21) When it is appropriate, the oblique solidus / will be used in the English version to denote a simple division.

(22) This replaces the equivalent 'quotientem'.

tatem quæ oritur dividendo 5 per 8, hoc est octavam partem numeri 5, et $\frac{a}{b}$ denotat quantitatem quæ oritur dividendo a per b: puta si a sit 15 et b 3, tum $\frac{a}{b}$ denotat 5. Et sic $\frac{ab-bb}{a+x}$ denotat quantitatem quæ oritur dividendo $ab-bb$ per $a+x$. Atq; ita in alijs. Hujusmodi autem quantitates fractiones dicuntur, parsq; superior Numerator, ac inferior Denominator.(23)

Aliquando Divisor quantitati divisæ, interjecto arcu, præfigitur. Sic ad designandam quantitatem quæ oritur ex divisione $\frac{axx}{a+x}$ per $a-b$, scribi potest $\overline{a-b})\frac{axx}{a+x}$.

Etsi multiplicatio per immediatam quantitatum conjunctionem denotari ‖[5] solet, ‖tamen numerus integer ante numerum fractum denotat summam utriusq;. Sic $3\frac{1}{2}$ denotat tria cum semisse.

(24)Ad designandam radicem alicujus quantitatis præfigi solet nota $\surd$ si radix sit quadratica, et $\surd 3$: vel $\surd$③ si sit cubica, et $\surd 4$: vel $\surd$④ si quadrato-quadratica, &c:(25) Sic $\surd aa$ denotat radicem quadraticam ex aa, hoc est a. Et $\surd\overline{ab-xx}$ denotat radicem quadraticam ex $ab-xx$. Et $\surd 3:\overline{abb-x^3}$ denotat radicem cubicam ex $abb-x^3$. Et $\surd 4:\frac{bbc^4+y^6}{cy}$ radicem quadrato-quadraticam ex $\frac{bbc^4+y^6}{cy}$.

Nonnulli(26) pro designanda quadratica potestate usurpant q, pro cubica c, pro quadrato-quadratica qq, pro quadrato-cubica qc, &c. Et ad hunc modum pro quadrato, cubo, et quadrato-quadrato ipsius A, scribitur A^q, A^c, A^{qq}, &c. Et pro radice cubica ex $abb-x^3$ scribitur $\surd c:\overline{abb-x^3}$. Alij alias notas adhibent sed quæ jam ferè exoleverunt.

Nota $=$(27) designat quantitates hinc inde æquales esse. Sic $x=b$ designat x æqualem esse b.

Nota :: significat quantitates hinc inde proportionales esse. Sic $a.b::c.d$,(28) significat esse a ad b ut c ad d. Et $a.b.e::c.d.f$(28) esse a, b et e inter se ut

(23) This last sentence is a late addition in the manuscript.

(24) In his revise (§2 below) Newton has inserted an additional paragraph at this point.

(25) The 'ring' notation for higher radicals was introduced by Stevin in his *Arithmetique* (Leyden, 1585) and afterwards popularized by Schooten in his 'Additamentum' on extracting binomial roots to the Latin editions of Descartes' *Geometrie* in 1649 and 1659; the notation using a colon as separatrix, a variant on Oughtred's '$\surd c$:' and '$\surd qq$:' (see the next paragraph) is seemingly Newton's present innovation. See Cajori's *History of Mathematical Notations* (note (11) above), **1**: 158, 192, 371–2.

(26) Following Oughtred (see Cajori's *History of Mathematical Notations*, **1**: 191); the equivalent upper-case notation of Q, C, QQ, ..., together with N ('numerus') for the unit had

eighth of the number 5; while $\frac{a}{b}$ denotes that which arises on dividing a by b—

if, say, a be 15 and b be 3, then it denotes 5. And similarly $\frac{ab-b^2}{a+x}$ denotes the

quantity which arises on dividing $ab-b^2$ by $a+x$. And likewise in other instances.
Quantities of this sort, however, are called fractions, the upper part the
numerator and the lower one the denominator.[23]

Sometimes the divisor is prefixed to the quantity divided, with an arc
inserted between. Thus, to designate the quantity which arises from the

division of $\frac{ax^2}{a+x}$ by $a-b$, there can be written $\overline{a-b})\frac{axx}{a+x}$.

Even though multiplication is usually denoted by an immediate joining of
quantities, an integer before a fraction none the less denotes the sum of each.
So $3\frac{1}{2}$ denotes three together with a half.

[24]To designate the root of some quantity the symbol $\sqrt{}$ is usually prefixed if
the root be square, $\sqrt{3}$: or $\sqrt{③}$ if it be a cube, $\sqrt{4}$: or $\sqrt{④}$ if it be a fourth one,
and so on.[25] Thus $\sqrt{aa}$ denotes the square root of a^2, that is, a, and $\sqrt{ab-xx}$
that of $ab-x^2$; while $\sqrt{3}:\overline{abb-x^3}$ denotes the cube root of ab^2-x^3, and
$\sqrt{4}:\frac{bbc^4+y^6}{cy}$ the fourth root of $\frac{b^2c^4+y^6}{cy}$.

Some people[26] use q to designate a quadratic (square) power, c for a cube,
qq for a square-square (fourth) power, and so on. And in the same way for the
square, cube and square-square of A is written A^q, A^c, A^{qq}, and so forth. And for
the cube root of ab^2-x^3 is written $\sqrt{c}:\overline{abb-x^3}$. Others employ still other
symbols, but these are now practically obsolete.

The symbol $=$[27] denotes that the quantities on its either side are equal.
Thus $x=b$ denotes that x is equal to b.

The symbol $::$ signifies that the quantities on its either side are proportional.
Thus $a.b::c.d$[28] signifies that a is to b as c to d; and $a.b.e::c.d.f$[28] that

appeared earlier in Alexander Anderson's edition of Viète's *De Emendatione Æquationum* (Paris,
1615) (see Cajori's *History*, **1**: 186). Both Viète and Oughtred invariably wrote their notations
for powers on the line immediately following the variable, and were copied in this by such
English mathematicians as Isaac Barrow in his Latin *Euclid* (Cambridge, 1655) and *Lectiones
Geometricæ* (London, 1670); the happy inspiration of raising them to be superscript in 'Cartesian' style was Newton's (compare II: 470 ff., III: 458 ff.) and is widely used by him in his
geometrical papers. In our English version we will henceforth employ the modern convention
$\sqrt[n]{}$ for the n-th root.

(27) Robert Recorde's symbol for equality, than which 'noe 2 thynges can be moare
equalle' (*The Whetstone of Witte* (London, 1557); see Cajori's *History*, **1**: 164–5), popularized
in England by Walter Warner (in his posthumous edition of Harriot's *Praxis*), Oughtred and
Richard Norwood (*ibid.*, **1**: 298).

(28) Newton's adaptation of Oughtred's notation (see Cajori's *History of Mathematical
Notations*, **1**: 196, 285–7), used by him universally in his mathematical papers from the time

sunt c, d et f inter se respectivè, vel esse a ad c, b ad d et e ad f in eadem ratione.

[1673]   DE ADDITIONE.

Lect 2   Numerorum, ubi non sunt admodum compositi, Additio per se manifesta est. Sic quod 7 et 9 seu $7+9$ faciunt 16, et quod $11+15$ faciunt 26 prima fronte patet. At in magis compositis opus peragitur scribendo numeros in serie descendente et summas columnarum sigillatim colligendo. Quemadmodum si numeri 1357 et 172 addendi sunt, scribe alterutrum 172 infra alterum 1357 ita ut hujus unitates 2 alterius unitatibus 7 subjiciantur, cæteriᴃ numeri   1357 numeris correspondentibus, nempe deni 7 denis 5 et centenus 1 centenis 3.   172 Tum incipiendo ad dextram, dic 2 et 7 faciunt 9 quem scribe infra. Item   ‾1529‾

‖[6]   7 et 5 faciunt 12, cujus posterio‖rem numerum 2 scribe infra, priorem vero 1 asserva proximis numeris 1 et 3 adjiciendum. Dic itaᴃ præterea 1 et 1 faciunt 2, cui 3 adjectus facit 5, et scribe 5 infra et manebit tantùm 1 prima figura superioris numeri, quæ etiam infra scribenda est, et sic habebitur summa 1529.

    Sic numeros $87899+23403+885+1920$, quo in unam summam redigantur, scribe in serie descendente ita ut unitates unam columnam, deni numeri aliam, centeni tertiam, milleni quartam constituant, et sic   87899 præterea. Deinde dic $5+3$ valent 8, et $8+9$ valent 17, scribeᴃ 7 infra   23403 et 1 adjice proximis numeris dicendo $1+8$ valent 9, $9+2$ valent 11, ac   1920 $11+9$ valent 20: Subscriptoᴃ 0, dic iterum ut ante $2+8$ valent 10,   885 $10+9$ valent 19, $19+4$ valent 23, et $23+8$ valent 31, adeoᴃ asservato   ‾114107‾ 3 subscribe 1 ut ante, et iterum dic $3+1$ valent 4, $4+3$ valent 7, et $7+7$ valent 14. Quare subscribe 4, denuoᴃ dic $1+2$ valent 3, et $3+8$ valent 11, quem ultimò subscribe et omnium summam habebis 114107.

    Ad eundem modum numeri decimales adduntur ut in annexo   630'953 paradigmate videre est.   51'0807

[29]In terminis Algebraicis Additio fit connectendo quantitates   305'27 addendas cum signis proprijs, et insuper uniendo quæ possunt   ‾987'3037‾ uniri. Sic a et b faciunt $a+b$; a et $-b$ faciunt $a-b$; $-a$ et $-b$ faciunt $-a-b$; $7a$ et $9a$ faciunt $7a+9a$; $-a\sqrt{ac}$ et $b\sqrt{ac}$ faciunt $-a\sqrt{ac}+b\sqrt{ac}$ vel $b\sqrt{ac}-a\sqrt{ac}$, nam perinde est quo ordine scribantur.

    Quantitates affirmativæ quæ ex parte specierum conveniunt, uniuntur[30]

(late 1660's) when he abandoned his youthful notation making use of the colon in place of the stop. In our English version we will henceforth use the modern equivalents $a:b=c:d$ and $a:b:e=c:d:f$.

(29) The following paragraph is a late insertion in the margin, replacing an equivalent passage cancelled on Newton's next page: 'Quæ nequeunt his modis uniri sufficit connectere signo $+$. Sic ad designandam summam quantitatum a & b scribimus tantum $a+b$' (Magnitudes impossible to unite by these means it is sufficient to connect with the sign $+$. Thus to designate the sum of the quantities a and b we write merely $a+b$).

a, *b* and *e* are to one another respectively as *c*, *d* and *f* are to each other, or that *a* to *c*, *b* to *d* and *e* to *f* are in the same ratio.

ADDITION

In the case of numbers which are not unduly complicated addition is self-evident. Thus it is clear at first glance that 7 and 9, that is, $7+9$, make 16 and that $11+15$ make 26. But in more complicated cases the operation is achieved by writing the numbers in a descending sequence and gathering the sums of the columns separately. For instance, if the numbers 1357 and 172 are to be added. Write the one, 172, below the other, 1357, so that the units, 2, of the latter lie beneath the units, 7, of the other and its remaining numbers beneath corresponding ones, namely the tens 7 below tens 5 and hundreds 1 below the hundreds 3; then, beginning on the right, say: 2 and 7 make 9, and write it underneath; likewise, 7 and 5 make 12, and write its latter number, 2, underneath, but keep the first, 1, for the next numbers 1 and 3 to be summed. Accordingly, say further: 1 and 1 make 2, and when to it 3 is added it makes 5; write 5 underneath and there will remain only 1, the first figure in the upper number, which must also be written beneath. And so the sum 1529 will be had.

Similarly, to reduce the numbers $87899 + 23403 + 885 + 1920$ to a single sum, write them in a descending sequence so that units form one column, tens a second, hundreds a third, thousands a fourth, and so on. Then say: the value of $5+3$ is 8 and of $8+9$ is 17, write 7 underneath and add 1 to the next numbers, saying: the value of $1+8$ is 9, of $9+2$ is 11 and of $11+9$ is 20. Again, having entered 0 below, say as before: the value of $2+8$ is 10, of $10+9$ is 19, of $19+4$ is 23 and of $23+8$ is 31; so, retaining 3, write 1 below as before, and once more say: the value of $3+1$ is 4, of $4+3$ is 7 and of $7+7$ is 14. Consequently, write 4 beneath, and yet again say: the value of $1+2$ is 3 and of $3+8$ is 11. Lastly, write this number below and you will have the total sum 114107.

In much the same way decimal numbers are added, as may be seen in the pattern of the example attached.

$$\begin{array}{r} 630{\cdot}953 \\ 51{\cdot}0807 \\ 305{\cdot}27 \\ \hline 987{\cdot}3037. \end{array}$$

[29]In algebraic expressions addition is performed by connecting quantities to be added with their appropriate signs and, besides, uniting what can be united. Thus *a* and *b* make $a+b$; *a* and $-b$ make $a-b$; $-a$ and $-b$ make $-a-b$; $7a$ and $9a$ make $7a+9a$; $-a\sqrt{ac}$ and $b\sqrt{ac}$ make $-a\sqrt{ac}+b\sqrt{ac}$, or $b\sqrt{ac}-a\sqrt{ac}$ (for the order in which they are to be written is of no consequence).

Positive quantities which concur in regard to their variables are united[30] by

---

(30) This replaces 'in unam summam colliguntur' (are collected into one sum). In seventeenth-century terminology such quantities are called 'communicant' (compare II: 309, note (29)).

addendo numeros præfixos quibus species multiplicantur. Sic $7a+9a$ faciunt $16a$, et $11bc+15bc$ faciunt $26bc$. Item $3\frac{a}{c}+5\frac{a}{c}$ faciunt $8\frac{a}{c}$, et $2\sqrt{ac}+7\sqrt{ac}$ faciunt $9\sqrt{ac}$, et $6\sqrt{ab-xx}+7\sqrt{ab-xx}$ faciunt $13\sqrt{ab-xx}$. Et ad eundem modum $6\sqrt{3}+7\sqrt{3}$ faciunt $13\sqrt{3}$.[31] Quinetiam $a\sqrt{ac}+b\sqrt{ac}$ faciunt $\overline{a+b}\sqrt{ac}$, additis nempe a et b tanquam si essent numeri multiplicantes $\sqrt{ac}$. Et sic $\overline{2a+3c}\sqrt{\frac{3axx-x^3}{a+x}}$ et $+3a\sqrt{\frac{3axx-x^3}{a+x}}$ faciunt $\overline{5a+3c}\sqrt{\frac{3axx-x^3}{a+x}}$ eo quod $2a+3c$ et $3a$ faciunt $5a+3c$.

‖[7]　　‖ Fractiones affirmativæ quarum idem est denominator, uniuntur[30] addendo numeratores. Sic $\frac{1}{5}+\frac{2}{5}$ faciunt $\frac{3}{5}$, et $\frac{2ax}{b}+\frac{3ax}{b}$ faciunt $\frac{5ax}{b}$, et $\frac{8a\sqrt{cx}}{2a+\sqrt{cx}}+\frac{17a\sqrt{cx}}{2a+\sqrt{cx}}$ faciunt $\frac{25a\sqrt{cx}}{2a+\sqrt{cx}}$, et $\frac{aa}{c}+\frac{bx}{c}$ faciunt $\frac{aa+bx}{c}$.

Negativæ quantitates eodem modo adduntur ac affirmativæ. Sic -2 et -3 faciunt -5; et $-\frac{4ax}{b}$ et $-\frac{11ax}{b}$ faciunt $-\frac{15ax}{b}$; $-a\sqrt{ac}$ et $-b\sqrt{ac}$ faciunt $\overline{-a-b}\sqrt{ac}$. Ubi verò negativa quantitas affirmativæ adjicienda est, oportet affirmativam negativa diminuere. Sic 3 et -2 faciunt 1; $\frac{11ax}{b}$ et $-\frac{4ax}{b}$ faciunt $\frac{7ax}{b}$; $-a\sqrt{ac}$ et $b\sqrt{ac}$ faciunt $\overline{b-a}\sqrt{ac}$. Et nota quod ubi negativa quantitas excedit[32] affirmativam, aggregatum erit negativum. Sic 2 et -3 faciunt -1; $-\frac{11ax}{b}$ & $\frac{4ax}{b}$ faciunt $-\frac{7ax}{b}$, ac $2\sqrt{ac}$ & $-7\sqrt{ac}$ faciunt $-5\sqrt{ac}$.

In additione aut plurium aut magis compositarum quantitatum convenit observare formam operationis supra in additione numerorum expositam. Quemadmodum si $17ax-14a+3$, et $4a+2-8ax$ et $7a-9ax$ addendæ sunt, dispono eas[33] in serie descendente ita scilicet[34] ut termini maxime affines stent in ijsdem columnis. Nempe numeri 3 et 2 in una columna, species $-14a$ et $4a$ et $7a$ in alia columna, atcɜ species $17ax$ et $-8ax$ et $-9ax$ in tertia. Dein terminos cujuscɜ columnæ sigillatim addo dicendo 2 et 3 faciunt 5 quod subscribo, dein $7a$ et $4a$ faciunt $11a$ et insuper $-14a$ facit $-3a$ quod iterum subscribo, denicɜ $-9ax$ et $-8ax$ faciunt $-17ax$ et insuper $17ax$ facit 0. Adeocɜ prodit summa $-3a+5$.

$$
\begin{array}{l}
17ax-14a+3 \\
-8ax\;\;+4a+2 \\
-9ax\;\;+7a \\
\hline
*\quad\;\;\; -3a+5
\end{array}
$$

Eadem methodo res in sequentibus exemplis absolvitur.

---

(31) Treating $\sqrt{3}$ as the 'communicant' *species*, that is.
(32) Understand in absolute magnitude.
(33) 'terminos eos' (the terms) was first written.

adding the prefixed numbers by which the variables are multiplied. Thus $7a + 9a$ make $16a$, and $11bc + 15bc$ make $26bc$. Likewise $3a/c + 5a/c$ make $8a/c$, and $2\sqrt{ac} + 7\sqrt{ac}$ make $9\sqrt{ac}$, and $6\sqrt{[ab - x^2]} + 7\sqrt{[ab - x^2]}$ make $13\sqrt{[ab - x^2]}$. And in much the same way $6\sqrt{3} + 7\sqrt{3}$ make $13\sqrt{3}$.[31] Indeed, $a\sqrt{ac} + b\sqrt{ac}$ make $(a + b)\sqrt{ac}$, when, to be sure, a and b are added just as though they were numbers multiplying $\sqrt{ac}$. And similarly

$$(2a + 3c)\sqrt{[(3ax^2 - x^3)/(a + x)]} \quad \text{and} \quad + 3a\sqrt{[(3ax^2 - x^3)/(a + x)]}$$

make $(5a + 3c)\sqrt{[(3ax^2 - x^3)/(a + x)]}$ for the reason that $2a + 3c$ and $3a$ make $5a + 3c$.

Positive fractions whose denominator is the same are united[30] by adding the numerators. Thus $\frac{1}{5} + \frac{2}{5}$ make $\frac{3}{5}$; and $2ax/b + 3ax/b$ make $5ax/b$,

$$8a\sqrt{cx}/(2a + \sqrt{cx}) + 17a\sqrt{cx}/(2a + \sqrt{cx}) \quad \text{make} \quad 25a\sqrt{cx}/(2a + \sqrt{cx})$$

and $a^2/c + bx/c$ make $(a^2 + bx)/c$.

Negative quantities are added in the same way as positive ones. Thus -2 and -3 make -5; and $-4ax/b$ and $-11ax/b$ make $-15ax/b$, $-a\sqrt{ac}$ and $-b\sqrt{ac}$ make $(-a - b)\sqrt{ac}$. When, indeed, a negative quantity is to be added to a positive one, the positive must be diminished by the negative one. Thus 3 and -2 make 1; $11ax/b$ and $-4ax/b$ make $7ax/b$, $-a\sqrt{ac}$ and $b\sqrt{ac}$ make $(b - a)\sqrt{ac}$. And note that when the negative quantity exceeds[32] the positive one, the total will be negative. Thus 2 and -3 make -1; $-11ax/b$ and $4ax/b$ make $-7ax/b$, while $2\sqrt{ac}$ and $-7\sqrt{ac}$ make $-5\sqrt{ac}$.

In the addition either of several quantities or of more compound ones it is convenient to keep to the working style displayed above in the addition of numbers. For instance, if $17ax - 14a + 3$, and $4a + 2 - 8ax$ and $7a - 9ax$ are to be added, I set them[33] out in descending sequence precisely in such a way[34] that terms most akin shall stand in the same columns. That is to say, the numbers 3 and 2 in one column, the variables $-14a$, $4a$ and $7a$ in a second one, and the variables $17ax$ and $-8ax$ in a third. Then I add the terms in each column separately, saying: 2 and 3 make 5, which I write beneath; then $7a$ and $4a$ make $11a$ and $-14a$ on top makes $-3a$, which I again write beneath; finally $-9ax$ and $-8ax$ make $-17ax$ and $17ax$ on top makes 0. Whence the resulting sum is $-3a + 5$.

$$
\begin{array}{ll}
17ax & -14a + 3 \\
-8ax & +4a + 2 \\
-9ax & +7a \\
\hline
0 & -3a + 5
\end{array}
$$

By the same method the matter is discharged in the following examples.

---

(34) Newton first continued: 'ut termini ejusdem ordinis (i.e. qui addendo uniri possunt vel saltem ex parte speciei alicujus conveniunt) stent in proprijs columnis' (that terms of the same order—that is, those which can be united by addition or at least concur in regard to some variable—shall stand in their own proper columns).

$$
\begin{array}{ll}
12x +7a & 11bc-7\sqrt{ac} \\
7x +9a & 15bc+2\sqrt{ac} \\
\hline
19x+16a\,. & 26bc-5\sqrt{ac}
\end{array}
\qquad
\begin{array}{l}
-\dfrac{4ax}{b}+6\sqrt{3}+\tfrac{1}{5} \\[4pt]
+\dfrac{11ax}{b}-7\sqrt{3}+\tfrac{2}{5}\,. \\[4pt]
\hline
\dfrac{7ax}{b}\;-\sqrt{3}+\tfrac{3}{5}
\end{array}
\qquad
\begin{array}{l}
-6xx +\tfrac{3}{7}x \\
5x^3 +\tfrac{5}{7}x \\
\hline
5x^3-6xx+\tfrac{8}{7}x\,.
\end{array}
$$

$\|$ [8]

$$
\begin{array}{l}
\| \; aay+2a^3-\dfrac{a^4}{2y} \\[4pt]
-2ayy \;-4aay \;+a^3 \\
y^3+2ayy \;-\tfrac{1}{2}aay \\
\hline
y^3 \quad * \quad -3\tfrac{1}{2}aay+3a^3-\dfrac{a^4}{2y}
\end{array}
\qquad
\begin{array}{l}
5x^4+2ax^3 \\
-3x^4-2ax^3 \\
-2x^4+5bx^3 \\
-4bx^3 \\
\hline
* \qquad bx^3+a^3\sqrt{aa+xx}-20a^3\sqrt{aa-xx}
\end{array}
\qquad
\begin{array}{l}
+8\tfrac{1}{4}a^3\sqrt{aa+xx} \\
-20a^3\sqrt{aa-xx} \\
-7\tfrac{1}{4}a^3\sqrt{aa+xx} \\
\end{array}
$$

[1673]

DE SUBDUCTIONE.

Lect 3   Numerorum non nimis compositorum inventio etiam Differentiæ per se patet. Quemadmodum quod 9 de 17 relinquunt 8. At in magis compositis Subductio fieri solet subscribendo numerum ablativum et sigillatim auferendo figuras inferiores de superioribus. Sic ad auferendum 63543 de 782579, subscripto 63543, dic 3 de 9 relinquit 6, quod scribe infra: Dein 4 de 7 relinquit 3 quod pariter scribe infra: Tum 5 de 5 relinquit 0 quod itidem subscribe: Postea 3 de 2 auferendum est, sed cum 3 sit majus, figura 1 a proxima figura 8 mutuò sumi debet, quæ una cum 2 faciat 12, a quo auferri potest 3, et restat 9, quod insuper subscribe: Adhæc cùm præter 6 etiam 1 de 8 auferendum sit, adde 1 ad 6, et summa 7 de 8 relinquet 1 quod etiam subscribe. Deniqꝫ cùm in inferiori numero nihil restet auferendum de superiori 7, subscribe etiam 7, et sic tandem habes differentiam 719036.

782579
63543
——
719036

Cæterùm omninò cavendum est ut figuræ numeri ablativi subscribantur in locis homogeneis; nempe unitates infra alterius numeri unitates, deni numeri infra denos, decimæ partes infra decimas, &c: sicut in Additione dictum est. Sic ad auferendum decimalem 0'63 ab integro 547, non dispone numeros hoc

$\|$ [9] $\|$ modo $\begin{array}{l}5\;47\\0'63\end{array}$ sed sic $\begin{array}{l}547\\0'63\end{array}$ ita nempe ut circulus[35] qui locum unitatum in decimali occupat, subjiciatur unitatibus alterius numeri. Tum, circulis in locis vacuis superioris numeri subintellectis, dic 3 de 0 auferendum esse, sed cùm nequeat, debet 1 de loco anteriori mutuo sumi ut 0 evadat 10 a quo 3 auferri potest et restabit 7, quod infra scribe. Dein illud 1 quod mutuò sumitur, adjectum 6 facit 7 et hoc de superiore 0 auferendum est, sed cùm nequeat, debet iterum 1 de loco anteriore sumi ut 0 evadet 10, et 7

547
0'63
——
546'37

---

(35) Literally, a 'circle' or 'round' (O), and hence a cipher.

$$12x \ +7a \qquad 11bc-7\sqrt{ac} \qquad -4ax/b+6\sqrt{3}+\tfrac{1}{5} \qquad -6x^2 \ +\tfrac{3}{7}x$$
$$\underline{7x \ +9a} \qquad \underline{15bc+2\sqrt{ac}} \qquad \underline{+11ax/b-7\sqrt{3}+\tfrac{2}{5}} \qquad \underline{5x^3 \ +\tfrac{5}{7}x}$$
$$19x+16a. \qquad 26bc-5\sqrt{ac}. \qquad 7ax/b \ -\sqrt{3}+\tfrac{3}{5}. \qquad 5x^3-6x^2+\tfrac{8}{7}x.$$

$$a^2y+2a^3-a^4/2y \qquad 5x^4+2ax^3$$
$$-2ay^2 \ -4a^2y \ +a^3 \qquad -3x^4-2ax^3 \qquad +8\tfrac{1}{4}a^3\sqrt{[a^2+x^2]}$$
$$\underline{y^3+2ay^2 \ -\tfrac{1}{2}a^2y} \qquad -2x^4+5bx^3 \qquad -20a^3\sqrt{[a^2-x^2]}$$
$$y^3 \quad 0 \ \ -3\tfrac{1}{2}a^2y+3a^3-a^4/2y. \qquad \underline{-4bx^3} \qquad \underline{-7\tfrac{1}{4}a^3\sqrt{[a^2+x^2]}}$$
$$0 \ \ +bx^3+a^3\sqrt{[a^2+x^2]}-20a^3\sqrt{[a^2-x^2]}.$$

SUBTRACTION

In the case of numbers not overly complicated, finding the difference, too, is self-evident: for instance, that 9 from 17 leaves 8. But in more complicated numbers subtraction is usually carried out by writing the subtrahend below and separately taking the lower figures from the upper ones. Thus to take 63 543 from 782 579, having written 63 543 below, say: 3 from 9 leaves 6, and write it beneath; next, 4 from 7 leaves 3, and this likewise write beneath; then 5 from 5 leaves 0, and similarly write it below; in sequel, 3 has to be taken from 2, but since 3 is the greater, a figure 1 should be borrowed from the next figure 8, which together with 2 shall make 12 and from this 3 can be taken with a remainder 9, which too write below; further, since, apart from 6, also 1 has to be taken from 8, add 1 to 6 and the sum 7 [taken] from 8 will leave 1, and this also write below; finally, since nothing remains in the lower number to be taken away from the upper one, 7, write 7 below also. And so at length you have the difference 719 036.

For the rest, every care is to be taken that the figures of the subtrahend be written below in places similar in sequence; namely, its units below the units of the other number, its tens below the number's tens, tenths below tenths, and so forth, as was prescribed in addition. Thus to take away the decimal 0·63 from the whole number 547, set the numbers out not in this manner $\begin{smallmatrix}5\ 47\\0\cdot63\end{smallmatrix}$, but in this $\begin{smallmatrix}547\\ \ \ 0\cdot63\end{smallmatrix}$; namely, so that the nought[35] which takes the place of units in the decimal shall lie below the units of the other number. Then, with noughts understood in the vacant places of the upper number, say: 3 has to be taken from 0, but, since it cannot, 1 should be borrowed from the preceding place so that 0 comes to be 10, and from this 3 can be taken with a remainder 7, which write beneath; next, the 1 which is borrowed, when added to 6, makes 7 and this has to be taken from the upper digit 0, but, since it cannot, 1 should again be borrowed from the preceding

$$\begin{array}{r} 547 \\ \underline{0\cdot63} \\ 546\cdot37. \end{array}$$

de 10 relinquet 3, quod similiter infra scribendum est. Tum illud 1 adjectum 0 facit 1, et hoc 1 de 7 relinquit 6, quod itidem subscribe. Deniĉp figuras etiam 54, siquidem de illis nihil ampliùs auferendum restat, subscribe, et habebis residuum 546'37.

Exercitationis gratia plura tum in integris tum in decimalibus numeris exempla subjecimus.

$$
\begin{array}{cccccc}
1673^{(36)} & 1673^{(36)} & 458074 & 35'72 & 46,5003^{(37)} & 308,7 \\
1541 & 1580 & 9205 & 14'32 & 3,078 & 25,74 \\
\hline
132 & 93 & 448869 & 21'4 & 43,4223 & 282,96
\end{array}
$$

Siquando major numerus$^{(38)}$ de minori auferendus est, oportet minorem de majore auferre et residuo præfigere negativum signum. Veluti si auferendum sit 1673 de 1541, e contra aufero 1541 de 1673 et residuo 132 præfigo signum$^{(39)}$ —.

In terminis Algebraicis Subductio fit connectendo quantitates cum signis omnibus quantitatis subducendæ mutatis, et insuper uniendo quæ possunt uniri perinde ut in Additione factum est. Sic $+7a$ de $+9a$ relinquit $+9a-7a$ sive $2a$; $-7a$ de $+9a$ relinquit $+9a+7a$ sive $16a$; $+7a$ de $-9a$ relinquit $-9a-7a$ sive $-16a$; et $-7a$ de $-9a$ relinquit $-9a+7a$ sive $-2a$. Sic $3\frac{a}{c}$ de $5\frac{a}{c}$ relinquit $2\frac{a}{c}$; $7\sqrt{ac}$ de $2\sqrt{ac}$ relinquit $-5\sqrt{ac}$; $\frac{2}{9}$ de $\frac{5}{9}$ relinquit $\frac{3}{9}$; $-\frac{4}{7}$ de $\frac{3}{7}$ relinquit $\frac{7}{7}$; $-\frac{2ax}{b}$ de $\frac{3ax}{b}$ relinquit $\frac{5ax}{b}$; $\frac{8a\sqrt{cx}}{2a+\sqrt{cx}}$ de $\frac{-17a\sqrt{cx}}{2a\times\sqrt{cx}}$ relinquit $\frac{-25a\sqrt{cx}}{2a+\sqrt{cx}}$; $\frac{aa}{c}$ de $\frac{bx}{c}$

‖[10] relinquit $\frac{bx-aa}{c}$; $a-b$ de $2a+b$ relinquit $2a+b-a+b$ sive ‖ $a+2b$; $3az-zz+ac$ de $3az$ relinquit $3az-3az+zz-ac$ sive $zz-ac$; $\frac{2aa-ab}{c}$ de $\frac{aa+ab}{c}$ relinquit $\frac{aa+ab-2aa+ab}{c}$ sive $\frac{-aa+2ab}{c}$; et $\overline{a-x}\sqrt{ax}$ de $\overline{a+x}\sqrt{ax}$ relinquit $\overline{a+x-a+x}\sqrt{ax}$ sive $2x\sqrt{ax}$. Et sic in alijs.

Cæterùm ubi quantitates pluribus terminis constant, operatio perinde ac in numeris institui potest. Id quod in sequentibus exemplis videre est.

$$
\begin{array}{cccc}
12x+7a & 15bc+2\sqrt{ac} & 5x^3+\frac{5}{7}x & \frac{11ax}{b}-7\sqrt{3}+\frac{2}{5} \\
7x+9a & -11bc+7\sqrt{ac} & 6xx-\frac{3}{7}x & \frac{4ax}{b}-6\sqrt{3}-\frac{1}{5} \\
\hline
5x-2a & 26bc-5\sqrt{ac} & 5x^3-6xx+\frac{8}{7}x & \frac{7ax}{b}-\sqrt{3}+\frac{3}{5}
\end{array}
$$

---

(36) Newton's unconscious choice of the number of the (Christian) year in which the example was concocted?

place so that 0 comes to be 10, and 7 from 10 will leave 3, which likewise is to be written beneath; then that 1 added to 0 makes 1, and this 1 [taken] from 7 leaves 6, which similarly write below; finally, seeing that nothing further remains to be taken from them, write the figures 54 below also. You will then have as the residue 546·37.

For practice's sake we append further examples both in whole numbers and in decimals:

$1673^{(36)}$	$1673^{(36)}$	458074	35·72	46·5003	308·7
1541	1580	9205	14·32	3·078	25·74
132.	93.	448869.	21·4 .	43·4223.	282·96.

Should ever the greater number$^{(38)}$ require to be taken from the less, the lesser one must be taken from the greater and a negative sign prefixed to the remainder. If, in instance, 1673 is to be taken away from 1541, I conversely take 1541 from 1673 and prefix the$^{(39)}$ sign — to the residue 132.

In algebraic terms subtraction is done by connecting quantities with all signs changed in the subtrahend and, moreover, uniting what can be united, exactly as happened in addition. Thus $+7a$ from $+9a$ leaves $+9a-7a$, that is, $2a$; $-7a$ from $+9a$ leaves $+9a+7a$, or $16a$; $+7a$ from $-9a$ leaves $-9a-7a$, or $-16a$; while $-7a$ from $-9a$ leaves $-9a+7a$, that is, $-2a$. So $3a/c$ from $5a/c$ leaves $2a/c$; $7\sqrt{ac}$ from $2\sqrt{ac}$ leaves $-5\sqrt{ac}$; $\frac{2}{9}$ from $\frac{5}{9}$ leaves $\frac{3}{9}$; $-\frac{4}{7}$ from $\frac{3}{7}$ leaves $\frac{7}{7}$; $-2ax/b$ from $3ax/b$ leaves $5ax/b$; $8a\sqrt{cx}/(2a+\sqrt{cx})$ from $-17a\sqrt{cx}/(2a+\sqrt{cx})$ leaves $-25a\sqrt{cx}/(2a+\sqrt{cx})$; a^2/c from bx/c leaves $(bx-a^2)/c$; $a-b$ from $2a+b$ leaves $2a+b-a+b$, that is, $a+2b$; $3az-z^2+ac$ from $3az$ leaves

$$3az-3az+z^2-ac, \quad \text{or} \quad z^2-ac;$$

$(2a^2-ab)/c$ from $(a^2+ab)/c$ leaves $(a^2+ab-2a^2+ab)/c$, that is, $(-a^2+2ab)/c$; and $(a-x)\sqrt{ax}$ from $(a+x)\sqrt{ax}$ leaves $(a+x-a+x)\sqrt{ax}$, or $2x\sqrt{ax}$. And likewise in other instances.

However, where quantities consist of several terms, the operation can be carried through exactly as in numbers. This may be seen in the following examples:

$12x+7a.$	$15bc+2\sqrt{ac}$	$5x^3 \quad +\frac{5}{7}x$	$11ax/b-7\sqrt{3}+\frac{2}{5}$
$7x+9a$	$-11bc+7\sqrt{ac}$	$6x^2 \quad -\frac{3}{7}x$	$4ax/b-6\sqrt{3}-\frac{1}{5}$
$5x-2a.$	$26bc-5\sqrt{ac}.$	$5x^3-6x^2+\frac{8}{7}x.$	$7ax/b \quad -\sqrt{3}+\frac{3}{5}.$

---

(37) Observe here that the separatrix comma (see note (11)) now drops onto the line, never again to rise to superscript position in the sequel.

(38) 'quantitas' (quantity) is cancelled.

(39) The superfluous adjective 'negativum' (negative) is deleted.

DE MULTIPLICATIONE.

[1673]

Lect 4   Numeri qui ex multiplicatione duorum quorumvis numerorum non majorum quàm 9 oriuntur, memoriter addiscendi sunt. Veluti quod 5 in 7 facit 35, quòdꝗ 8 in 9 facit 72, &c. Deinde majorum numerorum multiplicatio ad horum exemplorum normam instituetur.

Si 795 per 4 multiplicare oportet subscribe 4, ut vides. Dein dic, 4 in 5 facit 20, cujus posteriorem figuram 0 scribe infra 4, priorem vero 2 reserva in proximam operationem. Dic itaꝗ præterea 4 in 9 facit 36, cui adde præfatum 2 et fit 38, cujus posteriorem figuram 8 ut ante subscribe, et priorem 3 reserva. Deniꝗ dic 4 in 7 facit 28 cui adde prædictum 3 et fit 31. Eoꝗ pariter subscripto habebitur 3180 numerus qui prodit multiplicando totum 795 per 4.

$$\begin{array}{r} 795 \\ 4 \\ \hline 3180 \end{array}$$

Porrò si 9043 multiplicandus est per 2305, scribe alterutrum 2305 infra alterum 9043 ut ante, et multiplica superiorem 9043 per 5 pro more ostenso, et emerget 45215, dein per 0 et emerget 0000,[40] tertiò per 3 et emerget 27129, deniꝗ per 2 emerget 18086. Hosꝗ sic emergentes numeros, in serie descendente ita scribe ut ‖cujusꝗ inferioris ultima figura sit uno loco propior sinistræ quàm ultima superioris. Tandem hos omnes adde et orietur 20844115, numerus qui fit multiplicando totum 9043 per totum 2305.

‖[11]

$$\begin{array}{r} 9043 \\ 2305 \\ \hline 45215 \\ 00000 \\ 27129 \\ 18086 \\ \hline 20844115 \end{array}$$

Decimales numeri per integros vel per alios decimales perinde multiplicantur, ut vides in his exemplis.

72,4	50,18	3,9025
29	2,75	0,0132
651 6	2 5090	78050
1448	35 126	117075
2099,6	100 36	39025
	137,9950	0,05151300

Sed nota quod in prodeunte numero tot semper figuræ ad dextram pro decimalibus abscindi debent quot sunt figuræ decimales in utroꝗ numero multiplicante. Et si fortè non sint tot figuræ in prodeunte numero, deficientes loci circulis adimplendi sunt, ut hic fit in exemplo tertio.

Simplices termini Algebraici multiplicantur ducendo numeros in numeros et species in species, ac statuendo factum affirmativum si ambo factores sint affirmativi aut ambo negativi, et negativum si secus.[41] Sic 2*a* in 3*b* vel −2*a* in

---

(40) Correctly replacing each digit in the multiplicand by 0. An extra nought (omitted in all the printed editions) has crept into the corresponding line of the accompanying scheme.

MULTIPLICATION

Numbers which arise from the multiplication of any two numbers not greater than 9 are to be committed to memory: for instance, that 5 times 7 makes 35, and that 8 times 9 makes 72, and so on. Then the multiplication of greater numbers will be carried out on the pattern of these examples.

If 795 must be multiplied by 4, write 4 below, as you see. Then say: 4 times 5 makes 20, write its latter digit 0 beneath the 4, but retain the first, 2, till the next operation. Further, as a result say: 4 times 9 makes 36, to which add the previously mentioned 2 and it becomes 38; as before, write its latter figure 8 below, and keep the first digit 3. Say finally: 4 times 7 makes 28; to it add the previous 3 and it makes 31. And when this likewise is written below, there will be had the number 3180 which results from multiplying the total 795 by 4.

Furthermore, if 9043 is to be multiplied by 2305, write either one, 2305, below the other, 9043, as before, and in the fashion shown multiply the upper, 9043, by 5, when there emerges 45215; then by 0, when there emerges 0000;[40] thirdly by 3, when there emerges 27129; and finally by 2, when there emerges 18086. These numbers, as they emerge in this way, write in descending sequence so that the final figure of each lower one be one place nearer the left than the last digit of the one above it. Lastly, add these all up and there will appear the number, 20844115, which results on multiplying the total one, 9043, by the total other, 2305.

Decimal numbers are multiplied by integers or other decimals as you see in these examples:

72·4	50·18	3·9025
29	2·75	0·0132
651 6	2 5090	7 8050
1448	35 126	117 075
2099·6.	100 36	390 25
	137·9950.	0·0515 1300.

But note that in the resulting number so many figures ought always to be cut off on the right for decimals as there are decimal figures in both multiplying numbers. And if by chance there are not as many figures in the resulting number, the deficient places are to be filled in with noughts, as happens here in the third example.

Simple algebraic terms are multiplied by 'drawing' numbers into numbers and variables into variables, and then setting the product positive if both factors be positive or both negative, and negative otherwise.[41] Thus $2a$ times $3b$ and

(41) Namely, when the two factors are opposite in sign. It is surprising that Newton makes no attempt to justify this far from obvious fundamental rule.

$-3b$ facit $6ab$, vel $6ba$: nihil enim refert quo ordine ponantur. Sic etiam $2a$ in $-3b$ vel $-2a$ in $3b$ facit $-6ab$: et sic $2ac$ in $8bcc$ facit $16abccc$ sive $16abc^3$; et $7axx$ in $-12aaxx$ facit $-84a^3x^4$; et $-16cy$ in $31ay^3$ facit $-496acy^4$; et $-4z$ in $-3\sqrt{az}$ facit $12z\sqrt{az}$. Atꝗ ita 3 in -4 facit -12 et -3 in -4 facit 12.

Fractiones multiplicantur ducendo numeratores in numeratores ac denominatores in denominatores. Sic $\frac{2}{5}$ in $\frac{3}{7}$ facit $\frac{6}{35}$; et $\frac{a}{b}$ in $\frac{c}{d}$ facit $\frac{ac}{bd}$; et $2\frac{a}{b}$ in $3\frac{c}{d}$ facit

$6 \times \frac{a}{b} \times \frac{c}{d}$ seu $6\frac{ac}{bd}$; et $\frac{3acy}{2bb}$ in $\frac{-7cyy}{4b^3}$ facit $\frac{-21accy^3}{8b^5}$; et $\frac{-4z}{c}$ in $\frac{-3\sqrt{az}}{c}$ facit

$\frac{12z\sqrt{az}}{cc}$; et $\frac{a}{b}x$ in $\frac{c}{d}xx$ facit $\frac{ac}{bd}x^3$. Item 3 in $\frac{2}{5}$ facit $\frac{6}{5}$ ut pateat si 3 reducatur ad

formam fractionis $\frac{3}{1}$ adhibendo unitatem pro Denominatore. Et sic $\frac{15aaz}{cc}$ in $2a$

facit $\frac{30a^3z}{cc}$. Unde obiter nota quod $\frac{ab}{c}$ et $\frac{a}{c}b$ idem valent; ut et $\frac{abx}{c}$, $\frac{ab}{c}x$ & $\frac{a}{c}bx$,

‖[12] nec non $\frac{\overline{a+b}\sqrt{cx}}{a}$ et $\frac{a+b}{a}\sqrt{cx}$, ‖ & sic in alijs.[42]

Quantitates radicales ejusdem denominationis (hoc est, si sint ambæ radices quadraticæ, aut ambæ cubicæ, aut ambæ quadrato-quadraticæ &c) multiplicantur ducendo terminos in se invicem sub eodem signo radicali. Sic $\sqrt{3}$ in $\sqrt{5}$ facit $\sqrt{15}$, & $\sqrt{ab}$ in $\sqrt{cd}$ facit $\sqrt{abcd}$. Et $\sqrt{(3)}\ 5ayy$ in $\sqrt{(3)}\ 7ayz$ facit $\sqrt{(3)}\ 35aay^3z$. Et

*Vide Cap:
De Notatione[43]

$\sqrt{\frac{a^3}{c}}$ in $\sqrt{\frac{abb}{c}}$ facit $\sqrt{\frac{a^4bb}{cc}}$ hoc est $\frac{aab}{c}$* . Et $2a\sqrt{az}$ in $3b\sqrt{az}$ facit $6ab\sqrt{aazz}$ hoc est

$6aabz$. Et $\frac{3xx}{\sqrt{ac}}$ in $\frac{-2x}{\sqrt{ac}}$ facit $\frac{-6x^3}{\sqrt{aacc}}$ hoc est $\frac{-6x^3}{ac}$. Et $\frac{-4x\sqrt{ab}}{7a}$ in $\frac{-3dd\sqrt{5cx}}{10ee}$ facit

$\frac{12ddx\sqrt{5abcx}}{70aee}$.

Quantitates pluribus partibus constantes multiplicantur ducendo singulas unius partes in singulas alterius, perinde ut in multiplicatione numerorum ostensum est. Sic $c-x$ in a facit $ac-ax$, et $aa+2ac-bc$ in $a-b$ facit

$$a^3+2aac-aab-3bac+bbc.$$

Nam $aa+2ac-bc$ in $-b$ facit $-aab-2acb+bbc$, et in a facit $a^3+2aac-abc$, quorum summa est $a^3+2aac-aab-3abc+bbc$. Hujus multiplicationis specimen unà cum alijs consimilibus exemplis subjectum habes.

$aa+2ac-bc$	$a+b$	$a+b$
$a-b$	$a+b$	$a-b$
$-aab-2abc+bbc$	$ab+bb$	$-ab-bb$
$a^3+2aac-abc$	$aa\ +ab$	$aa+ab$
$a^3+2aac-aab-3abc+bbc$	$aa+2ab+bb$	$aa\ *\ -bb$

also $-2a$ times $-3b$ makes $6ab$, or $6ba$ (for the order in which they are put makes no difference). So also $2a$ times $-3b$, or $-2a$ times $3b$, makes $-6ab$. And thus $2ac$ times $8bc^2$ makes '$16abccc$', that is, $16abc^3$; $7ax^2$ times $-12a^2x^2$ makes $-84a^3x^4$; $-16cy$ times $31ay^3$ makes $-496acy^4$; and $-4z$ times $-3\sqrt{az}$ makes $12z\sqrt{az}$. So also 3 times -4 makes -12, and -3 times -4 makes 12.

Fractions are multiplied by 'drawing' numerators into numerators and denominators into denominators. Thus $\frac{2}{5}$ times $\frac{3}{7}$ makes $\frac{6}{35}$; a/b times c/d makes ac/bd; $2a/b$ times $3c/d$ makes $6(a/b) \times (c/d)$, that is, $6ac/bd$; $3acy/2b^2$ times $-7cy^2/4b^3$ makes $-21ac^2y^3/8b^5$; $-4z/c$ times $-3\sqrt{(az)}/c$ makes $12z\sqrt{(az)}/c^2$; and $(a/b)\,x$ times $(c/d)\,x^2$ makes $(ac/bd)\,x^3$. Likewise 3 times $\frac{2}{5}$ makes $\frac{6}{5}$, as may appear if 3 be reduced to its fractional form $\frac{3}{1}$ by employing unity for the denominator. And thus $15a^2z/c^2$ times $2a$ makes $30a^3z/c^2$. Hence note, by the way, that ab/c and $(a/c)\,b$ are equivalent; so also abx/c, $(ab/c)\,x$ and $(a/c)\,bx$, and equally $(a+b)\sqrt{(cx)}/a$ and $((a+b)/a)\sqrt{(cx)}$, and the like in other instances.

Radical quantities of the same denomination (if, that is, they be both square roots, or both cube ones, or both square-square [fourth] ones, and so on) are multiplied by drawing terms beneath the same radical sign into one another. Thus $\sqrt{3}$ times $\sqrt{5}$ makes $\sqrt{15}$, and $\sqrt{(ab)}$ times $\sqrt{(cd)}$ makes $\sqrt{(abcd)}$. And $\sqrt[3]{(5ay^2)}$ times $\sqrt[3]{(7ayz)}$ makes $\sqrt[3]{(35a^2y^3z)}$. And $\sqrt{(a^3/c)}$ times $\sqrt{(ab^2/c)}$ makes $\sqrt{(a^4b^2/c^2)}$, that is, a^2b/c.* And $2a\sqrt{(az)}$ times $3b\sqrt{(az)}$ makes $6ab\sqrt{(a^2z^2)}$, that is, $6a^2bz$. And $3x^2/\sqrt{(ac)}$ times $-2x\sqrt{(ac)}$ makes $-6x^3/\sqrt{(a^2c^2)}$, that is, $-6x^3/ac$. And $-4x\sqrt{(ab)}/7a$ times $-3d^2\sqrt{(5cx)}/10e^2$ makes $12d^2x\sqrt{(5abcx)}/70ae^2$.

*See the chapter, '*Notation*'.[43]

Quantities consisting of several parts are multiplied by drawing the individual parts of one into the individual portions of the other, exactly as was shown in the multiplication of numbers. Thus $c-x$ times a makes $ac-ax$, and $a^2+2ac-bc$ times $a-b$ makes $a^3+2a^2c-a^2b-3abc+b^2c$: for $a^2+2ac-bc$ times $-b$ makes $-a^2b-2abc+b^2c$, and times a it makes a^3+2a^2c-abc, the sum of which is $a^3+2a^2c-a^2b-3abc+b^2c$. Set out below you have a model for this multiplication, along with other closely similar examples:

$$
\begin{array}{ccc}
a^2+2ac-bc & a+b & a+b \\
a-b & a+b & a-b \\
\hline
-a^2b-2abc+b^2c & ab+b^2 & -ab-b^2 \\
a^3+2a^2c-abc & a^2\ +ab & a^2+ab \\
\hline
a^3+2a^2c-a^2b-3abc+b^2c. & a^2+2ab+b^2. & a^2\quad 0\quad -b^2.
\end{array}
$$

(42) This replaces the equivalent 'aliaq similia'.

(43) Uncritically following Whiston's 1707 *editio princeps* (page 20)—which replaces the preceding section 'De Notatione' by its 'corrected' version entitled 'De vocum quarundam & notarum significatione' (see note (9) above)—all printed editions illogically retain this marginal note unchanged.

$$yy + 2ay - \tfrac{1}{2}aa$$
$$yy - 2ay + aa$$
$$\overline{aayy + 2a^3y - \tfrac{1}{2}a^4}$$
$$-2ay^3 \quad -4aayy \quad +a^3y$$
$$y^4 + 2ay^3 \quad -\tfrac{1}{2}aayy$$
$$\overline{y^4 \quad * \quad -3\tfrac{1}{2}aayy + 3a^3y - \tfrac{1}{2}a^4}$$

$$\frac{2ax}{c} - \sqrt{\frac{a^3}{c}}$$
$$3a + \sqrt{\frac{abb}{c}}$$
$$\overline{\frac{2ax}{c}\sqrt{\frac{abb}{c}} - \frac{aab}{c}}$$
$$\frac{6aax}{c} - 3a\sqrt{\frac{a^3}{c}}$$
$$\overline{\frac{6aax}{c} - 3a\sqrt{\frac{a^3}{c}} + \frac{2ax}{c}\sqrt{\frac{abb}{c}} - \frac{aab}{c}.}$$

‖De Divisione.

‖[13]
[1673]
Lect 5

Divisio in numeris instituitur quærendo quot vicibus divisor in Dividendo continetur, totiesꝗ auferendo, et scribendo totidem unitates in Quoto. Idꝗ iteratò, si opus est, quamdiu divisor auferri potest. Sic ad dividendum 63 per 7, quære quoties 7 continetur in 63 et emerget 9 pro Quoto præcisè. Adeoꝗ $\frac{63}{7}$ valet 9. Insuper ad dividendum 371 per 7, præfige divisorem 7, et imprimis opus instituens in initialibus figuris Dividendi proximè majoribus Divisore, nempe in 37, dic quoties 7 continetur in 37? Resp: 5. Tum scripto 5 in Quoto, aufer 5×7 seu 35 de 37, et restabit 2, cui adnecte ultimam figuram Dividendi nempe 1, et fit 21 reliqua pars Dividendi in qua proximum opus instituendum est. Dic itaꝗ ut ante quoties 7 continetur in 21? Resp: 3. Quare scripto 3 in Quoto, aufer 3×7, seu 21 de 21 et restabit 0. Unde constat 53 esse numerum præcisè qui oritur ex divisione 371 per 7.

$$7)371(53$$
$$35$$
$$\overline{21}$$
$$21$$
$$\overline{0}$$

Atꝗ ita ad dividendum 4798 per 23, opus primò instituens in initialibus figuris 47 dic quoties 23 continetur in 47? Resp 2. Scribe ergo 2 in Quoto, et de 47 subduc 2×23 seu 46, restatꝗ 1, cui subjunge proximum numerum Dividendi, nempe 9, et fit 19 in subsequens opus. Dic itaꝗ quoties 23 continetur in 19? Resp 0. Quare scribe 0 in Quoto et de 19 subduc 0×23 seu 0 et restat 19, cui subjunge ultimum numerum 8, et fit 198 in proximum opus. Quamobrem dic ultimò quoties 23 continetur in 198, (id quod ex initialibus numeris 2 et 19 conjici potest, animadvertendo quoties 2 con-

$$23)4798(208,6086\ \&c^{(44)}$$
$$46$$
$$\overline{19}$$
$$00$$
$$\overline{198}$$
$$184$$
$$\overline{14{,}0}$$
$$13\ 8$$
$$\overline{20}$$
$$00$$
$$\overline{200}$$
$$184$$
$$\overline{160}$$

$$y^2 + 2ay - \tfrac{1}{2}a^2$$
$$y^2 - 2ay + a^2$$
$$\overline{}$$
$$a^2y^2 + 2a^3y - \tfrac{1}{2}a^4$$
$$-2ay^3 \quad -4a^2y^2 \quad +a^3y$$
$$\underline{y^4 + 2ay^3 \quad -\tfrac{1}{2}a^2y^2}$$
$$y^4 \quad 0 \quad -3\tfrac{1}{2}a^2y^2 + 3a^3y - \tfrac{1}{2}a^4.$$

$$2ax/c - \sqrt{(a^3/c)}$$
$$3a + \sqrt{(ab^2/c)}$$
$$\overline{}$$
$$2(ax/c)\sqrt{(ab^2/c)} - a^2b/c$$
$$6a^2x/c - 3a\sqrt{(a^3/c)}$$
$$\overline{}$$
$$6a^2x/c - 3a\sqrt{(a^3/c)} + 2(ax/c)\sqrt{(ab^2/c)} - a^2b/c.$$

DIVISION

Division is instituted in numbers by inquiring how many times the divisor is contained in the dividend, taking it away that number of times and entering an equal number of units in the quotient—and repeating this, if necessary, as long as the divisor can still be taken away. Thus, to divide 63 by 7, inquire how many times 7 is contained in 63 and there emerges 9, precisely, as the quotient, and accordingly the value of 63/7 is 9. Again, to divide 371 by 7, prefix the divisor 7 and, beginning work in the first instance with the initial figures of the dividend which are immediately greater than the divisor, namely with 37, say: how many times is 7 contained in 37? Reply: 5. Then, having written 5 in the quotient, take away 5 × 7, that is, 35, from 37 and there will remain 2: to this adjoin the last figure in the dividend, namely 1, and it becomes 21, the remaining part of the dividend, on which the next operation must be performed. As before, accordingly, say: how many times is 7 contained in 21? Reply: 3. Hence, having written 3 in the quotient, take away 3 × 7, that is, 21, from 21 and there will remain 0. It is therefore established that 53 is precisely the number which arises on dividing 371 by 7.

And so, to divide 4798 by 23, starting work first with the opening figures 47, say: how many times is 23 contained in 47? Reply: 2. Therefore enter 2 in the quotient and from 47 subtract 2 × 23, that is, 46, leaving 1: to this append the next number in the dividend, namely 9, and there comes 19 for the next-following stage. Accordingly, say: how many times is 23 contained in 19? Answer: 0. Consequently, write 0 in the quotient and from 19 subtract 0 × 23, that is, 0, and there remains 19: to this append the final number 8, and it becomes 198 at the next stage. In consequence of which, say lastly: how many times is 23 contained in 198 (which can be guessed from the initial numbers 2 and 19, on noticing how many

$$
\begin{array}{r}
23)\,4798\,(208{\cdot}6086\ldots^{(44)} \\
\underline{46} \\
19 \\
\underline{00} \\
198 \\
\underline{184} \\
14{\cdot}0 \\
\underline{13\ 8} \\
20 \\
\underline{00} \\
200 \\
\underline{184} \\
160
\end{array}
$$

---

(44) Since 7 × 23 = 161 the '6 &c' is very nearly '7'; the quotient is more accurately 208·6086956....

tinetur in 19.) ?[45] Resp 8. Quare scribe 8 in Quoto et de 198 subduc 8×23 seu
‖[14] 184, restabitcꝫ ‖ 14 adhuc dividendus per 23. Adeocꝫ Quotus erit $208\frac{14}{23}$. Quod
si hujusmodi fractio minùs placeat, possis Divisionem in Fractionibus decimalibus ultra ad libitum prosequi, semper adnectendo circulum numero
residuo. Sic residuo 14 adnecte 0, fitcꝫ 140. Tum dic quoties 23 sit in 140?
Resp 6. Scribe ergo 6 in Quoto et de 140 subduc 6×23 seu 138, et restabit 2,
cui adnecte 0 ut ante. Et sic opere ad arbitrium continuato emerget tandem
Quotus 208,6086 &c.

Ad eundem modum fractio decimalis 3,5218 per
fractionem decimalem 46,1 dividitur et prodit 0,07639
&c. Ubi nota[46] quod in Quoto tot figuræ pro decimalibus abscindendæ sunt quot sunt in ultimo dividuo plures
quam in divisore: ut in hoc exemplo quincꝫ quia sex
sunt in ultimo dividuo 0,004370 & una in Divisore
46,1.

```
46,1)3,5218(0,07639
     3 227
     ─────
     2948
     2766
     ─────
     1820
     1383
     ─────
     4370
```

Exempla plura lucis gratia subjunximus.

```
9043)20844115(2305.        72,4)2099,6(29[,]0.
     18086                      1448
     ─────                      ────
     27581                      651 6
     27129                      651 6
     ─────                      ─────
      45215                        0
      45215
     ──────
        0
```

```
50,18)137,995(2,75.        0,0132)0,051513(3,9025.
      100 36                       396
      ─────                        ───
      37 635                       1191
      35 126                       1188
      ──────                       ────
       2 5090                        330
       2 5090                        264
       ──────                        ───
          0                          660
                                     660
                                     ───
                                       0
```

---

(45) Since $[19/2] = 9$, the first 'responsum' will be 9, but this is evidently too great and
hence gives rise to Newton's modified 'Resp[onsum] 8' which follows.

(46) Newton first continued 'quod prima figura Quoti debet esse ejusdem ordinis cum
figura numeri dividendi quæ supereminet locum unitatum Divisoris ubi Divisor infra figuras
initiales se proximè majores collocatur. Sic scripto 46,1 infra 3,5218 ad hunc modum 3,5218

                                                  46,1

locus unitatum cadet infra 2 numerum centesimalem, et proinde 7, primus numerus quoti,
constitui debet in loco centesimali' (that the first figure in the quotient ought to be of the same

times 2 is contained in 19)?[45] Answer: 8. Hence write 8 in the quotient and from 198 subtract 8 × 23, that is, 184, and there will remain 14 still to be divided by 23. As a result, the quotient will be $208\frac{14}{23}$. But if a fraction of this sort is less to your liking, you might pursue the division in decimals as much further as you please, continually attaching a nought to the residual number. Thus to the residue 14 attach 0 and it becomes 140. Then say: how many 23's in 140? Answer: 6. So write 6 in the quotient and from 140 subtract 6 × 23, that is, 138, and 2 will remain: to this attach 0 as before. And, after the procedure has been carried on in this way at will, there will at length emerge as quotient 208·6086 ....

In much the same way the decimal fraction 3·5218 is divided by the decimal 46·1, yielding 0·07639 .... Here note[46] that in the quotient as many figures are to be cut off for decimals as there are more in the last divided number than in the divisor: in this example, for instance, five because there are six figures in the last dividend 0·004070 and one in the divisor 46·1.

$$
\begin{array}{r}
46\cdot1)3\cdot5218(0\cdot07639\ldots \\
3\ 227 \\
\hline
2948 \\
2766 \\
\hline
1820 \\
1383 \\
\hline
4370
\end{array}
$$

In elucidation we append several more examples:

$$
\begin{array}{r}
9043)20844115(2305. \\
18086 \\
\hline
27581 \\
27129 \\
\hline
45215 \\
45215 \\
\hline
0
\end{array}
\qquad
\begin{array}{r}
72\cdot4)2099\cdot6(29\cdot0. \\
1448 \\
\hline
651\ 6 \\
651\ 6 \\
\hline
0
\end{array}
$$

$$
\begin{array}{r}
50\cdot18)137\cdot995(2\cdot75. \\
100\ 36 \\
\hline
37\ 635 \\
35\ 126 \\
\hline
2\ 5090 \\
2\ 5090 \\
\hline
0
\end{array}
\qquad
\begin{array}{r}
0\cdot0132)0\cdot051513(3\cdot9025. \\
396 \\
\hline
1191 \\
1188 \\
\hline
330 \\
264 \\
\hline
660 \\
660 \\
\hline
0
\end{array}
$$

rank as the figure in the number to be divided which surmounts the units' place in the divisor when the divisor is placed beneath the initial figures next greater than it. Thus, when 46·1 has been written below 3·5218 in this manner..., its units' place 6 will fall beneath the hundredths' number 2, and consequently 7, the first number in the quotient, ought to be set in the hundredths' place).

In terminis Algebraicis Divisio fit resolvendo[47] quicquid per multiplica-

‖[15] tionem conflatur. Sic ab divis: per a dat b ‖ pro Quoto. $6ab$ div: per $2a$ dat $3b$, et div: per $-2a$ dat $-3b$. $-6ab$ div: per $2a$ dat $-3b$; et div: per $-2a$ dat $3b$. $16abc^3$ div: per $2ac$ dat $8bcc$. $-84a^3x^4$ div: per $-12aaxx$ dat $7axx$. Item $\frac{6}{35}$ divis: per $\frac{2}{5}$ dat $\frac{3}{7}$. $\frac{ac}{bd}$ div: per $\frac{a}{b}$ dat $\frac{c}{d}$. $\frac{-21accy^3}{8b^5}$ div: per $\frac{3acy}{2bb}$ dat $\frac{-7cyy}{4b^3}$. $\frac{6}{5}$ div: per 3 dat $\frac{2}{5}$; et vicissim $\frac{6}{5}$ div: per $\frac{2}{5}$ dat $\frac{3}{1}$ seu 3. $\frac{30a^3z}{2bb}$ div: per $2a$ dat $\frac{15aaz}{cc}$; et vicissim div: per $\frac{15aaz}{cc}$ dat $2a$. Item $\sqrt{15}$ div: per $\sqrt{3}$ dat $\sqrt{5}$. $\sqrt{abcd}$ div: per $\sqrt{cd}$ dat $\sqrt{ab}$. $\sqrt{a^3c}$ per $\sqrt{ac}$ dat $\sqrt{aa}$ seu a. $\sqrt{(3)}\,35aay^3z$ div: per $\sqrt{(3)}\,5ayy$ dat $\sqrt{(3)}\,7ayz$. $\sqrt{\frac{a^4bb}{cc}}$ div: per $\sqrt{\frac{a^3}{c}}$ dat $\sqrt{\frac{abb}{c}}$. $\frac{12ddx\sqrt{5abcx}}{70aee}$ div: per $\frac{-3dd\sqrt{5cx}}{10ee}$ dat $\frac{-4x\sqrt{ab}}{7a}$. Atcp ita $\overline{a+b}\sqrt{ax}$ div: per $a+b$ dat $\sqrt{ax}$, et vicissim div: per $\sqrt{ax}$ dat $a+b$. Et $\frac{a}{a+b}\sqrt{ax}$ div: per $\frac{1}{a+b}$ dat $a\sqrt{ax}$; vel div: per a dat $\frac{1}{a+b}\sqrt{ax}$ sive $\frac{\sqrt{ax}}{a+b}$; et vicissim div: per $\frac{\sqrt{ax}}{a+b}$ dat a. Cæterùm in hujusmodi resolutionibus omninò cavendum est ut quantitates sint ejusdem ordinis quæ ad invicem applicantur. Nempe ut numeri applicentur ad numeros, species ad species, radicales ad radicales, numeratores Fractionum ad Numeratores ac Denominatores ad Denomina- tores, nec non in Numeratoribus, Denominatoribus, et Radicalibus quantitates cujuscp generis ad quantitates homogeneas.

[1673]
Lect 6   Quod si quantitas dividenda nequeat sic per Divisorem resolvi, sufficit (ubi ambæ quantitates sint integræ) subscribere Divisorem cum lineola interjecta. Sic ad dividendum ab per c scribitur $\frac{ab}{c}$; et ad dividendum $\overline{a+b}\sqrt{cx}$ per a scribitur $\frac{\overline{a+b}\sqrt{cx}}{a}$ vel $\frac{a+b}{a}\sqrt{cx}$. Et sic $\sqrt{ax-xx}$ divis: per $\sqrt{cx}$ dat $\frac{\sqrt{ax-xx}}{\sqrt{cx}}$ sive $\sqrt{\frac{ax-xx}{cx}}$. Et $\overline{aa+ab}\sqrt{aa-2xx}$ divis: per $\overline{a-b}\sqrt{aa-xx}$ dat $\frac{aa+ab}{a-b}\sqrt{\frac{aa-2xx}{aa-xx}}$. Et $12\sqrt{5}$ div: per $4\sqrt{7}$ dat $3\sqrt{\frac{5}{7}}$.

Ubi verò fractæ sunt illæ quantitates, duc Numeratorem Dividendæ quanti-

‖[16] tatis in[48] Denomi‖natorem Divisoris ac Denominatorem in[48] Numeratorem, et factus prior erit Numerator, ac posterior Denominator Quoti. Sic ad dividendum $\frac{a}{b}$ per $\frac{c}{d}$ scribitur $\frac{ad}{bc}$, multiplicato scilicet a per d et b per c. Pariqp ratione $\frac{3}{7}$ div: per $\frac{5}{4}$ dat $\frac{12}{35}$. Et $\frac{3a}{4c}\sqrt{ax}$ divis: per $\frac{2c}{5a}$ dat $\frac{15aa}{8cc}\sqrt{ax}$; divis: autem

---

(47) Into its component parts, that is. Newton signifies that division is to be treated as inverse multiplication, and, to be sure, his following examples are the straightforward con- verses of those illustrating the preceding section 'De Multiplicatione'.

In algebraic quantities division is achieved by resolving[47] anything compounded by means of multiplication. Thus ab divided by a gives b for the quotient; $6ab$ divided by $2a$ gives $3b$, and divided by $-2a$ it gives $-3b$; $-6ab$ divided by $2a$ gives $-3b$, and divided by $-2a$ it gives $3b$; $16abc^3$ divided by $2ac$ gives $8bc^2$; and $-84a^3x^4$ divided by $-12a^2x^2$ gives $7ax^2$. Likewise $\frac{6}{35}$ divided by $\frac{2}{5}$ gives $\frac{3}{7}$; ac/bd divided by a/b gives c/d; $-21ac^2y^3/8b^5$ divided by $3acy/2b^2$ yields $-7cy^2/4b^3$; $\frac{6}{5}$ divided by 3 gives $\frac{2}{5}$, and inversely $\frac{6}{5}$ divided by $\frac{2}{5}$ yields 3; $30a^3z/2b^2$ divided by $2a$ yields $15a^2z/c^2$, and inversely when divided by $15a^2z/c^2$ it gives $2a$. Simiarly $\sqrt{15}$ divided by $\sqrt{3}$ gives $\sqrt{5}$; $\sqrt{(abcd)}$ divided by $\sqrt{(cd)}$ gives $\sqrt{(ab)}$; $\sqrt{(a^3c)}$ divided by $\sqrt{(ac)}$ yields $\sqrt{(a^2)}$, that is, a; $\sqrt[3]{(35a^2y^3z)}$ divided by $\sqrt[3]{(5ay^2)}$ yields $\sqrt[3]{(7ayz)}$; $\sqrt{(a^4b^2/c^2)}$ divided by $\sqrt{(a^3/c)}$ gives $\sqrt{(ab^2/c)}$; and

$$12d^2x\sqrt{(5abcx)}/70ae^2 \quad \text{divided by} \quad -3d^2\sqrt{(5cx)}/10e^2$$

yields $-4x\sqrt{(ab)}/7a$. And in this way $(a+b)\sqrt{(ax)}$ divided by $a+b$ gives $\sqrt{(ax)}$, and inversely when divided by $\sqrt{(ax)}$ it yields $a+b$. And $(a/(a+b))\sqrt{(ax)}$ divided by $1/(a+b)$ gives $a\sqrt{(ax)}$; or when divided by a it gives $(1/(a+b))\sqrt{(ax)}$, that is, $\sqrt{(ax)}/(a+b)$, and inversely when divided by $\sqrt{(ax)}/(a+b)$ it gives a. But in resolutions of this sort every precaution must be taken to see that quantities which divide one another are of the same order: namely, that numbers are divided into numbers, variables into variables, radicals into radicals, numerators of fractions into numerators and their denominators into denominators, and, generally, in numerators, denominators and radicals that quantities of any kind are divided into quantities of the same type.

If, however, the quantity to be divided cannot be resolved by means of a divisor in this fashion, it is sufficient (when both quantities are integral) to write the divisor below, with a short line intervening. Thus to divide ab by c, ab/c is written; while to divide $(a+b)\sqrt{(cx)}$ by a there is written $\dfrac{(a+b)\sqrt{(cx)}}{a}$ or $\dfrac{a+b}{a}\sqrt{(cx)}$. And so $\sqrt{(ax-x^2)}$ divided by $\sqrt{(cx)}$ gives $\dfrac{\sqrt{(ax-x^2)}}{\sqrt{(cx)}}$, that is, $\sqrt{\dfrac{ax-x^2}{cx}}$. And $(a^2+ab)\sqrt{(a^2-2x^2)}$ divided by $(a-b)\sqrt{(a^2-x^2)}$ gives $\dfrac{a^2+ab}{a-b}\sqrt{\dfrac{a^2-2x^2}{a^2-x^2}}$. While $12\sqrt{5}$ divided by $4\sqrt{7}$ yields $3\sqrt{\frac{5}{7}}$.

When, indeed, those quantities are fractional, multiply the numerator of the quantity to be divided into[48] the denominator of the divisor, and its denominator into[48] the numerator: the former product will then be the numerator of the quotient, the latter one its denominator. Thus to divide a/b by c/d, there is written ad/bc, where, of course, a is multiplied by d and b by c. Equally, $\frac{3}{7}$ divided by $\frac{5}{4}$ gives $\frac{12}{35}$. And $3a\sqrt{(ax)}/4c$ divided by $2c/5a$ yields $15a^2\sqrt{(ax)}/8c^2$; but when

---

(48) Newton originally wrote 'per' (by).

per $\dfrac{2c\sqrt{aa-xx}}{5a\sqrt{ax}}$ dat $\dfrac{15a^3x}{8cc\sqrt{aa-xx}}$. Et ad eundem modum $\dfrac{ad}{b}$ divis: per c $\left(\text{sive per } \dfrac{c}{1}\right)$

dat $\dfrac{ad}{bc}$. Et c $\left(\text{sive } \dfrac{c}{1}\right)$ divis: per $\dfrac{ad}{b}$ dat $\dfrac{bc}{ad}$. Et $\frac{3}{7}$ div: per 5 dat $\frac{3}{35}$. Et 3 div:

per $\frac{5}{4}$ dat $\frac{12}{5}$. Et $\dfrac{a+b}{c}\sqrt{cx}$ div: per a dat $\dfrac{a+b}{ac}\sqrt{cx}$. Et $\overline{a+b}\sqrt{cx}$ div: per $\dfrac{a}{c}$ dat

$\dfrac{ac+bc}{a}\sqrt{cx}$. Et $2\sqrt{\dfrac{axx}{c}}$ div: per $3\sqrt{cd}$ dat $\frac{2}{3}\sqrt{\dfrac{axx}{ccd}}$; Div: autem per $3\sqrt{\dfrac{cd}{x}}$ dat

$\frac{2}{3}\sqrt{\dfrac{ax^3}{ccd}}$. Et $\frac{1}{5}\sqrt{\dfrac{7}{11}}$ div: per $\frac{1}{2}\sqrt{\dfrac{3}{7}}$ dat $\frac{2}{5}\sqrt{\dfrac{49}{33}}$. Et sic in alijs.

Quantitas ex pluribus terminis composita dividitur applicando singulos ejus terminos ad Divisorem. Sic $aa+3ax-xx$ divisum per a dat $a+3x-\dfrac{xx}{a}$. At ubi Divisor etiam pluribus terminis constat, divisio perinde ac in Numeris institui debet. Sic ad dividendum $a^3+2aac-aab-3abc+bbc$ per $a-b$,[49] Dic quoties a continetur in a^3, nempe primus terminus Divisoris in primo Dividendi[50]? Resp: aa. Quare scribe aa in Quoto, et ablato $a-b$ in aa sive a^3-aab de Dividendo, restabit $2aac-3abc+bbc$ adhuc dividendum. Dic itaꝗ rursus quoties a continetur in $2aac$? Resp: $2ac$. Quare scribe etiam $2ac$ in Quoto, et ablato $a-b$ in $2ac$ sive $2aac-2abc$ de præfato Residuo, restabit etiamnum $-abc+bbc$. Quamobrem dic iterum quoties a continetur in $-abc$? Resp: $-bc$. Et proinde scribe $-bc$ in Quoto, et ablato denuò $+a-b$ in $-bc$ sive $-abc+bbc$ de novissimo Residuo, restabit nihil. Quod indicat Divisionem peractam esse, prodeunte Quoto $aa+2ac-bc$.

Cæterùm ut hujusmodi operationes ad formam quâ in Divisione numerorum usi sumus debitè reducantur, termini tùm dividendæ quantitatis tum Divisoris juxta dimensiones literæ[51] alicujus quæ ad hanc rem maximè idonea judicabitur in ordine disponendi sunt, ita nempe ut illi primum locum occupent ‖ in quibus litera ista est plurimarum dimensionum, ijꝗ secundum in quibus dimensiones ejus ad maximas proximæ sunt; et sic deinceps usꝗ ad terminos qui per literam istam non omninò multiplicantur, adeoꝗ ultimum locum occupabunt. Sic in allato exemplo si termini ordinentur juxta dimensiones literæ a, formam operis exhibebit adjunctum Diagramma:

‖[17]

$$
a-b \,\big)\, a^3 {\scriptstyle\begin{array}{l} +2aac \\ -aab \end{array}} -3abc+bbc \,\big(\, aa+2ac-bc.
$$

$$
\underline{\quad a^3\ -aab \quad}
$$

$$
\underline{0\ +2aac-3abc \quad}
$$

$$
\underline{2aac-2abc \quad}
$$

$$
\underline{0\quad -abc+bbc}
$$

$$
\underline{-abc+bbc}
$$

$$
0\qquad 0
$$

---

(49) The converse of Newton's first 'subjectum exemplum' on his page **13** above.

divided by $2c\sqrt{(a^2-x^2)}/5a\sqrt{(ax)}$ it gives $15a^3x/8c^2\sqrt{(a^2-x^2)}$. In much the same way ad/b divided by c (that is, by $c/1$) gives ad/bc; while c (or $c/1$) divided by ad/b yields bc/ad. And $\frac{3}{7}$ divided by 5 gives $\frac{3}{35}$; 3 divided by $\frac{5}{4}$ gives $\frac{12}{5}$. Again, $(a+b)\sqrt{(cx)}/c$ divided by a gives $(a+b)\sqrt{(cx)}/ac$; while $(a+b)\sqrt{(cx)}$ divided by a/c gives $(ac+bc)\sqrt{(cx)}/a$. And $2\sqrt{(ax^2/c)}$ divided by $3\sqrt{(cd)}$ yields $\frac{2}{3}\sqrt{(ax^2/c^2d)}$; but when divided by $3\sqrt{(cd/x)}$ it gives $\frac{2}{3}\sqrt{(ax^3/c^2d)}$. And $\frac{1}{5}\sqrt{\frac{7}{11}}$ divided by $\frac{1}{2}\sqrt{\frac{3}{7}}$ gives $\frac{2}{5}\sqrt{\frac{49}{33}}$. And the like in other examples.

A quantity composed of several terms is divided by 'applying' the divisor to its separate terms. Thus $a^2+3ax-x^2$ divided by a gives $a+3x-x^2/a$. But when the divisor too consists of several terms, division ought to be performed exactly as in numbers. So, to divide $a^3+2a^2c-a^2b-3abc+b^2c$ by $a-b$,[49] say: how many times is a contained in a^3, namely, the first term of the divisor in the first of the dividend? Answer: a^2. Therefore enter a^2 in the quotient, and when $a-b$ times a^2, that is, a^3-a^2b, is taken from the dividend there will remain $2a^2c-3abc+b^2c$ still to be divided. Accordingly, say again: how many times is a contained in $2a^2c$? Reply: $2ac$. Therefore enter also $2ac$ in the quotient, and when $a-b$ times $2ac$, or $2a^2c-2abc$, is taken from the above-mentioned residue there will yet remain $-abc+b^2c$. In consequence, once more say: how many times is a contained in $-abc$? Answer: $-bc$. Hence write $-bc$ in the quotient, and when again $a-b$ times $-bc$, or $-abc+b^2c$, is taken from the most recent residue, nothing will remain. This indicates that the division is finished, producing the quotient $a^2+2ac-bc$.

But to reduce operations of this sort appropriately to the pattern which we used in the division of numbers, the terms both of the quantity to be divided and of the divisor are to be set out in order according to the powers of some letter[51] which will be judged most fit for this purpose: namely, so that the primary position is held by those in which that letter is of most dimensions, the second by those whose dimensions are next greatest, and so on in sequence as far as the terms which are not multiplied at all by that letter and so will occupy the final place. Thus in the example adduced, if the terms be ordered according to the dimensions of the letter a, the layout of the work will be displayed by the appended scheme:

$$a-b \overline{)\, a^3+(2a^2c-a^2b)-3abc+b^2c \,\big(\, a^2+2ac-bc.}$$

$$
\begin{array}{llll}
\underline{a^3 \qquad\quad -a^2b} & & & \\
0 \ +2a^2c \quad 0 & & -3abc & \\
\underline{2a^2c} & & -2abc & \\
0 & & \underline{-abc+b^2c} & \\
& & -abc+b^2c & \\
& & \overline{0 \qquad 0} &
\end{array}
$$

---

(50) 'Dividendæ quantitatis' is replaced.

(51) More exactly, the algebraic quantity which it denotes.

Ubi videre est quod terminus a^3 sive a trium dimensionum occupat primum locum dividendæ quantitatis, terminicɜ $\begin{smallmatrix}2aac\\-aab\end{smallmatrix}$ in quibus a est duarum dimensionum secundum occupat, et sic præterea. Potuit etiam dividenda quantitas sic scribi $a^3 \begin{smallmatrix}+2c\\-b\end{smallmatrix} aa - 3bca + bbc$. Ubi termini secundum locum occupantes, uniuntur aggregando factores literæ juxta quam fit ordinatio. Et hoc modo si termini juxta dimensiones literæ b disponerentur, opus sicut in proximo Diagrammate institui deberet, cujus explicationem adnectere visum est.

$$-b+a \Big) cbb \begin{smallmatrix}-3ac\\-aa\end{smallmatrix} b \begin{smallmatrix}+a^3\\+2aac\end{smallmatrix} \Big(-cb \begin{smallmatrix}+2ac\\+aa\end{smallmatrix}.$$

$$\underline{cbb \;-ac\; b}$$

$$0 \begin{smallmatrix}-2ac\\-aa\end{smallmatrix} b \begin{smallmatrix}+a^3\\+2aac\end{smallmatrix}$$

$$\underline{\begin{smallmatrix}-2ac\\-aa\end{smallmatrix} b \begin{smallmatrix}+a^3\\+2aac\end{smallmatrix}}$$

$$0 \qquad 0$$

Dic quoties $-b$ continetur in cbb? Resp: $-cb$. Quare scripto $-cb$ in Quoto, aufer $-b+a$ in $-cb$ seu $bbc - abc$ et restabit in secundo loco $\begin{smallmatrix}-2ac\\-aa\end{smallmatrix} b$. Residuo huic adnecte, si placet, quantitates in ultimo loco, nempe $\begin{smallmatrix}a^3\\+2aac\end{smallmatrix}$, et dic iterum quoties $-b$ continetur in $\begin{smallmatrix}-2ac\\-aa\end{smallmatrix} b$? Resp: $\begin{smallmatrix}+2ac\\+aa\end{smallmatrix}$. Quare his in Quoto scriptis, aufer $-b+a$ in $\begin{smallmatrix}+2ac\\+aa\end{smallmatrix}$ seu $\begin{smallmatrix}-2ac\\-aa\end{smallmatrix} b \begin{smallmatrix}+2aac\\+a^3\end{smallmatrix}$ et restabit nihil. Unde constat divisionem peractam esse, prodeunte Quoto $-cb + 2ac + aa$ ut ante.

Atɜ ita si dividere oportet $aay^4 - aac^4 + yyc^4 + y^6 - 2y^4cc - a^6 - 2a^4cc - a^4yy$ per $yy - aa - cc$:[52] quantitates juxta literam y ad hunc modum ordino,

$$yy\begin{smallmatrix}-aa\\-cc\end{smallmatrix}\Big) y^6 \begin{smallmatrix}+aa\\-2cc\end{smallmatrix} y^4 \begin{smallmatrix}-a^4\\+c^4\end{smallmatrix} yy \begin{smallmatrix}-a^6\\-2a^4cc\\-aac^4\end{smallmatrix}.$$

Dein Divisionem ut in subjecto Diagrammate instituo. Adjiciuntur et alia ‖[18] exempla, de quibus insuper observandum est quòd ubi ‖ dimensiones literæ ad quam ordinatio fit, non in eadem ubicɜ progressione Arithmetica sed per saltum alicubi procedunt, locis vacuis substituitur nota $*$.

---

(52) This Cartesian example is probably borrowed from page 16 of Gerard Kinckhuysen's *Algebra ofte Stelkonst*, whose Latin translation was at this time (1673) in Newton's hands; compare II: 302 and Descartes' *Geometrie* (Leyden, 1637): 382.

Here it should be seen that the term a^3 (the third power of a, that is) occupies the first place in the dividend, the terms $2a^2c - a^2b$ in which a is of two dimensions occupy the second, and so forth. The dividend could also have been written this way: $a^3 + (2c - b) a^2 - 3bca + b^2c$, where the terms occupying the second place are united by bringing together factors of the letter according to which the ordering is made. And were the terms in this manner to be set out according to the dimensions of the letter b, the work ought to be performed as it is in the next array, an explanation of which we have thought fit to adjoin:

$$-b+a \Big) cb^2 + (-3ac - a^2)\, b + 2a^2c + a^3 \Big(-cb + 2ac + a^2.$$

$$\begin{array}{c} cb^2 \qquad\qquad -acb \\ \hline -(2ac + a^2)\, b + 2a^2c + a^3 \\ -(2ac + a^2)\, b + 2a^2c + a^3 \\ \hline 0 \qquad\qquad 0 \end{array}$$

Say: how many times is $-b$ contained in cb^2. Reply: $-cb$. Therefore, when $-cb$ is entered in the quotient, take away $-b+a$ times $-cb$, that is, $cb^2 - acb$, and there will be left in the second place $-(2ac + a^2)\, b$. Adjoin to this residue, please, the quantities in the final place—namely, $2a^2c + a^3$—and again say: how many times is $-b$ contained in $-(2ac + a^2)\, b$. Answer: $2ac + a^2$. Therefore, having written these in the quotient, take away $-b+a$ times $2ac + a^2$, that is,

$$-(2ac + a^2)\, b + 2a^2c + a^3,$$

and nothing will remain. It is hence established that the division is at an end, the resulting quotient being $-cb + 2ac + a^2$ as before.

And likewise, if it is necessary to divide

$$a^2y^4 - a^2c^4 + y^2c^4 + y^6 - 2y^4c^2 - a^6 - 2a^4c^2 - a^4y^2 \quad \text{by} \quad y^2 - a^2 - c^2,^{(52)}$$

I order the quantities according to the letter y in this manner:

$$y^2 - (a^2 + c^2) \Big) y^6 + (a^2 - 2c^2)\, y^4 - (a^4 - c^4)\, y^2 - (a^6 + 2a^4c^2 + a^2c^4).$$

Then I commence the division, as in the scheme below. Other examples, too, are added: regarding these it is further to be observed that, when the dimensions of the letter in line with which the ordering is made do not everywhere proceed in the same arithmetical progression but sometimes advance by a leap, substitution is made in the vacant places of the symbol '0'.

$$yy \begin{smallmatrix} -aa \\ -cc \end{smallmatrix}) y^6 \begin{smallmatrix} +aa \\ -2cc \end{smallmatrix} y^4 \begin{smallmatrix} -a^4 \\ +c^4 \end{smallmatrix} yy \begin{smallmatrix} -a^6 \\ -2a^4cc \\ -aac^4 \end{smallmatrix} \left(y^4 \begin{smallmatrix} +2aa \\ -cc \end{smallmatrix} yy \begin{smallmatrix} +a^4 \\ +aacc \end{smallmatrix} \right.$$

$$y^6 \begin{smallmatrix} -aa \\ -cc \end{smallmatrix} y^4$$

$$\overline{}$$

$$0 \begin{smallmatrix} +2aa \\ -cc \end{smallmatrix} y^4 \qquad\qquad a+b)\,aa \quad * \quad -bb\,(a-b.$$

$$\begin{smallmatrix} 2aa \\ -cc \end{smallmatrix} y^4 \begin{smallmatrix} -2a^4 \\ -aacc \end{smallmatrix} yy \qquad\qquad \overline{aa+ab}$$

$$\begin{smallmatrix} +c^4 \end{smallmatrix} \qquad\qquad\qquad 0 \; -ab$$

$$0 \begin{smallmatrix} +a^4 \\ +aacc \end{smallmatrix} yy \qquad\qquad\qquad -ab-bb$$

$$\begin{smallmatrix} a^4 \\ +aacc \end{smallmatrix} yy \begin{smallmatrix} -a^6 \\ -2a^4cc \\ -aac^4 \end{smallmatrix} \qquad\qquad 0 \qquad 0$$

$$\overline{}$$

$$0 \qquad 0$$

$$yy-2ay+aa)\,y^4 \quad * \quad -3\tfrac{1}{2}aayy+3a^3y-\tfrac{1}{2}a^4\,(yy+2ay-\tfrac{1}{2}aa.$$

$$\underline{y^4-2ay^3 \quad +aayy}$$

$$0+2ay^3-4\tfrac{1}{2}aayy$$

$$\underline{+2ay^3 \quad -4aayy+2a^3y}$$

$$0 \qquad -\tfrac{1}{2}aayy \quad +a^3y$$

$$\underline{-\tfrac{1}{2}aayy \quad +a^3y-\tfrac{1}{2}a^4}$$

$$0 \qquad\qquad 0 \qquad\qquad 0$$

$$aa+ab\surd 2+bb)\,a^4 \quad * \quad * \quad * \quad +b^4\,(aa-ab\surd 2+bb.^{(53)}$$

$$\underline{a^4+a^3b\surd 2 \;+aabb}$$

$$-a^3b\surd 2 \;-aabb$$

$$\underline{-a^3b\surd 2-2aabb-ab^3\surd 2}$$

$$+aabb+ab^3\surd 2$$

$$\underline{+aabb+ab^3\surd 2+b^4}$$

$$0 \qquad\qquad 0 \qquad\qquad 0$$

Aliqui Divisionem incipiunt ab ultimis terminis, sed eodem recidit si inverso terminorum ordine incipiatur a prioribus. Sunt et aliæ methodi dividendi[54] sed facillimam et commodissimam nosse sufficit.

---

(53) Compare this implied factorization of a^4+b^4 (a late marginal addition in Newton's manuscript) with his contemporary attempts to factorize the general binomial $1 \pm x^n$, n integral (IV: 205–8, especially 207). See also note (538).

$$y^2-(a^2+c^2)\Big)y^6+(a^2-2c^2)\,y^4 \qquad -(a^4-c^4)\,y^2\ -a^6-2a^4c^2-a^2c^4\Big(y^4+(2a^2-c^2)\,y^2+a^4+a^2c^2.$$

$$\frac{y^6\ -(a^2+c^2)\,y^4}{0+(2a^2-c^2)\,y^4}$$

$$\frac{(2a^2-c^2)\,y^4-(2a^4+a^2c^2-c^4)\,y^2}{0 \qquad\qquad +(a^4+a^2c^2)\,y^2}$$

$$\frac{(a^4+a^2c^2)\,y^2-(a^6+2a^4c^2+a^2c^4)}{0 \qquad\qquad 0}$$

$$a+b)a^2+0[a]-b^2(a-b.$$

$$\frac{a^2\ +ba}{-ba}$$

$$\frac{-ba-b^2}{0\qquad 0}$$

$$y^2-2ay+a^2)y^4+0[y^3]-3\tfrac12 a^2y^2+3a^3y-\tfrac12 a^4(y^2+2ay-\tfrac12 a^2.$$

$$\frac{y^4-2ay^3\ \ +a^2y^2}{0+2ay^3\ -4\tfrac12 a^2y^2}$$

$$\frac{2ay^3\ -4a^2y^2+2a^3y}{0\qquad -\tfrac12 a^2y^2\ +a^3y}$$

$$\frac{-\tfrac12 a^2y^2\ +a^3y-\tfrac12 a^4}{0\qquad\qquad 0\qquad\ 0}$$

$$a^2+ab\sqrt2+b^2)a^4\ \ +0\ \ \ \ +0\ \ \ +0\ \ +b^4(a^2-ab\sqrt2+b^2.^{(53)}$$

$$\frac{a^4+a^3b\sqrt2\ +a^2b^2}{-a^3b\sqrt2\ -a^2b^2}$$

$$\frac{-a^3b\sqrt2-2a^2b^2-ab^3\sqrt2}{+a^2b^2+ab^3\sqrt2}$$

$$\frac{a^2b^2+ab^3\sqrt2+b^4}{0\qquad\ 0\qquad\ 0}$$

Some people start division from the final terms, but there results the same if it be begun from the first ones with the order of the terms inverted. There are also other methods of dividing[54] but it is sufficient to be familiar with the easiest and most convenient.

---

(54) For instance, the variant on the galley method proposed by Kinckhuysen in his *Algebra* (see ii: 301); compare D. E. Smith, *History of Mathematics*, **2** (Boston/New York, 1925): 136–44. The present 'most commodious' method is, of course, that exemplified by Descartes in Book 3 of his *Geometrie* (see note (52) above).

[1673]

De extractione Radicum.[55]

Lect 7   Cùm numeri alicujus radix quadratica extrahi debet, is in locis alternis incipiendo ab unitate, punctis notandus est; Dein figura in Quoto seu Radice scribenda cujus quadratum figuræ vel figuris ante primum punctum aut æquale sit aut proximè minus. Et ablato illo quadrato, cæteræ radicis figuræ sigillatim invenientur dividendo residuum per duplum radicis eatinus extractæ, et singulis vicibus auferendo[56] e residuo illo factum a figura novissimè prodeunte & decuplo prædicti Divisoris figura illa aucti.

‖[19]   ‖ Sic ad extrahendam radicem ex 99856, imprimis nota eum[57] punctis ad hunc modum 9˙98˙56. Dein quære numerum cujus quadratum æquatur primæ figuræ 9, nempe 3; scribeꝗ in Quoto. Et de 9 ablato quadrato 3×3 seu 9, restabit 0; cui adnecte figuras ante proximum punctum, nempe 98 pro sequente opere. Tum neglecta ultima figura 8, dic quoties duplum 3 seu 6 continetur in priori 9? Resp: 1. Quare scripto 1 in Quoto, aufer factum 1×61 seu 61 de 98 et restabit 37, cui adnecte ultimas figuras 56, et fiet 3756 numerus in quo opus denuò institui debet. Quare et hujus ultima figura 6 neglecta, dic quoties duplum 31 seu 62 continetur in 375 (id quod ex initialibus figuris 6 et 37 conjici potest animadvertendo quoties 6 continetur in 37)? Resp 6. Et scripto 6 in Quoto aufer factum 6×626 seu 3756, et restabit nihil. Unde constat opus peractum esse prodeunte Radice 316.

```
9˙98˙56(316
9
─────
0 98
  61
─────
37 56
37 56
─────
    0
```

Atꝗ ita si radicem ex 22178791 extrahere oportet, imprimis facta punctatione quære numerum cujus quadratum, (siquidem id nequeat æquari) sit proximè minus figuris[58] 22 antecedentibus primum punctum, et invenies esse 4. Nam 5×5 sive 25 major est quàm 22, et 4×4 sive 16 minor. Quare 4 erit prima figura radicis. Et hac itaꝗ in Quoto scripta, de 22 aufer quadratum 4×4 seu 16, residuoꝗ 6 adjunge desuper proximas figuras 17, et habebitur 617, cujus divisione per duplum 4 elicienda est secunda figura

```
22˙17˙87˙91(4709,43637 &c
16
─────
 6 17
 6 09
───────
   8 87 91
   8 46 81
───────────
      41 10.00
      37 67 36
───────────
       3 42 6400
       2 82 5649
─────────────
         60 075100
         56 513196
─────────────
          3 56190400
          2 82566169
─────────────
            73624231
```

───

(55) Newton's first heading to this section was the more restrictive 'De extractione Radicis quadraticæ' (The extraction of a square root).

(56) Newton first continued 'tum quadratum figuræ novissimè prodeuntis tum factum ejus in duplum et prædictum Divisorem figura illa auctum' (both the square of the figure most recently ensuing and its product by two and the aforesaid divisor increased by that figure), forgetting that he had already divided the residue by twice the root 'thus far extracted'.

THE EXTRACTION OF ROOTS[55]

When the square root of some number ought to be extracted, it must be marked with points at every other place starting with the units; then in the quotient or root must be written the figure whose square is either equal to the figure or figures preceding the first point, or is the next less square. And after that square is taken away, the remaining figures in the root will be found one by one separately on dividing the residue at each stage by twice the root thus far extracted and taking away[56] from that residue the product of the figure which most recently results and ten times the aforesaid divisor increased by that figure.

So, to extract the root of 99 856, in the first instance mark it with points in this manner: 9˙98˙56. Then look for the number whose square is equal to the first figure 9—namely 3—and enter it in the quotient. Then, when its square (3 × 3 or 9) is taken away, there will remain 0: to this adjoin the figures, namely 98, preceding the next point in preparation for the work in sequel. Thereafter, neglecting the last figure 8, say: how many times is twice 3, that is, 6, contained in the former one, 9? Answer: 1. Therefore, having written 1 in the quotient, take away the product 1 × 61, that is, 61, from 98 and there will remain 37; to this adjoin the final figures 56, and it will become the number 3756 on which the operation has to be carried out afresh. Therefore, neglecting the last figure 6 of this also, say: how many times is twice 31, or 62, contained in 375? (This can be guessed from the initial figures 6 and 37 by noticing how many times 6 is contained in 37.) Reply: 6. Then, after 6 is written in the quotient, take away the product 6 × 626, that is, 3756, and nothing will be left. It is hence established that the task is finished, yielding the root 316.

And thus, if it proves necessary to extract the root of 22 178 791, first punctuate it and then (seeing that equality is impossible) seek the number whose square shall be next less than the figures[58] 22 preceding the first point, and you will find it to be 4. For 5 × 5, or 25, is larger than 22, and 4 × 4, that is, 16, is less. Therefore 4 will be the first figure in the root. Accordingly, when this is entered in the quotient, from 22 take away its square 4 × 4, or 16, and to the residue 6 adjoin from on top the next figures 17, and 617 will be obtained; by division of this by twice 4 the second figure in the root is to be

$$22\dot{}17\dot{}87\dot{}91\,(4709\cdot43637\,....$$
$$\underline{16}$$
$$6\ 17$$
$$...\quad...\quad...$$

---

(57) Misprinted 'cum' in Whiston's 1707 *editio princeps* (page 31)—a slip carried through into all subsequent printings. Compare note (2) above.

(58) 'quàm numero' (than the number) was first written.

radicis. Nempe, neglecta ultima figura 7, dic quoties 8 continetur in 61 ? Resp:
7. Quare scribe 7 in Quoto, et de 617 aufer factum 7 in 87 seu 609 et restabit 8,
cui adjunge proximas duas figuras 87, et habebitur 887, cujus divisione per
duplum 47 seu 94 elicienda est tertia figura. Utpote dic quoties 94 continetur
in 88 ? Resp: 0. Quare scribe 0 in quoto, adjungecɜ ultimas duas figuras 91, &
habebitur 88791 cujus divisione per duplum 470 seu 940 elicienda est ultima
figura. Nempe dic quoties 940 continetur in 8879 ? Resp: 9. Quare scribe 9 in
Quoto, et radicem habebis 4709.

‖[20]   ‖ Cæterùm cùm factus 9 × 9409 seu 84681 ablatus de 88791 relinquat 4110,
id indicio est numerum 4709 non esse radicem numeri 22178791 præcisè sed ea
paulo minorem existere. Et in hoc casu alijscɜ similibus si veram radicem magis
appropinquare placeat, prosequenda est operatio in decimalibus numeris
adnectendo ad residuum circulos duos in singulis operationibus. Sic residuo
4110 adnexis circulis, evadit 411000; cujus divisione per duplum 4709 seu 9418
elicietur figura prima decimalis, nimirum 4. Dein scripto 4 in Quoto, aufer
4 × 94184 seu 376736 de 411000 et restabit 34264. Atcɜ ita adnexis iterum
duobus circulis, opus pro lubitu continuari potest, prodeunte tandem radice
4709, 43637 &c.

Ubi verò radix ad medietatem aut ultra extracta est.[59] cæteræ figuræ per
divisionem solam obtineri possunt. Ut in hoc exemplo, si radicem ad uscɜ
novem figuras extrahere animus esset, postquam quincɜ priores 4709,4 extractæ
sunt, quatuor posteriores 3637 elici possent dividendo residuum 342[,]64 per
duplum 4709,4.[60]

Et ad hunc modum si radix ex 32976 ad uscɜ quincɜ figuras extrahi debet:
postquam figuræ punctis notantur, scribe 1 in Quoto,
utpote cujus quadratum 1 × 1 seu 1 maximum est quod in 3,
figurâ primum punctum antecedente, continetur. Ac de 3
ablato quadrato illo 1, restabit 2, dein huic 2 annexis
proximis figuris 29, Quære quoties duplum 1 seu 2 continetur in 22, et invenies quidem plusquam 10, sed nunquam
licet divisorem vel decies sumere, imò necɜ novies in hoc
casu, quia factus 9 × 29 sive 261 major est quàm 229 unde
deberet auferri. Quare pone tantum 8. Et perinde scripto 8
in Quoto, et ablato 8 × 28 sive 224 restabit 5. Huic insuper annexis figuris 76,

$$3\dot{\ }29\dot{\ }76(181,59$$
$$\underline{1}$$
$$2\ 29$$
$$2\ 24$$
$$\overline{5\ 76}$$
$$3\ 61$$
$$\overline{362)2\ 15(59}$$

---

(59) This replaces 'plusquam ad medietatem extrahitur' (…is extracted more than
halfway).

(60) In general, the error in approximating $\sqrt{[A^2+\epsilon]}$ by $A+\epsilon/2A$ is of order $\frac{1}{8}\epsilon^2/A^3$, which
is negligible in Newton's present terms.

elicited—namely, neglecting the final figure 7, say: how many times is 8 contained in 61? Answer: 7. Therefore write 7 in the quotient and from 617 take away the product of 7 times 87, that is, 609, when 8 will remain; to this adjoin the next two figures 87 and there will be had 887, and by division of this by twice 47, or 94, the third figure is to be determined. In consequence of which, say: how many times is 94 contained in 88? Reply: 0. Therefore write 0 in the quotient, attach the last pair of figures 91 and there will be had 88 791; by division of this by twice 470, that is, 940, the final figure is to be elicited— namely, say: how many times is 940 contained in 8879? Answer: 9. Therefore write 9 in the quotient, and you will obtain the root 4709.

But, since the product 9×9409, that is, 84 681, when taken from 88 791 still leaves 4110, this is confirmation that the number 4709 is not precisely the root of the number 22 178 791, being in fact a little less. And should you, in this and other similar cases, be inclined to approximate the exact root more closely, the operation must be pursued in decimals by adjoining two noughts to the residue at each repetition. Thus, on annexing noughts to the residue 4110, it becomes 411 000; and through division of this by twice 4709, or 9418, there will be derived the first decimal figure—that is to say, 4. Then, having entered 4 in the quotient, take away $4 \times 94 184$, or 376 736, from 411 000 and there will remain 34 264. And in this way, by further annexing pairs of noughts, the procedure can be continued as far as you please, producing at length the root 4709·43637 ....

When, indeed, the root has been extracted halfway or more,[59] the remaining figures can be obtained by division alone. As in this example, if it were your intention to extract the root as far as nine (significant) figures, after the first five 4709·4 have been extracted the latter four 3637 could be derived by dividing the residue 342·64 by twice 4709·4.[60]

And in this manner, if the root of 32 976 ought to be extracted to five (significant) figures, once the figures are marked with points, write 1 in the quotient, inasmuch as its square 1×1, that is, 1, is the greatest contained in 3, the figure preceding the first point. And when that square 1 is taken from 3 there will remain 2. Then, having appended to this 2 the next figures 29, ask how many times twice 1, or 2, is contained in 22 and to be sure you will find it to be more than 10, but it is never allowable to take a divisor even ten times, and in this case indeed not even nine times will do, because the product 9×29, that is, 261, is greater than 229 from which it is due to be taken away. Consequently, set it merely as 8 and then, having like-

$$
\begin{array}{r}
3{\cdot}29{\cdot}76(181{\cdot}59\ldots \\
1 \\
\hline
229 \\
224 \\
\hline
5\ 76 \\
3\ 61 \\
\hline
362)2\ 15(59\ldots
\end{array}
$$

wise entered 8 in the quotient and taken away 8×28, or 224, there will remain 5. Having in addition annexed to this the figures 76, ask how many times twice 18,

quære quoties duplum 18 seu 36 continetur in 57, et invenies 1, adeoꝗ scribe 1 in Quoto ac de 576 ablato 1×361 seu 361 restabit 215. Deniꝗ ad cæteras figuras eliciendas divide hunc 215 per duplum 181 seu 362 et exibunt figuræ 59, quibus etiam scriptis in Quoto, habebitur Radix 181,59.

    Eadem methodo radices etiam e decimalibus numeris extrahuntur. Sic ex 329,76 radix est 18,159. et ex 3,2976 radix est 1,8159. Et ex 0,032976 radix est 0,18159. Et sic præterea. Sed ex 3297,6 radix est 57,4247. Et ex 32,976 radix est 5,74247. Atꝗ ita ex 9,9856 radix est 3,16. Sed ex 0,99856 radix est 0,999279 &c. Quemadmodum e subjectis Diagrammis constare potest.[61]

‖[21]

‖32˙97˙,6 (57,4247 &c.	0,˙99˙85˙6(0,999279 &c.
25	81
7 97	18 85
7 49	17 01
48 60	1 84 60
45 76	1 79 01
1148) 2 84(247	1998) 5 59(279

Octob. 1674.   Extractionem radicis cubicæ et aliarum omnium regula generali compre-
Lect 1   hendam, praxi potiùs intellectu facili quàm expeditæ consulens, ne moram in eo quod rarò usu veniet, discentibus inferam. Nimirum tertia quæꝗ figura incipiendo ab unitate, primò punctis notanda est si radix sit cubica, aut unaquæꝗ quinta si sit quadrato-cubica, &c. Dein figura in Quoto scribenda est cujus maxima potestas (hoc est cubica si radix sit cubica, aut quadrato-cubica si radix sit quadrato-cubica &c) aut æquatur figuræ vel figuris ante primum punctum, aut proximè minor sit. Et ablata illa potestate, figura proxima elicietur dividendo residuum proxima numeri resolvendi figura auctum per potestatem Quoti pene-maximam ductam in[62] indicem maximæ potestatis, hoc est, per triplum Quadratum Quoti si radix sit cubica, aut per quintuplum quadrato-quadratum si radix sit quadrato-cubica &c. Rursusꝗ a numero resolvendo ablata maxima Quoti potestate, figura tertia invenietur dividendo residuum illud proxima numeri resolvendi figura auctum per potestatem Quoti pene-maximam ductam in[62] indicem maximæ potestatis. Et sic [in] infinitum.[63]

---

    (61) 'videre est' (may be seen) was first written.

    (62) The equivalent 'multiplicatam per' is replaced.

    (63) In general, Newton's procedure—in simplification of the cumbersome Viète method he had annotated as an undergraduate (see 1: 64–5)—will be to approximate the p-th root $A+E$ of the given number N (that is, the real root x of the equation $x^p - N = 0$) successively by choosing A such that $A^p \lessgtr N$ and then determining $E \approx (N-A^p)/pA^{p-1}$. The approach is analogous to the simplification of Viète's cumbrous general procedure for solving numerical equations which Newton introduced in his 'De Analysi' in 1669 (see II: 218–20, especially 218,

that is, 36, is contained in 57 and you will find it 1; so write 1 in the quotient, and when from 576 there is taken away 1 × 361, that is, 361, there will remain 215. To determine the remaining figures, finally, divide this residue 215 by twice 181, or 362, and there will ensue the figures [0·]59; when these, too, are entered in the quotient, the root 181·59 ... will be obtained.

By the same method roots are extracted from decimal numbers also. Thus of 329·76 the root is 18·159 ..., and of 3·2976 it is 1·8159 ...; while the root of 0·032976 is 0·18159 ..., and so forth. But of 3297·6 the root is 57·4247 ..., and of 32·976 it is 5·74247 .... And similarly of 9·9856 the root is 3·16, but of 0·99856 it is 0·999279 .... The details can be ascertained[61] from the appended arrays:

$$
\begin{array}{ll}
32\dot{}97\dot{}6(57\cdot4247\ldots & 0\dot{}99\dot{}85\dot{}6(0\cdot999279\ldots \\
25 & 81 \\
\hline
7\ 97 & 18\ 85 \\
7\ 49 & 17\ 01 \\
\hline
48\cdot60 & 1\ 84\ 60 \\
45\cdot76 & 1\ 79\ 01 \\
\hline
\end{array}
$$

114·8) 2·84([0·0]247 ...   1·998)   5 59([0·000]279 ...

The extraction of the cube root and of all others I shall comprehend in a general rule, concentrating on a technique easy to understand, rather than on one expeditious in performance, so as not to cause the student delay in what will come to be but rarely of use. Specifically, every third figure beginning from unity must first be marked with points if the root be cubic, or every fifth one if it be square-cube, and so on. Then in the quotient must be entered the figure whose greatest power (cubic, that is, if the root be cubic, or fifth if the root be square-cube, and so on) shall either be equal to the figure or figures in front of the first point or be the one next less. And, when that power is taken away, the next figure will be derived by dividing the residue, increased by the next figure in the number to be resolved, by the greatest-but-one power of the quotient multiplied into the index of the greatest power; that is, by triple the square of the quotient if the root be cube, or by five times its fourth power if the root be square-cube, and so on. On again taking away from the number to be resolved the greatest power of the quotient, the third figure will be found by dividing that residue, increased by the next figure in the number to be resolved, by the greatest-but-one power of the quotient multiplied into the index of the greatest power. And so indefinitely.[63]

note (45); and compare I: 66–70). In the present instance Viète's method of root-approximation would successively determine $E \approx (N-A^p)/((A+1)^p - A^p - 1)$, slightly more accurately but with considerably more labour. Newton's simplification represents, of course, the standard 'Newton–Raphson' iterative solution of the corresponding equation $x^p - N = 0$; compare IV: 665 note (24).

Sic ad extrahendam radicem cubicam ex 13312053, numerus ille primò punctis ad hunc modum 13˙312˙053 notandus est. Deinde in Quoto scribenda est illa figura 2 cujus cubus 8, siquidem æquari nequeat, proximè minor sit figuris 13 antecedentibus primum punctum. Et ablato illo cubo restabit 5, quod proxima numeri resolvendi figura 3 auctum, et per triplum quadratum quoti 2 divisum, quærendo nempe quoties 3 × 4 seu 12 continetur in 53, dat 4 pro secunda figura Quoti. Sed cùm Quoti 24 prodiret cubus 13824 major quàm qui auferri posset de figuris 13312 antecedentibus secundum punctum, scribi debet tantum 3 in

$$
\begin{array}{r}
13\text{˙}312\text{˙}053(237 \\ \hline
\text{aufer cub } \quad 8 \\
12)\text{restat} \quad 5\text{˙}3(4.\quad \text{aut 3.} \\ \hline
\text{aufer c.} \quad 12\ 167 \\
1587)\text{restat} \quad 1\text{˙}145\text{˙}0(7. \\ \hline
\text{aufer c.} \quad 13\ 312\ 053 \\
\text{restat} \quad 0
\end{array}
$$

Quoto. Tum Quotus 23 in charta aliqua seorsim per 23 multiplicatus dat
|| [22] quadratum 529, quod iterum per 23 multiplicatum dat || cubum 12167, et hic de 13312 ablatus relinquit 1145; quod proxima resolvendi numeri figura 0 auctum, et per triplum quadratum Quoti 23 divisum, quærendo nempe quoties 3 × 529 seu 1587 continetur in 11450, dat 7 pro tertia figura Quoti. Tum Quotus 237 per 237 multiplicatus dat quadratum 56169, quod iterum per 237 multiplicatum dat cubum 13312053, et hic de resolvendo numero ablatus relinquit nihil. Unde patet radicem quæsitam esse 237.

Atcp ita ad extrahendam radicem quadrato-cubicam ex 36430820, punctum ponitur ad quintam figuram,[64] et figura 3, cujus quadrato-cubus 243 proximè minor est figuris 364 antecedentibus punctum istud, scribitur in Quoto. Dein quadrato-cubo 243 de 364 ablato, restat 121 quod proxima resolvendi numeri figura 3 auctum et per quinquies quadrato-quadratum Quoti divisum, quærendo nempe quoties 5 × 81 seu 405 continetur in

$$
\begin{array}{r}
364\text{˙}30820(32,5 \\ \hline
243 \\
405)121\ 3(2 \\ \hline
335\ 54432 \\
5242880)28\ 76388,0(5
\end{array}
$$

1213, dat 2 pro secunda figura. Quotus ille 32 in se ter ductus efficit quadrato-quadratum 1048576, et hoc iterum in 32 ductum efficit quadrato-cubum 33554432; qui a numero resolvendo ablatus relinquit 2876388. Itacp 32 est integra pars radicis, sed non justa radix, et proinde si opus in decimalibus numeris prosequi animus est, residuum circulo auctum dividi debet per quinquies prædictum quadrato-quadratum Quoti, quærendo quoties 5 × 1048576 seu 5242880 continetur in 2876388,0, et prodibit tertia figura sive prima decimalis 5. Atcp ita auferendo quadrato-cubum Quoti 32,5 de numero resol-

---

(64) Understand 'a dextra' (from the right).

Thus, to extract the cube root of 13 312 053, that number must first be marked with points in this manner: 13˙312˙053. Next, in the quotient must be entered the figure 2 whose cube 8, seeing that it cannot be equal, shall be next less than the figures 13 preceding the first point. Then, after that cube has been taken away, there will remain 5: this, increased by the next figure 3 in the number to be resolved, and divided by triple the square of the quotient 2—namely, by asking how many times 3×4, that is, 12, is contained in 53—, yields 4 for the second figure in the quotient. But since 24 in the quotient would produce a cube, 13 824, greater than might be taken away from the figures 13 312 preceding the second point, there should be entered only 3 in the quotient. Then the quotient 23 multiplied (on some separate piece of paper) by 23 yields 529 as its square, and this, when again multiplied by 23, yields the cube 12 167, which, taken from 13 312, leaves 1145; this, increased by the next figure 0 in the number to be resolved, and divided by three times the square of the quotient 23—namely, by asking how many times 3×529, or 1587, is contained in 11 450—gives 7 for the third figure in the quotient. Then the quotient 237 multiplied by 237 yields 56 169 as its square, and, again multiplied by 237, it gives 13 312 053 as its cube: this, when taken from the number to be resolved, leaves nothing. It is hence evident that the required root is 237.

```
                 13˙312˙053(237
subtract cube     8  leaving
              12) 5˙3(4, or 3.
subtract cube   12 167 leaving
             1587)1˙145˙0(7.
subtract cube   13 312 053
              leaving    0
```

And likewise, to extract the square-cube (fifth) root of 36 430 820, a point is set at the fifth figure[64] and the figure 3, whose fifth power 243 is next less than the figures 364 preceding that point, is entered in the quotient. Then, after the fifth power 243 is taken from 364, there remains 121; this, increased by the next figure 3 in the resolvend and divided by five times the fourth power of the quotient—namely, by asking how many times 5×81, or 405, is contained in 1213—gives 2 for the second figure. The quotient 32, multiplied three times into itself, produces the fourth power 1 048 576, and this, again multiplied into 32, produces the fifth power 33 554 432; the latter, when taken from the resolvend, leaves 2 876 388. Accordingly, 32 is the integral part of the root, but not the exact root; in consequence, if you intend to pursue the work in decimals, the residue, increased by a nought, should be divided by five times the above-mentioned fourth power of the root (by inquiring how many times $5 \times 1 048 576$, or 5 242 880, is contained in 2 876 388˙0) and the third figure, that is, first decimal, 5 will result. And in this way, on taking the fifth power of the quotient 32˙5 from the resolvend and

```
            364˙30820(32˙5 ....
            243
       405)121 3(2
            335 54432
  5242880) 28 76388˙0(5
```

vendo ac dividendo residuum per quinquies quadrato-quadratum ejus, erui potest quarta figura. Et sic in infinitum.

Cùm radix quadrato-quadratica extrahenda est, oportet bis extrahere radicem quadraticam, eò quòd $\sqrt{④}$ valeat $\sqrt{(2 \times 2)}$.[65] Et cum radix cubo-cubica extrahenda est, oportet extrahere radicem cubicam et ejus radicis radicem quadraticam, eo quod $\sqrt{6:}$ valeat $\sqrt{2 \times 3:}$[65] Unde aliqui[66] radices hasce non cubo-cubicas sed quadrato-cubicas dixêre. Et idem in alijs radicibus quarum indices non sunt numeri primi observandum est.

‖ [23]
[1674]
Lect 2
E simplicibus quantitatibus Algebraicis extractio radicum ‖ ex ipsa Notatione patet. Quemadmodum quod $\sqrt{aa}$ sit a, et quod $\sqrt{aacc}$ sit ac, et quod $\sqrt{9aacc}$ sit $3ac$, et quod $\sqrt{49a^4xx}$ sit $7aax$. Atꝗ ita quod $\sqrt{\dfrac{a^4}{cc}}$ seu $\dfrac{\sqrt{a^4}}{\sqrt{cc}}$ sit $\dfrac{aa}{c}$, et quod $\sqrt{\dfrac{a^4bb}{cc}}$ sit $\dfrac{aab}{c}$, et quod $\sqrt{\dfrac{9aazz}{25bb}}$ sit $\dfrac{3az}{5b}$, et quod $\sqrt{\tfrac{4}{9}}$ sit $\tfrac{2}{3}$.[67] Et quod $\sqrt{③\dfrac{8b^6}{27a^3}}$ sit $\dfrac{2bb}{3a}$. Et quod $\sqrt{④aabb}$ sit $\sqrt{ab}$. Quinetiam quod $b\sqrt{aacc}$ seu b in $\sqrt{aacc}$ valeat b in ac sive abc. Et quod $3c\sqrt{\dfrac{9aazz}{25bb}}$ valeat $3c \times \dfrac{3az}{5b}$ sive $\dfrac{9acz}{5b}$. Et quod $\dfrac{a+3x}{c}\sqrt{\dfrac{4bbx^4}{81aa}}$ valeat $\dfrac{a+3x}{c} \times \dfrac{2bxx}{9a}$ sive $\dfrac{2abxx + 6bx^3}{9ac}$.

Hæc inquam patent siquidem propositas quantitates e radicibus in se ductis produci (ut aa ex a in a, $aacc$ ex ac in ac, $9aacc$ ex $3ac$ in $3ac$ &c) prima fronte constare potest. Ubi verò quantitates pluribus terminis constant, opus perinde ac in numeris absolvitur. Sic ad extrahendam radicem quadraticam ex $aa + 2ab + bb$, imprimis radicem primi termini aa nempe a scribe in Quoto. Et ablato ejus quadrato $a \times a$ restabit $2ab + bb$ pro elicienda reliqua parte radicis. Dic itaꝗ quoties duplum quoti seu $2a$ continetur in primo residui termino $2ab$? Resp: b. Adeoꝗ scribe b in Quoto, et ablato facto b in $\overline{2a+b}$ seu $2ab + bb$ restabit nihil. Quod indicat opus peractum esse, prodeunte radice $a+b$.[68]

$$aa + 2ab + bb\,(a+b$$
$$\underline{aa}$$
$$0$$
$$\overline{2ab + bb}$$
$$\overline{00}$$

Et sic ad extrahendam radicem ex $a^4 + 6a^3b + 5aabb - 12ab^3 + 4b^4$, imprimis pone in Quoto radicem primi termini a^4 nempe aa, et ablato ejus quadrato $aa \times aa$ seu a^4 restabit $6a^3b + 5aabb - 12ab^3 + 4b^4$ pro reliqua radice elicienda.

---

(65) That is, $\sqrt[2]{(\sqrt[2]{})}$ and $\sqrt[2]{(\sqrt[3]{})}$ respectively.

(66) Whom exactly Newton had in mind is not clear, though doubtless the distinction between $\sqrt[2+3]{}$ and $\sqrt[2\times3]{}$ was still not adequately grasped by the more plodding mathematicians of the period. In a letter (to John Kersey?) in the late 1660's claiming priority for Camillus Gloriosus 'in his first decade of exercises' (*Exercitationum Mathematicarum Decas Prima*, Naples, 1627) over Oughtred in the published use of the lower-case contractions q, c, qq, qc, ... for powers (compare note (26) above), John Collins went on to stress the superiority of the equivalent superscript Cartesian notation with the comment: 'Is not A^5 sooner wrote than

dividing the remainder by five times its fourth power, the fourth figure can be determined. And so on indefinitely.

When a square-square (fourth) root is to be extracted, you must twice extract a square root, inasmuch as $\sqrt[4]{\ }$ is equivalent to $\sqrt[2\times2]{\ }$.[65] And when a cube-cube (sixth) root is to be extracted, you must extract the cube root and then the square root of the root, seeing that $\sqrt[6]{\ }$ is equal in value to $\sqrt[2\times3]{\ }$.[65] For this reason certain people[66] have called these roots not cube-cube but square(d)-cube. And observe that the like holds in other roots whose indices are not primes.

In the case of simple algebraic quantities the extraction of roots is manifest from the very notation. For instance, $\sqrt{(a^2)}$ is a, $\sqrt{(a^2c^2)}$ is ac, $\sqrt{(9a^2c^2)}$ is $3ac$, and $\sqrt{(49a^4x^2)}$ is $7a^2x$. Similarly, $\sqrt{(a^4/c^2)}$, that is, $\sqrt{a^4}/\sqrt{c^2}$, is a^2/c, $\sqrt{(a^4b^2/c^2)}$ is a^2b/c, $\sqrt{(9a^2z^2/25b^2)}$ is $3az/5b$, and $\sqrt{(4/9)}$ is $2/3$. [67]Also $\sqrt[3]{(8b^6/27a^3)}$ is $2b^2/3a$; and $\sqrt[4]{(a^2b^2)}$ is $\sqrt{(ab)}$. Indeed $b\sqrt{(a^2c^2)}$, that is, b times $\sqrt{(a^2c^2)}$, is equivalent to b times ac, or abc; $3c\sqrt{(9a^2z^2/25b^2)}$ to $3c\times3az/5b$, or $9acz/5b$; and

$$((a+3x)/c)\sqrt{(4b^2x^4/81a^2)} \quad \text{to} \quad ((a+3x)/c)\times2bx^2/9a,$$

that is, $(2abx^2+6bx^3)/9ac$.

These reductions are, I say, obvious inasmuch as it can be accepted at first glance that the quantities proposed are the products of the roots multiplied into themselves (a^2, for example, that of a times a, a^2c^2 that of ac times ac, $9a^2c^2$ that of $3ac$ times $3ac$, and so on). But when the quantities consist of several terms, the task is accomplished exactly as in numbers. So, to extract the square root of $a^2+2ab+b^2$, first enter the root of the first term a^2, namely a, in the quotient; and, when its square $a\times a$ is taken away, there will remain $2ab+b^2$ for determining the remaining portion of the root. Accordingly, say: how many times is twice the quotient, that is, $2a$, contained in the first term $2ab$ of the residue? Answer: b.

$$
\begin{array}{l}
a^2+2ab+b^2(a+b. \\
\underline{a^2} \\
\ \ 0 \\
\hline
\ \ \ 2ab+b^2 \\
\hline
\ \ \ \ \ 0 \quad 0
\end{array}
$$

Therefore write b in the quotient and, when the product of b times $2a+b$, that is, $2ab+b^2$, is taken away, nothing will remain. This confirms that the work is finished, resulting in the root $a+b$.[68]

And thus, to extract the root of $a^4+6a^3b+5a^2b^2-12ab^3+4b^4$, in the first instance set in the quotient the root of the first term a^4, namely a^2, and when its square $a^2\times a^2$, or a^4, is taken away there will remain $6a^3b+5a^2b^2-12ab^3+4b^4$

Aqc? Let *A* be 2. The cube of 2 is 8, which squared is 64: one of the questions between Maghet[ɔ] Grisio and Camillus [in their textbooks Collins has just received from Italy] is whether $64[=(A^3)^2]=Acc$ or Aqc. The Cartesian method tells you it is A^6, and decides the doubt' (S. P. Rigaud, *Correspondence of Scientific Men of the Seventeenth Century*, **2** (Oxford, 1841): 477–82, especially 480).

(67) The next two examples are a late addition in the margin.
(68) The paradigm case of square-root extraction.

Dic itacp quoties $2aa$ continetur in $6a^3b$? Resp: $3ab$. Quare scribe $3ab$ in Quoto et ablato facto $3ab$ in $\overline{2aa+3ab}$ seu $6a^3b+9aabb$ restabit etiamnum

$$-4aabb-12ab^3+4b^4$$

pro opere persequendo. Adeocp dic iterum quoties duplum Quoti

nempe $2aa+6ab$ continetur in $-4aabb-12ab^3$, sive quod perinde est dic quoties ‖[24] duplum primi termini Quoti seu $2aa$ continetur in ‖ primo residui termino $-4aabb$? Resp: $-2bb$. Et proinde scripto $-2bb$ in Quoto, et ablato facto $-2bb$ in $2aa+6ab-2bb$ seu $-4aabb-12ab^3+4b^4$, restabit nihil Unde constat radicem esse $aa+3ab-2bb$.

$$
a^4+6a^3b+5aabb-12ab^3+4b^4(aa+3ab-2bb.
$$
$$
\frac{a^4}{0}
$$

$$
6a^3b+9aabb
$$
$$
\frac{0 \quad -4aabb}{\qquad -4aabb-12ab^3+4b^4}
$$
$$
0 \qquad 0 \qquad 0
$$

Atcp ita quantitatis $xx-ax+\tfrac14 aa$ radix est $x-\tfrac12 a$, et quantitatis

$$y^4+4y^3-8y+4 \quad \text{radix} \quad yy+2y-2,$$

et quantitatis $16a^4-24aaxx+9x^4+12bbxx-16aabb+4b^4$ radix $3xx-4aa+2bb$ ut e subjectis diagrammis constare potest.

$$
xx-ax+\tfrac14 aa(x-\tfrac12 a.
$$
$$
\frac{xx}{0}
$$
$$
\frac{-ax+\tfrac14 aa}{0 \qquad 0}
$$

$$
9x^4 \begin{matrix}-24aa \\ +12bb\end{matrix} xx \begin{matrix}+16a^4 \\ -16aabb \\ +4b^4\end{matrix} \left(3xx\begin{matrix}-4aa \\ +2bb\end{matrix}\right.
$$
$$
\frac{9x^4}{0}
$$
$$
\begin{matrix}-24aa \\ +12bb\end{matrix} xx \begin{matrix}+16a^4 \\ -16aabb \\ +4b^4\end{matrix}
$$
$$
0 \qquad 0
$$

$$
y^4+4y^3 \quad * \quad -8y+4(yy+2y-2.
$$
$$
\frac{y^4}{0}
$$
$$
\frac{4y^3+4yy}{-4yy}
$$
$$
\frac{-4yy-8y+4}{0 \qquad 0 \qquad 0}
$$

Si radicem cubicam ex $a^3+3aab+3abb+b^3$ oportet extrahere, operatio est hujusmodi. Extrahe radicem cubicam primi termini a^3 nempe a, et pone in Quoto. Tum ablato ejus cubo a^3 dic quoties triplum quadratum ejus seu $3aa$ continetur in proximo residui termino $3aab$? et prodit b. Quare scribe

$$
a^3+3aab+3abb+b^3(a+b.
$$
$$
\frac{a^3}{}
$$
$$
3aa) 0+3aab(b
$$
$$
\frac{a^3+3aab+3abb+b^3}{0 \qquad 0 \qquad 0 \qquad 0}
$$

for deriving the remaining root. Accordingly, say: how many times is $2a^2$ contained in $6a^3b$? Answer: $3ab$. Therefore write $3ab$ in the quotient and when the product $3ab(2a^2+3ab)$, or

$$6a^3b+9a^2b^2,$$

$$a^4+6a^3b+5a^2b^2+12ab^3+4b^4(a^2+3ab-2b^2.$$
$$\frac{a^4}{0}$$
$$\frac{6a^3b+9a^2b^2}{0\quad-4a^2b^2}$$
$$\frac{-4a^2b^2-12ab^3+4b^4}{0\qquad0\qquad0}$$

is taken away there will yet remain $-4a^2b^2-12ab^3+4b^4$ for pursuing the work. So say again: how many times is twice the quotient, namely $2a^2+6ab$, contained in $-4a^2b^2-12ab^3$? or, what is the same, say: how many times is twice the first term in the quotient, that is, $2a^2$, contained in the first term $-4a^2b^2$ of the residue? Answer: $-2b^2$. When in consequence $-2b^2$ is entered in the quotient and the product $-2b^2(2a^2+6ab-2b^2)$, that is, $-4a^2b^2-12ab^3+4b^4$, is taken away, nothing will remain. It is hence established that the root is $a^2+3ab-2b^2$.

And, likewise, of the quantity $x^2-ax+\frac{1}{4}a^2$ the root is $x-\frac{1}{2}a$, of the quantity y^4+4y^3-8y+4 the root is y^2+2y-2, and of

$$16a^4-24a^2x^2+9x^4+12b^2x^2-16a^2b^2+4b^4$$

the root is $3x^2-4a^2+2b^2$: as can be ascertained from the schemes appended.

$$x^2-ax+\frac{1}{4}a^2\Big(x-\frac{1}{2}a.\qquad 9x^4+(-24a^2+12b^2)x^2+16a^4-16a^2b^2+4b^4\Big(3x^2-4a^2+2b^2.$$
$$\frac{x^2}{0}\qquad\qquad\qquad\frac{9x^4}{0}$$
$$\frac{-ax+\frac{1}{4}a^2}{0\quad\;0}\qquad\qquad\frac{-(24a^2-12b^2)x^2+16a^4-16a^2b^2+4b^4}{0\qquad\qquad\qquad0}$$

$$y^4+4y^3+0[y^2]-8y+4\,(y^2+2y-2.$$
$$\frac{y^4}{0}$$
$$\frac{4y^3+4y^2}{-4y^2}$$
$$\frac{-4y^2\;-8y+4}{0\qquad0\qquad0}$$

If the cube root of $a^3+3a^2b+3ab^2+b^3$ must be extracted, the procedure is of this sort. Extract the cube root of the first term a^3, namely a, and set it in the quotient. Then, after its cube a^3 is taken away, say: how many times is triple its square, that is, $3a^2$, contained in the next term $3a^2b$ of the residue? The result is b. Therefore enter b also in the quotient, and

$$a^3+3a^2b+3ab^2+b^3(a+b.$$
$$\frac{a^3}{3a^2)\,0}+3a^2b(b$$
$$\frac{a^3+3a^2b+3ab^2+b^3}{0\qquad0\qquad0\qquad0}$$

etiam b in Quoto, et cubo Quoti $a+b$ ablato restabit nihil. Radix itacჳ est $a+b$.

Eodem modo radix cubica, si extrahatur ex $z^6+6z^5-40z^3+96z-64$, prodit $zz+2z-4$. Atcჳ ita in altioribus radicibus.

|| [25]
[1674]
Lect 3

||De Reductione Fractionum et Radicalium.

Præcedentibus operationibus inservit reductio fractarum et radicalium quantitatum, idcჳ vel ad minimos terminos vel ad eandem denominationem.

De Reductione Fractionum ad minimos terminos.

Fractiones ad minimos terminos reducuntur dividendo numeratores ac denominatores per maximam communem divisorem. Sic fractio $\frac{aac}{bc}$ reducitur ad simpliciorem $\frac{aa}{b}$ dividendo utrumcჳ aac et bc per c; et $\frac{203}{667}$[69] reducitur ad simpliciorem $\frac{7}{23}$ dividendo utrumcჳ 203 et 667 per 29; et $\frac{203aac}{667bc}$ reducitur ad $\frac{7aa}{23b}$ dividendo per $29c$. Atcჳ ita $\frac{6a^3-9acc}{6aa+3ac}$ evadit $\frac{2aa-3cc}{2a+c}$ dividendo per $3a$. Et $\frac{a^3-aab+abb-b^3}{aa-ab}$ evadit $\frac{aa+bb}{a}$ dividendo per $a-b$.

Et hac Methodo termini post Multiplicationem vel Divisionem plerumcჳ abbreviari possunt. Quemadmodum si multiplicare oportet $\frac{2ab^3}{3ccd}$ per $\frac{9acc}{bdd}$ vel id dividere per $\frac{bdd}{9aac}$, prodibit $\frac{18aab^3cc}{3bccd^3}$, et per reductionem $\frac{6aabb}{d^3}$. Sed in hujusmodi casibus præstat ante operationem concinnare terminos, dividendo per maximum communem divisorem quos postea dividere oporteret. Sic in allato exemplo si dividam $2ab^3$ et bdd per communem divisorem b, et $3ccd$ ac $9acc$ per communem divisorem $3cc$; emerget fractio $\frac{2abb}{d}$ multiplicanda per $\frac{3a}{dd}$ vel dividenda per $\frac{dd}{3a}$, prodeunte tandem $\frac{6aabb}{d^3}$ ut supra. Atcჳ ita $\frac{aa}{c}$ in $\frac{c}{b}$ evadit $\frac{aa}{1}$ in $\frac{1}{b}$ seu $\frac{aa}{b}$. Et $\frac{aa}{c}$ divis: per $\frac{b}{c}$ evadit aa div: per b, seu $\frac{aa}{b}$. Et $\frac{a^3-axx}{xx}$ in $\frac{cx}{aa+ax}$ evadit $\frac{a-x}{x}$ in $\frac{c}{1}$ seu $\frac{ac}{x}-c$.[70] Et 28 div: per $\frac{7}{3}$ evadit 4 div: per $\frac{1}{3}$, seu 12.[71]

---

(69) Newton's first example at this point was the fraction $\frac{126}{594}$ ($=\frac{7}{33}$), whose numerator and denominator share the more obvious common factor $18 = 2\times3\times3$.

(70) Implicitly invoking the prior simplification of $(a-x)/x$ to $a/x-1$.

when the cube of the quotient $a+b$ is taken away nothing will remain. The root is accordingly $a+b$.

In the same way, if the cube root of $z^6+6z^5-40z^3+96z-64$ be extracted, it proves to be z^2+2z-4. And similarly in the case of higher roots.

THE REDUCTION OF FRACTIONS AND RADICALS

Of service in the preceding operations is the reduction of fractions and radicals, and that either to their least terms or to the same denomination.

THE REDUCTION OF FRACTIONS TO LEAST TERMS

Fractions are reduced to least terms by dividing numerators and denominators by their greatest common divisor. Thus the fraction a^2c/bc is reduced to the simpler form a^2/b on dividing each of a^2c and bc by c; $\frac{203}{667}$ [69] is reduced to the simpler one $\frac{7}{23}$ on dividing each of 203 and 667 by 29; while $203a^2c/667bc$ is reduced to $7a^2/23c$ on dividing by $29c$. And likewise $(6a^3-9ac^2)/(6a^2+3ac)$ comes to be $(2a^2-3c^2)/(2a+c)$ on dividing by $3a$. And

$$(a^3-a^2b+ab^2-b^3)/(a^2-ab)$$

comes to be $(a^2+b^2)/a$ on dividing by $a-b$.

By this method terms can for the most part be shortened after multiplication and division. If, for instance, it is necessary to multiply $2ab^3/3c^2d$ by $9ac^2/bd^2$—or to divide it by $bd^2/9ac^2$—, there will result $18a^2b^3c^2/3bc^2d^3$ and so, by reduction, $6a^2b^2/d^3$. But in cases of this sort it is preferable to refine the terms before the operation, dividing through by the greatest common divisor which would subsequently need to be divided out. Thus, in the example quoted, if I should divide $2ab^3$ and bd^2 by the common divisor b, and $3c^2d$ and $9ac^2$ by their common divisor $3c^2$, there will emerge the fraction $2ab^2/d$ requiring to be multiplied by $3a/d^2$—or divided by $d^2/3a$—, with the eventual result $6a^2b^2/d^3$, as above. Similarly, $(a^2/c)\times(c/b)$ proves to be $(a^2/1)\times(1/b)$, that is, a^2/b. And $(a^2/c)\div(b/c)$ to be $a^2\div b$, or a^2/b. While $((a^3-ax^2)/x^2)\times(cx/(a^2+ax))$ comes to be

$$((a-x)/x)\times(c/1), \quad\text{or}\quad ac/x-c.\text{[70]}$$

And $28\div\frac{7}{3}$ to be $4\div\frac{1}{3}$, that is, 12.[71]

---

(71) At this place in his 1707 *editio princeps* (pages 42–51) Whiston uncritically introduced from the appended loose sheets of 'Correctiones' (ULC. Dd. 9.68: 259–65, reproduced in contracted form in §2 following) a draft section 'De inventione Divisorum', the revised, amplified version of which was inserted in the present text on his pages 173–87 below (as *soi-disant* 'lectiones' 1–5 of the 'Octob. 1682' series!). Though this sequence was retained by Newton in his 1722 revised edition, we keep to the page-order of the deposited copy, both as representing his maturer preference at the time of its composition and as necessitating no furtive change on Newton's following page 33, where the 'inventio divisorum' is referred to in the future tense, '...alibi satiùs docebitur' (illogically retained by Whiston in 1707, but changed by Newton in 1722 to 'eam priùs docuimus'). See also note (517).

Cæterùm maximus duorum numerorum communis divisor si prima fronte non innotescit, invenitur perpetua ablatione minoris de majore et reliqui de ‖[26] ablato. Nam quæsitus erit ‖ divisor qui tandem nihil relinquit.[72] Sic ad inveniendum maximum communem divisorem numerorum 203 et 667, aufer ter 203 de 667, et reliquum 58 ter de 203, et reliquum 29 bis de 58 restabitꝗ nihil: Quod indicat 29 esse divisorem quæsitum.

Haud secus in speciebus communis divisor, ubi compositus est, invenitur subducendo alterutram quantitatem, aut multiplicem ejus de altera: si modò et quantitates illæ et residuum juxta literæ alicujus dimensiones ut in Divisione ostensum est, ordinentur et qualibet vice concinnentur dividendo ipsas[73] per suos omnes divisores qui aut simplices sunt, aut singulos terminos instar simplicium dividunt. Sic ad inveniendum communem divisorem Numeratoris ac Denominatoris fractionis hujus $\dfrac{x^4-3ax^3-8aaxx+18a^3x-8a^4}{x^3-axx-8aax+6a^3}$, multiplica Denominatorem per x ut primus ejus terminus evadat idem cum primo termino numeratoris. Dein aufer, et restabit $-2ax^3+12a^3x-8a^4$, quod concinnatum dividendo per $-2a$ evadit $x^3-6aax+4a^3$. Hoc aufer de Denominatore et restabit $-axx-2aax+2a^3$. quod itidem per $-a$ divisum fit $xx+2ax-2aa$. Hoc autem per x multiplica ut ejus primus terminus evadat idem cum primo termino novissimi ablati $x^3-6aax+4a^3$, de quo auferendum est, et restabit $-2axx-4aax+4a^3$, quod per $-2a$ divisum fit etiam $xx+2aa-2aa$. Et hoc cùm idem sit ac superius residuum, proindeꝗ ablatum relinquat nihil, quæsitus erit divisor per quem fractio proposita, factâ Numeratoris ac Denominatoris divisione, reduci potest ad simpliciorem, nempe ad $\dfrac{xx-5ax+4aa}{x-3a}$.

Atꝗ ita si habeatur fractio $\dfrac{6a^5+15a^4b-4a^3cc-10aabcc}{9a^3b-27aabc-6abcc+18bc^3}$, termini ejus imprimis abbreviandi sunt dividendo numeratorem per aa ac Denominatorem per $3b$. Dein ablato bis $3a^3-9aac-2acc+6c^3$ de $6a^3+15aab-4acc-10bcc$, restabit $\begin{smallmatrix}15b\\+18c\end{smallmatrix}aa\begin{smallmatrix}-10bcc\\-12c^3\end{smallmatrix}$. Quod concinnatum dividendo terminum utrumꝗ per $5b+6c$ perinde ac si $5b+6c$ simplex esset quantitas evadit $3aa-2cc$. Hoc multiplicatum per a aufer de $3a^3-9aac-2acc+6c^3$ et secunda vice restabit $-9aac+6c^3$. quod itidem concinnatum per applicationem ad $-3c$, evadit etiam $3aa-2cc$ ut ante. ‖[27] Quare $3aa-2cc$ quæsitus est ‖ divisor. Quo invento, divide per eum partes fractionis propositæ et obtinebitur $\dfrac{2a^3+5aab}{3ab-9bc}$.

---

(72) Euclid's algorithm for determining the highest common factor of two numbers; see IV: 91 note (58).

(73) Newton originally wrote 'et reliquum qualibet vice concinnetur dividendo ipsum'

If, however, the greatest common divisor of two numbers is not discernible at first glance, it is found by continually taking away the lesser one from the greater and the remainder from what was taken away; for the divisor sought will be the one which eventually leaves nothing.[72] Thus, to find the greatest common divisor of the numbers 203 and 667, take three times 203 from 667, and three times the remainder 58 from 203, and then twice the remainder 29 from 58, and nothing will be left: this reveals that 29 is the required divisor.

With algebraic quantities it is no different. The common divisor, when it is compound, is ascertained by subtracting either quantity (or some multiple of it) from the other, provided that those quantities and the residue are, as was shown in 'DIVISION', ordered according to the dimensions of some letter and, at each and any stage, refined by dividing them[73] through by all their divisors which either are simple or, in the fashion of simple ones, divide their individual terms. Thus, to find the common divisor of numerator and denominator in the case of this fraction $(x^4 - 3ax^3 - 8a^2x^2 + 18a^3x - 8a^4)/(x^3 - ax^2 - 8a^2x + 6a^3)$, multiply the denominator by x so that its first term comes to be the same as the first term of the numerator; then take it away and there will remain $-2ax^3 + 12a^3x - 8a^4$, which, when refined by dividing through by $-2a$, comes to be $x^3 - 6a^2x + 4a^3$. Take this from the denominator and there will remain $-ax^2 - 2a^2x + 2a^3$, which, after a like division by $-a$, becomes $x^2 + 2ax - 2a^2$. Now multiply this by x so that its first term comes to be identical with that of the quantity most recently subtracted, $x^3 - 6a^2x + 4a^3$; from this it is to be taken, leaving $-2ax^2 - 4a^2x + 4a^3$, which, after division by $-2a$, also becomes $x^2 + 2ax - 2a^2$. Since this is the same as the preceding residue and hence, when taken away, will leave nothing, it will be the required divisor by which the proposed fraction, when division of its numerator and denominator is made, can be reduced to a simpler one: namely, $(x^2 - 5ax + 4a^2)/(x - 3a)$.

And similarly, if the fraction $\dfrac{6a^5 + 15a^4b - 4a^3c^2 - 10a^2bc^2}{9a^3b - 27a^2bc - 6abc^2 + 18bc^3}$ be had, its terms are first to be abbreviated by dividing the numerator by a^2 and the denominator by $3b$. Then, after twice $3a^3 - 9a^2c - 2ac^2 + 6c^3$ is taken from $6a^3 + 15a^2b - 4ac^2 - 10bc^2$, there will remain $(15b + 18c)\, a^2 - (10bc^2 + 12c^3)$. When this is refined by dividing each term by $5b + 6c$, just as though $5b + 6c$ were a simple quantity, it comes to be $3a^2 - 2c^2$. Having multiplied this by a, take it from $3a^3 - 9a^2c - 2ac^2 + 6c^3$ and there will remain $-9a^2c + 6c^3$ at the second stage. That, likewise refined through a division by $-3c$, comes also to be $3a^2 - 2c^2$, as before. Consequently $3a^2 - 2c^2$ is the divisor sought. Once it is found, divide the parts of the proposed fraction by it and there will be obtained $(2a^3 + 5a^2b)/(3ab - 9bc)$.

(and the residue, at each and any stage, refined by dividing it). The revision necessitated slight changes in the sequel also.

[1674]
Lect 4

Quod si divisor communis hoc pacto non inveniatur, certum est nullum omninò existere nisi forsan e terminis prodeat per quos Numerator ac Denominator fractionis abbreviantur.[74] Ut si habeatur fractio $\dfrac{aadd - ccdd - aacc + c^4}{4aad - 4acd - 2acc + 2c^3}$, ac termini ejus juxta dimensiones literæ d disponantur ita ut Numerator evadat

$\begin{matrix} aa \\ -cc \end{matrix} dd \begin{matrix} -aacc \\ +c^4 \end{matrix}$ ac Denominator $\begin{matrix} 4aa \\ -4cc \end{matrix} d \begin{matrix} -2acc \\ +2c^3 \end{matrix}$. Hos imprimis oportet abbreviare

dividendo utrumcg Numeratoris terminum per $aa - cc$ et utrumcg Denominatoris per $2a - 2c$, perinde ac si $aa - cc$ & $2a - 2c$ essent simplices quantitates. Atcg ita vice Numeratoris emerget $dd - cc$, et vice Denominatoris $2ad - cc$, ex quibus sic præparatis nullus communis divisor obtineri potest. Sed e terminis $aa - cc$ et $2a - 2c$ per quos Numerator ac Denominator abbreviati sunt, prodit ejusmodi divisor, nempe $a - c$, cujus ope fractio ad hanc $\dfrac{add + cdd - acc + c^3}{4ad - 2cc}$ reduci potest. Quod si necg termini $aa - cc$ & $2a - 2c$ communem divisorem habuissent, fractio proposita fuisset irreducibilis.[75]

Et hæc generalis est methodus inveniendi communes divisores, sed plerumcg expeditiùs inveniuntur quærendo omnes alterutrius quantitatis divisores primos, hoc est, qui per alios dividi nequeunt, ac dein tentando siqui alteram divident abscg residuo. Sic ad reducendum $\dfrac{a^3 - aab + aab - b^3}{aa - ab}$ ad minimos terminos, inveniendi sunt divisores quantitatis $aa - ab$, nempe a et $a - b$. Dein tentandum est an alteruter a vel $a - b$ dividet etiam $a^3 - aab + abb - b^3$ abscg residuo.[76]

Omnes autem cujusvis quantitatis primi divisores inveniuntur dividendo quantitatem illam per simplicissimam divisorem et Quotum perpetuò per simplicissimam divisorem ejus. Sic divisores quantitatis $9a^3b - 27aabc - 6abcc + 18bc^3$ inveniuntur dividendo ipsam imprimis per 3, et Quotum

$$3a^3b - 9aabc - 2abcc + 6bc^3$$

per b, et istud Quotum $3a^3 - 9aac - 2acc + 6c^3$ per $a - 3c$, prodeunte tandem Quoto $3aa - 2cc$ quod amplius dividi nequit. Adeocg divisores primi sunt 3, b, $a - 3c$, et $3aa - 2cc$.

De Reductione Fractionum ad communem Denominatorem.

∥[28]   ∥Fractiones ad communem Denominatorem reducuntur multiplicando terminos utriuscg per denominatorem alterius. Sic habitis $\dfrac{a}{b}$ et $\dfrac{c}{d}$, duc terminos

---

(74) A first, unfinished equivalent caution, cancelled at this point, reads '...nisi fortè aliquis reperiatur, qui singulos [? terminos Numeratoris et Denominatoris dividat]' (unless perchance some one be found [which shall divide] the individual [terms in the numerator and denominator]).   (75) This last sentence is a late addition in the manuscript.

But if a common divisor is not to be found by this means, it is certain that none at all exists, unless perhaps it ensues from the terms by which the numerator and denominator of the fraction are shortened.[74] Thus, if there be had the fraction $\frac{a^2d^2 - c^2d^2 - a^2c^2 + c^4}{4a^2d - 4acd - 2ac^2 + 2c^3}$, and its terms be arranged according to the powers of the letter d, so that the numerator comes to be $(a^2 - c^2)\,d^2 - (a^2c^2 - c^4)$ and the denominator $(4a^2 - 4c^2)\,d - (2ac^2 - 2c^3)$, it is first required to abbreviate these by dividing each term in the numerator by $a^2 - c^2$, and each in the denominator by $2a - 2c$, just as though $a^2 - c^2$ and $2a - 2c$ were simple quantities. In this way in the numerator's stead there will emerge $d^2 - c^2$ and in the denominator's place $2ad - c^2$; from these, however, when thus prepared, no common divisor can be obtained. But from the terms $a^2 - c^2$ and $2a - 2c$, by which the numerator and denominator have been shortened, there does result a divisor of this type, namely $a - c$; with its aid the fraction can be reduced to this:

$$(ad^2 + cd^2 - ac^2 + c^3)/(4ad - 2c^2).$$

If, however, the terms $a^2 - c^2$ and $2a - 2c$ had had no common divisor, the fraction proposed would have been irreducible.[75]

This method for finding common divisors is general; but they are for the most part located more speedily by searching out the prime divisors of one of the two quantities (those, that is, which cannot be divided by others) and then testing if any shall divide the other quantity without remainder. Thus, to reduce $(a^3 - a^2b + ab^2 - b^3)/(a^2 - ab)$ to its least terms, we have to find the divisors (namely, a and $a - b$) of the quantity $a^2 - ab$, and then test whether one or other of a or $a - b$ shall also divide $a^3 - a^2b + ab^2 - b^3$ without remainder.[76]

All prime divisors, however, of any quantity are found by dividing that quantity by its simplest divisor, and the quotient continually by its simplest divisor. Thus the divisors of the quantity $9a^3b - 27a^2bc - 6abc^2 + 18bc^3$ are found by dividing it first by 3, then the quotient $3a^3b - 9a^2bc - 2abc^2 + 6bc^3$ by b, and then that quotient $3a^3 - 9a^2c - 2ac^2 + 6c^3$ by $a - 3c$, producing at length a quotient, $3a^2 - 2c^2$, which cannot further be divided. As a result the prime divisors are 3, b, $a - 3c$ and $3a^2 - 2c^2$.

THE REDUCTION OF FRACTIONS TO A COMMON DENOMINATOR

Fractions are reduced to a common denominator by multiplying the terms of each by the denominator of the other. Thus, when a/b and c/d are had,

(76) Of course, $a - b$ is a divisor of the latter expression. Obeying Newton's dictate (in the draft 'correction' he had earlier inserted in preference to its revise; see note (71)) that he should proceed with his introduced text 'ut in pag 26 & 27 mutatis mutandis usꝗ ad verba: [absꝗ residuo]', Whiston in his 1707 *editio princeps* silently—and without any logical justification —omitted the final, following paragraph of this present section. It was not restored by Newton in 1722 and so has remained unpublished in any edition.

unius $\frac{a}{b}$ in d, et vicissim terminos alterius $\frac{c}{d}$ in b, et evadent $\frac{ad}{bd}$ et $\frac{bc}{bd}$, quarum communis est denominator bd. Atꝗ ita a et $\frac{ab}{c}$ sive $\frac{a}{1}$ et $\frac{ab}{c}$ evadunt $\frac{ac}{c}$ et $\frac{ab}{c}$. Ubi verò Denominatores communem habent divisorem, sufficit multiplicare alternè per Quotientes. Sic fractiones $\frac{a^3}{bc}$ et $\frac{a^3}{bd}$ ad hasce $\frac{a^3 d}{bcd}$ & $\frac{a^3 c}{bcd}$ reducuntur, multiplicando alternè per Quotientes c ac d ortos divisione denominatorum per communem divisorem b.

Hæc autem Reductio præcipuè usui est in Additione et Subductione fractionum, quæ si diversos habent denominatores, ad eundem reducendæ sunt antequam uniri possunt. Sic $\frac{a}{b} + \frac{c}{d}$ per reductionem evadit $\frac{ad}{bd} + \frac{bc}{bd}$, sive $\frac{ad + bc}{bd}$. Et $a + \frac{ab}{c}$ evadit $\frac{ac + ab}{c}$. Et $\frac{a^3}{bc} - \frac{a^3}{bd}$ evadit $\frac{a^3 d - a^3 c}{bcd}$ vel $\frac{d - c}{bcd} a^3$. Et $\frac{c^4 + x^4}{cc - xx} - cc - xx$ evadit $\frac{2x^4}{cc - xx}$. Atꝗ ita $\frac{2}{3} + \frac{5}{7}$ evadit $\frac{14}{21} + \frac{15}{21}$ sive $\frac{14 + 15}{21}$ hoc est $\frac{29}{21}$. Et $\frac{11}{6} - \frac{3}{4}$ evadit $\frac{22}{12} - \frac{9}{12}$ sive $\frac{13}{12}$. Et $\frac{3}{4} - \frac{5}{12}$ evadit $\frac{9}{12} - \frac{5}{12}$ sive $\frac{4}{12}$ hoc est $\frac{1}{3}$. Et $3\frac{4}{7}$ sive $\frac{3}{1} + \frac{4}{7}$ evadit $\frac{21}{7} + \frac{4}{7}$ sive $\frac{25}{7}$. Et $25\frac{1}{2}$ evadit $\frac{51}{2}$.

Fractiones ubi plures sunt gradatim uniri debent. Sic habito

$$\frac{aa}{x} - a + \frac{2xx}{3a} - \frac{ax}{a - x};$$

ab $\frac{aa}{x}$ aufer a et restabit $\frac{aa - ax}{x}$, huic adde $\frac{2xx}{3a}$ et prodibit $\frac{3a^3 - 3aax + 2x^3}{3ax}$, unde aufer deniꝗ $\frac{ax}{a - x}$ et restabit $\frac{3a^4 - 6a^3 x + 2ax^3 - 2x^4}{3aax - 3axx}$. Atꝗ ita si habeatur $3\frac{4}{7} - \frac{2}{3}$, imprimis aggregatum $3\frac{4}{7}$ inveniendum est nempe $\frac{25}{7}$ dein ab hoc auferendum $\frac{2}{3}$ et restabit $\frac{61}{21}$.

[1674] DE REDUCTIONE RADICALIUM AD MINIMOS TERMINOS.

Lect 5　　Radicalis, ubi totius radix extrahi nequit, plerumꝗ concinnatur extrahendo radicem divisoris alicujus.[77] Sic $\sqrt{aabc}$ extrahendo radicem divisoris aa fit $a\sqrt{bc}$. Et $\sqrt{48}$ extrahendo radicem divisoris 16 fit $4\sqrt{3}$. Et $\sqrt{48aabc}$ extrahendo radicem divisoris $16aa$ fit $4a\sqrt{3bc}$. Et $\sqrt{\dfrac{a^3 b - 4aabb + 4ab^3}{cc}}$ extrahendo radicem divisoris

‖[29] ‖ $\dfrac{aa - 4ab + 4bb}{cc}$ fit $\dfrac{a - 2b}{c}\sqrt{ab}$. Et $\sqrt{\dfrac{aaoomm}{ppzz} + \dfrac{4aam^3}{pzz}}$[78] extrahendo radicem divisoris $\dfrac{aamm}{ppzz}$ fit $\dfrac{am}{pz}\sqrt{oo + 4mp}$. Et $6\sqrt{\dfrac{75}{98}}$ extrahendo radicem divisoris $\dfrac{25}{49}$ fit

---

(77) Which understand to be a perfect square.

multiply the terms of one, a/b, into d and in sequel the terms of the other, c/d, into b; these will come to be ad/bd and bc/bd, having the common denominator bd. Likewise, a and ab/c, that is, $a/1$ and ab/c, come to be ac/c and ab/c. When, in fact, the denominators have a common divisor, it is enough to multiply alternately by the quotients. Thus the fractions a^3/bc and a^3/bd are reduced to these, a^3d/bcd and a^3c/bcd, by multiplying alternately by the quotients c and d which arise on division of the denominators by the common divisor b.

This reduction is, however, of especial use in the addition and subtraction of fractions; for, should they have differing denominators, they must be reduced to the same one before they can be united. Thus $a/b+c/d$ comes by reduction to be $ad/bd+bc/bd$, that is, $(ad+bc)/bd$. And $a+ab/c$ comes to be $(ac+ab)/c$. While $a^3/bc-a^3/bd$ comes to be $(a^3d-a^3c)/bcd$ or $((d-c)/bcd)\,a^3$. And $(c^4+x^4)/(c^2-x^2)-(c^2+x^2)$ comes to be $2x^4/(c^2-x^2)$. Similarly $\frac{2}{3}+\frac{5}{7}$ comes to be $\frac{14}{21}+\frac{15}{21}$ or $\dfrac{14+15}{21}$, that is, $\frac{29}{21}$. And $\frac{11}{6}-\frac{3}{4}$ comes to be $\frac{22}{12}-\frac{9}{12}$ or $\frac{13}{12}$. And $\frac{3}{4}-\frac{5}{12}$ comes to be $\frac{9}{12}-\frac{5}{12}$ or $\frac{4}{12}$, that is, $\frac{1}{3}$. While $3\frac{4}{7}$, that is, $\frac{3}{1}+\frac{4}{7}$, comes to be $\frac{21}{7}+\frac{4}{7}$ or $\frac{25}{7}$. And $25\frac{1}{2}$ comes to be $\frac{51}{2}$.

Where there are several fractions, they ought to be united in steps. Thus, having $a^2/x-a+2x^2/3a-ax/(a-x)$, from a^2/x take away a and there will remain $(a^2-ax)/x$; to this add $2x^2/3a$ and there will result

$$(3a^3-3a^2x+2x^3)/3ax;$$

from which, finally, take away $ax/(a-x)$ and there will result

$$(3a^4-6a^3x+2ax^3-2x^4)/(3a^2x-3ax^2).$$

Likewise, if there be had $3\frac{4}{7}-\frac{2}{3}$, you must first find the total of 3 and $\frac{4}{7}$, namely $\frac{25}{7}$, and then from this take away $\frac{2}{3}$, when there will remain $\frac{61}{21}$.

THE REDUCTION OF RADICALS TO LEAST TERMS

A radical, when its root cannot wholly be extracted, is commonly refined by extracting the root of some divisor[77] of it. Thus, by extracting the root of its divisor a^2, $\sqrt{[a^2bc]}$ becomes $a\sqrt{[bc]}$. And $\sqrt{48}$, on extracting the root of the divisor 16, becomes $4\sqrt{3}$. And $\sqrt{[48a^2bc]}$, on extracting the root of the divisor $16a^2$, becomes $4a\sqrt{[3bc]}$. While $\sqrt{[(a^3b-4a^2b^2+4ab^3)/c^2]}$, on extracting the root of the divisor $(a^2-4ab+4b^2)/c^2$, becomes $((a-2b)/c)\,\sqrt{[ab]}$. And

$$\sqrt{[a^2o^2m^2/p^2z^2+4a^2m^3/pz^2]},^{(78)}$$

on extracting the root of the divisor a^2m^2/p^2z^2, becomes $(am/pz)\sqrt{[o^2+4mp]}$. Also $6\sqrt{\frac{75}{98}}$, on extracting the root of the divisor $\frac{25}{49}$, becomes $\frac{30}{7}\sqrt{\frac{3}{2}}$, that is,

---

(78) An example borrowed from a text familiar to Newton, Descartes' analytical reduction of the Greek 3/4-line locus to a second-degree Cartesian curve in the second book of his *Geometrie* (Leyden, 1637): 330 [= *Geometria*, ₂1659: 31].

$\frac{30}{7}\sqrt{\frac{3}{2}}$, sive $\frac{30}{7}\sqrt{\frac{6}{4}}$, radicem$\not{q}$ denominatoris[79] adhuc extrahendo, fit $\frac{15}{7}\sqrt{6}$. Et sic $a\sqrt{\frac{b}{a}}$ sive $a\sqrt{\frac{ab}{aa}}$ extrahendo radicem denominatoris[79] fit $\sqrt{ab}$. Et $\sqrt{③}\;\overline{8a^3b+16a^4}$ extrahendo radicem cubicam divisoris $8a^3$ fit $2a\sqrt{③}:\overline{b+2a}$. Haud secus $\sqrt{④}\;a^3x$ extrahendo radicem [quadrato-]quadraticam[80] divisoris aa fit $\sqrt{a}$ in $\sqrt{④}\;ax$ vel extrahendo radicem quadrato-quadraticam divisoris a^4 fit $a\sqrt{④}\frac{x}{a}$. Atq$\not{q}$ ita $\sqrt{⑥}:a^7x^5$ convertitur in $a\sqrt{⑥}:ax^5$, vel in $ax\sqrt{⑥}:\frac{a}{x}$ vel in $\sqrt{ax}\times\sqrt{③}:aax$.

Cæterùm hæc reductio non tantùm concinnandis radicalibus inservit, sed et earum Additioni et Subductioni, si modò ex parte radicali conveniant[81] ubi ad formam simplicissimam reducuntur. Tunc enim uniri possunt, quod aliter non fit. Sic $\sqrt{48}+\sqrt{75}$ per reductionem evadit $4\sqrt{3}+5\sqrt{3}$ hoc est $9\sqrt{3}$. Et $\sqrt{48}-\sqrt{\frac{16}{27}}$ per reductionem evadit $4\sqrt{3}-\frac{4}{9}\sqrt{3}$ hoc est $\frac{32}{9}\sqrt{3}$. Et sic

$$\sqrt{\frac{4ab^3}{cc}}+\sqrt{\frac{a^3b-4aabb+4ab^3}{cc}}$$

extrahendo quicquid est rationale, evadit $\frac{2b}{c}\sqrt{ab}+\frac{a-2b}{c}\sqrt{ab}$ hoc est $\frac{a}{c}\sqrt{ab}$. Et $\sqrt{3}:\overline{8a^3b+16a^4}-\sqrt{3}:\overline{b^4+2ab^3}$ evadit $2a\sqrt{3}:\overline{b+2a}-b\sqrt{3}:\overline{b+2a}$ hoc est $\overline{2a-b}\sqrt{3}:\overline{b+2a}$.[82]

DE REDUCTIONE RADICALIUM AD EANDEM DENOMINATIONEM.

Cùm in radicalibus diversæ denominations[83] instituenda est multiplicatio vel divisio, oportet omnes ad eandem denominationem reducere, idq$\not{q}$ præfigendo signum radicale cujus index est minimus numerus quem earum indices dividunt absq$\not{q}$ residuo, et suffixas quantitates toties dempta[84] una vice in se ducendo quoties index ille jam major evaserit. Sic enim $\sqrt{ax}$ in $\sqrt{3}:aax$ evadit $\sqrt{6}:a^3a^3$ in $\sqrt{6}:a^4xx$ hoc est $\sqrt{6}:a^7x^5$. Et $\sqrt{a}$ in $\sqrt{4}:ax$ evadit $\sqrt{4}:aa$ in $\sqrt{4}:ax$ hoc est $\sqrt{4}:a^3x$. Et $\sqrt{6}$ in $\sqrt{4}:\frac{5}{6}$ evadit $\sqrt{4}:36$ in $\sqrt{4}:\frac{5}{6}$, hoc est $\sqrt{4}:30$. Eademq$\not{q}$ ratione $a\sqrt{bc}$ evadit $\sqrt{aa}$ in $\sqrt{bc}$ hoc est $\sqrt{aabc}$. Et $4a\sqrt{3bc}$ evadit $\sqrt{16aa}$ in $\sqrt{3bc}$ hoc est $\sqrt{48aabc}$.

Et $2a\sqrt{3}:\overline{b+2a}$ evadit $\sqrt{3}:8a^3$ in $\sqrt{3}:\overline{b+2a}$ hoc est $\sqrt{3}:\overline{8a^3b+16a^4}$. Atq$\not{q}$ ita $\frac{\sqrt{ac}}{b}$ fit $\frac{\sqrt{ac}}{\sqrt{bb}}$ sive $\sqrt{\frac{ac}{bb}}$. Et $\frac{6abb}{\sqrt{18ab^3}}$ fit $\frac{\sqrt{36aab^4}}{\sqrt{18ab^3}}$ sive $\sqrt{2ab}$. Et sic in alijs.[85]

---

(79) This replaces 'facti divisoris' (of the divisor [so] made).
(80) By a slip Newton wrote 'quadraticam' simply.
(81) Compare note (30).
(82) Newton has cancelled an abortive final sentence: 'Quod si post talem reductionem non conveniant ex parte radicali, non possunt uniri sed cu...' (But if after such reduction they should still not concur in their radical part, they cannot be united but...).

$\frac{30}{7}\sqrt{\frac{6}{4}}$, and by further extracting the root of the denominator[79] it comes to be $\frac{15}{7}\sqrt{6}$. And thus $a\sqrt{[b/a]}$, that is, $a\sqrt{[ab/a^2]}$, on extracting the root of the denominator[79] becomes $\sqrt{[ab]}$. And $\sqrt[3]{[8a^3b+16a^4]}$ on extracting the cube root of the divisor $8a^3$ becomes $2a\sqrt[3]{[b+2a]}$. No differently, $\sqrt[4]{[a^3x]}$ on extracting the square-square (fourth) root of the divisor a^2 becomes $\sqrt{a}\times\sqrt[4]{[ax]}$, or alternatively on extracting the square-square root of the divisor a^4 it becomes $a\sqrt[4]{[x/a]}$. And so $\sqrt[6]{[a^7x^5]}$ is converted to $a\sqrt[6]{[ax^5]}$, or to $ax\sqrt[6]{[a/x]}$, or to $\sqrt{ax}\times\sqrt[3]{[a^2x]}$.

For the rest, this reduction serves not merely to refine radicals but also to effect their addition and subtraction, provided they agree in their radical part[81] when they are reduced to simplest form: for then (and not otherwise) can they be united. Thus by reduction $\sqrt{48}+\sqrt{75}$ turns out to be $4\sqrt{3}+5\sqrt{3}$, that is, $9\sqrt{3}$. And $\sqrt{48}-\sqrt{\frac{16}{27}}$ by reduction comes to be $4\sqrt{3}-\frac{4}{9}\sqrt{3}$, that is, $\frac{32}{9}\sqrt{3}$. And thus $\sqrt{[4ab^3/c^2]}+\sqrt{[(a^3b-4aabb+4ab^3)/c^2]}$ on extracting whatever is rational proves to be $(2b/c)\sqrt{[ab]}+((a-2b)/c)\sqrt{[ab]}$, that is, $(a/c)\sqrt{[ab]}$. And

$$\sqrt[3]{[8a^3b+16a^4]}-\sqrt[3]{[b^4+2ab^3]}$$

proves to be $2a\sqrt[3]{[b+2a]}-b\sqrt[3]{[b+2a]}$, that is, $(2-b)\sqrt[3]{[b+2a]}$.[82]

THE REDUCTION OF RADICALS TO THE SAME DENOMINATION

When multiplication or division must be instituted in radicals of differing denomination,[83] it is necessary to reduce all to the same denomination—this, by prefixing a radical sign whose index is the least number which their indices divide without remainder, and then multiplying the quantities suffixed as many times but[84] one into themselves as the index has now become larger. For thus $\sqrt{[ax]}\times\sqrt[3]{[a^2x]}$ comes to be $\sqrt[6]{[a^3x^3]}\times\sqrt[6]{[a^4x^2]}$, that is, $\sqrt[6]{[a^7x^5]}$. And $\sqrt{a}\times\sqrt[4]{[ax]}$ comes to be $\sqrt[4]{[a^2]}\times\sqrt[4]{[ax]}$, that is, $\sqrt[4]{[a^3x]}$. And $\sqrt{6}\times\sqrt[4]{\frac{5}{6}}$ comes to be $\sqrt[4]{36}\times\sqrt[4]{\frac{5}{6}}$, that is, $\sqrt[4]{30}$. For the same reason $a\sqrt{[bc]}$ proves to be $\sqrt{[a^2]}\times\sqrt{[bc]}$, that is, $\sqrt{[a^2bc]}$. And $4a\sqrt{[3bc]}$ comes to be $\sqrt{[16a^2]}\times\sqrt{[3bc]}$, that is, $\sqrt{[48a^2bc]}$. And $2a\sqrt[3]{[b+2a]}$ comes to be $\sqrt[3]{[8a^2]}\times\sqrt[3]{[b+2a]}$, that is, $\sqrt[3]{[8a^3b+16a^4]}$. Similarly $\sqrt{[ac]}/b$ becomes $\sqrt{[ac]}/\sqrt{[b^2]}$, or $\sqrt{[ac/b^2]}$. And $6ab^2/\sqrt{[18ab^3]}$ becomes $\sqrt{[36a^2b^4]}/\sqrt{[18ab^3]}$, or $\sqrt{[2ab]}$. And so in other instances.[85]

---

(83) In their radical index ($\sqrt{}$, $\sqrt[3]{}$, $\sqrt[4]{}$, and so on).

(84) 'minùs' (less) is cancelled. Understand that the radical quantities are to be raised to the power whose number is that by which the radical index has been multiplied (so yielding an unchanged total quantity).

(85) Failing to realize that it is but a draft of the revised section incorporated into the present text on Newton's pages 207–10 following (as 'Lect. 9' of the 'Octob. 1682' series), Whiston in sequel in his 1707 *editio princeps* (pages 58–61) introduced from the concluding 'correctiones' (§2 below) an additional section 'De Reductione Radicaliũ ad simpliciores radicales per extractionem radicum', not questioning that Newton's instruction there that it be added 'In calce capitis *de Reductione radicalium*' held true. Though Newton in his own 1722

‖ [30]

‖De forma Æquationis.[86]

[1674]
Lect 6

Æquationes,[87] quæ sunt quantitatum aut sibi mutuò æqualium aut simul nihilo æquipollentium congeries, duobus præcipuè modis considerandæ veniunt: vel ut ultimæ conclusiones ad quas in Problematîs solvendis deventum est, vel ut media quorum ope finales æquationes acquirendæ sunt. Prioris generis æquatio ex unica tantum incognita quantitate cognitis involuta conflatur, modò Problema sit definitum[88] et aliquid certi quærendum innuat. Sed eæ posterioris generis involvunt plures quantitates incognitas, quæ ideò debent inter se comparari et ita connecti ut ex omnibus una tandem emergat æquatio nova cui inest unica[89] quam quærimus incognita quantitas admista cognitis. Quæ quantitas ut exinde facilius eliciatur, æquatio ista varijs plerumqʒ modis transformanda est donec evadat ea simplicissima quæ potest, atqʒ etiam similis alicui ex sequentibus gradibus, in quibus x designat quantitatem quæsitam ad cujus dimensiones termini, ut vides, ordinantur, et p, q, r, s alias quascunqʒ quantitates ex quibus determinatis et cognitis etiam x determinatur, et per methodos post explicandos investigari potest.

$$x = p.$$
$$xx = px + q.$$
$$x^3 = pxx + qx + r.$$
$$x^4 = px^3 + qxx + rx + s.$$
&c.

Vel

$$x - p = 0.$$
$$xx - px - q = 0.$$
$$x^3 - pxx - qx - r = 0.$$
$$x^4 - px^3 - qxx - rx - s = 0.$$
&c.

Ad horum normam itaqʒ termini æquationum secundum dimensiones incognitæ quantitatis in ordinem semper redigendi sunt ita ut primum locum occupent in quibus incognita quantitas est plurimarum dimensionum instar x, xx, x^3, x^4, & secundum locum in quibus ea est una dimensione minor instar p, px, pxx, px^3, & sic præterea. Et quod signa terminorum attinet, possunt omnibus modis se habere: imò et unus vel plures ex intermedijs terminis aliquando deesse. Sic $x^3 * - bbx + b^3 = 0$ vel $x^3 = bbx - b^3$, est æquatio tertij gradus, et

$$z^4 \begin{matrix} +a \\ -b \end{matrix} z^3 ** \begin{matrix} +ab^3 \\ -b^4 \end{matrix} = 0$$

æquatio quarti. Nam gradus æquationum æstimantur ex maxima dimensione quantitatis incognitæ nullo respectu ad quantitates cognitas habito, nec ad

---

revised edition made no effort to restore the passage to its alternative location at the end of the *Arithmetica* (immediately before the terminal appendix on the geometrical construction of equations), we are here—as in the preceding instance (see note (71)) of a revised section illogically suppressed by Whiston in favour of its preliminary draft—unwilling to tamper with the existing page-sequence of the manuscript we reproduce, one confirmed by Newton's own imposed marginal chronology, and we therefore ignore Whiston's insertion. Compare note (596).

  (86) A somewhat revised version of the corresponding introductory passage 'De Æquationibus' in Newton's additions to Kinckhuysen's *Algebra* (see II: 396–8).

THE FORM OF AN EQUATION[86]

Equations—the collections together of quantities[87] which are equal to one another or together effectively equal to nothing—come into consideration in two particular ways: either as the ultimate conclusions arrived at in solving problems, or as intermediaries with whose aid the final equations are obtained. An equation of the first kind is a conflation of but a single unknown quantity involved with known ones, provided that the problem is precise[88] and suggests that there is something definite to be ascertained. But those of the latter kind involve several unknown quantities, and these ought accordingly to be compared one with another and so connected as eventually to distil out of them all a single new equation containing the[89] unique unknown quantity we seek, intermingled with known ones. To determine this quantity more easily therefrom, that equation must be transformed—in differing ways, usually—till it comes to be the simplest possible, and so, also, similar to some one in the following grades, in which x denotes the quantity sought (according to whose dimensions the terms, as you see, are ordered) and $p, q, r, s, \ldots$ any other quantities in terms of which, when they are fixed and known, x also is determined and can, by methods to be elaborated later, be discovered:

$$
\begin{aligned}
x &= p. & x - p &= 0. \\
x^2 &= px + q. & \quad\text{Or}\quad \qquad x^2 - px - q &= 0. \\
x^3 &= px^2 + qx + r. & x^3 - px^2 - qx - r &= 0. \\
x^4 &= px^3 + qx^2 + rx + s. & x^4 - px^3 - qx^2 - rx - s &= 0. \\
\ldots \quad \ldots & & \ldots \quad \ldots
\end{aligned}
$$

On the pattern of these, in consequence, the terms of equations must always be reduced to order, following the dimensions of the unknown quantity, in such a way that first place is taken by ones in which the unknown is of most dimensions (as here $x, x^2, x^3, x^4, \ldots$), the second place by ones in which it is of one dimension less (as $p, px, px^2, px^3, \ldots$), and so forth. And, with respect to the signs of the terms, these can present themselves in all manner of ways; indeed, one or more of the intermediate terms may on occasion be missing. Thus

$$x^3 + 0[x^2] - b^2 x + b^3 = 0, \quad\text{or}\quad x^3 = b^2 x - b^3,$$

is an equation of third degree, and $z^4 + (a-b) z^3 + 0[z^2] + 0[z] + ab^3 - b^4 = 0$ one of the fourth: for the degree of an equation is reckoned from the greatest dimension of the unknown quantity in it, without paying regard to the known

---

(87) Newton first continued 'quæ ex quantitatibus aut sibi mutuò aut simul nihilo æqualibus constare possunt' (which can consist of quantities equal either mutually to one another or collectively to nothing).

(88) In modern terminology, 'well-formed'.

(89) 'illa' (that) is cancelled.

‖[31] intermedios terminos. Attamen ex defectu intermediorum terminorum æquatio plerumcp[90] fit multò simplicior et nonnun‖quam ad gradum inferiorem quodammodo deprimitur. Sic enim $x^4 = qxx + s$ æquatio secundi gradus censenda est, siquidem ea in duas secundi gradus æquationes resolvi potest. Nam supposito $xx = y$, et y pro xx in æquatione illa perinde scripto, ejus vice prodibit $yy = qy + s$, æquatio secundi gradus: cujus ope cum y inventa fuerit, æquatio $xx = y$ secundi etiam gradus, dabit x.

Atcp hæ sunt conclusiones ad quas Problemata deduci debent. Sed antequam eorum resolutionem[91] aggrediar, opus erit ut modos transformandi et in ordinem redigendi æquationes et ex medijs eliciendi finales æquationes abstractè doceam. Æquationis autem solitariæ reductionem in sequentibus regulis complectar.

DE CONCINNANDA ÆQUATIONE SOLITARIA.[92]

REG: 1. *Siquæ sunt quantitates quæ se mutuo destruere vel per additionem aut subductionem coalescere possunt, termini perinde minuendi sunt.* Veluti si habeatur $5b - 3a + 2x = 5a + 3x$, aufer utrincp $2x$ et adde $3a$ proditcp $5b = 8a + x$. Atcp ita $\frac{2ab + bx}{a} - b = a + b$, delendo æquipollentes $\frac{2ab}{a} - b = b$, evadit $\frac{bx}{a} = a$.

Ad hanc Regulam referri debet etiam ordinatio terminorum æquationis quæ fieri solet per translationem ad contrarias partes cum signo contrario. Ut si habita æquatione $5b = 8a + x$ desideretur[93] x; aufer utrincp $8a$, vel, quod eodem recidit, transfer $8a$ ad contrarias partes cum signo mutato, & prodibit $5b - 8a = x$. Eodem modo si habeatur $aa - 3ay = ab - bb + by$ ac desideretur[93] y, transpone $-3ay$ & $ab - bb$ eo ut ex una parte consistant termini multiplicati per y et ex altera reliqui termini, & prodibit $aa - ab + bb = 3ay + by$, unde y elicietur per Reg 5 sequentem, dividendo scilicet utramcp partem[94] per $3a + b$, prodibit enim $\frac{aa - ab + bb}{3a + b} = y$. Atcp ita æquatio $abx + a^3 - aax = aab - 2abx - x^3$ per debitam transpositionem et ordinationem evadit $x^3 = \begin{smallmatrix} aa \\ -3ab \end{smallmatrix} x \begin{smallmatrix} -a^3 \\ +abb \end{smallmatrix}$, vel $x^3 \begin{smallmatrix} -aa \\ +3ab \end{smallmatrix} x \begin{smallmatrix} +a^3 \\ -abb \end{smallmatrix} = 0$.

---

(90) In the case of a single missing term this is, of course, untrue since any listed term in an equation may be 'destroyed' by an appropriate linear transformation of the roots. The naïve belief that, by appropriate reduction, all the middle terms of a given equation might be removed was still paraded by François Du Laurens in his *Specimina Mathematica* (Paris, 1667); as we have seen (II: 420, note (101); III: 571, note (24)) Newton in summer 1670 gave John Collins pressing reasons why this was a false hope even in the cubic case.

(91) Understand 'particularem' (particular).

(92) A much amplified version of the corresponding chapter 'De transformatione æquationum, non mutatis radicibus' in Kinckhuysen's *Algebra* (II: 318–19), one which in 1670

quantities or to the intermediate terms. Nevertheless, through lacking intermediate terms an equation commonly[90] becomes much simpler and may not infrequently by some manner be depressed to a lower degree. So, for instance, $x^4 = qx^2 + s$ is to be thought of as an equation of second degree, seeing that it can be resolved into two second-degree equations: for, on supposing $x^2 = y$ and likewise writing y in place of x^2 in that equation, in its stead there will result $y^2 = qy + s$, an equation of second degree; and, once y has been discovered with its help, the equation $x^2 = y$ (also of second degree) will determine x.

These are the conclusions to which problems ought to be brought. But before I enter upon their[91] solution, it will be necessary for me to explain, in abstract fashion, ways of transforming equations and subduing them into order, and of deriving final equations from intermediate ones. The reduction of a lone equation, however, I shall embrace in the following rules.

THE REFINING OF A LONE EQUATION[92]

Rule 1. If there are any quantities which can mutually destroy one another or combine by addition or subtraction, the terms are correspondingly to be diminished. If, for example, there be had $5b - 3a + 2x = 5a + 3x$, take $2x$ away from either side and add $3a$, when there results $5b = 8a + x$. Similarly $(2ab + bx)/a - b = a + b$ comes, on deleting the equivalents $2ab/a - b = b$, to be $bx/a = a$.

To this rule ought also to be related the ordering of terms in an equation which usually is accomplished by a translation to the opposite side with a change of sign. For instance, if, when the equation $5b = 8a + x$ is had,[93] x were desired, take away $8a$ from either side or (what comes to the same) transpose $8a$ to the opposite side with its sign altered, and there will result $5b - 8a = x$. In the same way, if $a^2 - 3ay = ab - b^2 + by$ and [93]y be desired, transfer $-3ay$ and $ab - b^2$ with the aim that one side shall consist of terms multiplied by y and the other of the remaining terms, and there will result $a^2 - ab + b^2 = 3ay + by$: from this y will be derived by means of Rule 5 following—specifically, on dividing either side by $3a + b$, for there will then result $(a^2 - ab + b^2)/(3a + b) = y$. And thus the equation $abx + a^3 - a^2x = a^2b - 2abx - x^3$ through appropriate transposition and rearrangement comes to be $x^3 = (a^2 - 3ab)\,x - a^3 + a^2b$, or

$$x^3 - (a^2 - 3ab)\,x + a^3 - a^2b = 0.$$

Newton saw 'noe reason to change' (II: 400). The present 'Reg: 1' corresponds to Kinckhuysen's section 'Quomodo superflua æquationis eliminari possint, formacȝ ejus mutari, addendo vel subtrahendo'; 'Reg: 3' and 'Reg: 4' relate to Kinckhuysen's 'Quomodo fractiones alicujus æquationis mutentur in integra, et radicales in rationales, multiplicando'; while 'Reg: 2', 'Reg: 5', 'Reg: 6' and 'Reg: 7' are loosely contained in his final section 'Quomodo æquatio diminuatur, dividendo, et extrahendo radices'.

(93) 'valor' (the value of) is deleted.

(94) The equivalent 'utrobicȝ' was first written.

REG 2. *Siqua compareat quantitas per quam omnes æquationis termini multiplicantur, debent omnes per illam quantitatem dividi; vel si per eandem quantitatem omnes dividantur debent omnes per illam multiplicari.* Sic habito $15bb = 24ab + 3bx$, divide terminos omnes[95] per b & fit $15b = 24a + 3x$, deinde per 3 et fit $5b = 8a + x$. Vel habito $\frac{b^3}{ac} - \frac{bbx}{cc} = \frac{xx}{c}$ multiplica omnes per c et prodit $\frac{b^3}{a} - \frac{bbx}{c} = xx$.

REG: 3. *Siqua sit fractio irreducibilis in cujus denominatore reperiatur litera illa ad* ‖ [32] *cujus dimensiones æquatio ordinanda est,* ‖ *omnes æquationis termini per istum denominatorem, aut per aliquem divisorem ejus multiplicandi sunt.* Ut si æquatio $\frac{ax}{a-x} + b = x$ secundum x ordinanda sit, multiplicentur omnes ejus termini per $a - x$ denominatorem fractionis $\frac{ax}{a-x}$ siquidem x inibi reperiatur, & prodit

$$ax + ab - bx = ax - xx, \quad \text{seu} \quad ab - bx = -xx,$$

et facta utriusɋ partis translatione $xx = bx - ab$. Atɋ ita si habeatur $\frac{a^3 - abb}{2cy - cc} = y - c$ terminiɋ juxta y ordinandi sint, multiplicentur per denominatorem $2cy - cc$ vel saltem per divisorem $2y - c$ quo y tollatur e denominatore, et exurget $\frac{a^3 - abb}{c} = 2yy - 3cy + cc$ et ordinando $\frac{a^3 - abb}{c} - cc + 3cy = 2yy$. Ad eundem modum $\frac{aa}{x} - a = x$ multiplicando per x evadit $aa - ax = xx$, et $\frac{aabb}{cxx} = \frac{xx}{a+b-x}$ multiplicando primo per xx dein per $a + b - x$ evadit $\frac{a^3bb + aab^3 - aabbx}{c} = x^4$.

[1674]
Lect 7
REG: 4. *Sicui surdæ quantitati irreducibili litera illa involvatur ad cujus dimensiones æquatio ordinanda est, cæteri omnes termini ad contrarias partes cum signis mutatis transferendi sunt & utraɋ pars æquationis in se semel multiplicanda si radix quadratica sit, vel bis si sit cubica &c.* Sic ad ordinandam juxta x æquationem $\sqrt{aa - ax} + a = x$, transferatur a ad alteras partes fitɋ $\sqrt{aa - ax} = x - a$, et quadratis partibus, $aa - ax = xx - 2ax + aa$ seu $0 = xx - ax$ hoc est $x = a$.[96] Sic etiam

$$\sqrt{3} : \overline{aax + 2axx - x^3} - a + x = 0,$$

transponendo $-a + x$ evadit $\sqrt{3} : \overline{aax + 2axx - x^3} = a - x$, & partibus cubicè multiplicatis $aax + 2axx - x^3 = a^3 - 3aax + 3axx - x^3$ seu $xx = 4ax - aa$. Et sic $y = \sqrt{ay + yy - a\sqrt{ay - yy}}$ quadratis partibus evadit $yy = ay + yy - a\sqrt{ay - yy}$ et terminis debitè transpositis $ay = a\sqrt{ay - yy}$ seu $y = \sqrt{ay - yy}$, et partibus iterum quadratis $yy = ay - yy$, et transponendo denuo, $2yy = ay$ sive $2y = a$.

---

(95) This phrase replaces 'ubiɋ' (throughout).

Rule 2. If any quantity should appear as a multiplier of all terms in an equation, they ought all to be divided by that quantity; or should they all be divided by the same quantity, then they ought all to be multiplied by it. Thus, having $15b^2 = 24ab + 3bx$, divide all the terms[95] by b and there comes $15b = 24a + 3x$, and then by 3, when there comes $5b = 8a + x$. Alternatively, having $b^3/ac - b^2x/cc = x^2/c$, multiply throughout by c and there will prove to be $b^3/a - b^2x/c = x^2$.

Rule 3. If there be any irreducible fraction in whose denominator shall be found the letter according to whose dimensions the equation is to be ordered, all terms in the equation must be multiplied by that denominator or some divisor of it. For instance, if the equation $ax/(a-x) + b = x$ is to be ordered in sequence of x, let all its terms be multiplied by the denominator $a-x$ of the fraction $ax/(a-x)$ seeing that x is found therein, and there results $ax + ab - bx = ax - x^2$, that is $ab - bx = -x^2$, and, when transposition of the two sides is made, it is $x^2 = bx - ab$. And thus, if there be had $(a^3 - ab^2)/(2cy - c^2) = y - c$ and the terms have to be ordered according to y, let them be multiplied by the denominator $2cy - c^2$, or at least by $2y - c$, so as to remove y from the denominator, and there will arise

$$(a^3 - ab^2)/c = 2y^2 - 3cy + c^2,$$

and on ordering it $(a^3 - ab^2)/c - c^2 + 3cy = 2y^2$. In much the same way $a^2/x - a = x$ comes, on multiplying by x, to be $a^2 - ax = x^2$, and $a^2b^2/cx^2 = x^2/(a+b-x)$, on multiplying first by x^2 and then by $a+b-x$, comes to be

$$(a^3b^2 + a^2b^3 - a^2b^2x)/c = x^4.$$

Rule 4. If the letter according to whose dimensions the equation is to be ordered be involved in any irreducible surd quantity, all the remaining terms are to be transferred with signs changed to the opposite side and then either side of the equation multiplied into itself once if the root be square, twice if it be cube, and so on. Thus to order according to x the equation $\surd[a^2 - ax] + a = x$, transfer a to the opposite side and there comes $\surd[a^2 - ax] = x - a$; then, when the sides are squared, $a^2 - ax = x^2 - 2ax + a^2$ or $0 = x^2 - ax$, that is, $x = a$.[96] So also $\sqrt[3]{[a^2x + 2ax^2 - x^3]} - a + x = 0$ comes, on transposing $-a + x$, to be $\sqrt[3]{[a^2x + 2ax^2 - x^3]} = a - x$ and, when the sides are cubed, $a^2x + 2ax^2 - x^3 = a^3 - 3a^2x + 3ax^2 - x^3$, that is, $x^2 = 4ax - a^2$. And thus $y = \surd[ay + y^2 - a\surd(ay - y^2)]$, when its sides are squared, comes to be

$$y^2 = ay + y^2 - a\surd(ay - y^2)$$

and, with its terms appropriately transposed, $ay = a\surd(a^2 - y^2)$ or $y = \surd(a^2 - y^2)$; then, with its sides again squared, to be $y^2 = ay - y^2$ and, by a fresh transposition, $2y^2 = ay$ or $2y = a$.

---

(96) Or, of course, $x = 0$.

REG: 5. *Terminis secundum dimensiones literæ alicujus ope præcedentium regularum dispositis, si maxima ejusdem literæ dimensio per cognitam quamlibet quantitatem multiplicetur, debet tota æquatio per eandem dividi.* Sic $2y = a$ dividendo per 2 evadit $y = \frac{1}{2}a$,

et $\dfrac{bx}{a} = a$ dividendo per $\dfrac{b}{a}$ evadit $x = \dfrac{aa}{b}$. Et $\begin{array}{c} 2ac \\ -cc \end{array} x^3 \begin{array}{c} +a^3 \\ +aac \end{array} xx \begin{array}{c} -2a^3c \\ +aacc \end{array} x - a^3cc = 0$

dividendo per $2ac - cc$ evadit $x^3 \dfrac{\begin{array}{c} +a^3 \\ +aac \end{array} xx \begin{array}{c} -2a^3c \\ +aacc \end{array} x - a^3cc}{2ac - cc} = 0$, sive

$$x^3 \frac{+a^3 + aac}{2ac - cc} xx - aax - \frac{a^3c}{2a - c} = 0.$$

REG: 6. *Aliquando reductio institui potest dividendo æquationem per compositam*

‖ [33] *aliquam quantitatem.* Sic enim $y^3 = \begin{array}{c} -2c \\ +b \end{array} yy + 3bcy \,\|\, -bbc$, ad hanc $yy = -2cy + bc$

reducitur transferendo terminos omnes ad easdem partes hoc modo

$y^3 \begin{array}{c} +2c \\ -b \end{array} yy - 3bcy + bbc = 0$ et dividendo[97] per $y - b$ ut in capite de divisione

ostensum est: prodibit enim $yy + 2cy - bc = 0$. Ast hujusmodi divisorum inventio difficilis est et alibi satiùs docebitur.[98]

REG: 7. *Aliquando etiam reductio per extractionem radicis ex utraq; æquationis parte instituitur.* Quemadmodum si habeatur $xx = \frac{1}{4}aa - bb$, extracta utrobiq; radice prodit $x = \sqrt{\frac{1}{4}aa - bb}$.[99] Quod si habeatur $xx + aa = 2ax + bb$, transfer $2ax$ et exurget $xx - 2ax + aa = bb$, extractisq; partium radicibus $x - a = +\text{vel} - b$, seu $x = a \pm b$. Sic etiam habito $xx = ax - bb$, adde utrinq; $-ax + \frac{1}{4}aa$ et prodit $xx - ax + \frac{1}{4}aa = \frac{1}{4}aa - bb$, et extracta utrobiq; radice $x - \frac{1}{2}a = \pm\sqrt{\frac{1}{4}aa - bb}$ seu $x = \frac{1}{2}a \pm \sqrt{\frac{1}{4}aa - bb}$.

Et sic universaliter si sit $xx = . px . q$, erit $x = . \frac{1}{2}p \pm \sqrt{\frac{1}{4}pp . q}$.[100] Ubi $\frac{1}{2}p$ et q ijsdem signis ac p et q in æquatione priori afficienda sunt sed $\frac{1}{4}pp$ semper affirmativè ponendum. Estq; hoc exemplum regula ad cujus similitudinem æquationes omnes quadraticæ ad formam simplicium reduci possunt. E.g. Proposita æquatione $yy = \dfrac{2xxy}{a} + xx$, ad extrahendam radicem y confer

---

(97) 'hanc æquationē' (this equation) is cancelled.

(98) See Newton's pages 173–87 below. As we have earlier observed (note (71)), Whiston in his 1707 *editio princeps* of Newton's *Arithmetica* determined that a considerably shorter draft version of this section 'De inventione Divisorum' should be inserted on Newton's preceding page 25, but inconsistently retained the future tense of the present verb. The illogicality was none too happily ironed out by Newton himself in his 1722 revise, where the whole phrase is emended to read '... & eam prius docuimus' (and we have previously explained it).

(99) Strictly, '$x = \pm \sqrt{\frac{1}{4}aa - bb}$'. In the next example, comparably, Newton first deduced from '$xx - 2ax + aa = bb$' that '$x - a = b$, seu $x = a + b$'.

Rule 5. If, after the terms have, with the help of the preceding rules, been arranged according to the dimensions of some letter, the greatest dimension of this same letter should prove to be multiplied by any known quantity, the whole equation ought to be divided by it. Thus $2y = a$ comes, on dividing by 2, to be $y = \frac{1}{2}a$, and $bx/a = a$ on dividing by b/a comes to be $x = a^2/b$. While

$$(2ac - c^2)\, x^3 + (a^3 + a^2c)\, x^2 - (2a^3c - a^2c^2)\, x - a^3c^2 = 0$$

on dividing by $2ac - c^2$ comes to be

$$x^3 + ((a^3 + a^2c)\, x^3 - (2a^3c - a^2c^2)\, x - a^2c^2)/(2ac - c^2) = 0,$$

that is, $x^3 + ((a^3 + a^2c)/(2ac - c^2))\, x^2 - a^2x - a^3c/(2a - c) = 0$.

Rule 6. On occasion reduction can be effected by dividing the equation by some compound quantity. So, for instance, $y^3 = (b - 2c)\, y^2 + 3bcy - b^2c$ is reduced to this,

$$y^2 = -2cy + bc,$$

by transferring all terms to the same side in this manner,

$$y^3 - (b - 2c)\, y^2 - 3bcy + b^2c = 0,$$

and dividing[97] by $y - b$ as was shown in the chapter on 'Division': for there will then result $y^2 + 2cy - bc = 0$. But finding divisors of this sort is difficult and will more adequately be explained elsewhere.[98]

Rule 7. On occasion, too, reduction is achieved by extracting the root of either side of an equation. For example, if there be had $x^2 = \frac{1}{4}a^2 - b^2$, when the root is extracted there results $x = \sqrt{[\frac{1}{4}a^2 - b^2]}$.[99] And if there be had $x^2 + a^2 = 2ax + b^2$, transfer $2ax$ and there will arise $x^2 - 2ax + a^2 = b^2$; and, when the roots of the sides are extracted, it is $x - a = +$ or $-b$, that is, $x = a \pm b$. So also, when $x^2 = ax - b^2$ is had, add $-ax + \frac{1}{4}a^2$ to each side and there results $x^2 - ax + \frac{1}{4}a^2 = \frac{1}{4}a^2 - b^2$; then, when the root is extracted on either side, $x - \frac{1}{2}a = \pm\sqrt{[\frac{1}{4}a^2 - b^2]}$ or

$$x = \frac{1}{2}a \pm \sqrt{[\frac{1}{4}a^2 - b^2]}.$$

And so universally: if $x^2 = px + q$, then $x = \frac{1}{2}p \pm \sqrt{[\frac{1}{4}p^2 + q]}$.[100] In the latter $\frac{1}{2}p$ and q are to be invested with the same signs as p and q in the former equation, but $\frac{1}{4}p^2$ must always be set positive. This example is a rule, and on its analogy all quadratic equations can be reduced to the form of simple (linear) ones. For example, where the equation $y^2 = (2x^2/a)\, y + x^2$ is propounded, to extract its

---

(100) The points replace uncertain signs '$\pm$'. Newton evidently prefers this modified Cartesian notation (compare IV: 652, note (35)) to the more cumbrous equivalent '$\pm$' (and complementary '$\mp$') introduced on his page 3 above. In our English version we have, in modern style, absorbed the uncertainty in sign into the variables p and q, supposing that they may take on all values, both negative and positive.

$\dfrac{2xx}{a}$ cum p et xx cum q, hoc est scribe $\dfrac{xx}{a}$ pro $\frac{1}{2}p$ et $\dfrac{x^4}{aa}+xx$ pro $\frac{1}{4}pp \cdot q$, atqɜ orietur

$$y=\frac{xx}{a}+\sqrt{\frac{x^4}{aa}+xx}, \quad \text{vel} \quad y=\frac{xx}{a}-\sqrt{\frac{x^4}{aa}+xx}.$$

Eodem modo æquatio $yy=ay-2cy+aa-cc$ conferendo $a-2c$ cum p et $aa-cc$ cum q dabit $y=\frac{1}{2}a-c+{}^{(101)}\sqrt{\frac{5}{4}aa-ac}$. Quinetiam æquatio quadrato-quadratica $x^4=-aaxx+ab^3$ cujus termini impares[102] desunt, ope hujus regulæ evadit $xx=\frac{1}{2}aa+{}^{(101)}\sqrt{\frac{1}{4}a^4+ab^3}$, et extracta iterum radice $x=\sqrt{-\frac{1}{2}aa+\sqrt{\frac{1}{4}a^4+ab^3}}$. Et sic in alijs.

Suntqɜ hæ regulæ pro concinnanda æquatione solitaria, quarum usum cùm Analysta satis perspexerit, ita ut æquationem quamcunqɜ propositam secundum quamlibet literarum in ea complexarum disponere noverit, et ejusdem literæ si ea unius sit dimensionis, aut maximæ potestatis ejus si plurium, valorem elicere: haud difficilem sentiet comparationem plurium æquationum inter se quam pergo jam docere.[103]

[1674]
Lect 8

DE DUABUS PLURIBUSVE ÆQUATIONIBUS IN UNAM TRANSFORMANDIS UT INCOGNITÆ QUANTITATES EXTERMINENTUR.[104]

‖[34] Cum in alicujus problematis solutionem plures habentur æquationes ‖ statum quæstionis comprehendentes, quarum unicuiqɜ plures etiam incognitæ quantitates involvuntur: æquationes istæ (duæ per vices si modo sint plures duabus) sunt ita connectendæ ut una ex incognitis quantitatibus per singulas operationes tollatur et emergat æquatio nova. Sic habitis æquationibus $2x=y+5$, et $x=y+2$, demendo æqualia ex æqualibus prodibit $x=3$. Et sciendum est quod per quamlibet æquationem una quantitas incognita potest tolli, atqɜ adeo cum tot sunt æquationes quot quantitates incognitæ, omnes possunt ad unam deniqɜ reduci in qua unica manebit quantitas incognita. Sin quantitates incognitæ sint unâ plures quàm æquationes habentur tum in æquatione ultimò resultante duæ manebunt quantitates incognitæ, et si sint duabus plures quàm æquationes habentur tum in æquatione ultimò resultante manebunt tres, & sic præterea.

Possunt etiam duæ vel plures quantitates incognitæ per duas tantum æquationes fortasse tolli. Ut si sit $ax-by=ab-az$, et $bx+by=bb+az$: tum æqualibus[105] ad æqualia additis prodibit $ax+bx=ab+bb$, exterminatis utrisqɜ y et z. Sed ejusmodi casus vel arguunt vitium aliquod in statu quæstionis latere, vel

---

(101) Strictly, this should be '$\pm$'.

(102) That is, 'termini imparium dimensionum' (terms of odd dimension).

(103) 'ostendere' (...to reveal) was first written.

(104) This section is taken over without essential change by Newton from his 1670 'Observationes' on Kinckhuysen's *Algebra* (see II: 400–2).

root y compare $2x^2/a$ with p and x^2 with q, that is, write x^2/a in place of $\frac{1}{2}p$ and x^4/a^2+x^2 in place of $\frac{1}{4}p^2+q$, and there will then emerge $y = x^2/a+\sqrt{[x^4/a^2+x^2]}$ or $y = x^2/a-\sqrt{[x^4/a^2+x^2]}$. In the same way, on comparing $a-2c$ with p and a^2-c^2 with q the equation $y^2 = ay-2cy+a^2-c^2$ will yield

$$y = \tfrac{1}{2}a-c[\pm]\sqrt{[\tfrac{5}{4}a^2-ac]}.$$

To be sure, the quartic equation $x^4 = -a^2x^2+ab^3$, in which odd terms[102] are lacking, comes with the help of this rule to be

$$x^2 = -\tfrac{1}{2}a^2[\pm]\sqrt{[\tfrac{1}{4}a^4+ab^3]}$$

and then, after the root is again extracted, $x = [\pm]\sqrt{(-\tfrac{1}{2}a^2[\pm]\sqrt{[\tfrac{1}{4}a^4+ab^3]})}$. And so in other instances.

These are the rules for refining a lone equation. When the algebraist has familiarized himself with their use enough to know how to arrange any equation proposed according to any of the letters embraced in it and to derive the value of that letter, if it be of one dimension, or of its greatest power if it be of several, then he will feel no difficulty with the comparison of several equations one with another: a topic I now proceed to develop.[103]

TRANSFORMING TWO OR MORE EQUATIONS INTO ONE, SO THAT UNKNOWN QUANTITIES ARE ELIMINATED[104]

When, to solve some problem, several equations encompassing the circumstances of the question are obtained, each also involving several unknown quantities, those equations are to be connected (two by two should there be more than two) in such a way that at each separate operation one of the unknowns is removed and a fresh equation emerges. Thus, when the equations $2x = y+5$ and $x = y+2$ are had, by taking equals from equals there will result $x = 3$. And you should realize that by means of any equation one unknown quantity can be removed, and consequently that, when there are as many equations as unknowns, they can all ultimately be reduced to one in which a single unknown will remain. But, should the unknowns be one more than the number of equations obtained, then in the equation finally ensuing two unknown quantities will remain; and if they be two more than that number obtained, then in the equation finally ensuing three will remain; and so on.

It may also be, perhaps, that two or more unknowns can be removed by means of only two equations. As, if there be

$$ax-by = ab-az \quad \text{and} \quad bx+by = b^2+az,$$

then, when equals are added to equals there will result $ax+bx = ab+b^2$, each of y and z being eliminated. But cases of this sort argue that some deficiency is

---

(105) Newton first continued with 'simul' (as on II: 402).

calculum erroneum esse aut non satis artificiosum. Modus autem quo una quantitas incognita per singulas æquationes[106] tollatur ex sequentibus patebit.

EXTERMINATIO QUANTITATIS INCOGNITÆ PER ÆQUALITATEM VALORUM EJUS.[107]

Cùm quantitas tollenda unius est tantum dimensionis in utraꝗ æquatione, valor ejus in utraꝗ per regulas jam ante traditas quærendus est, et alter valor statuendus æqualis alteri.

Sic positis $a+x=b+y$ et $2x+y=3b$, ut exterminetur[108] y æquatio prima dabit $a+x-b=y$, et secunda dabit $3b-2x=y$. Est ergo $a+x-b=3b-2x$, sive ordinando $x=\dfrac{4b-a}{3}$.

Atꝗ ita $2x=y$, et $5+x=y$ dant $2x=5+x$ seu $x=5$.

‖[35]   Et $ax-2by=ab$ et $xy=bb$ dant $\dfrac{ax-ab}{2b}(=y)=\dfrac{bb}{x}$; sive ‖ordinando $xx-bx-\dfrac{2b^3}{a}=0$.

Item $\dfrac{bbx-aby}{a}=ab+xy$ et $bx+\dfrac{ayy}{c}=2aa$ tollendo x dant

$$\frac{aby+aab}{bb-ay}(=x)=\frac{2aac-ayy}{bc}:$$

et reducendo[109] $y^3-\dfrac{bb}{a}yy-\dfrac{-2aac-bbc}{a}y+bbc=0$.

Deniꝗ $x+y-z=0$ et $ay=xz$ tollendo z dant $x+y(=z)=\dfrac{ay}{x}$ sive $xx+xy=ay$.

Hoc idem quoꝗ perficitur subducendo alterutrum valorem quantitatis incognitæ ab altero et ponendo residuum æquale nihilo. Sic in exemplorum primo tolle $3b-2x$ ab $a+x-b$ et manebit $a+3x-4b=0$, sive $x=\dfrac{4b-a}{3}$.

EXTERMINATIO QUANTITATIS INCOGNITÆ SUBSTITUENDO PRO EA VALOREM SUUM.[110]

Cùm in altera saltem æquatione, tollenda quantitas unius tantum dimensionis existit, valor ejus in ea quærendus est et pro se in æquationem alteram substituendus. Sic propositis $xyy=b^3$ et $xx+yy=by-ax$; ut exterminetur x, prima

---

(106) 'operationes' (operations), evidently a momentary slip of Newton's pen (compare II: 402), is cancelled.

(107) This repeats the corresponding section in Newton's Kinckhuysen 'Observationes' (II: 402) all but word for word.

concealed in the statement of the question, or that the computation is either incorrect or insufficiently skilful. The way, however, in which one unknown quantity is to be removed by each of the equations[106] separately will be evident from the sequel.

ELIMINATION OF AN UNKNOWN QUANTITY BY EQUATION OF ITS VALUES[107]

Where the quantity to be removed is of but one dimension in each equation, its value in each is to be sought by the rules just now delivered above and the one value set equal to the other.

Thus, on setting $a+x = b+y$ and $2x+y = 3b$, in order to exterminate[108] y the first equation will give $a+x-b = y$ and the second, $3b-2x = y$. Therefore $a+x-b = 3b-2x$, that is, after ordering, $x = \frac{1}{3}(4b-a)$.

And thus $2x = y$ and $5+x = y$ yield $2x = 5+x$, or $x = 5$.

Also, $ax-2by = ab$ and $xy = b^2$ yield $(ax-ab)/2b\,[=y] = b^2/x$, and so, after ordering, $x^2-bx-2b^3/a = 0$.

Likewise, on removing x, $(b^2x-aby)/a = ab+xy$ and $bx+ay^2/c = 2a^2$ give $(aby+a^2b)/(b^2-ay)\,[=x] = (2a^2c-ay^2)/bc$ and, upon reduction,[109]

$$y^3 - (b^2/a)\,y^2 - (2ac+b^2c/a)\,y + b^2c = 0.$$

Finally, on removing z, $x+y-z = 0$ and $ay = xz$ give $x+y\,[=z] = ay/x$, that is, $x^2+xy = ay$.

The same is also accomplished by subtracting the one value of the unknown quantity from the other and setting the remainder equal to nothing. Thus in the first of the examples take off $3b-2x$ from $a+x-b$ and there will remain $a+3x-4b = 0$, or $x = \frac{1}{3}(4b-a)$.

ELIMINATION OF AN UNKNOWN QUANTITY BY SUBSTITUTING ITS VALUE IN ITS PLACE[110]

Where, in one or other of the equations at least, the quantity to be removed turns out to be of but one dimension, its value is to be ascertained and substituted in its place in the second equation. Thus, when

$$xy^2 = b^3 \quad \text{and} \quad x^2+y^2 = by-ax$$

---

(108) This replaces 'eliminetur' (eliminate), the reading on II: 402. A corresponding change is made by Newton several times in the sequel.

(109) As on II: 402, Newton first began to write 'redigend[o in ordinem]' (on reducing to order).

(110) An unchanged repeat of the equivalent section in the Kinckhuysen 'Observationes' (II: 402–4).

dabit $\frac{b^3}{yy}=x$: quare in secundam substituo $\frac{b^3}{yy}$ pro x et prodit $\frac{b^6}{y^4}+yy=by-\frac{ab^3}{yy}$, ac reducendo $y^6-by^5+ab^3yy+b^6=0$.

Propositis autem $ayy+aay=z^3$ et $yz-ay=az$, ut y tollatur, secunda dabit $y=\frac{az}{z-a}$. Quare pro y substituo $\frac{az}{z-a}$ in primam, proditꝗ

$$\frac{a^3zz}{zz-2az+aa}+\frac{a^3z}{z-a}=z^3.$$

Et reducendo, $z^4-2az^3+aazz-2a^3z+a^4=0$.

Pari modo propositis $\frac{xy}{c}=z$ et $cy+zx=cc$, ad z tollendum pro eo substituo $\frac{xy}{c}$ in æquationem secundam, et prodit $cy+\frac{xxy}{c}=cc$.

Cæterùm qui in hujus computationibus exercitatus fuerit sæpenumero contractiores modos percipiet quibus incognita quantitas exterminari possit. Sic habitis $ax=\frac{bbx-b^3}{z}$ et $x=\frac{az}{x-b}$ si æqualia multiplicentur æqualibus, prodibunt æqualia $axx=abb$ sive $x=b$. Sed casus ejusmodi particulares studiosis proprio Marte cum res tulerit investigandos linquo.

[1674]
Lect 9

EXTERMINATIO QUANTITATIS INCOGNITÆ QUÆ PLURIUM IN UTRAQUE ÆQUATIONE DIMENSIONUM EXISTIT.[111]

‖[36] Cum in neutra æquatione tollenda quantitas unius tantum di‖mensionis existit valor maximæ potestatis ejus in utraꝗ quærendus est; deinde si potestates istæ non sint eædem, æquatio potestatis minoris multiplicanda est per tollendam quantitatem aut per ejus quadratum aut cubum &c ut ea evadat ejusdem potestatis cum æquatione altera. Tum valores illarum potestatum ponendæ sunt æquales et æquatio nova prodibit ubi maxima potestas sive dimensio tollendæ quantitatis diminuitur. Et hanc operationem iterando quantitas illa tandem auferetur.

Quemadmodum si sit $xx+5x=3yy$ et $2xy-3xx=4$; ut x tollatur prima dabit $xx=-5x+3yy$ et secunda $xx=\frac{2xy-4}{3}$. Pono itaꝗ $3yy-5x=\frac{2xy-4}{3}$, et sic x ad unicam tantum dimensionem reducitur, adeoꝗ tolli potest per ea quæ paulo ante ostendi. Scilicet æquationem novissimam debite reducendo[112] prodit $9yy-15x=2xy-4$, sive $x=\frac{9yy+4}{2y+15}$. Hunc itaꝗ valorem pro x in aliquam ex æquationibus primò propositis (velut in $xx+5x=3yy$) substituo et oritur

are proposed, in order to eliminate x the first will give $b^3/y^2 = x$: I consequently substitute b^3/y^2 for x in the second and there results $b^6/y^4 + y^2 = by - ab^3/y^2$, and so, after reduction, $y^6 - by^5 + ab^3y^2 + b^6 = 0$.

Where, however, $ay^2 + a^2y = z^3$ and $yz - ay = az$ are proposed, to remove y the second will give $y = az/(z-a)$. I consequently substitute $az/(z-a)$ for y in the first and there results $a^3z^2/(z^2 - 2az + a^2) + a^3z/(z-a) = z^3$, and upon reduction $z^4 - 2az^3 + a^2z^2 - 2a^3z + a^4 = 0$.

Equally, when $xy/c = z$ and $cy + zx = c^2$ are proposed, to remove z in its stead I substitute xy/c in the second equation, and there results $cy + x^2y/c = c^2$.

For the rest, he who is practised in this sort of computation will frequently perceive more contracted ways in which the unknown quantity might be eliminated. Thus, when $ax = (b^2x - b^3)/z$ and $x = az/(x-b)$ are presented, if equals be multiplied by equals there will ensue the equals $ax^2 = ab^2$, that is, $x = b$. But particular cases of this sort I leave to be explored by the student on his own initiative as circumstance requires.

ELIMINATION OF AN UNKNOWN QUANTITY WHICH IS OF SEVERAL DIMENSIONS IN EACH EQUATION[111]

Where in neither equation is the quantity to be removed of but one dimension, the value of its greatest power in each is to be ascertained; next, if those powers should not be the same, the equation with the lesser power is to be multiplied by the quantity to be removed—or by its square or cube, and so on—so that it comes to be of the same power as the other equation. Then the values of those powers are to be set equal, and a fresh equation will result in which the greatest power, or dimension, of the quantity to be removed is diminished. And by repeating this operation that quantity will at length be completely taken away.

For instance, if there be $x^2 + 5x = 3y^2$ and $2xy - 3x^2 = 4$, so as to remove x the first will give $x^2 = -5x + 3y^2$ and the second $x^2 = \frac{1}{3}(2xy - 4)$. I accordingly set $3y^2 - 5x = \frac{1}{3}(2xy - 4)$, and in this way x is reduced to but a single dimension, and so can be removed by what I showed a little earlier. Specifically, by appropriately reducing[112] the most recent equation there ensues

$$9y^2 - 15x = 2xy - 4,$$

that is, $x = (9y^2 + 4)/(2y + 15)$. This value, accordingly, I substitute in place of x in one or other of the equations first proposed—say, in $x^2 + 5x = 3y^2$—and there

---

(111) An amplified revise of the corresponding section (II: 404–10) in Newton's Kinckhuysen 'Observationes'. His present 'Reg: 3' and 'Reg: 4' are additions in the style of the first two rules.

(112) Newton's earlier reading 'ordinando' (by ordering) is here cancelled.

$$\frac{81y^4+72yy+16}{4yy+60y+225}+\frac{45yy+20}{2y+15}=3yy.$$ Quam ut in ordinem redigatur multiplico per $4yy+60y+225$, & prodit

$$81y^4+72yy+16+90y^3+40y+675yy+300=12y^4+180y^3+675yy,$$

sive $69y^4-90y^3+72yy+40y+316=0$.

Præterea si sit $y^3=xyy+3x$, et $yy=xx-xy-3$: ut y tollatur multiplico posteriorem æquationem per y et fit $y^3=xxy-xyy-3y$ totidem dimensionum quot prior. Jam ponendo valores ipsius y^3 sibimet æquales habeo

$$xyy+3x=xxy-xyy-3y,$$

ubi y deprimitur ad duas dimensiones. Per hanc itaꝗ et simpliciorem ex æquationibus primo propositis $yy=xx-xy-3$ quantitas y prorsus tolli potest insistendo vestigijs prioris exempli.

Sunt et alij modi quibus hæc eadem absolvi possunt idꝗ sæpenumero contractiùs. Quemadmodum ex $yy=\dfrac{2xxy}{a}+xx$ & $yy=2xy+\dfrac{x^4}{aa}$ ut y deleatur, extrahe in utraꝗ radicem y sicut in Reg $7^{(113)}$ ostensum est, et prodibunt

$$y=\frac{xx}{a}+\sqrt{\frac{x^4}{aa}+xx}, \quad \text{et} \quad y=x+\sqrt{\frac{x^4}{aa}+xx}.$$

Jam hos ipsius y valores ponendo æquales habebitur $\dfrac{xx}{a}+\sqrt{\dfrac{x^4}{aa}+xx}=x+\sqrt{\dfrac{x^4}{aa}+xx}$,

‖[37] et rejiciendo æqualia $\sqrt{\dfrac{x^4}{aa}+xx}$, restabit $\dfrac{xx}{a}=x$, vel $xx=ax$ ‖ et $x=a.^{(114)}$

Porro ut ex æquationibus $x+y+\dfrac{yy}{x}=20$, et $xx+yy+\dfrac{y^4}{xx}=140$ tollatur x, aufer y de partibus æquationis primæ et restat $x+\dfrac{yy}{x}=20-y$, et partibus quadratis fit $xx+2yy+\dfrac{y^4}{xx}=400-40y+yy$, tollendoꝗ utrincꝗ yy restat $xx+yy+\dfrac{y^4}{xx}=400-40y$. Quare cum $400-40y$ & 140 ijsdem quantitatibus æquentur, erit $400-40y=140$, sive $y=6\frac{1}{2}.^{(115)}$ Et sic opus in plerisꝗ$^{(116)}$ alijs æquationibus contrahere liceat.

Cæterùm cùm quantitas exterminanda multarum dimensionum existit, ad eam ex æquationibus tollendam calculus maxime laboriosus nonnunquam requiritur: sed labor tunc plurimùm minuetur per exempla sequentia tanquam regulas adhibita.

---

(113) Of the preceding section 'De concinnanda Æquatione solitariâ', on Newton's page 33 above.

(114) There is a trivial extra solution $x = y = 0$; compare II: 407, (note 77).

arises $\dfrac{81y^4+72y^2+16}{4y^2+60y+225}+\dfrac{45y^2+20}{2y+15}=3y^2$. To reduce this to order I multiply it

by $4y^2+60y+225$, and there results

$$81y^4+72y^2+16+90y^3+40y+675y^2+300=12y^4+180y^3+675y^2,$$

that is, $69y^4-90y^3+72y^2+40y+316=0$.

Furthermore, if it be $y^3=xy^2+3x$ and $y^2=x^2-xy-3$, to remove y I multiply the latter equation by y and it becomes $y^3=x^2y-xy^2-3y$, of as many dimensions as the former. By now setting the values of y^3 equal to one another I have $xy^2+3x=x^2y-xy^2-3y$, in which y is lowered to two dimensions. By means of this, accordingly, and the simpler, $y^2=x^2-xy-3$, of the equations first proposed the quantity y can forthwith be removed by following in the steps of the first example.

There are also other ways in which this self-same reduction can be achieved, and then frequently in a more contracted manner. For instance, to delete y from $y^2=2x^2y/a+x^2$ and $y^2=2xy+x^4/a^2$, extract the root y in each (exactly as was shown in Rule 7)[113] and there will ensue $y=x^2/a+\sqrt{[x^4/a^2+x^2]}$ and $y=x+\sqrt{[x^4/a^2+x^2]}$. Now by setting these values of y equal there will be obtained $x^2/a+\sqrt{[x^4/a^2+x^2]}=x+\sqrt{[x^4/a^2+x^2]}$ and, after rejecting the equals $\sqrt{[x^4/a^2+x^2]}$, there will remain $x^2/a=x$, or $x^2=ax$ and $x=a$.[114]

Moreover, to remove x from the equations

$$x+y+y^2/x=20 \quad \text{and} \quad x^2+y^2+y^4/x^2=140,$$

take y away from the sides of the first equation and there remains

$$x+y^2/x=20-y;$$

with its sides squared it becomes $x^2+2y^2+y^4/x^2=400-40y+y^2$, and on removing y^2 from each side there remains $x^2+y^2+y^4/x^2=400-40y$. Consequently, since $400-40y$ and 140 are equal to the same quantities, there will be $400-40y=140$ or $y=6\frac{1}{2}$.[115]

And it may be permitted to contract the work in this manner in most[116] other equations.

When, however, the quantity to be eliminated proves to be of many dimensions, to remove it from the equations not infrequently requires an exceedingly laborious calculation. But the effort will then be very much diminished by applying the following examples as rules.

---

(115) This solution follows more neatly by evaluating

$$x-y+y^2/x = (x^2+y^2+y^4/x^2)/(x+y+y^2/x) = 7$$

and subtracting the quotient from the first equation.

(116) Does Newton just mean loosely 'a good many'?

REG: 1.

Ex $axx+bx+c=0$, *&* $fxx+gx+h=0$, *exterminato x prodit*

$$\overline{ah-bg-2cf}\times ah: +\overline{bh-cg}\times bf: +\overline{agg+cff}\times c=0.$$

REG: 2.

Ex $ax^3+bxx+cx+d=0$, *&* $fxx+gx+h=0$ *exterminato x prodit*

$$\overline{ah-bg-2cf}\times ahh: +\overline{bh-cg-2df}\times bfh: +\overline{ch-dg}\times\overline{agg+cff}:$$
$$+\overline{3agh+bgg+dff}\times df=0.$$

REG: 3.[117]

Ex $ax^4+bx^3+cxx+dx+e=0$, *&* $fxx+gx+h=0$ *exterminato x prodit*

$$\overline{ah-bg-2cf}\times ah^3: +\overline{bh-cg-2df}\times bfhh:$$
$$+\overline{agg+cff}\times\overline{chh-dgh+egg-2efh}+\overline{3agh+bgg+dff}\times dfh:$$
$$+\overline{2ahh+3bgh-dfg+eff}\times eff: \overline{-bg-2ah}\times efgg=0.$$

REG: 4.[117]

Ex $ax^3+bxx+cx+d=0$ *&* $fx^3+gxx+hx+k=0$ *exterminato x prodit*

$$\overline{ah-bg-2cf}\times\overline{adhh-achk}: +\overline{ak+bh-cg-2df}\times bdfh:$$
$$\overline{-ak+bh+2cg+3df}\times aakk: +\overline{cdh-ddg-cck+2bdk}\times\overline{agg+cff}:$$
$$\overline{+3agh+bgg+dff-3afk}\times ddf: \overline{-3ak-bh-cg+df}\times bcfk:$$
$$+\overline{bk-2dg}\times bbfk: \overline{-bbk-3adh-cdf}\times agk=0.$$

||[38]   Verbi gratia, ut ex æquationibus $xx+5x-3yy=0$, et || $3xx-2xy+4=0$ exterminetur x: in regulam primam si pro a, b, c; f, g, & h substituo 1, 5, $-3yy$; 3, $-2y$, & 4 respectivè. Et signis $+$ et $-$ probe observatis oritur

$$\overline{4+10y+18yy}\times 4: +\overline{20-6y^3}\times 15: +\overline{4yy-27yy}\times -3yy=0.$$

sive $16+40y+72yy+300-90y^3+69y^4=0$.[118]

---

(117) Much as in II: 408, note (78) and 409, note (79) we may encapsulate these two new Newtonian *regulæ* in modern determinant form as respectively

$$\begin{vmatrix} a & b & c & d & e & 0 \\ 0 & a & b & c & d & e \\ f & g & h & 0 & 0 & 0 \\ 0 & f & g & h & 0 & 0 \\ 0 & 0 & f & g & h & 0 \\ 0 & 0 & 0 & f & g & h \end{vmatrix}=0 \quad \text{and} \quad \begin{vmatrix} a & b & c & d & 0 & 0 \\ f & g & h & k & 0 & 0 \\ 0 & a & b & c & d & 0 \\ 0 & f & g & h & k & 0 \\ 0 & 0 & a & b & c & d \\ 0 & 0 & f & g & h & k \end{vmatrix}=0.$$

Rule 1

When x is eliminated from $ax^2 + bx + c = 0$ *and* $fx^2 + gx + h = 0$, *there results*

$$(ah - bg - 2cf)\,ah + (bh - cg)\,bf + (ag^2 + cf^2)\,c = 0.$$

Rule 2

When x is eliminated from $ax^3 + bx^2 + cx + d = 0$ *and* $fx^2 + gx + h = 0$, *there results*

$$(ah - bg - 2cf)\,ah^2 + (bh - cg - 2df)\,bfh + (ch - dg)(ag^2 + cf^2)$$
$$+ (3agh + bg^2 + df^2)\,df = 0.$$

Rule 3[117]

When x is eliminated from $ax^4 + bx^3 + cx^2 + dx + e = 0$ *and* $fx^2 + gx + h = 0$, *there results*

$$(ah - bg - 2cf)\,ah^3 + (bh - cg - 2df)\,bfh^2 + (ag^2 + cf^2)(ch^2 - dgh + eg^2 - 2efh)$$
$$+ (3agh + bg^2 + df^2)\,dfh + (2ah^2 + 3bgh - dfg + ef^2)\,ef^2 - (bg + 2ah)\,efg^2 = 0.$$

Rule 4[117]

When x is eliminated from $ax^3 + bx^2 + cx + d = 0$ *and* $fx^3 + gx^2 + hx + k = 0$, *there results*

$$(ah - bg - 2cf)(adh^2 - achk) + (ak + bh - cg - 2df)\,bdfh$$
$$+ (-ak + bh + 2cg + 3df)\,a^2k^2 + (cdh - d^2g - c^2k + 2bdk)(ag^2 + cf^2)$$
$$+ (3agh + bg^2 + df^2 - 3afk)\,d^2f + (-3ak - bh - cg + df)\,bcfk$$
$$+ (bk - 2dg)\,b^2fk - (b^2k + 3adh + cdf)\,agk = 0.$$

For example, to eliminate x from the equations $x^2 + 5x - 3y^2 = 0$ and $3x^2 - 2xy + 4 = 0$, if in the first rule $\begin{cases} \text{in place of } a,\ b, & c;\ f, & g, \text{ and } h \\ \text{I substitute } 1,\ 5, & -3y^2;\ 3, & -2y, \text{ and } 4 \end{cases}$ respectively, then when the signs $+$ and $-$ are properly observed there arises $(4 + 10y + 18y^2) \times 4 + (20 - 6y^3) \times 15 + (4y^2 - 27y^2) \times -3y^2 = 0$, that is,

$$16 + 40y + 72y^2 + 300 - 90y^3 + 69y^4 = 0.\text{[118]}$$

Newton's technique of reduction, illustrated in the instance of two cubics (Regula 4), is sketched in a somewhat later worksheet reproduced below as Appendix 1.

(118) Newton again (compare II: 408, note (80)) here fails to collect the constant terms as '316'.

Simili ratione ut y deleatur ex æquationibus $y^3 - xyy - 3x = 0$ et

$$yy + xy - xx + 3 = 0$$

in regulam secundam
proditɋ
$$\begin{array}{cccccccc} \text{pro} & a, & b, & c, & d; f, & g, & h, & \& \; x \\ \text{substituo} & 1, & -x, & 0, & -3x; 1, & x, & -xx+3, & \& \; y \end{array}$$
respectivè,

$$\overline{3 - xx + xx} \times \overline{9 - 6xx + x^4} : \overline{-3x + x^3 + 6x} \times \overline{-3x + x^3} : + 3xx \times xx :$$
$$+ \overline{9x - 3x^3 - x^3 - 3x} \times -3x = 0.$$

Tum delendo superflua & multiplicando, fit $27 - 18xx + 3x^4$, $-9xx + x^6$, $+3x^4$, $-18x^2 + 12x^4 = 0$, et ordinando $x^6 + 18x^4 - 45xx + 27 = 0$.[119]

Hactenus de unica incognita quantitate e duabus æquationibus tollenda. Quod si plures e pluribus tollendæ sunt, opus per gradus peragetur: ex æquationibus $ax = yz$, $x + y = z$ et $5x = y + 3z$, si quantitas y elicienda sit, imprimis tolle alterum quantitatum x aut z, puta x substituendo pro eo valorem ejus $\dfrac{yz}{a}$ (per æquationem primam inventum) in æquationem secundam ac tertiam. Quo pacto obtinebuntur $\dfrac{yz}{a} + y = z$ & $\dfrac{5yz}{a} = y + 3z$: E quibus deinde tolle z[120] ut supra.

De modo tollendi quantitates quotcunque surdas ex æquationibus.[121]

Huc referre licet quantitatum surdarum exterminationem fingendo eas literis quibuslibet æquales. Quemadmodum si sit $\sqrt{ay} - \sqrt{aa - ay} = 2a + \sqrt[3]{ayy}$, scribendo t pro $\sqrt{ay}$, v pro $\sqrt{aa - ay}$, & x pro $\sqrt[3]{ayy}$ habebuntur æquationes $t - v = 2a + x$, $tt = ay$, $vv = aa - ay$, & $x^3 = ayy$, ex quibus tollendo gradatim t v et x resultabit tandem æquatio libera ab omni asymmetria.

‖[39]   ‖**Quomodo Quæstio aliqua ad æquationem redigatur.**[122]

Octob.
1675
Lect: 1.

Postquam Tyro in æquationibus pro arbitrio transformandis & concinnandis aliquamdiu exercitatus fuerit, ordo exigit ut ingenij vires in quæstionibus ad

---

(119) Observe that Newton has accurately corrected his earlier text (see II: 409, note (81)). A remark on the fact that the given cubic and quadratic equations yield a sextic (in x or y) as their eliminant would here be in order; to be sure, by attending to the general pattern of his scheme of reduction it would have been well within Newton's power to have produced at this point a proof of his earlier generalization (compare II: 177, note (15)) that equations of m-th and n-th degrees yield as their eliminant an equation of mn-th degree—or equivalently, in his geometrical analogy, that the Cartesian curves defined by the given equations shall have 'soe many cut points as the rectangle of the curves dimensions'.

(120) Much as in his earlier 1670 text (II: 410) Newton here first concluded in more explicit fashion with 'per æqualitatem valorum ejus: nam earum prior per reductionem dat $z = \dfrac{ay}{a-y}$ & posterior $z = \dfrac{ay}{5y - 3a}$. Quare est $\dfrac{ay}{a-y} = \dfrac{ay}{5y - 3a}$. Hanc porro dividendo per ay prodit

For a like reason, to delete y from the equations

$$y^3 - xy^2 - 3x = 0 \quad \text{and} \quad y^2 + xy - x^2 + 3$$

in the second rule $\begin{cases} \text{in place of} \quad a, \quad b, c, \quad d; f, g, \quad h, \text{and } x \\ \text{I substitute } 1, -x, 0, -3x; 1, x, -x^2 + 3, \text{and } y \end{cases}$ respectively, and there results

$$(3 - x^2 + x^2)(9 - 6x^2 + x^4) + (-3x + x^3 + 6x)(-3x + x^3) + 3x^2 \times x^2$$
$$+ (9x - 3x^3 - x^3 - 3x) \times -3x = 0.$$

Then, on deleting superfluities and multiplying, there comes to be

$$27 - 18x^2 + 3x^4 | -9x^2 + x^6 | + 3x^4 | - 18x^2 + 12x^4 = 0$$

and, by ordering it, $x^6 + 18x^4 - 45x^2 + 27 = 0.$[119]

So much for removing a single unknown quantity from two equations. Should, however, several have to be removed from several, the work will be performed by stages. So, if the quantity y has to be elicited from the equations $ax = yz$, $x + y = z$ and $5x = y + 3z$, first remove one of the two quantities x or z—x, say—by substituting in its place in the second and third equation its value yz/a (found by means of the first equation). In this way there will be obtained $yz/a + y = z$ and $5yz/a = y + 3z$: from these then remove z[120] as above.

A WAY OF REMOVING ANY NUMBER OF SURD QUANTITIES
FROM EQUATIONS[121]

It is permissible to relate to this topic the elimination of surd quantities by conceiving them to be equal to arbitrary letters. If, for instance, there should be $\sqrt{(ay)} - \sqrt{(a^2 - ay)} = 2a + \sqrt[3]{(ay^2)}$, by writing t in place of $\sqrt{(ay)}$, v in place of $\sqrt{(a^2 - ay)}$ and x in place of $\sqrt[3]{(ay^2)}$ the equations $t - v = 2a + x$, $t^2 = ay$, $v^2 = a^2 - ay$ and $x^3 = ay^2$ will be obtained; on removing t, v and x from these by stages there will at length result an equation free from all irrationality.

HOW ANY QUESTION IS TO BE REDUCED TO AN EQUATION[122]

After the novice has exercised himself for some time in transforming and refining equations at will, the sequence demands that he test the strength of his ingenuity in reducing questions to an equation. Now when some question is

---

$\dfrac{1}{a-y} = \dfrac{1}{5y-3a}$, et per reductionem $y = \dfrac{2a}{3}$, (by equating its values: for, by reduction the first of these yields $z = ay/(a-y)$ and the latter $z = ay/(5y - 3a)$. Consequently $ay/(a - y) = ay/(5y - 3a)$. Further, division of this by ay produces $1/(a - y) = 1/(5y - 3a)$ and by reduction $y = \frac{2}{3}a$.

(121) An abridged version of the corresponding section in Newton's Kinckhuysen 'Observationes' (see II: 410).

(122) A slightly shortened repeat of Newton's revised opening to Kinckhuysen's 'Pars Tertia' (reproduced on II: 422–8).

æquationem redigendis tentet. Proposita autem aliqua Quæstione, Artificis ingenium in eo præsertim requiritur ut omnes ejus conditiones totidem æquationibus designet. Ad quod faciendum perpendet imprimis an propositiones sive sententiæ quibus enunciatur sint omnes aptæ quæ terminis algebraicis designari possint haud secus quàm conceptus nostri characteribus græcis vel latinis.[123] Et si ita, (ut solet in quæstionibus quæ circa numeros vel abstractas quantitates versantur) tunc nomina quantitatibus ignotis atcҙ etiam notis si opus fuerit imponat et sensum quæstionis sermone, ut ita loquar, analytico designet. Et conditiones ejus[124] ad algebraicos terminos sic translatæ tot dabunt æquationes, quot ei solvendæ sufficiunt.

Quemadmodum si *quærantur tres numeri continuè proportionales quorum summa sit 20 & quadratorum summa 140*: positis x y et z nominibus numerorum trium quæsitorum, Quæstio e latinis literis in algebraicas vertetur ut sequitur.

	Quæstio latine enunciata.	*Eadem algebraicè.*
	Quæruntur tres numeri his conditionibus	$x. y. z$?
VI. 17. Eucl.	Ut sint continuè proportionales,	$x.y::y.z$.[125] sive $xz=yy$.
	Ut omnium summa sit 20,	$x+y+z=20$.
	Et ut quadratorum summa sit 140.	$xx+yy+zz=140$.

Atcҙ ita quæstio deducitur ad æquationes $xz=yy$, $x+y+z=20$ &

$$xx+yy+zz=140,$$

quarum ope x y et z per regulas supradictas investigandi sunt.

Cæterum notandum est solutiones quæstionum eo magis expeditas et artificiosas utplurimum evadere quo pauciores incognitæ quantitates sub initio ponuntur. Sic in hac quæstione posito x pro primo numero et y pro secundo, erit $\frac{yy}{x}$ tertius continue proportionalis; quem proinde ponens pro tertio numero, ‖[40] quæstionem ‖ ad æquationes sic reduco.

	Quæstio latine enunciata.	*Eadem algebraicè.*
	Quæruntur tres numeri continue proportionales,	$x.y.\frac{yy}{x}$?
	Quorum summa sit 20,	$x+y+\frac{yy}{x}=20$.
	Et quadratorum summa 140.	$xx+yy+\frac{y^4}{xx}=140$.

---

(123) Understand 'in general verbal form'; in the sequel, correspondingly, we render Newton's 'latine' as 'verbally'.

propounded, the analyst's skill is especially required in endeavouring to denote all its conditions by an equal number of equations. To do this, let him first weigh carefully whether the statements or phrases by which it is enunciated are all suitable to be designated in algebraic terms in a way analogous to that in which we denote our concepts in Greek or Roman[123] characters. And if (as is usual with questions relating to numbers or abstract quantities) they are, then let him set names on the unknown quantities—and on the known ones, too, if need be—, symbolizing the meaning of the question in a language which is (so I might say) analytical. And, when its conditions[124] have been translated into algebraic terms in this way, they will give as many equations as suffice to solve it.

For instance, *let there be sought three numbers in continued proportion whose sum shall be 20 and the sum of their squares 140*: when x, y and z are set as the names of the three numbers sought, the question will be transliterated from verbal into algebraic form as follows:

The question verbally enunciated	*The same algebraically*
Three numbers are sought, subject to these conditions:	x? y? z?
that they be in continued proportion;	$x:y = y:z$,[125] that is, $xz = y^2$; Euclid, VI, 17.
that their total sum shall be 20;	$x+y+z = 20$;
and that the sum of their squares be 140.	$x^2+y^2+z^2 = 140$.

And thus the question is brought to the equations $xz = y^2$, $x+y+z = 20$ and $x^2+y^2+z^2 = 140$, by whose help x, y and z are to be found out following the rules cited above.

It should, however, be noted that solutions to questions turn out for the most part to be the more prompt and adroit, the fewer the unknown quantities which are supposed at the start. So in the present question, on putting x for the first number and y for the second, y^2/x will be the third continued proportional; consequently, setting this in place of the third number, I reduce the question to equations in this manner:

The question verbally enunciated	*The same algebraically*
Three numbers in continued proportion are wanted:	x? y? y^2/x?
their sum is to be 20;	$x+y+y^2/x = 20$,
and the sum of their squares 140.	$x^2+y^2+y^4/x^2 = 140$.

(124) Newton has cancelled his earlier equivalent phrase 'ejusꝗ status' (…its circumstances); compare II: 424. Below, correspondingly, 'translatus' has been emended to 'translatæ'.

(125) The marginal reference alongside justifies this algebraic 'translation' of the verbal concept of 'in continued proportion'.

Habentur itaq̃ æquationes $x+y+\dfrac{yy}{x}=20$ et $xx+yy+\dfrac{y^4}{xx}=140$ quarum reductione x et y determinandi sunt.[126]

Aliud exemplum accipe. *Mercator quidam nummos ejus triente quotannis adauget demptis 100$^{\overline{lib}}$ quas annuatim impendit in familiam, & post tres annos fit duplo ditior. Quæruntur nummi.*

latine.	*algebraicè.*
Mercator habet nummos quosdam	$x.$
Ex quibus anno primo expendit 100^{lib}	$x-100.$
Et reliquum adauget triente.	$x-100+\dfrac{x-100}{3}$ sive $\dfrac{4x-400}{3}.$
Annoq̃ secundo expendit 100^{lib}	$\dfrac{4x-400}{3}-100$ sive $\dfrac{4x-700}{3}.$
Et reliquum adauget triente	$\dfrac{4x-700}{3}+\dfrac{4x-700}{9}$ sive $\dfrac{16x-2800}{9}.$
Et sic anno tertio expendit 100^{lib}	$\dfrac{16x-2800}{9}-100$ sive $\dfrac{16x-3700}{9}.$
‖[41] ‖Et reliquo trientem similiter lucratus est	$\dfrac{16x-3700}{9}+\dfrac{16x-3700}{27}$ sive $\dfrac{64x-14800}{27}.$
Fitq̃ duplo ditior quam sub initio	$\dfrac{64x-14800}{27}=2x.$

Quæstio itaq̃ ad æquationem $\dfrac{64x-14800}{27}=2x$ redigitur cujus reductione eruendus est x. Nempe duc eam in 27 & fit $64x-14800=54x$. subduc $54x$ et restat $10x-14800=0$, seu $10x=14800$. Quare 1480^{lib} sunt nummi sub initio, ut et lucrum.

Vides itaq̃ quod ad solutiones quæstionum quæ circa numeros vel abstractas quantitatum relationes solummodo versantur, nihil aliud fere requiritur quàm ut e sermone Latino vel alio quovis in quo Problema proponitur, translatio fiat in sermonem (si ita loquar) Algebraicum, hoc est in characteres qui apti sunt ut nostros de quantitatum relationibus conceptus designent. Nonnunquam verò potest accidere quòd sermo quocum status quæstionis exprimitur ineptus videatur qui in Algebraicum possit verti; sed parvis mutationibus adhibitis & ad sensum potius quam verborum sonos[127] attendendo versio reddetur facilis.

---

(126) In sequel Newton has cancelled the detailed solution of this pair of simultaneous equations, evidently deciding at the last moment to delay it till later: indeed, the text of his following (arithmetical) Problema 14 (on Newton's pages 49–50) repeats this cancelled passage word for word except for its opening sentence: 'Utpote si *y* primò desideretur, ordi-

Accordingly, the equations $x+y+y^2/x = 20$ and $x^2+y^2+y^4/x^2 = 140$ are obtained, and by their reduction x and y are to be determined.[126]

Take another example. *A certain merchant each year increases his capital by a third, less £100 which he spends annually on his household, and after three years becomes twice as rich. What is his capital?*

verbally	*algebraically*
A merchant has a certain capital:	x.
of this the first year he spends £100,	$x-100$.
and increases the rest by a third;	$x-100+\frac{1}{3}(x-100)$, or $\frac{1}{3}(4x-400)$.
and the second year he spends £100,	$\frac{1}{3}(4x-400)-100$, or $\frac{1}{3}(4x-700)$.
and increases the rest by a third;	$\frac{1}{3}(4x-700)+\frac{1}{9}(4x-700)$, or $\frac{1}{9}(16x-2800)$.
and likewise the third year he spends £100,	$\frac{1}{9}(16x-2800)-100$, or $\frac{1}{9}(16x-3700)$.
making a similar profit of one-third on the rest;	$\frac{1}{9}(16x-3700)+\frac{1}{27}(16x-3700)$, or $\frac{1}{27}(64x-14800)$.
and comes to be twice as rich as at the start.	$\frac{1}{27}(64x-14800) = 2x$.

The question is accordingly reduced to the equation $\frac{1}{27}(64x-14800) = 2x$, and by its reduction x is to be determined. Namely, multiply it by 27 and it becomes $64x-14800 = 54x$; then subtract $54x$ and there remains $10x-14800 = 0$, that is, $10x = 14800$. Therefore the initial capital, and also the profit, is £1480.

You see, accordingly, that for the solution of questions whose preoccupation is merely with numbers, or the abstract relationships of quantities, almost nothing else is required than that a translation be made from the particular verbal language in which the problem is propounded into one (if I may call it so) which is algebraic; that is, into characters which are fit to symbolise our concepts regarding the relationships of quantities. Sometimes, indeed, it can happen that the language with which the circumstances of a question are expressed may appear unsuitable to be transmuted into an algebraic one; but, by introducing a few changes and paying heed to the meaning of words rather than to their phonetic form,[127] the change-over will be rendered an easy one.

---

nentur æquationes illæ secundum x ad hunc modum $xx \begin{smallmatrix}+y\\-20\end{smallmatrix} x+yy=0$. & $x^4 \begin{smallmatrix}+yy\\-140\end{smallmatrix} xx+y^4=0$.
Deinde in Reg: 3, pro a, b, c, d, e; f, g, & h substitue respectivè...' (Specifically, were y desired in the first instance, order those equations according to x in this manner:
$$x^2+(y-20)x+y^2 = 0, \quad \text{and} \quad x^4+(y^2-140)x^2+y^4 = 0.$$
Then, in Rule 3, in place of a, b, c, d, e; f, g and h substitute respectively...).
(127) Or, of course, their visual pattern.

Sic enim quælibet apud Gentes loquendi formæ propria habent Idiomata: quæ ubi obvenerint, translatio ex unis in alias non verbo tenus instituenda est sed **[1675]** ex sensu determinanda.[128] Cæterum ut hujusmodi Problemata hac methodo **Lect 2** ad æquationes redigendi familiaritatem convincam & illustram, & cùm Artes exemplis facilius quam præceptis addiscantur, placuit sequentium problema-tum[129] solutiones adjungere:

Prob: 1. Data duorum numerorum summa a et differentia quadratorum b, invenire numeros?

Sit eorum minor x et erit alter $a-x$, eorumcɢ quadrata xx et $aa-2ax+xx$: quorum differentia $aa-2ax$ supponitur b. Est itacɢ $aa-2ax=b$, indecɢ per reductionem $aa-b=2ax$ seu $\dfrac{aa-b}{2a}\left(=\tfrac{1}{2}a-\dfrac{b}{2a}\right)=x$.

Exempli gr. Si summa numerorum seu a sit 8, & quadratorum differentia seu b 16: erit $\tfrac{1}{2}a-\dfrac{b}{2a}(=4-1)=3=x$ et $a-x=5$. Quare numeri sunt 3 et 5.

Prob. 2. Invenire tres quantitates x y et z quarum paris cujuscɢ summa datur.

Si summa paris x et y sit a; paris x et z, b; ac paris y et z, c: pro determinandis tribus quæsitis $x y$ et z tres habebuntur æquationes $x+y=a$, $x+z=b$, & $y+z=c$. Jam ut incognitarum duæ puta y et z exterminentur, aufer x utrincɢ in prima et **‖[42]** secunda ‖ æquatione et emergent $y=a-x$ & $z=b-x$ quos valores pro y et z substitue in tertia et orietur $a-x+b-x=c$ et per reductionem $x=\dfrac{a+b-c}{2}$. Invento x æquationes superiores $y=a-x$ et $z=b-x$ dabunt y et z.[130]

Exempl. Si summa paris x et y sit 9, paris x et z 10, et paris y et z 13: tum in valoribus $x y$ et z scribe 9 pro a 10 pro b et 13 pro c et evadet $a+b-c=6$, adeocɢ $x\left(=\dfrac{a+b-c}{2}\right)=3$, $y(=a-x)=6$ & $z(=b-x)=7$.

Prob: 3. Quantitatem datam in partes quotcuncɢ dividere ut majores partes superent minimam per datas differentias.

Sit a quantitas[131] in quatuor ejusmodi partes dividenda, ejuscɢ prima atcɢ

---

(128) A rare, fleeting public glimpse of Newton's deep interest in the structure of language and its capacity to convey meaning. In his youth Newton had been much interested in phonetics and 'ye severall moodes of Speech, whereby wee expresse in what manner a pro-position is to bee understood', and in the possibility of creating 'an Universall Language' deduced 'from ye natures of things themselves' and freed from the 'divers & arbitrary Dialects' of each existing human language. (The text of Newton's 21-page draft 'Of An Universall Language', composed about 1660 and now in private possession in the United States, is reproduced by R. W. V. Elliott in his 'Isaac Newton's "Of An Universall Language"', *The Modern Language Review*, **52**, 1957: 1–18, especially 7–18; see also Elliott's 'Isaac Newton as Phonetician', *ibid.*, **49**, 1954: 5–12.) We have already reproduced the present paragraph on ii: 426–8 to help make up the content of a lost sheet of Newton's 'Observationes' on Kinckhuysen's *Algebra*.

As a parallel, to be sure, the various national forms of speech have their own peculiar idioms and, when these are met with, translation from one to another is not to be undertaken by a mere verbal transliteration, but must be determined from the sense.[128] However, so that I may develop an intimacy with this method of reducing problems of this sort to an equation and make it clear,—and since skills are more easily learnt by example than by precept—I have thought it right to append the solutions of the following problems:[129]

Problem 1. Given the sum a of two numbers and the difference b of their squares, to find the numbers?

Let the lesser of the two be x, and the other will be $a-x$, and their squares x^2 and $a^2-2ax+x^2$: of the last the difference a^2-2ax is taken to be b. Accordingly, $a^2-2ax=b$ and from this by reduction

$$a^2-b=2ax, \quad \text{or} \quad (a^2-b)/2a(=\tfrac{1}{2}a-b/2a)=x.$$

For example: if the sum of the numbers (that is, a) be 8 and the difference of their squares (that is, b) be 16, then $\tfrac{1}{2}a-b/2a(=4-1)=3=x$ and $a-x=5$. Hence the numbers are 3 and 5.

Problem 2. To find three quantities x, y and z, the sum of each pair of which is given.

If the sum of the pair x and y be a, that of the pair x and z be b, while that of the pair y and z is c, for determining the three magnitudes x, y and z sought three equations $x+y=a$, $x+z=b$ and $y+z=c$ will be obtained. Now to eliminate two of the unknowns, say y and z, take x away on each side in the first and second equations and there will emerge $y=a-x$ and $z=b-x$. Substitute these values in place of y and z in the third and there will arise $a-x+b-x=c$, and by reduction $x=\tfrac{1}{2}(a+b-c)$. Once x is found, the above equations $y=a-x$ and $z=b-x$ will yield y and z.[130]

Example. If the sum of the pair x and y be 9, of the pair x and z 10, and of the pair y and z 13, then in the values of x, y and z write 9 in place of a, 10 in place of b, and 13 in place of c and there will come to be $a+b-c=6$, and consequently $x(=\tfrac{1}{2}(a+b-c))=3$, $y(=a-x)=6$ and $z(=b-x)=7$.

Problem 3. To divide a given quantity into any number of parts so that the larger ones exceed the least by given differences.

Let a be a quantity[131] to be divided into four such parts, and let its first and

---

(129) The sixteen arithmetical problems which ensue are based in tone on the thirteen 'quæstiones, quæ æquationes quadraticas non excedunt' with which Kinckhuysen closed his 1661 *Algebra* (see II: 354–62). Indeed, Newton's cancelled 'Prob 13' is exactly Kinckhuysen's first 'quæstio' and his solution is here repeated from the one set down by him in the margin of Nicolaus Mercator's Latin translation of the *Algebra*; see II: 354, note (113).

(130) It would have been neater to find the semi-sum $s = x+y+z = \tfrac{1}{2}(a+b+c)$ of the three given equations and then compute at once $x = s-c$, $y = s-b$ and $z = s-a$.

(131) Understood in the sequel to be a geometrical line.

minima pars x, et super hanc excessus secundæ partis b, tertiæ partis c et quartæ partis d: et erit $x+b$ secunda pars, $x+c$ tertia pars et $x+d$ quarta pars, quarum omnium aggregatum $4x+b+c+d$ æquatur toti lineæ a. Aufer jam utrincß $b+c+d$ et restat $4x=a-b-c-d$ sive $x=\dfrac{a-b-c-d}{4}$.

Exempl. Proponatur linea 20 pedum sic in 4 partes distribuenda ut super primam partem excessus secundæ sit 2^{pedum} tertiæ $3^{\overline{\text{ped}}}$ & quartæ $7^{\overline{\text{ped}}}$. Et quatuor partes erunt $x\left(=\dfrac{a-b-c-d}{4}\ \text{sive}\ \dfrac{20-2-3-7}{4}\right)=2$, $x+b=4$, $x+c=5$ et $x+d=9$.

Eodem modo quantitas in plures partes ijsdem conditionibus dividitur.

Prob: 4. Viro cuidam nummos inter mendicantes distribuere volenti, desunt octo denarij quo minus det singulis tres denarios. Dat itacß singulis duos denarios et tres denarij supersunt. Quæritur numerus mendicantium.

Esto numerus mendicantium x et deerunt 8 denarij quo minus det omnibus $3x$ denarios; habet itacß $3x-8$ denarios: Ex his autem dat $2x$ denarios et reliqui denarij $x-8$ sunt tres. Hoc est $x-8=3$ seu $x=11$.

[1675]   *Prob. 5. Si Tabellarij*[132] *duo A et B 59 milliaribus distantes tempore matutino obviam*
Lect 3 *eant, quorum A conficit 7 milliaria in 2 horis et B 8 milliaria in 3 horis, ac B una hora serius iter instituit quam A: Quæritur longitudo itineris quod A conficiet antequam conveniet B.*

Dic longitudinem illam x et erit $59-x$ longitudo itineris B. Et cum A per-

‖[43] transeat 7^{mill} in 2^{hor}, pertransibit spatium ‖ x in $\dfrac{2x}{7}$ horis, eo quod sit 7^{mill}. 2^{hor} :: x^{mill}. $\dfrac{2x}{7}^{\text{hor}}$. Atcß ita cum $[B]$[133] pertranseat 8^{mill} in 3^{hor}, pertransibit spatium suum $59-x$ in $\dfrac{177-3x}{8}$ horis. Jam cum horum temporum differentia sit 1^{hor}; ut evadent æqualia adde differentiam illam breviori tempore nempe tempori $\dfrac{177-3x}{8}$, et emerget $1+\dfrac{177-3x}{8}=\dfrac{2x}{7}$. Et per reductionem $35=x$. Nam multiplicando per 8 fit $185-3x=\dfrac{16x}{7}$. Dein multiplicando etiam per 7 fit $1295-21x=16x$, seu $1295=37x$. Et dividendo denicß per 37, exoritur $35=x$. Sunt itacß 35^{mill} iter quod A conficiet antequam conveniet B.

Idem generaliùs.

Datis duorum mobilium A et B eodem cursu pergentium celeritatibus, una cum intervallo locorum ac temporum a quibus incipiunt moveri: determinare metam in qua conveniunt.

---

(132) In seventeenth-century parlance, 'post-boys', the carriers of express news and letters.

smallest one be x, the excess over this of the second part b, that of the third c, and that of the fourth d; then will $x+b$ be the second part, $x+c$ the third one and $x+d$ the fourth, while the total sum $4x+b+c+d$ is equal to the whole line a. Now take away $b+c+d$ from either side and there remains $4x = a-b-c-d$, that is, $x = \frac{1}{4}(a-b-c-d)$.

Example. Let it be proposed to apportion a 20-foot line into four parts in such a way that over the first the excess of the second shall be 4 feet, that of the third 3 feet and that of the fourth 7 feet: the four parts will then be $x(= \frac{1}{4}(a-b-c-d)$, that is, $\frac{1}{4}(20-2-3-7)) = 2$, $x+b = 4$, $x+c = 5$ and $x+d = 9$.

In the same manner a quantity is divided into several parts under the same conditions.

Problem 4. A man wanting to share a sum of money among a group of beggars is 8d. short of being able to give each of them 3d. So he gives each one 2d., and 3d. is left over. What is the number of beggars?

Let the number of beggars be x and he will be 8d. short of doling out altogether $3x$ pence, and he accordingly has $3x-8$ pence. Of this sum, however, he gives away $2x$ pence and the remaining $x-8$ pence are threepence. That is, $x-8 = 3$ or $x = 11$.

Problem 5. If two couriers[132] *A and B, 59 miles apart, set out one morning to meet each other, and of these A completes 7 miles in 2 hours and B 8 miles in 3 hours, while B starts his journey 1 hour later than A: how far a distance has A still to travel before he meets B?*

Call that distance x and then $59-x$ will be B's travelling distance. And since A covers 7 miles in 2 hours, he will cover the mileage x in $\frac{2}{7}x$ hours, because 7 miles: 2 hours = x miles: $\frac{2}{7}x$ hours. Similarly, since B[133] covers 8 miles in 3 hours, he will cover his mileage $59-x$ in $\frac{1}{8}(177-3x)$ hours. Now since these times have a difference of 1 hour, to make them equal add that difference to the shorter time, namely $\frac{1}{8}(177-3x)$, and there emerges $1+\frac{1}{8}(177-3x) = \frac{2}{7}x$, and by reduction $x = 35$. For on multiplying by 8 there comes $185-3x = \frac{16}{7}x$, and then, multiplying further by 7, $1295-21x = 16x$, or $37x = 1295$; and finally, on dividing by 37, there results $x = 35$. Accordingly, A has still to travel 35 miles before he meets B.

The same more generally

Given the speeds of two moving bodies A and B proceeding along the same path, together with the distance of the places and times from and at which they begin to move, to determine the point at which they meet.

---

(133) By a slip of Newton's pen the manuscript reads 'b', a reading allowed to pass into print by Whiston in 1707, but duly corrected by Newton in 1722.

Pone mobilis A eam esse celeritatem qua spatium c pertransire possit in tempore f, et mobilis B eam esse qua spatium d pertransire possit in tempore g; et locorum intervallum esse e, ac h temporum in quibus moveri incipiunt.

Cas: 1. Deinde si ambo ad easdem plagas tendant et A sit mobile quod sub initio motus longiùs distat a meta: pone distantiam illam esse x, indeqȝ aufer intervallum e et restabit $x-e$ pro distantia B a meta. Et cùm A pertranseat spatium c in tempore f, tempus in quo pertransibit spatium x erit $\frac{fx}{c}$, eo quod sit spatium c ad tempus f ut spatium x ad tempus $\frac{fx}{c}$. Atqȝ ita cùm B pertranseat spatium d in g, tempus in quo pertransibit spatium $x-e$ erit $\frac{gx-ge}{d}$. Jam cum horum temporum differentia supponatur h, ut ea evadant æqualia adde h breviori tempori nempe tempori $\frac{fx}{c}$ si modo B prius incipiat moveri, et evadet $\frac{fx}{c}+h=\frac{gx-ge}{d}$. Et per reductionem $\frac{cge+cdh}{cg-df}$ vel $\frac{ge+dh}{g-\frac{d}{c}f}=x$. Sin A prius moveri incipiat adde h tempori $\frac{gx-ge}{d}$ et evadet $\frac{fx}{c}=h+\frac{gx-ge}{d}$, et per reductionem $\frac{cge-cdh^{(134)}}{cg-df}=x$.

Cas 2. Quod si mobilia obviam eant, et x ut ante ponatur initialis[135] distantia mobilis A a meta, tum $e-x$ erit initialis[135] distantia ipsius B ab eadem meta; et $\frac{fx}{c}$ tempus in quo A conficiet[136] distantiam x, atqȝ $\frac{ge-gx}{d}$ tempus in quo B conficiet[136] distantiam suam $e-x$. Quorum temporum minori ut supra ‖[44] adde ‖ differentiam h, nempe tempori $\frac{fx}{c}$ si B prius incipiat moveri, et sic habebitur $\frac{fx}{c}+h=\frac{ge-gx}{c}$, et per reductionem $\frac{cge-cdh}{cg+df}=x$. Sin A prius incipiat moveri, adde h tempori $\frac{ge-gx}{d}$ et evadet $\frac{fx}{c}=h+\frac{ge-gx}{d}$, et per reductionem $\frac{cge+cdh}{cg+df}=x$.

Exempl. 1.[137] *Si Tabellarij duo A et B 59 milliaribus distantes tempore matutino obviam eant, quorum A conficit 7 milliaria in 2 horis et B 8 milliaria in 3 horis; & B una hora serius iter instituit quam A: quæritur iter quod A conficiet antequam conveniat B.*

Resp. 35^{mill}. Nam cùm obviam eant et A primò instituat iter, erit $\frac{cge+cdh}{cg+df}$ iter quæsitum. Et hoc, si scribatur 7 pro c, 2 pro f, 8 pro d, 3 pro g, 59 pro e et 1 pro h, evadet $\frac{7,3,59+7,8,1}{7,3+8,2}$; hoc est $\frac{1295}{37}$ sive 35.

Suppose the speed of the moving body A to be that with which it might cover a distance c in time f, and the speed of B that with which it might cover a distance d in time g; and take the distance of their starting places to be e, the difference in the times at which they begin to move h.

Case 1. If, then, both go the same way and A is the body which at the start of motion is further distant from the finishing point, suppose that distance to be x, from this take away the interval e and there will remain $x-e$ for B's distance from the finishing point. Again, since A covers the space c in time f, the time in which it will cover the space x will be $(f/c)\,x$, because space c to time f is as space x to time $(f/c)\,x$. Likewise, since B covers space d in g, the time in which it covers the space $x-e$ will be $(g/d)\,(x-e)$. Now since the difference of these times is supposed to be h, to make them equal add h to the shorter time, namely to $(f/c)\,x$ provided B begins to move first, and there will come to be

$$(f/c)\,x+h = (g/d)\,(x-e),$$

and by reduction $c(ge+dh)/(cg-df)$ or $(ge+dh)/(g-(d/c)f) = x$. But should A be the first to begin to move, add h to the time $(g/d)\,(x-e)$ and there will come $(f/c)\,x = h+(g/d)\,(x-e)$, and by reduction $x = c(ge-dh)/(cg-df)$.[134]

Case 2. But should the moving bodies go to meet each other, and x (as before) be supposed the initial[135] distance of the mobile A from the finishing point, then $e-x$ will be B's initial[135] distance from the same point; and so $(f/c)\,x$ the time in which A will complete[136] the distance x, and $(g/d)\,(e-x)$ the time in which B will complete[136] its distance $e-x$. To the lesser of these times add, as above, the difference h—namely, to $(f/c)\,x$ should B begin to move first, and in this way there will be obtained $(f/c)\,x+h = (g/d)\,(e-x)$, and by reduction $c(ge-dh)/(cg+df) = x$. But should A be the first to begin to move, add h to the time $(g/d)\,(e-x)$ and there will prove to be $(f/c)\,x = h+(g/d)\,(e-x)$, and by reduction $x = c(ge+dh)/(cg+df)$.

Example 1.[137] *If two couriers A and B, 59 miles apart, set out one morning to meet each other, and of these A completes 7 miles in 2 hours and B 8 miles in 3 hours, while B starts his journey 1 hour later than A: how far a distance has A still to travel before he meets B?* Answer: 35 miles. For since they go to meet each other and A begins his journey first, $c(ge+dh)/(cg+df)$ will then be the distance required. And this, if 7 be written for c, 2 for f, 8 for d, 3 for g, 59 for e and 1 for h, will prove to be $7(3\times59+8\times1)/(7\times3+8\times2)$, that is, $1295/37$ or 35.

---

(134) This replaces the more clumsy equivalent fraction $\dfrac{'ge-dh'}{g-\frac{d}{c}f}$.

(135) 'originalis' (original) was first written.

(136) 'transiget' (travel over) is cancelled.

(137) For some indeterminate reason Whiston inverted these two examples in sequence in his 1707 *editio princeps* and all subsequent editions retain the transposition.

Exempl: 2.[137] *Si quotidiè Sol unum gradum conficit et Luna tridecim,*[138] *et ad tempus aliquod, Sol sit in principio Cancri atqʒ post tres dies Luna in principio Arietis: quæritur locus conjunctionis proximè futuræ.* Resp: in $10\frac{3}{4}^{gr}$ Cancri. Nam cum ambo ad easdem plagas eant, et serior sit Epocha motûs lunæ quæ longius distat a meta: erit A Luna B Sol et $\dfrac{cge+cdh}{cg-df}$ longitudo itineris lunaris, quæ, si scribatur 13 pro c; 1 pro f, d ac g; 90[139] pro e; & 3 pro h; evadet $\dfrac{13,1,90+13,1,3}{13,1-1,1}$; hoc est $\dfrac{1209}{12}$ sive $100\frac{3}{4}$. Hos itaqʒ gradus adjice principio Arietis et prodibit $10\frac{3}{4}^{gr}$ Cancri.

Prob: 6. Data agentis alicujus potestate invenire quot ejusmodi agentes datum effectum (a) in dato tempore (b) producent.

Sit ea agentis potestas quæ effectum c producere potest in tempore d, et erit ut tempus d ad tempus b, ita effectus c quem agens iste producere potest in tempore d, ad effectum quem potest producere in tempore b, qui proinde erit $\dfrac{bc}{d}$. Deinde ut unius agentis effectus $\dfrac{bc}{d}$ ad omnium effectum a, ita agens iste unicus ad omnes agentes: adeoqʒ agentium numerus erit $\dfrac{ad}{bc}$.

Exempl. Si Scriba[140] *in 8 diebus 15 folia describere potest, quot ejusmodi scribæ requiruntur ad describendum 405 folia in 9 diebus?* Resp 24. Nam si substituantur 8 pro d, 15 pro c, 405 pro a et 9 pro b, numerus $\dfrac{ad}{bc}$ evadet $\dfrac{405,8}{9,15}$ hoc est $\dfrac{3240}{135}$, sive 24.

‖ [45]
[1675]
Lect 4

Prob 7. Datis plurium agentium viribus, tempus x determi‖nare in quo datum effectum (d) conjunctim producent.

Agentium A, B, C, vires ponantur quæ in temporibus e, f, g producant effectus a, b, c respectivè; et hæ in tempore x producent effectus $\dfrac{ax}{e}$, $\dfrac{bx}{f}$, $\dfrac{cx}{g}$. Quare est $\dfrac{ax}{e}+\dfrac{bx}{f}+\dfrac{cx}{g}=d$, et per reductionem $x=\dfrac{d}{\dfrac{a}{e}+\dfrac{b}{f}+\dfrac{c}{g}}$.

Exempl. Tres[141] *mercenarij opus aliquod certis temporibus perficere possunt viz: A semel in tribus septimanis B ter in octo septimanis, et C quinquies in duodecim septimanis. Quæritur quanto tempore simul absolvent?* Sunt itaqʒ agentium A B C vires quæ temporibus 3, 8, 12 producant effectus 1, 3, 5 respectivè: et quæritur tempus quo absolvent effectum 1. Quare pro a, b, c, d, e, f, g scribe 1, 3, 5, 1, 3, 8, 12, et proveniet $x=\dfrac{1}{\frac{1}{3}+\frac{3}{8}+\frac{5}{12}}$ sive $\frac{8}{9}^{sept}$, hoc est 6 dies $5\frac{1}{3}$ horæ tempus quo simul absolvent.

---

(138) As rounded off to the nearest whole number of degrees in the ecliptic, of course.

(139) Aries and Cancer are separated by Taurus and Gemini in the zodiac and hence are 3×30 degrees apart.

Example 2.[137] *If every day the sun completes one degree and the moon thirteen,*[138] *and at a certain time the sun is at the first point of Cancer, while 3 days later the moon is at the first point of Aries: at what position will their next conjunction occur?* Answer: at $10\frac{3}{4}°$ into Cancer. For since both move in the same direction, and this epoch in the moon's motion is the later the farther distant it is from the point of conjunction, then A will be the moon, B the sun and $c(ge+dh)/(cg-df)$ the length of the lunar journey; if 13 be written for c, 1 for f, d and g, 90[139] for e and 3 for h, this will come to be $13(1 \times 90 + 1 \times 3)/(13 \times 1 - 1 \times 1)$, that is, 1209/12 or $100\frac{3}{4}$. Add these degrees, accordingly, on to the first point of Aries and there will result $10\frac{3}{4}°$ into Cancer.

Problem 6. Given the power of some agent, to find how many agents of the same type will produce a given effect a in a given time b.

Let the agent's power be able to produce the effect c in the given time d, and then time d will be to time b as the effect c which that agent can produce in time d to the effect it can produce in time b—which, in consequence, will be bc/d. Next, as the effect bc/d of one agent to their total effect a, so that single agent to all the agents. As a result the number of agents will be ad/bc.

Example. If a scribe[140] *can copy out 15 sheets in 8 days, how many scribes of the same output are needed to copy 405 sheets in 9 days?* Answer: 24. For if substitution is made of 8 for d, 15 for c, 405 for a and 9 for b, the number ad/bc will come to be $405 \times 8/9 \times 15$, that is, 3240/135 or 24.

Problem 7. Given the forces of several agents, to ascertain the time x in which they shall jointly produce an effect d.

Let the forces of the agents A, B, C be supposed capable of producing the respective effects a, b, c in times e, f, g: then in the time x they will produce effects $(a/e)x$, $(b/f)x$, $(c/g)x$. Consequently, $(a/e)x + (b/f)x + (c/g)x = d$ and by reduction $x = d/(a/e + b/f + c/g)$.

Example. Three[141] *workmen can finish some job in certain times: namely, A once in three weeks, B three times in eight weeks, and C five times in twelve weeks. How great a time is required for them to complete it together?* The forces of the agents A, B, C are accordingly able to produce respective effects 1, 3, 5 in times 3, 8, 12, and the time is required in which they will achieve the effect 1. Consequently, in place of a, b, c, d, e, f, g write 1, 3, 5, 1, 3, 8, 12 and there will ensue $x = 1/(\frac{1}{3} + \frac{3}{8} + \frac{5}{12})$, that is, $\frac{8}{9}$ weeks or 6 days $5\frac{1}{3}$ hours is the time in which they will together complete it.

---

(140) That is, a professional longhand copier of documents of all kinds—a very necessary adjunct to commerce and scholarship in the days before the advent of rapid photocopying. An output of less than two sheets a day suggests either poor productivity or the necessity for complex penmanship!

(141) 'Quatuor' (Four) was first written: evidently Newton came rapidly to the conclusion that the extra 'agent' was a needless complication.

Prob 8. Dissimiles duarum pluriúmve rerum misturas ita componere ut res illæ commistæ datam inter se rationem acquirant.

Sit unius misturæ data quantitas $dA+eB+fC$, alterius eadem quantitas $gA+hB+kC$ et eadem tertiæ $lA+mB+nC$ ubi A, B et C denotent res mistas, et d, e, f, g, h, &c proportiones earundem in misturis:[142] et sit $pA+qB+rC$ mistura quam ex his tribus oportet componere; fingeꝗ x y et z numeros esse per quos si tres datæ misturæ respectivè multiplicentur, earum summa evadet $pA+qB+rC$.

Est itaꝗ $\left.\begin{array}{l} dxA+exB+fxC \\ +gyA+hyB+kyC \\ +lzA+mzB+nzC \end{array}\right\} = pA+qB+rC$, Adeoꝗ collatis terminis[143]

$$dx+gy+lz=p, \quad ex+hy+mz=q, \quad \text{et} \quad fx+ky+nz=r.$$

et[144] per reductionem

$$x=\frac{p-gy-lz}{d}=\frac{q-hy-mz}{e}=\frac{r-ky-nz}{f}.$$

Et rursus æquationes

$$\frac{p-gy-lz}{d}=\frac{q-hy-mz}{e} \quad \& \quad \frac{q-hy-mz}{e}=\frac{r-ky-nz}{f}$$

per reductionem dant $\dfrac{ep-dq+dmz-elz}{eg-dh}(=y)=\dfrac{fq-er+enz-fmz}{fh-ek}$. Quæ, si abbrevietur scribendo α pro $ep-dq$, β pro $dm-el$, γ pro $eg-dh$, δ pro $fq-er$, ζ pro $en-fm$ & θ pro $fh-ek$, evadet $\dfrac{\alpha+\beta z}{\gamma}=\dfrac{\delta+\zeta z}{\theta}$, et per reductionem $\dfrac{\theta\alpha-\gamma\delta}{\gamma\zeta-\beta\theta}=z$. Invento z pone $\dfrac{\alpha+\beta z}{\gamma}=y$ et $\dfrac{p-gy-lz}{d}=x$.

Exempl. Si tres sint metallorum colliquefactorum misturæ, quarum primæ pondo[145] *continet argenti* ℥ *12, æris* ℥ *1, & stanni* ℥ *3, secundæ pondo continet argenti.* ℥ *1, æris* ℥ *12, & stanni* ℥ *3, & tertiæ pondo continet æris* ℥ *14, stanni* ℥ *2 & argenti nihil;* ‖ [46] *sintꝗ* ‖ *hæ misturæ ita componendæ ut pondo compositionis contineat argenti* ℥ *4, æris* ℥ *9 & stanni* ℥ *3:* pro d, e, f; g, h, k; l, m, n; p, q, r scribe 12, 1, 3; 1, 12, 13; 0, 14, 2; 4, 9, 3 respectivè, et erit $\alpha(=ep-dq=1,4-12,9)=-104$, & $\beta(=dm-el=12,14-1,0)=168$, & sic $\gamma=-143$, $\delta=24$, $\zeta=-40$, & $\theta=33$. Adeoꝗ $z\left(=\dfrac{\theta\alpha-\gamma\delta}{\gamma\zeta-\beta\theta}=\dfrac{-3432+3432}{5720-5544}\right)=0$, $y\left(=\dfrac{\alpha+\beta z}{\gamma}=\dfrac{-104+0}{-143}\right)=\dfrac{8}{11}$, & $x\left(=\dfrac{p-gy-lz}{d}=\dfrac{4-\frac{8}{11}}{12}\right)=\dfrac{3}{11}$. Quare si misceantur $\frac{8}{11}$ partes pondo misturæ

---

(142) Newton first wrote '…cujuslibet quantitatem in data totius cujusꝗ misturæ quantitate' (the quantity of any in the given quantity of each total mixture).

(143) Understand the coefficients of A, B and C on each side.

(144) Newton proceeds to elaborate a rather cumbrous solution of the preceding trio of simultaneous equations which does not bring out the essential symmetry of the resulting values of x, y and z.

Problem 8. To compound dissimilar mixtures of two or more substances in such a way that the substances intermixed come to have a given proportion to one another.

Let a given quantity of one mixture be $dA + eB + fC$, the same quantity of a second $gA + hB + kC$ and the same of a third $lA + mB + nC$, in which A, B and C shall signify the substances mixed, and $d, e, f, g, h, \ldots$ the proportions of these in the mixtures;[142] also let $pA + qB + rC$ be the mixture it is necessary to compound from these three, and imagine that x, y and z are numbers such that, if the three given mixtures are respectively multiplied by them, their sum will come to be $pA + qB + rC$. Accordingly,

$$(dA + eB + fC)\,x + (gA + hB + kC)\,y + (lA + mB + nC)\,z = pA + qB + rC$$

and, when terms[143] are compared, $dx + gy + lz = p$, $ex + hy + mz = q$ and $fx + ky + nz = r$. Then[144] by reduction

$$x = (p - gy - lz)/d = (q - hy - mz)/e = (r - ky - nz)/f.$$

Again, the equations $(p - gy - lz)/d = (q - hy - mz)/e$ and

$$(q - hy - mz)/e = (r - ky - nz)/f$$

yield by reduction $\dfrac{ep - dq + dmz - elz}{eg - dh}\,(= y) = \dfrac{fq - er + enz - fmz}{fh - ek}$. If this be shortened by writing α in place of $ep - dq$, β for $dm - el$, γ for $eg - dh$, δ for $fq - er$, ζ for $en - fm$ and θ for $fh - ek$, it will come to be $(\alpha + \beta z)/\gamma = (\delta + \zeta z)/\theta$ and by reduction $z = (\theta\alpha - \gamma\delta)/(\gamma\zeta - \beta\theta)$. Once z is found, put $(\alpha + \beta z)/\gamma = y$ and $(p - gy - lz)/d = x$.

Example. Should there be three mixtures of metals molten together, the first of which contains per pound[145] *12oz. of silver, 1oz. of copper and 3 oz. of tin, the second per pound 1 oz. of silver, 12 oz. of copper and 3 oz. of tin, and the third per pound 14 oz. of copper, 2 oz. of tin and none of silver, and if these mixtures are to be so compounded that the amalgam contains per pound 4 oz. of silver, 9 oz. of copper and 3 oz. of tin:* in place of d, e, f; g, h, k; l, m, n; p, q, r write 12, 1, 3; 1, 12, 13; 0, 14, 2; 4, 9, 3 respectively, and there will be $\alpha\,(= ep - dq = 1 \times 4 - 12 \times 9) = -104$ and

$$\beta\,(= dm - el = 12 \times 14 - 1 \times 0) = 168,$$

and thus $\gamma = -143$, $\delta = 24$, $\zeta = -40$ and $\theta = 33$. Consequently

$$z\left(= \frac{\theta\alpha - \gamma\delta}{\gamma\zeta - \beta\theta} = \frac{-3432 + 3432}{5720 - 5544}\right) = 0, \quad y\left(= \frac{\alpha + \beta z}{\gamma} = \frac{-104 + 0}{-143}\right) = \frac{8}{11}$$

and $\qquad\qquad x\left(= \dfrac{p - qy - lz}{d} = \dfrac{4 - \frac{8}{11}}{12}\right) = \dfrac{3}{11}.$

Hence if $\frac{8}{11}$ parts per pound of the second mixture be mixed with $\frac{3}{11}$ parts per

---

(145) Avoirdupois, presumably, though the symbol '℥' is now usually reserved for 'ounce' in troy (apothecaries') weight, whose pound divides into 12 ounces only (and not the 16 Newton here requires).

secundæ, $\frac{3}{11}$ partes pondo tertiæ et nihil primæ aggregatum erit pondo continens quatuor uncias argenti, novem æris, et tres stanni.

Prob 9. Datis plurium ex ijsdem rebus misturarum pretijs & proportionibus mistorum inter se, pretium cujusvis e mistis determinare.

Cujusvis rerum A, B, C, misturæ $dA+gB+lC$ pretium esto p, misturæ $eA+hB+mC$ pretium q, & misturæ $fA+kB+nC$ pretium r; & rerum illarum A, B, C quærantur pretia x, y & z. Utpote pro rebus A, B, & C substitue earum pretia x, y & z, & exurgent æquationes $dx+gy+lz=p$, $ex+hy+mz=q$ & $fx+ky+nz=r$, ex quibus pergendo ut in præcedente Problemate, elicientur itidem $\dfrac{\theta\alpha-\gamma\delta}{\gamma\zeta-\beta\theta}=z$, $\dfrac{\alpha+\beta z}{\gamma}=y$, & $\dfrac{p-gy-lz}{d}=x$.

Exempl.[146] *Emit quidam 40 modios tritici 24 modios hordei et 20 modios avenæ simul 15*[libris] *12*[solidis]*; Deinde consimilis grani emit 26 modios tritici, 30 modios hordei et 50 modios avenæ simul 16 libris: Ac tertiò consimilis etiam grani emit 24 modios tritici 120 modios hordei et 100 modios avenæ simul 34*[lib]*. Quæritur quanti æstimandus sit modius cujusq̃ grani?* Resp: Modius tritici 5 solidis, hordei 3 solidis et avenæ 2 solidis. Nam pro d, g, l; e, h, m; f, k, n; p, q, &r scribendo respectivè 40, 24, 20; 26, 30, 50; 24, 120, 100; $15\frac{3}{5}$, 16, & 34; prodit $\alpha (=ep-dq=26,15\frac{3}{5}-40,16)=-234\frac{2}{5}$; & $\beta (=dm-el=40,50-26,20)=1480$. Atq̃ ita $\gamma=-576$, $\delta=-500$, $\zeta=1400$, & $\theta=-2400$. adeoq̃ $z\left(\dfrac{\theta\alpha-\gamma\delta}{\gamma\zeta-\beta\theta}=\dfrac{562560-28800}{-806400+355200}=\dfrac{274560}{2745600}\right)=\dfrac{1}{10}$.

$$y\left(=\frac{\alpha+\beta z}{\gamma}=\frac{-234\frac{2}{5}+148}{-576}\right)=\frac{3}{20}, \quad \& \quad x\left(=\frac{p-gy-lz}{d}=\frac{15\frac{3}{5}-\frac{18}{5}-2}{40}\right)=\frac{1}{4}.$$

|[47] Consistit itaq̃ modius tritici $\frac{1}{4}$[lb] seu 5 solidis, modius hordei || $\frac{3}{20}$[lb] seu 3 solidis & modius avenæ $\frac{1}{10}$[lb] seu 2 solidis.

[1675]
Lect: 5.
Prob 10. Datis et misturæ et mistorum gravitatibus specificis invenire proportionem mistorum inter se.

Sit e gravitas specifica misturæ $A+B$ cujus A gravitas specifica est a, et B gravitas b: & cum gravitas absoluta seu pondus componatur ex mole corporis et gravitate specifica, erit aA pondus ipsius A, bB pondus ipsius B et $eA+eB$ pondus aggregati $A+B$, adeoq̃ $aA+bB=eA+eB$, indeq̃ $aA-eA=eB-bB$ seu $e-b . a-e :: A . B$.

---

(146) In structure this example is identical with the eleventh 'quæstio' of Kinckhuysen's *Algebra* (II: 360); compare also the eleventh 'arithmetica propositio' in Frans van Schooten's *Exercitationum Mathematicarum Liber Primus. Continens Propositionum Arithmeticarum et Geometricarum Centuriam* (Leyden, 1657): 7. The various 'measures' of grain are presumably bushels.

pound of the third mixture and none of the first, the joint compound will have a content per pound of four ounces of silver, nine of copper and three of tin.

Problem 9. Given the prices of several mixtures of the same substances and the proportions of the things mixed together, to determine the price of any of the separate substances mixed.

Of any objects A, B, C let the price of the mixture $dA+gB+lC$ be p, that of the mixture $eA+hB+mC$ be q, and that of the mixture $fA+kB+nC$ be r; and let the prices x, y and z of the things A, B and C be sought. Correspondingly, in place of A, B and C substitute their prices x, y and z and there will arise the equations

$$dx+gy+lz = p, \quad ex+hy+mz = q \quad \text{and} \quad fx+ky+nz = r;$$

by proceeding from these as in the preceding problem there will in like manner be derived $z = (\theta\alpha-\gamma\delta)/(\gamma\zeta-\beta\theta)$, $y = (\alpha+\beta z)/\gamma$ and $x = (p-gy-lz)/d$.

Example.[146] *A person buys 40 measures of wheat, 24 measures of barley and 20 measures of oats for altogether £15. 12s.; then, of exactly similar grain, he buys 26 measures of wheat, 30 measures of barley and 50 measures of oats for altogether £16; and a third time, again of exactly similar grain, he buys 24 measures of wheat, 120 measures of barley and 100 measures of oats for altogether £34. Query: how much is a measure of each grain worth?* Answer: a measure of wheat is 5s., one of barley 3s. and one of oats 2s. For, in place of d, g, l; e, h, m; f, k, n; p, q and r, on entering respectively 40, 24, 20; 26, 30, 50; 24, 120, 100; $15\frac{3}{5}$, 16 and 34 there results $\alpha \, (= ep-dq = 26\times15\frac{3}{5}-40\times16) = -234\frac{2}{5}$ and

$$\beta \, (= dm-el = 40\times50-26\times20) = 1480,$$

and thus $\gamma = -576$, $\delta = -500$, $\zeta = 1400$ and $\theta = -2400$; so that

$$z \left(= \frac{\theta\alpha-\gamma\delta}{\gamma\zeta-\beta\theta} = \frac{562\,560-28\,800}{-806\,400+355\,200} = \frac{274\,560}{2\,745\,600}\right) = \frac{1}{10},$$

$$y \left(= \frac{\alpha+\beta z}{\gamma} = \frac{-234\frac{2}{5}+148}{-576}\right) = \frac{3}{20} \quad \text{and} \quad x \left(= \frac{p-gy-lz}{d} = \frac{15\frac{3}{5}-\frac{18}{5}-2}{40}\right) = \frac{1}{4}.$$

Accordingly, a measure of wheat costs £$\frac{1}{4}$, that is, 5s., a measure of barley £$\frac{3}{20}$ or 3s. and a measure of oats £$\frac{1}{10}$ or 2s.

Problem 10. Given the specific gravities both of a mixture and of the things intermixed, to find the relative proportions of the objects mixed.

Let e be the specific gravity of the mixture $A+B$, where a is the specific gravity of A and b that of B: then, since the absolute gravity, that is, weight, is compounded of the mass of a body and its specific gravity, aA will be the weight of A, bB the weight of B and $e(A+B)$ the weight of the aggregate $A+B$; so that $aA+bB = e(A+B)$ and thence $(a-e)A = (e-b)B$ or

$$(e-b):(a-e) = A:B.$$

Exempl.[147] Sit auri gravitas[148] ut 19, argenti ut $10\frac{1}{3}$, & Coronæ Hieronis ut 17; eritɋ $10.3(::e-b.a-e::A.B)::$moles auri in corona ad molem argenti, vel $190.31(::19\times10.10\frac{1}{3}\times3::a\times\overline{e-b}.b\times\overline{a-e})::$pondus auri in corona ad pondus argenti, et $221.31::$pondus coronæ ad pondus argenti.

Prob 11. Si boves a depascant pratum b in tempore c & boves d depascant pratum æque bonum e in tempore f, & gramen uniformiter crescat: quæritur quot boves depascent pratum simile g in tempore h.

Si boves a in tempore c depascant pratum b, tum per analogiam boves $\frac{e}{b}a$ in eodem tempore c, vel boves $\frac{ec}{bf}a$ in tempore f, vel boves $\frac{ec}{bh}a$ in tempore h depascent pratum e: puta si gramen post tempus c non cresceret. Sed cùm propter graminis incrementum boves d in tempore f depascant solummodo pratum e, ideò graminis in prato e incrementum illud per tempus $f-c$ tantum erit quantum per se sufficit pascendis bobus $d-\frac{eca}{bf}$ per tempus f, hoc est quantum sufficit pascendis bobus $\frac{df}{h}-\frac{eca}{bh}$ per tempus h. et in tempore $h-c$ per analogiam tantum erit incrementum quantum per se sufficit pascendis bobus $\frac{h-c}{f-c}$ in $\frac{df}{h}-\frac{eca}{bh}$ sive $\frac{bdfh-ecah-bdcf+aecc}{bfh-bch}$. Hoc incrementum adjice bobus $\frac{aec}{bh}$ et prodibit $\frac{bdfh-ecah-bdcf+ecfa}{bfh-bch}$ numerus boum quibus pascendis sufficit pratum e per tempus h. Adeoɋ per analogiam pratum g bobus $\frac{bdfgh-ecagh-bdcgf+ecfga}{befh-bceh}$ per idem tempus h pascendis sufficiet.[149]

Exempl. Si 12 boves depascant $3\frac{1}{3}$ jugera prati in 4 septimanis et 21 boves depascant 10 jugera consimilis prati in 9 septimanis; quæritur quot boves depascant 36 jugera in 18 septimanis? Resp: 36. Iste enim numerus invenietur substituendo in

$$\frac{bdfgh-ecagh-bdcgf+ecfga}{befh-bceh}$$

---

(147) The celebrated Archimedean problem of testing a gold object for purity, first narrated by Vitruvius (*De Architectura*, IX, 3) as set by King Hiero of Syracuse to Archimedes when he suspected that the 'gold' in a crown of his had been admixed with silver. Newton's assumption of the context of the problem as well known to his reader (or hearer) presupposes a familiarity with some work in which its background is more fully expounded—say, the fortyseventh 'arithmetica propositio' of Schooten's *Exercitationes Mathematicæ* (note (146) above): 30–2: 'Quo pacto ARCHIMEDES portionem argenti, coronæ aureæ permixtam deprehenderit.'

(148) Understand 'specifica' (specific), as before.

(149) It is assumed that the cows do not return to nibble at the new growth of grass in a meadow earlier eaten bare! With this restriction, if each single cow uniformly eats weekly a

Example.[147] Let the[148] gravity of gold be as 19, that of silver as $10\frac{1}{3}$ and that of Hiero's crown be as 17; then

mass of gold in the crown: mass of its silver $(= A:B = e-b:a-e) = 10:3$,

or

weight of gold in the crown: weight of its silver
$$(= a(e-b):b(a-e) = 19 \times 10:10\tfrac{1}{3} \times 3) = 190:31,$$

and   weight of crown: weight of silver $= 221:31$.

Problem 11. If cattle a should eat up a meadow b in time c, and cattle d an equally fine meadow e in time f, and if the grass grows at a uniform rate, how many cattle will eat up a similar meadow g in time h?

If a cattle in time c eat up the meadow b, then in proportion (e/b) a cattle in the same time c, or (ec/bf) a cattle in the time f, or (ec/bh) a cattle in the time h, will eat up the meadow e—if, that is, the grass were not to grow after time c. But since, because of the growth of the grass, d cattle in time f eat up only the meadow e, that growth of grass in the meadow e during time $f-c$ will consequently be sufficient enough to pasture $d-(ec/bf)$ a cattle during time f; that is, adequate to pasture $(d-eca/bf)$ (f/h) cattle during time h. And in time $h-c$, proportionately, the growth will be sufficient enough to pasture

$$(df/h-eca/bh)\ (h-c)/(f-c), \quad \text{that is,} \quad (bdfh-ecah-bdcf+aec^2)/bh(f-c)$$

cattle. Adjoin this increment to the aec/bh cattle and there will result

$$(bdfh-ecah-bdcf+ecfa)/bh(f-c)$$

as the number of cattle which the meadow e suffices to pasture during time h. Proportionately, therefore, the meadow g will suffice to pasture

$$g(bdfh-ecah-bdcf+ecfa)/beh(f-c)$$

cattle during the same time h.[149]

Example. If 12 cattle eat up $3\frac{1}{3}$ acres of meadow in 4 weeks and 21 cattle eat up 10 acres of exactly similar meadow in 9 weeks, how many cattle shall eat up 36 acres in 18 weeks? Answer: 36. To be sure that number will be found by substituting in

$$g(bdfh-ecah-bdcf+ecfa)/beh(f-c)$$

---

quantity α of grass which in one week grows by 2β of itself, then, given that a cows eat up a meadow b in c weeks, there follows $ac\alpha = b(1+\beta c)$ or $a = b(1+\beta c)/c\alpha$; similarly, given that d cows eat up a meadow e in f weeks, then $d = e(1+\beta f)/f\alpha$: hence to find the number x of cows which will, on these terms, eat up a meadow g in h weeks, we need only eliminate α and β from $x = g(1+\beta h)/h\alpha$ by means of the two previous equations. In comparison with this simple argument Newton's elaborate use of proportions seems heavy and contrived. We have traced no forerunners of this elegant variant on a traditional schoolboys' problem and are disposed to attribute it to Newton himself.

‖[48] numeros 12, 3[⅓],[150] 4, ‖ 21, 10, 9, 36, & 18 pro literis *a, b, c, d, e, f, g* & *h* respectivè. Sed solutio forte haud minus expedita erit si e primis principijs ad formam solutionis præcedentis literalis eruatur. Utpote si 12 boves in 4 septimanis depascant 3[⅓][150] jugera tum per analogiam 36 boves in 4 septimanis vel 16 boves in 9 septimanis vel 8 boves in 18 septimanis depascant 10 jugera: puta si gramen non cresceret. Sed cùm propter graminis incrementum 21 boves in 9 septimanis depascant solummodo 10 jugera, illud graminis in 10 jugeris per posteriores 5 septimanas incrementum tantum erit quantum per se sufficit excessui boum 21 supra 16, hoc est 5 bobus per 9 septimanas, vel quod perinde est $\frac{5}{2}$ bobus per 18 septimanas pascendis. Et in 14 septimanis (excessu 18 super 4 primas) incrementum illud graminis per analogiam tantum erit quantum sufficiat 7 bobus per 18 septimanas pascendis; est enim

$$5^{\text{sept}} . 14^{\text{sept}} :: \tfrac{5}{2} \text{ boves} . 7 \text{ boves}.$$

Quare 8 bobus quos 10 jugera sine incremento graminis pascere po[ssunt] per 18 septimanas adde hosce 7 boves quibus pascendis solum incrementum graminis sufficit, et summa erit 15 boves. Ac deniꝗ si 10 jugera 15 bobus per 18 septimanas pascendis sufficiant, tum per analogiam [36] jugera per idem tempus sufficient [54] bobus.

[1675] *Prob 12. Datis sphæricorum corporum in eadem recta moventium,*[151] *sibiꝗ occurrentium*
Lect 6. *magnitudinibus et motibus, determinare motus eorundem post reflexionem.*[152]

Hujus resolutio ex his dependet conditionibus, ut corpus utrumꝗ tantum reactione patiatur quantum agit in alterum,[153] et ut eadem celeritate post reflexionem recedant ab invicem qua ante accedebant. His positis sint corporum *A* et *B* celeritates *a* et *b* respectivè; & motus[154] (siquidem componantur ex mole

---

(150) The manuscript here has '3½', a slip of Newton's pen which Whiston allowed to pass into print in his *editio princeps* in 1707, subsequently to be caught by Newton himself in 1722.

(151) Corrected by Newton to 'motorum' in his 1722 revise.

(152) The classical problem of perfectly elastic impact between bodies moving uniformly in the same straight line. It is well known that Descartes, the first to attempt a systematic solution of the problem in the late 1630's (see R. Dugas, *La mécanique au XVIIᵉ siècle* (Paris, 1954): 159–62), came in the *Pars secunda* (§XL) of his *Principia Philosophiæ* (Paris, 1644) to formulate an inadequate *Lex tertia* of motion under impact, which in turn gave birth to a number of inexact rules for elastic 'shock' (compare Dugas' *Mécanique*: 183–4). Twelve years later Christiaan Huygens corrected Descartes' rules in a small tract (*Œuvres complètes*, **16**, 1929: 137–49) unpublished in his lifetime, making the crucial necessary allowance for conservation of 'motion' (momentum) in such impact; in subsequent years he made guarded disclosure of his findings to Sluse and, at London in 1661, accurately predicted for Wallis and Wren the results of an experiment on the collision of pendulums (Dugas, *Mécanique*: 286–7). In England, provoked in the middle 1660's by the new-found interest of fellow members of the Royal Society (particularly Wallis and Hooke) in both inelastic and elastic impact, Christopher Wren in December 1668 produced a 'hypothesis' regarding the latter, which he claimed to have formulated several years earlier; the following month he published a short paper 'Lex Naturæ de Collisione Corporum' in the *Philosophical Transactions* (**4**, No. 43 [for 11 January

the numbers 12, $3\frac{1}{3}$,[150] 4, 21, 10, 9, 36 and 18 for the letters a, b, c, d, e, f, g and h respectively. But the solution will perhaps be no less speedy if it be derived from first principles on the pattern of the preceding algebraic solution. Precisely, if 12 cattle in 4 weeks eat up $3\frac{1}{3}$[150] acres, then, proportionately, 36 cattle in 4 weeks, or 16 cattle in 9 weeks, or 8 cattle in 18 weeks should eat up 10 acres— if, that is, the grass were not to grow. But since, because of the growth of the grass, 21 cattle in 9 weeks eat up only 10 acres, that growth of grass in 10 acres during the latter 5 weeks will in itself be enough to pasture the excess of 21 over 16, that is, 5 cattle during 9 weeks, or, what is exactly the same, $\frac{5}{2}$ cattle during 18 weeks. And in 14 weeks (the excess of 18 over the first 4) the growth in the grass will proportionately be enough to pasture 7 cattle during 18 weeks; for 5 weeks : 14 weeks $= \frac{5}{2}$ cattle : 7 cattle. Consequently, to the 8 cattle which 10 acres can pasture without any growth of its grass add these 7 cattle which the growth of the grass by itself is sufficient to feed, and the total will be 15 cattle. Finally, if 10 acres suffice to pasture 15 cattle during 18 weeks, then, proportionately, 36 acres are sufficient during the same period for 54 cattle.

Problem 12. Given the sizes and motions of two spherical bodies moving in the same straight line and colliding with each other, to determine their motions after recoil.[152]

The solution of this depends on framing these stipulations: that each body shall suffer as much in reaction as it impresses in its action upon the other;[153] and that after recoil they shall depart from each other with the same speed as before they approached. With these suppositions, let the speeds of the bodies A and B be a and b respectively; and their 'motions'[154] (the compounds, namely,

---

1668/9]: 867–8; compare A. R. Hall, 'Mechanics and the Royal Society, 1668–70' [*British Journal for the History of Science*, **3**, 1966: 24–38], 30–3) with an essentially accurate 'calculus' for the impact of 'equal' and 'unequal' bodies. In readily admitting Wren's complete independence while claiming priority for his own investigations, Huygens rapidly smoothed over a potentially nasty international squabble (see Hall's 'Mechanics': 33–4; and Dugas' *Mécanique*: 289–93). How far Newton's own researches into collision were governed by either Huygens or his London contemporaries seems difficult to pinpoint, but the essential independence of his viewpoint is clear. His private Waste Book notes on the free and constrained motion of bodies (ULC. Add. 4004: 10^r–15^r/38^v/38^r, reproduced—with some omissions—in J. W. Herivel's *The Background to Newton's Principia* (Oxford, 1966): 133–82) reveal that already in (or soon after) January 1665 he had attained an exact insight into the problems of both elastic and inelastic collision between bodies moving not only in the same straight line but in different ones.

(153) Newton's third law of motion in his *Philosophiæ Naturalis Principia Mathematica* (London, 1687): 13: 'Actioni contrariam semper & æqualem esse reactionem.' Equivalently, in January 1665 Newton wrote in his Waste Book (ULC. Add. 4004: 10^v/11^r [= Herivel's *Background*: 142–3]) that, if two bodies moving uniformly in the same straight line 'reflect', then 'none of theire motion [momentum] shall bee lost.... For at their occursion [in the hypothesis of perfect elasticity] they presse equally uppon one another and [upon impact] they shall bee reflected, soe as to move as swiftly frome one another after ye reflection as they did to one another before it.'

(154) That is, 'momentum' in modern terminology.

et celeritate corporum) erunt aA & bB. Et si corpora ad easdem plagas tendant, & A celerius movens insequatur B, pone x decrementum motus aA & incrementum motus bB percussione exortum et post reflexionem motus erunt $aA - x$ et $bB + x$ et celeritates $\dfrac{aA - x}{A}$ ac $\dfrac{bB + x}{B}$ quarum differentia æquatur $a - b$

‖[49] differentiæ celeritatum ante reflexionem. Habetur itacȝ ‖ æquatio

$$\frac{bB + x}{B} \frac{-aA + x}{A} = a - b,$$

et inde per reductionem fit $x = \dfrac{2aAB - 2bAB}{A + B}$, quo pro x in celeritatibus $\dfrac{aA - x}{A}$ & $\dfrac{bB + x}{B}$ substituto, prodeunt $\dfrac{aA - aB + 2bB}{A + B}$ celeritas ipsius A, & $\dfrac{2aA - bA + bB}{A + B}$ celeritas ipsius B post reflexionem.

Quod si corpora obviam eant, tum signo ipsius b ubicȝ mutato, celeritates post reflexionem erunt $\dfrac{aA - aB - 2bB}{A + B}$ & $\dfrac{2aA + bA - bB}{A + B}$: quarum alterutra si forte negativa obvenerit, id arguit motum illum post reflexionem ad plagam dirigi ei contrariam ad quam A tendebat ante reflexionem. Id quod etiam de motu ipsius A in casu priori intelligendum est.

Exempl. Si corpora homogenea A trium librarum cum celeritatis gradibus 8, & B novem librarum cum celeritatis gradibus 2 ad easdem plagas tendant: tunc pro A, a, B et b scribe 3, 8, 9 & 2; et $\left(\dfrac{aA - aB + 2bB}{A + B}\right)$ evadet -1, ac $\left(\dfrac{2aA - bA + bB}{A + B}\right)$ 5. Recedet itacȝ A cum uno gradu celeritatis post reflexionem, & B cum quincȝ gradibus progredietur.

Prob 13. Vendit quis equum 144 florenis et pro quovis centenario florenorum quibus vendidit totidem lucratur quot ipsi florenis equus constitit. Quæritur de lucro.[155]

Pone equum illi constitisse x florenis & lucrum erit $144 - x$. Quare cum pro quovis centenario lucretur x florenos, hoc est, cùm pretium quo constitit equus sit ad lucrum ut 100 ad x, sive cum sit $x \,.\, 144 - x :: 100 \,.\, x$: erit (ponendo æqualitatem inter factos extremorum et mediorum) $xx = 14400 - 100x$,[156] et (radice per Reg 7 extracta) $x = 80$. Quare $64 (= 144 - x)$ est lucrum quæsitum.

Prob. 14. Invenire tres numeros continuè proportionales quorum summa sit 20 & quadratorum summa 140.[157]

---

(155) The first concluding 'quæstio' of Kinckhuysen's *Algebra* (**II**: 354). Newton's following solution is (compare note (129)) repeated word for word from his marginal insert in Mercator's Latin translation of Kinckhuysen. There seems no obvious reason for his subsequent cancellation of the whole problem in the present manuscript: his comparable simplification of Kinckhuysen's fourth 'quæstio' in 'Prob 15' below has been left intact.

of the mass and speed of the bodies) will be aA and bB. And if the bodies are directed the same way, with A—the swifter in motion—pursuing B, set x as the decrement in the 'motion' aA and (equal) increment in the 'motion' bB resulting from the impact, and after recoil their 'motions' will be $aA - x$ and $bB + x$, with their speeds $(aA - x)/A$ and $(bB + x)/B$: the difference between these equals the difference, $a - b$, of the speeds before recoil, and so the equation $(bB + x)/B - (aA - x)/A = a - b$ is had. From this there comes by reduction $x = 2(a - b) AB/(A + B)$, and when this is substituted for x in the speeds $(aA - x)/A$ and $(bB + x)/B$ there result

$$(a(A - B) + 2bB)/(A + B) \text{ and } (2aA - b(A - B))/(A + B)$$

as the respective speeds of A and B after recoil.

But if the bodies go to meet each other, then with the sign of b everywhere changed the speeds after recoil will be

$$(a(A - B) - 2bB)/(A + B) \quad \text{and} \quad (2aA + b(A - B))/(A + B).$$

If perchance one or other of these should happen to be negative, that shows that the motion after recoil is directed in the opposite sense to that in which A tended before impact. This should also be understood regarding A's motion in the former case.

Example. If the homogeneous bodies A (of three pounds weight and having 8 degrees of speed) and B (of nine pounds with 2 degrees of speed) tend in the same direction, then in place of A, a, B and b write 3, 8, 9 and 2, and then $(a(A - B) + 2bB)/(A + B)$ will come to be -1, $(2aA - b(A - B))/(A + B)$ to be 5. Accordingly, A will recede after recoil with one degree of speed, while B will proceed on its way with five degrees.

Problem 13. A man sells a horse for 144 florins and on the money for which he sold it his percentage profit is the same as the number of florins the horse cost him. What profit did he make ?[155]

Suppose the horse cost him x florins and the profit will then be $144 - x$. Accordingly, since the percentage profit is x florins—in other words, since the cost-price of the horse to the profit is as 100 to x, or, equivalently,

$$x : 144 - x = 100 : x$$

—then (on setting equality between the products of extremes and middles) $x^2 = 14400 - 100x$,[156] and (when the root is extracted by means of Rule 7) $x = 80$. Therefore $64 (= 144 - x)$ is the required profit.

Problem 14. To find three numbers in continued proportion, whose sum shall be 20 and the sum of their squares 140.[157]

---

(156) That is, $x^2 + 100x - 14400 = (x - 80)(x + 180) = 0$. For obvious reasons the negative root $x = -180$ is neglected in the sequel.

(157) Kinckhuysen's third 'quæstio' (II: 355).

Pone numerum primum x et secundum y eritꝗ tertius $\frac{yy}{x}$, adeoꝗ

$x+y+\frac{yy}{x}=20$ & $xx+yy+\frac{y^4}{xx}=140$. et per reductionem $xx\begin{smallmatrix}+y\\-20\end{smallmatrix}x+yy=0$ &

$x^4\begin{smallmatrix}+yy\\-140\end{smallmatrix}xx+y^4=0$. Jam ut exterminetur x, pro a, b, c, d, e, f, g & h in Reg: 3

substitue respective 1, 0, $yy-140$, 0, y^4, 1, $y-20$, & yy; et emerget

$$\overline{-yy+280}\times y^6:+\overline{2yy-40y}\times\overline{260y^4-40y^5}:+3y^4\times y^4:$$

$$\|-2yy\times\overline{y^6-40y^5+400y^4}:\ =0.$$

Et per multiplicationem $1600y^6-10400y^5=0$, seu $y=6\frac{1}{2}$. Id quod etiam breviùs
alia methodo sed minus obvia supra[158] inventum est. Porrò ut inveniatur x

substitue $6\frac{1}{2}$ pro y in æquatione $xx\begin{smallmatrix}+y\\-20\end{smallmatrix}x+yy=0$ et exurget $xx-13\frac{1}{2}x+42\frac{1}{4}=0$

seu $xx=13\frac{1}{2}x-42\frac{1}{4}$. Et extracta radice[159] $x=6\frac{3}{4}+$vel$-\sqrt{3\frac{5}{16}}$. Nempe $6\frac{3}{4}+\sqrt{3\frac{5}{16}}$
est maximus quæsitorum trium numerorum et $6\frac{3}{4}-\sqrt{3\frac{5}{16}}$ minimus. Nam x
alterutrum extremorum numerorum ambiguè designat, indeꝗ gemini pro-
deunt valores quorum alteruter potest esse x existente altero $\frac{yy}{x}$.

Idem aliter.[160] Positis numeris x, y, & $\frac{yy}{x}$ ut ante, erit $x+y+\frac{yy}{x}=20$, seu

$xx=\begin{smallmatrix}20\\-y\end{smallmatrix}x-yy$ et extracta radice $x=10-\frac{1}{2}y+\sqrt{100-10y-\frac{3}{4}yy}$ primus numerus.

Hunc et y aufer de 20 et restat $\frac{yy}{x}=10-\frac{1}{2}y-\sqrt{100-10y-\frac{3}{4}yy}$ tertius numerus

estꝗ summa quadratorum a tribus hisce numeris[161] $400-40y$, adeoꝗ
$400-40y=140$, sive $y=6\frac{1}{2}$. Invento medio numero $6\frac{1}{2}$, substitue eum pro y in
primo ac tertio numero supra invento et evadet primus $6\frac{3}{4}+\sqrt{3\frac{5}{16}}$ ac tertius
$6\frac{3}{4}-\sqrt{3\frac{5}{16}}$ ut ante.

*Prob 15. Invenire quatuor numeros continuè proportionales quorum duo medij simul
constituant 12 et duo extremi 20.*[162]

Sit x secundus numerus et erit $12-x$ tertius, $\frac{xx}{12-x}$ primus et $\frac{144-24x+xx}{x}$

quartus: adeoꝗ $\frac{xx}{12-x}+\frac{144-24x+xx}{x}=20$. Et per reductionem $xx=12x-30\frac{6}{7}$

seu $x=6+\sqrt{5\frac{1}{7}}$.[163] Quo invento cæteri numeri e superioribus dantur.

*Prob 16. Sunt numeri quidam in Arithmetica progressione quorum omnium præter
primum summa ducta in primum facit 76, & summa omnium excepto secundo ducta in*

---

(158) On Newton's page 37, as the marginal note affirms.
(159) Understand 'per Reg 7' (by Rule 7).

Suppose the first number to be x and the second one y, and the third will be y^2/x: in consequence, $x+y+y^2/x = 20$ and $x^2+y^2+y^4/x^2 = 140$, and by reduction $x^2+(y-20)x+y^2 = 0$ and $x^4+(y^2-140)x^2+y^4 = 0$. Now, to eliminate x, in place of a, b, c, d, e, f, g and h in Rule 3 substitute respectively 1, 0, y^2-140, 0, y^4, 1, $y-20$ and y^2, and there will emerge

$$(-y^2+280) \times y^6 + (2y^2-40y)(260y^4-40y^5)+3y^4 \times y^4$$

$$-2y^2 \times (y^6-40y^5+400y^4) = 0,$$

and so, by multiplication, $1600y^6-10400y^5 = 0$ or $y = 6\frac{1}{2}$. This has also been found above[158] in shorter manner by another but less obvious method. Further-   page 37
more, to find x substitute $6\frac{1}{2}$ in place of y in the equation $x^2+(y-20)x+y^2 = 0$ and there will ensue $x^2-13\frac{1}{2}x+42\frac{1}{4} = 0$, that is, $x^2 = 13\frac{1}{2}x-42\frac{1}{4}$, and, when the root is extracted,[159] $x = 6\frac{3}{4} +$ or $-\sqrt{3\frac{5}{16}}$; namely, $6\frac{3}{4}+\sqrt{3\frac{5}{16}}$ is the greatest of the three numbers sought, and $6\frac{3}{4}-\sqrt{3\frac{5}{16}}$ the least. For x ambiguously denotes either one of the extreme numbers, and from this twin values result, either one of which can be x with the other y^2/x.

The same another way.[160] With the numbers taken to be x, y and y^2/x as before, there will be $x+y+y^2/x = 20$, that is, $x^2 = (20-y)x-y^2$ and, when the root is extracted, the first number is $x = 10-\frac{1}{2}y+\sqrt{[100-10y-\frac{3}{4}y^2]}$. Take this and y away from 20 and there remains the third number

$$y^2/x = 10-\frac{1}{2}y-\sqrt{[100-10y-\frac{3}{4}y^2]}.$$

Now the sum of the squares of these three numbers[161] is $400-40y$, so that $400-40y = 140$, or $y = 6\frac{1}{2}$. Once the middle number $6\frac{1}{2}$ has been found, substitute it in place of y in the first and third number found above and the first will come to be $6\frac{3}{4}+\sqrt{3\frac{5}{16}}$, the third $6\frac{3}{4}-\sqrt{3\frac{5}{16}}$, as before.

Problem 15. To find four numbers in continued proportion, the two middle ones of which shall together make up 12 and the extremes 20.[162]

Let x be the second number, and then $12-x$ will be the third, $x^2/(12-x)$ the first and $(144-24x+x^2)/x$ the fourth; so that

$$x^2/(12-x)+(144-24x+x^2)/x = 20,$$

and by reduction $x^2 = 12x-30\frac{6}{7}$, that is, $x = 6+\sqrt{5\frac{1}{7}}$.[163] When this is found, the remaining numbers are yielded from what precedes.

Problem 16. Certain numbers are in arithmetical progression, and the sum of them all except the first when multiplied by the first makes 76, the sum of all apart from the second

---

(160) This variant is Kinckhuysen's solution on II: 355.
(161) Namely, $(10-\frac{1}{2}y+\sqrt{[100-10y-\frac{3}{4}y^2]})^2+y^2+(10-\frac{1}{2}y-\sqrt{[100-10y-\frac{3}{4}y^2]})^2$.
(162) Kinckhuysen's fourth concluding 'quæstio' (II: 356).
(163) Or, in Kinckhuysen's preferred form, $6+6\sqrt{\frac{1}{7}}$.

|| [51]

[1675]
Lect 7

secundum facit 175 summaꝗ omnium excepto tertio ducta in tertium facit 256: *quæritur quot sunt termini progressionis et qui?*[164]

Sit primus terminus x et differentia terminorum y eritꝗ progressio x, $x+y$, $x+2y$, &c cujus terminorum summa || existente z, habebuntur æquationes $zx-xx=76$, $zx+zy-xx-2xy-yy=175$, et $zx+2zy-xx-4xy-4yy=256$. Subduc primam a secunda ac tertia et restabunt

$$zy-2xy-yy=99 \ \& \ 2zy-4xy-4yy=180.$$

Ab harum primæ duplo subduc secundam et restabit $2yy=18$ seu $y=3$. In æquatione $zy-2xy-yy=99$ substitue 3 pro y et proveniet $3z-6x-9=99$, seu $z=2x+36$. Hoc pro z substitue in æquatione $zx-xx=76$ et orietur $xx+36x=76$ seu $x=2$, indeqꝗ $z(=2x+36)=40$. Quare progressio est 2, 5, 8, 11, 14; tot enim termini valent 40.[165]

Prob 16. Invenire quatuor numeros continuè proportionales quorum datur summa a et summa quadratorum b.[166]

Etsi desideratas quantitates utplurimùm immediatè quærere solemus, siquando tamen duæ obvenerint ambiguæ, hoc est quæ conditionibus omninò similibus præditæ sunt, (ut hic duo medij et duo extremi[167] numerorum quatuor proportionalium) præstat alias quantitates non ambiguas quærere per quas hæ determinantur, quemadmodum harum summam vel differentiam vel rectangulum.[168] Ponamus ergo summam duorū mediorum numerorum esse s et rectangulum r, et erit summa extremorum $a-s$, et rectangulum etiam r propter proportionalitatem. Jam ut ex his eruantur quatuor illi numeri, pone x primum et y secundum eritꝗ $s-y$ tertius et $a-s-x$ quartus, & rectangulum sub medijs $sy-yy=r$, indeqꝗ medij $y=\frac{1}{2}s+\sqrt{\frac{1}{4}ss-r}$ & $s-y=\frac{1}{2}s-\sqrt{\frac{1}{4}ss-r}$: Item rectangulum sub extremis $ax-sx-xx=r$, indeqꝗ extremi $x=\frac{a-s}{2}+\sqrt{\frac{ss-2as+aa}{4}-r}$ & $a-s-x=\frac{a-s}{2}-\sqrt{\frac{ss-2as+aa}{4}-r}$. Summa quadratorum ex hisce quatuor numeris est $2ss-2as+aa-4r$ quæ est $=b$. Ergo $r=\frac{1}{2}ss-\frac{1}{2}as+\frac{1}{4}aa-\frac{1}{4}b$, quo substituto pro r prodeunt quatuor numeri ut sequitur[169]

---

(164) This is essentially Kinckhuysen's sixth 'quæstio' (II: 356–8), slightly generalized and with new, simpler arithmetical quantities inserted. Again, there seems no valid reason for Newton to cancel this problem, other than personal dictate.

(165) A second solution -2, -5, -8, -11, -14 results on taking $y=-3$ and choosing the root $x=-2$ of the resulting equation $x^2-36x-76=0$. In sequel Newton began to pen a new paragraph 'Cæterùm numerus terminorum' (However, the number of terms...) but at once cancelled it.

(166) Evidently Newton's generalization of Kinckhuysen's third concluding 'quæstio' on II: 355.

(167) Newton first continued with 'termini numeri quæsiti' (numerical terms required).

when multiplied by the second makes 175, and the sum of all except the third when multiplied by the third makes 276: how many terms are there in the progression and what are they?[164]

Let the first term be x and the difference between the terms y, and the progression will be x, $x+y$, $x+2y$, ...; then, if the sum of the terms is z, the equations $zx-x^2 = 76$, $z(x+y)-(x^2+2xy+y^2) = 175$ and

$$z(x+2y)-(x^2+4xy+4y^2) = 256$$

will be obtained. Subtract the first from the second and third and there will remain $zy-2xy-y^2 = 99$ and $2zy-4xy-4y^2 = 180$. From twice the first of these subtract the second and there will remain $2y^2 = 18$, that is, $y = 3$. In the equation $zy-2xy-y^2 = 99$ substitute 3 in place of y and there will ensue $3z-6x-9 = 99$, or $z = 2x+36$. Substitute this for z in the equation $zx-x^2 = 76$ and there will arise $x^2+36x = 76$ or $x = 2$, and thence $z(= 2x+36) = 40$. In consequence, the progression is 2, 5, 8, 11, 14; for, of course, the (total) value of this many terms is 40.[165]

Problem 16. To find four numbers in continued proportion, the sum a of which is given, together with the sum b of their squares.[166]

Although we are apt most times to make a direct search for quantities desired, yet whenever two happen to be ambiguous, that is, circumscribed by exactly similar conditions,—as, here, the two middle and two extreme[167] of the four numbers in proportion—it is better to look for other, unambiguous quantities through which these are determined: their sum, for instance, or difference or product.[168] Let us suppose, therefore, that the sum of the two middle numbers is s and their product r, and the sum of the extremes will then be $a-s$ and their product (because of the proportionality) also r. Now to derive those four numbers from these latter ones, suppose x to be the first and y the second, and then $s-y$ will be the third and $a-s-x$ the fourth, with the product of the middles $sy-y^2 = r$, and in consequence the middles are

$$y = \tfrac{1}{2}s+\sqrt{[\tfrac{1}{4}s^2-r]} \quad \text{and} \quad s-y = \tfrac{1}{2}s-\sqrt{[\tfrac{1}{4}s^2-r]}.$$

Likewise, the product of the extremes is $ax-sx-x^2 = r$, and in consequence the extremes are $x = \tfrac{1}{2}(a-s)+\sqrt{[\tfrac{1}{4}(s^2-2as+a^2)-r]}$ and

$$a-s-x = \tfrac{1}{2}(a-s)-\sqrt{[\tfrac{1}{4}(s^2-2as+a^2)-r]}.$$

The sum of the squares of these four numbers is $2s^2-2as+a^2-4r$, which is hence equal to b. Therefore $r = \tfrac{1}{4}s^2-\tfrac{1}{2}as+\tfrac{1}{4}a^2-\tfrac{1}{4}b$, and when this is substituted for r the four numbers prove to be as follows:[169]

---

(168) A first introduction of Newton's 'rule of mates', soon to be elaborated in a scholium to his (geometrical) 'Prob 13' below.

(169) In sequence, the second, third, first and fourth respectively.

Duo medij $\begin{cases} \frac{1}{2}s + \sqrt{\frac{1}{4}b - \frac{1}{4}ss + \frac{1}{2}as - \frac{1}{4}aa}. \\ \frac{1}{2}s - \sqrt{\frac{1}{4}b - \frac{1}{4}ss + \frac{1}{2}as - \frac{1}{4}aa}. \end{cases}$

Duo extremi $\begin{cases} \dfrac{a-s}{2} + \sqrt{\frac{1}{4}b - \frac{1}{4}ss}. \\ \dfrac{a-s}{2} - \sqrt{\frac{1}{4}b - \frac{1}{4}ss}. \end{cases}$

∥ [52]   Restat tamen etiamnum inquirendus valor ipsius *s*. Quare ∥ ad abbreviandos terminos pro numeris hisce substitue

$$\frac{1}{2}s + p. \qquad \frac{a-s}{2} + q.$$
$$\&$$
$$\frac{1}{2}s - p. \qquad \frac{a-s}{2} - q.$$

et pone rectangulum sub secundo et quarto æquale quadrato tertij siquidem hæc problematis conditio nondum impleatur, eritɋ

$$\frac{as - ss}{4} - \frac{1}{2}qs + \frac{pa - ps}{2} - pq = \frac{1}{4}ss - ps + pp.$$

Pone etiam rectangulum sub primo et tertio æquale qua[dr]ato secundi, et erit $\dfrac{as - ss}{4} + \frac{1}{2}qs \dfrac{-pa + ps}{2} - pq = \frac{1}{4}ss + ps + pp.$ Harum æquationum priorem aufer e posteriori et restabit $qs - pa + ps = 2ps$, seu $qs = pa + ps$. Restitue jam

$$\sqrt{\frac{1}{4}b - \frac{1}{4}ss + \frac{1}{2}as - \frac{1}{4}aa}$$

in locum *p*, et $\sqrt{\frac{1}{4}b - \frac{1}{4}ss}$ in locum *q*, et habebitur

$$s\sqrt{\frac{1}{4}b - \frac{1}{4}ss} = \overline{a + s}\sqrt{\frac{1}{4}b - \frac{1}{4}ss + \frac{1}{2}as - \frac{1}{4}aa}.$$

Et quadrando $ss = -\dfrac{b}{a}s + \frac{1}{2}aa - \frac{1}{2}b$, seu $s = -\dfrac{b}{2a} + \sqrt{\dfrac{bb}{4aa} + \frac{1}{2}aa - \frac{1}{2}b}$, quo invento dantur quatuor numeri quæsiti e superioribus.

Prob 17. Si pensio librarum a per quinɋ annos proxime sequentes solvenda, ematur parata pecunia c, quæritur quanti æstimanda sit usura usuræ centum librarum per annum.

Pone $1 - x$ usuram usuræ pecuniæ *x* in anno, hoc est quod pecunia 1 post annum solvenda valeat *x* paratæ pecuniæ et per analogiam pecunia *a* post annum solvenda valebit *ax* paratæ pecuniæ, post duos annos *axx*, post tres ax^3, post quatuor ax^4 & post quinɋ ax^5. Adde jam hos quinɋ terminos et erit $ax^5 + ax^4 + ax^3 + axx + ax = c$, seu $x^5 + x^4 + x^3 + xx + x = \dfrac{c}{a}$, æquatio quinɋ dimensionum, cujus ope cum *x* per regulas post[170] docendas inventum

---

(170) No such rules for determining the roots of quintics or other higher general equations were in fact set down by Newton in his manuscript; doubtless they were intended to be introduced at the end of the present main text. The phrase passed without comment into Whiston's

the two middles $\begin{Bmatrix} \frac{1}{2}s + \sqrt{[\frac{1}{4}b - \frac{1}{4}s^2 + \frac{1}{2}as - \frac{1}{4}a^2]} \\ \frac{1}{2}s - \sqrt{[\frac{1}{4}b - \frac{1}{4}s^2 + \frac{1}{2}as - \frac{1}{4}a^2]} \end{Bmatrix}$;

the two extremes $\begin{Bmatrix} \frac{1}{2}(a-s) + \sqrt{[\frac{1}{4}b - \frac{1}{4}s^2]} \\ \frac{1}{2}(a-s) - \sqrt{[\frac{1}{4}b - \frac{1}{4}s^2]} \end{Bmatrix}$.

It yet remains, however, to seek out the value of s. So, to shorten terms, in place of these numbers substitute

$$\begin{Bmatrix} \frac{1}{2}s + p \\ \frac{1}{2}s - p \end{Bmatrix} \quad \text{and} \quad \begin{Bmatrix} \frac{1}{2}(a-s) + q \\ \frac{1}{2}(a-s) - q \end{Bmatrix}$$

and put the product of the second and the fourth equal to the square of the third, seeing that this condition of the problem is still not implemented: there will then be $\frac{1}{4}(as - s^2) - \frac{1}{2}qs + \frac{1}{2}(pa - ps) - pq = \frac{1}{4}s^2 - ps + p^2$. Put also the product of the first and third equal to the square of the second, and there will be $\frac{1}{4}(as - s^2) + \frac{1}{2}qs - \frac{1}{2}(pa - ps) - pq = \frac{1}{4}s^2 + ps + p^2$. Take the former of these equations away from the latter and there will be left $qs - pa + ps = 2ps$, or $qs = pa + ps$. Now restore $\sqrt{[\frac{1}{4}b - \frac{1}{4}s^2 + \frac{1}{2}as - \frac{1}{4}a^2]}$ in p's place and $\sqrt{[\frac{1}{4}b - \frac{1}{4}s^2]}$ in place of q, and there will be had $s\sqrt{[\frac{1}{4}b - \frac{1}{4}s^2]} = (a+s)\sqrt{[\frac{1}{4}b - \frac{1}{4}s^2 + \frac{1}{2}as - \frac{1}{4}a^2]}$ and, on squaring, $s^2 = -(b/a)s + \frac{1}{2}a^2 - \frac{1}{2}b$, that is, $s = -\frac{1}{2}b/a + \sqrt{[\frac{1}{4}b^2/a^2 + \frac{1}{2}a^2 - \frac{1}{2}b]}$. And when this is found, the four numbers sought are given from what precedes.

Problem 17. If an annuity of a pounds is to be paid during the five years immediately following, let its purchase price for ready money be c: what is the rate of compound interest on a hundred pounds per year?

Suppose $1 - x$ to be the compound interest on money x in a year, in other words, that money 1 paid back after a year shall be worth x in ready money, then, proportionately, the value in ready money of money a paid back after a year will be ax, after two years it will be ax^2, after three ax^3, after four ax^4, and after five ax^5. Now add these five terms, and there will be

$$ax^5 + ax^4 + ax^3 + ax^2 + ax = c \quad \text{or} \quad x^5 + x^4 + x^3 + x^2 + x = c/a;$$

then, when x has, with the help of this quintic equation, been found by means

1707 *editio princeps*, but in his 1722 revise Newton softened its baldness with the brief footnote: 'Nempe inveniendo figuras primas radicis per constructionem quamvis mechanicam et reliquas per methodum Vietæ' (Namely, by finding the first figures of the root by means of any mechanical construction and then the remaining ones by Viète's method). (The autograph draft of this footnote, differing only in the trivial change of 'Scilicet' for 'Nempe', exists on ULC. Add. 3960.7: 95.) There can be little doubt that the 'mechanical' constructions to which Newton refers are of the type of the 'linear' constructions of equations appended by him to the present manuscript (on Newton's pages 211–51 following); indeed, in his library copy of the 1707 printed edition he has cancelled the latter's heading 'Æquationum constructio linearis', replacing it by the title 'Inventio primarum figurarum Radicis' (The finding of the first figures of a root), and drastically abridging the following text to suit it; see note (615) below. By 'Viète's method' of resolving numerical equations he very possibly meant his own

fuerit,[171] pone $x \,.\, 1 :: 100 \,.\, y.$ et erit $y - 100$ usura usuræ centum librarum per annum.[172]

Atꝗ has in quæstionibus ubi solæ quantitatum proportiones absꝗ positionibus linearum considerandæ veniunt, instantias dedisse sufficiat: pergamus jam ad Problematum Geometricorum solutiones.

[1675]
Lect 8

QUOMODO QUÆSTIONES GEOMETRICÆ AD ÆQUATIONEM REDIGANTUR.[173]

‖ [53] Quæstiones Geometricæ eadem facilitate ijsdemꝗ legibus ad æquationes nonnunquam redigi possunt ac quæ de abstra‖ctis quantitatibus proponuntur. Ut si recta AB in extrema et media proportione secanda sit in C, hoc est ita ut BE quadratum maximæ partis sit æquale rectangulo BD sub tota et minore parte contento: posito $AB = a$, & $BC = x$ erit $AC = a - x$, et $xx = a$ in $a - x$; æquatio quæ per reductionem dat $x = -\frac{1}{2}a + \sqrt{\frac{5}{4}aa}$.[174]

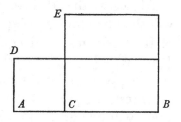

Sed in rebus Geometricis quæ frequentiùs occurrunt, a varijs linearum positionibus & relationibus complexis ita dependere solent ut egeant ulteriori inventione & artificio quo ad algebraicos terminos deduci possint. Et licet in hujusmodi casibus difficile sit aliquid præscribere, et cujusꝗ ingenium sibi debeat esse operandi norma: conabor tamen discentibus viam præsternere. Sciendum est itaꝗ quod quæstiones circa easdem lineas definito quolibet modo sibi invicem relatas possint variè proponi, ponendo alias atꝗ alias quærendas esse ex alijs atꝗ alijs datis. Sed de quibuscunꝗ tamen datis vel quæsitis instituitur quæstio, solutio ejus eadem plane methodo ex Analyseos serie perficietur, nulla omninò circumstantia variata præter fictas linearum species sive nomina quibus datas a quæsitis solemus distinguere.[175] Quemadmodum si quæstio sit de Isoscele CBD in circulum inscripto, cujus latera BC, BD et basis CD cum diametro circuli AB conferenda sunt: ea vel proponi potest de investigatione

improvement of the technique introduced into his *De Analysi* in 1669 (see II: 218–20) and first published by Wallis in his 1685 *Algebra* from the version sent by Newton to Leibniz in June 1676 (see IV: 667, note (35) and John Wallis, *A Treatise of Algebra, Both Historical and Practical* (London, 1685): chapter XCIV. 'A new Method of Extracting Roots in Simple and Affected Equations': 338–40), rather than the cumbrous method pioneered by Viète and refined by Harriot and Oughtred (compare I: 63–70).

(171) An inkblot partially obscures this word and the reading may just possibly be 'fuit'.

(172) Compare Newton's general solution, reproduced on IV: 203, of this problem of determining the value of an annuity whose purchase price and subsequent surrender value accumulates compound interest yearly. The present solution is obtained at once from the

of rules to be subsequently[170] taught, put $x:1 = 100:y$ and $y-100$ will be the compound interest on a hundred pounds per year.[172]

Let these instances presented suffice in the case of questions where the proportions alone of quantities, in divorce from the positions of lines, come into consideration: and now let us proceed to the solution of geometrical problems.

How geometrical questions are to be reduced to an equation[173]

Geometrical questions may not infrequently be reduced to equations with the same ease and by the same procedures as those propounded in regard to abstract quantities. If, for instance, AB has to be cut in extreme and mean proportion at C, that is, so that the square BE of the larger part shall be equal to the rectangle BD contained beneath the whole and the lesser part: on putting $AB = a$ and $BC = x$, there will be $AC = a-x$ and $x^2 = a(a-x)$, an equation which by reduction yields $x = -\frac{1}{2}a + \sqrt{(\frac{5}{4}a^2)}$.[174]

But in geometrical situations which more often occur they tend so to be dependent on a variety of positions of lines and their complex inter-relationships as to need further skilful manipulation to be reducible to algebraic terms. Even though in cases of this sort it is difficult to advance any firm principle and each individual's practised ingenuity ought to be the working criterion, I shall none the less attempt to smooth the way for the beginning student. It must accordingly be realized that questions regarding the same set of lines related to one another in any defined way can be variously proposed, by setting one or another group of them to be determined from this or that given one. Yet, whatever distribution between given and required magnitudes is affected in the question, its solution will be achieved manifestly by an approach born of the same sequence of analysis, with no variation in circumstance at all except in the variables which portray the lines, that is, in the names by which given lines are customarily distinguished from those sought.[175] For instance, should the question concern the isosceles triangle BCD inscribed in a circle, the sides BC, BD and base CD of which are to be compared with the circle's diameter AB, it can be proposed

---

preceding one (see IV: 202, note (2)) by setting $n = 5$ and making the substitutions $c \rightarrow x/(1-x)$ and $a/b \rightarrow c/a$, so yielding $x(1-x^5)/(1-x) = c/a$.

(173) This somewhat overlong section is essentially new, having no antecedents in Kinckhuysen's *Algebra* or Newton's 1670 'Observationes' thereon. The position adopted here by Newton has, of course, a deep if tenuously definable link with Descartes' attempt in the opening pages of his *Geometrie* to lay a firm basis for his development of 'analytical' geometry.

(174) This is Euclid, *Elements*, II, 11; compare III: 407, note (16). Since C is assumed to lie within AB, the negative root $x = -\frac{1}{2}(1 + \sqrt{5})a$ is rightly discarded.

(175) Doubtless Newton alludes to his practice in the sequel of allocating in Cartesian style 'names' $a, b, c, \ldots$ for knowns and $x, y, z, \ldots$ for unknowns.

diametri ex datis lateribus et basi, vel de investigatione basis ex datis lateribus et diametro,
vel deniq$\mathfrak{z}$ de investigatione laterum ex datis
basi et diametro: sed utcunq$\mathfrak{z}$ proponitur,
redigetur ad æquationem per eandem seriem
Analyseos. Nempe si quæratur diameter pono
$AB=x$, $CD=a$, & BC vel $BD=b$. Tum (ducta
AC) propter similia triangula ABC & CBE est
$AB.BC::BC.BE$, sive $x.b::b.BE$. Quare
$BE=\dfrac{bb}{x}$. Est et $CE=\frac{1}{2}CD$ sive $\frac{1}{2}a$: et propter

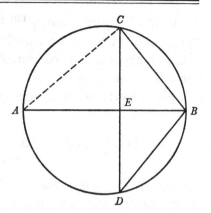

‖[54] angulum CEB rectum, $CE^q+BE^q=BC^q$, ‖ hoc est $\frac{1}{4}aa+\dfrac{b^4}{xx}=bb$. Quæ æquatio
per reductionem dabit quæsitum x.

Sin quæratur Basis, pono $AB=c$, $CD=x$ & BC vel $BD=b$. Tum (ducta AC)
propter sim. tri. ABC et CBE est $AB.BC::BC.BE$, sive $c.b::b.BE$. Quare
$BE=\dfrac{bb}{c}$. Est et $CE=\frac{1}{2}CD$ sive $\frac{1}{2}x$, et propter angulum CEB re[c]tum,
$CE^q+BE^q=BC^q$ hoc est $\frac{1}{4}xx+\dfrac{b^4}{cc}=bb$; æquatio quæ per reductionem dabit
quæsitum x.

Atq$\mathfrak{z}$ ita si latus BC vel BD quæratur pono $AB=c$, $CD=a$ & BC vel $BD=x$.
Et (AC ut ante ducta) propter similia triangula ABC et CBE est

$$AB.BC::BC.BE; \quad \text{sive} \quad c.x::x.BE.$$

Quare $BE=\dfrac{xx}{c}$. Est et $CE=\frac{1}{2}CD$ sive $\frac{1}{2}a$, et propter angulum CEB rectum est

$CE^q+BE^q=BC^q$, hoc est $\frac{1}{4}aa+\dfrac{x^4}{cc}=xx$; æquatio quæ per reductionem dabit
quæsitum x.

Vides itaq$\mathfrak{z}$ quod in unoquoq$\mathfrak{z}$ casu calculus quo pervenitur ad æquationem,
per omnia similis sit, et eandem æquationem pariat, excepto tantum quod
lineas alijs atq$\mathfrak{z}$ alijs literis designavi prout datæ vel quæsitæ ponuntur.[176] Ex
diversis quidem datis et quæsitis oritur diversitas in reductione æquationis in-
ventæ: nam æquationis $\frac{1}{4}aa+\dfrac{b^4}{xx}=bb$ alia est reductio ut obtineatur $x=\dfrac{2bb}{\sqrt{4bb-aa}}$

valor de AB, & æquationis $\frac{1}{4}xx+\dfrac{b^4}{cc}=bb$ alia reductio ut obtineatur $x=\dfrac{2b}{c}\sqrt{bb-cc}$

[!]valor de CD, & æquationis $\frac{1}{4}aa+\dfrac{x^4}{cc}=xx$ reductio longe alia ut obtineatur

$x=\sqrt{\frac{1}{2}cc\pm\frac{1}{2}c\sqrt{cc-aa}}$ valor de BC vel BD,[177] (perinde ut hæc $\frac{1}{4}aa+\dfrac{b^4}{cc}=bb$, ad

eliciendum c a vel b diversis modis reduci debet:) sed in harum æquationum
inventione nulla fuit diversitas. Et hinc est quod jubent ut nullum inter datas

in the form either of the determination of the diameter from the sides and base given, or of the determination of the base from the sides and diameter given, or finally of the determination of the sides from the base and diameter given: but, however it is proposed, it will be brought to an equation by the same sequence of analysis. Specifically, if the diameter be required, I put $AB = x$, $CD = a$ and BC (or BD) $= b$; then, when AC is drawn, because of the similar triangles ABC and CBE there is $AB:BC = BC:BE$ or $x:b = b:BE$, and therefore $BE = b^2/x$; also $CE = \frac{1}{2}CD$, that is, $\frac{1}{2}a$, and, because of the right angle $C\widehat{E}B$, $CE^2 + BE^2 = BC^2$ or $\frac{1}{4}a^2 + b^4/x^2 = b^2$: this equation by reduction will yield the x sought.

But if the base be required, I put $AB = c$, $CD = x$ and BC (or BD) $= b$; and then, when AC is drawn, because of the similar triangles ABC and CBE there is $AB:BC = BC:BE$ or $c:b = b:BE$, and therefore $BE = b^2/c$; also $CE = \frac{1}{2}CD$, that is, $\frac{1}{2}x$, and, because of the right angle $C\widehat{E}B$, $CE^2 + BE^2 = BC^2$ or

$$\tfrac{1}{4}x^2 + b^4/c^2 = b^2:$$

the equation which by reduction will yield the x sought.

And likewise, if the side BC or BD be sought, I put $AB = c$, $CD = a$ and BC (or BD) $= x$; and then, when AC is drawn as before, because of the similar triangles ABC and CBE there is $AB:BC = BC:BE$ or $c:x = x:BE$, and therefore $BE = x^2/c$; also $CE = \frac{1}{2}CD$, that is, $\frac{1}{2}a$, and, because of the right angle $C\widehat{E}B$, $CE^2 + BE^2 = BC^2$ or $\frac{1}{4}a^2 + x^4/c^2 = x^2$: the equation which by reduction will yield the x sought.

You see, as a result, that in each and every case the computation by which the equation is attained is alike in every detail and gives birth to the same equation, except only that I have represented the lines by different letters according as they are supposed to be given or sought.[176] From the variety, indeed, of ones given and sought there does ensue a diversity in the reduction of the equation once it is found: for the equation $\frac{1}{4}a^2 + b^4/x^2 = b^2$ requires one type of reduction to obtain $x = 2b^2/\sqrt{[4b^2 - a^2]}$ as the value of AB, the equation $\frac{1}{4}x^2 + b^4/c^2 = b^2$ a second sort to obtain $x = (2b/c)\sqrt{[c^2 - b^2]}$ as the value of CD, and the equation $\frac{1}{4}a^2 + x^4/c^2 = x^2$ a vastly different one to obtain $x = \sqrt{(\frac{1}{2}c^2 \pm \frac{1}{2}c\sqrt{[c^2 - a^2]})}$ as the value of BC or BD[177] (exactly as this, $\frac{1}{4}a^2 + b^4/c^2 = b^2$, has to be reduced in a variety of ways in order to ascertain c, a or b). No such diversity existed, however, in the discovery of these equations. Hence the injunction that no discrimination

---

(176) In the preceding, of course, the unknown x in turn replaces $AB = c$, $CD = a$ and $BC = BD = b$ in the basic equation $\frac{1}{4}a^2 + b^4/c^2 = b^2$.

(177) Given $AB = c$ and $CD = a$ there are evidently two positions of CD (perpendicular to AB) symmetrically placed round the circle's centre.

et quæsitas quantitates habeatur discrimen. Nam cùm eadem computatio cuicȝ casui datorum et quæsitorum competat, convenit ut sine discrimine concipiantur ‖[55] & conferantur quo rectiùs judicetur de modis computandi: vel potiùs ‖ convenit ut fingas quæstionem de ejusmodi datis et quæsitis propositam esse per quas arbitreris te posse ad æquationem facillimè pervenire.

Proposito igitur aliquo Problemate quantitates quas involvit confer, et nullo inter datas et quæsitas habito discrimine perpende quomodo aliæ ex alijs dependeant ut cognoscas quænam si assumantur,[178] syntheticè gradiendo, dabunt cæteras. Ad quod faciendum non opus est ut prima fronte de modo cogites quo aliæ ex alijs per calculum Algebraicum deduci possint, sed sufficit animadversio generalis quod possint directo nexu quomodocunȝ deduci. Verbi gratia, si quæstio sit de circuli diametro *AD* tribusȝ lineis *AB*, *BC*, et *CD* in semicirculo inscriptis et ex reliquis datis quæratur *BC*;[179] primo intuitu manifestum est diametrum *AD* determinare semicirculum, dein lineas *AB* et *CD* per inscriptionem determinare puncta *B* et *C* atȝ adeò quæsitum *BC*, idȝ nexu maximè directo; et quo pacto tamen *BC* ex his datis per Analysin eruatur non ita manifestum est. Hoc idem quoȝ de *AB* vel *CD* si ex reliquis datis quærerentur, intelligendum est. Quod si *AD* ex datis *AB*, *BC* et *CD* quæreretur æque patet id non fieri posse syntheticè siquidem punctorum *A* ac *D* distantia dependet ex angulis *B* et *C*, et illi anguli ex circulo cui datæ lineæ sunt inscribendæ, et ille circulus non datur ignota *AD* diametro. Rei igitur natura postulat ut *AD* non synthetice sed ex ejus assumptione quæratur ut ad data fiat regressus.

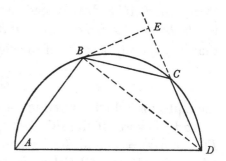

Cùm varios ordines quibus termini quæstionis sic evolvi possint perspexeris, e syntheticis quoslibet adhibe, assumendo lineas tanquam datas a quibus ad alias facillimus videtur progressus et ad ipsas vicissim difficillimus. Nam computatio utut per varia media possit incedere tamen ab istis lineis initium sumet; ac

---

(178) Newton first wrote 'ad arbitrium assumptæ' (assumed at will).

(179) This example—which will reappear several times in slightly variant form in following pages—is with little doubt based on the equivalent opening 'Problema. In semicirculo supra diametrum *AD* descripto quadrilatero *ABCD*. cognita sunt tria ejus latera *AB*, *BC*, & *CD*: oporteatque invenire diametrum seu quartum latus *AD*' set by Frans van Schooten in his *Appendix, De Cubicarum Æquationum Resolutione*, originally published in sequel to his *De Organica Conicarum Sectionum in Plano Descriptione, Tractatus* (Leyden, 1646) but more readily available to Newton in the reprint inserted by Schooten in his second Latin edition of Descartes' *Geometrie* (*Geometria*, **1** (Amsterdam, 1659): 345–68, especially 354, 362–5); compare J. E. Hofmann, *Frans van Schooten der Jüngere* [= *Boethius*, **2** (Wiesbaden, 1962)]: 14–16. The construction lines *BD*, *CE* and *BE* in Newton's present figure are explained on his page 59 below. On setting *AB* = *a*, *BC* = *b*, *CD* = *c* and *AD* = *x* Schooten determines the relation-

be made between quantities given and those sought. For, since the same computation fits each case of magnitudes given and sought, it is appropriate that they be conceived and compared without distinction so as to allow a more accurate decision regarding the methods of computation—or rather is it convenient that you should imagine the question to be propounded regarding magnitudes given and sought in this way, by which you think you are able most easily to arrive at an equation.

When, therefore, some problem is proposed, compare the quantities which it involves and, without placing any distinction between the given and the sought, ponder how one depends upon another so that you may find out those which, if they are assumed,[178] will by a sequence of synthetic steps yield the rest. To do this, there is no need for you, at first encounter, to reflect on the detailed manner in which one quantity can be deduced from another by algebraic computation—a general perception that such a deduction by direct interconnection can somehow be made is enough. For example, should the question concern the circle diameter AD and three lines AB, BC and CD inscribed in its semicircle, and if, given the rest, BC should be sought in terms of them:[179] at first inspection it is patent that the diameter AD determines the semicircle, and then that the lines AB and CD by their inscription in it determine the points B and C and hence the magnitude BC required—and this by an eminently direct chain of argument. The way, however, in which BC shall be derived from these given magnitudes by analysis is not so obvious. The same is also to be understood regarding AB or CD were they to be sought in terms of the rest when given. But if AD were to be sought in terms of given AB, BC and CD, it is equally evident that this cannot synthetically be done, inasmuch as the distance of the points A and D depends on the angles at B and C, and those angles on the (semi)circle in which the given lines are to be inscribed, while that circle is not given when the diameter AD is not known. The structure of the configuration therefore demands that AD be sought, not synthetically, but in pursuance to assuming it so that regress may be made to given magnitudes.

When you have glanced through the various sequences in which the terms of the question may in this way be unfolded, employ any of the synthetic approaches you like, assuming those lines as given from which advance to the others seems easiest and the return to them most difficult. For, although the computation can proceed by various means, it will take those lines as its starting

---

ship between these by dropping the perpendicular DE from D onto BC; it follows, because the triangles ECD, BAD are similar, that $CE = ac/x$, while (by Euclid, *Elements*, II, 13)

$$BD^2 = BC^2 + CD^2 + 2BC \times CE \quad \text{or} \quad x^2 - a^2 = b^2 + c^2 + 2abc/x:$$

whence $x^3 = (a^2 + b^2 + c^2)x + 2abc$.

|| [56] promptiùs perficietur fingendo quæstionem ejusmodi esse ac si de istis datis et quæsito aliquo ab istis facillimè prodituro institueretur quàm de || quæstione prout revera proponitur cogitando. Sic in exemplo jam allato si ex reliquis datis quæritur *AD*: cùm id[180] syntheticè fieri non posse percipiam, sed ab ipso tamen si modo daretur discursum ad alia directo nexu incedere, assumo *AD* tanquam datum et abinde computationem non secus incipio quam si revera daretur, et aliqua ex datis *AB BC* & *CD* quæreretur. Atcß hac methodo computationem ab assumptis ad cæteras quantitates eo more promovendo quo linearum relationes dirigunt, æquatio tandem inter duos ejusdem alicujus quantitatis valores semper obtinebitur, sive ex valoribus unus sit litera sub initio operis quantitati pro nomine imposita, & alter per computationem inventus, sive utercß per computationem diversimodè institutam inveniatur.

Cæterùm ubi terminos quæstionis sic in genere comparaveris plus artis et inventionis in eo requiritur ut advertas particulares istos nexus sive linearum relationes quæ computationi accommodantur. Nam quæ laxiùs perpendenti videantur immediatè & relatione proxima connecti, cùm illam relationem algebraicè designare volumus circuitum plerumcß quoad constructiones schematum de novo moliendas et computationem per gradus promovendam exigunt: quemadmodum de *BC* ex *AD AB* et *CD* colligendo constare potest. Per ejusmodi enim propositiones vel enunciationes solummodo gradiendum est quæ aptæ sunt ut terminis algebraicis designentur, quales præsertim ab Axiom. 19, Prop. 4. lib 6, & Prop 47 lib 1 Elem. scaturiunt.[181]

Imprimis itacß promovetur calculus per additionem vel subductionem linearum eo ut ex valoribus partium obtineatur valor totius, vel ex valoribus totius et unius partis obtineatur valor alterius.

Secundo promovetur ex linearum proportionalitate: ponimus enim (ut supra) factum a medijs terminis divisum per alterutrum extremorum esse || [57] valorem alterius. Vel quòd || perinde est, si valores omnium quatuor proportionalium priùs habeantur ponimus æqualitatem inter factos extremorum et factos mediorum. Linearum verò proportionalitas ex triangulorum similitudine maximè se prodit, quæ cùm ex æqualitate angulorum dignoscatur, in ijs comparandis Analysta[182] debet esse perspicax, atcß adeo non ignorabit Prop.5, 13, 15, 29 & 32 lib 1. Prop 4, 5, 6, 7 & 8 lib 6. et Prop 20, 21, 22, 27 ac 31 lib 3

---

(180) The repeated phrase 'ex reliquis datis' (from the rest given) is here cancelled.

(181) Newton is accordingly well aware of the primacy of these two làtter, essentially equivalent theorems (determining respectively that the corresponding sides of similar triangles are proportional, and that the square on the hypotenuse of a right triangle is equal to the sum of the squares of its two other sides) in establishing the Euclidean metric and thereby permitting the introduction of Cartesian 'analytical' measures of the basic line-lengths in the configuration under discussion.

(182) The practitioner of 'analysis speciosa', that is; compare the opening paragraph on Newton's page 1 above.

point, and will more readily be accomplished by imagining the question to be framed as if it related to these as given and some required magnitude most easily ready to result from them, rather than by pondering over the question as it is in fact proposed. Thus in the example just now introduced, if the rest are given and AD is sought in terms of them, since I perceive that this[180] cannot synthetically be done, but that from it, however, if only it were given, development to the others proceeds in a direct chain, I assume AD as given and begin the computation therefrom exactly as though it were given and one or other of the given lines AB, BC, CD were sought. By developing the computation on these lines from the quantities to the rest in the manner directed by the inter-relationships of the lines, an equation will at length always be obtained between two values of one and the same quantity, whether one of these values be the letter set as name on the quantity at the start of the work and the other that established by means of the computation, or whether each value be found by a computation instituted in a different way.

For the rest, when you have thus in general brought the terms of the question to uniformity, more skill and expertise is required in order that you may take proper notice of the particular interconnections or relationships of lines suited to the computation. For things which, on a rather loose consideration, seem to be joined in a direct and proximate relationship prove, when we want to denote that relationship in algebraic form, commonly to necessitate a devious argument in so far as constructions of figures have to be newly fabricated and the computation advanced by stages: the problem of gathering BC from AD, AB and CD can be cited as an instance in point. Naturally, you must progress only by means of propositions or enunciations of a kind suitable to be designated in algebraic terms—and in particular of the type which derive from Axiom 19, Proposition 4 of Book VI and Proposition 47 of Book I of the *Elements*.[181]

Thus, in the first place, the computation may be advanced through adding or subtracting lines, so as either to obtain from the values of the parts the value of the whole, or from the values of the whole and one part to obtain the value of the other.

Secondly, it may be advanced by means of the proportionality of lines: of course (as above) we suppose the product of the middle terms divided by either extreme to be the value of the other; or, equivalently, if the values of all four proportional magnitudes are previously had, we suppose there to be equality between the products of the extremes and the products of the middles. The proportionality of lines most readily, in fact, reveals itself as a consequence of similarity in triangles; and, since this is ascertained from the equality of angles, the algebraist[182] ought to be discerning in their comparison and hence familiar with Book I, Propositions 5, 13, 15, 29 and 32, Book VI, Propositions 4, 5, 6, 7 and 8, and Book III, Propositions 20, 21, 22, 27 and 31 of the *Elements*. To these

Elementorum. Quibus etiam referri potest Prop 3 lib 6 ubi ex proportionalitate linearum colligitur angulorum æqualitas et contra. Atცp idem aliquando præstant Prop 35 & 36 lib 3.

Tertiò promovetur per additionem vel subductionem quadratorum. In triangulis nempe rectangulis addimus quadrata minorum laterum ut obtineatur quadratum maximi, vel a quadrato maximi lateris subducimus quadratum unius e minoribus ut obtineatur quadratum alterius.

Atცp his paucis fundamentis (si adnumeretur Prop 1 lib 6 Elem. cum de superficiebus agitur, ut et aliquæ Propositiones ex lib 11 & 12 desumptæ cum agitur de solidis,) tota Ars Analytica quoad Geometriam rectilineam innititur.[183] Quinetiam ad solas linearum ex partibus compositiones et similitudines triangulorum possunt omnes Problematum difficultates reduci; adeo ut non opus sit alia Theoremata adhibere: quippe quæ omnia in hæc duo resolvi possunt, et proinde solutiones etiam quæ ex istis depromuntur. Inცp hujus rei instantiam subjunxi Problema de perpendiculo in basem obliquanguli trianguli demittendo sine adjumento Prop 47 lib 1 solutum. Etsi verò juvet simplicissima principia a quibus problematum solutiones dependent non ignorasse, et istis solis adhibitis posse quælibet solvere; expeditionis tamen gratia convenit ut non solùm Prop 47 lib 1 Elem. cujus usus est frequentissimus;[184] sed et alia etiam Theoremata nonnunquam adhibeantur.

Quemadmodum si perpendiculo in basem obliquanguli trianguli demisso, de segmentis basis ad calculum promovendum agatur; ex usu erit scire quod, Differentia quadratorum a lateribus æquetur duplo rectangulo sub basi et ‖[58] ‖distantia perpendiculi a medio basis.[185]

Si trianguli alicujus verticalis angulus bisecetur, computationi non solum inserviet quod basis secetur[186] in ratione laterum sed etiam quod differentia factorum a lateribus et a segmentis basis æquetur quadrato lineæ bisecantis angulum.[187]

Si de figuris in circulo inscriptis res est, Theorema[188] non raró subveniet quòd Inscripti cujuslibet quadrilateri factus a diagonijs æquetur summæ factorum a lateribus oppositis.

---

(183) Compare Newton's contemporary definition of 'simplex Geometria' in Proposition 8 of his 'Solutio Problematis Veterum de Loco solido' (IV: 302–4).

(184) In a similar vein Isaac Barrow spoke of 'Hoc nobilissimum, & utilissimum theorema ...Pythagoricum' (*Euclidis Elementorum Libri XV. breviter demonstrati* (Cambridge, 1655): 35).

(185) The familiar 'extensions of Pythagoras' theorem' (*Elements*, II, 12/13); compare III: 406.

(186) Understand by the bisector; this familiar theorem is *Elements*, VI, 3.

(187) A less familiar complement introduced by Pieter van Schooten in his *Tractatus de concinnandis Demonstrationibus geometricis ex Calculo Algebraïco* [=Descartes' *Geometria*, **2** (Amsterdam, 1661): 341–420]: 370; see I: 544, note (5). The theorem in its simple form was little known in the seventeenth century, but is cited by Leibniz, for instance, in 1689 in the equi-

may also be related Book VI, Proposition 3, in which the equality of the angles is gathered from the proportionality of the lines, and conversely so. And Book III, Propositions 35 and 36, will on occasion achieve the same end.

Thirdly, it may be advanced by adding or subtracting squares. In right-angled triangles, namely, we add the squares of the smaller sides to obtain the square of the largest one, or from the square of the largest side subtract the square of one of the smaller ones to obtain the square of the other.

And on these few foundations (if we count in *Elements*, Book VI, Proposition 1, when the question relates to surface, and also one or two propositions taken from Books XI and XII when it concerns solids) the whole art of (algebraic) analysis in so far as it regards the geometry of straight lines is supported.[183] Indeed, all difficulties in problems can be reduced to the compounding of lines from their parts and to the similarities of triangles alone—so much so that there is no need to employ any other theorems, seeing that they can all be resolved into the present two, and consequently the solutions which are drawn from them also. In instance of this assertion I have appended the problem of letting fall a perpendicular onto the base of an oblique triangle, solved without the assistance of Book I, Proposition 47. But though, in truth, it is well to be aware of the simplest principles on which the solutions of problems depend and to be able to solve any at will solely by applying these, it is nevertheless, for quickness' sake, convenient to employ from time to time not only *Elements*, Book I, Proposition 47—which is of service exceedingly often[184]—but also other theorems as well.

If, for example, when a perpendicular has been let fall onto the base of an oblique triangle, our concern should, to advance the calculation, be with the segments of the base, it will be of use to know that the difference of the squares of the sides is equal to twice the rectangle contained by the base and the distance of the perpendicular from the mid-point of the base.[185]

If the vertex angle of some triangle be bisected, it will be of service to the computation not only that the base is cut[186] in the ratio of the sides, but also that the difference of the products of the sides and of the base segments is equal to the square of the line bisecting the angle.[187]

If the matter relates to figures inscribed in a circle, assistance will not infrequently come from the theorem[188] that in any quadrilateral inscribed the product of the diagonals is equal to the sum of the products of opposite sides.

---

valent form 'in omni triangulo, si angulus ad verticem sit bisectus, rectam bisecantem esse ad rectam quæ potest differentiam inter potestates summæ laterum trianguli baseos, ut media proportionalis inter latera est ad laterum summam' (in §18 of the revise (?) of his 'Tentamen de Motuum Cœlestium Causis', first published by C. I. Gerhardt in *Leibnizens Mathematische Schriften*, **6** (Halle, 1860): 161–87, especially 174).

(188) Ptolemy's in the *Almagest* (*Syntaxis*, I, 10); see IV: 165, note (67).

Et hujusmodi plura inter exercendum observet Analysta, et in penum[189] fortè reservet; sed parciùs utatur si pari[190] facilitate aut non multò difficiliùs possit solutionem e simplicioribus computandi principijs extruere. Quamobrem ad tria primò proposita tanquam notiora, simpliciora, magis generalia, pauca, et omnibus tamen sufficientia animum præsertim advertat, & omnes difficultates ad ea præ cæteris reducere conetur.

[1675]
Lect 9

Sed ut hujusmodi Theoremata ad solvenda Problemata accommodari possint, schemata plerumꝗ sunt ultra construenda, idꝗ sæpissimè producendo aliquas ex lineis donec secent alias aut sint assignatæ longitudinis, vel ab insigniori quolibet puncto[191] ducendo lineas alijs parallelas aut perpendiculares, vel insigniora puncta conjungendo,[192] ut et aliter nonnunquam construendo, prout exigunt status Problematis et Theoremata quæ ad ejus solutionem adhibentur. Quemadmodum si duæ non concurrentes lineæ datos angulos cum tertia quadam efficiant, producimus fortè ut concurrentes constituant triangulum cujus anguli et proinde laterum rationes dantur. Vel si quilibet angulus detur aut sit alicui æqualis in triangulum sæpe complemus specie datum aut alicui simile, idꝗ vel producendo aliquas ex lineis in schemate vel subtensam aliter ducendo. Si triangulum sit obliquangulum, in duo rectangula sæpe resolvimus demittendo perpendiculum. Si de figuris multilateris

‖[59] agatur ‖ resolvimus in triangula ducendo lineas diagonales: Et sic in cæteris; ad hanc metam semper collimando ut schema in triangula vel data, vel similia, vel rectangula resolvatur. Sic in exemplo proposito[193] duco diagonium *BD* ut Trapezium *ABCD* in duo triangula, *ABD* rectangulum et *BDC* obliquangulum resolvatur. Deinde resolvo triangulum obliquangulum in duo rectangula demittendo perpendiculum a quolibet ejus angulo *B*, *C*, vel *D* in latus oppositum: quemadmodum a *B* in *CD* productam ad *E* ut huic perpendiculo *BE*

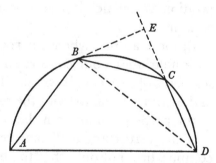

occurrat. Interea verò cum anguli *BAD* et *BCD* duos rectos (per 22.3 Elem) perinde ac *BCE* et *BCD* constituant; percipio angulos *BAD* et *BCE* æquales esse, adeoꝗ triangula *BCE* ac *DAB* similia. Atꝗ ita video computationem (assumendo *AD AB* et *BC* tanquam si *CD* quæreretur) ad hunc modum institui posse. viz: *AD* et *AB* (propter tri. *ABD* rect.) dant *BD. AD, AB, BD* et *BC* (propter sim. tri. *ABD* et *CEB*) dant *BE* et *CE. BD* et *BE* (propter triang. *BED* rect) dant *ED*; et *ED – EC* dat *CD*. Unde obtinebitur æquatio inter valorem de

---

(189) In the manner of the 'penus analytica' of Greek higher geometry which Pappus summarized in the seventh book of his *Mathematical Collection*; compare IV: 224, note (22).

(190) The equivalent 'æquali' is cancelled.

(191) 'ab invicem' (from one or another) was first written.

The algebraist may observe several others of this kind and perhaps store them away in his larder,[189] but let him use them rather sparingly if he can build up a solution on the basis of simpler computing principles with equal ease or without much more difficulty. And in consequence let him particularly direct his attention to the three first proposed as better known, simpler, more general, few in number and yet universally adequate, and to these in preference to others attempt to reduce all difficulties.

But, to be able to adapt theorems of this sort to solving problems, the figures must commonly be further built up—this most often by extending some of the lines till they intersect others or are of an assigned length, or from any more prominent point[191] drawing lines parallel or perpendicular to others, or joining[192] the more prominent points, and on occasion introducing still other constructions according as the circumstances of the problem and the theorems employed in its solution demand them. For instance, should two non-intersecting lines make given angles with a third, we may perhaps produce them so that by intersecting they form a triangle whose angles and hence sides are in given proportion. Or if any angle be given or equal to some other, we may often complete it into a triangle given in species or similar to some other: that either by extending some of the lines in the figure or otherwise drawing a line subtending it. If the triangle be scalene, we may often resolve it into two right-angled ones by letting fall a perpendicular. If it be a question of many-sided figures, we may resolve them into triangles by drawing their diagonal lines. And the like in other situations, always keeping in sight the goal of resolving figures into triangles which are either given or similar or right-angled. Thus in the example proposed[193] I draw the diagonal BD so as to resolve the quadrilateral $ABCD$ into two triangles, the right-angled one ABD and the obtuse one BDC; and then I resolve the obtuse triangle into two right-angled ones by letting fall a perpendicular from any of its corners B, C or D onto the opposite side—say from B onto CD extended to E to meet this perpendicular BE. Meanwhile indeed, since (by *Elements*, III, 22) the angles $B\widehat{A}D$ and $B\widehat{C}D$, like $B\widehat{C}E$ and $B\widehat{C}D$, make up two right angles, I perceive that the angles $B\widehat{A}D$ and $B\widehat{C}E$ are equal, and consequently that the triangles BCE and DAB are similar. And so I see that the computation can (on assuming AD, AB and BC with CD, as it were, to be sought) be instituted in this manner, namely: AD and AB (because of the right triangle ABD) give BD; AD, AB, BD and BC (because of the similar triangles ABD and CEB) give BE and CE; BD and BE (because of the right triangle BED) give ED; and $ED - EC$ gives CD. Hence an equation will be

(192) Originally 'connectendo' (connecting).

(193) On Newton's page 55 above; see note (179). For convenience we repeat the earlier figure.

CD sic inventum et litteram pro ea suffectam. Possumus etiam (& maximam partem satius est quàm opus in serie continuata nimis prosequi,) a diversis principijs computationem incipere, aut saltem diversis modis ad eandem quamlibet conclusionem promovere, ut duo tandem obtineantur ejusdem cujusvis quantitatis valores qui æquales ponantur. Sic *AD AB* et *BC* dant *BD BE* et *CE* ut priùs, deinde *CD* + *CE* dat *ED*, ac deniꝗ *BD* et *ED* (propter triang. rect. *BED*) dant *BE*. Potest etiam computatio hac lege optimè institui ut valores quantitatum investigentur quibus alia quæpiam relatio cognita intercedit, et illa deinde relatio æquationem dabit. Sic cùm relatio inter lineas *BD*, *DC*, *BC* et *CE* ex Prop 12 Lib 2 Elem. constet; nempe quòd sit

$$BD^q - BC^q - CD^q = 2CD \times CE:$$

|| [60] quæro BD^q ex assumptis *AD* et *AB*; ac *CE* ex assumptis *AD AB* et || *BC*. Et assumendo deniꝗ *CD* facio $BD^q - BC^q - CD^q = 2CD \times CE$. Ad hos modos et hujusmodi consilijs ductus, de serie Analyseos, deꝗ schemate propter eam construendo semper debes unà prospicere.

Ex his credo manifestum est quid sibi velint Geometræ cùm jubent putes factum esse quod quæris. Nullo enim inter cognitas et incognitas quantitates habito discrimine, quaslibet ad ineundum calculum assumere potes quasi omnes ex prævia solutione fuissent notæ et non ampliùs de solutione Problematis sed de probatione solutionis ageretur. Sic in primo ex tribus jam descriptis computandi modis, etsi fortè *AD* revera quæratur, fingo tamen *CD* quærendum esse quasi vellem probare an valor ejus ab *AD* derivatus quadret cum ejus quantitate priùs cognita. Sic etiam in duobus posterioribus modis pro meta non propono quantitatem aliquam quærendam esse sed æquationem e relationibus linearum utcunꝗ eruendam: Et in ejus rei gratiam assumo omnes *AD*, *AB*, *BC*, & *CD* tanquam notas, perinde ac si (quæstione priùs solutâ) de tentamine jam ageretur an conditionibus ejus hæ probe satisfaciant, quadrando cum quibuslibet æquationibus quas linearum relationes prodent. Opus quidem hâc ratione & consilijs prima fronte agressus sum sed cum ad æquationem deventum est sententiam muto et quantitatem desideratam per istius æquationis reductionem et solutionem quæro. Sic deniꝗ plures quantitates tanquam cognitas sæpenumerò assumimus quam in statu quæstionis exprimuntur. Hujusꝗ rei insignem in ultimo[194] sequentium problematum instantiam videre est ubi *a b* et *c* in æquatione $aa + bx + cxx = yy$ pro determinatione Sectionis Conicæ assumpsi, ut

---

(194) Understand 'Prob 42' on Newton's pages 106–9 below. Evidently he planned at this point in time to include only a portion of the geometrical problems which he came subsequently to incorporate in his text. Whiston in his 1707 *editio princeps* correctly replaced 'ultimo' by '42°' (42nd), which in turn—to take account of his renumbering of the problems in revise— Newton in 1722 further changed to '55°'.

obtained between the value of CD found in this way and the letter substituted
for it. We can also (and for the most part it is more satisfying to do so than to
pursue the work too far in a continuous sequence) begin the computation
from different principles, or at least advance it in different ways to the same
conclusion, no matter what, so that at length two values of the self-same quantity
are obtained, which may then be set equal. Thus AD, AB and BC give BD, BE
and CE as before; then $CD + CE$ gives ED; and finally BD and ED (because of
the right triangle BED) give BE. The computation can also very well be
instituted on this basis: that the values of quantities between which any other
known relationship holds are to be ascertained, and that relationship will then
yield an equation. Thus, since the relationship between the lines BD, DC, BC and
CE is determined by *Elements*, II, 12, namely that $BD^2 - BC^2 - CD^2 = 2CD \times CE$,
I seek BD^2 from the assumed lengths AD and AB, and CE from the assumed ones
AD, AB and BC; then finally, on assuming CD, I make

$$BD^2 - BC^2 - CD^2 = 2CD \times CE.$$

Guided in these ways and by advice of this sort, you ought always to keep a
watchful eye simultaneously on the sequence of the analysis and on the constructions in the figure necessary to it.

From these observations it is, I believe, clear what geometers mean when they
order you to conceive done what is sought. For, when no distinction is made
between known and unknown quantities, you can assume any of these you
please to begin the calculation from, as though all were known from a previous
solution and it were a question no longer of solving the problem but checking
that solution. Thus in the first of the three modes of computation just now
described, even if it chances in fact that AD is the one sought, I nonetheless
imagine that CD is to be sought as though I wished to test whether its value
derived from AD should square with its quantity as previously ascertained. So
also in the two latter modes I propose as the aim not that some quantity be
sought, but that an equation should somehow or other be elicited from the
relationships of the lines. And in acknowledgement of this goal I assume all of
AD, AB, BC and CD as known, just as if (with the question previously solved) it
were now a matter of testing whether they properly satisfy its conditions by
squaring with any equations produced from the relationships of the lines. This
is, to be sure, the reason and intention with which I approached the task on
first confronting it, but once an equation has been reached I change my tactics
and seek the quantity desired by reducing and solving that equation. So,
finally, we frequently assume more quantities as known than are expressed in
the circumstances of the question. Of this point an outstanding instance may be
seen in the last[194] of the following problems, where I have assumed a, b and c
in the equation $a^2 + bx + cx^2 = y^2$ for determining a conic section and also still

et alias etiam lineas *r*, *s*, *t*, *v* de quibus Problema prout proponitur nihil innuit. Nam quaslibet quantitates assumere licet quarum ope possibile sit ad æquationes pervenire: hoc solum cavendo ut ex illis tot æquationes obtineri possint ‖[61] ‖quot assumptæ sunt quantitates revera incognitæ.

Postquam de computandi methodo constat et ornatur schema, quantitatibus quæ computationem ingredientur (hoc est ex quibus assumptis aliarum valores derivandi sunt donec tandem ad æquationem perveniatur) nomina impone, delegendo quæ problematis omnes conditiones involvunt & operi præ cæteris accommodatæ videntur & conclusionem (quantum possis conjicere[)] simpliciorem reddent, sed non plures tamen quàm proposito sufficiunt. Itaqȝ pro quantitatibus quæ ex aliarum vocabulis facilè deduci possint, propria vocabula vix tribuas. Sic ex tota linea et ejus partibus, ex tribus lateribus trianguli rectanguli, & ex tribus vel quatuor proportionalibus unum aliquod minimùm sine nomine permittere solemus, eo quòd valor ejus e reliquorum nominibus facilè derivari possit. Quemadmodum in exemplo jam allato si dicam $AD = x$ et $AB = a$ ipsum BD nulla litera designo quòd sit tertium latus trianguli rectanguli ABD et proinde valeat $\sqrt{xx - aa}$. Dein si dicam $BC = b$, cùm triangula DAB et BCE sint similia et inde lineæ $AD . AB :: BC . BE$ proportionales, quarum tribus AD, AB, et BC imposita sunt nomina; eapropter quartam CE

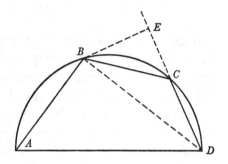

sine nomine permitto & ejus vice valorem $\dfrac{ab}{x}$ ex hac proportionalitate detectum usurpo. Atqȝ ita si DC vocetur c, ipsi DE nomen non assigno quod ex partibus ejus DC et CE sive c et $\dfrac{ab}{x}$, valor $c + \dfrac{ab}{x}$ prodeat.

Cæterùm dum de his moneo, Problema ad æquationem penè redactum est. Nam postquam literæ pro speciebus principalium linearum præscriptæ sunt, nihil aliud agendum restat quàm ut ex istis speciebus valores aliarum linearum juxta methodum præconceptam eruantur, donec modo quovis proviso in æquationem coeant. Et in hoc casu nihil restare video nisi ut per triangula rectangula BCE et BDE dupliciter eliciam BE. Nempe est $BC^q - CE^q$ $\left(\text{sive } bb - \dfrac{aabb}{xx}\right) = BE^q$, ut et $BD^q - DE^q$ $\left(\text{sive } xx - aa - cc - \dfrac{2abc}{x} - \dfrac{aabb}{xx}\right) = BE^q$.

‖[62] Et hinc $\left(\text{utrobiqȝ deleto } \| \dfrac{aabb}{xx}\right)$ æquationem habebo $bb = xx - aa - cc - \dfrac{2abc}{x}$: quæ reducta fit $x^3 = \genfrac{}{}{0pt}{}{+\,aa}{+\,bb}\,x + 2abc.$[195]
$+cc$

other lines r, s, t, v regarding which the problem as it is proposed intimates nothing. For it is permissible to assume any quantities at all with whose aid it may be possible to attain an equation, with the sole proviso that it shall be possible to obtain from them the same number of equations as in fact there are assumed unknown quantities.

After the method of computing is established and the figure decked out, on the quantities which will enter into the computation (those, that is, from which, when assumed, the values of the others are to be derived till at length an equation is attained) set names, choosing those which involve all the conditions of the problem and seem, more than the others, suited to the work, rendering the conclusion—as far as an advance guess is possible—simpler, but which are yet no more in number than is sufficient for the purpose. Accordingly, you should refrain from assigning separate appellations to quantities which may easily be deduced in terms of the designations set on others. Thus, in the case of a whole line and its parts, the three sides of a right-angled triangle and three or four quantities in proportion, we usually allow one at least—no matter which— to go nameless so as to be able easily to derive its value in terms of the others' names. In the example just now adduced, for instance, if I call $AD = x$ and $AB = a$, I do not designate BD with a letter because it is the third side of the right-angled triangle ABD and consequently its value is $\sqrt{[x^2 - a^2]}$. If I then call $BC = b$, since the triangles DAB and BCE are similar and as a result the lines AD, AB, BC, BE are in proportion $AD:AB = BC:BE$, and since names have been put on three of these, AD, AB and BC, in consequence I allow the fourth to go without a name and in its place employ the value ab/x disclosed by this proportion. Likewise, if DC be called c I assign no name to DE because from its parts DC and CE, that is, c and ab/x, there results $c + ab/x$ as its value.

But, even as I stress these points, the problem is all but brought to an equation. For, once letters have been set on the algebraic magnitudes of the principal lines, nothing else remains to be done except to determine from those magnitudes, following a preconceived method, the values of the other lines involved till at length in any way planned they combine in an equation. And in the present case I see that nothing remains but to elicit BE, by means of the right-angled triangles BCE and BDE, in two ways. Specifically,

$$BC^2 - CE^2 \text{ (or } b^2 - a^2 b^2/x^2) = BE^2,$$

and also $BD^2 - DE^2$ (or $x^2 - a^2 - c^2 - 2abc/x - a^2 b^2/x^2$) $= BE^2$; and hence (after $[-]a^2 b^2/x^2$ is deleted on each side) I shall have the equation

$$b^2 = x^2 - a^2 - c^2 - 2abc/x,$$

which when reduced becomes $x^3 = (a^2 + b^2 + c^2)\, x + 2abc$.[195]

---

(195) This is effectively Schooten's deduction of the cubic in his tract *De Cubicarum Æquationum Resolutione*; see note (179).

Cùm vero de solutione Problematis hujus plures modos etsi non multùm dissimiles in præcedentibus recensuerim quorum iste de Prop. 12. Lib. 2, Elem. desumptus sit cæteris quodammodo concinnior; eundem placet etiam subjungere. Sit itaꝗ $AD=x$, $AB=a$, $BC=b$, & $CD=c$, eritꝗ $BD^q=xx-aa$, et $CE=\dfrac{ab}{x}$ ut priùs. Hisce dein speciebus in Theorema

$$BD^q - BC^q - CD^q = 2CD \times CE$$

substitutis orietur $xx-aa-bb-cc=\dfrac{2abc}{x}$; et facta reductione $x^3 = \begin{array}{l} aa \\ +bb \\ +cc \end{array} x + 2abc.$

ut ante.

Sed ut pateat quanta sit in solutionum inventione varietas, et proinde quod in eas incidere prudenti Geometræ non sit admodum difficile: visum fuit plures adhuc modos hoc idem perficiendi docere. Atꝗ equidem ducto Diagonio si vice perpendiculi BE a puncto B in latus DC supra demissi demittatur perpendiculum a puncto D in latus BC vel a puncto C in latus BD, quo obliquangulum triangulum BCD in duo rectangula utcunꝗ resolvatur, ijsdem fermè quas jam descripsi methodis ad æquationem pervenire licet: sunt [e]t alij modi ab istis satis differentes;

Quemadmodum si diagonij duo AC et BD ducantur, dabitur BD ex assumptis AD et AB, ut et AC ex assumptis AD et CD; deinde per notum Theorema de figuris quadrilateris in circulo inscriptis,[196] nempe quòd sit $AD \times BC + AB \times CD = AC \times BD$ obtinebitur æquatio. Stantibus itaꝗ linearum AD, AB, BC, CD vocabulis x, a, b, c; erit $BD=\sqrt{xx-aa}$ & $AC=\sqrt{xx-cc}$ per 47.1 Elem. Et his linearum speciebus in Theorema jam recensitum substitutis, exibit $xb+ac=\sqrt{xx-cc}\times\sqrt{xx-aa}$. Cujus æquationis partibus deniꝗ quadratis et reductis[197] obtinebitur iterum $x^3 = \begin{array}{l} aa \\ +bb \\ +cc \end{array} x + 2abc.$

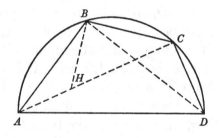

[1675]
Lect 10

‖[63]

Cæterùm ut pateat etiam quo pacto solutiones ex isto Theoremate petitæ possint inde ad solas triangulorum similitudines redigi: erigatur BH ipsi BC perpendicularis et occurrens AC in H, et fient triangula BCH, BDA similia propter an‖gulos ad B rectos et ad C ac D (per 21.3[198] Elem) æquales; ut et triangula BCD BHA similia propter æquales angulos tum ad B (ut pateat demendo communem angulum DBH a duobus rectis,) tum ad D ac A (per

(196) Ptolemy's; see note (188).

However, since in the preceding pages I have reviewed several ways—admittedly not greatly dissimilar—pertinent to the solution of this problem, and of these that taken from *Elements*, II, 12, is somewhat neater than the rest, it is agreeable to append this one also. Accordingly, let $AD = x$, $AB = a$, $BC = b$ and $CD = c$, and then $BD^2 = x^2 - a^2$ and $CE = ab/x$, as previously. Next, when these magnitudes are substituted in the theorem

$$BD^2 - BC^2 - CD^2 = 2CD \times CE$$

there will arise $x^2 - a^2 - b^2 - c^2 = 2abc/x$; and, when reduction is made,

$$x^3 = (a^2 + b^2 + c^2)\, x + 2abc,$$

as before.

But to make clear how great a variety there is in finding solutions and in consequence that it is not overly difficult for the practised geometer to light upon them, I have thought it appropriate to explain still further ways of accomplishing this same result. And to be sure, if, when the diagonal is drawn, instead of dropping the above perpendicular BE from the point B onto the side DC a perpendicular be let fall from the point D onto the side BC or from the point C onto the side BD in order that the obtuse triangle BCD may somehow be resolved into two right-angled ones, the equation is attainable by almost the same methods as I have already described. There are also other methods appreciably different from these.

For example, if the two diagonals AC and BD be drawn, BD will be given in consequence of AD and AB assumed, and likewise AC from AD and CD assumed; then, by the known theorem on quadrilateral figures inscribed in a circle,[196] namely, that $AD \times BC + AB \times CD = AC \times BD$, an equation will be obtained. Accordingly, with the designations x, a, b, c of the lines AD, AB, BC, CD standing, there will be $BD = \sqrt{[x^2 - a^2]}$ and $AC = \sqrt{[x^2 - c^2]}$ by *Elements*, I, 47. And when these algebraic quantities of the lines are substituted in the theorem just now cited there will result $xb + ac = \sqrt{[x^2 - c^2]} \times \sqrt{[x^2 - a^2]}$. Finally, on squaring the sides of this equation and reducing,[197] there will once more be obtained $x^3 = (a^2 + b^2 + c^2)\, x + 2abc$.

However, to make clear also how solutions derived from that theorem can therefrom be reduced to similarities between triangles alone, raise BH perpendicular to BC, meeting AC in H, and the triangles BCH, BDA will come to be similar because of their angles at B being right and those at C and D (by *Elements*, III, 21)[198] equal; and likewise the triangles BCD, BHA will be similar since they have equal angles both at B (as may be patent on taking away the common angle $D\widehat{B}H$ from the two right angles) and also at D and A (by

---

(197) Namely, by removing an unwanted factor x and reordering.
(198) Asserting that angles subtended by a chord in the same circle-arc are equal.

21.3 Elem.) Videre est itaꝗ quod ex proportionalitate $BD.AD::BC.HC$ detur HC, ut et AH ex proportionalitate $BD.CD::AB.AH$. Unde cùm sit $AH+HC=AC$, habebitur æquatio.[199] Stantibus ergo præfatis linearum vocabulis x, a, b, c, nec non ipsarum AC et BD valoribus $\sqrt{xx-cc}$ & $\sqrt{xx-aa}$: prima proportionalitas dabit $HC=\dfrac{ac}{\sqrt{xx-aa}}$, et secunda dabit $AH=\dfrac{bx}{\sqrt{xx-aa}}$. Unde propter $AH+HC=AC$ erit $\dfrac{bx+ac}{\sqrt{xx-aa}}=\sqrt{xx-cc}$; æquatio quæ (multiplicando per $\sqrt{xx-aa}$ et quadrando) reducetur ad formam in præcedentibus sæpius descriptam.

Adhæc ut magis pateat quanta sit solvendi copia, producantur BC et AD donec conveniant in F; et fient triangula ABF et CDF similia, quippe quorum angulus ad F communis est et anguli ABF et CDF (dum complent ang CDA ad duos rectos per 13. 1 & 22. 3 Elem.) æquales. Quamobrem si præter quatuor terminos de quibus instituitur quæstio, daretur AF, proportio $AB.AF::CD.CF$

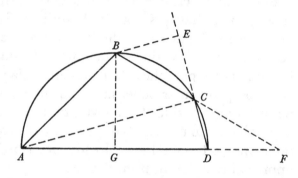

daret CF. Item $AF-AD$ daret DF, et proportio $CD.DF::AB.BF$ daret BF: unde (cum sit $BF-CF=BC$) emergeret æquatio. Sed cùm duæ quantitates incognitæ AD ac DF tanquam datæ assumantur, restat alia æquatio invenienda. Demitte ergo BG in AF ad rectos angulos, et proportio $AD.AB::AB.AG$. dabit AG: quo habito, Theorema e 13. 2 Elem. petitum, nempe quod sit $BF^q+2FAG=AB^q+AF^q$, dabit æquationem alteram. Stantibus ergo a, b, c, x ut priùs, et dicto $AF=y$: erit (insistendo vestigijs Theoriæ jam excogitatæ) $\dfrac{cy}{a}=CF$.

$y-x=DF$. $\dfrac{\overline{y-x}\times a}{c}=BF$. Indéꝗ $\dfrac{\overline{y-x}\times a}{c}-\dfrac{cy}{a}=b$, æquatio prima. Erit etiam $\dfrac{aa}{x}=AG$, adeoꝗ $\dfrac{aayy-2aaxy+aaxx}{cc}+\dfrac{2aay}{x}=aa+yy$, æquatio secunda. Quæ duæ

‖[64] per reductionem dabunt ‖ æquationem desideratam. Nempe valor ipsius y per æquationem priorem inventus est $\dfrac{abc+aax}{aa-cc}$, qui in secundam, $\Big($sic si placet ordinatam $\dfrac{aa-cc}{cc}y+\dfrac{aa-bb}{y}=\dfrac{2aa}{x}+\dfrac{2bc}{a}[!]\Big)^{(200)}$ substitutus dabit æquationem ex quâ rectè disposita fiet $x^3=\begin{smallmatrix}aa\\+bb\\+cc\end{smallmatrix}x+2abc$ ut ante.

Elements, III, 21). It may accordingly be seen that from the proportion $BD:AD = BC:HC$ there will be given HC, and likewise AH is yielded by the proportion $BD:CD = AB:AH$. Whence, since $AH+HC = AC$, an equation will be had.[199] With the above-stated designations x, a, b, c of the lines continuing to stand, therefore, and the values $\sqrt{[x^2-c^2]}$ and $\sqrt{[x^2-a^2]}$ of AC and BD still holding, the first proportion will give $HC = ac/\sqrt{[x^2-a^2]}$ and the second, $AH = bx/\sqrt{[x^2-a^2]}$. Hence, because $AH+HC = AC$, there will be $(bx+ac)/\sqrt{[x^2-a^2]} = \sqrt{[x^2-c^2]}$: an equation which (on multiplying through by $\sqrt{[x^2-a^2]}$ and squaring) will be reduced to the form somewhat frequently written out in the preceding pages.

Furthermore, to make yet clearer how richly profuse the possible solutions are, let BC and AD be extended till they meet in F; and the triangles ABF, CDF will come to be similar, seeing that the angle at F is common, while the angles $\widehat{ABF}$ and $\widehat{CDF}$ (in so far as, by *Elements*, I, 13, and III, 22, they supplement the angle $\widehat{CDA}$) are equal. Consequently, if—apart from the four terms regarding which the question is framed—AF were given, the proportion

$$AB:AF = CD:CF$$

would yield CF. Similarly, $AF-AD$ would give DF, and the proportion $CD:DF = AB:BF$ would yield BF; and from these (since $BF-CF = BC$) an equation would emerge. But since two unknown quantities AD and DF are assumed as given, another equation remains to be found. Let fall, therefore, BG at right angles to AF, and the proportion $AD:AB = AB:AG$ will give AG; and, once this is had, a theorem derived from *Elements*, II, 13—namely, that $BF^2+2AF \times AG = AB^2+AF^2$—will give this other equation. With a, b, c, x standing as before and on calling $AF = y$, there will therefore (on retracing the steps of the procedure just now devised) be $CF = (c/a)\,y$, $DF = y-x$,

$$BF = (a/c)\,(y-x)$$

and thence $(a/c)\,(y-x) - (c/a)\,y = b$, the first equation. Also there will be $AG = a^2/x$, and in consequence $(a^2/c^2)\,(y^2-2xy+x^2)+2a^2y/x = a^2+y^2$, the second equation. These two will, by reduction, give the equation desired. Namely, the value of y found by the former equation is $(abc+a^2x)/(a^2-c^2)$, and this, when substituted in the second (ordered, if you think fit, in this manner: $((a^2-c^2)/c^2)\,y+(a^2-b^2)/y = 2a^2/x+2bc/a[!])$,[200] will yield an equation from which, after suitable rearrangement, will come $x^3 = (a^2+b^2+c^2)\,x+2abc$, as before.

---

(199) Effectively, in lieu of accepting the truth of Ptolemy's theorem, Newton proves it *ab initio* in Ptolemaic style!

(200) This erroneous parenthesis, which occupies a line by itself in the manuscript, was omitted by Whiston from his 1707 *editio princeps* and is accordingly to be found in no subsequent edition. On correctly reducing these results $(a^2-c^2)y = a(x^2-c^2)(2a/x-a/y)$.

Atꝗ ita si *AB* ac *DC* producantur donec sibi mutuò occurrant, solutio haud aliter se habebit, nisi forte futura sit paulo facilior. Quare aliud hujus rei specimen e fonte multùm dissimili petitum potiùs subjungam, quærendo nempe aream quadrilateri propositi, idꝗ dupliciter. Duco igitur diagonium *BD* ut in duo triangula quadrilaterum resolvatur. Dein usurpatis linearum vocabulis *x*, *a*, *b*, *c* ut ante, invenio $BD = \sqrt{xx - aa}$ indeꝗ $\frac{1}{2}a\sqrt{xx - aa}\,(= \frac{1}{2}AB \times BD)$ aream trianguli *ABD*. Porro demisso *BE* perpendiculariter in *CD*, erit (propter similia triangula *ABD*, *BCE*) $AD . BD :: BC . BE$, et proinde $BE = \frac{b}{x}\sqrt{xx - aa}$. Quare etiam $\frac{bc}{2x}\sqrt{xx - aa}\,(= \frac{1}{2}CD \times BE)$ erit area trianguli *BCD*. Hasce jam areas addendo orietur $\frac{ax + bc}{2x}\sqrt{xx - aa}$ area totius quadrilateri. Non secus ducendo diagonium *AC* et quærendo areas triangulorum *ACD* et *ACB*, easꝗ addendo, rursus obtinebitur[201] area quadrilateri $\frac{cx + ba}{2x}\sqrt{xx - cc}$. Quare ponendo hasce areas æquales et utrasꝗ multiplicando per 2*x*, habebitur $\overline{ax + bc}\sqrt{xx - aa} = \overline{cx + ba}\sqrt{xx - cc}$, æquatio quæ quadrando ac dividendo per $aax - ccx$ redigetur ad formam sæpius inventam $x^3 = \begin{matrix} aa \\ +bb \\ +cc \end{matrix} x + 2abc.$

Ex his constare potest quanta sit solvendi copia et obiter quòd alij modi sint alijs multò concinniores. Quapropter si in primas de solutione Problematis alicujus cogitationes modus computationi male accommodatus inciderit, relationes linearum iterum evolvendæ sunt donec modum quàm poteris idoneum et elegantem machinatus fueris. Nam quæ leviori curæ se offerunt laborem satis molestum plerumꝗ parient si ad opus adhibeantur. Sic in Problemate de quo ‖ [65] agitur, nil ‖ difficilius foret[202] in sequentem modum quàm in aliquem e præcedentibus incidere. Demissis nempe *BR* et *CS* ad *AD* normalibus ut et *CT* ad *BR*, figura resolvetur in triangula rectangula. Et videre est quod

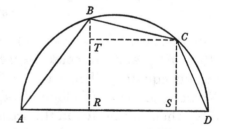

AD et *AB* dant *AR*, *AD* et *CD* dant *SD*, $AD - AR - SD$ dant *RS* vel *TC*. Item *AB* et *AR* dant *BR*, *CD* et *SD* dant *CS* vel *TR*, & $BR - TR$ dant *BT*. Deniꝗ *BT* ac *TC* dant *BC*, unde obtinebitur æquatio. Siquis autem hoc modo computationem aggressus fuerit is in terminos Algebraicos profusiores quàm ulli præcedentium incidet et ad finalem æquationem ægriùs reducibiles.[203]

---

(201) Simply by interchanging $AB = a$ and $CD = c$.

And likewise, if AB and DC be produced till they meet each other, there will be obtained a solution differing in no way except, perhaps, that it may prove to be a little easier. In consequence, I shall in preference append a further specimen solution of the problem derived from a greatly different source, seeking the area, namely, of the quadrilateral proposed, and that in two forms. I therefore draw the diagonal BD so as to resolve the quadrilateral into two triangles, and I then find, employing the designations x, a, b, c for the lines as before, that $BD = \sqrt{[x^2 - a^2]}$ and thence that the area $(\frac{1}{2}AB \times BD)$ of the triangle ABD is $\frac{1}{2}a\sqrt{[x^2 - a^2]}$. Moreover, on letting fall BE perpendicularly onto CD, there will (because the triangles DAB, BCE are similar) be $AD:BD = BC:BE$ and consequently $BE = b\sqrt{[x^2 - a^2]}/x$. Hence also $\frac{1}{2}bc\sqrt{[x^2 - a^2]}/x$ $(=\frac{1}{2}CD \times BE)$ will be the area of the triangle BCD. By now adding these areas there will result $\frac{1}{2}(ax + bc)\sqrt{[x^2 - a^2]}/x$ as the area of the whole quadrilateral. In like manner, on drawing the diagonal AC and ascertaining the areas of the triangles ACD and ACB, and then adding them, the area of the quadrilateral will again be obtained[201] as $\frac{1}{2}(cx + ba)\sqrt{[x^2 - c^2]}/x$. Whence, on setting these areas equal and multiplying both by $2x$, there will be had

$$(ax + bc)\sqrt{[x^2 - a^2]} = (cx + ba)\sqrt{[x^2 - c^2]},$$

an equation which, after squaring and dividing through by $(a^2 - c^2)\,x$ will be reduced to the form all too often now found, $x^3 = (a^2 + b^2 + c^2)\,x + 2abc$.

On this basis we may be convinced of the rich profusion of solutions possible, and agree, incidentally, that some ways are much neater than others. Because of this, if in your first thoughts on the solution of some problem you should chance upon an approach ill-suited to computation, the relationships of the lines should further be evolved till you have devised as effective and elegant a way as you can. For what suggests itself upon a shallow consideration will generally give birth to an offspring troublesome enough if it be applied to the task in hand. Thus, in the context of the problem which is of present concern it would be[202] no more difficult to chance upon the following way than on any of the preceding ones. Namely, on letting fall the normals BR and CS to AD and likewise CT to BR, the figure will be resolved into right triangles. It may be seen that AD and AB give AR, AD and CD give SD, $AD - AR - SD$ give RS (or TC). Similarly, AB and AR give BR, CD and SD give CS (or TR), and $BR - TR$ give BT. Finally, BT and TC give BC, and from this an equation will be obtained. Should anyone, however, attack the computation by this approach, he will fall in with algebraic terms more profuse than any in the preceding and not so easily reducible to the final equation.[203]

---

(202) Newton first wrote 'est' (is) in an all but identical, cancelled phrase.

(203) It is evident that Newton has not attempted the solution by this approach, for he rather overstates its complexity. On taking (as before) $AB = a$, $BC = b$, $CD = c$ and $AD = x$,

Et hæc de solutione problematum in rectilinea Geometria: nisi forte operæ pretium fuerit annotasse præterea quòd cùm anguli sive positiones linearum per angulos expressæ[204] statum quæstionis ingrediuntur, angulorum vice debent adhiberi lineæ aut linearum proportiones, tales nempe quæ ab angulis datis possunt per calculum Trigonometricum derivari[,] a quibus inventis[205] anguli quæsiti per eundem calculum prodeunt; hoc est quæ se mutuo determinant: cujus rei plures instantias videre est in sequentibus.

Octob. 1676
Lect: 1

[206]Quod ad Geometriam circa lineas curvas attinet, illæ designari solent vel describendo eas per motum localem[207] rectarum, vel adhibendo æquationes indefinite exprimentes relationem rectarum certa aliqua lege dispositarum & ad curvas desinentium.[208] Idem fecerunt Veteres per sectiones[209] Solidorum, sed minùs commodè. Computationes verò quæ curvas primo modo descriptas respiciunt haud secus quàm in præcedentibus peraguntur. Quemadmodum si

AκC sit curva linea[210] descripta per *κ* verticale punctum normæ *Aκφ*, cujus unum crus *Aκ* per punctum *A* positione datum liberè dilabitur, dum alterum *κφ* datæ longitudinis super rectam ‖*AD* positione datam promovetur, et quæratur punctum *C* in quo recta quævis *CD* positione data hanc

‖[66]

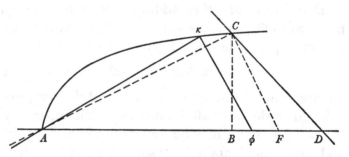

curvam secabit: duco rectas *ACF* quæ normam in positione quæsita referant, et relatione linearum (sine aliquo dati et quæsiti discrimine aut respectu ad curvam) considerata, percipio dependentiam cæterarum a *CF* et qualibet harum quatuor *BC*, *BF*, *AF* et *AC* syntheticam[211] esse; quarum duas itaq ut *CF=a* et *CB=x* assumo, et inde computum ordiendo statim lucratus sum $BF=\sqrt{aa-xx}$ &

$AB=\dfrac{xx}{\sqrt{aa-xx}}$ propter ang: rectum *CBF*, lineasq *BF*.*BC*::*BC*.*AB* continuò

---

at once $AR = a^2/x$, $SD = c^2/x$, $BR = (a/x)\sqrt{[x^2-a^2]}$ and $CS = (c/x)\sqrt{[x^2-c^2]}$, whence the equation $BC^2 = (AD-AR-SD)^2+(BR-CS)^2$ yields on reduction

$$x^4 - (a^2+b^2+c^2)x^2 + 2a^2c^2 - 2ac\sqrt{[(x^2-a^2)(x^2-c^2)]} = 0$$

or $(\sqrt{[(x^2-a^2)(x^2-c^2)]}-ac)^2 = b^2x^2$ and so $\sqrt{[(x^2-a^2)(x^2-c^2)]} = ac \pm bx$; accordingly, on squaring and reordering, $x(x^3 - (a^2+b^2+c^2)x \mp 2abc) = 0$. (The dubious sign of the last term corresponds to choice of B and C on the same or opposite sides of the diameter *AD*.)

(204) 'determinatæ' (determined) was first written.

(205) Newton has deleted 'vicissim' (in turn).

(206) Before proceeding to the detailed solution of his set of geometrical problems Newton inserts a short précis of Cartesian analytical geometry, making no attempt to designate the

This will do for the solution of problems in the geometry of straight lines, unless perhaps it should be worth-while to contribute the further remark that, when angles or positions of lines expressed[204] by means of angles enter the context of the question, in place of the angles there ought to be introduced lines or ratios of lines, specifically of a type which can be derived from given angles by trigonometrical calculation, and from which, when found, the angles required[205] result by the same calculation—that is, which mutually determine each other. Several instances of this kind may be observed in the sequel.

[206]With regard to the subject of geometrical curves, these are usually represented either by describing them through the local motion[207] of straight lines or by employing equations indefinitely expressing the relationship of straight lines arranged in some fixed order and terminating at the curves.[208] The Ancients attained the same end by means of[209] sections of solids, but less conveniently so. Computations, indeed, which relate to curves described in the first manner are effected no differently than in the preceding. For instance, if $A\kappa C$ be the curve[210] described by the vertex point κ of a right-angled rule $A\kappa\phi$, one leg $A\kappa$ glides freely through the point A given in position, while the second one $\kappa\phi$ of given length advances along the straight line AD given in position, and if there should be required the point C in which any straight line CD given in position cuts this curve: I draw straight lines AC/CF which mark the rule in its required position, and, having considered the inter-relationship of the lines (without any distinction between given and sought or any regard to the curve itself), I perceive that the dependence of the rest from CF and any one of these four, BC, BF, AF and AC, is synthetic.[211] Accordingly, I assume two of these—$CF = a$ and $CB = x$, say—and, beginning the calculation therefrom, I immediately gain $BF = \sqrt{[a^2 - x^2]}$ and $AB = x^2/\sqrt{[a^2 - x^2]}$ because the angle $C\widehat{BF}$ is right and the lines BF, BC, AB are in the continued proportion

$$BF : BC = BC : AB.$$

basic coordinate systems in terms of which a given equation defines a corresponding curve. (As we have seen (III: 120–46), in his 1671 tract he distinguished nine such 'modes' of coordinates.) It is difficult to conceive that his contemporary reader would have been much enlightened by so sketchy a discussion of this complex and fundamental topic. Or does Newton assume a familiarity with Descartes' *Geometrie* and the several available commentaries upon it?

(207) See III: 70, note (80) and compare note (14).

(208) Disposed in some fixed coordinate-system in terms of which the curves are defined by the equations, that is.

(209) In context, understand 'planas' (plane). No doubt Newton has primarily in mind the Apollonian definition of a conic as the general plane section of a cone.

(210) The (first quadrant of the) kappa curve (Gutschoven quartic) whose area Newton had earlier evaluated in his 1671 tract; see III: 268, note (602).

(211) Representable, that is, in a synthetic (deductive) sequence.

proportionales. Porro ex data positione CD datur AD quam itacg dico b, datur etiam ratio BC ad BD quam pono d ad e & fit $BD = \dfrac{ex}{d}$ et $AB = b - \dfrac{ex}{d}$. Est ergo $b - \dfrac{ex}{d} = \dfrac{xx}{\sqrt{aa - xx}}$, æquatio quæ (quadrando partes et multiplicando per $aa - xx$ &c) reducetur ad hanc formam

$$x^4 = \frac{2bdex^3 \begin{smallmatrix} -bbdd \\ +aaee \end{smallmatrix} xx - 2aabdex + aabbdd}{dd + ee};$$

unde demum e datis a, b, d, et e erui debet x per regulas post[212] tradendas, & intervallo isto x sive BC acta ipsi AD parallela recta secabit CD in quæsito puncto C.[213]

Quod si non descriptiones Geometricæ sed æquationes pro curvis designandis adhibeantur, computationes eo pacto faciliores & breviores evadent in quantum ejusmodi æquationes ipsis lucro cedunt. Quemadmodum si datæ Ellipseos ACE intersectio C cum recta CD positione data quæratur: pro Ellipsi designanda sumo notam aliquam æquationem ei propriam ut

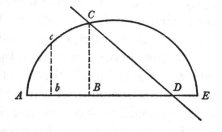

$rx - \dfrac{r}{q}xx = yy$[214] ubi x indefinite ponitur pro qualibet axis parte Ab vel AB, et y pro perpendiculo bc vel BC ad curvam terminato; r vero et q[215] dantur ex data specie Ellipsis.

Cùm itacg CD positione detur dabitur et AD quam dic a et erit $BD = a - x$, dabitur etiam an‖gulus ADC et inde ratio BD ad BC quam dic 1 ad e, et erit ‖[67]

$BC(y) = ea - ex$, cujus quadratum $eeaa - 2eeax + eexx$ æquabitur $rx - \dfrac{r}{q}xx$. Indecg per reductionem orietur $xx = \dfrac{2aeex + rx - aaee}{ee + \dfrac{r}{q}}$, seu $x = \dfrac{aee + \frac{1}{2}r \pm e\sqrt{ar + \dfrac{rr}{4ee} - \dfrac{aar}{q}}}{ee + \dfrac{r}{q}}$.

Quinetiam etsi Curva per Descriptionem Geometricam vel per sectionem solidi designetur, potest tamen inde æquatio obtineri quæ naturam Curvæ definiet, adeocg huc omnes Problematum quæ circa eam proponuntur difficultates reduci.

Sic in exemplo priori si AB dicatur x et BC y, tertia proportionalis BF erit $\dfrac{yy}{x}$, cujus quadratum una cum quadrato BC æquatur CF^q, hoc est $\dfrac{y^4}{xx} + yy = aa$; sive $y^4 + xxyy = aaxx$. Estcg hæc æquatio qua Curvæ $A\kappa C$ unumquodcg punctum C

---

(212) Newton here probably refers to the general algebraic determination of the roots of a quartic equation outlined on his pages 205–7 below, rather than to any approximate geometrical construction or numerical approximation; but see note (170).

Moreover, from CD given in position there is given AD, which I accordingly call b; the ratio of BC to BD is also given, and this I set as d to e. There then comes $BD = (e/d)\,x$ and $AB = b - (e/d)\,x$, and therefore

$$b - (e/d)\,x = x^2/\sqrt{[a^2 - x^2]},$$

an equation which will (on squaring its sides, multiplying by $a^2 - x^2$ and so on) be reduced to this form $x^4 = (2bdex^3 + (a^2e^2 - b^2d^2)\,x^2 - 2a^2bdex + a^2b^2d^2)/(d^2 + e^2)$. From this, lastly, x has to be elicited in terms of the given a, b, d, e by means of rules to be delivered subsequently,[212] and then a straight line drawn parallel to AD at that distance x (that is, BC) will intersect CD in the required point C.[213]

But if equations—and not geometrical descriptions—should be employed for representing curves, the computations involved will prove easier and shorter in as much as equations of the sort are a positive gain to them. For instance, if the intersection C of the given ellipse ACE with the straight line CD given in position be required, in order to represent the ellipse I take some known equation appropriate to the purpose, such as $rx - (r/q)\,x^2 = y^2$,[214] in which x is indefinitely put for any portion Ab or AB of the axis and y for the perpendicular bc or BC terminated at the curve, while r and q,[215] to be sure, are given from the given species of the ellipse. Accordingly, since CD is given in position, there will be given AD also: call this a, and then $BD = a - x$. The angle $\widehat{ADC}$ is given also, and thereby the ratio of BD to BC: this I call 1 to e, and then BC (or y) $= ea - ex$, the square of which $e^2a^2 - 2e^2ax + e^2x^2$ will equal $rx - (r/q)\,x^2$. And thence by reduction there will arise

$$x^2 = \frac{(2ae^2 + r)\,x - a^2e^2}{e^2 + r/q}, \quad \text{that is,} \quad x = \frac{ae^2 + \tfrac12 r \pm e\sqrt{[ar + \tfrac14 r^2/e^2 - a^2r/q]}}{e^2 + r/q}.$$

Even though, indeed, a curve be designated by means of a geometrical description or by a solid's section, an equation can yet be obtained therefrom which shall define the nature of the curve, and in consequence all difficulties in problems propounded concerning it reduced to that equation.

Thus, in the former example, if AB be called x and BC y, the third proportional BF will be y^2/x, the square of which together with that of BC is equal to CF^2; that is, $y^4/x^2 + y^2 = a^2$, or $y^4 + x^2y^2 = a^2x^2$. This is the equation by which any arbitrary point C of the curve $A\kappa C$ corresponding to any length AB of the

(213) To be precise, there will be two or four such intersections. In effect, where $AB = y$, Newton constructs the meets of the kappa curve $y = x^2/\sqrt{[a^2 - x^2]}$ and the straight line $y = b - (e/d)\,x$.

(214) Descartes' canonical equation in Book 2 of his *Geometrie*; here, of course, r denotes the ellipse's *latus rectum* and q its main axis (*latus transversum*). As we have seen, Newton made much use of this equation in his early mathematical researches into tangent and curvature problems: see, for instance, **1**: 252, 417.

(215) See the previous note.

unicuiꝗ basis longitudini *AB* congruens (adeoꝗ ipsa Curva) definitur, et e qua proinde solutiones Problematum quæ de hâc curva proponuntur petere liceat.

Ad eundem fere modum cum curva non datur specie sed determinanda proponitur, possis pro arbitrio æquationem fingere quæ naturam ejus generaliter contineat, & hanc pro ea designanda tanquam si daretur assumere ut ex ejus assumptione quomodocunꝗ perveniatur ad æquationes ex quibus assumpta tandem determinentur: Cujus rei exempla habes in ultimis sequentium problematum quæ in pleniorem illustrationem hujus doctrinæ & exercitium discentium congessi, quæꝗ jam pergo tradere.[216]

[1676]
Lect: 2.

Prob: 1.[217]

Data recta terminata BC a cujus extremitatibus duæ rectæ BA, CA ducuntur in datis angulis ABC ACB: invenire AD altitudinem concursus A supra datam BC.

Sit $BC=a$, & $AD=y$; & cum angulus *ABD* detur, dabitur (ex tabula sinuum vel tangentium) ratio inter lineas *AD* et *BD* quam pone ut *d* ad *e*. Est ergo $d \cdot e :: AD(y) \cdot BD$. Quare $BD=\frac{ey}{d}$. Similiter propter datum angulum *ACD* dabitur ratio inter *AD* ac *DC* quam pone ut *d* ad *f* et

‖[68] erit $DC=\frac{fy}{d}$. At $BD+DC=BC$, hoc est $\frac{ey}{d}+\frac{fy}{d}=a$. Quæ ‖ reducta multiplicando utramꝗ partem æquationis per *d* ac dividendo per $e+f$ evadit $y=\frac{ad}{e+f}$.

Prob. 2.[218]

Cujuslibet Trianguli ABC datis lateribus AB AC & Basi BC quam perpendiculum AD ab angulo verticali secat in D: invenire segmenta BD ac DC.

---

(216) We have already observed (note (194)) that at a fairly late stage in penning the preceding text Newton refers to Problem 42 as his 'last' geometrical example and we may hence infer that the remaining nineteen problems were added in afterthought. Moreover, Problem 23 is a still later addition (on Newton's page 257) to the text some time after the initial state of the present manuscript was completed. Eight of the opening fifteen problems repeat corresponding 'geometrick ones' (see II: 428–44)—though not in the same order—which Newton had earlier, in his 1670 'Observationes' on Kinckhuysen's *Algebra*, suggested to Collins might be subjoined in republication 'After the Authors Algebraick problems' (II: 428). It is difficult not to conclude that the remainder, drawn in large part (and with some revision) from contemporary printed sources rather than his own fertile imagination, were added piecemeal at whim, without system or regard to any grading by relative difficulty. The present sequence of the following sixty-one problems was retained unchanged by Whiston in his 1707 *editio princeps*, but Newton himself soon after—in an evident attempt at a more logical

base—and hence the curve itself—is defined, and from which it is consequently permissible to derive the solutions of problems proposed about this curve.

In much the same way, almost, when a curve is not given in species but it is proposed to determine it, you might imagine an equation at will which is to contain its nature generally, and then assume this to represent it as though it were given in order that from its assumption equations might be somehow or other attained, from which in turn the quantities assumed may at length be determined. You have examples of this technique in the final problems of the following set, which I have collected together to afford a fuller illustration of the present doctrine and an exercise for students. These I now proceed to elaborate.[216]

Problem 1[217]

Given the determinate straight line BC from whose extremities two straight lines BA, CA are drawn at given angles $\widehat{ABC}$, $\widehat{ACB}$, to find the height AD of their meeting point A above the given line BC.

Let $BC = a$ and $AD = y$; and then, since the angle $\widehat{ABD}$ is given, there will be given (from a table of sines or tangents) the ratio between the lines AD and BD. Take this to be d to e, and hence $d:e = AD$ (or y):BD, so that $BD = (e/d)\,y$. Similarly, because the angle $\widehat{ACD}$ is given, there will be given the ratio between AD and DC. Take this to be d to f, and there will be $DC = (f/d)\,y$. But $BD + DC = BC$, that is, $(e/d)\,y + (f/d)\,y = a$: which, when reduced on multiplying both sides of the equation by d and dividing through by $e + f$, comes to be $y = ad/(e+f)$.

Problem 2[218]

In any triangle ABC given the sides AB, AC and the base BC, which the perpendicular AD from the vertex angle shall intersect in D, to find the segments BD and DC.

---

regrouping—introduced into his library copy of that edition a marginal renumbering which was carried through without deviation in his 1722 revise, and is accordingly that of all subsequent editions. The following table conveys the essence of the reordering (1707 numbers on the left, their 1722 renumberings on the right):

1–3	1–3	23	12	33	25	50	13
4–6	8–10	24–6	37–9	34–41	40–7	51	11
7–11	17–21	27/28	26/27	42	55	52–7	56–61
12/13	23/24	29–31	14–16	43–5	48–50	58–61	51–4.
14–22	28–36	32	22	46–9	4–7		

The reader will make up his own mind regarding the success and effectiveness of this recasting.

(217) Repeated, with trivial amplification, from Newton's Kinckhuysen 'Observationes' (II: 432).

(218) A repeat of Newton's second Kinckhuysen problem (see II: 430).

Sit $AB=a$, $AC=b$, $BC=c$,   & $BD=x$,   eritϙ $DC=c-x$. Jam cùm $AB^q-BD^q\,(aa-xx)=AD^q$; & $AC^q-DC^q\,(bb-cc+2cx-xx)=AD^q$: erit

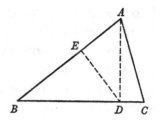

$$aa-xx=bb-cc+2cx-xx;$$

quæ per reductionem fit $\dfrac{aa-bb+cc}{2c}=x$.

Cæterum ut pateat omnes omnium Problematum difficultates per solam linearum proportionalitatem sine adminiculo Prop: 47 primi Elementorum, licet non absϙ circuitu, enodari posse: placuit sequentem hujus solutionem ex abundanti subjungere. A puncto D in latus AB demitte DE normalem, et stantibus jam positis linearum nominibus, erit

$$AB.BD::BD.BE.$$
$$a.\quad x::\quad x.\ \tfrac{xx}{a}.$$
et $BA-BE\left(a-\dfrac{xx}{a}\right)=EA.$

nec non $EA.AD::AD.AB$, adeoϙ $EA\times AB(aa-xx)=AD^q$. Et sic ratiocinando circa triangulum ACD invenietur iterum $AD^q=bb-cc+2cx-xx$.[219] Unde obtinebitur ut ante $x=\dfrac{aa-bb+cc}{2c}$.

Prob. 3.[220]

Trianguli rectanguli ABC perimetro & area datis invenire hypotenusam BC.

Esto perimeter a, area bb, $BC=x$, et $AC=y$, eritϙ $AB=\sqrt{xx-yy}$. Unde rursus perimeter $(BC+CA+AB)$ est $x+y+\sqrt{xx-yy}$, & area $(\tfrac{1}{2}AC\times AB)$ est $\tfrac{1}{2}y\sqrt{xx-yy}$. Adeoϙ $x+y+\sqrt{xx-yy}=a$, & $\tfrac{1}{2}y\sqrt{xx-yy}=bb$.

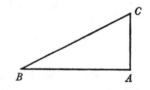

Harum æquationum posterior dat $\sqrt{xx-yy}=\dfrac{2bb}{y}$, quare scribo $\dfrac{2bb}{y}$ pro $\sqrt{xx-yy}$ in æquatione priori ut asymmetria tollatur & prodit $x+y+\dfrac{2bb}{y}=a$, sive multiplicando per y et ordinando $yy=ay-xy-2bb$. Porrò ex partibus æquationis prioris aufero $x+y$ & restat $\sqrt{xx-yy}=a-x-y$, cujus partes quadrando ut asymmetria rursus tollatur, prodit $xx-yy=aa-2ax-2ay+xx+2xy+yy$, quæ in ordinem redacta et per 2 divisa fit $yy=ay-xy+ax-\tfrac{1}{2}aa$. Deniϙ ponendo

---

(219) Newton here has first copied and then cancelled the phrase

'Quare $aa-xx=bb-cc+2cx-xx$'

(Therefore $a^2-x^2=b^2-c^2+2cx-x^2$) from his Kinckhuysen notes (compare II: 430).

Let $AB = a$, $AC = b$, $BC = c$ and $BD = x$, and there will be $DC = c-x$. Now, since $AB^2 - BD^2$ (or a^2-x^2) $= AD^2$ and $AC^2 - DC^2$ (or $b^2-c^2+2cx-x^2$) $= AD^2$, there will be $a^2-x^2 = b^2-c^2+2cx-x^2$, which by reduction becomes

$$x = \tfrac{1}{2}(a^2-b^2+c^2)/c.$$

However, to make it clear that every difficulty in every problem can be un-ravelled through the proportionality of lines alone without the assistance of Proposition 47 of the first book of the *Elements*, though not without some circuitousness, I have thought fit to adjoin the following solution of the present problem as an extra. From the point D onto the side AB let fall the normal DE and then, with the designations already set on the lines still standing, there will be $AB(a):BD(x) = BD(x):BE$, or $BE = x^2/a$ and so

$$EA \text{ (or } BA-BE) = a-x^2/a;$$

and also $EA:AD = AD:AB$, so that $AD^2 = (EA \times AB$ or$)$ a^2-x^2. And by reasoning in this way about the triangle ACD there will again be found $AD^2 = b^2-c^2+2cx-x^2$.[219] There will hence be obtained, as before,

$$x = \tfrac{1}{2}(a^2-b^2+c^2)/c.$$

Problem 3[220]

Given the perimeter and area of the right-angled triangle ABC, to find its hypotenuse BC.

Let the perimeter be a, the area b^2, $BC = x$ and $AC = y$, and there will be $AB = \surd[x^2-y^2]$; from this, again, the perimeter $(BC+CA+AB)$ is

$$x+y+\surd[x^2-y^2]$$

and the area $(\tfrac{1}{2}AC \times AB)$ is $\tfrac{1}{2}y\surd[x^2-y^2]$, so that $x+y+\surd[x^2-y^2] = a$ and $\tfrac{1}{2}y\surd[x^2-y^2] = b$.

The latter of these equations gives $\surd[x^2-y^2] = 2b^2/y$, and accordingly I write $2b^2/y$ in place of $\surd[x^2-y^2]$ in the former equation to free it from irrationality and there results $x+y+2b^2/y = a$, and so, on multiplying by y and ordering, $y^2 = ay-xy-2b^2$. Further, from the sides of the first equation I take away $x+y$ and there remains $\surd[x^2-y^2] = a-x-y$, from which, on squaring its sides so that irrationality may again be removed, there results

$$x^2-y^2 = a^2-2ax-2ay+x^2+2xy+y^2;$$

and, when this is reduced to order and divided by 2, there comes

$$y^2 = ay-xy+ax-\tfrac{1}{2}a^2.$$

---

(220) Newton's opening problem in his Kinckhuysen 'Observationes' (II: 428–30), here reproduced without change.

æqualitatem inter duos valores ipsius yy, habeo $ay-xy-2bb=ay-xy+ax-\frac{1}{2}aa$,

‖ [69] ‖ quæ reducta fit $\frac{1}{2}a-\dfrac{2bb}{a}=x$.

Idem aliter.

Esto $\frac{1}{2}$ perimeter $=a$, area $=bb$, & $BC=x$, eritſ̧ $AC+AB=2a-x$. Jam cùm sit $xx\ (BC^q)=AC^q+AB^q$, & $4bb=2AC\times AB$, erit

$$xx+4bb=AC^q+AB^q+2AC\times AB=\text{quadrato ex } AC+AB$$
$$=\text{quadrato ex } 2a-x=4aa-4ax+xx.$$

Hoc est $xx+4bb=4aa-4ax+xx$; quæ reducta fit $a-\dfrac{bb}{a}=x$.

[1676]
Lect 3

Prob: 4.[221]

Trianguli cujuscunſ̧ ABC, datis area perimetro & uno angulorum A, cætera determinare.

Esto perimeter $=a$ & area $=bb$, & ab ignotorum angulorum alterutro C ad latus oppositum AB demitte perpendiculum CD et propter angulum A datum erit AC ad CD in data ratione, puta d ad e. Dic ergo $AC=x$ & erit $CD=\dfrac{ex}{d}$, per quam divide duplam aream et prodibit $\dfrac{2bbd}{ex}=AB$. Adde

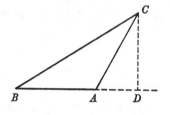

$AD\left(\text{nempe } \sqrt{AC^q-CD^q},\ \text{sive } \dfrac{x}{d}\sqrt{dd-ee}\right)$ et emerget $BD=\dfrac{2bbd}{ex}+\dfrac{x}{d}\sqrt{dd-ee}$: cujus quadrato adde CD^q et orietur $\dfrac{4b^4dd}{eexx}+xx+\dfrac{4bb}{e}\sqrt{dd-ee}=BC^q$. Adhæc a perimetro aufer AC & AB, et restabit $a-x-\dfrac{2bbd}{ex}=BC$, cujus quadratum $aa-2ax+xx-\dfrac{4abbd}{ex}+\dfrac{4bbd}{e}+\dfrac{4b^4dd}{eexx}$ pone æquale quadrato prius invento & neglectis æquipollentibus erit $\dfrac{4bb}{e}\sqrt{dd-ee}=aa-2ax-\dfrac{4abbd}{ex}+\dfrac{4bbd}{e}$. Et hæc, assumendo $4af$ pro datis terminis $aa+\dfrac{4bbd}{e}-\dfrac{4bb}{e}\sqrt{dd-ee}$, et reducendo, evadit $xx=2fx-\dfrac{2bbd}{e}$, sive $x=f\pm\sqrt{ff-\dfrac{2bbd}{e}}$.

Eadem æquatio prodijsset etiam quærendo crus AB, nam crura AB et AC similiter se habent ad omnes conditiones problematis. Quare si AC ponatur $f-\sqrt{ff-\dfrac{2bbd}{e}}$ erit $AB=f+\sqrt{ff-\dfrac{2bbd}{e}}$, & vicissim: atſ̧ horum summa $2f$ subducta de perimetro relinquit tertium latus $BC=a-2f$.

Finally, on setting equality between the two values of y, I have

$$ay - xy - 2b^2 = ay - xy + ax - \tfrac{1}{2}a^2,$$

which after reduction becomes $x = \tfrac{1}{2}a - 2b^2/a$.

The same another way

Let the semi-perimeter $= a$, the area $= b^2$ and $BC = x$, and then

$$AC + AB = 2a - x.$$

Now, since x^2 (or BC^2) $= AC^2 + AB^2$ and $4b^2 = 2AC \times AB$, there will be

$$x^2 + 4b^2 = AC^2 + AB^2 + 2AC \times AB = (AC + AB)^2 = (2a - x)^2 = 4a^2 - 4ax + x^2;$$

that is, $x^2 + 4b^2 = 4a^2 - 4ax + x^2$, which after reduction becomes $x = a - b^2/a$.

Problem 4[221]

In any triangle ABC given the area, perimeter and one of the angles A, to determine its other elements.

Let the perimeter $= a$ and area $= b^2$, and from either one of the unknown angles C let fall the perpendicular CD onto the opposite side AB; then, because angle A is given, AC will be to CD in a given ratio, say d to e. Call, therefore, $AC = x$ and there will be $CD = (e/d)\,x$; by this divide twice the area and there will result $AB = 2b^2 d/ex$. Add AD (namely, $\sqrt{[AC^2 - CD^2]}$ or $x\sqrt{[d^2 - e^2]}/d$, and there will emerge $BD = 2b^2 d/ex + x\sqrt{[d^2 - e^2]}/d$: add CD^2 to its square and there will ensue $4b^4 d^2/e^2 x^2 + x^2 + 4b^2\sqrt{[d^2 - e^2]}/e = BC^2$. Furthermore, from the perimeter take away AC and AB and there will remain $BC = a - x - 2b^2 d/ex$, whose square $a^2 - 2ax + x^2 - 4ab^2 d/ex + 4b^2 d/e + 4b^4 d^2/e^2 x^2$ put equal to the square previously found and, with equivalents neglected, there will be

$$4b^2\sqrt{[d^2 - e^2]}/e = a^2 - 2ax - 4ab^2 d/ex + 4b^2 d/e.$$

On taking $4af$ in place of the given terms $a^2 + 4b^2 d/e - 4b^2\sqrt{[d^2 - e^2]}/e$ and reducing, this comes to be $x^2 = 2fx - 2b^2 d/e$, that is, $x = f \pm \sqrt{[f^2 - 2b^2 d/e]}$.

The same equation would have resulted also from seeking the 'leg' AB, for the *crura* AB and AC are similarly involved in all the defining circumstances of the problem. Hence, if AC be put (equal to) $f - \sqrt{[f^2 - 2b^2 d/e]}$, then

$$AB = f + \sqrt{[f^2 - 2b^2 d/e]},$$

and conversely so; while their sum $2f$ subtracted from the perimeter leaves the third side $BC = a - 2f$.

---

(221) We have already reproduced the text of this problem in the second volume as our conjectural restoration of Newton's (lost) third Kinckhuysen problem; see II: 430–2, especially 431, note (123).

Prob: 5.[222]

Datis altitudine basi et summa laterum invenire triangulum.

Sit altitudo $CD=a$, basis AB dimidium $=b$, laterum semisumma $=c$ & semidifferentia $=z$: eritq majus latus puta $BC=c+z$, & minus $AC=c-z$. Subduc CD^q de BC^q & AC^q et exibit hinc $BD=\sqrt{cc+2cz+zz-aa}$ & inde $AD=\sqrt{cc-2cz+zz-aa}$. Subduc etiam AB de BD et exibit iterum

$$AD=\sqrt{cc+2cz+zz-aa}-2b.$$

‖[70] Qua‖dratis jam valoribus AD et ordinatis terminis, orietur

$$bb+cz=b\sqrt{cc+2cz+zz-aa}.$$

Rursusq quadrando et redigendo in ordinem obtinebitur

$$cczz-bbzz=bbcc-bbaa-b^4. \quad\text{Et}\quad z=b\sqrt{1-\frac{aa}{cc-bb}}.$$

Unde dantur latera.

Prob. 6.

Datis basi AB, summa laterum AC+BC, & angulo verticali C, determinare latera.

Sit basis $=a$, semisumma laterum $=b$, & semidifferentia $=x$ eritq majus latus $BC=b+x$ & minus $AC=b-x$. Ab alterutro ignotorum angulorum A ad latus oppositum BC demitte perpendiculum AD & propter angulum C datum dabitur ratio AC ad CD puta d ad e, & proinde erit $CD=\dfrac{eb-ex}{d}$. Est etiam per 11. 2. Elem.

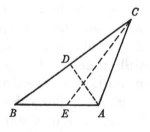

$\dfrac{AC^q-AB^q+BC^q}{2BC}$ hoc est $\dfrac{2bb+2xx-aa}{2b+2x}=CD$; adeoq

habetur æquatio inter valores CD. Et hæc reducta fit $x=\sqrt{\dfrac{daa+2ebb-2dbb}{2d+2e}}$.

Unde dantur latera.

Si anguli ad basin quærerentur conclusio foret concinnior[:] utpote ducatur EC datum angulum bisecans et basi occurrens in E; et erit

$$AB \cdot AC+BC(::AE \cdot AC)::\sin:\text{ang}:ACE \cdot \sin:\text{ang}:AEC.$$

Et ab angulo AEC ejusq complemento BEC si subducatur dimidium anguli C relinquentur anguli ABC & BAC.

---

(222) This and the following problem are of a type discussed by William Oughtred in his *Clavis Mathematicæ denuo limata, Sive potius fabricata* (Oxford, ₃1652): Cap. xix, Probl. xv: 90–1. If this was Newton's source, his present algebraic solutions are his own.

Problem 5[(222)]

Given the altitude, base and sum of the sides, to find the triangle.

Let the altitude $CD = a$, half the base $AB = b$, the half-sum of the (inclined) sides $= c$ and their half-difference $= z$: the greater side will then be, say, $BC = c+z$ and the lesser one $AC = c-z$. Subtract CD^2 from BC^2 and AC^2 and from the former will result $BD = \sqrt{[c^2+2cz+z^2-a^2]}$, from the latter

$$AD = \sqrt{[c^2-2cz+z^2-a^2]}.$$

Subtract also AB from BD and there will result a second time

$$AD = \sqrt{[c^2+2cz+z^2-a^2]}-2b.$$

When the values of AD are now squared and the terms ordered, there will arise $b^2+cz = b\sqrt{[c^2+2cz+z^2-a^2]}$. And, on again squaring and reducing to order, there will be obtained $(c^2-b^2)z^2 = b^2(c^2-a^2-b^2)$, and so

$$z = b\sqrt{[1-a^2/(c^2-b^2)]}.$$

From this the sides are given.

Problem 6

Given the base AB, the sum of the sides $AC+BC$ and the vertex angle C, to determine the sides.

Let the base $= a$, the half-sum of the sides $= b$ and their semi-difference $= x$, and then the greater side will be $BC = b+x$, the lesser one $AC = b-x$. From either one, A, of the unknown angles let fall the perpendicular AD onto the opposite side BC, and because the angle C is given there will be given the ratio of AC to CD, say d to e; whence $CD = e(b-x)/d$. Also, by *Elements*, II, 11 $CD = (AC^2-AB^2+BC^2)/2BC$, that is, $(2b^2+2x^2-a^2)/2(b+x)$; and in consequence there is had an equation between the values of CD. This, when reduced, becomes $x = \sqrt{[(a^2d+2b^2e-2b^2d)/2(d+e)]}$; and from this the sides are given.

Were the angles at the base required, the deduction would be neater. Specifically, let EC be drawn bisecting the given angle and meeting the base in E; there will then be $AB:AC+BC(=AE:AC) = \sin \widehat{ACE}:\sin \widehat{AEC}$, and if from the angle $\widehat{AEC}$ and its supplement $\widehat{BEC}$ there be taken the angle C there will remain the angles $\widehat{ABC}$ and $\widehat{BAC}$.

Prob 7.

Datis Parallelogrammi cujuscunqȝ lateribus AB, BD, DC & AC, & una linea diagonali BC, invenire alteram diagonalem AD.

Sit *E* concursus diagonalium & ad diagonalem *BC* demitte normalem *AF*, et per 13. 2. Elem. erit $\dfrac{AC^q - AB^q + BC^q}{2BC} = CF$, atqȝ etiam

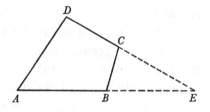

$\dfrac{AC^q - AE^q + EC^q}{2EC} = CF.$ Quare cùm sit $EC = \tfrac{1}{2}BC$, & $AE = \tfrac{1}{2}AD$, erit

$$\frac{AC^q - AB^q + BC^q}{2BC} = \frac{AC^q - \tfrac{1}{4}AD^q + \tfrac{1}{4}BC^q}{BC},$$

et facta reductione $AD = \sqrt{2AC^q + 2AB^q - BC^q}.$[223]

Unde obiter in quolibet parallelogrammo, summa quadratorum laterum æquatur summæ quadratorum diagonalium.[224]

Prob 8.

Datis Trapezij ABCD angulis perimetro et area, determinare latera.

Latera duo quælibet *AB* ac *DC* produc donec concurrant in *E*, sitqȝ $AB = x$ & $BC = y$ & propter angulos omnes datos dantur rationes *BC* ad *CE* et *BE* quas pone *d* ad *e* & *f* et erit $CE = \dfrac{ey}{d}$

& $BE = \dfrac{fy}{d}$ adeoqȝ $AE\| = x + \dfrac{fy}{d}.$ Dantur etiam rationes *AE* ad *AD* ac *DE* quas pone *g* & *h* ad *d*, et erit $AD = \dfrac{dx + fy}{g}$ & $ED = \dfrac{dx + fy}{h}$, adeoqȝ $CD = \dfrac{dx + fy}{h} - \dfrac{ey}{d}$, & summa omnium laterum $x + y + \dfrac{dx + fy}{g} + \dfrac{dx + fy}{h} - \dfrac{ey}{d}$; quæ, cùm detur, esto *a*, & abbrevientur etiam termini scribendo $\dfrac{p}{r}$ pro dato $1 + \dfrac{d}{g} + \dfrac{d}{h}$, & $\dfrac{q}{r}$ pro dato $1 + \dfrac{f}{g} + \dfrac{f}{h} - \dfrac{e}{d}$, & habebitur æquatio $\dfrac{px + qy}{r} = a.$

---

(223) Whence $AD^2 + BC^2 = (2AB^2 + 2AC^2$ or) $AB^2 + BD^2 + DC^2 + AC^2$, Newton's immediately following corollary.

(224) The modern reader, familiar with the theorem relating the length of a median of a triangle to that of its sides, may wonder why Newton insists on so apparently trivial a corollary. However, the median theorem was not at all well known in the seventeenth century, and in

Problem 7

In any parallelogram, given the sides AB, BD, DC and AC along with one diagonal BC, to find the other diagonal AD.

Let E be the meeting point of the diagonals, and onto the diagonal BC let fall the normal AF; then by *Elements*, II, 13 there will be

$$CF = (AC^2 - AB^2 + BC^2)/2BC \quad \text{and also} \quad CF = (AC^2 - AE^2 + EC^2)/2EC.$$

Consequently, since $EC = \tfrac{1}{2}BC$ and $AE = \tfrac{1}{2}AD$, there will be

$$(AC^2 - AB^2 + BC^2)/2BC = (AC^2 - \tfrac{1}{4}AD^2 + \tfrac{1}{4}BC^2)/BC$$

and so, when reduction is made, $AD = \sqrt{[2AC^2 + 2AB^2 - BC^2]}$.[223]

Hence note, incidentally, that in any parallelogram the sum of the squares of the sides is equal to the sum of the squares of the diagonals.[224]

Problem 8

Given the angles, perimeter and area of the quadrilateral ABCD, to determine its sides.

Extend any two sides AB and DC till they meet in E, and let $AB = x$ and $BC = y$; and, because all the angles are given, the ratios of BC to CE and BE are given: put these as d to e and f, and there will be $CE = (e/d)\,y$ and $BE = (f/d)\,y$, so that $AE = x + fy/d$. The ratios of AE to AD and DE are also given: put these to be g and h to d, and then $AD = (dx + fy)/g$ and $ED = (dx + fy)/h$, so that $CD = (dx + fy)/h - ey/d$, and the total sum of the sides is

$$x + y + (dx + fy)/g + (dx + fy)/h - ey/d.$$

Since this is given, let it be a; and its terms may also be shortened by writing p/r in place of the given quantity $1 + d/g + d/h$ and q/r in place of the given

$$1 + f/g + f/h - e/d.$$

There will then be had the equation $(px + qy)/r = a$.

---

particular is not listed in the compendia of elementary geometry (notably Barrow's 1655 *Euclid* and Oughtred's *Clavis Mathematicæ*; see I: 12, 22–3) with which Newton was acquainted —indeed, its early history is more than a little obscure. (It is first known in an extant text when cited by Pappus (from Euclid?) in his *Mathematical Collection*, VII, 122, as his first lemma for the third locus of Book 2 of Euclid's lost treatise on *Plane Loci*. Tropfke, in his *Geschichte der Elementar-Mathematik*, 4 (Berlin/Leipzig, ₂1923): 151, records that Christopher Clavius included it, with a citation of Pappus, in his edition of Euclid (Rome, 1574) as a gloss on *Elements*, II, 12/13, but can give no seventeenth-century reference other than its repeat in Clavius' *Opera*, 1 (Mainz, 1612): 101.) A geometrical proof of the present result, essentially a transliteration of the above algebraic one, is given—without back reference by Newton—in Problem 44 below; see note (370).

Adhæc propter datos omnes angulos datur ratio BC^q ad triangulum BCE[225] quam pone m ad n et erit triang. $BCE = \dfrac{n}{m}yy$. Datur etiam ratio AE^q ad triangulū ADE[225] quam pone m ad d, et erit triang. $ADE = \dfrac{ddxx + 2dfxy + ffyy}{dm}$. Quare cùm area AC, quæ est horum triangulorum differentia, detur, esto bb, & erit $\dfrac{ddxx + 2dfxy + ffyy - dnyy}{dm} = bb$. Atꝗ ita habentur duæ æquationes ex quarum reductione omnia determinantur. Nempe superior æquatio dat $\dfrac{ra - qy}{p} = x$, & scribendo $\dfrac{ra - qy}{p}$ pro x in inferiori, provenit

$$\frac{drraa - 2dqray + dqqyy}{ppm} + \frac{2afry - 2fqry}{pm} + \frac{ffyy - dnyy}{dm} = bb.$$

Et abbreviatis terminis scribendo s pro dato $\dfrac{dqq}{pp} - \dfrac{2fq}{p} + \dfrac{ff}{d} - n$, & st pro dato $-\dfrac{adqr}{pp} + \dfrac{afr}{p}$, ac stv pro dato $bbm - \dfrac{drraa}{pp}$, oritur $yy = 2ty + tv$ seu $y = t + \sqrt{tt + tv}$.[226]

[1676]
Lect 5

Prob: 9.[227]

Piscinam ABCD perambulatorio ABCDEFGH datæ areæ & ejusdem ubiꝗ latitudinis circundare.

Esto perambulatorij latitudo x et ejus area aa. Et a punctis A, B, C, D ad lineas EF, FG, GH & HE demissis perpendicularibus AK, BL, BM, CN, CO, DP, DQ, AI, perambulatorium dividetur in quatuor trapezia IK, LM, NO, PQ & in quatuor parallelogramma AL, BN, CP, DI, latitudinis x, & ejusdem longitudinis cum lateribus dati trapezij. Sit ergo summa laterum $(AB + BC + CD + DA) = b$ & erit summa parallelogrammorum $= bx$.

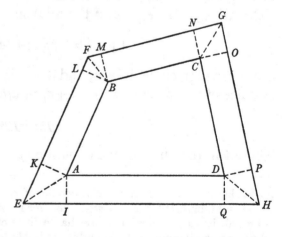

Porrò ductis AE, BF, CG, DH; cum sit $AI = AK$ erit

ang $AEI = $ ang. $AEK = \frac{1}{2}IEK$ sive $\frac{1}{2}DAB$.

Datur ergo ang. AEI et proinde ratio ipsius AI ad IE quam pone d ad e et erit $IE = \dfrac{ex}{d}$. Hanc duc in $\frac{1}{2}AI$ sive $\frac{1}{2}x$ et fiet area trianguli $AEI = \dfrac{exx}{2d}$. Sed

Further, because all angles are given, there is given the ratio of BC^2 to the triangle BCE:[225] take this to be m to n, and then triangle $BCE = (n/m)\,y^2$. There is also given the ratio of AE^2 to the triangle ADE:[225] take this to be m to d, and then triangle $ADE = (d^2x^2 + 2dfxy + f^2y^2)/dm$. Consequently, since the area $A[B]C[D]$—the difference of these triangles—is given, let it be b^2, and there will be $(d^2x^2 + 2dfxy + f^2y^2 - dny^2)/dm = b^2$. And in this way two equations are obtained, and from their reduction everything is determined. Namely, the former equation yields $(ra - qy)/p = x$ and on replacing x by $(ra - qy)/p$ in the latter there ensues

$$d(a^2r^2 - 2arqy + q^2y^2)/p^2m + 2fr(a - q)\,y/pm + (f^2 - dn)\,y^2/dm = b^2.$$

And, when the terms are shortened by writing s in place of the given quantity $dq^2/p^2 - 2fq/p + f^2/d - n$, st in place of the given $-adqr/p^2 + afr/p$ and stv in place of the given $b^2m - a^2dr^2/p^2$, there arises $y^2 = 2ty + tv$, that is, $y = t + \sqrt{[t^2 + tv]}$.[226]

Problem 9[227]

To surround the fish-pond ABCD with the promenade ABCDEFGH of given area and everywhere of the same width.

Let the width of the walk be x and its area a^2. And from the points A, B, C, D to the lines EF, FG, GH and HE let fall the perpendiculars AK, BL, BM, CN, CO, DP, DQ, AI, so dividing the walk into the four quadrilaterals IK, LM, NO, PQ and the four rectangles AL, BN, CP, DI of width x and of the same length as the sides of the given quadrilateral [$ABCD$]. Let, therefore, the sum of its sides $(AB + BC + CD + DA) = b$ and then the sum of the rectangles $= bx$.

Moreover, when AE, BF, CG, DH are drawn, since $AI = AK$ there will be $\widehat{AEI} = \widehat{AEK} = \tfrac{1}{2}\widehat{IEK}$, that is, $\tfrac{1}{2}\widehat{DAB}$. Angle $\widehat{AEI}$ is therefore given and so also, in consequence, the ratio of AI to IE: put this to be d to e, and then $IE = (e/d)\,x$. Multiply this by $\tfrac{1}{2}AI$ (or $\tfrac{1}{2}x$) and the area of the triangle will come to be

---

(225) That is, $2BC:CE\sin\hat{C}$ and $2AE:DE\sin\hat{E}$ respectively.

(226) Since BC cannot be negative, Newton has tacitly discarded the second root of the preceding quadratic in y.

(227) This reworking of Schooten's 'fish-pond' problem is adapted from the equivalent generalization earlier introduced by Newton into his Kinckhuysen 'Observationes' (see II: 432–4).

propter æquales angulos & latera, triangula AEI et AEK sunt æqualia, adeoqʒ

‖[72] trapezium $IK(=2\,\text{triang. }AEI)=\dfrac{exx}{d}$. Simili modo ponendo ‖ $BL.LF::d.f$, &

$CN.NG::d.g$, & $DP.DH::d.h$ (nam illæ etiam rationes dantur ex datis

angulis B, C, ac D) habebitur trapezium $LM=\dfrac{fxx}{d}$, $NO=\dfrac{gxx}{d}$, & $PQ=\dfrac{hxx}{d}$.

Quamobrem $\dfrac{exx}{d}+\dfrac{fxx}{d}+\dfrac{gxx}{d}+\dfrac{hxx}{d}$ sive $\dfrac{pxx}{d}$ scribendo p pro $e+f+g+h$, erit

æquale trapezijs quatuor $IK+LM+NO+PQ$ & proinde $\dfrac{pxx}{d}+bx$ æquabitur

toti perambulatorio[228] aa. Quæ æquatio dividendo omnes terminos per $\dfrac{p}{d}$

& extrahendo radicem ejus evadet $x=\dfrac{-db+\sqrt{bbdd+4aapd}}{2p}$.[229] Latitudine

Perambulatorij sic inventa facile est ipsum describere.

Prob: 10.

A dato puncto C rectam lineam CF ducere quæ cum alijs duabus positione datis rectis AE et AF triangulum datæ magnitudinis AEF comprehendet.

Age CD parallelam AE, et CB ac EG
perpendiculares AF sitqʒ $AD=a$, $CB=b$,
$AF=x$ & trianguli AEF area cc, & propter proportionales

     $DF.AF(::DC.AE)::CB.EG$,

hoc est $a+x.x::b.\dfrac{bx}{a+x}=EG$. Hanc duc

in $\tfrac{1}{2}AF$ & emerget $\dfrac{bxx}{2a+2x}$ quantitas areæ AEF quæ proinde æquatur cc. Atqʒ

adeo æquatione ordinata est $xx=\dfrac{2ccx+2cca}{b}$ seu $x=\dfrac{cc+\sqrt{c^4+2ccab}}{b}$.[230]

Nihil secus recta per datum punctum ducitur quæ triangulum vel trapezium
quodvis in data ratione secabit.[231]

[1676]
Lect 6
Prob. 11.

Punctum C in data recta linea DF determinare a quo ad alia duo positione data puncta A Vide
Pro[b] 39[232]
et B ductæ rectæ AC & BC datam habeant differentiam.

---

(228) ‘zonæ’ (belt) was first written, as in Newton's earlier text at this point (II: 434).
(229) The suppressed negative root $x=\tfrac{1}{2}(-bd-\sqrt{[b^2d^2+4a^2pd]})/p$ is, of course, the width
of a ‘promenade’ within[!] the perimeter of the fish-pond.

$= \frac{1}{2}(e/d) x^2$. But because their angles and sides are equal the triangles AEI and AEK are equal, so that quadrilateral $IK (= 2 \times \text{triangle } AEI) = ex^2/d$. In a similar manner, by putting $BL:LF = d:f$, $CN:NG = d:g$ and $DP:DH = d:h$ (for these ratios also are given from the given angles B, C and D) there will be had quadrilateral $LM = fx^2/d$, quadrilateral $NO = gx^2/d$ and quadrilateral $PQ = hx^2/d$. As a result $(e+f+g+h) x^2/d$—or px^2/d on writing p in place of $e+f+g+h$—will be equal to the four quadrilaterals $IK+LM+NO+PQ$, and consequently $(p/d) x^2 + bx$ will be equal to the total promenade[228] a^2. On dividing all its terms by p/d and then extracting its root, this equation will prove to be $x = \frac{1}{2}(-bd + \sqrt{[b^2 d^2 + 4a^2 pd]})/p$.[229] Once the width of the walk is found in this way, it is easy to mark it out.

Problem 10

From the given point C to draw a straight line CF which shall, with two other straight lines AE and AF given in position, embrace a triangle AEF of given size.

Draw CD parallel to AE, and CB, EG perpendicular to AF, and let $AD = a$, $CB = b$, $AF = x$ and the area of the triangle AEF be c^2. Then, because of the proportionals $DF:AF (= DC:AE) = CB:EG$ or $(a+x):x = b:EG$, $EG = bx/(a+x)$. Multiply this by $\frac{1}{2}AF$ and there will emerge $\frac{1}{2}bx^2/(a+x)$ as the magnitude of the area AEF. This is consequently equal to c^2, and accordingly, when the equation is ordered, it is $x^2 = 2(c^2 x + c^2 a)/b$ and so

$$x = (c^2 + \sqrt{[c^4 + 2abc^2]})/b.^{(230)}$$

In exactly the same way a straight line is drawn through a given point which shall cut any triangle or quadrilateral in a given ratio.[231]

Problem 11

To determine the point C in the given straight line DF, such that two straight lines AC and BC drawn from it to two other points A and B given in position shall have a given difference. See Problem 39.[232]

---

(230) Read ' $x = \dfrac{cc \pm \sqrt{c^4 + 2ccab}}{b}$ ', so correctly yielding the two values of AF (and hence positions of the transversal CEF) possible. Compare the figure alongside.

(231) Namely, where EAF is an appropriate corner of the triangle or quadrilateral, we need to cut off a triangle EAF whose area is in the desired ratio to that of the total figure.

(232) On Newton's pages 101–2. The inference is that the solution of the present problem can be made to yield the construction of a circle passing

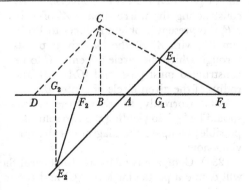

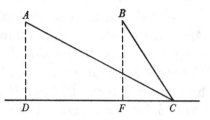

A datis punctis ad datam rectam demitte perpendiculares AD et BF, et dic $AD=a$, $BF=b$, $DF=c$ & $DC=x$, & erit $AC=\sqrt{aa+xx}$, $FC=x-c$, & $BC=\sqrt{bb+xx-2cx+cc}$. Sit jam data harum differentia d, existente AC majori quam BC erit $\sqrt{aa+xx}-d=\sqrt{bb+xx-2cx+cc}$. et quadratis partibus $aa+xx+dd-2d\sqrt{aa+xx}=bb+xx-2cx+cc$. factaꝗ reductione et abbreviandi causa pro datis $aa+dd-bb-cc$ scripto $2ee$, emerget $ee+cx=d\sqrt{aa+xx}$. Iterumꝗ quadratis partibus $e^4+2ceex+ccxx=ddaa+ddxx$. et æquatione reducta $xx=\dfrac{2eecx+e^4-aadd}{dd-cc}$, seu $x=\dfrac{eec+\sqrt{e^4dd-aad^4+aaddcc}}{dd-cc}$. [233]

|| [73] Haud secus problema resolvitur si linearum AC et BC summa vel quadratorum summa aut differentia, vel proportio vel || rectangulum vel angulus ab ipsis comprehensus detur: Vel etiam si vice rectæ DC, circumferentia circuli, aut alia quævis curva linea adhibeatur, modò calculus (in hoc ultimo præsertim casu) referatur ad lineam conjungentem puncta A et B. [234]

Prob. 12. [235]

Punctum Z determinare a quo ad quatuor positione datas rectas lineas FA, EB, FC, GD, si aliæ quatuor lineæ ZA, ZB, ZC, & ZD in datis angulis ducantur, duarum e ductis ZA et ZB summa [236] *& aliarum duarum ZC & ZD rectangulum* [236] *detur.*

---

through two given points to touch a second given circle: in fact, if A, A' are the two given points, we need only locate the centre C of the first circle in the perpendicular bisector DF of AA' such that $|CA-CB|$ is equal to the radius of the given circle of centre B. (There will of course be two such points C, corresponding to $C_1A-C_1B=E_1B$ and $C_2B-C_2A=E_2B$.) Conversely, by constructing the mirror-image A' of A in DF, the present problem reduces to Problem 39, with A, A' the two given points through which the circle of centre C to be constructed must pass, and $|CA-CB|$ the radius of the given circle of centre B.

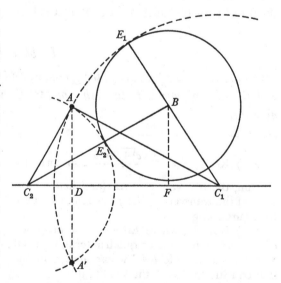

(233) Correctly, the sign of the radical should be '$\pm$', so yielding the two solutions possible; compare the diagram of the previous note.

(234) Compare IV: 246–54. In general the given relationship between the bipolars AC, BC will define a particular locus (C) and the meet of this locus with the given line or curve DC

From the given points to the given straight line let fall the perpendiculars AD and BF, and call $AD = a$, $BF = b$, $DF = c$ and $DC = x$; then

$$AC = \sqrt{[a^2 + x^2]}, \quad FC = x - c \quad \text{and} \quad BC = \sqrt{[b^2 + x^2 - 2cx + c^2]}.$$

Let now the given difference of these lines be d, and then—where AC is greater than BC—there will be $\sqrt{[a^2 + x^2]} - d = \sqrt{[b^2 + x^2 - 2cx + c^2]}$, and, with the sides squared, $a^2 + x^2 + d^2 - 2d\sqrt{[a^2 + x^2]} = b^2 + x^2 - 2cx + c^2$. When reduction is made and $2e^2$ written, for brevity's sake, in place of the given quantities

$$a^2 + d^2 - b^2 - c^2,$$

there will emerge $e^2 + cx = d\sqrt{[a^2 + x^2]}$ and so, with the sides again squared, $e^4 + 2ce^2x + c^2x^2 = d^2a^2 + d^2x^2$; that is, when the equation is reduced,

$$x^2 = (2ce^2x + e^4 - a^2d^2)/(d^2 - c^2)$$

or $x = (ce^2 + \sqrt{[d^2e^4 - a^2d^4 + a^2c^2d^2]})/(d^2 - c^2)$.[233]

The problem is resolved in much the same way if of the lines AC and BC the sum, or sum or difference of the squares, or ratio or product or included angle be given; or, indeed, if instead of the straight line DC a circle's circumference or any other curved line be employed, provided that the calculation (in this last case especially) be related to the line joining the points A and B.[234]

Problem 12[235]

To determine a point Z such that, if from it to four straight lines FA, EB, FC, GD given in position there be drawn four other lines ZA, ZB, ZC and ZD at given angles, the [product] of two of those drawn, ZA and ZB, and the [sum] of the two others, ZC and ZD, are given.

---

will yield corresponding points C which solve the problem; choice of a rectangular Cartesian system of coordinates in which AB is an axis and its mid-point the origin will evidently much simplify the analytical determination of the locus (C)—which will be a hyperbola when (as in the present problem) $AC - BC$ is given, an ellipse for $AC + BC$ given, a Fermatian circle for $AC^2 + BC^2$ given but a straight line when $AC^2 - BC^2$ is given, an Apollonian circle when AC/BC is given, a Cassini oval for the product $AC \times BC$ given, and a Euclidean circle for $\widehat{ACB}$ given (see IV: 248, note (62)).

(235) A determinate variant on Descartes' version of the classical *locus ad quatuor lineas* (see IV: 219–20) in which the locus condition $ZA \times ZB/ZC \times ZD = $ constant is replaced by the particular conditions $ZA \times ZB = g^2$ and $ZC + ZD = f$, so yielding but two points Z (the meets of the corresponding hyperbola and straight line which these revised Newtonian conditions now define). It is interesting to notice that in the worked 'Prob. 11' of his revised 'Arithmeticæ Universalis Liber Primus' (see **2**, [1]: note (82)) Newton will return to an instance of the classical curve, there seeking the meets of a 3-line locus with a straight line.

(236) In his *editio princeps* in 1707 Whiston has correctly interchanged these two words to accord with Newton's following text.

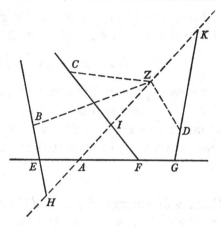

E lineis elige aliquam positione datam *FA* ut et positione non datam *ZA*, quæ ad illam ducitur, ex quarum longitudinibus punctum *Z* determinetur,[237] & cæteras positione datas lineas produc donec his, si opus est etiam productis, occurrant, ut vides. Dictisꝗ *EA*=*x* & *AZ*=*y* propter angulos trianguli *AEH* datos dabitur ratio

AE ad *AH* quam pone *p* ad *q* et erit $AH = \dfrac{qx}{p}$.

Adde *AZ*, fitꝗ $ZH = y + \dfrac{qx}{p}$. Et inde cum propter datos angulos trianguli *HZB* detur ratio *HZ* ad *BZ* si ea ponatur *n* ad *p* habebitur $ZB = \dfrac{py + qx}{n}$.

Præterea si data *EF* dicatur *a* erit *AF*=*a*−*x*, indeꝗ, si propter datos angulos trianguli *AFI* statuatur *AF* ad *AI* in ratione *p* ad *r* evadet $AI = \dfrac{ra - rx}{p}$. Hanc aufer ab *AZ* et restabit $IZ = y - \dfrac{ra - rx}{p}$. Et propter datos angulos trianguli *ICZ* si ponatur *IZ* ad *ZC* in ratione *m* ad *p* evadet $ZC = \dfrac{py - ra + rx}{m}$.

Ad eundem modum si ponatur *EG*=*b*. *AG*.*AK*::*p*.*s*, & *ZK*.*ZD*::*l*.*p*.[238] obtinebitur $ZD = \dfrac{sb - sx - py}{l}$.

Jam ex statu quæstionis si duarum *ZC* et *ZD* summa $\dfrac{py - ra + rx}{m} + \dfrac{sb - sx - py}{l}$ ponatur æqualis dato alicui *f* & aliarum duarum rectangulum $\dfrac{pyy + qxy}{n}$ æquale *gg* habebuntur duæ æquationes pro determinandis *x* et *y*. Per posteriorem fit

‖[74] $x = \dfrac{ngg - pyy}{qy}$ & hunc ipsius *x* valorem scribendo pro ‖ eo in priori æquatione,

evadet $\dfrac{py - ra}{m} + \dfrac{rngg - rpyy}{mqy} + \dfrac{sb - py}{[l]} - \dfrac{sngg - spyy}{[l]qy} = f$. Et reducendo

$$yy = \frac{a[l]qry - bmqsy + f[l]mqy + ggmns - gg[l]nr}{[l]pq - [l]pr - mpq + mps}.$$

Et abbreviandi causa scripto 2*h* pro $\dfrac{a[l]qr - bmqs + f[l]mq}{[l]pq - [l]pr - mpq + mps}$, & *kk* pro

---

(237) That is, by the pair of Cartesian coordinates *EA* and *AZ*, departing from origin *E* and including the fixed angle $\widehat{EAZ}$.

From the lines select one, *FA*, given in position and that, *ZA*, not given in position, which is drawn to it, in order to determine the point *Z* from their lengths,[237] and extend the remaining lines given in position till they meet these—also produced, if need be—as you see. Then, on calling $EA = x$ and $AZ = y$, because the angles of the triangle *AEH* are given, there will be given the ratio of *AE* to *AH*. Put this to be p to q, and there will be $AH = (q/p)\,x$; then add *AZ* and there comes $ZH = y + qx/p$. Since now (because the angles in the triangle *HZB* are given) the ratio of *HZ* to *ZB* is given, if this be set as n to p, there will thence be obtained $ZB = (py + qx)/n$.

Further, if the given line *EF* be called a, there will be $AF = a - x$ and therefrom, if (because the angles of the triangle *AFI* are given) *AF* be settled to be to *AI* in the ratio p to r, there will come $AI = r(a - x)/p$. Take this away from *AZ* and there will remain $IZ = y - (ra - rx)/p$. And because the angles in the triangle *ICZ* are given, if *IZ* be set to be to *ZC* in the ratio m to p, there will prove to be $ZC = (py - ra + rx)/m$.

In much the same way, if there be put $EG = b$, $AG:AK = p:s$ and

$$ZK:ZD = l:p,\text{[238]}$$

there will be obtained $ZD = (sb - sx - py)/l$.

Now if, in agreement with the terms of the question, the sum

$$(py - ra + sx)/m + (sb - sx - py)/l$$

of the two lines *ZC* and *ZD* be set equal to some given quantity f, and the product $(py^2 + qxy)/n$ of the two others equal to g^2, there will be had two equations for determining x and y. By the latter there comes $x = (ng^2 - py^2)/qy$, and on writing this value of x in its place in the former equation there will result

$$(py - ra)/m + r(ng^2 - py^2)/mqy + (sb - py)/l - s(ng^2 - py^2)/lqy = f,$$

and after reduction

$$y^2 = ((alr - bms + flm)\,qy + (ms - lr)\,g^2 n)/p(lq - lr - mq + ms).$$

And when, to shorten this, $2h$ is written for

$$(alr - bms + flm)\,q/p(lq - lr - mq + ms)$$

---

(238) Unnecessarily 'corrected' by Whiston in 1707 to read '$AG.AK::l.s$, & $ZK.ZD::p.l$', an interchange of l and p which further required changing the following equation to

$$`ZD = \frac{sb - sx - ly}{p}`.$$

The alteration was retained by Newton in his 1722 revise, but in our edited text we have preferred to make the few changes of 'p' into 'l' which retention of the manuscript reading necessitates in the sequel.

$$\frac{ggmns - gg[l]nr}{[l]pq - [l]pr - mpq + mps} \text{ fiet } yy = 2hy + kk, \text{ sive } y = h \pm \sqrt{hh + kk}. \text{ Cujus æquationis}$$

ope cum y innotescit, æquatio $yy + \dfrac{qxy}{p} = \left[\dfrac{ggn}{p}\right]^{(239)}$ dabit x. Quod sufficit ad determinandum punctum Z.

Ad eundem fere modum punctum determinatur a quo ad plures vel pauciores positione datas rectas totidem aliæ rectæ ducantur ea lege ut aliquarum summa vel differentia vel contentum detur aut æquetur cæterarum summæ vel differentiæ vel contento vel ut alias quaslibet habeant assignatas conditiones.[240]

[1676]
Lect 7

Prob: 13.[241]

Angulum rectum EAF data recta EF subtendere quæ transibit per datum punctum C a lineis rectum angulum comprehendentibus æquidistans.

Quadratum *ABCD* compleatur et linea *EF* bisecetur in *G*. Tum dic *CB* vel *CD* esse a, *EG* vel *FG* esse b, & *CG* esse x; eritꝗ $CE = x - b$ & $CF = x + b$. Dein cum $CF^q - BC^q = BF^q$, erit $BF = \sqrt{xx + 2bx + bb - aa}$. Deniꝗ propter similia triangula *CDE*, *FBC*, est $CE \cdot CD :: CF \cdot BF$, sive

$$x - b \cdot a :: x + b \cdot \sqrt{xx + 2bx + bb - aa}.$$

Unde $ax + ab = \overline{x - b}\sqrt{xx + 2bx + bb - aa}$. Cujus æquationis utraꝗ parte quadrata, & prodeuntibus terminis in ordinem redactis, prodit $x^4 = \begin{array}{l} 2aa \\ +2bb \end{array} xx \begin{array}{l} +2aabb \\ -b^4 \end{array}$. Et extracta radice sicut fit in æquationibus quadraticis, prodit $xx = aa + bb + \sqrt{a^4 + 4aabb}$. Adeoꝗ $x = \sqrt{aa + bb + \sqrt{a^4 + 4aabb}}$.[242] *CG* sic inventa dat *CE* vel *CF*, quæ determinando punctum *E* vel *F* problemati satisfacit.

Idem aliter.

Sit $CE = x$, $CD = a$, et $EF = b$, eritꝗ $CF = x + b$ et $BF = \sqrt{xx + 2bx + bb - aa}$. Et

||[75] proinde cum sit $CE \cdot CD :: \| CF \cdot BF$, sive $x \cdot a :: x + b \cdot \sqrt{xx + 2bx + bb - aa}$, erit

---

(239) The manuscript illogically reads 'x'. Whiston has cancelled Newton's whole equation and inserted in the margin alongside '$\dfrac{ngg - pyy}{qy} = x$', a correction which passed into his 1707 *editio princeps* and was subsequently sanctioned by Newton in 1722.

(240) Since each of *ZA*, *ZB*, *ZC* and *ZD* are linear functions of x and y, the degree $(\alpha + \beta + \gamma + \delta)$ of a given algebraic defining *relatio* $\phi(ZA, ZB, ZC, ZD) = 0$ will be that of the highest term(s) $k \cdot ZA^\alpha \cdot ZB^\beta \cdot ZC^\gamma \cdot ZD^\delta$ contained in it. Thus the classical 4-line locus defined (see note (235)) by $ZA \times ZB = \lambda \cdot ZC \times ZD$ will be of second degree, and hence a conic.

and k^2 for $(ms-lr)\,g^2n/p(lq-lr-mq+ms)$, there will come $y^2 = 2hy+k^2$, that is, $y = h\pm\sqrt{[h^2+k^2]}$. When, with the aid of this equation y is ascertained, the equation $y^2+(q/p)\,xy = g^2n/p^{(239)}$ will yield x. This suffices to determine the point Z.

In almost the same manner a point is determined from which to more or fewer straight lines given in position an equal number of other straight lines may be drawn under the restriction that the sum, difference or product of some of them be given, or be equal to the sum, difference or product of the remainder, or fulfil any other assigned conditions.[240]

Problem 13[241]

To subtend the right angle $\widehat{EAF}$ by the given straight line EF which shall pass through the given point C equidistant from the lines comprising the right angle.

Let the square $ABCD$ be completed and the line EF bisected in G. Then call CB or CD a, EG or FG b, and CG x; then will there be $CE = x-b$ and $CF = x+b$. Next, since $CF^2 - BC^2 = BF^2$, there will be $BF = \sqrt{[x^2+2bx+b^2-a^2]}$. Finally, because of the similar triangles CDE, FBC, there is $CE:CD = CF:BF$ or $(x-b):a = (x+b):\sqrt{[x^2+2bx+b^2-a^2]}$. Hence

$$ax+ab = (x-b)\sqrt{[x^2+2bx+b^2-a^2]}.$$

When each side of this equation is squared and the resulting terms reduced to order, there results $x^4 = 2(a^2+b^2)\,x^2+2a^2b^2-b^4$. And when the root is extracted just as it is done in quadratic equations, there ensues

$$x^2 = a^2+b^2+\sqrt{[a^4+4a^2b^2]}.$$

Consequently $x = \sqrt{(a^2+b^2+\sqrt{[a^4+4a^2b^2]})}$.[242] When CG has been thus found it gives CE or CF, and these, by determining the point E or F, satisfy the problem.

The same another way

Let $CE = x$, $CD = a$ and $EF = b$, and there will be $CF = x+b$ and

$$BF = \sqrt{[x^2+2bx+b^2-a^2]}.$$

Consequently, since $CE:CD = CF:BF$ or $x:a = (x+b):\sqrt{[x^2+2bx+b^2-a^2]}$,

---

(241) Repeated without essential alteration from Newton's 1670 'Observationes' on Kinckhuysen's *Algebra* (see II: 434–8). Observe, however, that the lines CA and GK perpendicular to it are now explicitly shown in the present figure.

(242) More accurately '$x = \pm\sqrt{aa+bb\pm\sqrt{a^4+4aabb}}$', yielding four or two real solutions according as $(a^2+b^2)^2 \gtrless a^4+4a^2b^2$, that is, as $b^2 \gtrless 2a^2$; compare II: 436, note (129).

$ax + ab = x\sqrt{xx + 2bx + bb - aa}$. Hujus æquationis partibus quadratis & terminis in ordinem redactis, prodibit $x^4 + 2bx^3 \begin{smallmatrix} +bb \\ -2aa \end{smallmatrix} xx - 2aabx - aabb = 0$, æquatio biquadratica cujus radicis investigatio difficilior est quam in priori casu. Sic autem investigari potest. Pone $x^4 + 2bx^3 \begin{smallmatrix} +bb \\ -2aa \end{smallmatrix} xx - 2aabx + a^4 = aabb + a^4$, et extracta utrobiꝗ radice erit $xx + bx - aa = \pm a\sqrt{aa + bb}$.[243]

Ex his occasionem nactus sum tradendi regulam de electione terminorum ad ineundum calculum. Scilicet cùm duorum terminorum talis obvenit affinitas sive similitudo relationis ad cæteros terminos quæstionis, ut oporteret æquationes per omnia similes ex utrovis adhibito produci, aut ambos si simul adhiberentur easdem in æquatione finali dimensiones & eandem omninò formam (signis forte + & − exceptis) habituros esse; (id quod facilè prospicitur:) tunc neutrum adhibere convenit sed eorum vice tertium quemvis eligere qui similem utriꝗ relationem gerit, puta semisummam vel semidifferentiam, vel medium proportionale forsan, aut quamvis aliam quantitatem utriꝗ indifferenter & sine compare relatam. Sic in præcedente problemate cùm viderem lineam *EF* pariter ad utramꝗ *AB* et *AD* referri (quod patebit si ducas itidem *EF* in angulo *BAH*,) atꝗ adeò nulla ratione suaderi possem cur *ED* potius quàm *BF* vel *CE* potius quàm *CF* pro quærenda quantitate adhiberetur: vice punctorū [*E*][244] et *F* unde hæc ambiguitas proficiscitur sumpsi (in solutione priori) intermedium *G* quod parem relationem ad utramꝗ linearum *AB* et *AD* observat. Deinde ab hoc *G* non demisi perpendiculum ad *AF* pro quærenda quantitate quia potui eadem ratione demisisse ad *AD*. Et eapropter in neutrum *CB* vel *CD* demisi, sed institui *CG* quærendum esse quod nullum admittit compar: & sic æquationem biquadraticam obtinui sine terminis imparibus.[245]

---

(243) These two last sentences (replacing a lengthier equivalent on II: 436) are a late addition to the text.

(244) The manuscript (and Whiston's 1707 *editio princeps*) reads '*C*', a slip of the pen caught by Newton in his 1722 revise.

(245) Newton elaborates, as earlier in his 1670 Kinckhuysen 'Observationes' (II: 436–8), the 'rule of mates' he has already adumbrated in his present text (see note (168)) in working his sixteenth arithmetical problem. A century later Lagrange (in his '[Suite des] Réflexions sur la Résolution algébrique des Équations' (*Nouveaux Mémoires de l'Académie Royale des Sciences et Belles Lettres [de Berlin]. Année MDCCLXXI* (Berlin, 1773: 138–253): §115: 250–3, especially 253) could see in this Newtonian simplification little more than a *Deus ex machina* dependent on a prior knowledge of the result: 'De là on voit qu'on seroit parvenu d'abord à une équation du quatrième degré sans puissances impaires de l'inconnue, si on eût pris pour inconnue la ligne [*CG*]. ...mais on doit avouer, ce me semble, qu'un tel choix de l'inconnue est assés peu naturel, & que ce n'est, pour ainsi dire, qu'après coup qu'on peut le faire; du moins il me paroit que le principe d'où Newton le fait dépendre n'a pas toute l'évidence que l'on est en droit d'exiger dans ces sortes de matières.' Lagrange, intent on establishing general methods of solution of problems to which the answer is not previously known, is right in his criticism:

there will be $ax + ab = x\sqrt{[x^2 + 2bx + b^2 - a^2]}$. When the sides of this equation are squared and its terms reduced to order, there will result

$$x^4 + 2bx^3 + (b^2 - 2a^2)\,x^2 - 2a^2bx - a^2b^2 = 0,$$

a quartic equation the determination of whose root is more difficult than in the preceding case. It can, however, be determined in this way. Set

$$x^4 + 2bx^3 + (b^2 - 2a^2)\,x^2 - 2a^2bx + a^4 = a^2b^2 + a^4$$

and, when the root is extracted on either side there will be

$$x^2 + bx - a^2 = \pm a\sqrt{[a^2 + b^2]}.^{(243)}$$

Let me take this occasion to hand on a rule dealing with the selection of the terms to begin the calculation with. Specifically, when two terms chance to have such an affinity or similarity of relationship to the remaining terms of the question that, were either employed, the equations resulting therefrom would of necessity be similar in every respect or, if both were simultaneously employed, they would necessarily have the same dimensions in the final equation and a wholly identical form—as is easy to anticipate—, then it is convenient to employ neither but rather to choose in their place some third term which bears a similar relationship to each: their half-sum, say, or half-difference, or maybe their mean proportional or any other quantity related in no different manner to each and without a fellow. Thus, in the preceding problem, since I saw that the line *EF* was related equally to each of *AB* and *AD* (as will be evident if you draw *EF* correspondingly in the angle $\widehat{BAH}$) and consequently that no reason could possibly be urged why I should employ *ED* rather than *BF* or *CE* rather than *CF* for ascertaining the quantity, instead of the points $E^{(244)}$ and *F* which give rise to this ambiguity I took (in the first solution) their mid-point *G*, which maintains an equal relationship to each of the lines *AB* and *AD*. Next, I did not drop a perpendicular from this point *G* to *AF* in order to seek the quantity because I could with equal reason have let it fall to *AD*. For the same reason I let it fall neither to *CB* nor to *CD*, but fixed on *CG* as the quantity to be determined because it admits of no fellow; and in this way I obtained a quartic equation lacking its odd terms.$^{(245)}$

the successful application of Newton's simplifying technique does demand a narrow pre-knowledge of the number and type of solutions possible, and this can come only after considerable previous (geometrical or analytical) familiarity with a given problem. On the other hand, at the relatively sophisticated stage of 'composing' the solution for elegant, public presentation (an approach already seen as old-fashioned and without essential point by the school of algebraic analysis which Lagrange typified) the awareness that coupled solutions exist can, by appropriately revised choice of variable in one or other of Newton's suggested ways, considerably simplify and shorten the deduction of the solutions possible. Newton will further elaborate and exemplify his concept of a general 'rule of mates' in a contemporary manuscript sheet which, we judge, was meant to form part of his revised 'Arithmeticæ Universalis Liber Primus' (and is accordingly reproduced as **2**, [2] below).

Potui etiam (animadverso quod punctum G jaceat in peripheria circuli centro A, radio EG descripti) demisisse GK perpendiculum in diagonalem AC & quæsivisse AK vel CK, (quippe quæ similem etiam utriꝗ AB et AD relationem ‖[76] gerunt:) atꝗ ita in æquationem quadraticam $yy=\frac{1}{2}ey+\frac{1}{2}bb$ ‖ incidissem posito $AK=y$,(246) $AC=e$ & $EG=b$. Et AK sic invento erigendum fuisset perpendiculum KG præfato circulo occurrens in G, per quod CF transiret.

Ad hanc regulam animum advertens, in Prob: 5 & 6 ubi trianguli latera germana BC et AC determinandæ erant, quæsivi potius semidifferentiam quam alterutrum eorum. Sed regulæ hujus utilitas e sequenti Problemate magis elucescet.

[1676]
Lect 8

Prob: 14.(247)

Rectam DC datæ longitudinis in datam Conicam sectionem DAC sic inscribere ut ea per punctum G positione datum transeat.

Sit AF axis Curvæ(248) & a punctis D G et C ad hunc demitte normales DH, GE, et CB. Jam ad determinandam positionem rectæ DC puncti D aut C inventio proponi potest: sed cùm hæc sint germana & adeo paria ut ad alterutrum determinandum operatio similis evasura esset, sive quærerem CG, CB, aut AB, sive comparia DG DH aut AH: eapropter de tertio aliquo puncto prospicio quod utrumꝗ D et C similiter respectet, & unà determinet. Et hujusmodi video esse punctum F.

Jam sit $AE=a$, $EG=b$, $DC=c$, $EF=z$, & præterea cùm relatio inter AB et BC habeatur in æquatione quam suppono pro Conica sectione determinanda datam esse, sit $AB=x$ & $BC=y$ & erit $FB=x-a+z$. & propter $GE.EF::CB.FB$ erit iterum $FB=\frac{yz}{b}$. Ergo $x-a+z=\frac{yz}{b}$.

His ita præparatis tolle x per æquationem quæ curvam designat. Quemadmodum si Curva sit Parabola per æquationem $rx=yy$ designata, scribe $\frac{yy}{r}$ pro x ‖[77] & orietur $\frac{yy}{r}-a+z=\frac{yz}{b}$, et extracta radice, $y=\frac{rz}{2b}\pm\sqrt{\frac{rrzz}{4bb}+ar-rz}$. ‖ Unde

---

(246) Presupposing that the positive sense of y is from A in the direction of C.

I could also (observing that the point G lies in the circumference of the circle described on centre A and with radius EG) have let fall the perpendicular GK to the diagonal AC and sought AK or CK (seeing that they too bear a similar relationship to each of AB and AD); and so have fallen upon the quadratic equation $y^2 = \frac{1}{2}ey + \frac{1}{2}b^2$ on putting $AK = y$,[246] $AC = e$ and $EG = b$. Once AK has been found in this way, the perpendicular KG would have to be erected, meeting the above-mentioned circle in G, through which CF was to pass.

Paying heed to this rule, in Problems 5 and 6 where the closely related sides BC and AC of the triangle were to be determined, I sought their half-difference rather than either individually. But the usefulness of this rule will be more transparent from the following problem.

Problem 14[247]

To inscribe the straight line DC of given length in the given conic DAC such that it shall pass through the point G given in position.

Let AF be the axis of the curve[248] and from the points D, G and C let fall to this the normals DH, GE and CB. Now, to determine the position of the straight line DC, finding the points D or C may be proposed; but since these are mates and so closely fellow that to determine either one a similar sequence of operation would ensue, whether I were to seek CG, CB or AB or their fellows DG, DH or AH, I accordingly look out for some third point which shall regard each of D and C similarly and determine them jointly. And I see that F is a point of this sort.

Now let $AE = a$, $EG = b$, $DC = c$, $EF = z$; further, since the relationship between AB and BC is had in an equation which I suppose given for determining the conic section, let $AB = x$ and $BC = y$, and there will be $FB = x - a + z$. Again, because $GR:EF = CB:FB$, it will be $FB = yz/b$. Therefore

$$x - a + z = yz/b.$$

With this preparation so made, remove x by means of the equation which designates the curve. For instance, if the curve be a parabola denoted by the equation $rx = y^2$, write y^2/r in place of x and there will arise $y^2r - a + z = yz/b$ and, when the root is extracted, $y = \frac{1}{2}rz/b \pm \sqrt{[\frac{1}{4}r^2z^2/b^2 + ar - rz]}$. It is conse-

---

(247) A repeat of the first half (II: 438–40) of Newton's seventh geometrical example in his Kinckhuysen 'Observationes', omitting the sketch solution of the equivalent problem in the case of an ellipse or hyperbola, and also the concluding paragraphs (II: 442) 'De electione terminorum ad ineundum calculum' and regarding the construction of a tangent to the parabola which results 'DC evanescente'.

(248) As earlier (see II: 438) Newton first began: 'Sit *AF* recta quævis ad quam Curva simplicissimè refertur' (Let *AF* be the straight line to which the curve is most simply referred). He now assumes as evident that the simplest such line is the parabola's main diameter.

patet $\sqrt{\dfrac{rrzz}{bb}+4ar-4az}$ esse differentiam gemini valoris y, id est linearum $+BC$ & $-DH$, adeoç (demisso CK in CB normali) valere CK. Est autem $FG.GE::DC.CK$, hoc est $\sqrt{bb+zz}\,.\,b::c\,.\,\sqrt{\dfrac{rrzz}{bb}+4ar-4az}$. ducendoç quadrata extremorum et mediorum in invicem et facta ordinando orietur

$$z^4=\dfrac{4bbrz^3\begin{smallmatrix}-4abbr\\-bbrr\end{smallmatrix}zz+4b^4rz\begin{smallmatrix}-4ab^4r\\+b^4cc\end{smallmatrix}}{rr},$$

æquatio quatuor tantū dimensionum, quæ ad octo dimensiones ascendisset si quæsivissem CG vel CB aut AB.[249]

Prob. 15.[250]

Datum angulum per datum numerum multiplicare vel dividere.

In angulo quovis FAG inscribe lineas AB, BC, CD, DE, &c ejusdem cujusvis longitudinis, et erunt triangula ABC, 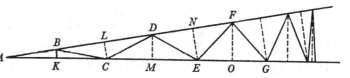 BCD, CDE, DEF, &c isoscelia: adeoç per 32. 1 Elem. erit

$$\text{ang}:CBD=\text{ang}:A+ACB=2\,\text{ang}:A,\ \&$$

$$\text{ang}:DCE=\text{ang}:A+ADC=3\,\text{ang}:A.\ \&$$

$$\text{ang}:EDF=\text{ang}:A+AED=4\,\text{ang}:A,$$

& ang:$FEG=5$ ang:A, & sic deinceps. Positis jam AB, BC, CD &c radijs æqualium circulorum, perpendicula BK, CL, DM &c demissa in AC, BD, CE &c erunt sinus istorum angulorum, & AK, BL, CM, DN &c sinus complementorum ad rectum. Vel positâ AB diametro illæ AK, BL, CM &c erunt chordæ. Sit ergo $AB=2r$ & $AK=x$, dein sic operare.

$AB.AK::AC.AL.$ Et $\left.\begin{matrix}AL-AB\\[2pt]\dfrac{xx}{r}-2r\end{matrix}\right\}=BL$, Duplicatio.

$2r\ .\ x\ ::\ 2x\,.\,\dfrac{xx}{r}\,.$

$AB.AK::AD(2AL-AB).AM.$ Et $\left.\begin{matrix}AM-AC\\[2pt]\dfrac{x^3}{rr}-3x\end{matrix}\right\}=CM$, Triplicatio.

$2r\ .\ x\ ::\ \dfrac{2xx}{r}-2r\ .\ \dfrac{x^3}{rr}-x.$

$AB.AK::AE(2AM-AC).AN\ .$ Et $\left.\begin{matrix}AN-AD\\[2pt]\dfrac{x^4}{r^3}-\dfrac{4xx}{r}+2r\end{matrix}\right\}=DN$, Quadruplicatio.

$2r\ .\ x\ ::\ \dfrac{2x^3}{rr}-4x\ .\ \dfrac{x^4}{r^3}-\dfrac{2xx}{r}\,.$

$AB.AK::AF(2AN-AD).AO.$ Et $\left.\begin{matrix}AO-AE\\[2pt]\dfrac{x^5}{r^4}-\dfrac{5x^3}{rr}+5x\end{matrix}\right\}=EO$, Quintuplicatio.

$2r\ .\ x\ ::\ \dfrac{2x^4}{r^3}-\dfrac{6xx}{r}+2r\ .\ \dfrac{x^5}{r^4}-\dfrac{3x^3}{rr}+x.$

quently evident that $\sqrt{[r^2z^2/b^2+4(ar-rz)]}$ is the difference of the twin value of y, that is, of the lines $+BC$ and $-DH$, and therefore (on letting fall the normal CK onto CB) that it is the value of CK. However, $FG:GE = DC:CK$, that is, $\sqrt{[b^2+z^2]}:b = c:\sqrt{[r^2z^2/b^2+4ar-4az]}$, and on multiplying the squares of the extremes and middles into each other and ordering the products there will come to be $z^4 = (4b^2rz^3-(4ab^2r+b^2r^2)\,z^2+4b^4rz-4ab^4r+b^4c^2)/r^2$, an equation of only four dimensions, which would have risen to eight dimensions had I sought CG, CB or AB.[249]

Problem 15[250]

To multiply or divide a given angle by a given number.

In any angle $F\widehat{A}G$ inscribe the lines AB, BC, CD, DE, ... of the same arbitrary length, and the triangles ABC, BCD, CDE, DEF, ... will then be isosceles; consequently, by *Elements*, I, 32, there will be $C\widehat{B}D = \hat{A}+A\widehat{C}B = 2\hat{A}$,

$$D\widehat{C}E = \hat{A}+A\widehat{D}C = 3\hat{A}, \quad E\widehat{D}F = \hat{A}+A\widehat{E}D = 4\hat{A}, \quad F\widehat{E}G = 5\hat{A},$$

and so on. With AB, BC, CD, ... now supposed to be the radii of equal circles, the perpendiculars BK, CL, DM, ... let fall to AC, BD, CE, ... will be the sines of those angles and AK, BL, CM, DN, ... their complementary sines (cosines). Or, if AB be supposed the diameter, those lines AK, BL, CM, ... will be their chords. Let therefore $AB = 2r$ and $AK = x$, and then work this way:

$$\left.\begin{array}{l} AB:AK = AC:AL \\ 2r \,:\, x \;=\; 2x:x^2/r \end{array}\text{ and }\begin{array}{l} AL-AB \\ x^2/r-2r \end{array}\right\} = BL, \text{ duplication.}$$

$$\left.\begin{array}{l} AB:AK = AD(2AL-AB):\quad AM \\ 2r \,:\, x \;=\; \quad 2x^2/r-2r \quad :x^3/r^2-x \end{array}\text{ and }\begin{array}{l} AM-MC \\ x^3/r^2-3x \end{array}\right\} = CM, \text{ triplication.}$$

$$\left.\begin{array}{l} AB:AK = AE(2AM-AC):\quad AN \\ 2r \,:\, x \;=\; \quad 2x^3/r^2-4x \quad :x^4/r^3-2x^2/r \end{array}\text{ and }\begin{array}{l} AN-AD \\ x^4/r^3-4x^2/r+2r \end{array}\right\} = DN, \text{ quadruplication.}$$

$$\left.\begin{array}{l} AB:AK = AF(2AN-AD)\,:\quad\quad AO \\ 2r \,:\, x \;=\; 2x^4/r^3-6x^2/r+2r:x^5/r^4-3x^3/r^2+x \end{array}\text{ and }\begin{array}{l} AO-AE \\ x^5/r^4-5x^3/r^2+5x \end{array}\right\} = EO, \text{ quintuplication.}$$

---

(249) In sequel Newton has cancelled an unfinished added remark: 'Simplicius[!] autem evasisset ponendo $CK = z$ et hanc quærendo. Prodijsset enim' (It would, however, have proved to be simpler on setting $CK = z$ and seeking this. For there would have ensued...). In fact, change of variable from EF to $CK = DK \times EG/EF$ is equivalent to the transformation $z \to b\sqrt{[c^2-z^2]}/z$ which yields the new equation $z^4-r(4a-r)\,z^2+4brz\sqrt{[c^2-z^2]}-c^2r^2 = 0$, effectively of eighth degree and so no simpler at all; this may much more fussily be derived *ab initio* by eliminating y between

$$(y^2/r-a)/(y^2/r-(y-z)^2/r) = (y-b)/z \quad \text{and} \quad ((y^2-(y-z)^2)/r)^2+z^2 = c^2,$$

the equations which now determine the problem.

(250) An essentially unaltered repeat of the final problem in Newton's earlier Kinckhuysen 'Observationes' (II: 444; compare I: 484).

Et sic deinceps. Quod si velis angulum in aliquot partes dividere, pone q pro

‖[78] BL, CM, DN, &c, et habebis $xx-2rr=qr$ ‖ ad bisectionem, $x^3-3rrx=qrr$ ad trisectionem, $x^4-4rrxx+2r^4=qr^3$ ad quadrisectionem, $x^5-5rrx^3+5r^4x=qr^4$ ad quinquisectionem &c.

[1676]
Lect 9

Prob: 16.

Cometæ in linea recta BD uniformiter progredientis positionem cursûs ex tribus observationibus determinare.[251]

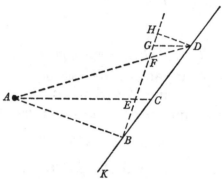

Sit A oculus spectatoris, B locus Cometæ in primâ observatione, C in secunda ac D in tertia: & quærenda erit inclinatio lineæ BD ad lineam AB. Ex observationibus itaq̃ dantur anguli BAC BAD adeoq̃ si BH ducatur ad AB normalis & occurrens AC et AD in E & F, ex assumpto utcunq̃ AB dabuntur BE et BF tangentes nempe præfatorum angulorum respectu radij AB. Sit ergo $AB=a$, $BE=b$, & $BF=c$. Porro ex datis observationum intervallis dabitur ratio BC ad BD quæ si ponatur b ad e et agatur DG parallela AC, cum sit BE ad BG in eadem ratione et BE dicta fuerit b erit $BG=e$, adeoq̃ $GF=e-c$. Adhæc si demittatur DH normalis ad BG propter triangula ABF et DHF similia et similiter secta lineis AE ac DG, erit $FE.AB::FG.HD$, hoc est $c-b\,.\,a::e-c\,.\,\dfrac{ae-ac}{c-b}=HD$.

Erit etiam $FE.FB::FG.FH$, hoc est $c-b\,.\,c::e-c\,.\,\dfrac{ce-cc}{c-b}=FH$: cui adde BF sive c et fit $BH=\dfrac{ce-cb}{c-b}$. Quare est $\dfrac{ce-cb}{c-b}$ ad $\dfrac{ae-ac}{c-b}$ (sive $ce-cb$ ad $ae-ac$, vel $\dfrac{ce-cb}{e-c}$ ad a) ut BH ad HD, hoc est ut tangens anguli HDB sive ABK ad radium.

Quare cum a supponatur esse radius erit $\dfrac{ce-cb}{e-c}$ tangens anguli ABK, adeoq̃

---

(251) Without broaching the vast literature on comets in the early and middle seventeenth century, we may note that it was Kepler in his *De Cometis* (Augsburg, 1619) who first publicly announced the hypothesis that comets travel in straight-line paths, though not everywhere with uniform speed, and that the apparent curvature of their paths is but an optical illusion. (See especially his opening *Assumpta* (p. 8): 'II. Cometam non secus ac Trajectionem aliquam ferri per spacia mundi in lineâ rectâ, in directum continuè' and 'III. Trajectionem Cometæ initio æquabili celeritate procedere, inde paulatim increscere...lege...ordinatâ'; and his later statement (p. 97) that 'Olim existimavi [quod Cometæ incipiunt curvare itinera sua] esse reale quippiam, itaque de causis Physicis philosophabar: sed est mera visus deceptio'.) The narrower hypothesis that a comet moves rectilinearly 'through equal spaces in equal times'— over an appreciable portion of its path at least—was taken up by Christopher Wren (then

And so forth. But should you wish to divide an angle into some number of parts, put q in place of BL, CM, DN, ... and you will have $x^2 - 2r^2 = qr$ in the case of bisection, $x^3 - 3r^2x = qr^2$ in that of trisection, $x^4 - 4r^2x^2 + 2r^4 = qr^3$ in that of quadrisection, $x^5 - 5r^2x^3 + 5r^4x = qr^4$ in that of quinquisection, and so on.

Problem 16

To determine the position of a comet proceeding uniformly in the straight line BD from three observations of its course.[251]

Let A be the viewer's eye, B the location of the comet at the first observation, C that of the second and D that of the third; and there will need to be found out the slope of the line BD to the line AB. Now from the observations the angles $\widehat{BAC}$, $\widehat{BAD}$ are given, and hence if BH be drawn normal to AB, meeting AC and AD in E and F, from arbitrarily assuming AB there will be given BE and BF: namely, the tangents (with respect to radius AB) of the above-cited angles. Let then $AB = a$, $BE = b$ and $BF = c$. From the given intervals between the observations, moreover, there will be given the ratio of BC to BD; if this be set as b to e, and DG be drawn parallel to AC, since BE is to BG in the same ratio while BE has just been called b, there will be $BG = e$ and accordingly $GF = e - c$. If, further, DH be let fall at right angles to BG, because the triangles ABF and DHF are similar and similarly divided by the lines AE and DG, therefore $FE:AB = FG:HD$, that is, $(c-b):a = (e-c):HD$ or $HD = a(e-c)/(c-b)$. Also there will be $FE:FB = FG:FH$, that is, $(c-b):c = (e-c):FH$ or

$$FH = c(e-c)/(c-b).$$

To this add BF, that is, c, and there comes $BH = c(e-b)/(c-b)$. Hence

$$c(e-b)/(c-b):a(e-c)/(c-b)$$

(that is, $c(e-b):a(e-c)$ or $c(e-b)/(e-c):a$) is the ratio of BH to HD—in other words, as the tangent of angle $\widehat{HDB}$ (or $\widehat{ABK}$) to its radius. Hence, when a be taken to be the radius, $c(e-b)/(e-c)$ will be the tangent of the angle $\widehat{ABK}$, so

Savilian Professor of Astronomy at Oxford) 'circiter 1661 aut 1662' because of its mathematical convenience and subsequently (during 1678–81) adopted by Hooke, Wallis and Huygens; see Hooke's *Cometa, or, Remarks about Comets* [= *Lectures and Collections made by Robert Hooke* (London, 1678): 1–80]: 40–2; John Wallis, 'De Cometarum Distantiis Investigandis' (Bodleian. MS Don. d. 45: 283v–280r; printed as the first of his three *Exercitationes* appended to the second edition of Jeremiah Horrocks' *Opera Posthuma* (London, 1678): 1–9 [= Wallis' *Opera Mathematica*, **2** (Oxford, 1693): 455–62]); and Christiaan Huygens' *Œuvres complètes*, **19** (The Hague, 1937): 283–304. Newton's present further specialization of the hypothesis is wildly implausible in that it requires the observer (presumably placed on the moving earth) to remain at rest at A as the comet uniformly traverses the straight line BCD. In his 'Prob: 52' following (on his pages 121–4) Newton's presents Wren's more realistic determination of a comet's path (from four sightings) when the observer is himself allowed to move, with the Earth, round the Sun.

facta resolutione erit ut $e-c$ ad $e-b$ (sive GF ad GE) ita c (sive tangens anguli BAF) ad tangentem anguli ABK.

Dic itaꝗ ut tempus inter primam et secundam observationem ad tempus inter primam ac tertiam, ita tangens anguli BAE ad quartam proportionalem. Dein ‖[79]‖ ut differentia inter illam quartam proportionalem et tangentem anguli BAF ad differentiam inter eandem quartam proportionalem et tangentem anguli BAE, ita tangens anguli BAF ad tangentem anguli ABK.[252]

Prob. 17.

Radijs a puncto lucido ad sphæricam superficiem refringentem divergentibus, invenire concursus singulorum refrac-torum cum axe sphæræ per punctum illud lucidum tran-seunte.

Sit A punctum illud lucidum, & BV sphæra, cujus axis AD, centrum

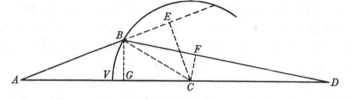

C, et vertex V, sitꝗ AB radius incidens & BD refractus ejus, ac demissis ad radios istos perpendicularibus CE & CF, ut et BG perpendiculari ad AD, actaꝗ BC, dic $AC=a$, VC vel $BC=r$, $CG=x$, et $CD=z$, eritꝗ $AG=a-x$, $BG=\sqrt{rr-xx}$, $AB=\sqrt{aa-2ax+rr}$, et propter sim. tri. ABG et ACE, $CE=\dfrac{a\sqrt{rr-xx}}{\sqrt{aa-2ax+rr}}$. Item $GD=z+x$, $BD=\sqrt{zz+2zx+rr}$, & propter sim. tri. DBG ac DCF, $CF=\dfrac{z\sqrt{rr-xx}}{\sqrt{zz+2zx+rr}}$. Præterea cùm ratio sinuum incidentiæ & refractionis adeoꝗ CE ad CF detur pone illam rationem esse a ad f, et erit

$$\frac{fa\sqrt{rr-xx}}{\sqrt{aa-2ax+rr}}=\frac{az\sqrt{rr-xx}}{\sqrt{zz+2zx+rr}}$$

ac multiplicando in crucem dividendoꝗ per $a\sqrt{rr-xx}$, erit

$$f\sqrt{zz+2zx+rr}=z\sqrt{aa-2ax+rr},$$

et quadrando ac redigendo terminos in ordinem $zz=\dfrac{2ffxz+ffrr}{aa-2ax+rr-ff}$. Deniꝗ pro dato $\dfrac{ff}{a}$ scribe p, & q pro dato $a+\dfrac{rr}{a}-p$, et erit $zz=\dfrac{2pxz+prr}{q-2x}$ ac $z=\dfrac{px+\sqrt{ppxx-2prrx+pqrr}}{q-2x}$.[253] Inventum est itaꝗ z hoc est longitudo CD adeoꝗ punctum quæsitum D quo refractus BD concurrit cum axe. Q.E.F.

that, when resolution is made, it will be as $e-c$ to $e-b$ (or $GF:GE$), so is c (or the tangent of the angle $B\widehat{A}F$) to the tangent of the angle $A\widehat{B}K$.

Accordingly, say: as the time between the first and second observations to the time between the first and the third, so let the tangent of the angle $B\widehat{A}E$ be to a fourth proportional; then, as the difference between that fourth proportional and the tangent of the angle $B\widehat{A}F$ to the difference between the same fourth proportional and the tangent of the angle $B\widehat{A}E$, so is the tangent of the angle $B\widehat{A}F$ to the tangent of the angle $A\widehat{B}K$.[252]

Problem 17

Where rays diverge from a luminous point onto a refracting spherical interface, to find the meets of individual refracted rays with the axis of the sphere passing through that luminous point.

Let A be that luminous point and BV the sphere, having axis AD, centre C and vertex V; also let AB be an incident ray and BD its refraction. Then, having let fall the perpendiculars CE and CF to those rays, and likewise BG perpendicular to AD, and drawing BC, call $AC = a$, VC or $BC = r$, $CG = x$ and $CD = z$, and there will be $AG = a-x$, $BG = \sqrt{[r^2-x^2]}$, $AB = \sqrt{[a^2-2ax+r^2]}$; also, because of the similar triangles ABG and ACE,

$$CE = a\sqrt{[r^2-x^2]}/\sqrt{[a^2-2ax+r^2]}.$$

Likewise $GD = z+x$, $BD = \sqrt{[z^2+2zx+r^2]}$ and, because of the similar triangles DBG and DCF, $CF = z\sqrt{[r^2-x^2]}/\sqrt{[z^2+2zx+r^2]}$. Furthermore, since the ratio of the sines of incidence and refraction is given, and therefore that of CE to CF, set that ratio to be a to f and there will be

$$fa\sqrt{[r^2-x^2]}/\sqrt{[a^2-2ax+r^2]} = az\sqrt{[r^2-x^2]}/\sqrt{[z^2+2zx+r^2]};$$

then, on multiplying cross-wise and dividing through by $a\sqrt{[r^2-x^2]}$, there will come $f\sqrt{[z^2+2zx+r^2]} = z\sqrt{[a^2-2ax+r^2]}$ and, by squaring and reducing the terms to order, $z^2 = (2f^2xz+f^2r^2)/(a^2+r^2-f^2-2ax)$. Finally, in place of f^2/a given write p, and then q in place of $a+r^2/a-p$ given, and there will be $z^2 = (2pxz+pr^2)/(q-2x)$ and so $z = (px+\sqrt{[p^2x^2-2pr^2x+pqr^2]})/(q-2x)$.[253] Accordingly, z is found—that is, the length of CD and hence the required point D in which the refracted ray BD encounters the axis. As was to be done.

---

(252) Newton's synthetic composition of the preceding algebraic analysis. Observe that even in the present simplified hypothesis of cometary motion three observations serve to determine only the inclination of the comet's rectilinear path, not its absolute position.

(253) A light revise of Newton's equivalent earlier solution (reproduced on III: 534) of this problem of elementary geometrical optics: for simplicity he here effectively makes the substitutions $e^2/d^2 \to p/a$ and $1+r^2/a^2-e^2/d^2 = q/a$.

Posui hic incidentes radios divergentes esse et in Medium densius incidere. sed mutatis mutandis Problema perinde resolvitur ubi convergunt vel incidunt e densiori Medio in rarius.[254]

|| [80]

|| *Prob: 18.*[255]

[1676]
Lect 10

Si conus plano quolibet secetur, invenire figuram sectionis.

Sit *ABC* conus circulari basi *BC* insistens; *IEM* ejus sectio quæsita; *KILM* alia quælibet sectio parallela basi, et occurrens priori sectioni in *HI*; & *ABC* tertia sectio perpendiculariter bisecans priores duas in *EH* & *KL* & conum in triangulo *ABC*. Et producto *EH* donec occurrat ipsi *AK* in *D*, actisq *EF* ac *DG* parallelis *KL* et occurrentibus *AB* et *AC* in *F* ac *G* dic *EF=a*, *DG=b*, *ED=c*, *EH=x*, et *HI=y*: et propter sim. tri. *EHL*, *EDG*, erit $ED.DG::EH.HL=\dfrac{bx}{c}$. Dein propter sim. tri. *DEF*, *DHK*, erit *DE.EF::DH* ($c-x$ in Fig. 1, et $c+x$ in Fig. 2). $HK=\dfrac{ac \mp ax}{c}$. Deniq cùm sectio *KIL* sit parallela basi adeoq circularis, erit $HK \times HL=HI^q$, hoc est $\dfrac{ab}{c}x \mp \dfrac{ab}{cc}xx=yy$, æquatio quæ exprimit relationem inter *EH(x)* & *HI(y)* hoc est inter axem et ordinatim applicatam sectionis *EIDM*, quæ æquatio cùm sit ad Ellipsin in Fig. 1, & ad Hyperbolam in Fig. 2[256] patet sectionem illam perinde Ellipticam vel Hyperbolicam esse.

Quod si *ED* nullibi occurrat *AK*, ipsi parallela existens, tunc erit *HK=EF(a)*, & inde $\dfrac{ab}{c}x(HK \times HL)=yy$, æquatio ad Parabolam.[256]

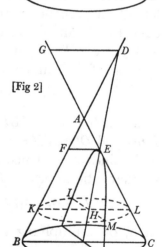

[Fig 1]

[Fig 2]

Prob: 19.

Si recta XY circa axem AB, ad distantiā CD in data inclinatione ad planum DCB convolvatur, & solidum[257] *PQRVTS ista convolutione generatum secetur plano quolibet INQLK: invenire figuram Sectionis.*

---

(254) See III: 514–16 for Newton's previous discussion of these other cases.

(255) Compare 1: 40, 42, 44 for this Schootenian variant on Apollonius' classical derivation of the defining 'symptom' of the conic from its definition as the general plane section of a circular cone.

(256) Newton, in asserting the identity of the second-degree Cartesian locus defined by the *æquatio* $rx \mp (r/q)x^2 = y^2$ with the classical Apollonian conic, ought really to have distinguished the *latus rectum* r and *latus transversum* (main axis) q in terms of which the latter's defining

Here I have supposed that the rays are divergent and incident into a denser medium, but with appropriate changes the problem is correspondingly resolved when they converge or are incident from a denser medium into a rarer one.[254]

Problem 18[255]

If a cone be cut by any plane, to find the figure of the section.

Let *ABC* be a cone standing on the circular base *BC*, *IEM* the section of it sought, *KILM* any other section parallel to the base and meeting the former section in *HI*, and *ABC* a third section perpendicularly bisecting the two previous ones in *EH* and *KL* and the cone in the triangle *ABC*. And, having extended *EH* till it meets *AK* in *D* and also drawn *EF* and *DG* parallel to *KL* and meeting *AB* and *AC* in *F* and *G*, call $EF = a$, $DG = b$, $ED = c$, $EH = x$ and $HI = y$. Then, because of the similar triangles *EHL*, *EDG*, there will be $ED:DG = EH:HL$ and so $HL = bx/c$. Again, because of the similar triangles *DEF*, *DHK*, there will be $DE:DF = DH$ ($c-x$ in Figure 1, $c+x$ in Figure 2): *HK*, and so $HK = a(c \mp x)/c$. Finally, since the section *KIL* is parallel to the base and hence circular, there will

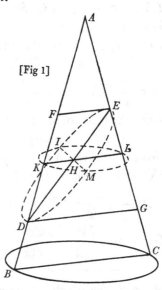

[Fig 1]

be $HK \times HL = HI^2$, that is, $(ab/c)\,x \mp (ab/c^2)\,x^2 = y^2$, an equation expressing the relationship between $EH(x)$ and $HI(y)$—that is, between the axis and (general) ordinate of the section *EIDM*. Since this equation belongs to an ellipse in Figure 1 and to a Hyperbola in Figure 2,[256] it is clear that the section is likewise elliptical or hyperbolical.

But should *ED* nowhere meet *AK*, proving to be parallel to it, then will there be $HK = EF$ (or a) and thence $(ab/c)\,x$ (or $HK \times HL$) $= y^2$, an equation to a parabola.[256]

Problem 19

If the straight line XY should revolve round the axis AB at the distance CD, maintaining a given slope to the plane DCB, and the solid[257] PQRVTS generated by that revolution be cut by any plane INQLK, to find the figure of the section.

'sympton' is stated. Alternatively, if he implicitly accepts (in modern style) that the ellipse, parabola and hyperbola are the three (non-degenerate) species of the general second-degree Cartesian locus, a reference to Descartes' not altogether rigorous reduction (in Book 2 of his *Geometrie*; compare IV: 219, note (9)) of the general second-degree defining equation to its three canonical forms could well here have been made by Newton.

(257) A hyperboloid of revolution of one sheet, a shape not unfamiliar in round wicker baskets of the period—indeed, John Wallis told Sluse on 10 September 1669 that it was

‖ [81] ‖

Esto *BHQ* vel *GHO* inclinatio axis *AB* ad planum sectionis & *L* quilibet concursus rectæ *XY* cum plano illo. Age *DF* parallelam *AB*, et ad *AB*, *DF* & *HO* demitte perpendiculares *LG*, *LF*, *LM*, ac junge *FG* & *MG*.[258] Dictisqß *CH*=*b*, *HM*=*x*, & *ML*=*y*; et propter datum angulum *GHO* posito *MH*.*HG*::*d*.*e*:

erit $\dfrac{ex}{d}$=*HG*, & *b*+$\dfrac{ex}{d}$=*GC* vel *FD*. Adhæc propter angulum datum *LDF* (nempe inclinationem rectæ *XY* ad planum *GCDF*) posito *FD*.*FL*::*g*.*h*,

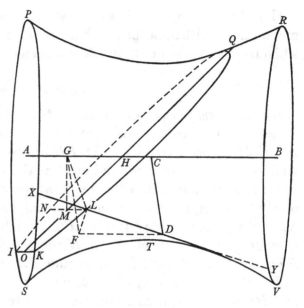

erit $\dfrac{hb}{g}+\dfrac{hex}{dg}$=*FL*, cujus quadrato adde *FG*^{q} (*DC*^{q} seu *aa*) & emerget

$$GL^q=aa+\frac{hhbb}{gg}+\frac{2hhbex}{dgg}+\frac{hheexx}{ddgg}.$$ Hinc aufer $MG^q\left(HM^q-HG^q \text{ seu } xx-\frac{ee}{dd}xx\right)$

observing this which had inspired Christopher Wren to analyse the surface mathematically: 'quum viderit aliquando in officina, inter alia, venalem corbem quendam vimineum rotundum, ex viminibus tantum rectis contextum, situ obliquo positis (credo, et decussatis) cujus superficies lateralis, Cylindrum extrinsecus excavatum, exhiberet (ea forma, qua salina solent apud nostrates confieri, vel trochlearum orbiculi;) animadvertit, uno ex viminibus illis circa Cylindri axem circumducto, manente situ illo ad axem obliquo, descriptum iri superficiem illam concavo-convexam, adeoque Torno posse confici cylindroides ejusmodi, per aciem Dolabræ rectam, obliquo ad Cylindri axem situ positam, cujus sectio per axem, foret ea linea curva' (Royal Society MS W. 1.95, quoted in *The Correspondence of Henry Oldenburg*, **6** (Madison, 1969): 238). With some slight help from Wallis (*ibid.*: 237–8) Wren was then able to give elementary geometrical proof of the hyperboloid's property—anticipated some four years earlier by Newton and applied in an equivalent optical context (see I: 561)—that all plane sections of it parallel to a section through the axis are hyperbolas (or, in the case of a tangential section, a pair of generating lines), publishing the result in his 'Generatio Corporis Cylindroidis Hyperbolici, elaborandis Lentibus Hyperbolicis accommodati' (*Philosophical Transactions*, **4**, 1669, No. 48 [for 21 June 1669]: 961–2). Duly inspired by Wren's article (communicated to him almost at once by Oldenburg), Sluse soon after attacked and solved the problem of determining the hyperboloid's general plane section, writing back to Oldenburg on 10 March 1670 (N.S.) that 'non parallelogrammum tantùm et triangulum in eo [solido hyperbolico] secari posse; sed parabolam quoque et Ellipsin, et Hyperbolam pluribus modis' (C. Le Paige, 'Correspondance de René-François de Sluse' [=*Bullettino di Bibliografia e Storia di Scienze Matematiche e Fisiche*, **17**, 1884: 427–554/603–59]: 642). Two years later, in May 1672, Richard Towneley told Collins that 'now a long time since' Sluse had written to him, proposing 'the solution of Dr. Wren's problem more generally'—namely (and apparently so only) by using a skew generator to define the hyperboloid (see Sluse's letter to Oldenburg of 6/16 August 1669,

Let $\widehat{BHQ}$ or $\widehat{GHO}$ be the slope of the axis AB to the plane of section and L any meet of the straight line XY with that plane. Draw DF parallel to AB, and to

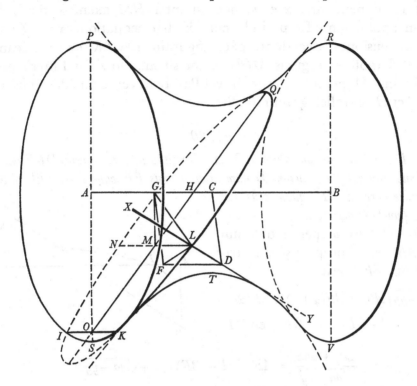

AB, DF and HO let fall the perpendiculars LG, LF and LM, and then join FG and MG.[258] Calling now $CH = b$, $HM = x$ and $ML = y$, because the angle $\widehat{GHO}$ is given I put $MH:HG = d:e$, and there will be $HG = (e/d)\,x$ and GC or $FD = b+(e/d)\,x$. Again, because the angle $\widehat{LDF}$ (namely, the slope of the straight line XY to the plane $GCDF$) is given, I put $FD:FL = g:h$ and there will be $FL = hb/g+(he/dg)\,x$. To the square of this add FG^2 (that is, DC^2 or a^2) and there will emerge $GL^2 = a^2+h^2b^2/g^2+2(h^2be/dg^2)\,x+(h^2e^2/d^2g^2)\,x^2$. From this take off $MG^2(= HM^2-HG^2$ or $x^2-(e^2/d^2)\,x^2)$ and there will remain (ML^2 or)

---

reproduced in *The Correspondence of Henry Oldenburg*, **6**: 178–82)—and adding that 'the hyperbolical cylindroid might so be cut as to give all the sections both of cone and cylinder,...and proposed to me the finding them' (S. P. Rigaud, *Correspondence of Scientific Men of the Seventeenth Century*, **1** (Oxford, 1841): 191). Towneley's analytical solution of Sluse's challenge problem (*ibid.*: 192) and the excerpts he quoted for Collins of Sluse's extension to it (*ibid.*: 193–4), both making use of Fermatian variables *a* and *e* in the latter's usual style, are not essentially different from Newton's present determination of the 'figura [planæ] Sectionis'; how much, however, Newton knew of his predecessors' solutions is not known and we may well wish to conclude that he here presents a wholly independent resolution of the problem.

(258) Since Newton's somewhat distorted figure is not easy to visualize spatially, we have added a slightly augmented modern redrawing of it in the English version.

& restabit $\dfrac{aagg+hhbb}{gg}+\dfrac{2hhbe}{dgg}x+\dfrac{hhee-ddgg+eegg}{ddgg}xx(=ML^q)=yy$: æquatio quæ exprimit relationem inter x et y, hoc est inter HM axem sectionis & ML ordinatim applicatam. Et proinde cùm in hac æquatione x et y ad duas tantùm dimensiones ascendant, patet figuram $INQLK$ esse conicam sectionem.[259] Utpote si angulus MHG major sit angulo LDF Ellipsis erit hæc figura, si minor Hyperbola, si æqualis vel Parabola vel (coincidentibus insuper punctis C et H) parallelogrammum.[260]

Octob. 1677
Lect. 1.

Prob. 20

Si ad AF erigatur perpendiculum AD datæ longitudinis, & normæ DEF crus unum ED continuò transeat per punctum D dum alterum crus EF æquale AD dilabatur super AF: invenire curvam HIC quam crus EF medio ejus puncto C describit.

Sit EC vel $CF=a$, perpendiculum $CB=y$, $AB=x$, & propter similia triangula FBC, FEG, erit

$$BF(\sqrt{aa-yy})\,.\,BC+CF(y+a)::EF(2a).$$
$$EG+GF(AG+GF)\text{ seu }AF.$$

Quare

$$\frac{2ay+2aa}{\sqrt{aa-yy}}(=AF=AB+BF)=x+\sqrt{aa-yy}.$$

Jam multiplicando per $\sqrt{aa-yy}$ fit $2ay+2aa=aa-yy+x\sqrt{aa-yy}$ seu $2ay+aa+yy=x\sqrt{aa-yy}$, et quadrando partes divisas per $\sqrt{a+y}$[261] ac ordinando prodit $y^3+3ayy\genfrac{}{}{0pt}{}{+3aa}{+xx}y\genfrac{}{}{0pt}{}{+a^3}{-axx}=0$.[262]

Idem aliter.

In BC cape hinc inde BI, & CK æquales CF et age KF, HI, HC, ac DF quarum HC ac DF occurrant ipsis AF et $\parallel IK$ in M et N et in HC demitte normalem IL. Eritꝗ angulus $K=\frac{1}{2}BCF=\frac{1}{2}EGF=GFD=AMH=MHI=CIL$; adeoꝗ tri-

---

(259) Evidently (compare note (256)) the preceding equation may be reduced to the form $rx-(r/q)x^2=y^2$, $rx=y^2$ or $rx+(r/q)x^2=y^2$ (defining an ellipse, parabola or hyperbola respectively) according as the coefficient of x^2 is less than, equal to or greater than zero; that is, as $e/d=\cos\widehat{MHG}$ is less than, equal to or greater than $g/\sqrt{[g^2+h^2]}=\cos\widehat{LDF}$. Newton's following criterion for distinguishing the conic species is an immediate corollary. (It is understood that both $\widehat{MHG}$ and $\widehat{LDF}$ are taken to be acute.)

(260) That is, a pair of parallel generating lines (terminated by the perpendicular endfaces of the hyperboloid). The hyperbolic case also degenerates into a pair of (intersecting) generators when the section plane is tangent to the hyperboloid.

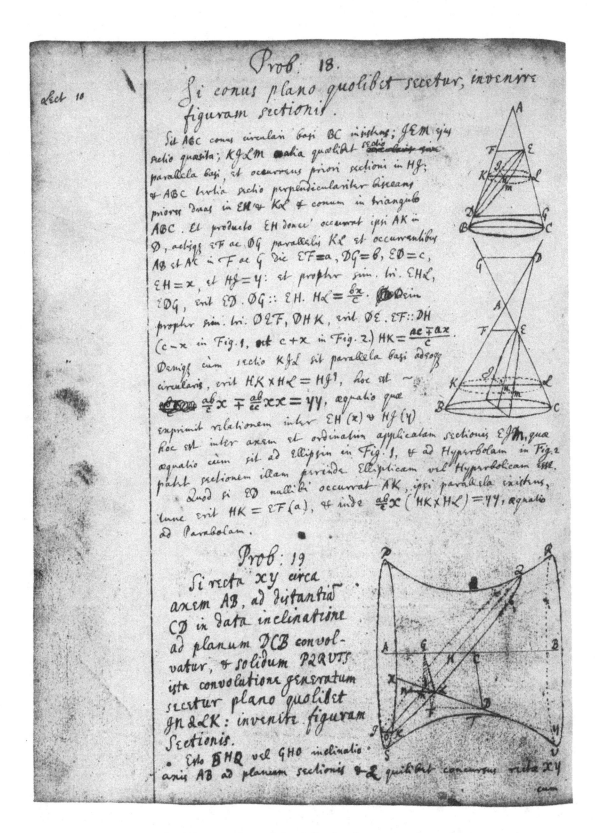

Prob: 18.

Si conus plano quolibet secetur, invenire figuram sectionis.

Sit ABC conus circularis basi BC insistens; JEM sic sectio quæsita; KJLM alia qualibet sectio parallela basi, et occurrens priori sectioni in HJ; & ABC tertia sectio perpendiculariter bisecans priores duas in EH & KL & conum in triangulo ABC. Et producto EH donec' occurrat ipsi AK in D, actisq; EF ac DG parallelis KL et occurrentibus AB et AK in F ac G dic EF=a, DG=b, ED=c, EH=x, et HJ=y: et propter sim. tri. EHL, EDG, erit ED. DG :: EH. HL = $\frac{bx}{c}$. Dein propter sim. tri. DEF, DHK, erit DE. EF :: DH (c−x in Fig.1, et c+x in Fig. 2.) HK = $\frac{ac \mp ax}{c}$. Deniq; cum sectio KJL sit parallela basi adeoq; circularis, erit HK × HL = HJ², hoc est $\frac{ab}{c}x \mp \frac{ab}{cc}xx = yy$, æquatio quæ exprimit relationem inter EH (x) & HJ (y) hoc est inter axem et ordinatim applicatam sectionis EJM, quæ æquatio cum sit ad Ellipsin in Fig.1, & ad Hyperbolam in Fig.2 patet sectionem illam perinde Ellipticam vel Hyperbolicam esse.

Quod si ED nullibi occurrat AK, ipsi parallela existens, tunc erit HK = EF (a), & inde $\frac{ab}{c}x$ (HK × HL) = yy, æquatio ad Parabolam.

Prob: 19

Si recta xy circa axem AB, ad distantiam CD in data inclinatione ad planum DCB convolvatur, & solidum PQRVSS ista convolutione generatum secetur plano quolibet JN&LK: invenire figuram Sectionis.

Esto BHQ vel GHO inclinatio axis AB ad planum sectionis & L quilibet concursus rectæ xy

Plate I.   Plane sections of the right circular cone and one-sheet hyperboloid of revolution (**1, 2, §1**).

$y^2 = (a^2g^2 + h^2b^2)/g^2 + 2(h^2be/dg^2)\,x + ((h^2e^2 - d^2g^2 + e^2g^2)/d^2g^2)\,x^2$: an equation which expresses the relationship between x and y, that is between the axis HM of the section and its ordinate ML. In consequence, since x and y in this equation rise but to two dimensions, it is evident that the figure $INQLK$ is a conic.[259] Precisely, if the angle $\widehat{MHG}$ be greater than angle $\widehat{LDF}$ this will be an ellipse, if less it will be a hyperbola, if equal to it the figure will be either a parabola or (if in addition the points C and H coincide) a rectangle.[260]

Problem 20

If to AF there be erected the perpendicular AD of given length, and one leg ED of the right-angled rule DEF shall continually pass through the point D while the other leg EF, equal to AD, slides along on AF, to find the curve HIC which the leg EF describes at its mid-point C.

Let EC or $CF = a$, the perpendicular $CB = y$, $AB = x$, and then, because of the similar triangles FBC, FEG, there will be

$$BF(\sqrt{[a^2 - y^2]}):BC + CF(y + a) = EF(2a):(EG + GF \text{ or } AG + GF, \text{ that is) } AF.$$

Hence $2a(y + a)/[a^2 - y^2] = (AF = AB + BF \text{ or}) \; x + \sqrt{[a^2 - y^2]}$. On now multiplying through by $\sqrt{[a^2 - y^2]}$ there comes $2ay + 2a^2 = a^2 - y^2 + x\sqrt{[a^2 - y^2]}$ or $a^2 + 2ay + y^2 = x\sqrt{[a^2 - y^2]}$; and so, on dividing through by $\sqrt{[a + y]}$,[261] squaring the sides and ordering, there results $y^3 + 3ay^2 + (3a^2 + x^2)\,y + a^3 - ax^2 = 0$.[262]

The same otherwise

In BC, on either side of it, take BI and CK equal to CF and draw KF, HI, HC and DF; of these let HC and DF meet AF and IK in M and N, and onto HC let fall the normal IL. Then will

$$\text{angle } \hat{K} = \tfrac{1}{2}\widehat{BCF} = \tfrac{1}{2}\widehat{EGF} = \widehat{GFD} = \widehat{AMH} = \widehat{MHI} = \widehat{CIL},$$

---

(261) Clearly, the equation $y = -a$ yields the particular solution of the line through H parallel to AF: the locus of (C) when the *norma DEF* is set below the base line AF. Newton did not at once notice the multiplying factor $\sqrt{[y + a]}$ but straightforwardly squared the previous equation to derive '$y^4 + 4ay^3 \genfrac{}{}{0pt}{}{+6aa}{+aa}\,yy + 4a^3y \genfrac{}{}{0pt}{}{+a^4}{-a[a]xx} = 0$', only then deducing the final cubic 'dividendo totam per $y + a$' (by dividing through by $y + a$).

(262) That is, $(y + a)^3 + x^2(y + a) - 2ax^2 = 0$, a cissoid of vertex H and asymptote parallel to AF and distant AH from it; compare II: 61, note (58) and III: 270, 'Exempl: 3'. Newton's present elegant construction of the cissoid as the locus of a point on a moving 'angle', one side of which is constrained to slide through a fixed pole while the end-point of the other slides along a fixed line, here appears for the first time; we have, however, already noticed its reappearance a few years afterwards in Newton's 'Geometria Curvilinea' (compare IV: 481, note (157)).

angula rectangula *KBF*, *FBN*, *HLI* &
ILC similia. Dic ergo *FC*=*a*. *HI*=*x* &
IC=*y*; & erit

　　BN(2*a*−*y*).*BK*(*y*)::*LC*.*LH*

　　　　　　::*CI*^{q}(*yy*).*HI*^{q}(*xx*)

adeoꝗ 2*axx*−*yxx*=*y*³. Ex qua æquatione facilè colligitur hanc curvam esse
Cissoidem Veterum ad circulum cujus
centrum sit *A* ac radius *AH* pertinen-
tem.[263]

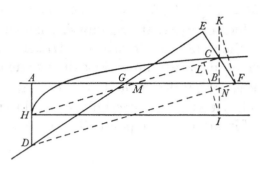

Prob. 21.[264]

　　*Si datæ longitudinis recta ED angulum datum EAD subtendens ita moveatur ut termini
ejus D et E anguli istius latera AD et AE perpetim contingant: proponatur Curvam FCG
determinare quam punctum quodvis C in recta ista ED
datum describit.*

　　A dato puncto *C* age *CB* parallelam *EA* et dic
AB=*x*. *BC*=*y*. *CE*=*a* & *CD*=*b*, et propter similia triangula *DCB*, *DEA* erit *EC*.*AB*::*CD*.*BD*.

hoc est *a*.*x*::*b*.*BD*=$\dfrac{bx}{a}$. Præterea demisso per-

pendiculo *CH*, propter datum angulum *DAE* vel
DBC adeoꝗ datam rationem laterum trianguli

rectanguli *BCH* sit *a*.*e*::*BC*.*BH* et erit *BH*=$\dfrac{ey}{a}$.

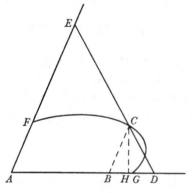

Aufer hanc de *BD* et restabit *HD*=$\dfrac{bx-ey}{a}$. Jam

in triangulo *BCH* propter angulum rectum *BHC* est *BC*^{q}−*BH*^{q}=*CH*^{q} hoc est

$yy-\dfrac{eeyy}{aa}$=*CH*^{q}. Similiter in triangulo *CDH* propter angulum *CHD* rectum, est

CD^q−*CH*^{q}=*HD*^{q}, hoc est

$$bb-yy+\frac{eeyy}{aa}\left(=HD^q=\overline{\frac{bx-ey}{aa}}^{\text{quad}}\right)=\frac{bbxx-2bexy+eeyy}{aa}.$$

Et per reductionem $yy=\dfrac{2be}{aa}xy+\dfrac{aabb-bbxx}{aa}$: Ubi cùm incognitæ quantitates

sint duarum tantum dimensionum patet curvam esse Conicam sectionem.

Præterea extracta radice fit $y=\dfrac{bex\mp b\sqrt{eexx-aaxx+a^4}}{aa}$. Ubi in termino radicali

‖[83]　coefficiens ipsius *xx* est *ee*−*aa*. Atqui erat *a*.*e*::*BC*.*BH* ‖ et *BC* necessariò major
est linea quam *BH*, nempe Hypotenusa trianguli rectanguli major latere; ergo
a ⎾ *e* et *ee*−*aa* negativa est quantitas, atꝗ adeo curva erit Ellipsis.[265]

so that the right triangles *KBF*, *FBN*, *HLI* and *ILC* are similar. Call, therefore, *FC* = *a*, *HI* = *x* and *IC* = *y*, and there will be

$$BN(2a-y):BK(y) = LC:LH = CI^2(y^2):HI^2(x^2),$$

so that $2ax^2 - yx^2 = y^3$. From this equation it is readily gathered that the present curve is the Ancients' cissoid, relating to a circle whose centre is *A* and radius *AH*.[263]

Problem 21[264]

If the straight line ED of given length and subtending the given angle $\widehat{EAD}$ move in such a way that its end-points D and E shall perpetually be in contact with the sides AD and AE of that angle, let it be proposed to determine the curve FCG which any point C given in that line ED shall describe.

From the given point *C* draw *CB* parallel to *EA* and call *AB* = *x*, *BC* = *y*, *CE* = *a* and *CD* = *b*. Then, because of the similar triangles *DCB*, *DEA*, there will be *EC*: *AB* = *CD*: *BD*, that is, *a*: *x* = *b*: *BD* or *BD* = *bx*/*a*. Moreover, on letting fall the perpendicular *CH*, because the angle $\widehat{DAE}$ is given and hence so is the proportion of the sides of the right triangle *BCH*, let *BC*: *BH* = *a*: *e* and there will be *BH* = *ey*/*a*. Take this away from *BD* and there will remain *HD* = (*bx*−*ey*)/*a*. Now in the triangle *BCH*, because the angle $\widehat{BHC}$ is right, there is $CH^2 = BC^2 - BH^2$, that is, $CH^2 = y^2 - e^2y^2/a^2$. Similarly in the triangle *CDH*, because the angle $\widehat{CHD}$ is right, there is $HD^2 = CD^2 - CH^2$, that is, $b^2 - y^2 + e^2y^2/a^2 (= HD^2$ or $(bx-ey)^2/a^2) = (b^2x^2 - 2bexy + e^2y^2)/a^2$, and by reduction $y^2 = 2(be/a^2) xy + b^2(a^2 - x^2)/a^2$: here, since the unknown quantities have but two dimensions, it is evident that the curve is a conic section. Moreover, when the root is extracted, there comes $y = (bex \mp b\sqrt{[(e^2 - a^2) x^2 + a^4]})/a^2$. The coefficient of x^2 within the radical is here $e^2 - a^2$. But there was *a*: *e* = *BC*: *BH* and *BC* is necessarily a larger line than *BH*—the hypotenuse of a right triangle being, namely, greater than a side—and therefore *a* > *e* and $e^2 - a^2$ is a negative quantity, so that the curve will be an ellipse.[265]

---

(263) If the parallel to *AF* through *C* meets *HA* in α and the circle of centre *A* and radius *AH* in β, the derived equation $y^2 = x\sqrt{[y(2a-y)]}$ yields the classical defining 'symptom' $H\alpha^2 = \alpha C \times \alpha\beta$. Newton's essential simplification in the present *idem aliter* is, of course, to make the substitution $y \to y - a$ in the preceding Cartesian equation of the locus (see note (262)).

(264) Newton noted (and improved upon) this Schootenian locus in his last undergraduate year (see I: 32, [*g*]), but had then determined it to be an ellipse by more traditional geometrical means.

(265) Evidently of centre *A*, with $y = (be/a^2) x$ the diameter (of length

$$2\sqrt{[(a^4 + (2a+b) be^2)/(a^2 - e^2)]})$$

conjugate to *AF* (*x* = 0). It is readily shown (compare I: 32, note (25)) that the circumradius of the triangle *ADE* has the constant value $\frac{1}{2}(a+b)\sqrt{[a^2 - e^2]}/a$.

[1677]
Lect 2

Prob. 22.[266]

Si norma EBD ita moveatur ut ejus crus unum EB continuò subtendat angulum rectum EAB dum terminus alterius cruris BD describat curvam aliquam lineam FDG:[267] *invenire lineam istam FDG quam punctum D describit.*

A puncto D ad latus AC demitte perpendiculum DC & dictis $AC = x$ et $DC = y$, atqɜ $EB = a$ et $BD = b$: in triangulo BDC propter angulum rectum ad C est

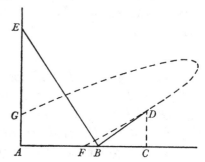

$$BC^q = BD^q - DC^q = bb - yy.$$

Ergo $BC = \sqrt{bb - yy}$ et $AB = x - \sqrt{bb - yy}$. Præterea propter similia triangula BEA. DBC est $BD . DC :: EB . AB$. hoc est $b . y :: a . x - \sqrt{bb - yy}$.

Quare $bx - b\sqrt{bb - yy} = ay$, sive $bx - ay = b\sqrt{bb - yy}$. et partibus quadratis ac debitè reductis $yy = \dfrac{2abxy + b^4 - bbxx}{aa + bb}$. et extracta radice $y = \dfrac{abx \pm bb\sqrt{aa + bb - xx}}{aa + bb}$.

Unde[268] patet iterum Curvam esse Ellipsin.

Hæc ita se habent ubi anguli EBD et EAB recti sunt, sed si anguli isti sunt alterius cujusvis magnitudinis, dummodo sint æquales,[269] sic procedendum erit. Demittatur DC perpendicularis ad AC ut ante, et agatur DH constituens angulum DHA æqualem angulo HAE puta obtusum, dictisqɜ $EB = a$. $BD = b$. $AH = x$ & $HD = y$, propter similia triangula EAB BHD erit

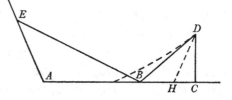

$BD . DH :: EB . AB$. hoc est $b . y :: a . AB = \dfrac{ay}{b}$. Aufer hanc de AH et restabit

‖[84] $BH = x - \dfrac{ay}{b}$. Præterea in triangulo DHC ‖ propter omnes angulos datos adeoqɜ datam rationem laterum assume DH ad HC in ratione quavis data puta b ad

---

(266) Schooten's extension of the preceding problem; compare I: 32, [g].

(267) Again an ellipse, as the following analysis shows. In Whiston's 1707 *editio princeps* Newton's accurate manuscript figure was crudely reproduced, with the ellipse arc *FDG* reduced to two broken straight lines capped by a small circle-arc at their 'vertex'; in subsequent editions all attempt, in either of the accompanying diagrams, to represent the ellipse arcs by other than undeviatingly straight lines is abandoned.

(268) Since the coefficient of x^2 within the radical is -1; compare note (259), and also IV: 219, note (9).

(269) As Newton well knew (see I: 32, note (25)) this restriction is unnecessary—indeed, it may here be made only to abbreviate the analysis. In general, if D is an arbitrary fixed point in the plane of the moving line BE, then the circumcentre O of triangle ABE (in the perpendi-

Problem 22[(266)]

If the right-angled rule EBD should move in such a way that one of its legs EB shall continually subtend the right angle $E\widehat{A}B$ while the end-point of its other leg BD describes some curved line FDG,[(267)] *to find that line FDG which the point D describes.*

From the point D to the side AC let fall the perpendicular DC and call $AC = x$ and $DC = y$, with $EB = a$ and $BD = b$. Then in the triangle BDC, because the angle at C is right, there is $BC^2 = BD^2 - DC^2 = b^2 - y^2$ and therefore

$$BC = \sqrt{[b^2 - y^2]}, \quad \text{so that} \quad AB = x - \sqrt{[b^2 - y^2]}.$$

Moreover, because the triangles BEA, DBC are similar, there is

$$BD:DC = EB:AB, \quad \text{that is,} \quad b:y = a:(x - \sqrt{[b^2 - y^2]}).$$

Hence $bx - b\sqrt{[b^2 - y^2]} = ay$, or $bx - ay = b\sqrt{[b^2 - y^2]}$, and so, when the sides are squared and duly ordered, $y^2 = (2abxy + b^2(b^2 - x^2))/(a^2 + b^2)$ and on extracting the root $y = (abx \pm b^2\sqrt{[a^2 + b^2 - x^2]})/(a^2 + b^2)$. It is hence[(268)] evident that the curve is an ellipse.

This is the situation when the angles $E\widehat{B}D$ and $E\widehat{A}B$ are (each) right, but if those angles are of any other size—provided they are equal[(269)]—this will be the procedure. Let fall DC perpendicular to AC as before, and draw DH forming an angle $D\widehat{H}A$ equal to the angle $H\widehat{A}E$, obtuse say. Then on calling $EB = a$, $BD = b$, $AH = x$ and $HD = y$, because the triangles EAB, BHD are similar, there will be $BD:DH = EB:AB$, that is, $b:y = a:AB$ or $AB = ay/b$. Take this from AH and there will remain $BH = x - ay/b$. In the triangle DHC, moreover, because all the angles are given and so consequently is the proportion of the sides, assume DH to be to HC in any given ratio, say b to e, and since DH is y

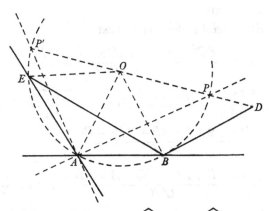

cular bisector of BE such that (exterior) angle $E\widehat{O}B = 2 \times E\widehat{A}B$) is a second point fixed in that moving plane; further, where P, P' are the meets of OD with that circumcircle, the loci of P and P' are a pair of perpendicular lines through A. At once the locus of D (now conceived as a fixed point in the line PP' of constant length, whose end-points travel in the perpendicular lines AP, AP') is an ellipse of axes coincident with AP, AP'.

e, & cum DH sit y erit $HC[=]\frac{ey}{b}$, et $HB \times HC = \frac{exy}{b} - \frac{aeyy}{bb}$. Deniꝗ per

12. 2. Elem. in triangulo BHD est $BD^q = BH^q + DH^q + 2BH \times HC$, hoc est

$$[bb]^{(270)} = xx - \frac{2axy}{b} + \frac{aayy}{bb} + yy + \frac{2exy}{b} - \frac{2aeyy}{bb}.$$

Et extracta radice $x = \dfrac{ay - ey \pm \sqrt{eeyy - bbyy + [b^4]}}{b}$. Ubi cum b sit major $e^{(271)}$ hoc

est $ee - bb$ negativa quantitas patet iterum curvam esse Ellipsin.

|| [257]
                     || *Prob. 23.*[(272)]

Trianguli cujusvis rectilinei datis lateribus et basi, invenire segmenta basis, perpendiculum, aream et angulos.

Trianguli ABC
dentur latera AC,
BC & basis AB.
Biseca AB in I et
in ea[(273)] utrinꝗ
producta cape AF

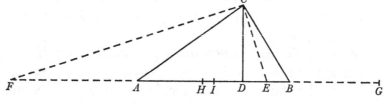

et AE æquales AC, atꝗ BG et BH æquales BC. Junge CE, CF et a C ad basem demitte perpendiculum CD. Et erit

$$AC^q - BC^q = AD^q + CD^q - CD^q - BD^q = AD^q - BD^q$$
$$= AD + BD \times AD - BD = AB \times 2DI.$$

Ergo $\dfrac{AC^q - BC^q}{2AB} = DI$. Et $2AB . AC + BC :: AC - BC . DI$. Quod est Theorema[(274)]

pro determinandis segmentis basis.

De IE hoc est de $AC - \frac{1}{2}AB$ aufer DI et restabit

$$DE = \frac{BC^q - AC^q + 2AC \times AB - AB^q}{2AB}.$$

hoc est $= \dfrac{BC + AC - AB \times BC - AC + AB}{2AB}$ sive $= \dfrac{HE \times EG}{2AB}$. Aufer DE de FE sive

$2AC$ et restabit $FD = \dfrac{AC^q + 2AC \times AB + AB^q - BC^q}{2AB}$

hoc est $= \dfrac{AC + AB + BC \times AC + AB - BC}{2AB}$ sive $= \dfrac{FG \times FH}{2AB}$.

Et cum sit CD medium proportionale inter DE ac DF, CE medium proportionale inter DE et EF, ac CF medium proportionale inter DF et EF: erit

$$CD = \frac{\sqrt{FG \times FH \times HE \times EG}}{2AB}.$$

there will be $HC = ey/b$, while $HB \times HC = (e/b) xy - (ae/b^2) y^2$. Finally, by *Elements*, ii, 12, in the triangle BHD there is $BD^2 = BH^2 + DH^2 + 2BH \times HC$, that is, $b^2 = x^2 - 2(a/b) xy + (a^2/b^2) y^2 + y^2 + 2(e/b) xy - 2(ae/b^2) y^2$. And when the root x is extracted it is $x = ((a-e) y \pm \sqrt{[(e^2-b^2) y^2 + b^4]})/b$. Since b is here greater than e,[271] in other words $e^2 - b^2$ is a negative quantity, it is evident that the curve is again an ellipse.

Problem 23[272]

In any rectilinear triangle, given the sides and base, to find its base segments, altitude, area and angles.

In the case of the triangle ABC let there be given the sides AC, BC and the base AB. Bisect AB in I and in it,[273] produced either way, take AF and AE equal to AC, and also BG and BH equal to BC. Join CE, CF and from C let fall to the base the perpendicular CD. Then will there be

$$AC^2 - BC^2 = (AD^2 + CD^2) - (CD^2 + BD^2) = AD^2 - BD^2$$
$$= (AD + BD)(AD - BD) = AB \times 2DI.$$

Therefore $DI = (AC^2 - BC^2)/2AB$ and $2AB : (AC + BC) = (AC - BC) : DI$. This is the[274] theorem for determining the base segments.

From IE (that is, from $AC - \frac{1}{2}AB$) take away DI and there will remain $DE = (BC^2 - AC^2 + 2AC \times AB - AB^2)/2AB$, that is,

$$(BC + AC - AB)(BC - AC + AB)/2AB \quad \text{or} \quad HE \times EG/2AB.$$

Take away DE from FE (or $2AC$) and there will remain

$$FD = (AC^2 + 2AC \times AB + AB^2 - BC^2)/2AB,$$

that is, $(AC + AB + BC)(AC + AB - BC)/2AB$ or $FG \times FH/2AB$. Then, since CD is a mean proportional between DE and DF, CE one between DE and EF, and CF one between DF and EF, there will be $CD = \sqrt{[FG \times FH \times HE \times EG]}/2AB$,

---

(270) Newton wrote '*aa*' but Whiston has corrected the manuscript reading (and that of his *editio princeps* in 1707) to read '*bb*'; likewise, he has corrected a similar slip '*bbaa*' on Newton's part in the next equation to the somewhat cumbrous form '*bbbb*'.

(271) Since $DH > HC$.

(272) We have already, in reproducing the preliminary draft of this problem (in 1, §3), observed that Newton's insertion of this late addition in its present place encourages little trust in the marginal chronology imposed on the manuscript of his Lucasian 'lectures'. In terms of its context his chosen location is equally odd: certainly, it would have been much more logical to set it with Problem 51 (on Newton's pages 120–1 below), whose content it repeats and augments. For a historical sketch of this 'Heronian' (more probably Archimedean) theorem, see 1, §3: note (1). Compare also note (392).

(273) Understand 'rectâ' (that straight line).

(274) 'primum' (first) is cancelled.

$$CE=\sqrt{\dfrac{AC\times HE\times EG}{AB}},\text{ et }CF=\sqrt{\dfrac{AC\times FG\times FH}{AB}}.\text{ Duc }CD\text{ in }\tfrac12AB\text{ et habebitur}$$

area $=\tfrac14\sqrt{FG\times FH\times HE\times EG}$. Pro angulo vero A determinando prodeunt Theoremata multiplicia, viz[275]

1. $2AB\times AC\,.\,HE\times EG(::AC\,.\,DE)::$radius ad sinum versum anguli A.[276]
2. $2AB\times AC\,.\,FG\times FH(::AC\,.\,FD)::$rad. cosin: vers: A.[277]
3. $2AB\times AC\,.\,\sqrt{FG\times FH\times HE\times EG}(::AC\,.\,CD)::$rad. sin A.
4. $\sqrt{FG\times FH}\,.\,\sqrt{HE\times EG}(::CF\,.\,CE)::$rad. tang: $\tfrac12A$.
5. $\sqrt{HE\times EG}\,.\,\sqrt{FG\times FH}(::CE\,.\,CF)::$rad. cotang $\tfrac12A$.
6. $2\sqrt{AB\times AC}\,.\,\sqrt{HE\times EG}(::FE\,.\,CE)::$rad. sin $\tfrac12A$.
7. $2\sqrt{AB\times AC}\,.\,\sqrt{FG\times FH}(::FE\,.\,FC)::$rad. cosin $\tfrac12A$.[278]

‖[85]　　　　　　　　　　　　　‖*Prob. 24.*[279]

[1677]
Lect 3　　*In dato angulo PAB actis utcunꝗ rectis BD, PD in data ratione hac semper lege ut BD sit parallela AP, et PD terminetur ad punctum P in recta AP datum: invenire locum puncti D.*

　　Age CD parallelam AB et DE perpendicularem AP ac dic $AP=a$, $CP=x$, & $CD=y$;[280] sitꝗ BD ad PD in ratione d ad e et erit AC vel $BD=a-x$, & $PD=\dfrac{ea-ex}{d}$. Sit insuper propter datum angulum DCE ratio CD ad CE, d ad f, et erit $CE=\dfrac{fy}{d}$, et $EP=x-\dfrac{fy}{d}$. Atqui

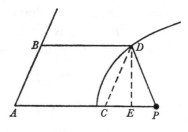

propter angulos ad E rectos est $CD^q-CE^q(=ED^q)=PD^q-EP^q$ hoc est

$$yy-\frac{ffyy}{dd}=\frac{eeaa-2eeax+eexx}{dd}-xx+\frac{2fxy}{d}-\frac{ffyy}{dd}.$$ Ac deletis utrobiꝗ $-\dfrac{ffyy}{dd}$,

terminisꝗ ritè dispositis $yy=\dfrac{2fxy}{d}+\dfrac{eeaa-2eeax+eexx-ddxx}{dd}$. Et extracta radice

$$y=\frac{fx}{d}\pm\sqrt{\dfrac{eeaa-2eeax{+ee\atop-dd}xx+ff}{dd}}.$$ Ubi cùm x et y in æquatione penultima non

---

(275) In sequence these are entered in the manuscript in the ranking 1, 2, 7, 3, 6, 4, 5 but are here rearranged to accord with Newton's numbering. It is understood that all the trigonometrical functions are the products of the radius and their modern, dimensionless equivalents.

(276) Namely, $R(1-\cos A)=2R\sin^2\tfrac12A$.

(277) That is, $R(1+\cos A)=2R\cos^2\tfrac12A$.

(278) On the preceding history of these half-angle formulas see J. Tropfke, *Geschichte der Elementar-Mathematik*, **5** (Berlin/Leipzig, ₂1923): 82–4.

$CE = \sqrt{[AC \times HE \times EG / AB]}$ and $CF = \sqrt{[AC \times FG \times FH / AB]}$. Multiply CD into $\frac{1}{2}AB$ and the area will be had $= \frac{1}{4}\sqrt{[FG \times FH \times HE \times EG]}$. Now, to be sure, there result several theorems for determining the angle A, namely:[275]

1. $2AB \times AC : HE \times EG = (AC : DE \text{ or}) \text{ radius} : \text{versin}\,A$.[276]

2. $2AB \times AC : FG \times FH = (AC : FD \text{ or}) \text{ radius} : \text{vercos}\,A$.[277]

3. $2AB \times AC : \sqrt{[FG \times FH \times HE \times EG]} = (AC : CD \text{ or}) \text{ radius} : \sin A$.

4. $\sqrt{[FG \times FH]} : \sqrt{[HE \times EG]} = (CF : CE \text{ or}) \text{ radius} : \tan \frac{1}{2}A$.

5. $\sqrt{[HE \times EG]} : \sqrt{[FG \times FH]} = (CE : CF \text{ or}) \text{ radius} : \cot \frac{1}{2}A$.

6. $2\sqrt{[AB \times AC]} : \sqrt{[HE \times EG]} = (FE : CE \text{ or}) \text{ radius} : \sin \frac{1}{2}A$.

7. $2\sqrt{[AB \times AC]} : \sqrt{[FG \times FH]} = (FE : FC \text{ or}) \text{ radius} : \cos \frac{1}{2}A$.[278]

Problem 24[279]

Where the straight lines BD, PD in given ratio are drawn randomly in the given angle $\widehat{PAB}$, but universally subject to the restriction that BD be parallel to AP and that PD terminate at the point P given in the straight line AP, to find the locus of the point D.

Draw CD parallel to AB and DE perpendicular to AP, and call $AP = a$, $CP = x$ and $CD = y$;[280] also, let BD be to PD in the ratio d to e, and there will be AC or $BD = a - x$ and $PD = e(a-x)/d$. In addition, because the angle $\widehat{DCE}$ is given, let the ratio of CD to CE be d to f, and there will be $CE = fy/d$ and $EP = x - fy/d$. However, because the angles at E are right, there is

$$CD^2 - CE^2 (= ED^2) = PD^2 - EP^2,$$

that is, $y^2 - f^2 y^2 / d^2 = e^2(a^2 - 2ax + x^2)/d^2 - x^2 + 2fxy/d - f^2 y^2 / d^2;$

and, when $-f^2 y^2 / d^2$ is deleted on both sides and the terms duly rearranged, $y^2 = 2fxy/d + (e^2 a^2 - 2e^2 ax + (e^2 - d^2)\,x^2)/d^2$ and, with the root extracted,

$$y = fx/d \pm \sqrt{[(e^2 a^2 - 2e^2 ax + (e^2 - d^2 + f^2)\,x^2)/d^2]}.$$

Here, since (in the last equation but one) x and y rise at most to two dimensions,

---

(279) In the manuscript this comes after 'Prob. 25' following, which therefore belongs to the preceding 'Lect 2': one more inconsistency in the marginal chronology!

(280) Newton chooses oblique Cartesian coordinates PC, CD to determine the locus-point D. In hindsight it would be much simpler to draw PA' and DB' perpendicular to AB and then DE' perpendicular to PA', and hence choose the rectangular coordinates PE' and $E'D$ to define D. At once, since $B'D = BD\sin\widehat{PAB}$, the ratio of PD to $B'D$ is given and the problem reduces to verifying that the classical focus-directrix defining property of a general conic (Pappus, *Mathematical Collection*, VII, 235–7; compare IV: 243, note (38)) yields a Cartesian equation of second degree in the corresponding analytical equivalent.

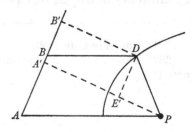

nisi ad duas dimensiones ascendant, locus puncti D erit Conica sectio, eaꝗ Hyperbola Parabola vel Ellipsis prout $ee-dd+ff$ (coefficiens ipsius xx in æquatione posteriori) sit majus, æquale vel minus nihilo.[281]

‖ [84 *rursum*]   ‖ *Prob.* 25.[282]

Rectis duabus VE et VC positione datis & ab alia recta PE circa polum positione datum P vertente sectis utcunꝗ in C et E: si recta intercepta CE dividatur in partes CD, DE rationem datam habentes, proponatur invenire locum puncti D.

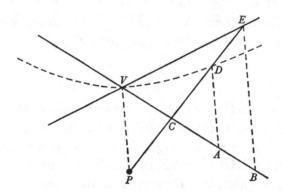

Age VP eiꝗ parallelas DA EB occurrentes VC in A et B. Dic $VP=a$. $VA=x$ & $AD=y$ & cum detur ratio CD ad DE vel conversè CD ad CE hoc est ratio DA ad $EB_{[,]}$ sit ista ratio d ad e et erit $EB=\frac{ey}{d}$. Præterea cùm detur angulus EVB adeoꝗ ratio EB ad VB, sit ista ratio e ad f et erit $VB=\frac{fy}{d}$. Deniꝗ propter similia triangula CEB, CDA, CPV, est $EB\,.\,CB::DA\,.\,CA::VP\,.\,VC$. et componendo $EB+VP\,.\,CB+VC::DA+VP\,.\,CA+VC$. hoc est $\frac{ey}{d}+a\,.\,\frac{fy}{d}::y+a\,.\,x$. Ductisꝗ extremis et medijs in se $eyx+dax=fyy+fay$. Ubi cùm indefinitæ quantitates x et y non nisi ad duas dimensiones ascendant, sequitur curvam VD in qua punctum D perpetim reperitur esse conicam sectionem eamꝗ Hyperbolam[283] quia una ex indefinitis quantitatibus x est unius tantùm dimensionis et in termino exy multiplicatur per alteram indefinitam quantitatem y.

---

(281) Compare note (259).
(282) This is set in the manuscript as the concluding part of 'Lect 2' preceding; see note (279).

the locus of the point D will be a conic—and a hyperbola, parabola or ellipse according as $e^2 - d^2 + f^2$ (the coefficient of x in the last equation) is greater than, equal to, or less than zero.[281]

Problem 25[282]

Where the two straight lines VE and VC given in position are arbitrarily cut in C and E by another straight line PE as it turns round the pole P given in position, if the intercepted straight line CE be divided into parts CD, DE having a given ratio (to one another), let it be proposed to find the locus of the point D.

Draw VP and, parallel to it, DA and EB meeting VC in A and B. Call $VP = a$, $VA = x$ and $AD = y$; and then, since the ratio of CD to DE is given, or *convertendo* that of CD to CE, that is, the ratio of DA to EB, let that ratio be d to e and there will be $EB = ey/d$. Moreover, since the angle $E\widehat{V}B$ is given, and hence the ratio of EB to VB, let that ratio be e to f and there will be $VB = fy/d$. Finally, because the triangles CEB, CDA, CPV are similar, there is

$$EB:CB = DA:CA = VP:VC,$$

and by compounding

$$(EB + VP):(CB + VC) = (DA + VP):(CA + VC);$$

that is, $(ey/d + a):fy/d = (y+a):x$, and, when extremes and middles are multiplied into each other, $exy + adx = fy^2 + afy$. Here, since the indefinite quantities x and y rise at most to two dimensions, it follows that the curve VD in which the point D is perpetually found is a conic—in fact a hyperbola,[283] seeing that one of the indefinite quantities, x, is merely of one dimension and in the term exy is multiplied by the other indefinite quantity y.

---

(283) Namely, with asymptotes $\alpha\beta$ ($y = -ad/e$) parallel to VC, and $\alpha\gamma$ ($ex - fy = af(1-d/e)$) parallel to VE respectively. Newton has not drawn the second branch of the hyperbola (D)—manifestly through $P(0, -a)$—which resolves the problem when CE lies in the exterior angle at V.

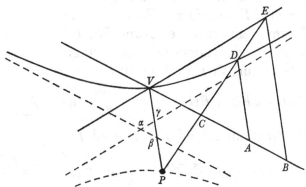

‖ [85 *rursum*]

‖ *Prob. 26.*[(284)]

Si rectæ duæ AC, BC a duobus positione datis punctis A et B in data quavis ratione ad tertium quodvis punctum C ducantur: invenire locum puncti concursûs C.

‖ [86] ‖ Junge AB et ad hanc demitte normalem CD dictisꝗ $AB=a$, $AD=x$, $DC=y$: erit $AC=\sqrt{xx+yy}$. $BD=x-a$ et $BC(=\sqrt{BD^q+CD^q})=\sqrt{xx-2ax+aa+yy}$. Jam cùm detur ratio AC ad BC sit ista d ad e[(285)] et extremis et medijs in se ductis erit $e\sqrt{xx+yy}=d\sqrt{xx-2ax+aa+yy}$. et per reductionem $\sqrt{\dfrac{ddaa-2ddax}{ee-dd}-xx}=y$.

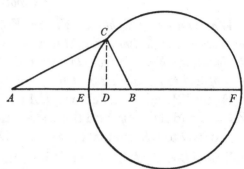

Ubi cum xx sit negativum et sola unitate affectum atꝗ etiam angulus ADC rectus, patet[(286)] curvam in quâ punctum C locatur esse circulum. Nempe in recta AB cape puncta E et F ita ut sint $d.e::AE.BE::AF.BF$, et erit EF circuli hujus diameter.

Et hinc e converso patet hoc Theorema quod in circuli cujusvis diametro EF infinitè producta datis utcunꝗ duobus punctis A et B hac lege ut sit $AE.AF::BE.BF$, et a punctis hisce actis duabus rectis AC, BC concurrentibus ad circulum in puncto quovis C: erit AC ad BC in data ratione AE ad BE.

[1677]
Lect 4

Prob. 27.

Invenire punctum D a quo tres rectæ DA, DB, DC ad totidem alias positione datas rectas AE, BF, CF perpendiculariter demissæ, datam inter se rationem obtineant.

E rectis positione datis producatur una puta BF ut et ejus perpendicularis BD donec reliquis AE et CF ‖ [87] occurrant, BF quidem in E et F, BD autem ‖ in H et G. Jam sit $EB=x$ et $EF=a$ eritꝗ $BF=a-x$. Cum autem propter datam positionem rectarum EF, EA, et FC, anguli E et F adeoꝗ rationes laterum triangulorum EBH & FBG dentur: sit EB ad BH ut d ad e et

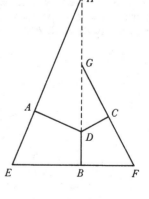

erit $BH=\dfrac{ex}{d}$, et $EH(=\sqrt{EB^q+BH^q})=\sqrt{xx+\dfrac{eexx}{dd}}$. hoc est $=\dfrac{x}{d}\sqrt{dd+ee}$.[(287)] Sit etiam BF ad BG ut d ad f et erit $BG=\dfrac{fa-fx}{d}$ et $FG(=\sqrt{BF^q+BG^q})=\sqrt{aa-2ax+xx\dfrac{+ffaa-2ffax+ffxx}{dd}}$.

Problem 26[284]

If two straight lines AC, BC be drawn in any given ratio from two points A and B, given in position, to any third point C, to find the locus of their meeting point C.

Join AB and to this let fall the normal CD; then, on calling $AB = a$, $AD = x$ and $DC = y$, there will be $AC = \sqrt{[x^2+y^2]}$, $BD = x - a$ and

$$BC \text{ (or } \sqrt{[BD^2 + CD^2]}) = \sqrt{[x^2 - 2ax + a^2 + y^2]}.$$

Now, since the ratio of AC to BC is given, let it be d to e[285] and, when extremes and middles have been multiplied together, there will be

$$e\sqrt{[x^2+y^2]} = d\sqrt{[x^2 - 2ax + a^2 + y^2]}$$

and by reduction $y = \sqrt{[(a^2d^2 - 2ad^2x)/(e^2 - d^2) - x^2]}$. Since x^2 is here negative and with a unit coefficient too, while, only, the angle $A\widehat{D}C$ is right, it is clear[286] that the curve in which the point C is located is a circle. Specifically, in the straight line AB take points E and F such that $AE:BE = AF:BF = d:e$ and EF will be a diameter of this circle.

It is hence clear that this converse theorem holds: given, in the infinitely extended diameter EF of any circle, two random points A restricted by $AE:AF = BE:BF$, when from these points two straight lines AC, BC are drawn meeting at the circle in any point C, then will AC be to BC in the given ratio AE to BE.

Problem 27

To find a point D such that, when three straight lines DA, DB, DC are let fall perpendicularly to a corresponding number of other straight lines AE, BF, CF given in position, a given ratio obtains between them.

Of the lines given in position extend one, BF say, and also its perpendicular, BD, till they meet the remaining lines AE and CF: BF, namely, in E and F, but BD in H and G. Now let $EB = x$ and $EF = a$, and there will be $BF = a - x$. However, because the position of the lines EF, EA and FC is given, the angles E and F, and hence the ratios of the sides of the triangles EBH and FBG are given: in consequence, let EB be to BH as d to e and there will be $BH = ex/d$ and EH (or $\sqrt{[EB^2 + BH^2]}$) $= \sqrt{[x^2 + e^2x^2/d^2]}$, that is, $x\sqrt{[d^2+e^2]}/d$.[287] Also let BF be to BG as d to f, and there will be $BG = f(a-x)/d$ and

$$FG \text{ (or } \sqrt{[BF^2 + BG^2]}) = \sqrt{[a^2 - 2ax + x^2 + f^2(a^2 - 2ax + x^2)/d^2]},$$

---

(284) The problem of Apollonius' circle; see IV: 315, note (76).

(285) Since $AC/CB = AE/EB > 1$, by implication $d > e$.

(286) Compare Descartes' *Geometrie* [= *Discours de la Methode* (Leyden, 1637): 297–413]: 328–9.

hoc est $=\dfrac{a-x}{d}\sqrt{dd+ff}$.[287] Præterea dicatur $BD=y$ et erit $HD=\dfrac{ex}{d}-y$, et

$GD=\dfrac{fa-fx}{d}-y$: adeoqʒ cùm sit $AD\,.\,HD(::EB\,.\,EH)::d\,.\,\sqrt{dd+ee}$, &

$$DC\,.\,GD(::BF\,.\,FG)::d\,.\,\sqrt{dd+ff},$$

erit $AD=\dfrac{ex-dy}{\sqrt{dd+ee}}$, et $DC=\dfrac{fa-fx-dy}{\sqrt{dd+ff}}$. Deniqʒ ob datas rationes linearum BD,

AD, DC, sit $BD\,.\,AD::\sqrt{dd+ee}\,.\,h-d$, et erit $\dfrac{hy-dy}{\sqrt{dd+ee}}(=AD)=\dfrac{ex-dy}{\sqrt{dd+ee}}$, sive

$hy=ex$.[288] Sit etiam $BD\,.\,DC::\sqrt{dd+ff}\,.\,k-d$ et erit

$$\frac{ky-dy}{\sqrt{dd+ff}}(=DC)=\frac{fa-fx-dy}{\sqrt{dd+ff}},$$

sive $ky=fa-fx$.[288] Est itaqʒ $\dfrac{ex}{h}(=y)=\dfrac{fa-fx}{k}$, & per reductionem $\dfrac{fha}{ek+fh}=x$.

Quare cape $EB\,.\,EF::h\,.\,\dfrac{ek}{f}+h$, dein $BD\,.\,EB::e\,.\,h$, & habebitur punctum quæsitum D.

Prob. 28.

Invenire punctum D, a quo tres rectæ DA, DB, DC ad data tria puncta A, B, C ductæ, datam inter se rationem obtineant.

E datis tribus punctis junge duo quævis puta A et C, et a tertio B ad lineam

‖[88] con‖jungentem AC demitte perpendiculum BE
ut et perpendiculum DF a puncto quæsito D:
dictisqʒ $AE=a$, $AC=b$, $EB=c$, $AF=x$, & $FD=y$
erit $AD^q=xx+yy$. $FC=b-x$.

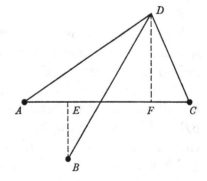

$$CD^q(=FC^q+FD^q)=bb-2bx+xx+yy.$$

$EF=x-a$, ac

$$BD^q(=EF^q+\overline{EB+FD}{}^{\text{quad}})$$
$$=xx-2ax+aa+cc+2cy+yy.$$

Jam cùm sit AD ad CD in data ratione sit ista ratio d ad e, et erit $CD=\dfrac{e}{d}\sqrt{xx+yy}$. Cùm etiam

sit AD ad BD in data ratione sit ista ratio d ad f et erit $BD=\dfrac{f}{d}\sqrt{xx+yy}$.

Adeoqʒ est $\dfrac{eexx+eeyy}{dd}(=CD^q)=bb-2bx+xx+yy$, &

$$\frac{ffxx+ffyy}{dd}(=BD^q)=xx-2ax+aa+cc+2cy+yy.$$ [289]

---

(287) More shortly, $EH = EB\sec(\tan^{-1}[e/d])$ and $FG = BF\sec(\tan^{-1}[f/d])$.

that is, $(a-x)\sqrt{[d^2+f^2]}/d$.[287] Moreover, call $BD=y$ and there will be $HD=ex/d-y$ and $GD=f(a-x)$ $d-y$; consequently, since

$$AD:HD\,(=EB:EH)=d:\sqrt{[d^2+e^2]}$$

and
$$DC:GD\,(=BF:FG)=d:\sqrt{[d^2+f^2]},$$

there will be $AD=(ex-dy)/\sqrt{[d^2+e^2]}$ and $DC=(f(a-x)-dy)/\sqrt{[d^2+f^2]}$. Finally, because the lines BD, AD, DC have given ratios, let

$$BD:AD=\sqrt{[d^2+e^2]}:(h-d)$$

and there will be $(hy-dy)/\sqrt{[d^2+e^2]}=(AD\text{ or})$ $(ex-dy)/\sqrt{[d^2+e^2]}$ and so $hy=ex$.[288] Let also $BD:DC=\sqrt{[d^2+f^2]}:(k-d)$ and there will be

$$(ky-dy)/\sqrt{[d^2+f^2]}=(DC\text{ or})\ (f(a-x)-dy)/\sqrt{[d^2+f^2]}$$

and so $ky=f(a-x)$.[288] Accordingly, $ex/h(=y)=f(a-x)/k$ and by reduction $x=afh/(ek+fh)$. Therefore take $EB:EF=h:(ek/f+h)$ and then

$$BD:EB=e:h,$$

and the point D required will be obtained.

Problem 28

To find a point D such that when three straight lines DA, DB, DC are drawn to the three given points A, B, C, a given proportion obtains between them.

Of the three given points join any two, say A and C, and from the third, B, to the joining line AC drop the perpendicular BE, and likewise the perpendicular DF from the point D sought; then, on calling $AE=a$, $AC=b$, $EB=c$, $AF=x$ and $FD=y$, there will be $AD^2=x^2+y^2$, $FC=b-x$,

$$CD^2\ (\text{or }FC^2+FD^2)=b^2-2bx+x^2+y^2,$$

$EF=x-a$ and BD^2 (or $EF^2+(EB+FD)^2)=x^2-2ax+a^2+c^2+2cy+y^2$. Now, since AD is to CD in given ratio, let that ratio be d to e, and there will be $CD=e\sqrt{[x^2+y^2]}/d$. Since AD is also to BD in a given ratio, let that ratio be d to f and there will be $BD=f\sqrt{[x^2+y^2]}/d$. Hence

$$e^2(x^2+y^2)/d^2\,(=CD^2)=b^2-2bx+x^2+y^2$$

and $f^2(x^2+y^2)/d^2(=BD^2)=x^2-2ax+a^2+c^2+2cy+y^2$.[289] If in these, for

---

(288) The outright computation of AD and DC is here something of a red herring; since the ratios of AD, BD and DC are given, the quadrilaterals $ADBE$ and $CDBF$ are given in species, whence it follows that the ratios $BD:EB=e:h$ and $BD:BF=f:k$ are given. In sequel Newton merely constructs the point D as the meet of ED ($y=(e/h)\,x$) and FD ($y=(f/k)\,(a-x)$).

(289) These are, of course, the Cartesian equations of the Apollonius circles (see Problem 26 above) defined by the constancy of the ratios $AD/CD=d/e$ and $AD/BD=d/f$: it is immediately obvious that their (two) intersections will satisfy the conditions of the problem.

In quibus si abbreviandi causa pro $\dfrac{dd-ee}{d}$ scribatur p, & q pro $\dfrac{dd-ff}{d}$, emerget

$bb-2bx+\dfrac{p}{d}xx+\dfrac{p}{d}yy=0$. & $aa+cc-2ax+2cy+\dfrac{q}{d}xx+\dfrac{q}{d}yy=0$. Per priorem est

$\dfrac{2bqx-bbq}{p}=\dfrac{q}{d}xx+\dfrac{q}{d}yy$. Quare in posteriori pro $\dfrac{q}{d}xx+\dfrac{q}{d}yy$ scribe $\dfrac{2bqx-bbq}{p}$, et

orietur $\dfrac{2bqx-bbq}{p}+aa+cc-2ax+2cy=0$.[(290)] Iterum abbreviandi causa scribe

m pro $a-\dfrac{bq}{p}$ & $2cn$ pro $\dfrac{bbq}{p}-aa-cc$, et erit $2mx+2cn=2cy$, terminisꝗ per $2c$

divisis $\dfrac{mx}{c}+n=y$. Quamobrem in æquatione $bb-2bx+\dfrac{p}{d}xx+\dfrac{p}{d}yy=0$ pro yy

scribe quadratum de $\dfrac{mx}{c}+n$ & habebitur

$$bb-2bx+\dfrac{p}{d}xx+\dfrac{pmm}{dcc}xx+\dfrac{2pmn}{dc}x+\dfrac{pnn}{d}=0.$$

Ubi denuò si abbreviandi causa $\dfrac{b}{r}$ scribatur pro $\dfrac{p}{d}+\dfrac{pmm}{dcc}$, & $\dfrac{sb}{r}$ pro $b-\dfrac{pmn}{dc}$

||[89] habebitur $xx=2sx-rb$. Et extracta radice $x=s\pm\sqrt{ss-rb}$. Invento x ||æquatio

$\dfrac{mx}{c}+n=y$ dabit y, et ex datis x & y hoc est AF et FB determinatur punctum

quæsitum D.

<table>
<tr><td>[1677]
Lect 5</td><td></td></tr>
</table>

Prob 29.

Invenire Triangulum ABC cujus tria latera AB, AC, BC & perpendiculum DC sunt in Arithmetica progressione.

    Dic $AC=a$. $BC=x$ et erunt $DC=2x-a$ et $AB=2a-x$.
Erunt etiam $AD(=\sqrt{AC^q-DC^q})=\sqrt{4ax-4xx}$ &

$$BD(=\sqrt{BC^q-DC^q})=\sqrt{4ax-3xx-aa}.$$

Atꝗ adeo rursus $AB=\sqrt{4ax-4xx}+\sqrt{4ax-3xx-aa}$.
Quare $2a-x=\sqrt{4ax-4xx}+\sqrt{4ax-3xx-aa}$, sive

$$2a-x-\sqrt{4ax-4xx}=\sqrt{4ax-3xx-aa}.$$

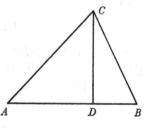

et partibus quadratis $4aa-3xx\overline{-4a+2x}\sqrt{4ax-4xx}=4ax-3xx-aa$, sive

$$5aa-4ax=\overline{4a-2x}\sqrt{4ax-4xx}.$$

Et partibus iterum quadratis ac terminis ritè dispositis

$$16x^4-80ax^3+144aaxx-104a^3x+25a^4=0.$$

Hanc æquationem divide per $2x-a$[(291)] et orietur

$$8x^3-36axx+54aax-25a^3=0,$$

brevity's sake, p be written in place of $(d^2-e^2)/d$ and q in place of $(d^2-f^2)/d$, there will emerge $b^2-2bx+(p/d)\,(x^2+y^2) = 0$ and

$$a^2+c^2-2ax+2cy+(q/d)\,(x^2+y^2) = 0.$$

By the former there is $(2bx-b^2)\,q/p = (q/d)\,(x^2+y^2)$. Therefore in the latter in place of $(q/d)\,(x^2+y^2)$ write $2bqx/p-b^2q/p$ and there will arise

$$a^2+c^2-2ax+2cy+2bqx/p-b^2q/p = 0.^{(290)}$$

Again for brevity's sake, write m for $a-bq/p$ and $2cn$ for $-a^2-c^2+b^2q/p$, and there will come $2mx+2cn = 2cy$ and so, with the terms divided by $2c$, $mx/c+n=y$. In consequence, in the equation $b^2-2bx+(p/d)\,(x^2+y^2) = 0$ in place of y^2 write the square of $mx/c+n$ and there will be had

$$b^2-2bx+(p/d)\,x^2+(pm^2/dc^2)\,x^2+2(pmn/dc)\,x+pn^2/d = 0.$$

If in this, for brevity's sake once more, b/r be written in place of $p/d+pm^2/dc^2$ and sb/r in place of $b-pmn/dc$, there will be obtained $x^2 = 2sx-rb$ and so, when the root is extracted, $x = s\pm\sqrt{[s^2-rb]}$. Once x is found, the equation $mx/c+n = y$ will give y, and from x and y, that is, AF and FB, given the point D sought is determined.

Problem 29

To determine a triangle ABC whose three sides AB, AC, BC and altitude DC are in arithmetical progression.

Call $AC = a$ and $BC = x$, and there will then be $DC = 2x-a$ and $AB = 2a-x$; also AD (or $\sqrt{[AC^2-DC^2]}$) $= \sqrt{[4ax-4x^2]}$ and

$$BD \text{ (or } \sqrt{[BC^2-DC^2]}) = \sqrt{[4ax-3x^2-a^2]},$$

so that again $AB = \sqrt{[4ax-4x^2]}+\sqrt{[4ax-3x^2-a^2]}$. Therefore

$$2a-x = \sqrt{[4ax-4x^2]}+\sqrt{[4ax-3x^2-a^2]},$$

or $2a-x-\sqrt{[4ax-4x^2]} = \sqrt{[4ax-3x^2-a^2]}$ and, on squaring the sides,

$$4a^2-3x^2-(4a-2x)\,\sqrt{[4ax-4x^2]} = 4ax-3x^2-a^2,$$

or $5a^2-4ax = (4a-2x)\,\sqrt{[4ax-4x^2]}$, and so, with the sides again squared and the terms duly rearranged, $16x^4-80ax^3+144a^2x^2-104a^3x+25a^4 = 0$. Divide this equation by $2x-a^{(291)}$ and there will result $8x^3-36ax^2+54a^2x-25a^3 = 0$,

---

(290) This is, in fact, the equation of the radical axis of the two preceding Apollonius circles (see note (289)).

(291) The root $x = \frac{1}{2}a$ corresponds to the trivial particular solution for which point C lies in AB with $AB = \frac{3}{2}a$, $AC = a$, $BC = \frac{1}{2}a$ and $DC = 0$.

æquatio cujus resolutione dabitur x ex assumpto utcunꝗ a. Habitis a et x constitue triangulum cujus latera erunt $2a-x$, a, & x & perpendiculum in latus $2a-x$ demissum erit $2x-a$.

Si posuissem differentiam laterum trianguli esse d & perpendiculum esse x: opus evasisset aliquanto concinnius, prodeunte tandem æquatione

$$x^3 = 24ddx - 48d^3.^{(292)}$$

Prob 30.

‖ [90]   *Invenire Triangulum ABC cujus tria latera AB, AC, BC, & perpen‖diculum CD sunt in geometrica progressione.*

Dic $AC=x$ & $BC=a$ et erit $AB=\dfrac{xx}{a}$ & $CD=\dfrac{aa}{x}$. Est et

$$AD\left(=\sqrt{AC^q-CD^q}\right)=\sqrt{xx-\frac{a^4}{xx}} \quad \& \quad BD\left(=\sqrt{BC^q-DC^q}\right)=\sqrt{aa-\frac{a^4}{xx}}:$$

adeoꝗ $\dfrac{xx}{a}\,(=AB)=\sqrt{xx-\dfrac{a^4}{xx}}+\sqrt{aa-\dfrac{a^4}{xx}}$, sive $\dfrac{xx}{a}-\sqrt{aa-\dfrac{a^4}{xx}}=\sqrt{xx-\dfrac{a^4}{xx}}$. Et

partibus æquationis quadratis $\dfrac{x^4}{aa}-\dfrac{2xx}{a}\sqrt{aa-\dfrac{a^4}{xx}}+aa-\dfrac{a^4}{xx}=xx-\dfrac{a^4}{xx}$, hoc est

$x^4-aaxx+a^4=2aax\sqrt{xx-aa}$. Et partibus iterum quadratis

$$x^8-2aax^6+3a^4x^4-2a^6xx+a^8=4a^4x^4-4a^6xx.$$

Hoc est $x^8-2aax^6-a^4x^4+2a^6xx+a^8=0$. Divide hanc æquationem per

$$x^4-aaxx-a^4 \quad \text{et orietur} \quad x^4-aaxx-a^4[=0].^{(293)}$$

Quare est $x^4=aaxx+a^4$ et extracta radice $x=a\sqrt{\tfrac{1}{2}+\sqrt{\tfrac{5}{4}}}.^{(294)}$ Cape ergo a sive BC cujusvis longitudinis & fac $BC.AC::AC.AB::1.\sqrt{\tfrac{1}{2}+\sqrt{\tfrac{5}{4}}}$; et trianguli ABC ex his lateribus constituti perpendiculum DC erit ad latus BC in eadem ratione.

Idem aliter.

Cùm sit $AB.AC::BC.DC$ dico angulum ACB rectum esse. Nam si negas age CE constituentem angulum ECB rectum. Sunt ergo triangula BCE, DBC similia per 8. vi. Elem. adeoꝗ $EB.EC::BC.DC$. hoc est $EB.EC::AB.AC$. Age AF perpendicularem CE & propter parallelas $AF_{[,]}$ BC erit $EB.EC::AE.FE::AB.FC$. Ergo per 9. v

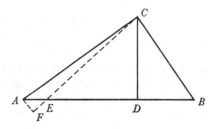

---

(292) Understand 'assumpto d' (when d is assumed). In his 1722 revise Newton correctly altered the last term to '$+48d^3$'.

an equation whose resolution will yield x in terms of any a assumed. Having obtained a and x, set up the triangle whose sides shall be $2a-x$, a and x, and the perpendicular let fall onto the side $2a-x$ will be $2x-a$.

If I had put the difference of the triangle's sides to be d and the altitude x, the work would have proved to be somewhat neater, yielding at length the equation $x^3 = 24d^2x - 48d^3$.[(292)]

Problem 30

To find a triangle ABC whose three sides AB, AC, BC and altitude CD are in geometrical progression.

Call $AC = x$ and $BC = a$ and there will be $AB = x^2/a$ and $CD = a^2/x$. Also AD (or $\sqrt{[AC^2 - CD^2]}$) $= \sqrt{[x^2 - a^4/x^2]}$ and

$$BD \text{ (or } \sqrt{[BC^2 - DC^2]}) = \sqrt{[a^2 - a^4/x^2]},$$

so that $x^2/a (= AB) = \sqrt{[x^2 - a^4/x^2]} + \sqrt{[a^2 - a^4/x^2]}$; that is,

$$x^2/a - \sqrt{[a^2 - a^4/x^2]} = \sqrt{[x^2 - a^4/x^2]}$$

and, when the sides of the equation are squared,

$$x^4/a^2 - 2(x^2/a)\sqrt{[a^2 - a^4/x^2]} + a^2 - a^4/x^2 = x^2 - a^4/x^2,$$

whence $x^4 - a^2x^2 + a^4 = 2a^2x\sqrt{[x^2 - a^2]}$ and so, when the sides are again squared, $x^8 - 2a^2x^6 + 3a^4x^4 - 2a^6x^2 + a^8 = 4a^4x^4 - 4a^6x^2$, that is,

$$x^8 - 2a^2x^6 - a^4x^4 + 2a^6x^2 + a^8 = 0.$$

Divide this equation by $x^4 - a^2x^2 - a^4$ and there will arise $x^4 - a^2x^2 - a^4 = 0$.[(293)] Consequently $x^4 = a^2x^2 + a^4$ and when the root is extracted $x = a\sqrt{[\frac{1}{2} + \sqrt{\frac{5}{4}}]}$.[(294)] Therefore take a, that is, BC, of any length and make

$$BC:AC = AC:AB = 1:\sqrt{[\tfrac{1}{2}(1 + \sqrt{5})]},$$

and the triangle ABC formed from these sides will have the perpendicular DC to the side BC in the same ratio.

The same another way

Since $AB:AC = BC:DC$, I assert that the angle $\widehat{ACB}$ is right. For, should you deny it, draw CE forming the right angle $\widehat{ECB}$. Accordingly, the triangles BCE, BDC are similar by *Elements*, VI, 8 and hence $EB:EC = BC:DC$, that is, $EB:EC = AB:AC$. Draw AF perpendicular to CE and because of the parallels AF, BC there will be $EB:EC = AE:FE = AB:FC$. Therefore by *Elements*, V, 9,

---

(293) A roundabout way of saying 'Extrahe radicem quadraticam et orietur

$$x^4 - aaxx - a^4 = 0 \text{'}$$

(Extract the square root and there will arise $x^4 - a^2x^2 - a^4 = 0$).

(294) On taking $\alpha = \sqrt{[\frac{1}{2}(1 + \sqrt{5})]}$ it follows that $AB = \alpha^2a$, $AC = \alpha a$, $BC = a$, $CD = \alpha^{-1}a$ (in geometrical progression) with $AD = a$ and $DB = \alpha^{-2}a$.

Elem. est $AC=FC$, hoc est Hypotenusa trianguli rectanguli æqualis lateri
‖[91] contra 19. I. Elem. Non est ergo angulus ‖ ECB rectus, et proinde ipsum ACB
rectum esse oportet.[295] Est itacp $AC^q+BC^q=AB^q$. Sed est $AC^q=AB\times BC$.
Ergo $AB\times BC+BC^q=AB^q$, et extracta radice $AB=\frac{1}{2}BC+\sqrt{\frac{5}{4}}BC^q$. Quamobrem cape $BC.AB::1.\dfrac{1+\sqrt5}{2}$, & AC mediam proportionalem inter BC et

AB, & triangulo ex his lateribus constituto erunt $AB.AC.BC.DC$ continuè
proportionales.

[1677]
Lect 6

Prob. 31.

*Super data basi AB triangulum ABC constituere, cujus vertex C erit ad rectam EC
positione datam, basis autem medium existet
arithmeticum inter latera.*

Basis AB bisecetur in F, & producatur donec rectæ EC positione datæ
occurrat in E, & ad ipsam demittatur
perpendicularis CD: dictiscp $AB=a$,
$FE=b$, & $BC-AB=x$, erit $BC=a+x$,
$AC=a-x$. Et per 13. II. Elem.

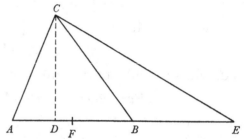

$$BD\left(=\frac{BC^q-AC^q+AB^q}{2AB}\right)=2x+\tfrac{1}{2}a.$$

Adeocp $FD=2x$, $DE=b+2x$, & $CD(=\sqrt{CB^q-BD^q})=\sqrt{\frac{3}{4}aa-3xx}$. Sed propter
datas positiones rectarum CE et AB datur angulus CED adeocp et ratio DE
ad CD, quæ si ponatur d ad e dabit analogiam $d.e::b+2x.\sqrt{\frac{3}{4}aa-3xx}$. Unde,
multiplicatis extremis et medijs in se, oritur æquatio $eb+2ex=d\sqrt{\frac{3}{4}aa-3xx}$,
cujus partibus quadratis et rite dispositis, fit $xx=\dfrac{\frac{3}{4}ddaa-eebb-4eebx}{4ee+3dd}$. et radice
extracta $x=\dfrac{-2eeb+d\sqrt{3eeaa-3eebb+\frac{9}{4}ddaa}}{4ee+3dd}$.[296] Dato autem x, datur $BC=a+x$
et $AC=a-x$.

---

(295) This involved, indirect proof scarcely seems necessary, for at once, since the ratio
$AB:AC = BC:DC$ is the cosecant of $\widehat{ABC}$, the angle $\widehat{ACB}$ must be right. But Newton is
manifestly concerned to give Euclidean proof of an unusual test for the similarity (and hence
equiangularity) of two triangles: one to be found neither in the *Elements*, nor in any other
text accessible to him.

there is $AC = FC$, that is, the hypotenuse of a right-angled triangle equal to a side, contrary to *Elements*, I, 19. The angle $\widehat{ECB}$ is therefore not right, and in consequence $\widehat{ACB}$ itself must be right.[295] Accordingly $AC^2 + BC^2 = AB^2$. But $AC^2 = AB \times BC$ and therefore $AB \times BC + BC^2 = AB^2$, and, when the root is extracted, $AB = \frac{1}{2}BC + \sqrt{[\frac{5}{4}BC^2]}$. Wherefore take $BC:AB = 1:\frac{1}{2}(1+\sqrt{5})$ and AC the mean proportional between BC and AB, then, once the triangle is formed from these sides, AB, AC, BC, DC will be in continued proportion.

Problem 31

On the given base AB to set up a triangle ABC whose vertex C shall be on a straight line EC given in position, while its base prove to be the arithmetic mean between the sides.

Let the base AB be bisected in F and extended till it meets the straight line EC given in position, and to it let fall the perpendicular CD. On calling $AB = a$, $FE = b$ and $BC - AB = x$, there will then be $BC = a+x$, $AC = a-x$ and by *Elements*, II, 13, BD (or $(BC^2 - AC^2 + AB^2)/2AB) = 2x + \frac{1}{2}a$, so that $FD = 2x$, $DE = b + 2x$ and CD (or $\sqrt{[CB^2 - BD^2]}) = \sqrt{[\frac{3}{4}a^2 - 3x^2]}$. But, because the lines CE and AB are given in position, the angle $\widehat{CED}$ is given and hence also the ratio of DE to CD; if this be put to be d to e, it will yield the proportion

$$d:e = (b+2x):\sqrt{[\tfrac{3}{4}a^2 - 3x^2]}.$$

From this, on multiplying extremes and middles into each other, there results the equation $be + 2ex = d\sqrt{[\frac{3}{4}a^2 - 3x^2]}$ and, when its sides are squared and duly rearranged, there comes $x^2 = (\frac{3}{4}a^2d^2 - b^2e^2 - 4be^2x)/(4e^2 + 3d^2)$ and with the root extracted $x = (-2be^2 + d\sqrt{[3a^2e^2 - 3b^2e^2 + \frac{9}{4}a^2d^2]})/(4e^2 + 3d^2)$.[296] Given x, however, there is given $BC = a+x$ and $AC = a-x$.

---

(296) Strictly, the sign attached to the radical should be ' $\pm$ ', so allowing for the two positions of C in the construction line CE. These (shown in the accompanying figure as C, C') will evidently be the intersections of CE with the ellipse drawn on foci A, B with main axis equal to $(AC+BC =) 2AB$.

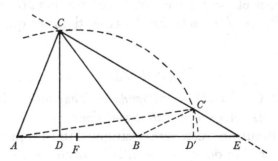

|| [92]

|| *Prob. 32.*

Datis positione tribus rectis AD, AE, BF, quartam DF ducere cujus partes DE EF prioribus interceptæ, datarum erunt longitudinum.

Ad *BF* demitte perpendicularem *EG* ut et obliquam *EC* parallelam *AD*, et rectis tribus positione datis concurrentibus in *A*, *B*, et *H*, dic $AB=a$. $BH=b$. $AH=c$. $ED=d$. $EF=e$, & $HE=x$. Jam propter similia triangula *ABH*, *ECH*, est $AH.AB::HE.EC=\dfrac{ax}{c}$,

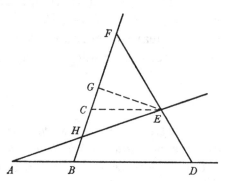

& $AH.HB::HE.CH=\dfrac{bx}{c}$. Adde *HB* et fit

$CB=\dfrac{bx+bc}{c}$. Insuper propter similia triangula *FEC*, *FDB*, est $ED.CB::EF.CF=\dfrac{ebx+ebc}{dc}$. Deniqȝ per 12 & 13. II.

Elem. est $\dfrac{EC^q-EF^q}{2FC}+\tfrac{1}{2}FC(=CG)=\dfrac{HE^q-EC^q}{2CH}-\tfrac{1}{2}CH$. hoc est

$$\dfrac{\dfrac{aaxx}{cc}-ee}{\dfrac{2ebx+2ebc}{dc}}+\dfrac{ebx+ebc}{2dc}=\dfrac{xx-\dfrac{aaxx}{cc}}{\dfrac{2bx}{c}}-\dfrac{bx}{2c}.$$

sive $\dfrac{aadxx-eedcc}{ebx+ebc}+\dfrac{ebx}{d}+\dfrac{ebc}{d}=\dfrac{ccx-aax-bbx}{b}$. Hic abbreviandi causa pro $\dfrac{cc-aa-bb}{b}-\dfrac{eb}{d}$ scribe *m* et erit $\dfrac{aadxx-eedcc}{ebx+ebc}+\dfrac{ebc}{d}=mx$, ac terminis omnibus multiplicatis per $x+c$ fiet $\dfrac{aadxx-eedcc}{eb}[+]\dfrac{ebcx}{d}+\dfrac{ebcc}{d}=mxx+mcx$. Iterum pro $\dfrac{aad}{eb}-m$ scribe *p*, pro $mc[-]\dfrac{ebc}{d}$ scribe $2pq$, & pro $[-]\dfrac{ebcc}{d}+\dfrac{eedcc^{(297)}}{eb}$ scribe *prr* et evadet $xx=2qx+rr$ et $x=q\pm\sqrt{qq+rr}$. Invento *x* sive *HE* age *EC* parallelam

|| [93] *AB*, & cape $FC.BC::e.d$, & acta *FED* || conditionibus quæstionis satisfaciet.

[1677]
Lect 7

Prob. 33.

Ad Circulum centro C radio CD descriptum ducere Tangentem DB cujus pars PB inter rectas positione datas AP, AB sita sit datæ longitudinis.

A centro *C* ad alterutram rectarum positione datarum puta *AB* demitte normalem *CE* eamqȝ produc donec Tangenti *DB* occurrat in *H*. Ad eandem *AB* demitte etiam normalem *PG*, &[298] dictis $EA=a$, $EC=b$, $CD=c$, $BP=d$, & $PG=x$,

Problem 32

Given three straight lines AD, AE, BF in position, to draw a fourth, DF, whose parts DE, EF intercepted by the previous ones shall be of given lengths.

To *BF* let fall the perpendicular *EG* and also the oblique line *EC* parallel to *AD*; then, where the three lines given in position meet in *A*, *B* and *H*, call *AB = a*, *BG = b*, *AH = c*, *ED = d*, *EF = e* and *HE = x*. Now, because the triangles *ABH*, *ECH* are similar, it is $AH:AB = HE:EC$ or $EC = ax/c$, and $AH:HB = HE:CH$ or $CH = bx/c$. Add *HB* and it becomes $CB = b(x+c)/c$. In addition, because the triangles *FEC*, *FDB* are similar, it is $ED:CB = EF:CF$ or $CF = be(x+c)/cd$. Finally, by *Elements*, II, 12/13 there is

$$\tfrac{1}{2}(EC^2 - EF^2)/FC + \tfrac{1}{2}FC(= CG) = \tfrac{1}{2}(HE^2 - EC^2)/CH - \tfrac{1}{2}CH,$$

that is,

$$\tfrac{1}{2}(a^2x^2/c^2 - e^2)/(be(x+c)/cd) + \tfrac{1}{2}be(x+c)/cd = \tfrac{1}{2}(x^2 - a^2x^2/c^2)/(bx/c) - \tfrac{1}{2}bx/c$$

or $d(a^2x^2 - c^2e^2)/be(x+c) + be(x+c)/d = (c^2 - a^2 - b^2)\,x/b$. Here, for brevity's sake, in place of $(c^2 - a^2 - b^2)/b - be/d$ write *m* and there will be

$$d(a^2x^2 - c^2e^2)/be(x+c) + bce/d = mx;$$

this, when all its terms are multiplied by $x+c$, will become

$$d(a^2x^2 - c^2e^2)/be + bce(x+c)/d = mx(x+c).$$

Again, in place of $a^2d/be - m$ write *p*, for $cm - bce/d$ write $2pq$ and for

$$-bc^2e/d + c^2de^2/be^{(297)}$$

write pr^2, and there will come to be $x^2 = 2qx + r^2$ and so $x = q \pm \sqrt{[q^2 + r^2]}$. Once *x*, that is, *HE*, is found, draw *EC* parallel to *AB* and take $FC:BC = e:d$, then *FED* when drawn will satisfy the conditions of the question.

Problem 33

To the circle described on centre C and with radius CD to draw a tangent DB whose section PB lying between the straight lines AP, AB given in position shall be of given length.

From the centre *C* to one or other of the lines given in position, say *AB*, let fall the normal *CE* and extend it till it meets the tangent *DB* in *H*. To the same line *AB* let fall also the normal *PG*, and[298] call $EA = a$, $EC = b$, $CD = c$, $BP = d$

---

(297) Newton has failed to notice the common factor *e* in the numerator and denominator of this fraction. The unreduced form is retained in all printed editions.

(298) 'junge *CD*' (join *CD*) is cancelled.

propter similia triangula PGB, CDH erit $GB(\sqrt{dd-xx}) \cdot PB :: CD \cdot CH = \dfrac{cd}{\sqrt{dd-xx}}$.

Adde EC et fiet $EH = b + \dfrac{cd}{\sqrt{dd-xx}}$.

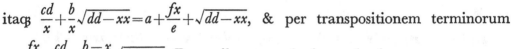

Porrò est

$$PG \cdot GB :: EH \cdot EB = \frac{b}{x}\sqrt{dd-xx} + \frac{cd}{x}.$$

Adhæc propter angulum PAG datum datur ratio PG ad AG qua posita e ad f erit $AG = \dfrac{fx}{e}$. Adde EA et BG & habebitur denuò $EB = a + \dfrac{fx}{e} + \sqrt{dd-xx}$. Est

itacɜ $\dfrac{cd}{x} + \dfrac{b}{x}\sqrt{dd-xx} = a + \dfrac{fx}{e} + \sqrt{dd-xx}$, & per transpositionem terminorum

$a + \dfrac{fx}{e} - \dfrac{cd}{x} = \dfrac{b-x}{x}\sqrt{dd-xx}$. Et partibus æquationis quadratis

$$aa + \frac{2afx}{e} - \frac{2acd}{x} + \frac{ffxx}{ee} - \frac{2cdf}{e} + \frac{ccdd}{xx} = \frac{bbdd}{xx} - bb - \frac{2bdd}{x} + 2bx + dd - xx.$$

$$\begin{matrix} +2aef \\ -2bee \end{matrix}x^3 \quad \begin{matrix} +aaee \\ +bbee \\ -ddee \end{matrix}xx \quad \begin{matrix} +2bddee \\ -2acdee \end{matrix}x \quad \begin{matrix} +ccddee \\ -bbddee \end{matrix}$$

Et per debitam reductionem $x^4 \dfrac{\begin{matrix} -2cdef \end{matrix}}{ee+ff} = 0.$[299]

|| [94]

|| *Prob. 34.*[300]

Si punctum lucidum A, radios versus refringentem superficiem planam CD ejiciat: invenire radium AC cujus refractus CB impinget in datum punctum B.

A puncto isto lucido ad refringens planum demitte perpendiculum AD, et cum eo utrinꝗ producto concurrat refractus radius BC in E, & perpendiculum a puncto B demissum in F, & agatur BD, dictisꝗ $AD = a$, $DB = b$, $BF = c$, $DC = x$, statue rationem sinuum incidentiæ et refractionis hoc est sinuum

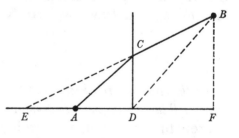

angulorum CAD, CED esse d ad e, et cùm EC & AC (ut notum est) sint in eadem ratione,[301] & AC sit $\sqrt{aa+xx}$ erit $EC = \dfrac{d}{e}\sqrt{aa+xx}$. Præterea est

$$ED\left(=\sqrt{EC^q - CD^q}\right) = \sqrt{\frac{ddaa + ddxx}{ee} - xx}, \quad \& \quad DF = \sqrt{bb - cc},$$

and $PG = x$. Then, because the triangles PGB, CDH are similar there will be GB (or $\sqrt{[d^2-x^2]}):PB = CD:CH$ and so $CH = cd/\sqrt{[d^2-x^2]}$. Add EC and there will come $EH = b+cd/\sqrt{[d^2-x^2]}$. Moreover, $PG:GB = EH:EB$ and so

$$EB = (b\sqrt{[d^2-x^2]}+cd)/x.$$

Further, because the angle $\widehat{PAG}$ is given, there is given the ratio of PG to AG, and, when this is put to be e to f, there will be $AG = fx/e$. Add EA and BG, and there will be obtained afresh $EB = a+fx/e+\sqrt{[d^2-x^2]}$. Accordingly, there is $cd/x+b\sqrt{[d^2-x^2]}/x = a+fx/e+\sqrt{[d^2-x^2]}$ and, by transposition of the terms, $a+fx/e-cd/x = (b-x)\sqrt{[d^2-x^2]}/x$, and then, on squaring the sides of the equation

$$a^2+2afx/e-2acd/x+f^2x^2/e^2-2cdf/e+c^2d^2/x^2$$
$$= b^2d^2/x^2-b^2-2bd^2/x+2bx+d^2-x^2.$$

Hence, by due reduction,

$$x^4 + \frac{2e(af-be)\,x^3+(e^2(a^2+b^2-d^2)-2cdef)\,x^2+2de^2(bd-ac)\,x+(c^2-b^2)\,d^2e^2}{e^2+f^2} = 0.^{(299)}$$

Problem 34[300]

If a luminous point A should throw out rays in the direction of the plane interface CD, to find the ray AC whose refraction CB will strike the given point B.

From that luminous point to the refracting plane let fall the perpendicular AD, and let its extension either way be met by the refracted ray BC in E and by the perpendicular dropped from point B to it in F, then draw BD and, calling $AD = a$, $DB = b$, $BF = c$, $DC = x$, set the ratio of the sines of incidence and refraction—that is, the sines of angles $\widehat{CAD}$ and $\widehat{CED}$—to be d to e. Since EC and AC (as is known) are in the same ratio[301] and AC is $\sqrt{[a^2+x^2]}$, there will be $EC = d\sqrt{[a^2+x^2]}/e$. Furthermore,

$$ED \text{ (or } \sqrt{[EC^2-CD^2]}) = \sqrt{[d^2(a^2+x^2)/e^2-x^2]} \quad \text{and} \quad DF = \sqrt{[b^2-c^2]},$$

---

(299) It will be evident geometrically that, depending on the length of PB and the position of PA and PB, all four roots of this quartic may be real, or just two, or none at all.

(300) The Lower–Sluse problem of constructing the point of refraction at a plane interface between two media; for Newton's earlier equivalent solution—with the resulting quartic equation constructed in Cartesian style by the intersections of a circle and parabola—see III: 450–3.

(301) Namely, $CD/AC:CD/EC$.

atɋ $EF = \sqrt{bb-cc} + \sqrt{\dfrac{ddaa+ddxx}{ee} - xx}$. Deniɋ propter similia triangula ECD, EBF, est $ED.DC::EF.FB$, et ductis extremorum et mediorum valoribus in se

$$c\sqrt{\dfrac{ddaa+ddxx}{ee} - xx} = x\sqrt{bb-cc} + x\sqrt{\dfrac{ddaa+ddxx}{ee} - xx}. \text{ sive}$$

$$\overline{c-x}\sqrt{\dfrac{ddaa+ddxx}{ee} - xx} = x\sqrt{bb-cc}.$$

Et partibus æquationis quadratis & rite dispositis

$$x^4 - 2cx^3 \dfrac{\begin{array}{l}+ddcc\\+ddaa\,xx - 2ddaacx + ddaacc\\-eebb\end{array}}{dd-ee} = 0.$$

[1677]
Lect 8

‖ [95]

Prob. 35.[302]

Invenire locum verticis trianguli D cujus basis AB datur, et anguli ad ‖*basem DAB, DBA datam habent differentiam.*

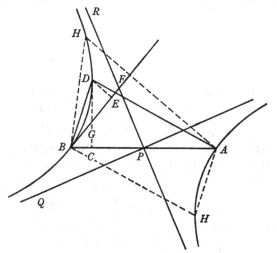

Ubi angulus ad verticem, sive (quod perinde est) ubi summa angulorum ad basem datur, docuit Euclides[303] locum verticis esse circumferentiam circuli; proposuimus igitur inventionem loci ubi differentia angulorum ad basem datur. Sit angulus DBA major angulo DAB, sitɋ ABF eorum data differentia, rectâ BF occurrente AD in F. Insuper ad BF demittatur normalis DE, ut et ad AB normalis DC occurrens BF in G. Dictisɋ $AB = a$, $AC = x$, & $CD = y$, erit $BC = a-x$.

Jam in triangulo BCG cum dentur omnes anguli dabitur ratio laterum BC et GC. Sit ista d ad a et erit $CG = \dfrac{aa-ax}{d}$. Aufer hanc de DC sive y et restabit $DG = \dfrac{dy-aa+ax}{d}$. Præterea propter similia triangula $BGC\ DGE$ est $BG.BC::DG.DE$. Est autem in triangulo BGC, $a.d::CG.BC$, adeoɋ $aa.dd::CG^q.BC^q$ et componendo $aa+dd.dd::BG^q.BC^q$. et extractis radicibus

---

(302) This problem is briefly enunciated in Newton's contemporary notes on geometrical loci (see IV: 246, note (58)) and again in Problem 7 of his 'Quæstionum solutio Geometrica' (IV: 252).

so that $EF = \sqrt{[b^2 - c^2]} + \sqrt{[d^2(a^2 + x^2)/e^2 - x^2]}$. Finally, because the triangles ECD, EBF are similar, there is $ED:DC = EF:FB$ and so, when extremes and middles are multiplied into one another,

$$c\sqrt{[d^2(a^2 + x^2)/e^2 - x^2]} = x\sqrt{[b^2 - c^2]} + x\sqrt{[d^2(a^2 + x^2)/e^2 - x^2]}$$

or $(c - x)\sqrt{[d^2(a^2 + x^2)/e^2 - x^2]} = x\sqrt{[b^2 - c^2]}$; and with the sides of the equation squared and duly rearranged,

$$x^4 - 2cx^3 + \frac{(a^2d^2 + c^2d^2 - b^2e^2)\, x^2 - 2a^2cd^2x + a^2c^2d^2}{d^2 - e^2} = 0.$$

Problem 35[302]

To find the locus of the vertex D of a triangle whose base AB is given and in which the angles $D\widehat{A}B$, $D\widehat{B}A$ at the base have a given difference.

When the angle at the vertex is given—or (what is effectively the same thing) when the sum of the angles at the base is—, Euclid[303] has taught that the locus of the vertex is a circle's circumference; we have therefore proposed finding the locus when the difference of the angles at the base is given. Let the angle $D\widehat{B}A$ be greater than the angle $D\widehat{A}B$ and let $A\widehat{B}F$, where the straight line BF meets AD in F, be their difference. In addition, to BF let fall the normal DE, and likewise to AB the normal DC meeting BF in G. Then, on calling $AB = a$, $AC = x$ and $CD = y$, there will be $BC = a - x$. Now in the triangle BCG, since all its angles are given, there will be given the ratio of the sides BC and GC. Let that be d to a and there will be $CG = a(a - x)/d$. Take this from DC, that is, y, and there will remain $DG = (dy - a^2 + ax)/d$. Moreover, because the triangles BGC DGE are similar, it is $BG:BC = DG:DE$. But, in the triangle BGC, $CG:BC = a:d$, so that $CG^2:BC^2 = a^2:d^2$ and, by compounding,

$$BG^2:BC^2 = (a^2 + d^2):d^2$$

---

(303) In the margin alongside Whiston has added the correct reference to '**III**. 29. Euclid:' (*Elements*, **III**, 29) and this is incorporated in all the printed editions. Newton's present variant on a theorem in Euclid is in keeping with the widespread contemporary desire to improve on classical mathematics. In the middle 1690's, for example, Johann Bernoulli thought to improve on 'decantatissima illa proprietas circuli demonstrata prop. 35 & 36 lib. 3 Euclid' (that the product of the intercepts cut off an arbitrary line through a fixed point by a given circle is constant) by proposing the more general problem: 'Invenire & construere curvam...ita ut ex puncto dato...ducta quavis recta...secante curvam in [duobus] punctis, rectangulum sub segmentis...sit semper æquale eidem constanti dato' ('Supplementum Defectus Geometriæ Cartesianæ circa Inventionem Locorum', *Acta Eruditorum* (June 1696): 264–7). As we shall see in the eighth volume, on the evening of 29 January 1696/7 Newton was able speedily to solve a related generalization, to the case where the sum of the n-th powers is to be constant, within minutes of encountering it—and its companion problem of the brachistochrone—in his copy of the *Programma* containing these challenges which was circulated by Bernoulli early in January.

$\sqrt{aa+dd} \cdot d (::BG \cdot BC) :: DG \cdot DE$. Ergo $DE = \dfrac{dy-aa+ax}{\sqrt{aa+dd}}$. [304] Adhæc cùm ang

ABF sit differentia angulorum BAD et ABD, adeoჳ anguli BAD et FBD æquentur, similia erunt triangula rectangula CAD et EBD, & proinde latera proportionalia $DA \cdot DC :: DB \cdot DE$. Sed est $DC = y$. $DA (=\sqrt{AC^q + DC^q}) = \sqrt{xx+yy}$.

$DB (=\sqrt{BC^q + DC^q}) = \sqrt{aa-2ax+xx+yy}$, & supra erat $DE = \dfrac{dy-aa+ax}{\sqrt{aa+dd}}$.

‖[96] Quare est $\sqrt{xx+yy} \cdot y :: \sqrt{aa-2ax+xx+yy} \cdot \dfrac{dy-aa+ax}{\sqrt{aa+dd}}$. ‖ Et extremorum et mediorum quadratis in se ductis

$aayy - 2axyy + xxyy + y^4$

$$= \frac{ddxxyy + ddy^4 - 2aadxxy - 2aady^3 + 2adyx^3 + 2adxy^3 + a^4xx + a^4yy - 2a^3x^3 - 2a^3xyy + aax^4 + aaxxyy}{aa+dd}.$$

Duc omnes terminos in $aa+dd$ et prodeuntes redige in debitum ordinem et

orietur $x^4 \begin{matrix} -2a \\ +\frac{2d}{a}y \end{matrix} x^3 \begin{matrix} -2dy \\ +aa \end{matrix} xx \begin{matrix} +\frac{2d}{a}y^3 \\ +\frac{2dd}{a}yy \end{matrix} x \begin{matrix} -ddyy \\ -2dy^3 \\ -y^4 \end{matrix} = 0$. Divide hanc æquationem per

$xx - ax \begin{matrix} +dy \\ +yy \end{matrix}$, & orietur $xx \begin{matrix} -a \\ +2\frac{d}{a}y \end{matrix} x \begin{matrix} -yy \\ -dy \end{matrix} = 0$. Duæ itaჳ prodierunt æquationes in

solutione hujus Problematis. Prior $xx - ax \begin{matrix} +dy \\ +yy \end{matrix} = 0$ est ad circulum, [305] locum

nempe puncti D ubi angulus FBD sumitur ad alias partes rectæ BF quàm in figura describitur, existente angulo ABF summa angulorum DAB DBA ad

basem, adeoჳ angulo ADB ad verticem dato. Posterior $xx \begin{matrix} -a \\ +2\frac{d}{a}y \end{matrix} x \begin{matrix} -yy \\ -dy \end{matrix} = 0$ est

ad Hyperbolam, [306] locum puncti D ubi ang. FBD situm obtinet a recta BF quem in Figura descripsimus, hoc est ita ut angulus ABF sit differentia angulorum DAB, DBA ad basem. Hyperbolæ autem hæc est determinatio.

---

(304) Readily recognized by modern eyes as the perpendicular distance from $D(x, y)$ onto $BG(a(x-a)+dy=0)$; the general formula for computing such a distance was, however, systematized only by S. F. Lacroix in his *Traité élémentaire de trigonométrie rectiligne et sphérique et l'application de l'algèbre à la géométrie* (Paris, 1798): §66: 79. (See Tropfke's *Geschichte der Elementar-Mathematik*, **6**: 124; and compare C. B. Boyer, *History of Analytical Geometry* (New York, 1956): 213.)

(305) Of centre $(\frac{1}{2}a, -\frac{1}{2}d)$, vertically below P, and passing through A, B, as is geometrically evident.

and so, when the roots are extracted, $(BG:BC$ or$)$ $DG:DE = \sqrt{[a^2+d^2]}:d$. Therefore $DE = (dy - a^2 + ax)/\sqrt{[a^2+d^2]}$.[304] Further, since angle $\widehat{ABF}$ is the difference of the angles $\widehat{BAD}$ and $\widehat{ABD}$, and hence the angles $\widehat{BAD}$ and $\widehat{FBD}$ are equal, the right triangles CAD and EBD will be similar and consequently their sides in proportion $DA:DC = DB:DE$. Here, however, $DC = y$,

$$DA \text{ (or } \sqrt{[AC^2+DC^2]}) = \sqrt{[x^2+y^2]},$$

DB (or $\sqrt{[BC^2+DC^2]}) = \sqrt{[a^2-2ax+x^2+y^2]}$, while above there was

$$DE = (dy - a^2 + ax)/\sqrt{[a^2+d^2]}.$$

There is in consequence

$$\sqrt{[x^2+y^2]}:y = \sqrt{[a^2-2ax+x^2+y^2]}:(dy-a^2+ax)/\sqrt{[a^2+d^2]};$$

and, when the squares of extremes and middles are multiplied into one another,

$$a^2y^2 - 2axy^2 + x^2y^2 + y^4$$
$$= \frac{d^2x^2y^2 + d^2y^4 - 2a^2dx^2y - 2a^2dy^3 + 2adyx^3 + 2adxy^3 + a^4x^2 + a^4y^2 - 2a^3x^3 - 2a^3xy^2 + a^2x^4 + a^2x^2y^2}{a^2+d^2}.$$

Multiply all the terms into $a^2 + d^2$ and reduce those resulting to due order, and there will ensue

$$x^4 + 2(dy/a - a)\,x^3 + (-2dy + a^2)\,x^2 + 2(d/a)\,(y^3 + dy^2)\,x - (y^4 + 2dy^3 + d^2y^2) = 0.$$

Divide this equation by $x^2 - ax + (y^2 + dy)$ and there will ensue

$$x^2 + (2dy/a - a)\,x - (y^2 + dy) = 0.$$

Accordingly, two equations will result in solution of this problem. The first, $x^2 - ax + y^2 + dy = 0$, defines a circle:[305] namely, the locus of point D when the angle $\widehat{FBD}$ is taken on the other side of the straight line BF from that depicted in the figure, the angle $\widehat{ABF}$ here being the sum of the angles $\widehat{DAB}$, $\widehat{DBA}$ at the base, so that the vertex angle $\widehat{ADB}$ is given. The latter equation,

$$x^2 + (2dy/a - a)\,x - (y^2 + dy) = 0,$$

is that of a hyperbola,[306] the locus of point D when the angle $\widehat{FBD}$ maintains its position on the side of the line BF which we have depicted in the figure—such that, in other words, the angle $\widehat{ABF}$ is the difference of the angles $\widehat{DAB}$, $\widehat{DBA}$ at the base. The hyperbola, however, is fixed in location as follows. Bisect AB in P,

---

(306) Having perpendicular asymptotes PQ, PR defined by the equations

$$x = ((d \pm \sqrt{[d^2+a^2]})/a)\,y + \tfrac{1}{2}a$$

and with centre $P(\tfrac{1}{2}a, 0)$: these Newton constructs in the sequel, making use of the corollary

$$\tan \widehat{BPQ} = -(\sqrt{[d^2+a^2]} - d)/a = -(a/\sqrt{[d^2+a^2]})/(1 + d/\sqrt{[d^2+a^2]})$$
$$= -\sin \widehat{FBC}/(1 + \cos \widehat{FBC}) = -\tan \tfrac{1}{2}\widehat{FBC}.$$

Biseca *AB* in *P*. Age *PQ* constituentem angulum *BPQ* æqualem
dimidio anguli *ABF*. Huic erige normalem *PR*, et erunt *PQ, PR* Asymp-
toti hujus Hyperbolæ, & *B* punctum
per quod Hyperbola transibit.

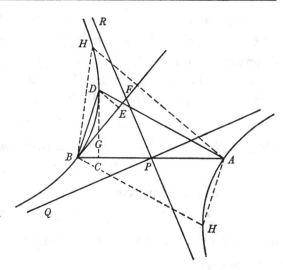

Et hinc prodit tale Theorema.
Hyperbolæ rectangulæ diametro
quavis *AB* ducta, & a terminis ejus
ad Hyperbolæ puncta duo quævis *D*
et *H* ductis rectis *AD, BD, AH,*
BH: hæ rectæ angulos *DAH, DBH*
ad terminos diametri constituent
æquales.[307]

[1677]
Lect 9

‖[97]

Idem brevius.

Ad Prob. 13 Regulam[308] de commoda terminorum ‖ ad ineundum calculum
electione tradidi, ubi obvenit ambiguitas in electione. Hic differentia angulorum
ad basem eodem modo se habet ad utrumcȝ angulum, & in constructione
schematis æque potuit addi ad angulum minorem *DAB* ducendo ab *A* rectam
ipsi *BF* parallelam,[309] ac substrahi ab angulo majori *DBA* ducendo rectam *BF*.
Quamobrem nec addo nec substraho, sed dimidium ejus uni angulorum addo,
alteri substraho. Deinde cùm etiam ambiguum sit utrum *AC* vel *BC* pro
termino indefinito cui ordinatim applicata *DC*
insistit adhibeatur, neutrum adhibeo sed biseco
AB in *P* & adhibeo *PC*: vel potiùs actâ *MPQ*
constituente hinc inde angulos *APQ, BPM*
æquales dimidio differentiæ angulorum ad
basem, ita ut ea cum rectis *AD, BD* constituat
angulos *DQP, DMP* æquales; ad *MQ* demitto
normales *AR, BN, DO* et adhibeo *DO* pro
ordinatim applicata ac *PO* pro indefinita linea
cui insistit.[310] Voco itacȝ *PO=x, DO=y, AR*
vel *BN=b*, et *PR* vel *PN=c*. Et propter similia
triangula *BNM, DOM* erit *BN . DO :: MN . MO*. et dividendo

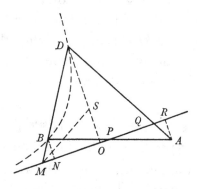

$$DO - BN(y-b) \cdot DO(y) :: MO - MN(ON \text{ sive } c-x) \cdot MO.$$

Quare $MO = \dfrac{cy - xy}{y - b}$. Similiter ex altera parte propter similia triangula *ARQ*,
DOQ, erit *AR . DO :: RQ . QO* et componendo

$$DO + AR(y+b) \cdot DO(y) :: QO + RQ(OR \text{ sive } c+x) \cdot QO.$$

draw PQ forming an angle $B\widehat{P}Q$ equal to half the angle $A\widehat{B}F$ and on it erect the normal PR: then PQ, PR will be the hyperbola's asymptotes and B a point through which the hyperbola shall pass.

A theorem derives from this to the effect that: if any diameter AB of a rectangular hyperbola be drawn, and from its end-points to any two points D and H of the hyperbola straight lines AD, BD, AH, BH are extended, then these lines shall form at the end-points of the diameter equal angles $D\widehat{A}H$, $D\widehat{B}H$.[307]

The same a shorter way

At the close of Problem 13 I set down a rule[308] regarding the convenient choice of terms to begin a calculation with when some ambiguity in that choice presents itself. In the present instance the difference of the angles at the base is related in the same manner to either angle and in the construction of the scheme could equally well have been added to the lesser angle $D\widehat{A}B$, by drawing from A a straight line parallel to BF,[309] rather than subtracted from the greater angle $D\widehat{B}A$ by drawing the line BF. Consequently, I neither (wholly) add or subtract it, but add its half to one of the angles and subtract it from the other. Next, since it is also ambiguous whether AC or BC is to be employed for the indefinite base-line on which the ordinate DC is stood, I employ neither but bisect AB in P and employ PC—or better still, having drawn MPQ forming on its either side angles $A\widehat{P}Q$, $B\widehat{P}M$ equal to half the difference of the angles at the base, so making equal angles $D\widehat{Q}P$, $D\widehat{M}P$ with the lines AD, BD, I then to MQ let fall the normals AR, BN, DO and employ DO for the ordinate, with PO as the indefinite line on which it stands.[310] I accordingly call $PO = x$, $DO = y$, AR or $BN = b$ and PR or $PN = c$. Then, because the triangles BNM, DOM are similar, there will be $BN:DO = MN:MO$ and *divisim*

$$DO - BN(y-b):DO(y) = MO - MN(ON \text{ or } c-x):MO,$$

whence $MO = (c-x)y/(y-b)$. Similarly, because on the other hand the triangles ARQ, DOQ are similar, there will be $AR:DO = RQ:QO$ and by compounding $DO + AR(y+b):DO(y) = QO + RQ(OR \text{ or } c+x):QO$, whence

---

(307) Compare Charles Taylor, *An Introduction to the Ancient and Modern Geometry of Conics* (Cambridge, 1881): 172, note ∗. This elegant property of the hyperbola would appear to be Newton's present discovery.

(308) See note (245).

(309) Understand 'et ad oppositas partes rectæ *AB*' (and on the opposite side of the line *AB*).

(310) The line *MS* (bisector of $D\widehat{M}Q$) relates to Problem 36 following.

Quare $QO = \dfrac{cy+xy}{y+b}$. Deniq propter æquales angulos DMQ DQM æquantur

MO et $QO_{[,]}$ hoc est $\dfrac{cy-xy}{y-b} = \dfrac{cy+xy}{y+b}$. Divide omnia per y & multiplica per

denominatores, & orietur $cy+cb-xy-xb = cy-cb+xy-xb$, sive $cb=xy$,[311]
notissima æquatio ad Hyperbolam.

Quinetiam locus puncti D sine calculo Algebraico prodire potuit. Est
enim ex superioribus $DO-BN . ON :: DO . MO(QO) :: DO+AR . OR$. Hoc est

||[98] $\| DO-BN . DO+BN :: ON . OR$ et mixtim

$$DO . BN :: \frac{ON+OR}{2}(NP) . \frac{OR-ON}{2}(OP).$$

Adeoq $DO \times OP = BN \times NP$.[312]

Prob. 36.[313]

*Locum verticis trianguli invenire cujus Basis datur & angulorum ad Basem unus dato
angulo differt a duplo alterius.*

In schemate novissimo superioris Problematis[314] sit ABD triangulum istud,
AB basis bisecta in P, APQ vel BPM dimidium[315] anguli dati quo angulus
DBA excedit duplum anguli DAB: et angulus DMQ erit duplus anguli DQM.
Ad MQ demitte perpendicula AR, BN, DO, & angulum DMQ biseca rectâ
MS occurrente DO in S, et erunt triangula DOQ SOM similia, adeoq
$OQ . OM :: OD . OS$ & dividendo $OQ-OM . OM :: SD . OS ::$ (per 3. vi. Elem.)
$DM . OM$. Quare (per 9. v Elem) $OQ-OM = DM$. Dictis jam $PO = x$. $OD = y$.

AR vel $BN = b$. & PR vel $PN = c$ erit ut in superiori Problemate $OM = \dfrac{cy-xy}{y-b}$

& $OQ = \dfrac{cy+xy}{y+b}$ adeoq $OQ-OM = \dfrac{[-]2bcy+2xyy}{yy-bb}$.[316] Pone jam

$$DO^q + O[M]^q = D[M]^q,\text{[317]}$$

---

(311) This follows less fussily *componendo et dividendo* in the form $xy/b = cy/y$ (or c).

(312) Robert Simson in his *Sectionum Conicarum Libri Quinque* (Edinburgh, ₂1750): Appendix,
Propositio 1: 221 gives a still more elegant proof that the locus (D) is a hyperbola by locating
point E instantaneously in AB such that $\widehat{BED} = \widehat{ABD} - \widehat{BAD}$, constant: at once, $\widehat{BDE} = \widehat{BAD}$
and so the triangles ADE, DBE are similar, whence $ED^2 = (AE \times BE$ or$) OE^2 - OB^2$. Newton
might well have noticed, too, that the locus (D) can be generated by an 'organic' construction
in which the fixed angles $\widehat{DAB}$, $\widehat{DBA}$ rotate round the poles A, B and is therefore a conic—in
fact, a rectangular hyperbola since it has points at infinity in the perpendicular directions PQ
and PR.

(313) In line with the one preceding, this problem also is briefly enumerated in Newton's
contemporary notes on 'linear' loci; see iv: 248, note (60).

(314) For convenience we have introduced a more accurate illustrating figure in our
English version.

$QO = (c+x)\,y/(y+b)$. Finally, because the angles $D\widehat{M}Q$, $D\widehat{Q}M$ are equal, so also are MO and QO; that is, $(c-x)\,y/(y-b) = (c+x)\,y/(y+b)$. Divide throughout by y and multiply by the denominators, and there will then ensue

$$cy + cb - xy - xb = cy - cb + xy - xb,$$

that is, $cb = xy$[311]—a very well-known equation to the hyperbola.

To be sure, the locus of the point D could have been derived without recourse to an algebraic computation. For, by the above,

$$(DO-BN):ON = DO:MO \text{ (or } QO) = (DO+AR):OR$$

and therefore $(DO-BN):(DO+BN) = ON:OR$, and *mixtim*

$$DO:BN = (\tfrac{1}{2}(OR+ON) \text{ or}) \; NP:(\tfrac{1}{2}(OR-ON) \text{ or}) \; OP;$$

hence $DO \times OP = BN \times NP$.[312]

Problem 36[313]

To find the locus of the vertex of a triangle whose base is given and one of whose base angles differs by a given angle from twice the other.

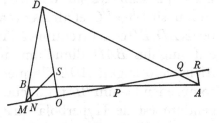

In the most recent diagram in the preceding problem[314] let ABD be that triangle, AB the base bisected in P, $A\widehat{P}Q$ or $B\widehat{P}M$ a half[!][315] of the given angle by which angle $D\widehat{B}A$ exceeds twice the angle $D\widehat{A}B$: whence the angle $D\widehat{M}Q$ will be twice the angle $D\widehat{Q}M$. To MQ let fall the perpendiculars AR, BN, DO and bisect angle $D\widehat{M}Q$ by the straight line MS meeting DO in S; the triangles DOQ, SOM will then be similar, so that $OQ:OM = OD:OS$ and *divisim* $(OQ-OM):OM = SD:OS = $ (by *Elements*, VI, 3) $DM:OM$. Consequently (by *Elements*, V, 9) $OQ-OM = DM$. On now calling $PO = x$, $OD = y$, AR or $BN = b$ and PR or $PN = c$, there will, as in the previous problem, be $OM = (c-x)\,y/(y-b)$ and $OQ = (c+x)\,y/(y+b)$, so that

$$OQ - OM = 2(-bc+xy)\,y/(y^2-b^2).\text{[316]}$$

Now set
$$DO^2 + OM^2 = DM^2,\text{[317]}$$

(315) Read 'triens' (a third). Newton's momentary slip passed into print in Whiston's 1707 *editio princeps*, but was caught by its author in his library copy of the edition and publicly amended in 1722.

(316) Newton in his manuscript carelessly set the numerator as '$2bcy+2xyy$': to compensate, an occasional interchange of '$+b$' and '$-b$' is necessitated in the sequel. Due correction was made in 1722.

(317) The manuscript, subsequently corrected by Whiston to the version reproduced, reads '$DO^q+OQ^q=DQ^q$' in manifest error.

hoc est $yy + \dfrac{cc - 2cx + xx}{yy - 2by + bb} yy = \dfrac{4bbcc[-]8bcxy + 4xxyy}{y^4 - 2bbyy + b^4} yy$.[318] et per debitam

reductionem orietur tandem

$$y^4 * \begin{matrix} +cc \\ -2bb \\ -2cx \\ -3xx \\ -bb \\ +cc \end{matrix}\, yy \begin{matrix} [+]2bxx \\ [+]4bcx \\ [+]2bcc \\ \\ \end{matrix}\, y \begin{matrix} +b^4 \\ -3bbcc \\ -2bbcx \\ [+]bbxx \end{matrix} = 0.$$ Divide omnia

per $y[-]b$, & evadet

$$y^3 [+]byy \begin{matrix} -bb \\ +cc \\ -2cx \\ -3xx \end{matrix}\, y \begin{matrix} [-]b^3 \\ [+]3bcc \\ [+]2bcx \\ [-]bxx \end{matrix} = 0.$$ Quare punctum D est ad

Curvam trium dimensionum,[319] quæ tamen evadit Hyperbola ubi angulus *BPM* statuitur nullus, sive angulorum ad basem unus *DAB* duplus alterius *DBA*. Tunc enim *BN*, sive b evanescente, æquatio fiet $yy = 3xx + 2cx - cc$.[320]

Ex hujus autem æquationis constructione, tale elicitur Theorema. Si centro C, Asymptotis *CS*, *CT*, angulum ‖*SCT* 120 graduum continentibus describatur Hyperbola quævis *DV* cujus semiaxes sint *CV*, *CA*: produc *CV* ad B ut sit $VB = VC$, et ab A et B actis utcunq̃ rectis *AD*, *BD* concurrentibus ad Hyperbolam erit angulus *BAD* dimidium anguli *ABD*, triens verò anguli *ADE*[321] quem recta *AD* comprehendit cum *BD* producta. Hoc intelligendum est de Hyperbola quæ transit per punctum *V*. Quod si ab ijsdem punctis A et B actæ rectæ *Ad Bd* conveniant ad conjugatam Hyperbolam quæ transit per A: tunc

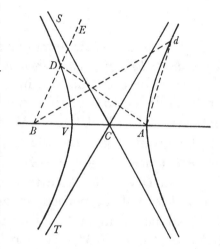

externorum angulorum trianguli ad basem, ille ad B erit duplus alterius ad A.

Prob. 37.[322]

Lect 10

Circulum per data duo puncta describere qui rectam positione datam continget.

Sunto A et B puncta data et *EF* recta positione data et requiratur circulum *ABE* per ista puncta describere qui contingat rectam istam *FE*. Junge *AB* et

---

(318) Whence, on multiplying through by $(y+b)^2(y-b)/y^2$, there results

$$(y+b)^2(y-b) = (4(xy - bc)^2 - (x-c)^2(y+b)^2)/(y-b) = (x+c)(b(x-3c) + (3x-c)y).$$

Newton's following 'due reduction' unnecessarily introduces an extra factor $y - b$ which has then—not a little mysteriously—to be divided out at the next stage.

(319) A general tridiametral cubic whose three asymptotes

$$y - \tfrac{1}{3}b = 0 \quad \text{and} \quad y - \tfrac{1}{3}b = \pm\sqrt{3}.(x + \tfrac{1}{3}c)$$

that is,

$$y^2 + (c^2 - 2cx + x^2)\,y^2/(y^2 - 2by + b^2) = 4(b^2c^2 - 2bcxy + x^2y^2)\,y^2/(y^4 - 2b^2y^2 + b^4),^{(318)}$$

and by appropriate reduction there will at length arise

$$y^4 + (-2b^2 + c^2 - 2cx - 3x^2)\,y^2 + 2b(x^2 + 2cx + c^2)\,y + b^2(b^2 - 3c^2 - 2cx + x^2) = 0.$$

Divide throughout by $y - b$ and there will result

$$y^3 + by^2 + (-b^2 + c^2 - 2cx - 3x^2)\,y - b(b^2 - 3c^2 - 2cx + x^2) = 0.$$

Consequently the point D is on a curve of third degree.[319] This, however, proves to be a hyperbola when the angle $B\widehat{P}M$ is set to be zero—that is, when one of the base angles, $D\widehat{A}B$, is twice the other, $D\widehat{B}A$; for then, with $(BN$ or) b vanishing, the equation will become $y^2 = 3x^2 + 2cx - c^2.$[320]

From the construction of this equation, indeed, a theorem to this purpose is derived. If, with centre C and asymptotes CS, CT containing an angle $S\widehat{C}T$ of 120 degrees, there be described any hyperbola DV whose semi-axes shall be CV and CA, extend CV to B so that $VB = CV$ and then, when there are drawn from A and B arbitrary straight lines AD, BD meeting at the hyperbola, the angle $B\widehat{A}D$ will be half the angle $A\widehat{B}D$ and, indeed, a third of the angle $A\widehat{D}E$[321] which the line AD comprehends with BD produced. Understand this of the hyperbolic branch which passes through point V. If, however, straight lines Ad, Bd drawn from the same points A and B intersect on the conjugate branch which passes through A, then of the triangle's exterior base angles that at B will be twice the other at A.

Problem 37[322]

To describe a circle through two given points which shall touch a straight line given in position.

Let A and B be the given points, EF the straight line given in position and let it be required to describe a circle ABE through those points so as to touch that

concur in the point $(-\tfrac{1}{3}c, \tfrac{1}{3}b)$; there are two hyperbolic branches and a main curve which 'snakes' round the asymptote $y - \tfrac{1}{3}b = \sqrt{3}.(x + \tfrac{1}{3}c)$; compare II: 49.

(320) Paired with the axis $AB(y = 0)$. This hyperbolic locus is classical, occurring in Pappus' *Mathematical Collection*, IV, 34. As Pappus shows geometrically (compare T. L. Heath, *A History of Greek Mathematics*, 1 (Oxford, 1921): 241–3), it is readily established from the present Cartesian equation that the hyperbola has centre $C(-\tfrac{1}{3}c, 0)$ in AB such that $BC = 2CA$, passing through points $A(-c, 0)$ and $V(\tfrac{1}{3}c, 0)$, where V is the midpoint of BC; less obviously, since the hyperbola has eccentricity $\sqrt{[3+1]} = 2$, the second base-point $B(c, 0)$—distant $2VC$ from the centre C—is a focus.

(321) That is, $A\widehat{B}D + B\widehat{A}D$.

(322) We have reproduced in the fourth volume the contemporary 'Quæstionum solutio Geometrica' (IV: 254–8) where Newton has given concise synthetic proofs of this and the three following problems. This corresponds to 'Prob 1' on IV: 254.

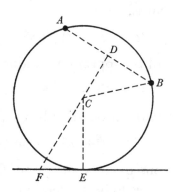

eam biseca in D. Ad D erige normalem DF occurrentem rectæ FE in F, et circuli centrum incidet in hanc novissime ductam DF puta in C. Junge ergo CB et ad FE demitte CE normalem, eritq E punctum contactûs, ac CB CE æquales inter se utpote radij circuli quæsiti. Jam cùm puncta A, B, D et F dentur esto $DB=a$, ac $DF=b$ et ad determinandum centrum circuli quæratur DC quam ideo dic x. Jam in triangulo CDB propter angulum ad D rectum est $\sqrt{DB^q+DC^q}$ hoc est $\sqrt{aa+xx}=CB$. Est et $DF-DC$ sive $b-x=CF$. Et in triangulo

|| [100] rectangulo CFE cum dentur anguli, dabitur ratio laterum CF et CE, sit || ista

d ad e, et erit $CE=\frac{e}{d}\times CF$ hoc est $=\frac{eb-ex}{d}$. Pone jam CB et CE (radios nempe

circuli quæsiti) æquales inter se, & habebitur æquatio $\sqrt{aa+xx}=\frac{eb-ex}{d}$. Cujus

partibus quadratis et multiplicatis per dd oritur $aadd+ddxx=eebb-2eebx+eexx$

sive $xx=\dfrac{-2eebx\begin{smallmatrix}+eebb\\-aadd\end{smallmatrix}}{dd-ee}$. et extracta radice $x=\dfrac{-eeb+d\sqrt{eebb+eeaa-ddaa}}{dd-ee}$.[323]

Inventa est ergo longitudo DC adeoq centrum C quo circulus per puncta A et B describendus est ut contingat rectam FE.

Octob 1678
Lect 1

Prob. 38.

Circulum per datum punctum describere qui rectas duas positione datas continget.

Esto datum punctum A, et sint EF FG rectæ duæ positione datæ et AEG circulus quæsitus easdem contingens ac transiens per punctum istud A. Rectâ CF bisecetur angulus EFG et centrum circuli in ipsa reperietur. Sit istud C, et ad EF et FG demissis perpendiculis CE CG, erunt E ac G puncta contactus. Jam in triangulis CEF CGF, cum anguli ad E et G sint recti et anguli ad F semisses sint anguli EFG, dantur omnes anguli adeoq ratio laterum CF et CE vel CG. Sit ista d ad e et si ad determinandum centrum circuli quæsiti C, assumatur $CF=x$, erit CE vel

Resolvitur ut Pro[b]. 37. Nam dato puncto A datur et aliud punctum I.[324]

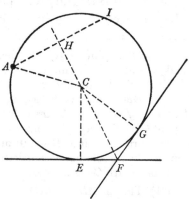

$CG=\frac{ex}{d}$. Præterea ad FC demitte normalem AH et cum punctum A detur

line *FE*. Join *AB* and bisect it in *D*, at *D* erect the normal *DF* meeting the straight line *FE* in *F* and the circle's centre will fall in this most recently drawn line *DF*, say at *C*. So join *CB* and onto *FE* let fall the normal *CE*; *E* will then be the point of contact, with *CB*, *CE*—being the radii of the circle sought—equal to each other. Now since the points *A*, *B*, *D* and *F* are given, let $DB = a$ and $DF = b$, while to determine the centre let *DC* be sought and hence call it x. Then in the triangle *CDB*, because the angle at *D* is right, *CB* (that is,

$$\sqrt{[DB^2 + DC^2]}) = \sqrt{[a^2 + x^2]}.$$

Also $CF = (DF - DC \text{ or}) \; b - x$. And since in the right-angled triangle *CFE* the angles are given, there will be given the ratio of the sides *CF* and *CE*; let it be d to e, and there will be $CE = (e/d) \, CF$, that is, $e(b - x)/d$. Now set *CB* and *CE* (the radii, namely, of the circle sought) equal to one another, and there will be had the equation $\sqrt{[a^2 + x^2]} = e(b - x)/d$. When the sides of this are squared and multiplied by d^2, there arises $a^2 d^2 + d^2 x^2 = e^2(b^2 - 2bx + x^2)$, that is,

$$x^2 = (-2be^2 x + b^2 e^2 - a^2 d^2)/(d^2 - e^2),$$

and with the root extracted $x = (-be^2 + d\sqrt{[(a^2 + b^2) \, e^2 - a^2 d^2]})/(d^2 - e^2)$.[323] The length of *DC* is therefore found and hence, too, the centre *C* on which the circle is to be described through the points *A* and *B* in order to touch the straight line *FE*.

Problem 38

To describe a circle through a given point which shall touch two straight lines given in position.

This is resolved as Problem 37 is. For, given the point *A*, the other point *I* is also given.[324]

Let the given point be *A*, with *EF*, *FG* the two straight lines given in position, and *AEG* the required circle touching them and passing through the point *A*. Bisect the angle $\widehat{EFG}$ by the straight line *CF* and the circle's centre will be found in it. Let that be *C* and, on letting fall to *EF* and *FG* the perpendiculars *CE* and *CG*, then will *E* and *G* be the points of contact. Now in the triangles *CEF*, *CGF*, since the angles at *E* and *G* are right and those at *F* halves of the angle $\widehat{EFG}$, all the angles are given and hence so too is the ratio of the sides *CF* and *CE* or *CG*. Let that be d to e and if, to determine the centre *C* of the circle sought, you should assume $CF = x$, there will be *CE* or $CG = ex/d$. Moreover, onto *FC* let fall the normal *AH* and then, since the point *A* is given,

(323) The sign preceding the radical should be ' $\pm$ ', so correctly presaging the two solutions possible.

(324) Namely, the mirror-image of *A* in *FC*, the bisector of $\widehat{EFG}$. This reduction is employed by Newton in 'Prob 2' on IV: 254.

dabuntur etiam rectæ *AH* et *FH*. Dicantur istæ *a* et *b* et ab *FH* sive *b* ablato *FC* sive *x* restabit *CH* = *b* − *x*. Cujus quadrato *bb* − 2*bx* + *xx* adde quadratum ‖ [101] ipsius *AH*, sive *aa* ‖ et summa *aa* + *bb* − 2*bx* + *xx* erit *AC*^q per 47. 1 Elem. siquidem angulus *AHC* ex Hypothesi sit rectus. Pone jam radios circuli *AC* et *CG* inter se æquales₍ₛ₎ hoc est pone æqualitatem inter eorum valores, vel inter quadrata eorum et habebitur æquatio $aa + bb - 2bx + xx = \frac{eexx}{dd}$. Aufer

utrobiꝗ *xx*, et mutatis omnibus signis erit $-aa - bb + 2bx = xx - \frac{eexx}{dd}$. Duc

omnia in *dd* ac divide per *dd* − *ee* et evadet $\frac{-aadd - bbdd + 2bddx}{dd - ee} = xx$. Cujus

æquationis extracta radix est $x = \frac{bdd - d\sqrt{eebb + eeaa - ddaa}}{dd - ee}$. [325] Inventa est itaꝗ

longitudo *FC* adeoꝗ punctum *C* quod centrum est circuli quæsiti.

Si inventus valor *x* sive *FC* auferatur de *b* sive *HF* restabit

$$HC = \frac{-eeb + d\sqrt{eebb + eeaa - ddaa}}{dd - ee}$$

eadem æquatio quæ[326] in priori problemate prodijt, ad determinandam longitudinem *DC*.

[1678] Lect 2

Prob 39.[327]

Vide Pro[b]. 11.[328]

Circulum per data duo puncta describere qui alium circulum positione datum continget.

Sint *A*, *B* puncta data, *EK* circulus positione et magnitudine datus, *F* centrum ejus, *ABE* circulus quæsitus per puncta *A* et *B* transiens ac tangens alterum circulum in *E*, et *C* centrum ejus. Ad *AB* product[a]m demitte perpendicula *CD* et *FG* et age *CF* secantem circulos in puncto contactus *E*, ac age etiam *FH* ‖ [102] paral‖lelam *DG* et occurrentem *CD* in *H*. His constructis dic *AD* vel *DB* = *a*, *DG* vel *HF* = *b*, *GF* = *c* & *EF*

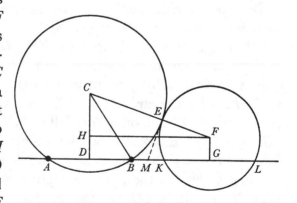

(radium nempe circuli dati) = *d*, atꝗ *DC* = *x*: et erit *CH* (= *CD* − *FG*) = *x* − *c*, et *CF*^q (= *CH*^q + *HF*^q) = *xx* − 2*cx* + *cc* + *bb*, atꝗ *CB*^q (= *CD*^q + *DB*^q) = *xx* + *aa*, adeoꝗ *CB* vel $CE = \sqrt{xx + aa}$. Huic adde *EF* et habebitur $CF = d + \sqrt{xx + aa}$ cujus

---

(325) Compare note (323); the radical sign should again be ' ± ', so yielding two real solutions.

the straight lines AH and FH will be given also. Let these be called a and b, and when from FH or b there is taken FC or x there will remain $CH = b - x$. To its square $b^2 - 2bx + x^2$ add the square of AH, that is, a^2, and (by *Elements*, I, 47) the sum $a^2 + b^2 - 2bx + x^2$ will be AC^2, seeing that the angle $A\widehat{H}C$ is, by hypothesis, right. Now put the circle's radii AC and CG equal to one another—that is, set equality between their values, or rather between (those of) their squares—and there will be had the equation $a^2 + b^2 - 2bx + x^2 = e^2 x^2 / d^2$. Take x^2 away from either side and, when all the signs are changed there will be $x^2(1 - e^2/d^2) = 2bx - a^2 - b^2$. Multiply throughout by d^2 and divide by $d^2 - e^2$, and it will come to be $x^2 = (2bd^2 x - (a^2 + b^2)\, d^2)/(d^2 - e^2)$. And after the root of this equation is extracted it is $x = (bd^2 - d\sqrt{[(a^2 + b^2)\, e^2 - a^2 d^2]})/(d^2 - e^2)$.[325] The length of FC is accordingly found and hence so too is the point C which is the centre of the required circle.

If the value of FC or x found be taken from HF or b, there will be left

$$HC = (-be^2 + d\sqrt{[(a^2 + b^2)\, e^2 - a^2 d^2]})/(d^2 - e^2),$$

the same value which resulted in the previous problem for determining the length of DC.

Problem 39[327]

See Problem 11.[328]

To describe a circle through two given points which shall touch another circle given in position.

Let A, B be the given points, EK the circle given in position and magnitude, F its centre, ABE the required circle passing through the points A and B and touching the other circle in E, with C its centre. Onto AB produced let fall the perpendiculars CD and FG, then draw CF cutting the circles in their point E of contact and draw also FH parallel to DG, meeting CD in H. Once these are constructed, call AD or $DB = a$, DG or $HF = b$, $GF = c$ and EF (the radius, namely, of the given circle) $= d$, and also $DC = x$; there will then be

$$CH(\text{or } CD - FG) = x - c,$$

and so $CF^2(\text{or } CH^2 + HF^2) = x^2 - 2cx + c^2 + b^2$, while

$$CB^2(\text{or } CD^2 + DB^2) = x^2 + a^2,$$

so that CB or $CE = \sqrt{[x^2 + a^2]}$. Add EF to this and there will be obtained

---

(326) We would expect here 'isdem valor qui' and so we render Newton's phrase in our English version.

(327) Compare Newton's synthetic revise in 'Prob 3' of his contemporary 'Quæstionum solutio Geometrica' (IV: 256).

(328) This (see note (232)) can be made to yield a variant construction of the present problem.

quadratum $dd+aa+xx+2d\sqrt{xx+aa}$ æquatur valori ejusdem CF^q prius invento, nempe $xx-2cx+cc+bb$. Aufer utrobiꝗ xx et restabit

$$dd+aa+2d\sqrt{xx+aa}=cc+bb-2cx.$$

Aufer insuper $dd+aa$ et habebitur $2d\sqrt{xx+aa}=cc+bb-dd-aa-2cx$. Jam abbreviandi causa pro $cc+bb-dd-aa$ scribe $2gg$, et habebitur

$$2d\sqrt{xx+aa}=2gg-2cx$$

sive $d\sqrt{xx+aa}=gg-cx$. Et partibus æquationis quadratis, erit

$$ddxx+ddaa=g^4-2ggcx+ccxx.$$

Utrinꝗ aufer $ddaa$ et $ccxx$ et restabit $ddxx-ccxx=g^4-ddaa-2ggcx$. Et partibus æquationis divisis per $dd-cc$, habebitur $xx=\dfrac{g^4-ddaa-2ggcx}{dd-cc}$. atꝗ per

extractionem radicis affectæ $x=\dfrac{-ggc+\sqrt{g^4dd-d^4aa+ddaacc}}{dd-cc}$ [329].

Inventa igitur x sive longitudine DC, biseca AB in D et ad D erige perpendiculum $DC=\dfrac{-ggc+d\sqrt{g^4-aadd+aacc}}{dd-cc}$. Dein centro C per punctum A vel B describe circulum ABE; nam hic continget alterum circulum EK et transibit per utrumꝗ punctum A, B. Q.E.F.

[1678]
Lect 3

Pro[b] 40.[330]

Circulum per datum punctum describere qui datum circulum & rectam lineam positione datam continget.

‖[103] ‖Sit circulus iste describendus BD, ejus centrum C, punctum per quod describi debet B, recta quam continget AD, punctum contactus D, circulus quem continget GEM, ejus centrum F, & punctum contactus E. Junge CB, CD, CF et CD erit perpendicularis ad AD, atꝗ CF secabit circulos in puncto contactus E. Produc CD ad Q ut sit $DQ=EF$ et per Q age QN parallelam AD. Deniꝗ a B et F ad AD et QN demitte perpendicula BA, FN, et a C ad AB et FN perpendicula CK, CL. Et cùm sit $BC=CD$

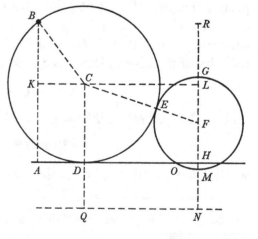

vel AK, erit $BK(=AB-AK)=AB-BC$, adeoꝗ $BK^q=AB^q-2AB\times BC+BC^q$. Aufer hoc de BC^q et restabit $2AB\times BC-AB^q$ pro quadrato de CK. Est itaꝗ $AB\times\overline{2BC-AB}=CK^q$. Et eodem argumento erit $FN\times\overline{2FC-FN}=CL^q$ atꝗ

$CF = d + \sqrt{[x^2+a^2]}$, whose square, $d^2+a^2+x^2+2d\sqrt{[x^2+a^2]}$, is equal to the value of the same quantity CF^2 previously found, namely $x^2-2cx+c^2+b^2$. Take off x^2 from each side and there will be left $d^2+a^2+2d\sqrt{[x^2+a^2]} = c^2+b^2-2cx$. Take away in addition d^2+a^2 and there will be had

$$2d\sqrt{[x^2+a^2]} = c^2+b^2-d^2-a^2-2cx.$$

Now, for brevity's sake, in place of $c^2+b^2-d^2-a^2$ write $2g^2$ and there will be had $2d\sqrt{[x^2+a^2]} = 2g^2-2cx$ or $d\sqrt{[x^2+a^2]} = g^2-cx$. And when the sides of this equation are squared there will be $d^2(x^2+a^2) = g^4-2g^2cx+c^2x^2$. Take away on each side a^2d^2 and c^2x^2, and there will remain $(d^2-c^2)x^2 = -2cg^2x+g^4-a^2d^2$; and, with the sides of the equation divided by d^2-c^2, there will be had

$$x^2 = (-2cg^2x+g^4-a^2d^2)/(d^2-c^2)$$

and so, by extraction of the 'affected' root,

$$x = (-cg^2+d\sqrt{[g^4-(d^2-c^2)a^2]})/(d^2-c^2).^{(329)}$$

Therefore when x, that is, the length of DC, is found, bisect AB in D and at D erect the perpendicular $DC = (-cg^2+d\sqrt{[g^4-(d^2-c^2)a^2]})/(d^2-c^2)$, then with centre C and through the point A or B describe the circle ABE; for this will touch the other circle EK and pass through each of the points A and B. As was to be done.

Problem 40[330]

To describe a circle through a given point which shall touch a given circle and a straight line given in position.

Let the circle which must be described be BD, C its centre, B the point through which it has to be described, AD the straight line which it is to touch and D the point of contact, GEM the circle it is to touch, F its centre and E the point of contact. Join CB, CD, CF and CD will be perpendicular to AD, while CF will cut the circles in their point E of contact. Extend CD to Q so that $DQ = EF$ and through Q draw QN parallel to AD. Finally, from B and F to AD and QN let fall the perpendiculars BA, FN and from C to AB and FN the perpendiculars CK, CL. Then, since $BC = CD$ (or AK), there will be

$$BK = (AB-AK \text{ or}) \ AB-BC$$

and hence $BK^2 = AB^2-2AB\times BC+BC^2$. Take this from BC^2 and there will remain $2AB\times BC-AB^2$ for the square of CK. Accordingly

$$AB(2BC-AB) = CK^2.$$

And by the same argument there will be $FN(2FC-FN) = CL^2$; hence

---

(329) Once more Newton omits the full sign ' $\pm$ ' before the radical which allows the double solution possible; compare notes (323) and (325).

(330) Compare the equivalent synthetic construction given by Newton in 'Prob 4' on IV: 256–8.

adeo $\frac{CK^q}{AB}+AB=2BC$ et $\frac{CL^q}{FN}+FN=2FC$. Quamobrem si pro AB, CK, FN, KL,

et CL scribas a, y, b, c, & $c-y$, erit $\frac{yy}{2a}+\frac{1}{2}a=BC$ et $\frac{cc-2cy+yy}{2b}+\frac{1}{2}b=FC$. De FC

aufer BC et restabit $EF=\frac{cc-2cy+yy}{2b}+\frac{1}{2}b-\frac{yy}{2a}-\frac{1}{2}a$. Jam si puncta ubi FN

producta secat rectam AD et circulum GEM notentur literis H, G, et M et in HG producta capiatur $HR=AB$, cum sit $HN(=DQ=EF)=GF$ addendo FH utrinq erit $FN=GH$, adeoq $AB-FN(=HR-GH)=GR$, et $AB-FN+2EF$ hoc est $a-b+2EF=RM$, et $\frac{1}{2}a-\frac{1}{2}b+EF=\frac{1}{2}RM$. Quare cum supra fuerit

$EF=\frac{cc-2cy+yy}{2b}+\frac{1}{2}b-\frac{yy}{2a}-\frac{1}{2}a$ si hoc scribatur pro EF[331] habebitur

[104]
$$\tfrac{1}{2}RM=\|\frac{cc-2cy+yy}{2b}-\frac{yy}{2a}.$$

Dic ergo RM d, et erit $d=\frac{cc-2cy+yy}{b}-\frac{yy}{a}$. Duc omnes terminos in a et b et

orietur $abd=acc-2acy+ayy-byy$. Aufer utrinq $acc-2acy$ et restabit

$$abd-acc+2acy=ayy-byy.$$

Divide per $a-b$ et orietur $\frac{abd-acc+2acy}{a-b}=yy$. et extracta radice

$$y=\frac{ac}{a-b}\pm\sqrt{\frac{aabd-abbd+abcc}{aa-2ab+bb}}.$$

Quæ conclusiones sic abbreviari possunt. Pone $c.b::d.e$, dein $a-b.a::c.f$ et erit $fe-fc+2fy=yy$ sive $y=f\pm\sqrt{ff+fe-fc}$. Invento y sive KC vel AD, cape $AD=f\pm\sqrt{ff+fe-fc}$, ad D erige perpendiculum $DC(=BC)=\frac{KC^q}{2AB}+\frac{1}{2}AB$, et centro C, intervallo CB vel CD describe circulum BDE, nam hic transiens per datum punctum B, tanget rectam AD in D et circulum GEM in E. Q.E.F.

Hinc circulus etiam describi potest qui duos datos circulos et rectam positione datam contin-get. Sint enim circuli dati RT, SV, eorum centra B, F, & recta positione data PQ. Centro F radio $FS-BR$ describe circulum EM. A puncto B ad rectam PQ demitte perpendiculum BP et producto eo ad A ut sit $PA=BR$ per A age AH parallelam PQ, & circulus describatur qui transeat

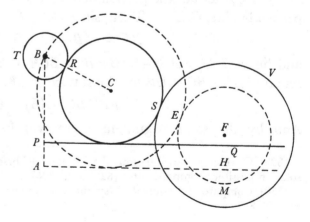

$2BC = CK^2/AB + AB$ and $2FC = CL^2/FN + FN$. Consequently, if in place of AB, CK, FN, KL and CL you write a, y, b, c and $c-y$, there will be

$$BC = \tfrac{1}{2}(y^2/a + a) \quad \text{and} \quad FC = \tfrac{1}{2}((c^2 - 2cy + y^2)/b + b).$$

From FC take BC and there will remain $EF = \tfrac{1}{2}((c^2 - 2cy + y^2)/b + b - y^2/a - a)$. Now if the points in which FN produced intersects the straight line AD and the circle GEM be marked by the letters H, G and M, while in HG produced there be taken $HR = AB$, since HN (or $DQ = EF$) $= GF$, by adding FH on either side there will be $FN = GH$, and so $AB - FN$ (or $HR - GH$) $= GR$ and RM (that is, $AB - FN + 2EF$) $= a - b + 2EF$, or $\tfrac{1}{2}RM = \tfrac{1}{2}(a-b) + EF$. Therefore, since above there was $EF = \tfrac{1}{2}((c^2 - 2cy + y^2)/b + b - y^2/a - a)$, if this now be written in place of EF[331] there will be had $\tfrac{1}{2}RM = \tfrac{1}{2}((c^2 - 2cy + y^2)/b - y^2/a)$. So call $RM = d$ and then $d = (c^2 - 2cy + y^2)/b - y^2/a$. Multiply all the terms by a and b and there will ensue $abd = ac^2 - 2acy + ay^2 - by^2$. Take away $ac^2 - 2acy$ from each side and there will remain $abd - ac^2 + 2acy = (a-b)y^2$, and after division by $a-b$ there will arise $y^2 = (2acy + a(bd - c^2))/(a-b)$; then, when the root is extracted, $y = (ac \pm \sqrt{[a^2bd - ab^2d + abc^2]})/(a-b)$. These conclusions may be shortened in this manner. Put $c:b = d:e$ and $(a-b):a = c:f$, and there will be

$$y^2 = 2fy + f(e-c),$$

that is, $y = f \pm \sqrt{[f(f+e-c)]}$. Having once found y, that is, KC or AD, take $AD = f \pm \sqrt{[f(f+e-c)]}$, at D erect the perpendicular

$$DC \text{ (or } BC) = \tfrac{1}{2}(KC^2/AB + AB)$$

and then with centre C and radius CB or CD describe the circle BDE; for this—passing through the given point B—will touch the line AD at D and the circle GEM at E. As was to be done.

A circle can, from this, also be described which shall touch two given circles and a straight line given in position. For let the given circles be RT, SV, their centres B, F, and the straight line given in position PQ. With centre F and radius $FS - BR$ describe the circle EM; then from point B to the line PQ let fall the perpendicular BP and, having extended it to A so that $PA = BR$, through A draw AH parallel to PQ, and describe a circle to pass through the point B and

---

(331) Newton first wrote 'adde $\tfrac{1}{2}a - \tfrac{1}{2}b$ ad utramcg partem et' (add $\tfrac{1}{2}(a-b)$ to each side and). It is curious that Newton does not at once set EF (the radius of the given circle) equal to some constant $\alpha[= 2(-a+b+d)]$ in order to derive his following quadratic in y.

per punctum B tangatꝗ rectam AH et circulum EM. Sit ejus centrum $C_{[,]}$ junge BC secantem circulum RT in R, et eodem centro C, radio verò CR descriptus circulus RS tanget circulos $RT\ SV$ et rectam PQ, ut ex constructione manifestum est.[332]

|| *Prob 41.*[333]

[1678]
Lect 4

Circulum describere qui per datum punctum transibit et alios duos positione et magnitudine datos circulos continget.

Esto punctum datum A sintꝗ circuli positione et magnitudine dati TIV, RHS, centra eorum C et B, circulus describendus AIH, centrum ejus D, & puncta contactus I et H. Junge AB, AC, AD, DB, secetꝗ AB producta circulum RHS in punctis R et S et AC producta circulum TIV in T et V. Et a punctis D et C demissis perpendiculis DE ad AB et DF ad AC occurrente[334] AB in G, atꝗ CK ad AB; in triangulo ADB erit $AD^q - DB^q + AB^q = 2AE \times AB$, per 13. 2 Elem. Sed $DB = AD + BR$,

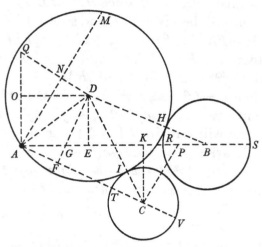

adeoꝗ $DB^q = AD^q + 2AD \times BR + BR^q$. Aufer hoc de $AD^q + AB^q$ et restabit $AB^q - 2AD \times BR - BR^q$ pro $2AE \times AB$. Est [e]t[335]

$$AB^q - BR^q = \overline{AB - BR} \times \overline{AB + BR} = AR \times AS.$$

Quare $AR \times AS - 2AD \times BR = 2AE \times AB$. et $\dfrac{AR \times AS - 2AB \times AE}{BR} = 2AD$. Et

simili ratiocinio in triangulo ADC emerget iterum $2AD = \dfrac{TAV - 2CAF}{CT}$.[336]

Quare $\dfrac{RAS - 2BAE}{BR} = \dfrac{TAV - 2CAF}{CT}$. et $\dfrac{TAV}{CT} - \dfrac{RAS}{BR} + \dfrac{2BAE}{BR} = \dfrac{2CAF}{CT}$. et

$$\overline{\dfrac{TAV}{CT} - \dfrac{RAS}{BR} + \dfrac{2BAE}{BR}} \times \dfrac{CT}{2AC} = AF.$$

Unde cum sit $AK.AC::AF.AG$, erit $AG = \overline{\dfrac{TAV}{CT} - \dfrac{RAS}{BR} + \dfrac{2BAE}{BR}} \times \dfrac{CT}{2AK}$. Aufer

---

(332) This reduction of the tangency of two circles and a line to the problem of drawing a circle through a given point to touch a given circle and line is, of course, due to François Viète in his *Apollonius Gallus* (Paris, 1600): Problema VII [= *Francisci Vietæ Opera Mathematica* (Leyden, 1646): 325–46, especially 332–3], as Newton well knew (compare I: 21).

touch both the straight line AH and the circle EM. Let its centre be C, join BC cutting the circle RT in R, and the circle RS described on the same centre C but with radius CR will touch the circles RT, SV and the straight line PQ: as is manifest from the construction.[332]

Problem 41[333]

To describe a circle which shall pass through a given point and touch two other circles given in position and magnitude

Let the given point be A, the circles given in position and magnitude TIV, RHS, their centres C and B, the circle to be described AIH, its centre D, and the points of contact I and H. Join AB, AC, AD, DB and let AB produced cut the circle RHS in the points R and S, and AC produced the circle TIV in T and V. Then, on letting fall from points D and C the perpendiculars DE to AB, DF to AC (meeting AB in G) and CK to AB, in the triangle ADB there will be

$$AD^2 - DB^2 + AB^2 = 2AE \times AB,$$

by *Elements*, II, 13. But $DB = AD + BR$, and therefore

$$DB^2 = AD^2 + 2AD \times BR + BR^2.$$

Take this from $AD^2 + AB^2$ and there will remain $AB^2 - 2AD \times BR - BR^2$ for $2AE \times AB$. In addition $AB^2 - BR^2 = (AB - BR)(AB + BR) = AR \times AS$. Hence $AR \times AS - 2AD \times BR = 2AE \times AB$ and $2AD = (AR \times AS - 2AB \times AE)/BR$. And, by similar reasoning, in the triangle ADC there will again emerge

$$2AD = (AT \times AV - 2AC \times AF)/CT.\text{[336]}$$

Consequently $(AR \times AS - 2AB \times AE)/BR = (AT \times AV - 2AC \times AF)/CT$ and so $AT \times AV/CT - AR \times AS/BR + 2AB \times AE/BR = 2AC \times AF/CT$, or

$$AF = (AT \times AV/CT - AR \times AS/BR + 2AB \times AE/BR)\,CT/2AC.$$

Hence, since $AK : AC = AF : AG$, there will be

$$AG = (AT \times AV/CT - AR \times AS/BR + 2AB \times AE/BR)\,CT/2AK.$$

---

(333) Observe Newton's sudden switch from Cartesian to classical (geometrical) analysis, almost as though this were a concluding 'Prob 5' of his contemporary 'Quæstionum solutio Geometrica' (see note (322) above).

(334) Though the overwritten original is not clear it may be that 'occurrens' was written in over 'occurrente', so replacing it, rather than being replaced by it. We reproduce Whiston's reading, printed in his 1707 *editio princeps* and subsequently sanctioned by Newton in 1722.

(335) The manuscript reads 'Est at'. Whiston's 1707 *editio princeps* has 'Est &', an emendation accepted by Newton in 1722.

(336) Observe how Newton, having previously worked with two geometrical variables AD and AE, now eliminates one in order to continue with the single unknown AE.

hoc de AE sive $\dfrac{2KAE}{CT} \times \dfrac{CT}{2AK}$ et restabit

$$GE = \frac{RAS}{BR} - \frac{TAV}{CT} - \frac{2BAE}{BR} + \frac{2KAE}{CT} \times \frac{CT}{2AK}.$$

Unde cum sit $KC.AK::GE.DE$, erit

$$DE = \frac{RAS}{BR} - \frac{TAV}{CT} - \frac{2BAE}{BR} + \frac{2KAE}{CT} \times \frac{CT}{2KC}.$$

‖[106]　In AB cape AP quæ sit ad AB ut ‖ CT ad BR, et erit $\dfrac{2PAE}{CT} = \dfrac{2BAE}{BR}$ adeoꝗ

$\dfrac{2PK \times AE}{CT} = \dfrac{2BAE}{BR} - \dfrac{2KAE}{CT}$, adeoꝗ $DE = \dfrac{RAS}{BR} - \dfrac{TAV}{CT} - \dfrac{2PK \times AE}{CT} \times \dfrac{CT}{2KC}$. Ad

AB erige ergo perpendiculum $AQ = \dfrac{RAS}{BR} - \dfrac{TAV}{CT} \times \dfrac{CT}{2KC}$, et in eo cape

$QO = \dfrac{PK \times AE}{KC}$, et erit $AO = DE$.[337] Junge DO, DQ, CP, et triangula DOQ

CKP erunt similia, quippe quorum anguli ad O et K sunt recti & latera
$(KC.PK::AE$ vel $DO.QO)$ proportionalia. Anguli ergo OQD, KPC æquales
sunt & proinde QD perpendicularis est ad CP. Quamobrem si agatur AN
parallela CP et occurrens QD in N, angulus ANQ erit rectus et triangula AQN,

PCK similia, adeoꝗ $PC.KC::AQ.AN$. Unde cum AQ sit $\dfrac{RAS}{BR} - \dfrac{TAV}{CT} \times \dfrac{CT}{2KC}$,

AN erit $\dfrac{RAS}{BR} - \dfrac{TAV}{CT} \times \dfrac{CT}{2PC}$. Produc AN ad M ut sit $NM = AN$ et erit $AD = DM$

adeoꝗ circulus quæsitus transibit per punctum M.[338] Cum ergo punctum M
datum sit, ex his sine ulteriori Analysi talis emergit Problematis resolutio.

　　In AB cape AP quæ sit ad AB ut CT ad BR, junge CP eiꝗ parallelam age

AM quæ sit ad $\dfrac{RAS}{BR} - \dfrac{TAV}{CT}$ ut CT ad PC: et ope Prob 39 per puncta A et M

describe circulum $AIHM$ qui tangat alterutrum circulorum TIV, RHS, et idem
circulus tanget utrumꝗ. Q.E.F.

　　Et hinc circulus etiam describi potest qui tres circulos positione et magnitudine datos continget. Sunto trium datorū circulorum radij A. B. C & centra
D, E, F. Centris E et F radijs $B \pm A$, $C \pm A$ describantur duo circuli, et tertius

---

(337) Whence the slope $QO/AE = PK/KC$ of QD is fixed. Since Q (fixed by the preceding
value of AQ) is determined in position, this serves to determine the line QN passing through
the centre D of the required circle.

(338) A neat reduction (to Problem 39 preceding) which obviates the need to construct the
centre D or points H and I of contact directly.

Take this from AE or $(2AK \times AE/CT)\, CT/2AK$ and there will remain

$$GE = (AR \times AS/BR - AT \times AV/CT - 2AB \times AE/BR$$
$$+ 2AK \times AE/CT)\, CT/2AK.$$

Hence, since $KC:AK = GE:DE$, there will be

$$DE = (AR \times AS/BR - AT \times AV/CT - 2AB \times AE/BR$$
$$+ 2AK \times AE/CT)\, CT/2KC.$$

In AB take AP to be to AB as CT to BR and there will be

$$2AP \times AE/CT = 2AB \times AE/BR,$$

so that $2AB \times AE/BR - 2AK \times AE/CT = 2PK \times AE/CT$ and consequently

$$DE = (AR \times AS/BR - AT \times AV/CT - 2PK \times AE/CT)\, CT/2KC.$$

To AB erect therefore the perpendicular

$$AQ = (AR \times AS/BR - AT \times AV/CT)\, CT/2KC$$

and in it take $QO = PK \times AE/KC$, and there will be $AO = DE$.[337] Join DO, DQ, CP and the triangles DOQ, CKP will be similar, seeing that their angles at O and K are right and their sides in proportion $(KC:PK = (AE \text{ or})\, DO:QO)$. The angles $O\widehat{Q}D$, $K\widehat{P}C$ are therefore equal and consequently QD is perpendicular to CP. As a result, if AN be drawn parallel to CP, meeting QD in N, the angle $A\widehat{N}Q$ will be right and the triangles AQN, PCK similar, so that

$$PC:KC = AQ:AN.$$

Hence, since AQ is $(AR \times AS/BR - AT \times AV/CT)\, CT/2KC$, AN will be

$$(AR \times AS/BR - AT \times AV/CT)\, CT/2PC.$$

Extend AN to M so that $NM = AN$ and there will be $AD = DM$, and hence the required circle will pass through the point M.[338] Since, therefore, the point M is given, from these results a resolution of the problem emerges, without further analysis, to this effect;

In AB take AP to be to AB as CT to BR, join CP and parallel to it draw AM, which shall be to $AR \times AS/BR - AT \times AV/CT$ as CT to PC; then with the aid of Problem 39 through the points A and M describe the circle $AIHM$ to touch one or other of the circles TIV, RHS, and this same circle will touch both. As was to be done.

And hence a circle can also be described which shall touch three circles given in position and magnitude. Let the radii of the three given circles be A, B and C, their centres D, E and F. With centres E and F and radii $B \pm A$, $C \pm A$ let two circles be described, and also a third circle to touch these and pass through the

circulus qui hosce tangat transeatꝗ per punctum [*D*].[339] Sit hujus radius *G* et centrum *H* et eodem centro *H* radio $G \pm A$[340] descriptus circulus continget tres primos circulos ut fieri oportuit.[341]

Prob. 42.[342]

Erectis alicubi terrarum tribus baculis ad Horizontale planum in punctis ‖ *A, B, et C perpendicularibus, quorum is qui in A sit sex pedum, qui in B octodecim pedum & qui in C octo pedum, existente linea AB triginta trium pedum: Contingit quodam die extremitatem umbræ baculi A transire per puncta B et C, baculi autem B per A et C ac baculi C per punctum A.*[343] *Quæritur declinatio Solis et elevatio Poli, sive dies locusꝗ ubi hæc evenerint?*

Quoniam umbra baculi cujusꝗ descripsit Conicam sectionem, sectionem nempe Coni radiosi[344] cujus vertex est baculi summitas: fingam *BCDEF* esse

---

(339) The manuscript reads '*A*', a slip of Newton's pen passed by Whiston into his 1707 *editio princeps* but caught soon after by Newton in his library copy of it and duly corrected in his 1722 edition.

(340) We would expect '$G \mp A$' to balance the preceding augments/decrements.

(341) Analytical proof of Newton's construction is not difficult and considerably more enlightening to modern eyes untrained in the subtleties of classical analysis using geometrical variables. If we impose a Cartesian system of coordinates in which *A* is the origin, $AE = x$, $ED = y$ and set $B(a, 0)$, $C(b, -c)$ as the centres of the two given circles of respective radii *r* and *s*, then the condition $DB^2 = DA^2 + AB^2 - 2AB \times AE$ yields

$$(\sqrt{[x^2+y^2]}+r)^2 = x^2+y^2+a^2-2ax \quad \text{or} \quad 2r\sqrt{[x^2+y^2]} = (a^2-r^2)-2ax;$$

similarly, $DC^2 = DA^2 + AC^2 - 2AC \times AF$ gives, after reduction,

$$2s\sqrt{[x^2+y^2]} = (b^2+c^2-s^2)-2(bx-cy):$$

whence $2(br-as)x - 2cry = r(b^2+c^2-s^2)-s(a^2-r^2)$, that is, $(\alpha-b)x+c(y-\beta) = 0$ where $AP = as/r = \alpha$ (so that $KP = \alpha-b$) and also $AQ = \frac{1}{2}(a^2-r^2)s/cr - \frac{1}{2}(b^2+c^2-s^2)/c = \beta$. This last equation determines $D(x, y)$ to be on a line of given slope $-(\alpha-b)/c$ through $Q(0, \beta)$, and from this Newton's reduction to Problem 39 follows immediately.

(342) This problem has an interesting history. First proposed as a *Problema Astronomicum et Geometricum* (The Hague, 1638) published in a broadsheet challenge to his fellow geometers by the Dutch mathematical professor Jan Stampioen de Jonge, the question was brought to Descartes' attention by a young Utrecht surveyor Jacob van Waessenaer, who had earlier come into conflict with Stampioen over a companion *Questie aem de Batavische Ingenieurs* and now sought help in outwitting his adversary (see C. Adam and P. Tannery, *Œuvres de Descartes*, **2** (Paris, 1898): 611–13; and compare II: 316, note (41)). Descartes subsequently permitted his solution to appear anonymously in the third, concluding section of Waessenaer's hard-hitting *Den On-wissen Wis-konstenaer I. I. Stampioënius ontdeckt door sijne ongegronde Weddinge ende mis-lucte Solutien van sijne eygene Questien...* (Leyden, 1640): five years later he wrote (to Haestrecht?) that 'pour remarquer l'industrie de bien demesler les équations, ie n'en sçache point de plus propre que celle des trois bâtons' (Clerselier, *Lettres de M. Descartes*, **3** (Paris, 1667): 458–60 [= *Œuvres de Descartes*, **4**, 1901: 227–32], especially 459). In 1649 his solution was incorporated by Frans van Schooten in an extended *addendum* to the Latin version of Descartes' *Geometrie* (*Geometria*, ₁1649: 295–323), reappearing a decade later in lightly revised form in the second

point D.[339] Let the radius of this be G and its centre H, and a circle described with the same centre H and radius $G \pm A$[340] will then touch the first three circles: as was required to happen.[341]

Problem 42[342]

When, somewhere on Earth, three staves are erected perpendicular to the horizontal plane at points A, B and C—that at A being 6 feet long, that at B 18 feet and that at C 8 feet, with the line AB 33 feet in length—, it happens on a certain day that the tip of stave A's shadow passes through the points B and C, that of stave B, however, through A and C, and that of stave C through the point A.[343] What is the sun's declination and the polar elevation? in other words, on what day and at what place do these events occur?

Since the shadow cast by each stave describes a conic—viz. a (plane) section of the light-cone[344] whose vertex is the stave's tip—, I shall imagine that

edition (*Geometria*, ₂1659: 369–89). This last version is with little doubt the source for Newton's present reworking of the problem: to point the divergences between the two (structurally identical) solutions a précis of the Cartesian computation is given at appropriate points in following footnotes.

(343) Whence it must pass through B also, as Descartes proved; see note (346). As quoted by Descartes to Haestrecht(?) in 1645 (see previous note), the original version of the problem read: '...Et una atque eadem die extremitas umbræ solaris quam facit baculus A transit per puncta B & C, extremitas umbræ baculi B per A & C, et ex consequenti [!] etiam baculi C, per A & B.'

(344) Literally, the 'cone of rays', PRQ in the accompanying diagram, which the tip R of any 'stave' AR will cast as its sun-shadow as the earth rotates diurnally around its polar axis

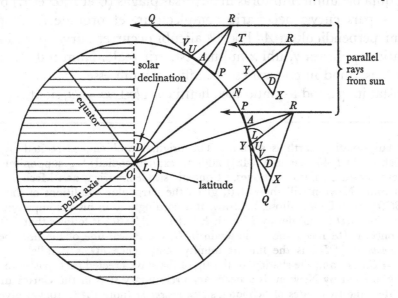

NO. It is assumed that on any given day the sun's declination to the equator (equal to the complement $R\widehat{X}Y$ of half the vertex angle of the light-cone) is constant, so that the light-cone is effectively circular and therefore the (plane) shadow it casts at any point A on the surface

hujusmodi curvam (sive ea
sit Hyperbola, Parabola vel
Ellipsis)[345] quam umbra
baculi A eo die descripsit,
ponendo AD, AE, AF ejus
umbras fuisse cum BC, BA,
CA respectivè fuerunt umbræ
baculorum B & C.[346] Et
præterea fingam PAQ esse
lineam Meridionalem[347] sive
axem hujus curvæ ad quem
demissæ perpendiculares BM,
CH, $DK_{[,]}$ EN et FL sunt

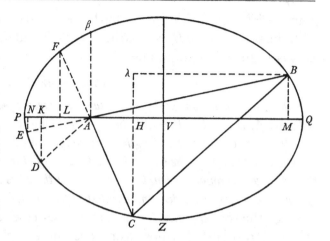

ordinatim applicatæ. Has vero ordinatim applicatas indefinitè designabo litera
y, & axis partes interceptas AM, AH, AK, AN et AL litera x. Fingam deniçz
æquationem $aa \perp bx \perp cxx = yy$[348] ipsarum x et y relationem (i.e. naturam Curvæ)
designare, assumendo aa, b, & c tanquam cognitas ut ex Analysi tandem inveniantur. Ubi incognitas quantitates x et y duarum tantum dimensionum posui
quia æquatio est ad Conicam sectionem, et ipsius y dimensiones impares omisi
quia ipsa est ordinatim applicata ad axem. Signa autem ipsorum b et c quia
indeterminata sunt designavi notula $\perp$ quam indifferenter pro $+$ aut $-$ usurpo
‖ [108] et ejus oppositum $\top$ pro signo contrario.[349] At signum qua‖drati aa affirmativum
posui, quia baculum A umbras in adversas plagas (C et F, D et E) projicientem
concava pars curvæ necessariò complectitur, et proinde si ad punctum A
erigatur perpendiculum $A\beta$ hoc[350] alicubi occurret curvæ puta in β, hoc est,
ordinatim applicata y, ubi x nullum est, erit reale. Nam inde sequitur quadratum ejus, quod in eo casu est aa, affirmativum esse.

Constat itaçz quod æquatio hæc fictitia $aa \perp bx \perp cxx = yy$, sicut terminis super-

---

of the daily-rotating earth is a conic. The original problem, as recited by Descartes to
Haestrecht(?) in 1645 (see note (343)) adds an explicit codicil: '...supponimus illas umbras
describere accurate conicas sectiones, ut quæstio sit Geometrica, non Mechanica'.

(345) Since Newton will show that, given the particular conditions of the problem, this
conic *BCDEF* will be an ellipse, he draws it accordingly so in his accompanying figure.

(346) Since AD is parallel to BC with $EA:AB = AD:BC$, it follows that ED is parallel to
AC; at once, if DA meets the conic again in G, the Pascal-line of the inscribed Leibnizian
hexagrammum BCFGDE is the line at infinity, and hence FG is parallel to EAB, so that
$CB:AG = CA:AF$ and the shadow of the tip of the stave erected on C passes through B. This
corollary, unused by Newton, is a necessary preliminary step in the Cartesian solution (see
note (351)): the two states of Schooten's *Additamentum* (note (342) above) give variant syn-
thetic/analytical proofs of the result (*Geometria*, ₁1649: 296–9 = ₂1659: 370–1), each effectively
employing the projective definition of a conic implicit in Apollonius, *Conics*, III, 17 (compare
IV: 225). In essence, if (as before) DA meets the ellipse in G and GH, drawn parallel to FC,

BCDEF is the curve of this class (be it a hyperbola, parabola or ellipse)[345] which the shadow of stave *A* describes on that particular day, supposing that *AD*, *AE*, *AF* were its shadows when *BC*, *BA*, *CA* respectively were those of the staves *B* and *C*.[346] I shall further imagine that *PAQ* is the meridional line;[347] that is, the (main) axis of this curve, to which, when they are let fall, the perpendiculars *BM*, *CH*, *DK*, *EN* and *FL* are ordinately applied. These ordinates, indeed, I shall indefinitely designate by the letter *y*, and the intercepted portions *AM*, *AH*, *AK*, *AN* and *AL* by the letter *x*. Finally, I shall imagine that the equation $a^2 \pm bx \pm cx^2 = y^2$ [348] denotes the relationship of the *x*'s and *y*'s (that is, the nature of the curve), assuming a^2, *b* and *c* as known so that they may at length be ascertained from the analysis. I have here set the unknown quantities to be of but two dimensions because the equation defines a conic, and I have omitted odd dimensions of *y* because it is ordinate to the axis. Because, however, the signs of *b* and *c* are indeterminate, I have denoted them by the little symbol '⊥' ($\pm$) which I employ without distinction in place of + or −, using its opposite '⊤' ($\mp$) in place of the contrary sign.[349] But the sign of the square a^2 I have set as positive, because the stave *A*, casting its shadows in opposite directions (both to *C* and *F*, and to *D* and *E*) is necessarily enclosed within the hollow region of the curve, and in consequence, if a perpendicular *Aβ* be erected at the point *A*, it[350] will meet the curve at some point, say in *β*; in other words, when *x* is zero, the ordinate *y* will be real—from which, of course, it follows that its square, a^2 in the present case, is positive.

It may therefore be agreed that this fictitious equation $a^2 \pm bx \pm cx^2 = y^2$,

---

meets the conic again in *H* and also *AB*, *BC* in *I* and *K*, then by Apollonius' proposition $DA \times AG/CK \times KB = FA \times AC/GK \times KH$ or $FA \times KB = DA \times KH$, and

$$FA \times AC/GI \times IH = EA \times AB/EI \times IB \quad \text{or} \quad FA \times CB/AG \times IH = DA \times CB/DG \times KB;$$

whence

$$\frac{AG}{FA} = \frac{DG \times KB}{DA \times IH} = \frac{(DA+AG)\,KB}{DA \times IK + FA \times KB} = \frac{KB}{IK},$$

and so *FG* is parallel to *EAB*. Orthogonal projection of the ellipse into a circle, preserving the ratios of lines drawn in the same direction, would yield a simpler elementary proof: perhaps Newton had this in mind when, about the time he penned the present manuscript, he inserted a brief, somewhat mysterious note in his Waste Book (ULC. Add. 4004: 96ᵛ) that 'The Problem in Schooten *de tribus baculis* may be solved more easily by supposing yᵉ Ellipsis to be a circle first & then reducing it to yᵉ desired [Ellipsis]'.

(347) The North–South line.

(348) The conic's equation can be put in this simple form because the 'meridional line' *PAQ* manifestly is the main axis of the shadow cast at *A* by the light-cone; compare note (344).

(349) This notation is evidently Newton's *ad hoc* concoction. We may or may not agree with Florian Cajori's supposition (*A History of Mathematical Notations*, **1** (Chicago, 1928): 245) that 'These signs appear to be the + with half the vertical stroke excised'.

(350) Newton first wrote equivalently 'ordinatim applicetur *Aβ* hæc' (...the ordinate *Aβ* be applied..., it), understanding that the ordination angle is right.

fluis non refertur sic neq restrictior est quam ut ad omnes hujus problematis conditiones se extendat, Hyperbolam Ellipsin vel Parabolam quamlibet designatura prout ipsorum *aa*, *b*, *c* valores determinabuntur aut nulli forte reperientur. Quid autem valent, quibusq signis *b* et *c* debent affici, et inde quænam sit hæc curva ex sequenti Analysi constabit.

Analyseos pars prior.[351]

Cùm umbræ sint ut altitudines baculorum erit

$$BC.AD::AB.AE(::18.6)::3.1.$$

Item $CA.AF(::8.6)::4.3$. Quare nominatis $AM=r$, $MB=s$, $AH=t$ & $HC=\perp v$. ex similitudine triangulorum *AMB*, *ANE* & *AHC*, *ALF* erunt $AN=-\dfrac{r}{3}$. $NE=-\dfrac{s}{3}$. $AL=-\dfrac{3t}{4}$ et $LF=\top\dfrac{3v}{4}$: quorum signa signis ipsarum *AM*, *MB*, *AH*, *HC* contraria posui quia tendunt ad contrarias plagas respectu puncti *A* a quo ducuntur axisv́e *PQ* cui insistunt. His autem pro *x* et *y* in æquatione fictitia $aa\perp bx\perp cxx=yy$, respectivè scriptis,

r et s dabunt $aa\perp br\perp crr=ss$.

$-\dfrac{r}{3}$ & $-\dfrac{s}{3}$ dabunt $aa\top\dfrac{br}{3}\perp\frac{1}{9}crr=\frac{1}{9}ss$.

t & $\perp v$ dabunt $aa\perp bt\perp ctt=vv$.

$-\frac{3}{4}t$ & $\top\frac{3}{4}v$ dabunt $aa\top\frac{3}{4}bt\perp\frac{9}{16}ctt=\frac{9}{16}vv$.

Jam e prima[352] harum exterminando ss ut obtineatur r, prodit $\dfrac{2aa}{\perp b}=r$. Unde

---

(351) This represents only a slight variation on Descartes' approach (*Geometria*, ₁1649: 301–10 = ₂1659: 371–9). Having shown that *DA* meets the conic again in *G* such that *FG*

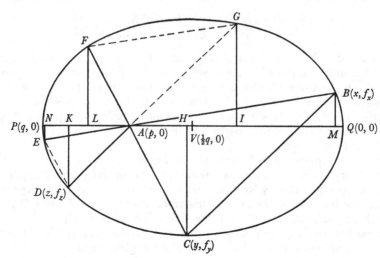

neither reckoning with superfluous terms nor being too restricted to extend to all the conditions of the present problem, is ready to express any hyperbola, ellipse or parabola according as the values of a^2, b and c shall be determined, or perhaps found to be zero. But what these values are, what signs ought to be affixed to b and c, and in consequence what the species of the curve is will be settled from the following analysis.

First part of the analysis[351]

Since the shadows are proportional to the heights of their staves, there will be $BC:AD = AB:AE = (18:6$ or$)$ $3:1$. Likewise $CA:AF = (8:6$ or$)$ $4:3$. Therefore, on naming $AM = r$, $MB = s$; $AH = t$ and $HC = \pm v$, from the similarity of the triangles AMB, ANE and AHC, ALF there will be $AN = -\frac{1}{3}r$, $NE = -\frac{1}{3}s$; $AL = -\frac{3}{4}t$ and $LF = \mp\frac{3}{4}v$. (I have set their signs contrary to those of AM, MB, AH, HC because they tend in the opposite direction with regard to the point A from which they are drawn or the axis PQ to which they are ordinate.) And, when these are entered in the fictitious equation $a^2 \pm bx \pm cx^2 = y^2$ in place of x and y respectively,

$$r \text{ and } s \text{ will yield } a^2 \pm br \pm cr^2 = s^2,$$

$$-\tfrac{1}{3}r \text{ and } -\tfrac{1}{3}s \text{ will yield } a^2 \mp \tfrac{1}{3}br \pm \tfrac{1}{9}cr^2 = \tfrac{1}{9}s^2,$$

$$t \text{ and } \pm v \text{ will yield } a^2 \pm bt \pm ct = v^2, \text{ and}$$

$$-\tfrac{3}{4}t \text{ and } \mp\tfrac{3}{4}v \text{ will yield } a^2 \mp \tfrac{3}{4}bt \pm \tfrac{9}{16}ct^2 = \tfrac{9}{16}v^2.$$

On now eliminating s^2 from the first[352] of these in order to obtain r, there results

---

is parallel to EAB (see note (346)), from the given ratios $BA:AE = CB:DA = 3:1$ and $CA:AF = CB:AG = 4:3$ Descartes deduces that $DA:AG = 4:9$; then he assumes the conic $BCDEFG$ to be an ellipse of main axis $QP = q$ and *latus rectum r*, setting its Cartesian equation, referred to origin $Q(0, 0)$, vertex $P(q, 0)$ and centre $V(\frac{1}{2}q, 0)$, to be effectively $f_a^2 = ra - (r/q)\,a^2$, where its general point (a, f_a) is referred to abscissa a and (perpendicular) ordinate f_a. On putting $QA = p$, $QM = x$, $QH = y$ and $QK = z$ there is $QN = \frac{1}{3}(4p - x)$, $QL = \frac{1}{4}(7p - 3y)$ and $QI = \frac{1}{4}(13p - 9z)$; and thence from the equations $\frac{1}{3}f_x = f_{\frac{1}{3}(4p-x)}$, $\frac{3}{4}f_y = f_{\frac{1}{4}(7p-3y)}$ and $\frac{9}{4}f_z = f_{\frac{1}{4}(13p-9z)}$ there comes respectively $x = p(4p - 3q)/(2p - q)$, $y = \frac{1}{3}p(7p - 4q)/(2p - q)$ and $z = \frac{1}{9}p(13p - 4q)/(2p - q)$; while from the proportion $(BM + HC):KD = (CB:DA$ or$)\,3:1$ there results $f_x + f_y = 3f_z$, which after reduction (and on putting $v = 7/16\sqrt{3}$) gives

$$(p - \tfrac{1}{2}q)^2(p^2 - pq + (\tfrac{1}{4} - v^2)\,q^2) = 0.$$

Hence, since A and V do not coincide (and so $p \neq \frac{1}{2}q$), there results $QA = p = (v + \frac{1}{2})\,q$, $AP = (-v + \frac{1}{2})\,q$ and so $VA = p - \frac{1}{2}q = vq$, wherefore $QM = x = (2v + \frac{1}{2} - \frac{1}{4}v^{-1})\,q$ and $MA = (-v + \frac{1}{4}v^{-1})\,q$. This agrees with Newton's following computation using the variant defining equation $y^2 = a^2 + bx - cx^2$ of the ellipse referred to V as origin, since on substituting $c = \frac{143}{196}b^2/a^2$ Newton's expression for the length of the main axis $PQ = q = 2\sqrt{[\frac{1}{4}b^2/c^2 + a^2/c]}$ yields $q = \frac{224}{143}\sqrt{3}.a^2/b$ and hence $v = VA/q = 7/16\sqrt{3}$, as above.

(352) In his 1722 revise Newton added the necessary complement 'et secunda' (and second).

patet $\perp b$ esse affirmativum.[353] Item e tertia et quarta exterminando vv ut

||[109] obtineatur t prodit $\dfrac{aa}{3b}=t$. Et scrip||tis insuper $\dfrac{2aa}{b}$ pro r in prima, et $\dfrac{aa}{3b}$ pro t in

tertia, oriuntur $3aa\perp\dfrac{4a^4c}{bb}=ss$ & $\frac{4}{3}aa\perp\dfrac{a^4c}{9bb}=vv$.

Porro demissa $B\lambda$ perpendiculari in CH, erit

$$BC.AD(::3.1)::B\lambda.AK::C\lambda.DK.$$

Quare cùm sit $B\lambda(=AM-AH=r-t)=\dfrac{5aa}{3b}$, erit $AK=\dfrac{5aa}{9b}$, vel potius[354]

$=-\dfrac{5aa}{9b}$. Item cum sit $C\lambda(=CH\perp BM=v\perp s)=\sqrt{\dfrac{4aa}{3}\perp\dfrac{a^4c}{9bb}}\perp\sqrt{3aa\perp\dfrac{4a^4c}{bb}}$, erit

$DK(=\frac{1}{3}C\lambda)=\sqrt{\dfrac{4aa}{27}\perp\dfrac{a^4c}{81bb}}\perp\sqrt{\dfrac{1}{3}aa\perp\dfrac{4a^4c}{9bb}}$. Quibus in æquatione $aa+bx\perp cxx=yy$,

pro AK ac DK sive x et y respectivè scriptis, prodit

$$\dfrac{4aa}{9}\perp\dfrac{25a^4c}{81bb}=\dfrac{13}{27}aa\perp\dfrac{37a^4c}{81bb}\perp2\sqrt{\dfrac{4aa}{27}\perp\dfrac{a^4c}{81bb}}\times\sqrt{\dfrac{aa}{3}\perp\dfrac{4a^4c}{9bb}}.$$

Et per reductionem $-bb\perp4aac$[355]$=2\sqrt{36b^4\perp51aabbc+4a^4cc}$, & partibus quad-

ratis iterumcȝ reductis exit $0=143b^4\perp196aabbc$ sive $\dfrac{-143bb}{196aa}=\perp c$. Unde constat

$\perp c$ negativam esse, ade ocȝ æquationem fictitiam $aa\perp bx\perp cxx=yy$ hujus esse

formæ $aa+bx-cxx=yy$, et ideo curvam quam designat Ellipsin esse. Ejus vero

centrum et axes duo[356] sic eruuntur.

Ponendo $y=0$, sicut in Figuræ verticibus P et Q contingit, habebitur

$aa+bx=cxx$, et extracta ra-

dice, $x=\dfrac{b}{2c}\pm\sqrt{\dfrac{bb}{4cc}+\dfrac{aa}{c}}=\dfrac{AQ.}{AP.}$

Adeocȝ sumpto $AV=\dfrac{b}{2c}$, erit

V centrum Ellipsis et VQ vel

$VP\left(\sqrt{\dfrac{bb}{4cc}+\dfrac{aa}{c}}\right)$ semiaxis maxi-

mus. Si porro ipsius AV valor

$\dfrac{b}{2c}$ pro x in æquatione

$aa+bx-cxx=yy$

scribatur, fiet $aa+\dfrac{bb}{4c}=yy$.

Quare est $aa+\dfrac{bb}{4c}=VZ^q$ hoc est quadrato semiaxis minimi. Denicȝ in valoribus

$r = 2a^2/\pm b$: from which it is clear that $\pm b$ is positive.[353] Likewise, on eliminating v^2 from the third and fourth in order to obtain t, there results $\frac{1}{3}a^2/b = t$. And, when in addition $2a^2/b$ is written in place of r in the first and $\frac{1}{3}a^2/b$ in place of t in the third, there arise $3a^2 \pm 4a^4c/b^2 = s^2$ and $\frac{4}{3}a^2 \pm \frac{1}{9}a^4c/b^2 = v^2$.

Moreover, when $B\lambda$ is let fall perpendicularly to CH, there will be

$$BC : AD \, (= 3 : 1) = B\lambda : AK = C\lambda : DK.$$

Consequently, since $B\lambda$ (or $AM - AH = r - t$) $= \frac{5}{3}a^2/b$, there will be $AK = \frac{5}{9}a^2/b$, or rather[354] $-\frac{5}{9}a^2/b$. Likewise, since

$$C\lambda \,(\text{or } CH \pm BM = v \pm s =)\, \sqrt{[\tfrac{4}{3}a^2 \pm \tfrac{1}{9}a^4c/b^2]} \pm \sqrt{[3a^2 \pm 4a^4c/b^2]},$$

there will be DK (or $\frac{1}{3}C\lambda$) $= \sqrt{[\tfrac{4}{27}a^2 \pm \tfrac{1}{81}a^4c/b^2]} \pm \sqrt{[\tfrac{1}{3}a^2 \pm \tfrac{4}{9}a^4c/b^2]}$. And when these are entered in the equation $a^2 + bx \pm cx^2 = y^2$ in place of AK and DK, that is, x and y respectively, there results

$$\tfrac{4}{9}a^2 \pm \tfrac{25}{81}a^4c/b^2 = \tfrac{13}{27}a^2 \pm \tfrac{37}{81}a^4c/b^2 \pm 2\sqrt{[\tfrac{4}{27}a^2 \pm \tfrac{1}{81}a^4c/b^2]} \times \sqrt{[\tfrac{1}{3}a^2 \pm \tfrac{4}{9}a^4c/b^2]}$$

and after reduction $-b^2 \pm 4a^2c$[355] $= 2\sqrt{[36b^4 \pm 51a^2b^2c + 4a^4c^2]}$; after the sides are squared and again reduced there comes out $0 = 143b^4 \pm 196a^2b^2c$ or $\pm c = -\frac{143}{196}b^2/a^2$. It is accordingly established that $\pm c$ is negative, and hence that the fictitious equation $a^2 \pm bx \pm cx^2 = y^2$ is of this form, $a^2 + bx - cx^2 = y^2$: in consequence the curve it denotes is an ellipse. Its centre, however, and its two axes[356] are derived in this manner.

On setting $y = 0$ (as happens at the vertices P and Q of the curve) there will be had $a^2 + bx = cx^2$ and, when the root is extracted,

$$x = \tfrac{1}{2}b/c \pm \sqrt{[\tfrac{1}{4}b^2/c^2 + a^2/c]} = \begin{Bmatrix} AQ \\ AP \end{Bmatrix}$$

and hence, when there is taken $AV = \frac{1}{2}b/c$, V will be the ellipse's centre and VQ or VP ($\sqrt{[\tfrac{1}{4}b^2/c^2 + a^2/c]}$) the semi-major-axis. If, furthermore, the value $\frac{1}{2}b/c$ of AV be written in place of x in the equation $a^2 + bx - cx^2 = y^2$, there will come $a^2 + \frac{1}{4}b^2/c = y^2$ and consequently

$$VZ^2 \text{ (that is, the square of the semi-minor-axis)} = a^2 + \tfrac{1}{4}b^2/c.$$

---

(353) Whence the conic's defining equation assumes the form '$aa + bx \perp cxx = yy$'.

(354) Since AK is a negative abscissa.

(355) Newton rightly corrected this term to read '$\top 4aac$' ($\mp 4a^2c$) in his 1722 revise.

(356) 'diametri duæ' (two diameters) was first written.

ipsarum AV, VQ, VZ jam inventis scripto $\dfrac{143bb}{196aa}$ pro c, exeunt $\dfrac{98aa}{143b}=AV$,

$\dfrac{112aa\sqrt{3}}{143b}=VQ$, & $\dfrac{8a\sqrt{3}}{\sqrt{143}}=VZ$.

[1678]
Lect 6

Analyseos pars altera.[357]

Supponatur jam baculum puncto A insistens esse AR et erit RPQ planum meridionale ac $RPZQ$ conus radiosus cujus vertex est R. Sit insuper TXZ planum secans Horizontem in VZ ut et meridionale planum

‖[110] in TVX, quæ sectio sit ad axem ‖ mundi[358] conive perpendicularis,

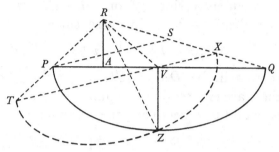

et ipsum planum TXZ erit ad eundem axem perpendiculare et conum secabit in peripheria circuli TZX quæ ab ejus vertice pari ubiꝗ intervallo RX, RZ, RT distabit. Quamobrem si PS ipsi TX parallela ducatur, fiet $RS=RP$ propter æquales RX, RT; nec non $SX=XQ$ propter æquales PV, VQ. Unde est RX vel

$RZ\left(=\dfrac{RS+RQ}{2}\right)=\dfrac{RP+RQ}{2}$. Deniꝗ ducatur RV et cùm VZ perpendiculariter

insistat plano RPQ, (sectio utiꝗ existens planorum eidem perpendiculariter insistentium) fiet triangulum RVZ rectangulum ad V.

---

(357) Having established the proportion of AV to the semi-axes $PV = VQ$ and knowing the height of the stave $AR = 6$, Newton now determines the triangle PRQ in absolute dimension from the condition that R shall be the vertex of the right circular cone whose plane section through PQ (perpendicular to the plane of the triangle PRQ) is the ellipse PZQ of semi-axes $PV = VQ$ and VZ, defined (in terms of the coefficients a and b of the ellipse's Cartesian equation $y^2 = a^2 + bx - \frac{143}{196}(b^2/a^2)x^2$) as in the preceding 'Analyseos pars prior'. The required solar declination $D = R\widehat{X}T$ and latitude $L = R\widehat{U}P$ (see the augmented perspective sketch inserted in the English version to clarify Newton's jejune figure, and compare note (344)) are then immediately derivable as the complements of the semi-sum $X\widehat{R}Y$ and semi-difference $A\widehat{R}Y$ of the vertex angles $A\widehat{R}Q$ and $A\widehat{R}P$. Descartes' approach (*Geometria*, $_1$1649: 310–18 $=_2$1659: 379–84) is not essentially different. In the terms of note (351), on putting $\alpha = 2v + \frac{1}{2} - \frac{1}{4}v^{-1}$, $\beta = -v + \frac{1}{4}v^{-1}$ $(v = 7/16\sqrt{3})$ so that $QM = \alpha q$ and $MA = \beta q$, and also for convenience setting $AR(= 6) = c$ and $AB(= 33) = kc$ (where $k = 5\frac{1}{2}$) he draws the cone's axis RU, meeting PQ, XT and the parallel to it through P in U, Y and Y' respectively, and then takes $AU = (1-f)vq$, so that $PU = (\frac{1}{2}-fv)q$ and $RU = n$, where $n^2 = c^2 + (f-1)^2v^2q^2$. At once UY (or $AU \times UV/RU$) $= f(1-f)v^2q^2/n$ and so $RY = (c^2 + (1-f)v^2q^2)/n$;

$$UY' \text{ (or } AU \times PU/RU) = (\tfrac{1}{2}-fv)(1-f)vq^2/n$$

and therefore $RY' = (c^2 + (1-f)(v^2 - \frac{1}{2}v)q^2)/n$; YV (or $AR \times UV/RU$) $= cfvq/n$; and

$$YX \text{ (or } AR \times PV/RU) = \tfrac{1}{2}cq/n.$$

Finally, when in the values of AV, VQ, VZ just now found $\frac{143}{196}b^2/a^2$ is written in place of c, there results $AV = \frac{98}{143}a^2/b$, $VQ = \frac{112}{143}a^2\sqrt{3}/b$ and $VZ = 8a\sqrt{3}/\sqrt{143}$.

Second part of the analysis[357]

Now suppose that the stave standing on the point A is AR: RPQ will then be the meridional plane and $RPZQ$ the light-cone whose vertex is R. In addition let TXZ be a plane cutting the horizontal in VZ and also the meridional plane in TVX; let this line of section be perpendicular to the world's axis[358]—and so to that of the cone—and the plane TXZ will itself be perpendicular to the same axis, cutting the cone in the circumference TZX of a circle at distances RX, RZ, RT from its vertex everywhere of equal length. In consequence, if PS be drawn parallel to TX, there will come to be $RS = RP$ because RX, RT are equal, and also $SX = XQ$ because PV, VQ are equal; hence

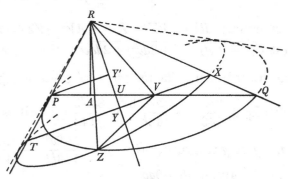

$$RX \text{ or } RZ = (\tfrac{1}{2}(RS+RQ) \text{ or) } \tfrac{1}{2}(RP+RQ).$$

Finally, let RV be drawn and then, since VZ stands perpendicularly on the plane RPQ—being, of course, the line of section of planes standing perpendicularly to it—, the triangle RVZ will prove to be right-angled at V.

Hence, since $RT/(RT-RY) = TX/TV$, there follows $c^2 = (f-1)(v^2-1/4f)q^2$ and so $n^2 = (1-1/f)(f^2v^2-\tfrac{1}{4})q^2$. Further

$$(VZ^2 \text{ or) } \tfrac{1}{4}qr = (YX^2 - YV^2 \text{ or) } c^2(\tfrac{1}{4}-f^2v^2)q^2/n^2 = c^2f/(1-f) = \tfrac{1}{4}(1-4fv^2)q^2$$

and consequently $r = (1-4fv^2)q$. Finally, where $\gamma = \alpha - \alpha^2$,

$$(MB^2 \text{ or) } \gamma qr = (AB^2 - MA^2 \text{ or) } k^2c^2 - \beta^2q^2$$

and therefore $\gamma(1-4fv^2) = \gamma r/q = k^2(f-1)(v^2-1/4f) - \beta^2$, whence

$$(k^2+4\gamma)v^2f^2 - (k^2(v^2+\tfrac{1}{4})+\beta^2+\gamma)f + \tfrac{1}{4}k^2 = 0:$$

a quadratic whose root f (in the interval $[0, 1]$) is readily determinable in terms of the known coefficients. On back-tracking, q is determined as the (positive) root of

$$k^2c^2/(\beta^2+\gamma(1-4fv^2));$$

and from this are obtained the solar declination $\tan^{-1}[(c^2+(1-f)v^2q^2)/\tfrac{1}{2}cq]$ and the latitude $\tan^{-1}[c/(1-f)vq]$.

(358) The polar axis NO in the figure of note (344).

Dictis jam $RA=d$, $AV=e$, VP vel $VQ=f$, & $VZ=g$ erit $AP=f-e$, & $RP=\sqrt{ff-2ef+ee+dd}$. Item $AQ=f+e$ & $RQ=\sqrt{ff+2ef+ee+dd}$: adeoqჳ $RZ\left(=\dfrac{RP+RQ}{2}\right)=\dfrac{\sqrt{ff-2ef+ee+dd}+\sqrt{ff+2ef+ee+dd}}{2}$. Cujus quadratum

$$\frac{dd+ee+ff}{2}+\tfrac{1}{2}\sqrt{f^4-2eeff+e^4+2ddff+2ddee+d^4}$$

est æquale $(RV^q+VZ^q=RA^q+AV^q+VZ^q=)\ dd+ee+gg$. Jam reductione facta est $\sqrt{f^4-2eeff+e^4+2ddff+2ddee+d^4}=dd+ee-ff+2gg$, et partibus quadratis ac in ordinem redactis $ddff=ddgg+eegg-ffgg+g^4$, sive $\dfrac{ddff}{gg}=dd+ee-ff+gg$.

Deniqჳ 6, $\dfrac{98aa}{143b}$, $\dfrac{112aa\sqrt{3}}{143b}$ & $\dfrac{8a\sqrt{3}}{\sqrt{143}}$ (valoribus ipsorum AR, AV, VQ & VZ) pro d, e, f ac g restitutis, oritur $36-\dfrac{196a^4}{143bb}+\dfrac{192aa}{143}=\dfrac{36,14,14aa}{143bb}$ et inde per reductionem $\dfrac{49a^4+36,49aa}{48aa+1287}=bb$.

In primo schemate est $AM^q+MB^q=AB^q$, hoc est $rr+ss=33\times33$. Erat autem $r=\dfrac{2aa}{b}$ & $ss=3aa-\dfrac{4a^4c}{bb}$, unde $rr=\dfrac{4a^4}{bb}$ & (substituto $\dfrac{143bb}{196aa}$ pro c) $ss=\dfrac{4aa}{49}$. Quare $\dfrac{4a^4}{bb}+\dfrac{4aa}{49}=33\times33$,[359] et inde per reductionem iterum resultat $\dfrac{4,49a^4}{53361-4aa}=bb$. Ponendo igitur æqualitatem inter duo bb, & dividendo

‖[111]   utramqჳ partem æquationis per 49 fit $\dfrac{a^4+36aa}{48aa+1287}=\dfrac{4a^4}{53361-4aa}$. ‖ Cujus partibus in crucem multiplicatis ordinatis ac divisis per $49[aa]$, exit

$$4a^4=981aa+274428,^{(360)}$$

cujus radix aa est $\dfrac{981+\sqrt{1589625}}{8}=280|2254144$.

Supra inventum fuit $\dfrac{4,49a^4}{53361-4aa}=bb$, sive $\dfrac{14aa}{\sqrt{53361-4aa}}=b$. unde $AV\left(\dfrac{98aa}{143b}\right)$ est $\dfrac{7\sqrt{53361-4aa}}{143}$, & VP vel $VQ\left(\dfrac{112aa\sqrt{3}}{143b}\right)$ est $\tfrac{8}{143}\sqrt{160083-12aa}$. Hoc est substituendo $280|2254144$ pro aa ac terminos in decimales numeros reducendo, $AV=11|188297$ & VP vel $VQ=22|147085$. Adeoqჳ $AP(PV-AV)=10|958788$ et $AQ(AV+VQ)=33|335382$.

Deniqჳ si $\tfrac{1}{6}AR$ sive 1 ponatur Radius, erit $\tfrac{1}{6}AQ$ sive $5|555897$ tangens anguli ARQ 79${}^{\text{gr}}$. 47′. 48″ & $\tfrac{1}{6}AP$ sive $1|826465$ tangens anguli ARP 61${}^{\text{gr}}$. 17′. 52″. Quorum angulorum semisumma 70${}^{\text{gr}}$. 32′. 50″ est complementum declinationis

On now calling $RA = d$, $AV = e$, VP or $VQ = f$ and $VZ = g$, there will be $AP = f-e$ and $RP = \sqrt{[f^2-2ef+e^2+d^2]}$; similarly, $AQ = f+e$ and

$$RQ = \sqrt{[f^2+2ef+e^2+d^2]},$$

so that RZ (or $\frac{1}{2}(RP+RQ)$) $= \frac{1}{2}(\sqrt{[f^2-2ef+e^2+d^2]}+\sqrt{[f^2+2ef+e^2+d^2]})$. The square, $\frac{1}{2}(d^2+e^2+f^2)+\frac{1}{2}\sqrt{[f^4-2e^2f^2+e^4+2d^2f^2+2d^2e^2+d^4]}$, of this is equal to RV^2+VZ^2 (or $RA^2+AV^2+VZ^2$) $= d^2+e^2+g^2$. When reduction is now made it is $\sqrt{[f^4-2e^2f^2+e^4+2d^2f^2+2d^2e^2+d^4]} = d^2+e^2-f^2+2g^2$, and so, with the sides squared and reduced to order, $d^2f^2 = d^2g^2+e^2g^2-f^2g^2+g^4$, that is $d^2f^2/g^2 = d^2+e^2-f^2+g^2$. Finally, when $6, \frac{98}{143}a^2/b, \frac{112}{143}a^2\sqrt{3}/b$ and $8a\sqrt{3}/\sqrt{143}$ (the values of AR, AV, VQ and VZ) are restored in place of d, e, f and g, there ensues

$$36-\tfrac{196}{143}a^4/b^2+\tfrac{192}{143}a^2 = 36\times\tfrac{196}{143}a^2/b^2$$

and from this by reduction $b^2 = 49(a^4+36a^2)/(48a^2+1287)$.

In the first scheme it is $AM^2+MB^2 = AB^2$, that is, $r^2+s^2 = 33^2$. Previously, however, $r = 2a^2/b$ and $s^2 = 3a^2-4a^4c/b^2$, whence $r^2 = 4a^4/b^2$ and (when $\frac{143}{196}b^2/a^2$ is substituted for c) $s^2 = \frac{4}{49}a^2$. Hence $4a^4/b^2+\frac{4}{49}a^2 = 33^{2(359)}$ and from this by reduction there again results $b^2 = 4\times49a^4/(53361-4a^2)$. Therefore, on setting equality between the two b^2's and dividing both sides of the equation by 49 there comes $(a^4+36a^2)/(48a^2+1287) = 4a^4/(53361-4a^2)$. When the sides of this are multiplied cross-wise, then ordered and divided by $49a^2$, there comes out $4a^4 = 981a^2+274428,^{(360)}$ whose root a^2 is $\frac{1}{8}(981+\sqrt{1589625}) = 280\cdot2254144\dots$.

Above there was found $b^2 = 4\times49a^4/(53361-4a^2)$, that is,

$$b = 14a^2/\sqrt{[53361-4a^2]}; \quad \text{whence} \quad AV \text{ (or } \tfrac{98}{143}a^2/b)$$

is $\frac{7}{143}\sqrt{[53361-4a^2]}$ and VP or VQ $(\frac{112}{143}a^2\sqrt{3}/b)$ is $\frac{8}{143}\sqrt{[160083-12a^2]}$. That is, on substituting $280\cdot2254144$ in place of a^2 and reducing these terms to decimals, $AV = 11\cdot188297\dots$ and VP or $VQ = 22\cdot147085\dots$, so that

$$AP \text{ (or } PV-AV) = 10\cdot958788\dots \quad \text{and} \quad AQ \text{ (or } AV+VQ) = 33\cdot335382\dots.$$

Finally, if $\frac{1}{6}AR$ or 1 be put as the radius, then $\frac{1}{6}AQ$ or $5\cdot555897\dots$ is the tangent of $\widehat{ARQ} = 79°\,47'\,48''$ and $\frac{1}{6}AP$ or $1\cdot826465\dots$ the tangent of

$$\widehat{ARP} = 61°\,17'\,52''.$$

The half-sum $70°\,32'\,50''$ of these angles is the complement of the solar declina-

---

(359) Namely, $1089 = 53361/49$.

(360) A slip of Newton's pen, passed by Whiston into the 1707 *editio princeps* but changed by its author in his 1722 revise to the correct constant term '39204'. The root computed in sequel is accurate, and so the succeeding calculation is unaffected.

solis & semidifferentia 9^{gr}. 14′. 58″ complementum latitudinis Loci. Proinde declinatio solis erat 19^{gr}. 27′. 10″ & Latitudo loci 80^{gr}. 45′. 2″.[361] Quæ erant invenienda.

Octob 1679
Lect 1

Prob 43.

Si ad extremitates fili DAE circa paxillum[362] *A labentis appendantur pondera duo D et E quorum pondus E labitur per lineam obliquam BG: invenire locum ponderis E ubi pondera hæc in æquilibrio consistunt.*

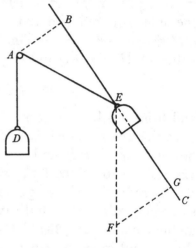

Puta factum et ipsi *AD* age parallelam *EF* quæ sit ad *AE* ut pondus *E* ad pondus *D*. Et a punctis *A* et *F* ad lineam *BG* demitte perpendicula *AB*, *FG*. Jam cùm pondera ex Hypothesi sint ut lineæ *AE*, *EF*, exponantur pondera per lineas istas, pondus *D* per lineam *AE* et pondus ‖[112] *E* per lineam *EF*. Ergo Corpus[363] ‖ *E* proprij ponderis vi directa *EF* tendit versus *F* et vi obliqua *EG* tendit versus *G*. Et idem Corpus *E* ponderis *D* vi directa *AE* trahitur versus *A* et vi obliqua *BE* trahitur versus *B*. Cum itacɜ pondera se mutuo sustineant in æqulibrio, vis qua pondus *E* trahitur versus *B* æqualis esse debet vi contrariæ qua tendit versus *G*,[364] hoc est *BE* æqualis esse debet ipsi *EG*. Jam vero datur ratio *AE* ad *EF* ex Hypothesi, & propter datum angulum *FEG* datur etiam ratio *FE* ad *EG* cui *BE* æqualis est. Ergo datur ratio *AE* ad *BE*. Datur etiam *AB* longitudine. Et inde triangulum *ABE* et punctum *E* facile dabitur. Nempe dic $AB = a$, $BE = x$ & erit $AE = \sqrt{aa + xx}$.

---

(361) Descartes, using only five decimal places in his computations, finds the slightly less accurate values 'grad. 19, & 27 min.' and 'grad. 80, & 45 min.' respectively (*Geometria*, $_1$1649: 322–3 = $_2$1659: 388–9). In the Northern hemisphere this places point *A* within the Arctic circle (in the latitude of Spitzbergen) at a time near to summer solstice when the sun does not set in the sky (so indeed yielding an elliptical gnomon shadow over the horizontal plane)—'in Zonis frigidis, cùm Sol non occidit' in Descartes' phrase (*Geometria*, $_1$1649: 295 = $_2$1659: 369). The problem is, of course, hopelessly theoretical and Delambre, practical astronomer that he was, savagely reviewed Newton's present solution—a deal less cumbrous computationally than Descartes' version!—in his *Histoire de l'Astronomie au dix-huitième siècle* (Paris, 1827): 37, 42: '. . .jamais aucun hasard n'en pourra fournir les données. . . . Son analyse occupe quatre pages [in the 1732 third edition available to Delambre]; ses équations sont hérissées de radicaux, et en se bornant même à ce qui est indispensable, le calcul est encore d'une longuer énorme et bien inutile. . . . Newton, apparemment pour montrer les ressources de son analyse, se complaît à accumuler les difficultés au lieu de les écarter. Il est évident qu'il n'a en vue que les géomètres et nullement les astronomes, qui savent fort bien que ce n'est pas dans les ouvrages d'analyse qu'il faut chercher les solutions des problèmes usuels de Trigonométrie. . . . [Mais] si ce problème n'a aucune utilité réelle, il a du moins le mérite d'être l'un des plus extraordinaires qui aient jamais été proposées.' In consequence of

tion and their half-difference 9° 14′ 58″ the complement of the latitude of the location. In consequence, the solar declination was 19° 27′ 10″ and the place's latitude 80° 45′ 2″:[361] as we were required to find.

Problem 43

If at the extremities of a thread DAE slipping round a pin[362] A there be attached two weights D and E, of which weight E slides along the oblique line BG, to find the position of weight E when these weights are at rest in balance.

Suppose it done, and parallel to AD draw EF, which is to be to AE as weight E to weight D. And from the points A and F to the line BG let fall the perpendiculars AB, FG. Now since the weights are, by hypothesis, as the lines AE, EF, let the weights be represented by those lines: weight D by the line AE and weight E by the line EF. In consequence, the body[363] E by the direct pull EF of its own weight tends towards F and so EG is the oblique pull upon it towards G. And the same body E is drawn by the direct pull AE of weight D towards A, and so the oblique pull drawing it towards B is BE. Accordingly, since the weights mutually hold one another in balance, the pull by which weight E is drawn towards B ought to be equal to the contrary pull by which it tends towards G;[364] that is, BE ought to be equal to EG. Now, in fact, the ratio of AE to EF is given by hypothesis, and, because the angle $\widehat{FEG}$ is given, there is also given the ratio of EF to EG and its equal BE: the ratio of AE to BE is therefore given. Also, AB is given in length. And from these the triangle ABE and so the point E will easily be determined. Namely, call $AB = a$, $BE = x$ and there will be

$$AE = \surd[a^2 + x^2].$$

amending $\widehat{ARP}$ to be '61^{gr}. 17′. 57″' Newton in his 1722 revise refined his present solution to read 'declinatio solis . . . 19^{gr}. 27′. 8″ & Latitudo loci 80^{gr}. 45′. 4″.'

(362) Literally 'spike' (supposed to be a fulcrum offering no frictional resistance to the weighted thread passing round it).

(363) A stray catchword 'sive' (or) at this place was, despite its illogicality, reproduced by Whiston in his 1707 *editio princeps*, but subsequently deleted by Newton in his 1722 revise.

(364) Stevin's celebrated principle of equilibrium between inclined weights, which in the next Problem Newton will, as Stevin, make the basis of his discussion of a balanced 'triangle' of static 'forces'. The principle was first enunciated by Stevin in Proposition 27 of his *Beghinselen der Weeghconst* (Leyden, 1586) and elaborated in a supplement (*Byvough der Weeghconst*) added to its second edition in his collected *Wisconstighe Ghedachtenissen* (Leyden, 1605–8), whose 'Eerste Deel' [= *The Principal Works of Simon Stevin*, **1** (Amsterdam, 1955): 528–49] applies the principle to a variety of loadings of an inextensible stretched string. It is improbable that Newton ever read Stevin's work, but he was undoubtedly familiar with Pardies' contemporary extension of Stevin's researches in his *Statique, ou La Science des Forces* (Paris, 1673): 110–43. As we have already observed (see III: 391, note (8)), he had received a copy of Pardies' 'little but ingenious tract' as a gift from Collins in September 1673, soon after its publication.

Sit insuper AE ad BE in data ratione d ad e et erit $e\sqrt{aa+xx}=dx$. et partibus æquationis quadratis & reductis, $eeaa=ddxx-eexx$ sive $\dfrac{ea}{\sqrt{dd-ee}}=x$. Inventa est igitur longitudo BE quæ determinat locum ponderis E. Q.E.F.

Quod si pondus utrumcʒ per lineam obliquam descendat, computum sic institui potest. Sint $CD\ BE$ obliquæ lineæ positione datæ per quas pondera ista D et E descendunt. A paxillo A ad has lineas demitte perpendicula AC, AB, ijscʒ productis occurrant in punctis G et H lineæ $EG\ DH$ a ponderibus perpendiculariter ad Horizontem erectæ, et vis qua pondus E conatur descendere juxta lineam perpendicularem hoc est tota gravitas ipsius E erit ad vim qua pondus idem conatur descendere juxta lineam obliquam BE ut GE ad BE atcʒ vis qua conatur juxta lineam

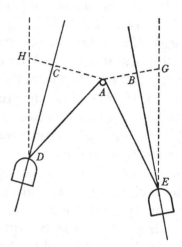

‖ [113] istam obliquam ‖ BE descendere erit ad vim qua conatur juxta lineam AE descendere hoc est ad vim qua filum AE distenditur ut BE ad AE. Adeocʒ gravitas ipsius E erit ad tensionem fili AE ut GE ad AE. Et eadem ratione gravitas ipsius D erit ad tensionem fili AD ut HD ad AD. Sit itacʒ fili totius $DA+AE$ longitudo c, sitcʒ pars ejus $AE=x$ & erit altera pars $AD=c-x$.[(365)] Et quoniam est $AE^q-AB^q=BE^q$ & $AD^q-AC^q=CD^q$ sit insuper $AB=a$ & $AC=b$, et erit $BE=\sqrt{xx-aa}$ & $CD=\sqrt{xx-2cx+cc-bb}$. Adhæc cum triangula BEG, CDH dentur specie, sit $BE.EG::f.E$ et $CD.DH::f.g$ et erit $EG=\dfrac{E}{f}\sqrt{xx-aa}$, et $DH=\dfrac{g}{f}\sqrt{xx-2cx+cc-bb}$. Quamobrem cùm sit $GE.AE::$pond$:E.$tens$:AE.$ et

$$HD.AD::\text{pond}:D.\text{tens}:AD,$$

---

(365) Newton first inadequately concluded: 'Et quoniam est

$$AE^q-AB^q(BE^q)+BG^q=GE^q, \quad \text{et} \quad AD^q-AC^q+HC^q=HD^q$$

dic $BG^q-AB^q=m$ & $HC^q-AC^q=n$, et erit $xx+m=GE^q$ et $xx-2cx+cc+n=HD^q$. Quare cùm sit $GE.AE::$pond$:E.$ tens$:AE$ et $HD.AD::$pond$:D.$ tens$:AD$, erit

$$\frac{Ex}{\sqrt{xx+m}}=\text{tens }AE=\text{tens }AD=\frac{Dc-Dx}{\sqrt{xx-2cx+cc+n}}.$$

Cujus æquationis reductione provenit $\begin{array}{l}DD\\-EE\end{array}x^4\begin{array}{l}-2DDc\\+2EEc\end{array}x^3\begin{array}{l}+DDcc\\-EEcc\\-EEn\\+DDm\end{array}xx-2DDcmx+DDccm=0.$

In addition, let AE be to BE in the given ratio d to e, and there will be

$$e\sqrt{[a^2+x^2]} = dx;$$

then, with the sides of the equation squared and reduced, $e^2a^2 = (d^2-e^2)\,x^2$ and so $x = ea/\sqrt{[d^2-e^2]}$. The length of BE is consequently found, and this in turn determines the position of the weight E. As was to be done.

But if each weight should descend along an oblique line, the calculation can be set out in this way. Let CD, BE be the oblique lines given in position along which the weights D and E descend. From the pin A let fall to these lines the perpendiculars AC, AB and let these, when produced, be met in the points G and H by lines EG, DH erected from the weights perpendicularly to the horizon. Then the pull by which the weight E endeavours to descend following a vertical path—the total gravity of E, that is—will be to the pull by which the same weight endeavours to descend following the oblique line BE as GE to BE, while the pull by which it endeavours to descend following that oblique line BE will be to the pull by which it endeavours to descend following the line AE—the force, that is, by which the thread AE is stretched tight—as BE to AE; consequently, E's gravity will be to the tension in the thread AE as GE to AE. And by the same reasoning D's gravity will be to the tension in the thread AD as HD to AD. So let the length of the whole thread $DA+AE$ be c and its part $AE = x$, and then its other part will be $AD = c-x$.[365] And, since $AE^2-AB^2 = BE^2$ and

$$AD^2-AC^2 = CD^2,$$

let there in addition be $AB = a$ and $AC = b$, and there will be $BE = \sqrt{[x^2-a^2]}$ and $CD = \sqrt{[x^2-2cx+c^2-b^2]}$. Further, since the triangles BEG and CDH are given in species, let $BE:EG = f:E$ and $CD:DH = f:g$, so that there will be $EG = (E/f)\sqrt{[x^2-a^2]}$ and $DH = (g/f)\sqrt{[x^2-2cx+c^2-b^2]}$. As a result, since $GE:AE = $ weight $E:$ tension in AE and $HD:AD = $ weight $D:$ tension in AD,

---

sive $x^4 - 2cx^3 \dfrac{+cc}{\underset{\overline{DD-EE}}{+DDm-EEn}} xx - \dfrac{2DDcm}{DD-EE}\,x + \dfrac{DDccm}{DD-EE} = 0$.' (Then, seeing that

$$AE^2-AB^2 \text{ (or } BE^2) + BG^2 = GE^2 \quad \text{and} \quad AD^2-AC^2+HC^2 = HD^2,$$

call $BG^2-AB^2 = m$ and $HC^2-AC^2 = n$ and there will be $GE^2 = x^2+m$ and

$$HD^2 = x^2-2cx+c^2+n.$$

Consequently, since $GE:AE = $ weight $E:$ tension AE and $HD:AD = $ weight $D:$ tension AD, there will be $Ex/\sqrt{[x^2+m]} = $ tension $AE = $ tension $AD = D(c-x)/\sqrt{[x^2-2cx+c^2+n]}$. And on reduction of this equation there results....)

& tensiones istæ æquentur inter se, erit

$$\frac{Ex}{\frac{E}{f}\sqrt{xx-aa}} = \text{tens}:AE = \text{tens}:AD = \frac{Dc-Dx}{\frac{g}{f}\sqrt{xx-2cx+cc-bb}}.$$

Cujus æquationis reductione provenit $gx\sqrt{xx-2cx+cc-bb}=\overline{Dc-Dx}\sqrt{xx-aa}$

sive $\begin{matrix} gg \\ -DD \end{matrix}x^4 \begin{matrix} -2gg \\ +2DD \end{matrix}x^3 \begin{matrix} +ggcc \\ -ggbb \\ -DDcc \\ +DDaa \end{matrix}xx-2DDcaax+DDccaa=0.$

Si casum desideras quo hoc Problema per Regulam et circinum[366] construi queat, pone pondus D ad pondus E ut ratio $\frac{BE}{EG}$ ad rationem $\frac{CD}{DH}$, et evadet

$g=D$, adeoq̃ vice præcedentis æquationis habebitur hæc $\begin{matrix} aa \\ -bb \end{matrix}xx-2aacx+aacc=0.$

sive $x=\frac{ac}{a+b}$. [367]

‖ [114]   ‖ *Prob: 44.*[368]

[1679]
Lect 2

Si ad filum DACBF circa paxilla duo A, B, labile appendantur tria pondera D, E, F; D et F ad extremitates fili et E ad medium ejus punctum C inter pagos positum: ex datis ponderibus et situ pagorum invenire situm puncti C ad quod medium pondus appenditur ubi pondera consistunt in æquilibrio.

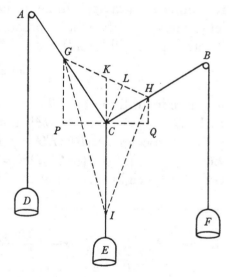

Cum tensio fili AC æquetur tensioni fili AD et tensio fili BC tensioni fili BF, tensiones filorum AC, BC, EC erunt ut pondera D, F, E. In eadem ponderum ratione cape partes filorum CG, CH, CI. Produc IC donec ea occurrat GH in K et erit $GK=KH$ et $CK=\frac{1}{2}CI$, adeoq̃ C centrum gravitatis trianguli GHI.[369] Nam per C agatur ipsi CE perpendiculare PQ et huic a punctis G et H perpendicularia GP, HQ. Et si vis qua filum

---

(366) Resolvable algebraically, that is, by a (linear or) quadratic equation.

(367) The alternative solution '$vel\ x=\frac{ac}{a-b}$' is cancelled in sequel.

(368) Stevin's principle for the equilibrium of static 'tensions' (see note (364)) is now applied to the classical case of a loaded string supported at its two ends. Newton's inspiration may well have been Pardies' problem 'Des poids suspendus au milieu d'une corde attachée par les deux bouts' (*La Statique, ou La Science des Forces Mouvantes* (Paris, 1673): §§LXVI/

and because those tensions are equal to each other, there will be

$$Ex/(E/f) \sqrt{[x^2 - a^2]} = \text{tension in } AE = \text{tension in } AD$$
$$= D(c - x)/(g/f) \sqrt{[x^2 - 2cx + c^2 - b^2]}.$$

From reduction of this equation there ensues

$$gx \sqrt{[x^2 - 2cx + c^2 - b^2]} = D(c - x) \sqrt{[x^2 - a^2]},$$

that is,

$$(g^2 - D^2) x^4 - 2(g^2 - D^2) x^3 + (g^2(c^2 - b^2) + (a^2 - c^2) D^2) x^2 - 2a^2cD^2x + a^2c^2D^2 = 0.$$

Should you desire a case in which this problem is constructible by ruler and compasses,[366] put weight D to be to weight E as the ratio BE/EG to the ratio CD/DH and there will prove to be $g = D$, and hence instead of the preceding equation there will be had this one, $(a^2 - b^2) x^2 - 2a^2cx + a^2c^2 = 0$, that is, $x = ac/(a + b)$.[367]

Problem 44.[368]

If to the thread DACBF freely slidable round the two pins A, B there be attached three weights D, E and F (D and F at the extremities of the thread, E at a central point C of it positioned between the spikes), from the weights and location of the spikes given to find the location of the point C at which the middle weight is appended when the weights are at rest in balance.

Since the tension in the thread AC is equal to that in the thread AD and the tension in the thread BC to that in the thread BF, the tensions in the threads AC, BC and EC will be as the weights D, F and E. In the same ratio of the weights take the parts CG, CH and CI of the threads. Extend IC till it meets GH in K and there will be $GK = KH$ and $CK = \frac{1}{2}CI$, so that C is the centroid of the triangle GHI.[369] For through C draw PQ perpendicular to CE and, from the points G and H, GP and HQ perpendicular to this. Then if the pull by which the thread

---

LXVII: 110–15), but his solution differs considerably in constructional detail. We may, however, point to yet one more parallel here with Christiaan Huygens, who on 20 December 1688 began to pen a series of notes on the problem of the loaded string—published only thirty years ago (in Huygens' *Œuvres complètes*, **19**: 60–2)—in which the 'centre de gravité' C of the triangle GHI formed by the balanced 'tensions' CG, CH, CI is constructed in a way entirely analogous to Newton's present one. Still more remarkably, Huygens' extension of his solution of the problem of the loaded string in September 1690 (*ibid.*: 66–8) to construct the defining differential equation of the catenary exactly mirrors a hitherto unpublished Newtonian extension of the present problem (ULC. Add. 3965.13: 375ʳ, reproduced as Appendix 2 below) to the same purpose some twenty years afterwards. Although Huygens met Newton on a visit to London in mid-June 1689, it seems unlikely that the problem of the freely hanging chain—not to be thrown out by Jakob Bernoulli as a public challenge, 'Invenire, quam curvam referat funis laxus & inter duo puncta fixa libere suspensus' (*Acta Eruditorum*, May 1690: 219), till the next spring—arose in their conversation.

(369) The Stevin triangle of the 'forces' proportional to and acting in the direction of CG, CH and CI respectively; see note (364).

AC vi ponderis *D* trahit punctum *C* versus *A* exponatur per lineam *GC*, vis qua filum istud trahet idem punctum versus *P* exponetur per lineam *CP* et vis qua trahit illud versus *K* exponetur per lineam *GP*. Et similiter vires quibus filum *BC* vi ponderis *F* trahit idem punctum *C* versus *B*, *Q* & *K* exponentur per lineas *CH*, *CQ*, *HQ*; et vis quâ filum *CE* vi ponderis *E* trahit punctum illud *C* versus *E* exponetur per lineam *CI*. Jam cùm punctum *C* viribus æquipollentibus sustineatur in æquilibrio, summa virium quibus fila *AC* et *BC* simul trahunt punctum *C* versus *K* æqualis erit vi contrariæ qua filum *EC* trahit punctum illud versus *E*, hoc est summa *GP+HQ* æqualis erit ipsi *CI*: Et vis qua filum *AC* trahit punctum *C* versus *P* æqualis erit vi contrariæ qua filum *BC* trahit idem punctum *C* versus *Q*[,] hoc est linea *PC* ‖ æqualis lineæ *CQ*. Quare

‖ [115]

cùm *PG*, *CK* et *QH* parallelæ sint, erit etiam $GK = KH$, et $CK\left(=\dfrac{GP+HQ}{2}\right)=\tfrac{1}{2}CI$.

Quod erat ostendendum. Restat itaqʒ triangulum *GCH* determinandum cujus latera *GC* et *HC* dantur una cum linea *CK* quæ a vertice *C* ad medium basis ducitur. Demittatur itaqʒ a vertice *C* ad basem *GH* perpendiculum *CL* et erit $\dfrac{GC^q - CH^q}{2GH} = KL = \dfrac{GC^q - KC^q - GK^q}{2GK}$. Pro $2GK$ scribe *GH*, et rejecto com[mu]ni divisore *GH*, et ordinatis terminis, erit $GC^q - 2KC^q + CH^q = 2GK^q$, sive $\sqrt{\tfrac{1}{2}GC^q - KC^q + \tfrac{1}{2}CH^q} = GK$.[370] Invento GK[371] vel *KH*, dantur simul anguli *GCK KCH* sive *DAC*, *FBC*. Quare a punctis *A*, et *B* in datis istis angulis *DAC*, *FBC* duc lineas *AC*, *BC* concurrentes in puncto *C* et istud *C* erit punctum quod quæritur.

Cæterùm quæstiones omnes quæ sunt ejusdem generis non semper opus est per Algebram[372] sigillatim solvere, sed ex solutione unius plerumqʒ consectatur solutio alterius. Ut si jam proponeretur hæc quæstio.

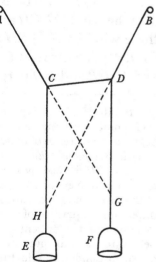

Filo ACDB in datas partes AC, CD, DB diviso, et extremitatibus ejus ad paxilla duo A, B positione data ligatis, si ad puncta divisionum C ac D appendantur pondera duo E et F: ex dato pondere F et situ punctorum C ac D, cognoscere pondus E.

Ex præcedentis Problematis solutione satis facilè colligetur hæcce solutio hujus. Produc lineas *AC*, *BD*, donec occurrant lineis *DF*, *CE* in *G* & *H*: et erit pondus *E* ad pondus *F* ut *DG* ad *CH*.[373]

***

(370) This result has already been established in Newton's preceding Problem 7; see note (224).

(371) 'ejusqʒ duplo *GH*' (and its double *GH*) is cancelled.

(372) In context Newton ought here to add 'vel per Geometriam' (or by geometry)!

AC through the force of the weight *D* draws the point *C* towards *A* be represented by the line *GC*, the pull by which that thread will drag the same point towards *P* will be represented by the line *CP* and the pull by which it drags it towards *K* by the line *GP*. Similarly, the pull with which the thread *BC* through the force of the weight *F* draws the same point *C* towards *B*, *Q* and *K* will be expressed by the lines *CH*, *CQ* and *HQ*; and the pull by which the thread *CE* through the force of the weight *E* draws that point *C* towards *E* will be represented by the line *CI*. Now since the point *C* is held in equilibrium by counter-balancing forces, the resultant sum of the pulls by which the threads *AC* and *BC* together draw point *C* towards *K* will be equal to the contrary pull by which the thread *EC* drags that point towards *E*—in other words, the sum of *GP* + *HQ* will be equal to *CI*; and the pull by which the thread *AC* draws the point *C* towards *P* will be equal to the contrary pull by which the thread *BC* drags the same point *C* towards *Q*—whence the line *PC* is equal to the line *CQ*. Consequently, since *PG*, *CK* and *QH* are parallel, there will also be *CK* = *KH* and *CK* (or $\frac{1}{2}(GP + HQ)$) = $\frac{1}{2}CI$: as was to be shown. It accordingly remains to determine the triangle *GCH* whose sides *GC* and *HC* are given, along with the line *CK* which is drawn from its vertex *C* to the mid-point of the base. So let fall the perpendicular *CL* from the vertex *C* to the base *GH* and there will be $(GC^2 - CH^2)/2GH = KL = (GC^2 - KC^2 - GK^2)/2GK$. In place of 2*GK* write *GH* and, on rejecting the common divisor *GH* and ordering the terms, there will be $GC^2 - 2KC^2 + CH^2 = 2GK^2$, or $GK = \sqrt{[\frac{1}{2}GC^2 - KC^2 + \frac{1}{2}CH^2]}$.[370] After *GK*[371] or *KH* is found, the angles $\widehat{GCK}$ and $\widehat{KCH}$—that is, $\widehat{DAC}$ and $\widehat{FBC}$—are given at once. In consequence, from the points *A* and *B* at those given angles $\widehat{DAC}$, $\widehat{FBC}$ draw lines *AC*, *BC* meeting in the point *C*, and that *C* will then be the point required.

There is, however, no need always to solve all questions which are of the same type by a separate algebraic[372] argument; instead, a solution of one commonly yields the solution of a second as its corollary. If, for instance, this question were now proposed:

> *If, where a thread ACDB is divided into given parts AC, CD, DB and its extremities are tied to two pins A, B given in position, two weights E and F are attached at the points C and D of division, given the weight of F and the position of the points C and D, to ascertain from these the weight of E.*

From the solution of the preceding problem there is easily enough gathered the following solution of the present one: Extend the lines *AC*, *BD* till they meet the lines *DF*, *CE* in *G* and *H*, and then the weight of *E* will be to the weight of *F* as *DG* to *CH*.[373]

---

(373) For, since the 'force' triangles *CDG* and *CDH* are each in balance, it follows respectively that tension of *CE*:tension of *CD* = *DG*:*CD* and tension of *CD*:tension of *DF* = *CD*:*CH*, whence weight of *E*:weight of *F* = tension of *CE*:tension of *DF* = *DG*:*CH*. As a corollary

Et hinc obiter patet ratio componendi stateram ex solis filis, qua pondus corporis cujusvis E ex unico dato pondere F cognosci potest.[374]

‖[116]

‖*Prob 45.*

[1679]
Lect 3

Lapide in puteum decidente, ex sono lapidis fundum percutientis altitudinem putei cognoscere.[375]

Sit altitudo putei x, et si lapis motu uniformiter accelerato descendat per spatium quodlibet datum a in tempore dato b, et sonus motu uniformi transeat per idem spatium datum a in tempore dato d, lapis descendet per spatium x in tempore $b\sqrt{\dfrac{x}{a}}$, sonus autem qui fit a lapide in fundum putei impingente ascendet per idem spatium x in tempore $\dfrac{dx}{a}$. Ut enim sunt spatia gravibus decidentibus descripta, ita sunt quadrata temporum descensus. Vel ut radices spatiorum hoc est ut $\sqrt{x}$ & $\sqrt{a}$ ita sunt ipsa tempora. Et ut spatia x et a per quæ sonus transeunt ita sunt tempora transitus. Ex horum temporum $b\sqrt{\dfrac{x}{a}}$ & $\dfrac{dx}{a}$ summa conflatur tempus a lapide demisso ad sonus reditum. Hoc tempus ex observatione cognosci potest. Sit ipsum t et erit $b\sqrt{\dfrac{x}{a}}+\dfrac{dx}{a}=t$. Ac $b\sqrt{\dfrac{x}{a}}=t-\dfrac{dx}{a}$. Et partibus quadratis $\dfrac{bbx}{a}=tt-\dfrac{2tdx}{a}+\dfrac{ddxx}{aa}$. Et per reductionem $xx=\dfrac{2adt+abb}{dd}x-\dfrac{aatt}{dd}$. Et extracta radice $x=\dfrac{adt+\frac{1}{2}abb}{dd}-\dfrac{ab}{2dd}\sqrt{bb+4dt}$.[376]

Prob. 46.[377]

Dato trianguli rectanguli perimetro et perpendiculo, invenire triangulum.

Trianguli ABC sit C rectus angulus et CD perpendiculum inde ad basem AB demissum. Detur $AB+BC+CA=a$ et $CD=b$. Pone basem $AB=x$, &

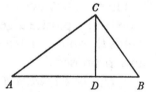

(one added explicitly by Newton some thirty years later in the extension to the present problem reproduced in Appendix 2 following; compare note (368)), if the length CD is now supposed to be the vanishing increment ds (of weight $d(w_s)$, equal to that of weight F) of arc $AC = s$, bounded by the point $C(x, y)$, whose instantaneous slope at C is dy/dx, then the difference in 'tension' at points C and D is proportional to $DG-CH$, that is, to the increment $d(dy/dx)$ of slope at D as compared with that at C; accordingly $d(w_s) \propto d(dy/dx)$, the defining differential equation of the loaded string. Where the load is uniformly distributed horizontally ($w_s \propto x$), Huygens' 1646 result in his 'De Catena pendente' (*Œuvres complètes*, **11**: 37–44) that the simple suspension bridge is a parabola, $y = ax^2+bx+c$, is immediate; equally, for the simple catenary ($w_s \propto s$) the resulting differential equation $s = k \cdot dy/dx$ yields eventually the defining Cartesian

Hence, by the way, is plain a method of constructing a balance from threads alone, by which the weight of any body E can be found out from but a single given weight F.[374]

Problem 45

Where a stone drops down a well, from the sound of the stone striking the bottom to ascertain the depth of the well.[375]

Let the depth of the well be x, and then, if the stone in uniformly accelerated motion should descend through any given space a in time b, while sound passes with uniform motion through the same given space a in time d, the stone will descend through the space x in time $b\sqrt{[x/a]}$, but the sound made by the stone hitting the well's bottom will ascend through the same space x in time dx/a. (Of course, spaces described by heavy bodies as they fall are as the squares of the times of descent; alternatively, the (square) roots of the spaces—that is, $\sqrt{x}$ and $\sqrt{a}$—are as the times themselves. The spaces x and a through which sound passes are as the times of passage.) From the sum of these times $b\sqrt{[x/a]}$ and dx/a is made up the time from the dropping of the stone to the returning echo: this time-span can be found out by observation. Let it be t and there will be

$$b\sqrt{[x/a]} + dx/a = t, \quad \text{and so} \quad b\sqrt{[x/a]} = t - dx/a.$$

Then, on squaring the sides, $b^2x/a = t^2 - 2dtx/a + d^2x^2/a^2$ and by reduction $x^2 = 2a(dt + \tfrac{1}{2}b^2)\,x/d^2 - a^2t^2/d^2$, and so, when the root is extracted,

$$x = a(dt + \tfrac{1}{2}b^2)/d^2 - \tfrac{1}{2}ab\sqrt{[b^2 + 4dt]}/d^2.[376]$$

Problem 46[377]

Given the perimeter and altitude of a right-angled triangle, to find the triangle.

In the triangle ABC let C be a right angle and CD the perpendicular let fall from there to the base, and let there be given $AB + BC + CA = a$ and $CD = b$.

---

equation $\log[(y + \sqrt{[y^2 - k^2]})/k] = x/l$ of the curve. (See C. Truesdell, *The Rational Mechanics of Flexible or Elastic Bodies, 1638–1788* [= *Leonhardi Euleri Opera Omnia* (2) **11**. 2 (Zurich, 1960): 1–435]: 44–6, 64–75.)

(374) Namely, by hanging the given weight F and unknown weight E from the points D and C respectively, and then measuring the ratio of CH to DG which is that of their weights.

(375) As the following text makes clear, it is assumed that the stone falls with constant acceleration, but that the sound of its impact echoes back at a uniform speed.

(376) It would have been a good deal less tedious to put $\sqrt{[x/a]} = y$, so solving $dy^2 + by - t = 0$ and thence evaluating $x/a = (\tfrac{1}{2}(-b + \sqrt{[b^2 + 4dt]})/d)^2$. The second root of Newton's quadratic, resolving the condition $-b\sqrt{[x/a]} + dx/a = t$, is evidently inadmissible.

(377) This and the three following Problems 47–9 are of the type discussed by Oughtred in Problemata IX–XIII of 'Cap. XIX. Exempla Æquationis Analyticæ, pro Theorematibus inveniendis, Problematibusꝗ solvendis' of his *Clavis Mathematicæ Denuo Limata, sive potius Fabricata* (Oxford, ₃1652): 83–8. On that model Newton makes various choice of the two conditions necessary to determine a right triangle.

‖[117] erit laterum[378] summa $a-x$. Pone laterum[378] differentiam y, et ‖erit majus latus $AC=\dfrac{a-x+y}{2}$; minus $BC=\dfrac{a-x-y}{2}$. Jam ex natura trianguli rectanguli[379] est $AC^q+BC^q=AB^q$ hoc est $\dfrac{aa-2ax+xx+yy}{2}=xx$. Est et $AB.AC::BC.DC$, adeoœ $AB\times DC=AC\times BC$ hoc est $bx=\dfrac{aa-2ax+xx-yy}{4}$. Per priorem æquationem est $yy=xx+2ax-aa$. Per posteriorem $yy=xx-2ax+aa-4bx$. Adeoœ $xx+2ax-aa=xx-2ax+aa-4bx$. Et per reductionem $4ax+4bx=2aa$. sive $x=\dfrac{aa}{2a+2b}$.

Geometricè sic. In omni triangulo rectangulo: ut est summa perimetri et perpendiculi ad perimetrum, ita dimidium perimetri ad basem.

Aufer $2x$ de a et restabit $\dfrac{ab}{a+b}$ excessus laterum super basem.[380] Unde rursus.

In omni triangulo rectangulo Ut summa perimetri et perpendiculi ad perimetrum ita perpendiculum ad excessum laterum super basem.

[1679]
Lect 4.

Prob: 47.

Datis trianguli rectanguli basi AB et summa perpendiculi et laterū $CA+CB+CD$, invenire triangulum.

Esto $CA+CB+CD=a$, $AB=b$, $CD=x$ & erit $AC+CB=a-x$. Pone $AC-CB=y$ et erit $AC=\dfrac{a-x+y}{2}$ & $CB=\dfrac{a-x-y}{2}$. Est autem $AC^q+CB^q=AB^q$ hoc est $\dfrac{aa-2ax+xx+yy}{2}=bb$. Est et $AC\times CB=AB\times CD$, hoc est

$$\dfrac{aa-2ax+xx-yy}{4}=bx.$$

Quibus comparatis fit $2bb-aa+2ax-xx=yy=aa-2ax+xx-4bx$. Et per reductionem $xx=2ax+2bx-aa+bb$. Et $x=a+b-\sqrt{2ab+2bb}$.[381]

Geometrice sic. In omni triangulo rectangulo de summa perimetri et perpendiculi aufer mediam proportionalem inter eandem summam et duplum basis et restabit perpendiculum.

‖[118]

‖*Idem aliter.*

Sit $CA+CB+CD=a$, $AB=b$, & $AC=x$ et erit $BC=\sqrt{bb-xx}$, &

$$CD=\dfrac{x\sqrt{bb-xx}}{b}.$$

---

(378) The sides *CA*, *CB* only (without the hypotenuse *AB*) are understood.
(379) Understand 'per 47.1 Elem' (by *Elements*, I, 47).

Put the base $AB = x$ and the sum of the sides[378] will be $a - x$. Put the difference of the sides[378] to be y, and there will be the greater side $AC = \frac{1}{2}(a - x + y)$, the lesser one $BC = \frac{1}{2}(a - x - y)$. Now by the nature of a right triangle[379] it is $AC^2 + BC^2 = AB^2$, that is, $\frac{1}{2}(a^2 - 2ax + x^2 + y^2) = x^2$. Also $AB : AC = BC : DC$ and hence $AB \times DC = AC \times BC$, that is, $bx = \frac{1}{4}(a^2 - 2ax + x^2 - y^2)$. By the first equation there is $y^2 = x^2 + 2ax - a^2$, by the latter $y^2 = x^2 - 2ax + a^2 - 4bx$, so that $x^2 + 2ax - a^2 = x^2 - 2ax + a^2 - 4bx$, and by reduction $4ax + 4bx = 2a^2$ or

$$x = \tfrac{1}{2}a^2/(a+b).$$

In geometrical terms, thus: In every right-angled triangle the sum of the perimeter and the altitude is to the perimeter as half the perimeter to the base.

Take away $2x$ from a and there will remain $ab/(a+b)$ for the excess of the sides over the base. Hence again: In every right-angled triangle the sum of the perimeter and altitude is to the perimeter as the altitude to the excess of the sides over the base.

Problem 47

Given in any right-angled triangle the base AB and the sum of the altitude and the sides $CA + CB + CD$, to find the triangle.

Let $CA + CB + CD = a$, $AB = b$, $CD = x$ and there will be $AC + CB = a - x$. Put $AC - CB = y$ and there will be $AC = \frac{1}{2}(a - x + y)$ and $CB = \frac{1}{2}(a - x - y)$. However, $AC^2 + CB^2 = AB^2$, that is, $\frac{1}{2}(a^2 - 2ax + x^2 + y^2) = b^2$. Also

$$AC \times CB = AB \times CD,$$

that is, $\frac{1}{4}(a^2 - 2ax + x^2 - y^2) = bx$. On comparing these there comes

$$2b^2 - a^2 + 2ax - x^2 = y^2 = a^2 - 2ax + x^2 - 4bx,$$

and by reduction

$$x^2 = 2(a+b)\,x - a^2 + b^2, \quad \text{and so} \quad x = a + b - \sqrt{[2a(a+b)]}.\text{[381]}$$

In geometrical terms, thus: In every right-angled triangle from the sum of the perimeter and the altitude take away the mean proportional between that same sum and twice the base and there will remain the altitude.

The same another way

Let $CA + CB + CD = a$, $AB = b$ and $AC = x$, and there will be

$$BC = \sqrt{[b^2 - x^2]} \quad \text{and} \quad CD = x\sqrt{[b^2 - x^2]}/b;$$

---

(380) Newton first wrote 'Aufer x de a et restabit $\dfrac{aa + 2ab}{2a + 2b}$ pro summa laterum' (Take away x from a and there will remain $\frac{1}{2}(a^2 + 2ab)/(a+b)$ for the sum of the sides). This produces, in place of the proportion $(a+b) : a = b : ([a-x] - x)$ stated verbally in sequel, the more cumbrous equivalent (say) $(a+b) : \frac{1}{2}a = (a+2b) : (a-x)$.

(381) The quadratic's second root does not yield an allowable solution $(x < a)$.

et $x+CB+CD=a$ sive $CB+CD=a-x$ atȝ adeo $\dfrac{b+x}{b}\sqrt{bb-xx}=a-x$. et quadratis partibus atȝ multiplicatis per bb, fiet

$$-x^4-2bx^3+2b^3x+b^4=aabb-2abbx+bbxx.$$

Qua æquatione per transpositionem partium ad hunc modum ordinata

$$x^4+2bx^3\begin{matrix}+3bb\\+2ab\end{matrix}xx\begin{matrix}+2b^3\\+2abb\end{matrix}x\begin{matrix}+b^4\\+2ab^3\\+aabb\end{matrix}=\begin{matrix}2bb\\+2ab\end{matrix}xx\begin{matrix}+4b^3\\+4abb\end{matrix}x\begin{matrix}+2b^4\\+2ab^3\end{matrix}$$

extracta utrobiȝ radice, orietur $xx+bx+bb+ab=\overline{x+b}\sqrt{2ab+2bb}$.[382] Et extracta iterum radice $x=-\tfrac{1}{2}b+\sqrt{\tfrac{1}{2}bb+\tfrac{1}{2}ab}\pm\sqrt{b\sqrt{\tfrac{1}{2}bb+\tfrac{1}{2}ab}-\tfrac{1}{4}bb-\tfrac{1}{2}ab}$.

Constructio Geometrica.[383]

Cape igitur $AB=\tfrac{1}{2}b$. $BC=\tfrac{1}{2}a$, $CD=\tfrac{1}{2}AB$, AE mediam proportionalem inter b et AC, & EF hinc inde mediam proportionalem inter b et DE, et erunt BF, BF duo latera trianguli.

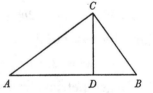

[1679]
Lect. 5.

Prob 48.[384]

Datis in triangulo rectangulo ABC summa laterum AC+BC & perpendiculo CD invenire triangulum.

Sit $AC+BC=a$, $CD=b$, $AC=x$ & erit $BC=a-x$, $AB=\sqrt{aa-2ax+2xx}$. Est et $CD.AC::BC.AB$. Ergo rursus $AB=\dfrac{ax-xx}{b}$. Quare $ax-xx=b\sqrt{aa-2ax+2xx}$ et partibus quadratis et ordinatis

$$x^4-2ax^3\begin{matrix}+aa\\-2bb\end{matrix}xx+2abbx-aabb=0.$$

Adde ad utramȝ partem $aabb+b^4$ & fiet $x^4-2ax^3\begin{matrix}+aa\\-2bb\end{matrix}xx+2abbx+b^4=aabb+b^4$.

---

(382) Newton will elaborate this technique of factorizing a quartic by means of 'surd' divisors on his pages 187–92 below. The present quadratic results more simply perhaps by setting $CD=y$, when $x(a-x-y)=by$ and $x^2+(a-x-y)^2=b^2$, so that at once

$$(a-y)^2=2by+b^2 \quad\text{and so}\quad y=a+b-\sqrt{[2(ab+b^2)]};$$

on substituting this in the first equation there accordingly ensues

$$x^2+(b-\sqrt{[2(ab+b^2)]})\,x+b(a+b-\sqrt{[2(ab+b^2)]})=0.$$

and, again, $x + CB + CD = a$ or $CB + CD = a - x$. Hence

$$(b+x)\sqrt{[b^2-x^2]}/b = a-x$$

and, when the sides are squared and multiplied by b^2, there will come

$$-x^4 - 2bx^3 + 2b^3x + b^4 = a^2b^2 - 2ab^2x + b^2x^2.$$

After this equation is ordered by transposition of its sides in this manner:

$$x^4 + 2bx^3 + (2ab+3b^2)\,x^2 + (2ab^2+2b^3)\,x + b^2(b^2+2ab+a^2)$$
$$= (2ab+2b^2)\,(x^2+2bx+b^2)$$

and the root is extracted on either side, there will arise

$$x^2 + bx + ab + b^2 = (x+b)\sqrt{[2ab+2b^2]}.^{(382)}$$

And, when the root is extracted a second time,

$$x = -\tfrac{1}{2}b + \sqrt{[\tfrac{1}{2}(ab+b^2)]} \pm \sqrt{(b\sqrt{[\tfrac{1}{2}(ab+b^2)]} - \tfrac{1}{2}ab - \tfrac{1}{4}b^2)}.$$

Geometrical construction[(383)]

Take, therefore, $AB = \tfrac{1}{2}b$, $BC = \tfrac{1}{2}a$, $CD = \tfrac{1}{2}AB$, AE the mean proportional between b and AC, and EF (taken either way) the mean proportional between b and DE, and the two BF's will be the two sides of the triangle.

Problem 48[(384)]

Given in a right-angled triangle ABC the sum of the sides $AC + BC$ and the altitude CD, to find the triangle.

Let $AC + BC = a$, $CD = b$, $AC = x$ and there will be $BC = a - x$,

$$AB = \sqrt{[a^2-2ax+2x^2]}.$$

Also $CD:AC = BC:AB$ and therefore again $AB = x(a-x)/b$. Consequently $ax - x^2 = b\sqrt{[a^2-2ax+2x^2]}$ and, with the sides squared and ordered,

$$x^4 - 2ax^3 + (a^2-2b^2)\,x^2 + 2ab^2x - a^2b^2 = 0.$$

To each side add $a^2b^2 + b^4$ and there will come

$$x^4 - 2ax^3 + (a^2-2b^2)\,x^2 + 2ab^2x + b^4 = a^2b^2 + b^4,$$

---

(383) Newton constructs $BF = x$ by making successively

$$AD = \tfrac{1}{4}b + \tfrac{1}{2}a, \quad AE = \sqrt{[\tfrac{1}{2}b(a+b)]} \quad \text{and} \quad EF = \sqrt{[b\,.\,(AE-AD)]}$$

with, finally, $BF = -AB + AE \pm EF$.

(384) This is Problema x of Oughtred's Caput xix (*Clavis* (note (377)): 84–5): 'Datis summa laterum trianguli rectanguli,... & perpendiculari ab angulo recto in hypotenusam...: invenire tum hypotenusam, tum triangulum ipsum.' Newton's following *Idem aliter* is the algebraic equivalent of Oughtred's geometrical solution.

‖[119] Et extracta utrobiꝗ radice $xx - ax - bb = -b\sqrt{aa + bb}$.[385] ‖Et radice iterum extracta $x = \frac{1}{2}a \pm \sqrt{\frac{1}{4}aa + bb - b\sqrt{aa + bb}}$.

Constructio Geometrica.[386]

Cape $AB = BC = \frac{1}{2}a$. Ad C erige perpendiculum $CD = b$. Produc DC ad E ut sit $DE = DA$. Et inter CD et CE cape medium proportionale CF. Centroꝗ F radio BC descriptus circulus GH secet rectam BG in G et H et erunt BG et BH latera duo trianguli.

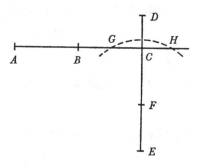

Idem aliter.[387]

Sit $AC + BC = a$. $AC - BC = y$. $AB = x$ ac $DC = b$ et erit $\frac{a+y}{2} = AC$, $\frac{a-y}{2} = BC$.

$\frac{aa+yy}{2} = AC^q + BC^q = AB^q = xx$. $\frac{aa-yy}{4b} = \frac{AC \times BC}{DC} = AB = x$. Ergo

$$2xx - aa = yy = aa - 4bx. \quad \text{et} \quad xx = aa - 2bx$$

et extracta radice $x = -b + \sqrt{bb + aa}$. Unde in superiori constructione est CE Hypotenusa trianguli quæsiti.

Data autem basi[388] et perpendiculo tam in hoc quam in superiore Problemate, triangulum sic expeditè construitur. Fac parallelogrammum CG cujus latus CE erit basis trianguli[,] latus alterum CF perpendiculum. Et super CE describe semicirculum secantem latus oppositum FG in H. Age CH, EH et erit CHE triangulum quæsitum.

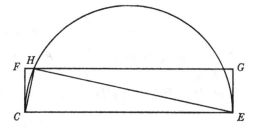

Prob. 49.

In triangulo rectangulo, datis summa laterum, et summa perpendiculi et basis invenire Triangulum.

Sit laterum AC et BC summa a, basis AB et perpendiculi CD summa b, latus $AC = x$, basis $AB = y$, et erit $BC = a - x$. $CD = b - y$.

$$aa - 2ax + 2xx = AC^q + BC^q = AB^q = yy.$$

$ax - xx = AC \times BC = AB \times CD = by - yy = by - aa + 2ax - 2xx$. & $by = aa - ax + xx$.

---

(385) It is again simpler (compare note (382)) to set $AB = y$, thence deriving

$$y^2 = x^2 + (a-x)^2 \quad \text{and} \quad by = x(a-x),$$

and, when the root is extracted on each side, $x^2 - ax - b^2 = -b\sqrt{[a^2 + b^2]}$.[385]
And, after the root is once more extracted, $x = \frac{1}{2}a \pm \sqrt{(\frac{1}{4}a^2 + b^2 - b\sqrt{[a^2 + b^2]})}$.

Geometrical construction[386]

Take $AB = BC = \frac{1}{2}a$, at C erect the perpendicular $CD = b$, extend DC to E so that $DE = DA$, and between CD and CE take the mean proportional CF; then let the circle GH described on centre F and with radius BC cut the straight line BG in G and H, and BG, BH will be the two sides of the triangle.

The same another way[387]

Let $AC + BC = a$, $AC - BC = y$, $AB = x$ and $DC = b$, and there will be $AC = \frac{1}{2}(a+y)$, $BC = \frac{1}{2}(a-y)$, $\frac{1}{2}(a^2 + y^2) = AC^2 + BC^2 = AB^2 = x^2$ and

$$(a^2 - y^2)/4b = AC \times BC/DC = AB = x.$$

Therefore $2x^2 - a^2 = y^2 = a^2 - 4bx$ and so $x^2 = -2bx + a^2$, and with the root extracted $x = -b + \sqrt{[a^2 + b^2]}$. Hence in the above construction CE is the hypotenuse of the triangle sought.

Given, however, the base[388] and altitude both in this problem and the preceding one, the triangle is speedily constructed as follows. Build the rectangle CG, whose side CE shall be the triangle's base, the other side CF its altitude, then on CE describe a semicircle cutting the opposite side FG in H; draw CH, EH and CHE will then be the required triangle.

Problem 49

In a right-angled triangle, given the sum of the sides and the sum of the altitude and base, to find the triangle.

Let the sum of the sides AC and BC be a, the sum of the base AB and altitude CD b, the side $AC = x$, the base $AB = y$, and there will be $BC = a - x$, $CD = b - y$, $a^2 - 2ax + 2x^2 = AC^2 + BC^2 = AB^2 = y^2$,

$$x(a - x) = AC \times BC = AB \times CD = y(b - y) = by - a^2 + 2ax - 2x^2,$$

---

and so $y^2 + 2by = a^2$ or $y = -b + \sqrt{[a^2 + b^2]}$ (as in the following *Idem aliter*); on substituting this, it follows at once that $x^2 - ax + b(-b + \sqrt{[a^2 + b^2]}) = 0$.

(386) Newton successively constructs $DA = \sqrt{[a^2 + b^2]}$, $CE = (b - DA$ or) $b - \sqrt{[a^2 + b^2]}$, $GC = \sqrt{[(\frac{1}{2}a)^2 + b \times CE]}$ and finally $BG = \frac{1}{2}a \pm GC$.

(387) See note (384).

(388) More precisely, understand that this is constructed from the given elements according to Newton's preceding directions.

Hujus quadratum $a^4 - 2a^3x + 3aaxx - 2ax^3 + x^4$ pone æquale yy in bb hoc est ‖[120] æquale $aabb \| - 2abbx + 2bbxx$. et ordinata æquatione fiet

$$x^4 - 2ax^3 {+3aa \atop -2bb} xx {-2a^3 \atop +2abb} x {+a^4 \atop -aabb} = 0.$$

As utramɋ partem æquationis adde $b^4 - aabb$ et fiet

$$x^4 - 2ax^3 {+3aa \atop -2bb} xx {-2a^3 \atop +2abb} x {+a^4 \atop -2aabb \atop +b^4} = b^4 - aabb.$$

Et extracta utrobiɋ radice $xx - ax + aa - bb = -b\sqrt{bb - aa}$.[389] Et radice iterum extracta $x = \frac{1}{2}a \pm \sqrt{bb - \frac{3}{4}aa - b\sqrt{bb - aa}}$.

Const[r]uctio Geometrica.

Cape R mediam proportionalem inter $b + a$ et $b - a$, et S mediam proportionalem inter R et $b - R_{[,]}$ et T mediam proportionalem inter $\frac{1}{2}a + S$ & $\frac{1}{2}a - S$, et erunt $\frac{1}{2}a + T$ & $\frac{1}{2}a - T$ latera trianguli.[390]

Prob 50.

Datum angulum CBD recta data CD subtendere ita ut si a termino istius rectæ D ad punctum A in recta CB producta datum agatur AD, fuerit angulus ADC æqualis angulo ABD.

Dicatur $CD = a$, $AB = b$, $BD = x$ et erit $BD . BA :: CD . DA = \frac{ab}{x}$. Demitte perp. DE.

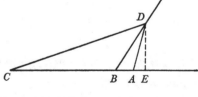

Erit $\quad BE = \dfrac{BD^q - AD^q + BA^q}{2BA} = \dfrac{xx - \dfrac{aabb}{xx} + bb}{2b}$.

Ob datum angulum DBA pone $BD . BE :: b . e$ et habebitur iterum $BE = \frac{ex}{b}$.

Ergo $xx - \dfrac{aabb}{xx} + bb = 2ex$. et $x^4 - 2ex^3 + bbxx - aabb = 0$.[391]

---

(389) Once more (compare notes (382) and (385)) it is simpler to compute the auxiliary equation $y^2 - 2by + a^2 = 0$ from the given conditions $y^2 = x^2 + (a - x)^2$ and $y(b - y) = x(a - x)$, then substituting its root $y = b - \sqrt{[b^2 - a^2]}$ in the latter.

and so $by = a^2 - ax + x^2$. The latter's square, $a^4 - 2a^3x + 3a^2x^2 - 2ax^3 + x^4$, put equal to $b^2 \times y^2$, that is, $b^2(a^2 - 2ax + 2x^2)$, and when the equation is ordered there will come $x^4 - 2ax^3 + (3a^2 - 2b^2)x^2 - (2a^3 - 2ab^2)x + a^4 - a^2b^2 = 0$. To each side of the equation add $b^4 - a^2b^2$ and there will prove to be

$$x^4 - 2ax^3 + (3a^2 - 2b^2)x^2 - (2a^3 - 2ab^2)x + a^4 - 2a^2b^2 + b^4 = b^4 - a^2b^2,$$

and, with the root extracted, $x^2 - ax + a^2 - b^2 = -b\sqrt{[b^2 - a^2]}$.[389] And, when the root is once more extracted, $x = \tfrac{1}{2}a \pm \sqrt{(b^2 - \tfrac{3}{4}a^2 - b\sqrt{[b^2 - a^2]})}$.

Geometrical construction

Take R to be the mean proportional between $b+a$ and $b-a$, S the mean proportional between R and $b-R$, and T the mean proportional between $\tfrac{1}{2}a + S$ and $\tfrac{1}{2}a - S$: then $\tfrac{1}{2}a + T$ and $\tfrac{1}{2}a - T$ will be the triangle's sides.[390]

Problem 50

To subtend the given angle $C\widehat{B}D$ by the given straight line CD in such a way that, if AD be drawn from the end-point D of that line to the given point A in the straight line CB produced, the angle $A\widehat{D}C$ will have become equal to the angle $A\widehat{B}D$.

Call $CD = a$, $AB = b$, $BD = x$ and there will be

$$(BD:BA = CD:DA \text{ or) } DA = ab/x.$$

Let fall the perpendicular DE and there will be

$$BE \text{ (or } \tfrac{1}{2}(BD^2 - AD^2 + BA^2)/BA) = \tfrac{1}{2}(x^2 - a^2b^2/x^2 + b^2)/b.$$

Since the angle $D\widehat{B}A$ is given, put $BD:BE = b:e$ and there will again be had $BE = ex/b$. Consequently $x^2 - a^2b^2/x^2 + b^2 = 2ex$, and

$$x^4 - 2ex^3 + b^2x^2 - a^2b^2 = 0.\text{[391]}$$

---

(390) It is evident that Newton constructs $R = \sqrt{[b^2 - a^2]}$,

$$S^2 = (R(b-R) \text{ or) } b\sqrt{[b^2 - a^2]} + a^2 - b^2 \quad \text{and} \quad T = \sqrt{[\tfrac{1}{4}a^2 - S^2]},$$

whence $x = \tfrac{1}{2}a \pm T$.

(391) Newton does not here attempt to solve this general quartic equation, reserving his sketch of Descartes' method of so doing till pages 205–6 below.

Octob 1680
Lect. 1.

Prob. 51.[392]

Datis trianguli lateribus invenire angulos.

Dentur latera $AB=a$. $AC=b$. $BC=c$. [&] quæratur angulus A. Demisso ad AB perpendiculo CD quod angulo isti opponitur, erit imprimis

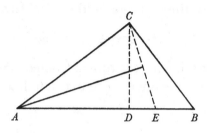

$$bb-cc=AC^q-BC^q=AD^q-BD^q$$
$$=\overline{AD+BD}\times\overline{AD-BD}$$
$$=AB\times\overline{2AD-AB}=2AD\times a-aa.$$

||[121] Adeoq $\frac{1}{2}a+\dfrac{bb-cc}{2a}=AD$.[393] Unde prodit || hocce primum Theorema. Ut AB, ad $AC+BC$ ita $A[C]-BC$ ad quartam proportionalem N. $\dfrac{AB+N}{2}=AD$. Ut AC ad AD ita radius ad cosinum anguli A.

Adhæc

$$DC^q=AC^q-AD^q=\frac{2aabb+2aacc+2bbcc-a^4-b^4-c^4}{4aa}^{[394]}$$

$$=\frac{a+b+c\times a+b-c\times a-b+c\times +a+b+c}{4aa}.$$

Unde multiplicatis numeratoris & denominatoris radicibus per b conflatur hocce Theorema secundum. Ut $2ab$ ad medium proportionale inter

$$a+b+c\times a+b-c \quad \text{et} \quad a-b+c\times -a+b+c$$

ita radius ad sinum anguli A.

Insuper in AB cape $AE=AC$, et age CE, er erit angulus ECD æqualis dimidio anguli A.[395] Aufer AD de AE et restabit

$$DE=b-\tfrac{1}{2}a-\frac{bb-cc}{2a}=\frac{cc-aa+2ab-bb}{2a}=\frac{c+a-b\times c-a+b}{2a}.$$

Unde $DE^q=\dfrac{c+a-b\times c+a-b\times c-a+b\times c-a+b}{4aa}$. Et hinc confit Theorema

tertium quartumq, vizt. Ut $2ab$ ad $c+a-b\times c-a+b$ (ita AC ad DE) ita radius ad sinum versum anguli A.[396] Et, Ut medium proportionale inter $a+b+c$ & $a+b-c$ ad medium proportionale inter $c+a-b$ & $c-a+b$ (ita CD ad DE) ita radius ad tangentem dimidij anguli A, vel dimidij cotangens ad radium.

---

(392) As we have seen (note (272)) this, despite its number, precedes in time the more sophisticated equivalent 'Prob. 23' which Newton subsequently inserted at an earlier place in the present manuscript. Significantly, when Newton came to change the order of his geometrical problems for his revised 1722 edition (see note (216)), he placed these two 'Heronian'

Problem 51[392]

Given the sides of a triangle, to find its angles.

Let the given sides be $AB = a$, $AC = b$ and $BC = c$, and let the angle $\hat{A}$ be sought. On letting fall to AB the perpendicular CD which lies opposite to that angle, there will in the first place be

$$b^2 - c^2 = AC^2 - BC^2 = AD^2 - BD^2 = (AD + BD)(AD - BD)$$
$$= AB(2AD - AB) = a(2AD - a),$$

so that $AD = \frac{1}{2}a + \frac{1}{2}(b^2 - c^2)/a$.[393] Hence there results this first theorem: Take AB to $AC + BC$ to be as $AC - BC$ to the fourth proportional N; then

$$AD = \tfrac{1}{2}(AB + N)$$

and AC is to AD as the radius to the cosine of angle $\hat{A}$.

Further,

$$DC^2 = AC^2 - AD^2 = (2a^2b^2 + 2a^2c^2 + 2b^2c^2 - a^4 - b^4 - c^4)/4a^2 \text{ [394]}$$
$$= (a + b + c)(a + b - c)(a - b + c)(-a + b + c)/4a^2.$$

Hence, when the roots of numerator and denominator are multiplied by b, there is produced this second theorem: As $2ab$ to the mean proportional between $(a + b + c)(a + b - c)$ and $(a - b + c)(-a + b + c)$, so is the radius to the sine of angle $\hat{A}$.

Moreover, in AB take $AE = AC$ and draw CE, and the angle $E\widehat{CD}$ will then be equal to half the angle $\hat{A}$.[395] Take AD from AE and there will remain

$$DE = b - \tfrac{1}{2}a - \tfrac{1}{2}(b^2 - c^2)/a = (c^2 - a^2 + 2ab - b^2)/2a = (c + a - b)(c - a + b)/2a,$$

whence $DE^2 = (c + a - b)^2 (c - a + b)^2/4a^2$. And from this there comes out a third theorem, and a fourth, namely: As $2ab$ to $(c + a - b)(c - a + b)$, so is (AC to DE or) the radius to the versed sine of angle $\hat{A}$;[396] and: As the mean proportional between $a + b + c$ and $a + b - c$ is to the mean proportional between $a - b + c$ and $-a + b + c$, so is (CD to DE or) the radius to the tangent of half the angle $\hat{A}$, or the cotangent of half that to the radius.

---

problems together but inverted their sequence, renumbering them as '11' and '12' respectively.

(393) It would have been simpler and more elegant to derive this value 'per 12.2 Elem' (by *Elements*, II, 12).

(394) The numerator is the expansion of

$$(2ab)^2 - (a^2 + b^2 - c^2)^2 = ((a + b)^2 - c^2)(-(a - b)^2 + c^2).$$

(395) For, where the bisector of $\hat{A}$ (drawn by Newton in his figure, but not marked) meets CE in F, the triangles DCE and FAE are readily shown to be similar.

(396) That is, radius $\times (1 - \cos \hat{A})$; compare note (276).

Præterea est $CE^q = CD^q + DE^q = \dfrac{2abb + bcc - baa - b^3}{a} = \dfrac{b}{a} \times \overline{c+a-b} \times \overline{c-a+b}$.

Unde Theorema quintum et sextum.[397] Ut medium proportionale inter $2a$ et $2b$ ad medium proportionale inter $c+a-b$ & $c-a+b$, vel ut 1 ad medium proportionale inter $\dfrac{c+a-b}{2a}$ & $\dfrac{c-a+b}{2b}$ (ita AC ad $\frac{1}{2}CE$ vel CE ad DE) ita radius ad sinum[398] dimidij anguli A. Et ut medium proportionale inter $2a$ et $2b$ ad medium proportionale inter $a+b+c$ et $a+b-c$ (ita CE ad CD) ita radius ad cosinum dimidij anguli A.

Si præter angulos desideretur etiam area trianguli duc CD^q in $\frac{1}{4}AB^q$ et radix vizt $\frac{1}{4}\sqrt{a+b+c} \times \overline{a+b-c} \times \overline{a-b+c} \times \overline{-a+b+c}$ erit area illa quæsita.[399]

Prob. 52.

[1680]
Lect 2
‖[122]

E Cometæ motu uniformi rectilineo per Cœlum trajicientis locis quatuor ‖ observatis, distantiam a terra, motusǫ determinationem, in Hypothesi Copernicanæa[400] colligere.

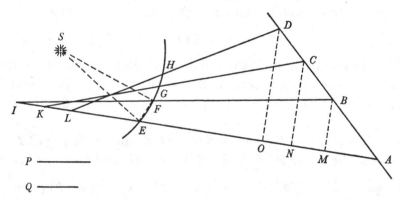

Si e centro Cometæ in locis quatuor observatis, ad planum Eclipticæ demittantur totidem perpendicula: sintǫ A, B, C, D puncta in plano illo in quæ perpendicula incidunt. Per puncta illa agatur recta[401] AD, et hæc secabitur a perpendiculis in eadem ratione cum linea quam Cometa motu suo describit, hoc est, ita ut sit AB ad AC ut tempus inter primam et secundam observationem ad tempus inter primam ac tertiam, et AB ad AD ut tempus illud inter primam

---

(397) Newton first wrote 'quintum, sextum et septimum' (a fifth, sixth and seventh).

(398) This replaces 'duplam chordam' (twice the chord), that is, four times the 'sinus' of $\frac{1}{2}\hat{A}$.

(399) For historical details regarding this classical Archimedean theorem, see 1, §3: note (1) preceding.

(400) Understand by this seventeenth-century *terminus technicus* the more accurate modified hypothesis, introduced by Kepler in his *Astronomia Nova ΑΙΤΙΟΛΟΓΗΤΟΣ, seu Physica Cœlestis, tradita Commentariis De Motibus Stellæ Martis* (Prague, 1609), in which the Earth— exactly as the other solar planets—is borne round the Sun, set at a focus, in an ellipse of small eccentricity, with its instantaneous speed determined according to the areal law: in the case of the Earth it is usually enough, because of its negligible eccentricity, to take its orbit to be

observatis, distantiam a terra, motusque determinationem, in Hypothesi Copernicea colligere.

Si è centro Cometæ in locis quatuor observatis, ad planum Eclipticæ demittantur totidem
perpendicula: sintque A, B, C, D puncta in plano

illo in qua perpendicula incidunt. Per puncta
illa agatur recta AD, et hæc secabitur a perpendiculis in eadem ratione cum linea quam
Cometa motu suo describit, hoc est, ita ut sit
AB ad AC ut tempus inter primam et secundam
observationem ad tempus inter primam ac tertiam,
et AB ad AD ut tempus illud inter primam et
secundam observationem ad tempus inter primam
et quartam. Ex observationibus itaque dantur
rationes linearum AB, AC, AD ad invicem.

Insuper in eodem Eclipticæ plano sit S sol
EH arcus lineæ Eclipticæ in qua terra movetur.
~~Et arcus~~ E, F, G, H loca quatuor terræ temporibus observationum, E locus primus, F secundus,
G tertius H quartus. Jungantur AE, BF, CG, DH
et producantur donec tres posteriores priorem secent
in J, K et L, BF in J, CG in K DH in L. Et erunt
anguli AJB, AKC, ALD differentia longitudinum
observatarum Cometæ AJB differentia longitudinum loci primi Cometæ et secundi AKC differentia longitudinum loci primi ac tertij et ALD
differentia longitudinum loci primi et quarti. Dantur itaque
ex observationibus anguli AJB, AKC, ALD.

Junge SE, SF, EF et ob data puncta
S, E, F

Plate II.   A rectilinear cometary orbit constructed from four
angular observations (1, 2, §1).

Furthermore, there is

$$CE^2 = CD^2 + DE^2 = (2ab^2 + bc^2 - a^2b - b^3)/a = (b/a)(a-b+c)(-a+b+c).$$

Whence a fifth and a sixth[397] theorem: As the mean proportional between $2a$ and $2b$ to the mean proportional between $a-b+c$ and $-a+b+c$, or as 1 to the mean proportional between $(a-b+c)/2a$ and $(-a+b+c)/2b$, so is (AC to $\frac{1}{2}CE$, or CE to DE, that is) the radius to the sine[398] of half the angle $\hat{A}$; and: As the mean proportional between $2a$ and $2b$ to the mean proportional between $a+b+c$ and $a+b-c$, so is (CE to CD or) the radius to the cosine of half the angle $\hat{A}$.

If besides the angles there were also desired the triangle's area, multiply CD^2 into $\frac{1}{4}AB^2$ and the root, namely $\frac{1}{4}\sqrt{[(a+b+c)(a+b-c)(a-b+c)(-a+b+c)]}$ will be the area required.[399]

Problem 52

From four observed positions of a comet crossing the sky with a uniform rectilinear motion, to gather its distance from the Earth and the direction of its motion, supposing the 'Copernican hypothesis'.[400]

If from the comet's centre in its four observed positions an equal number of perpendiculars be let fall to the plane of the ecliptic, let A, B, C, D be the points in that plane onto which the perpendiculars fall. Through those points draw the straight line[401] AD and this will be cut by the perpendiculars in the same ratio as the line which the comet describes in its motion—that is, such that AB is to AC as the time between the first and second observations is to the time between the first and third, and that AB is to AD as that time between the first and second

---

a uniformly traversed circle centred on the Sun (as Newton will do in the worksheet calculations reproduced in Appendix 3 below). However, the solution to the present problem does not require that the terrestrial orbit be specified absolutely, but only that the timed sightings IE, IF, KG and LH be given.

We have already noticed (see note (251)) that the present assumption of the cometary path as being exactly a straight line traversed uniformly is a simplification of Kepler's 'assumptum' in his *De Cometis* (Augsburg, 1619) introduced by Christopher Wren in the early 1660's. Wren's ensuing geometrical construction of the derived 'Problema. Datis quatuor lineis utcunque ductis (quarum nec tres sunt parallelæ neque ab eodem puncto ductæ) quintam ducere quæ à quatuor primo datis in tres partes secetur ratione & positione datas'—first printed by Robert Hooke in his *Cometa, or, Remarks about Comets* [= *Lectures and Collections made by Robert Hooke* (London, 1678): 1–80]: 41–2, but earlier revamped by John Wallis in an improved form (Bodleian. MS Don. d. 45: 283ᵛ–280ʳ) communicated by him in March 1677 to John Collins and published by the latter the following year as the first exercise 'De Cometarum Distantiis Investigandis' appended to his second edition of Jeremiah Horrocks' *Opera Posthuma* (London, 1678) and reissued in Caput cv of Wallis' Latin *Algebra* [= *Opera Mathematica*, **2** (Oxford, 1693): 1–482, especially 455–62]—is essentially that now presented by Newton.

(401) Evidently, since the comet's path is taken to be rectilinear, its orthogonal projection on the ecliptic will also be a straight line.

et secundam observationem ad tempus inter primam et quartam. Ex observationibus itacɜ dantur rationes linearum *AB, AC, AD* ad invicem.

Insuper in eodem Eclipticæ plano sit *S* sol, *EH* arcus lineæ Eclipticæ[402] in qua terra movetur, *E, F, G, H* loca quatuor terræ temporibus observationum, *E* locus primus, *F* secundus, *G* tertius[,] *H* quartus. Jugantur *AE, BF, CG, DH* et producantur donec tres posteriores priorem secent in *I, K* et *L, BF* in *I, CG* in *K*[,] *DH* in *L*. Et erunt anguli *AIB, AKC, ALD* differentiæ longitudinum observatarum Cometæ[,] *AIB* differentia longitudinum loci primi Cometæ et secundi *AKC* differentia longitudinum loci primi ac tertij et *ALD* differentia longitudinū loci primi et quarti. Dantur itacɜ ex observationibus anguli *AIB, AKC, ALD*.

‖ [123]   Junge *SE, SF, EF* et ob data puncta ‖ *S, E, F*, datumcɜ angulum *ESF*, dabitur angulus *SEF*. Datur etiam angulus *SEA* utpote differentia longitudinis Cometæ et Solis tempore observationis primæ. Quare si complementum ejus ad duos rectos, nempe angulū *SEI*, addas angulo *SEF*, dabitur angulus *IEF*. Trianguli igitur *IEF* dantur anguli cum latere *EF*, adeocɜ datur etiam latus *IE*. Et simili argumento dantur *KE* et *LE*.[403] Dantur igitur positione lineæ quatuor *AI, BI, CK, DL*, adeocɜ Problema huc redit, Ut lineis quatuor positione datis, quintam inveniamus quæ ab his in data ratione secabitur.[404]

Demissis ad *AI* perpendiculis *BM, CN, DO*, ob datum angulum *AIB* datur ratio *BM* ad *MI*. Est et *BM* ad *CN* in data ratione *BA* ad *CA*, et ob datum angulum *CKN* datur ratio *CN* ad *KN*. Quare datur etiam ratio *BM* ad *KN*: et inde ratio *CN* ad *KN*. Quare datur etiam ratio *BM* ad *KN*: et inde ratio quocɜ *BM* ad *MI−KN*, hoc est ad *MN+IK*. Cape *P* ad *IK* ut est *AB* ad *BC*, et cum sit *MA* ad *MN* in eadem ratione, erit etiam *P+MA* ad *IK+MN* in eadem ratione hoc est in ratione data. Quare datur ratio *BM* ad *P+MA*. Et simili argumento si capiatur *Q* ad *IL* in ratione *AB* ad *BD*, dabitur ratio *BM* ad *Q+MA*. Et proinde ratio *BM* ad ipsorum *P+MA* & *Q+MA* differentiam quocɜ dabitur. At differentia illa, nempe *P−Q* vel *Q−P*, datur. Et proinde[405]

---

(402) Accurately, an ellipse with the Sun set at a focus; see note (400).

(403) Newton has deleted 'atcɜ horum etiam differentiæ *IK* et *IL*' (and of these, also, the differences *IK* and *IL*).

(404) The reduction is Wren's; see note (400).

(405) The crucial assumption is made at this point that the magnitudes P and Q are different in quantity. On setting $IK = a$, $IL = b$, $AB:BC:CD = 1:\alpha:\beta$, $IM/MB = \lambda$, $KN/NC = \mu$ and $LO/OD = \nu$, at once $P = a/\alpha$ and $Q = b/(\alpha+\beta)$; also,

$$\text{since } IK/P = NM/MA = (IM-KN)/(P+MA)$$

$$\text{and } IL/Q = OM/MA = (IM-LO)/(Q+MA)$$

$$\text{it follows that } BM/(P+MA) = BM \times IK/P(IM-KN) = \alpha/(\lambda-(1+\alpha)\mu)$$

and $BM/(Q+MA) = BM \times IL/Q(IM-LO) = (\alpha+\beta)/(\lambda-(1+\alpha+\beta)\nu)$. Hence, if $P = Q$,

observations is to the time between the first and fourth. From the observations, accordingly, are given the ratios of the lines AB, AC, AD to one another.

Moreover, in the same plane of the ecliptic let S be the Sun, $\overset{\frown}{EH}$ the arc of the ecliptic curve[402] in which the Earth moves, and E, F, G, H the four positions of the Earth at the times of observation: E the first position, F the second, G the third and H the fourth. Join AE, BF, CG, DH and extend them till the three latter lines intersect the first in I, K and L: BF in I, CG in K, DH in L. Then will the angles $\widehat{AIB}$, $\widehat{AKC}$, $\widehat{ALD}$ be the differences in observed longitude of the comet: $\widehat{AIB}$ the difference in longitude between the first position of the comet and its second, $\widehat{AKC}$ the difference in longitude between its first position and its third, and $\widehat{ALD}$ the difference in longitude between its first position and its fourth. Accordingly, the angles $\widehat{AIB}$, $\widehat{AKC}$, $\widehat{ALD}$ are given from the observations.

Join SE, SF, EF and because the points S, E, F are given, and also the angle $\widehat{ESF}$, the angle $\widehat{SEF}$ will be given. The angle $\widehat{SEA}$ is also given, being, namely, the difference in longitude between the comet and the Sun at the time of the first observation. Consequently, if you add its supplement—the angle $\widehat{SEI}$, to be precise—to the angle $\widehat{SEF}$, the angle $\widehat{IEF}$ will be given. In the triangle IEF, therefore, the angles together with the side EF are given, and hence the side IE is also given. And by a similar reasoning KE and LE[403] are given. The four lines AI, BI, CK, DL are therefore given in position, and in consequence the problem reduces to this: Given four lines in position, we are to find a fifth which shall be cut by these in a given ratio.[404]

When the perpendiculars BM, CN, DO are let fall to AI, because the angle $\widehat{AIB}$ is given, so is the ratio of BM to MI. Also BM is to CN in the given ratio BA to CA, and, because the angle $\widehat{CKN}$ is given, so is the ratio CN to KN. Consequently the ratio of BM to KN is also given; and from this the ratio, too, of BM to $MI - KN$, that is, $MN + IK$. Take P to IK as AB is to BC and, since MA is to MN in the same ratio, there will also be $P + MA$ to $IK + MN$ in the same ratio— a given one, that is. Consequently the ratio of BM to $P + MA$ is given. And by a similar reasoning, if Q be taken to IL in the ratio AB to BD, the ratio of BM to $Q + MA$ will be given. As a result, the ratio of BM to the difference of $P + MA$ and $Q + MA$ will also be given. But that difference, namely, $P - Q$ (or $Q - P$), is

---

then $a:b = \alpha:(\alpha+\beta) = (\lambda-(1+\alpha)\mu):(\lambda-(1+\alpha+\beta)\nu)$: in geometrical terms, if LD meets IB, KC in m, n respectively—when it is readily shown that

$$Lm/Ln = [IL/(\lambda-\nu)]/[KL/(\mu-\nu)],$$

this implies the dual condition $IK/IL = BC/BD$ and $AB/BC = Lm/mn$—whence the parallels RS, TS (through C and B) to LA, LD respectively meet in a point S of the straight line In, and, conversely, any point S of In will determine a line $ABCD$ which solves the problem when $P = Q$. The importance of this particular, indeterminate case is that it holds very nearly true in physical reality: for, to the approximation that the comet's path may be accurately assumed

dabitur *BM*. Dato autem *BM*, simul dantur *P+MA*, & *MI*, & inde *MA*, *ME*, *AE* et angulus *EAB*.[406]

His inventis, erige ad *A* lineam plano Eclipticæ perpendicularem, quæ sit ad lineam *EA* ut tangens latitudinis Cometæ in observatione prima ad radium, et istius perpendicularis terminus erit locus centri Cometæ in observatione prima. Unde datur distantia Cometæ a Terra tempore illius observationis. Et eodem ‖[124] modo si e puncto *B* erigatur perpendicularis quæ ‖ sit ad lineam *BF* ut tangens latitudinis Cometæ in observatione secunda ad radium, habebitur locus centri Cometæ in observatione illa secunda. Et acta linea a loco primo ad locum secundum ea est in qua Cometa per Cœlum trajicit.[407]

---

to be the uniformly traversed straight line *ABCD*, it follows that the Earth will very nearly travel over a correspondingly rectilinear orbit *EFGH* at an equivalently uniform rate; accordingly, *EF*:*FG*:*GH* = *AB*:*BC*:*CD* and the parallels *rs*, *ts* (through *G* and *F*) will meet on

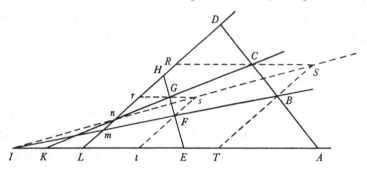

In, whose general point *S* will yield an infinity of possible positions for the line *ABCD*. Though, as we saw in the previous volume (IV: 270–3, especially 273, note (13)), Newton was already aware of this indeterminate instance in 1665, he seems never to have realized that it imposes an ineliminable block to the successful application of the present problem to the accurate construction of cometary orbits; while during 1681–2 he made several elaborate attempts to make the construction viable (compare Appendix 3 below), he had abandoned the Keplerian hypothesis of rectilinear cometary paths altogether by late 1684, henceforward (as will be seen in the sixth volume) referring all his computations to parabolic paths. In yet one more parallel between the mathematical development of Newton and Christiaan Huygens, it was in fact the latter who, in a note (*Œuvres complètes*, **19**: 297–300) dated '16 febr. 1681' but unpublished till the present century, both independently repeated Newton's 1665 proof that the 'center of motion' of two uniformly moving bodies is a straight line, and then went on to remark (*ibid.*: 300) that 'L'on peut conclure...que si les observations d'une comète de 20 ou 30 jours peuvent estre representees ou a peu pres en la faisant aller egalement dans une ligne droite, il y aura une infinitè d'autres telles lignes, dans les quelles estant supposee aller d'un mouvement egal les mesmes observations seront representees a peu pres de mesme; car quoyque la terre dans son orbite n'aille point en ligne droite...l'arc de cercle qu'elle parcourt dans 20 ou 30 jours ne s'eloigne pas beaucoup d'une ligne droite, et son mouvement ne differe pas sensiblement du mouvement egal.' More than sixty years later Roger Boscovich's attention was drawn to this crucial point when a young student at Rome. In the early 1740's Zanotti, 'un des premiers Astronomes de l'Italie', having computed the position of a comet according to the simplified Wrennian scheme, found to his astonishment that the computed location was diametrically opposite to its real position in the sky, but could not resolve the paradox: thus

given. And as a consequence[405] BM will be given. Given BM, however, $P+MA$ and MI are at once given, and from these MA, ME, AE and the angle $\widehat{EAB}$.[406]

Once these are found, erect at A a line perpendicular to the plane of the ecliptic, which is to be to the line EA as the tangent of the comet's latitude in the first observation, and the end-point of that perpendicular will be the position of the comet's centre at the first observation. Hence there is given the comet's distance from the Earth at the time of that observation. And in the same way, if from point B there be erected a perpendicular which shall be to the line BF as the tangent of the comet's latitude in the second observation, there will be had the position of the comet's centre at that second observation. And a (straight) line drawn from the first position to the second one is that in which the comet crosses the sky.[407]

---

prodded, Boscovich independently established that the problem became effectively indeterminate when applied to terrestrial sightings of solar comets, publishing a somewhat cumbrous proof in his doctoral *Dissertatio de Cometis, Habita…in Collegio Romano Anno 1746 mense Septembri die 5* (Rome, 1746) [= *Rogeri Josephi Boscovich Opera pertinentia ad Opticam, et Astronomiam…*, **3** (Venice, 1785): 316–68, especially 329–32]. Fifteen years afterwards, taking advantage— as we ourselves have done above—of the elegant construction of the general, determinate proposition given in Problem XXXVIII of Thomas Simpson's *Elements of Geometry; with their Application…to the Construction of a great Variety of Geometrical Problems* (London, ₁1747), Boscovich added to Castiglione's overblown commented edition of the *Arithmetica* an improved 'Observatio in Problema [52 →] LVI' (*Arithmetica Universalis…. Cum Commentario Johannis Castillionei*, **2** (Amsterdam, 1761): 124–30), in which he for the first time enunciated the dual geometrical condition for indeterminacy stated above, showing both its necessity and sufficiency and concluding with the wry remark that 'In hoc problemate illud est dolendum maxime, quod unus ex casibus indeterminatis ibi potissimum accidat, ubi ipsum problema maxime usum habere posset & pro quo in primis fuerat consideratum' (*ibid.*: 130).

(406) More explicitly, on assuming (as in the previous note) that $IK = a$, $IL = b$,

$$AB:BC:CD = 1:\alpha:\beta, \quad IM/MB = \lambda, \quad KN/NC = \mu \quad \text{and} \quad LO/OD = \nu,$$

and now setting $AM = x$, $MB = y$, the ensuing equalities $IM-KN = \alpha(P+AM)$ and $IM-LO = (\alpha+\beta)(Q+AM)$ produce the respective equations $(\lambda-(1+\alpha)\mu)y = a+\alpha x$ and $(\lambda-(1+\alpha+\beta)\nu)y = b+(\alpha+\beta)x$, from which

$$x = \frac{(\mu-\lambda)(b-a)-(\nu-\mu)a(1+\alpha+\beta)}{(\lambda-\mu)\beta-(\mu-\nu)\alpha(1+\alpha+\beta)} \quad \text{and} \quad y = \frac{a(\alpha+\beta)-b\alpha}{(\lambda-\mu)\beta-(\mu-\nu)\alpha(1+\alpha+\beta)}.$$

The point $B(x, y)$—and hence the line $ABCD$—is evidently indeterminate when

$$a:(b-a) = \alpha:\beta = (\lambda-\mu):(\mu-\nu)(1+\alpha+\beta),$$

or equivalently (see note (405)) $a:b = \alpha:(\alpha+\beta) = (\lambda-(1+\alpha)\mu):(\lambda-(1+\alpha+\beta)\nu)$.

(407) Two worksheets, in which Newton about spring 1681 attempted to compute the position of the 1680/1 comet under the present Wrennian supposition that it travelled, during late December and early January at least, uniformly in a rectilinear path, are reproduced in Appendix 3 following, along with an algebraic solution—essentially equivalent to Wallis' reworking of Wren's geometrical construction (see note (400))—which, despite Newton's present marginal 'dating' of his 'Prob. 52' as 'Octob. 1680', may well precede the geometrical analysis he here gives.

[1680]
Lect 3

Prob. 53.[408]

Si angulus datus CAD circa punctum angulare A positione datum, et angulus datus CBD circa punctum angulare B positione datum ea lege circumvolvantur ut crura AD, BD ad rectam positione datam EF sese semper intersecent: invenire lineam illam curvam quam reliquorum crurum AC, BC intersectio C describit.

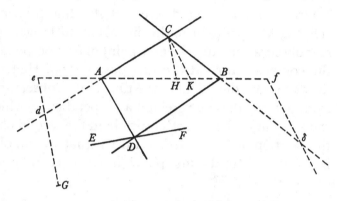

Produc *CA* ad *d* ut sit *Ad*=*AD*, et *CB* ad *δ* ut sit *Bδ*=*BD*. Fac angulum *Ade* æqualem angulo *ADE* et angulum *Bδf* æqualem angulo *BDF*, et produc *AB* utrincꝫ donec ea occurrat *de* et *δf* in *e* et *f*. Produc etiam *ed* ad *G* ut sit *dG*=*δf*, et a puncto *C* ad lineam *AB*, ipsi *ed* parallelam age *CH* et ipsi *fδ* parallelam *CK*. Et concipiendo lineas *eG fδ* immobiles manere dum anguli *CAD CBD* lege præscripta circa polos *A* et *B* volvantur, semper erit *Gd* æqualis *fδ*, et triangulum *CHK* dabitur specie.[409] Dic itaꝫ *Ae*=*a*. *eG*=*b*. *Bf*=*c*. *AB*=*m*. *BK*=*x* et

‖[125] *CK*=*y*. Et erit *BK*.*CK*::‖*Bf*.*fδ*. Ergo $f\delta = \dfrac{cy}{x} = Gd$. Aufer hoc de *Ge* et restabit $ed = b - \dfrac{cy}{x}$. Cum detur specie triangulum *CKH*, pone *CK*.*CH*::*d*.*e*; & *CH*.*HK*::*d*.*f* et erit $CH = \dfrac{ey}{d}$ et $HK = \dfrac{fy}{d}$. adeoꝗ $AH = m - x - \dfrac{fy}{d}$. Est autem *AH*.*HC*::*Ae*.*ed*, hoc est $m - x - \dfrac{f}{d}y \cdot \dfrac{ey}{d} :: a \cdot b - \dfrac{cy}{x}$. Ergo ducendo media et extrema in se fiet $mb - \dfrac{mcy}{x} - bx + cy - \dfrac{bf}{d}y + \dfrac{cfyy}{dx} = \dfrac{aey}{d}$. Duc omnes terminos in *dx*, eosꝗ in ordinem redige et fiet $fcyy \begin{subarray}{l} +dc \\ -ae \\ -fb \end{subarray} xy - dcmy - bdxx + bdmx = 0$. Ubi cùm incognitæ quantitates *x* et *y* ad duas tantùm dimensiones ascendunt patet curvam lineam quam punctum *C* describit esse Conicam Sectionem. Pone $\dfrac{ae + fb - dc}{c} = 2p$ et fiet $yy = \dfrac{2pxy}{f} + \dfrac{dm}{f}y + \dfrac{bd}{fc}xx - \dfrac{bdm}{fc}x$. et extracta radice $y = \dfrac{p}{f}x + \dfrac{dm}{2f} \pm \sqrt{\dfrac{pp}{ff}xx + \dfrac{bd}{fc}xx + \dfrac{pdm}{ff}x - \dfrac{bdm}{fc}x + \dfrac{ddmm}{4ff}}$. Unde colligitur Curvam

---

(408) The organic construction of a general conic from a rectilinear *directrix*; compare
II: 106, 118 ff. and especially 134–48. In his earlier researches Newton had at one point (see

Problem 53[408]

If the given angle $C\widehat{A}D$ revolves round the angular point A given in position, and the given angle $C\widehat{B}D$ round the angular point B given in position, with the restriction that their legs AD, BD shall always intersect on the straight line EF given in position, to find the curved line which the intersection C of their remaining legs AC, BC describes.

Extend CA to d so that $Ad = AD$, and CB to δ so that $B\delta = BD$. Make angle $A\widehat{d}e$ equal to the angle $A\widehat{D}E$ and angle $B\widehat{\delta}f$ equal to the angle $B\widehat{D}F$, and extend AB each way till it meets de and δf in e and f. Extend also ed to G so that $dG = \delta f$, and from the point C as far as the line AB draw CH parallel to ed and CK parallel to $f\delta$. Then, on conceiving that the lines eG, $f\delta$ remain stationary while the angles $C\widehat{A}D$, $C\widehat{B}D$ by the restriction imposed revolve round the poles A and B, Gd will always be equal to $f\delta$ and the triangle CHK ever given in species.[409] So call $Ae = a$, $eG = b$, $Bf = c$, $AB = m$, $BK = x$ and $CK = y$. There will then be $BK:CK = Bf:f\delta$, and therefore $f\delta = Gd = cy/x$. Take this from Ge and there will remain $ed = b - cy/x$. Since the triangle CKH is given in species, put $CK:CH = d:e$ and $CH:HK = d:f$, and then $CH = ey/d$ and $HK = fy/d$, so that $AH = m - x - fy/d$. However, $AH:HC = Ae:ed$, that is,

$$(m - x - (f/d)\,y) : (e/d)\,y = a : (b - cy/x).$$

Therefore, on multiplying middles and extremes into one another, there will come to be $bm - cmy/x - bx + cy - (bf/d)\,y + cfy^2/dx = (ae/d)\,y$. Multiply all the terms by dx, reduce them to order and there will come

$$cfy^2 - (ae + bf - cd)\,xy - cdmy - bdx^2 + bdmx = 0.$$

Since the unknown quantities here rise but to two dimensions, it is evident that the curved line which the point C describes is a conic. Set $(ae + bf - cd)/c = 2p$ and there will come to be $y^2 = 2(p/f)\,xy + (dm/f)\,y + (bd/cf)\,x^2 - (bdm/cf)\,x$, and, with the root extracted,

$$y = (p/f)\,x + \tfrac{1}{2}dm/f \pm \sqrt{[(p^2/f^2 + bd/cf)\,x^2 + (p/f^2 - b/cf)\,dmx + \tfrac{1}{4}d^2m^2/f^2]}.$$

II: 152–5) made a crude, unsatisfactory approach to an analytical discussion of the organic construction in terms of a non-standard coordinate system, leaving his computations unfinished when they appeared to lead nowhere. He now successfully achieves a reduction of the construction to standard oblique Cartesian coordinates.

(409) For its exterior angles at H and K are equal to the given ones $A\widehat{e}d = A\widehat{E}D$ (where $DE = de$) and $B\widehat{f}\delta = B\widehat{F}D$ (where $DF = \delta f$) respectively, and hence all its angles are given. Since the equality $Gd = f\delta$ between the general segments of the fixed lines Gde and $f\delta$ measured from the respective fixed points G and f determines the pair of points d, δ to be in 1,1 correspondence 'per simplicem Geometriam', it follows as an immediate corollary to Proposition 12 of Newton's contemporary 'Solutio Problematis Veterum de Loco solido' (IV: 282–320, especially 308–10) that the locus (C) of the meets of lines dA, δB drawn through the fixed poles A, B is a conic passing through A and B. Compare also D. T. Whiteside, 'Patterns of Mathematical Thought in the later Seventeenth Century' (*Archive for History of Exact Sciences*, **1**, 1961: 179–388): 307–8.

Hyperbolam esse si sit $\frac{bd}{fc}$ affirmativum, vel negativum et non majus quam $\frac{pp}{ff}$;

Parabolam si sit $\frac{bd}{fc}$ negativum et æquale $\frac{pp}{ff}$; Ellipsin vel circulum si sit $\frac{bd}{fc}$ et

negativum et majus quam $\frac{pp}{ff}$.[(410)] Q.E.I.

[1680]
Lect 4

Prob. 54.

Parabolam describere quæ per data quatuor puncta transibit.

Sint puncta illa data A, B, C, D. Junge AB et eam biseca in E. Et per E age rectam aliquam VE, quam concipe diametrum esse Parabolæ, puncto V existente vertice ejus.[(411)] Junge AC ipsiꝗ AB parallelam age DG occurrentem AC in G. Dic $AB=a$.

‖[126] ‖$AC=b$. $AG=c$. $GD=d$. In AC cape AP cujusvis longitudinis et a P age PQ parallelam AB, et concipiendo Q punctum esse Parabolæ: dic $AP=x$, $PQ=y$, et æquationem quamvis ad Parabolam assume quæ relationem inter AP et PQ exprimat. Ut quod sit $y=e+fx\pm\sqrt{gg+hx}$.[(412)]

Jam si ponatur AP sive $x=0$, puncto P incidente in ipsum A, fiet PQ sive $y=0$ ut et $=-AB$. Scribendo autem in æquatione assumpta 0 pro x fiet $y=e\pm\sqrt{gg}$ hoc est $=e\pm g$. Quorum valorum ipsius y major $e+g$ est $=0_{[,]}$ minor $e-g=-AB$ sive $-a$. Ergo $e=-g$ et $e-g$ hoc est $-2g=-a$ sive $g=\frac{1}{2}a$. Atꝗ adeo vice æquationis assumptæ habebitur hæc $y=-\frac{1}{2}a+fx\pm\sqrt{\frac{1}{4}aa+hx}$.

Adhæc si ponatur AP sive $x=AC$ ita ut punctum P incidat in C fiet iterum $PQ=0$. Pro x igitur in æquatione novissima scribe AC sive b, et pro y 0, et fiet $0=-\frac{1}{2}a+fb+\sqrt{\frac{1}{4}aa+hb}$. sive $\frac{1}{2}a-fb=\sqrt{\frac{1}{4}aa+hb}$ et partibus quadratis $-afb+ffbb=hb$. sive $ffb-fa=h$. Atꝗ ita vice assumptæ æquationis habebitur isthæc $y=-\frac{1}{2}a+fx\pm\sqrt{\frac{1}{4}aa+ffbx-fax}$.

---

(410) This is exactly Descartes' reduction of the general second-degree Cartesian defining equation to standard form in Book 2 of his *Geometrie* (compare IV: 219, note (9)), and repeats its several minor inadequacies. Thus, when the lines Gde and $f\delta$ prove to be parallel, then (HC and so) $f=0$ and the locus to be reduced is $(dc-ae)xy-bdx^2-dcmy+bdmx=0$, that of a hyperbola. Again, when the directrix EF passes through either A or B, then

$$4(p^2/f^2+bd/fc)\,d^2m^2/4f^2 = (pdm/f^2-bdm/fc)^2, \quad \text{that is,} \quad c(d+2p)=bf \quad \text{and so} \quad ae=0;$$

hence $a=0$ or $e=0$ and the locus (C) reduces to a corresponding straight line through A or B (paired with the base-line AB). A much modified version of the present proof, there referred

It is hence gathered that the curve is a hyperbola if bd/cf is positive, or negative and not greater than p^2/f^2; a parabola if bd/cf is negative and equal to p^2/f^2; an ellipse (or circle) if bd/cf is both negative and greater than p^2/f^2.[410] As was to be found.

Problem 54

To describe a parabola which shall pass through four given points.

Let those given points be A, B, C, D. Join AB and bisect it in E. Then through E draw some straight line VE and suppose it to be a diameter of the parabola, the point V being its vertex.[411] Join AC and parallel to AB draw DG meeting AC in G. Call $AB = a$, $AC = b$, $AG = c$, $GD = d$. In AC take AP of any length and from P draw PQ parallel to AB; then, conceiving Q to be a point on the parabola, call $AP = x$, $PQ = y$ and assume any defining equation of the parabola to express the relationship between AP and PQ. Let there, for instance, be
$y = e + fx \pm \sqrt{[g^2 + hx]}$.[412]

If now AP (or x) be set $= 0$, there will, with the point P falling at A, be $PQ = 0$ and also $-AB$; while, on writing 0 for x in the equation assumed, it will become $y = e \pm \sqrt{g^2}$, that is, $e \pm g$: the larger of these values, $e + g$, is then $= 0$, the lesser one $e - g = -AB$, that is, $-a$. Therefore $e = -g$ and ($e - g$, that is) $-2g = -a$, or $g = \frac{1}{2}a$. Hence, instead of the equation assumed there will be had this one:
$y = -\frac{1}{2}a + fx \pm \sqrt{[\frac{1}{4}a^2 + hx]}$.

Further, if AP (or x) be set $= AC$—so that the point P falls at C—, there will again prove to be $PQ = 0$. Consequently, in place of x in the most recent equation write AC (or b) and 0 in place of y, and there will come $0 = -\frac{1}{2}a + fb + \sqrt{[\frac{1}{4}a^2 + hb]}$, that is, $\frac{1}{2}a - bf = \sqrt{[\frac{1}{4}a^2 + bh]}$, and, when the sides are squared, $-abf + b^2f^2 = bh$ or $h = bf^2 - af$. And hence instead of the equation assumed there will now be had this: $y = -\frac{1}{2}a + fx \pm \sqrt{[\frac{1}{4}a^2 + (bf^2 - af)x]}$.

---

to perpendicular Cartesian coordinates departing from one of the fixed poles as origin, was subsequently presented by Colin Maclaurin in his *Geometria Organica: sive Descriptio Linearum Curvarum Universalis* (London, 1720): Pars Prima, Sectio I. 'De Descriptione Curvarum primi Generis seu Linearum Ordinis secundi', Propositio 1: 1–3; in it Newton's open appeal to the geometrical 1,1 correspondence of points which lies at the heart of his *constructio organica* has become heavily cloaked with irrelevant detail. Maclaurin justifies his variation of Newton's elegant argument 'quoniam nostra [Analysis] in sequentibus calculis usus non erit exigui & æquationem ad Ordinatas axi normales exhibeat' (*ibid.*: 3).

(411) Newton first continued: 'Ipsi AB parallelas age CG, FD secantes $EF[!]$ in F et G' (Parallel to AB draw CG, FD cutting [VE] in F and G). The manuscript figure bears no traces of such a construction, and it is not easy to see what he intended thereby.

(412) Assuming this Cartesian defining equation for the required parabola, whose main diameter VEI (conjugate to the direction of the ordinates PQ) accordingly has the equation $y = e + fx$, Newton now proceeds to evaluate the coefficients e, f, g and h from the conditions that it shall pass through $A(0, 0)$, $B(0, -a)$, $C(b, 0)$ and $D(c, -d)$.

Insuper si ponatur AP sive $x = AG$ sive c fiet PQ sive $y = -GD$ sive $-d$. Quare pro x et y in æquatione novissima scribe c et $-d$ et fiet

$$-d = -\tfrac{1}{2}a + fc - \sqrt{\tfrac{1}{4}aa + ffbc - fac}.$$

|| [127] sive $\tfrac{1}{2}a - d - fc = [-]\sqrt{\tfrac{1}{4}aa + ffbc - fac}$.[413] Et partibus || quadratis

$$-ad - fac + dd + 2dcf + ccff = ffbc - fac.$$

Et æquatione ordinata et reducta $ff = \dfrac{2d}{b-c}f + \dfrac{dd - ad}{bc - cc}$. Pro $b - c$ hoc est pro

GC scribe k et æquatio illa fiet $ff = \dfrac{2d}{k}f + \dfrac{dd - ad}{kc}$. Et extracta radice

$$f = \frac{d}{k} \pm \sqrt{\frac{ddc + ddk - adk}{kkc}}.\ ^{(414)}$$

Invento autem f, æquatio ad Parabolam, vizt $y = -\tfrac{1}{2}a + fx \pm \sqrt{\tfrac{1}{4}aa + ffbx - fax}$, plenè determinatur: cujus itaꝗ constructione Parabola etiam determinabitur. Constructio autem ejus hujusmodi est.[415] Ipsi BD parallelam age CH occurrentem DG in H. Inter DG ac DH cape mediam proportionalem DK, et ipsi CK parallelam age EI bisecantem AB in E et occurrentem DG in I. Dein produc IE ad V ut sit $EV \cdot EI :: EB^q \cdot DI^q - EB^q$, et erit V vertex, VE diameter et $\dfrac{BE^q}{VE}$ latus rectum Parabolæ quæsitæ.

[1680]
Lect 5

Prob 55.

Conicam Sectionem per data quinꝗ puncta describere.[416]

Sint puncta ista A, B, C, D, E. Junge AC, BE se mutuò secantes in H. Age DI parallelam BE et occurrentem AC in I. Item EK parallelam AC et occurrentem DI productæ in K. Produc ID ad F et EK ad G ut sit

$$AHC \cdot BHE :: AIC \cdot FID :: EKG \cdot FKD,$$

et erunt puncta F ac G in conica sectione ut notum est. Hoc tamen observare debebis, quod si punctum H cadit inter puncta omnia A, C & B, E vel extra ea

---

(413) The sign of the radical is taken negative since in Newton's figure as drawn

$$\tfrac{1}{2}a - d - fc\ (\text{or } EA - DG - [IG - EA]) = -DI.$$

(414) In sequel Newton will construct this by setting off $DH = d + (d-a)k/c$ in DG—whence $GH : GC = (DG - BA) : AG$ and so CH is parallel to DB—and then making

$$DK = \sqrt{[DG \times DH]}.$$

In consequence KC has slope $GK/GC = f$ and is therefore parallel to the main diameter VEI, of Cartesian equation $y = -\tfrac{1}{2}a + fx$. The corresponding 'vertex' V is determined at once from the proportion $VE : VI = EB^2 : ID^2$.

Moreover, if AP (or x) be set $= AG$ (or c), there will come to be

$$PQ \text{ (or } y) = -GD \text{ (or } -d).$$

Consequently, in place of x and y in the newest equation write c and $-d$, and there will come $-d = -\frac{1}{2}a + fc - \sqrt{[\frac{1}{4}a^2 + (bf^2 - af)\,c]}$ or

$$\tfrac{1}{2}a - d - cf = -\sqrt{[\tfrac{1}{4}a^2 + bcf^2 - acf]}.^{(413)}$$

Then, with the sides squared, $-ad - acf + d^2 + 2cdf + c^2f^2 = bcf^2 - acf$ and, when the equation is ordered and reduced, $f^2 = 2df/(b-c) + d(d-a)/c(b-c)$. In place of $b-c$ (that is, GC) write k and that equation will become

$$f^2 = 2df/k + d(d-a)/ck,$$

and, with the root extracted, $f = d/k \pm \sqrt{[d(cd+dk-ak)/ck^2]}.^{(414)}$ Once f is found, however, the equation to the parabola, namely

$$y = -\tfrac{1}{2}a + fx \pm \sqrt{[\tfrac{1}{4}a^2 + (bf^2 - af)\,x]},$$

is fully determined; and, correspondingly, by its construction the parabola will also be determined. Its construction, however, is effected in this manner: Parallel to BD draw CK meeting DG in H, between DG and DH take the mean proportional DK, and parallel to CK draw EI bisecting AB in E and meeting DG in I; then extend IE to V so that $VE:EI = EB^2:(DI^2 - EB^2)$ and V will be the vertex, VE its diameter and BE^2/VE the *latus rectum* of the required parabola.

Problem 55

To describe a conic section through five given points.[416]

Let those points be A, B, C, D, E. Join AC, BE, mutually intersecting in H, draw DI parallel to BE and meeting AC in I, and likewise EK parallel to AC and meeting DI produced in K; then extend ID to F and EK to G so that

$$AH \times HC : BH \times HE = AI \times IC : FI \times ID = EK \times KG : FK \times KD,$$

and—as is known—the points F and G will be on the conic. You ought, however, to notice this: if the point H falls between each pair of points A, C and

---

(415) Newton first continued: 'Inter d et $d + \dfrac{dk-ak}{c}$ cape mediam proportionalem M, dein dic $d \pm M.d :: AE.AF$' (Between d and $(d-a)k/c$ take the mean proportional M, then say $(d \pm M):d = AE:AF$). This evidently determines the point F (in CA continued) to be in the diameter VEI: the manuscript figure, correspondingly, has a deleted portion in which CA and IEV are extended in broken line to their meet at F.

(416) Newton's following synthetic construction is essentially that of his contemporary 'Veterum Loca solida restituta' (IV: 274–82, especially 278), where his strong debt to Pappus (*Mathematical Collection*, VIII, 13) is explicitly acknowledged. An analytical solution in parallel to that of the preceding problem, making use of the general (second-degree) Cartesian defining equation of a conic, is given in a scholium on Newton's pages 130–2 following.

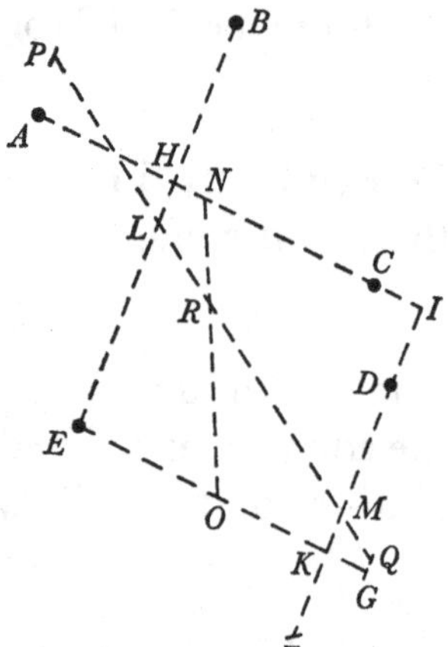

omnia, punctum I cadere debebit vel inter puncta omnia A, C & F, D vel extra ea omnia, et punctum K

|| [128] inter omnia D, F & E, G vel extra ea || omnia. At si punctum H cadit inter duo puncta A, C et extra alia duo B, E vel inter illa duo B, E et extra altera duo A, C, debebit punctum I cadere inter duo punctorum A, C & F, D et extra alia duo eorum & similiter punctum K debebit cadere inter duo punctorum D, F & E, G et extra alia duo eorum. Id quod fiet capiendo IF, KG ad hanc vel illam partem punctorum I, K, pro exigentia problematis. Inventis punctis F ac G, biseca AC EG in N et O; item BE, FD in L et M. Junge NO, LM se mutuo secantes in R et erunt LM et NO diametri conicæ sectionis, R centrum ejus et BL, FM ordinatim applicatæ ad diametrum LM. Produc LM hinc inde si opus est ad P et Q ita ut sit $BL^q.FM^q::PLQ.PMQ$, et erunt P et Q vertices. Fac $PLQ.LB^q::PQ.T.$ et erit T latus rectum. Quibus cognitis cognoscitur Figura.

Restat tantum ut doceamus quomodo LM hinc inde producenda sit ad P et Q ita ut fiat $BL^q.FM^q::PLQ.PMQ$. Nempe PLQ sive $PL\times LQ$ est $PR-LR\times PR+LR$, nam PL est $PR-LR$ et LQ est $RQ+LR$ seu(417) $PR+LR$. Porro $PR-LR\times PR+LR$ multiplicando fit PR^q-LR^q. Et ad eundem modum PMQ est $PR+RM\times PR-RM$ seu PR^q-RM^q. Ergo

$$BL^q.FM^q::PR^q-LR^q.PR^q-RM^q,$$

et dividendo $BL^q-FM^q.FM^q::RM^q-LR^q.PR^q-RM^q$. Quamobrem cum dentur BL^q-FM^q, FM^q, & RM^q-LR^q dabitur PR^q-RM^q. Adde datum RM^q et dabitur summa PR^q, adeoq; et latus ejus PR, cui QR æqualis est.

[1680] Lect 6

Prob. 56.(418)

Conicam sectionem describere quæ transibit per quatuor data puncta, et in uno istorum punctorum continget rectam positione datam.

|| [129] Sint puncta quatuor data A, B, C, D, et || recta positione data AE quam conica sectio contingat in puncto A. Junge duo quævis(419) puncta D C, et DC, producta si opus est, occurrat tangenti in E. Per quartum punctum B ipsi DC age

(417) Since R is the conic's centre, and so $PR = RQ$.

(418) The particular case of 'Prob 55' preceding in which two of the given points coincide. An analytical solution follows on Newton's page 132.

(419) Other than A, it is understood.

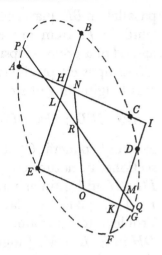

B, E, or outside each pair, then the point I ought to fall either between each pair of points A, C and F, D, or outside each pair, and the point K between each pair D, F and E, G, or outside each pair; but if the point H falls between the two points A, C and outside the other two B, E, or between the latter two B, E and outside the other two A, C, point I ought to fall between two of the points A, C and F, D and outside the other pair, and similarly the point K ought to fall between one of the pairs of points D, F and E, G and outside the other pair of these. This will be accomplished by taking IF and KG on one or other side of the points I, K, depending on the conditions of the problem. When the points F and G are found, bisect AC and EG in N and O, and likewise BE and FD in L and M, then join NO, LM, mutually intersecting in R, and LM, NO will be diameters of the conic, R its centre and BL, FM ordinates to the diameter LM. Extend LM, either way if need be, to P and Q so that $BL^2 : FM^2 = PL \times LQ : PM \times MQ$, and then P and Q will be vertices. Make $PL \times LQ : LB^2 = PQ : T$ and T will be the *latus rectum*. Once these are known, the figure is known.

It but remains to tell how LM is to be extended each way to P and Q so that there comes to be $BL^2 : FM^2 = PL \times LQ : PM \times MQ$. Precisely,

$$\text{`}PLQ\text{'} \text{ or } PL \times LQ$$

is $(PR - LR)(PR + LR)$—for PL is $PR - LR$ and LQ is $RQ + LR$, that is,[417] $PR + LR$—, while on multiplying it out $(PR - LR)(PR + LR)$ becomes $PR^2 - LR^2$; and in much the same way

$$PM \times MQ \text{ is } (PR + RM)(PR - RM) \text{ or } PR^2 - RM^2.$$

Therefore $BL^2 : FM^2 = (PR^2 - LR^2) : (PR^2 - RM^2)$ and *divisim*

$$(BL^2 - FM^2) : FM^2 = (RM^2 - LR^2) : (PR^2 - RM^2).$$

Consequently, since $BL^2 - FM^2$, FM^2 and $RM^2 - LR^2$ are given, so will be $PR^2 - RM^2$. Add the given square RM^2 and their sum PR^2 will be given, and hence also its side PR, to which QR is equal.

Problem 56[418]

To describe a conic section which shall pass through four given points, and in one of those points shall touch a straight line given in position.

Let the four given points be A, B, C, D, the straight line given in position AE, and its point of contact with the conic A. Join any two[419] points D, C and let DC, produced if necessary, meet the tangent in E. Through the fourth point B

parallelam *BF* quæ occurrat eidem tangenti in *F*. Item tangenti parallelam age *DI* quæ occurrat ipsi *BF* in *I*. In *FB*, *DI*, si opus est productis, cape *FG*, *HI* ejus longitudinis ut sit

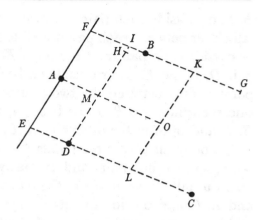

$$AE^q \cdot CED :: AF^q \cdot BFG :: DIH \cdot BIG.$$

et erunt puncta *G* et *H* in Conica sectione, ut notum est: si modò capias *FG*, *IH* ad legitimas partes punctorum *F* et *I*, juxta regulam in superiore Problemate traditam. Biseca *BG*, *DC*, *DH* in *K*, *L* et *M*. Junge *KL*, *AM* se mutuo secantes in *O*, et erit *O* centrum, *A* vertex, & *HM* ordinatim applicata ad semidiametrum *AO*. Quibus cognitis cognoscitur figura.

Prob. 57.[420]

Conicam sectionem describere quæ transibit per tria data puncta et in duobus istorum punctorum continget rectas positione datas.

Sint puncta illa data *A*, *B*, *C*. Tangentes *AD*, *BD* ad puncta *A* et *B*. *D* communis intersectio tangentium. Biseca *AB* in *E*. Age *DE* et produc eam donec in *F* occurrat *CF* actæ parallelæ *AB*: Et erit *DF* diameter, et *AE*, *CF* ordinatim applicatæ ad diametrum. Produc *DF* ad *O* et in *DO* cape *OV* mediam proportionalem inter *DO* et *EO*,[421] ea lege ut sit etiam

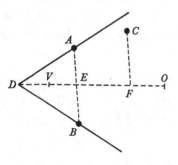

‖ [130] $$AE^q \cdot CF^q :: VE \times VO + OE. \| VF \times VO + OF:$$

et erit *V* vertex et *O* centrum figuræ. Quibus cognitis Figura simul cognoscitur. Est autem $VE = VO - OE$ adeoꝗ

$$VE \times \overline{VO + OE} = \overline{VO - OE} \times \overline{VO + OE} = VO^q - OE^q.$$

Præterea quia *VO* media proportionalis est inter *DO* et *EO* erit $VO^q = DOE$ adeoꝗ $VO^q - OE^q = DOE - OE^q = DEO$. Et simili argumento erit

$$VF \times \overline{VO + OF} = VO^q - OF^q = DOE - OF^q.$$

---

(420) The particular case of 'Prob 55' in which two pairs of points are coincident. Newton gives an equivalent analytical solution on his pages 132–4 below.

parallel to *DC* draw *BF* to meet the same tangent in *F*; likewise, parallel to the tangent draw *DI* to meet *BF* in *I*. In *FB*, *DI* (produced if necessary) take *FG*, *HI* of such a length that

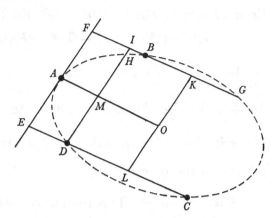

$$AE^2 : CE \times ED = AF^2 : BF \times FG$$
$$= DI \times IH : BI \times IG,$$

and the points *G* and *H* will, as is known, be on the conic, provided only that you take *FG*, *IH* on their legitimate sides of the points *F* and *I*, following the rule delivered in the previous problem. Bisect *BG*, *DC*, *DH* in *K*, *L* and *M*, and join *KL*, *AM* mutually intersecting in *O*; then will *O* be the centre, *A* a vertex and *HM* ordinate to the semi-diameter *AO*. And once these are known, the figure is known.

Problem 57[(420)]

To describe a conic which shall pass through three given points and in two of those points shall touch straight lines given in position.

Let those given points be *A*, *B*, *C* with *AD*, *BD* the tangents at the points *A* and *B*, and *D* the common intersection of the tangents. Bisect *AB* in *E*, draw *DE* and extend it till it meets *CF*, drawn parallel to *AB*, in *F*: *DF* will then be a diameter, with *AE*, *CF* ordinate to that diameter. Extend *DF* to *O* and in *DO* take *VO* to be the mean proportional between *DO* and *EO*,[(421)] with the added restriction that $AE^2 : CF^2 = VE(VO + EO) : VF(VO + FO)$, and

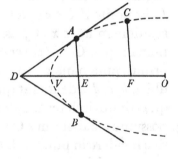

then *V* will be a vertex and *O* the figure's centre. And when these are known, the figure is at once known. Here, however, $VE = VO - EO$, so that

$$VE(VO + EO) = (VO - EO)(VO + EO) = VO^2 - EO^2;$$

further, because *VO* is a mean proportional between *DO* and *EO*, there will be $VO^2 = DO \times EO$, and hence $VO^2 - EO^2 = DO \times EO - EO^2 = DE \times EO$. And by a similar argument there will be

$$VF(VO + FO) = VO^2 - FO^2 = DO \times EO - FO^2.$$

---

(421) Since *AB* is the polar of *D* with respect to the conic, the points *D*, *E* are inverse with regard to the centre *O* and so $VO^2 = DO \times EO$.

Ergo $AE^q . CF^q :: DEO . DOE - OF^q$. Est $OF^q = EO^q - 2FEO + FE^q$. adeoq̧

$$DOE - OF^q = DOE - OE^q + 2FEO - FE^q = DEO + 2FEO - FE^q.$$

et $AE^q . CF^q :: DEO . DEO + 2FEO - FE^q :: DE . DE + 2FE - \dfrac{FE^q}{EO}$. Datur ergo

$DE + 2FE - \dfrac{FE^q}{EO}$. Aufer hoc de dato $DE + 2FE$ et restabit $\dfrac{FE^q}{EO}$ datum. Sit illud

N et erit $\dfrac{FE^q}{N} = EO$ adeoq̧ dabitur EO. Dato autem EO simul datur VO medium

proportionale inter DO et EO.

[1680]
Lect 7
     (422)Hoc modo per Theoremata quædam Apollonij(423) satis expedite resol-
vuntur hæc problemata: quæ tamen sine istis Theorematibus per Algebram
solam resolvi possent. Ut si proponatur primum(424) trium novissimorum Problematum: sint puncta quinq̧ data A, B, C, D, E,
per quæ Conica sectio transire debet. Junge duo
quævis $A\,C$ et alia duo $B\,E$ rectis se secantibus
in H. Ipsi BE parallelam age DI occurrentem
AC in I: ut et aliam quamvis rectam KL
occurrentem AC in K et conicæ sectioni in L.
Et finge Conicam sectionem datam esse, ita ut
cognito puncto K simul cognoscatur punctum
L. Et posito $AK = x$ et $KL = y$ ad exprimendam
relationem inter x et y assume quamvis æquationem quæ Conicas sectiones generaliter exprimit, puta hanc $a + bx + cxx + dy + exy + yy = 0$

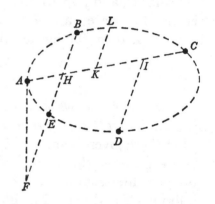

ubi a, b, c, d, e denotant quantitates determinatas cum signis suis,(425) x vero et y
quantitates indeterminatas. Si jam quantitates determinatas a, b, c, d, e invenire
‖[131] possumus, habebimus Conicam sectionem. Fin‖gamus ergo punctum L successive incidere in puncta A, C, B, E, D et videamus quid inde sequetur.(426) Si ergo
punctum L incidit in punctum A, erit in eo casu AK et KL hoc est x et y nihil.
Proinde æquationis omnes termini præter a evanescent, et restabit(427) $a = 0$.
Quare delendum est a in æquatione illa, et cæteri termini

$$bx + cxx + dy + exy + yy \text{ erunt} = 0.$$

Porrò si L incidit in C erit AK seu $x = AC$ et LK seu $y = 0$. Pone ergo $AC = f$ et
substituendo f pro x et 0 pro y æquatio ad curvam $bx + cxx + dy + exy + yy = 0$
evadet $bf + cff = 0$. seu(428) $b = -cf$. Et in æquatione illa scripto $-cf$ pro b

---

     (422) In this scholium to the three preceding Problems 55–7 Newton gives an alternative
Cartesian reduction of their respective requirements.

     (423) Notably the fundamental propositions in *Conics*, III, 17/18.

     (424) 'Prob 55' on Newton's pages 127–8 above.

Therefore $AE^2:CF^2 = DE \times EO:(DO \times EO - FO^2)$. Now

$$FO^2 = EO^2 - 2EF \times EO + EF^2$$

and hence

$$DO \times EO - FO^2 = DO \times EO - EO^2 + 2EF \times EO - EF^2 = (DE + 2EF)\,EO - EF^2,$$

and so

$$AE^2:CF^2 = DE \times EO:((DE + 2EF)\,EO - EF^2) = DE:(DE + 2EF - EF^2/EO).$$

Therefore $DE + 2EF - EF^2/EO$ is given. Take this from the given quantity $DE + 2EF$ and there will remain EF^2/EO, given. Let that be N and then $EO = EF^2/N$, so that EO will be given. But when EO is given, there is at once given VO, the mean proportional between DO and EO.

[422]In this way by means of certain theorems of Apollonius[423] these problems are resolved speedily enough: they might also, however, be resolved without the aid of those theorems by algebra alone. Thus, if the first[424] of the three most recent problems be proposed, let A, B, C, D, E be the five given points through which the conic must pass. Join any two A, C and any other two B, E by straight lines intersecting in H; parallel to BE draw DI meeting AC in I, and also any other line KL meeting AC in K and the conic in L; then conceive the conic to be given, so that when the point K is known the point L shall at once be known. On setting $AK = x$ and $KL = y$, to represent the relationship between x and y assume any equation which represents conics in a general fashion, this one, say: $a + bx + cx^2 + dy + exy + y^2 = 0$, where a, b, c, d, e denote determinate quantities with their signs,[425] x and y, however, indeterminate ones. If we can now find the determinate quantities a, b, c, d, e, we shall have the conic. Let us conceive, therefore, that the point L falls successively at the points A, C, B, E, D and see what will follow from that.[426] If, then, the point L falls at the point A, in this case AK and KL, that is, x and y, will be zero; consequently all terms in the equation except a will disappear and there will remain[427] $a = 0$, and as a result a in that equation must be deleted: the rest of the terms $bx + cx^2 + dy + exy + y^2$ will then be $= 0$. Moreover, if L falls at C, then AK or x will be $= AC$ and KL or $y = 0$; so put $AC = f$ and, on substituting f for x and 0 for y, the equation $bx + cx^2 + dy + exy + y^2 = 0$ to the curve will come to be $bf + cf^2 = 0$, and so[428] $b = -cf$: with $-cf$ written in place of b in it, the equation will come to be

(425) That is, undetermined constants either positive or negative. Newton ever feels he must acknowledge this departure from contemporary convention, in which unknowns can only take on positive values; indeed, he at once proceeds to assume the older convention in sequel.

(426) Newton proceeds to determine the constants a, b, c, d, e by requiring the conic to pass through the Cartesian points $A(0, 0)$, $B(g, h)$, $C(f, 0)$, $D(m, n)$ and $E(g, -k)$.

(427) 'solum' (only) is deleted here.

(428) Since points A and C are distinct, $f \neq 0$.

evadet $-cfx+cxx+dy+exy+yy=0$. Adhæc si punctum L incidit in punctum B, erit AK seu $x=AH$ et KL seu $y=BH$. Pone ergo $AH=g$ et $BH=h$ et perinde scribe g pro x et h pro y, et æquatio $-cfx+cxx$ &c $[=0]$ evadet

$$-cfg+cgg+dh+egh+hh=0.$$

Quod si punctum L incidit in E erit $AK=AH$ seu $x=g$ et KL seu $y=HE$. Pro HE ergo scribe $-k$ cum signo negativo quia HE jacet ad contrarias partes lineæ AC, et substituendo g pro x et $-k$ pro y, æquatio $-cfx+cxx$ &c $[=0]$ evadet $-cfg+cgg-dk-egk+kk=0$. Aufer hoc de superiori æquatione

$$-cfg+cgg+dh+egh+hh[=0]$$

et restabit $dh+egh+hh+dk+egk-kk=0$. Divide hoc per $h+k$[429] et fiet $d+eg+h-k=0$. Hoc ductum in h aufer de $-cfg+cgg+dh+egh+hh=0$ et restabit $-cfg+cgg+hk=0$ seu $\dfrac{hk}{-gg+fg}=c$. Deniꝗ si punctum L incidit in punctum D erit AK seu $x=AI$ et KL seu $y=ID$. Quare pro AI scribe m et pro ID n[430] et perinde pro x et y substitue m et n et æquatio $-cfx+cxx$ &c $[=0]$ evadet $-cfm+cmm+dn+emn+nn=0$. Hoc divide per n et fiet

$$\frac{-cfm+cmm}{n}+d+em+n=0.$$

Aufer $d+eg+h-k=0$ et restabit $\dfrac{-cfm+cmm}{n}+em-eg+n-h+k=0$. sive

‖[132] $\dfrac{cmm-cfm}{n}+n-h+k=eg-em$. ‖ Jam verò ob data puncta A, B, C, D, E dantur AC, AH, AI, BH, EH, DI, hoc est f, g, m, h, k, n. Atꝗ adeo per æquationem $\dfrac{hk}{fg-gg}=c$ datur c. Dato autem c, per æquationem $\dfrac{cmm-cfm}{n}+n-h+k=eg-em$ datur $eg-em$. Divide hoc datum per datum $g-m$[431] et emerget datum e. Quibus inventis æquatio $d+eg+h-k=0$ seu $d=k-h-eg$[432] dabit d. Et his cognitis simul determinatur æquatio ad quæsitam Conicam sectionem $cfx=cxx+dy+exy+yy$. Et ex ea æquatione per methodum Cartesij[433] determinabitur Conica sectio.

[1680]
Lect 8 Quod si[434] quatuor A, B, C, E et positio rectæ AF quæ tangit Conicam sectionem ad unum istorum punctorum A daretur, posset Conica sectio sic facilius determinari. Inventis ut supra æquationibus $cfx=cxx+dy+exy+yy$,

---

(429) Similarly, since points B and E are distinct, $h \neq -k$.

(430) As Newton has drawn his figure (and in line with the convention according to which unknowns have only positive values, used by him just now in setting $HE = -k$), this should strictly be '$-n$'. Of course, the point $D(m, n)$ is in general position.

(431) Since D coincides with neither B nor E, $g \neq m$.

$-cfx+cx^2+dy+exy+y^2=0$. Further, if the point L falls at the point B, then AK or $x=AH$ and KL or $y=HB$; so put $AH=g$ and $HB=h$ and, correspondingly, write g for x and h for y, and the equation $-cfx+cx^2\ldots=0$ will come to be $-cfg+cg^2+dh+egh+h^2=0$. But if the point L falls at E, then $AK=AH$ or $x=g$ and KL or $y=HE$; in place of HE, therefore, write $-k$ (with a negative sign because HE lies on the opposite side of the line AC) and, on substituting g for x and $-k$ for y, the equation $-cfx+cx^2\ldots=0$ will come to be $-cfg+cg^2-dk-egk+k^2=0$. Take this from the previous equation

$$-cfg+cg^2+dh+egh+h^2=0$$

and there will be left $dh+egh+h^2+dk+egk-k^2=0$; divide this by $h+k$[429] and it will become $d+eg+h-k=0$. Multiply this by h and take it from

$$-cfg+cg^2+dh+egh+h^2=0$$

and there will remain $-cfg+cg^2+hk=0$, that is, $c=hk/(f-g)g$. If, finally, the point L falls at the point D, then AK or $x=AI$ and KL or $y=ID$; consequently, for AI write m and for $ID\,n$,[430] and, correspondingly, in place of x and y substitute m and n: the equation $-cfx+cx^2\ldots=0$ will come to be

$$-cfm+cm^2+dn+emn+n^2=0.$$

Divide this by n and it will become $cm(-f+m)/n+d+em+n=0$. Take away $d+eg+h-k=0$ and there will remain $cm(-f+m)/n+em-eg+n-h+k=0$, or $cm(m-f)/n+n-h+k=e(g-m)$. Now, indeed, because the points A, B, C, D, E are given, there are given AC, AH, AI, HB, EH, DI, that is, f, g, m, h, k, n; and hence by the equation $c=hk/(f-g)g$ there is given c. But, once c is given, by the equation $cm(m-f)/n+n-h+k=e(g-m)$ there is given $e(g-m)$; divide this by the given magnitude $g-m$[431] and there will emerge e, given. And, when these are found, the equation $d+eg+h-k=0$, that is, $d=k-h-eg$,[432] will give d. As soon as these are ascertained, the equation $cfx=cx^2+dy+exy+y^2$ to the required conic is at once determined. And from its equation the conic section will—by Descartes' method[433]—itself be determined.

But should[434] four points A, B, C, E and the position of a straight line AF touching the conic at one of those four, A, be given, the conic might be more easily determined in this manner. Having found, as above, the equations

---

(432) The equivalent equation '$\dfrac{k-h-d}{g}=e$' is cancelled.

(433) In Book 2 of his *Geometrie*; compare IV: 219, note (9) and also note (410) above.

(434) The case of Problem 56 preceding. Here, in modification of his previous supposition, Newton assumes that the conic is to pass through $A(0,0)$, $B(g,h)$, $C(f,0)$ and $E(g,-k)$, being touched at A by the line AF of slope p/g.

$d=k-h-eg$, et $c=\dfrac{hk}{fg-gg}$, concipe tangentem

AF occurrere rectæ EH in F, dein punctum L moveri per perimetrum figuræ CDE donec incidat in punctum A: et ultima ratio ipsius LK ad AK erit ratio FH ad AH, ut contemplanti figuram constare potest.[435] Dic ergo $FH=p$ et in hoc casu ubi LK est ad AK in ultima ratione erit $p.g::y.x$, sive $\dfrac{gy}{p}=x$. Quare pro x in æquatione $cfx=cxx+dy+exy+yy$, scribe

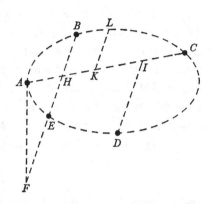

$\dfrac{gy}{p}$ et orietur $\dfrac{cfgy}{p}=\dfrac{cggyy}{pp}+dy+\dfrac{egyy}{p}+yy$. Divide omnia per y et emerget $\dfrac{cfg}{p}=\dfrac{cggy}{pp}+d+\dfrac{egy}{p}+y$. Jam quia supponitur punctum L incidere in punctum A, adeoq KL seu y infinite parvum vel nihil esse, dele terminos qui per y multiplicantur et restabit $\dfrac{cfg}{p}=d$. Quare fac $\dfrac{hk}{fg-gg}=c$, dein $\dfrac{cfg}{p}=d$, deniq $\dfrac{k-h-d}{g}=e$, et inventis c, d et e, æquatio $cfx=cxx+dy+exy+yy$ determinabit conicam sectionem.

Si deniq[436] tria tantum puncta A, B, C dentur una cum positione duarum rectarum AT, CT quæ tangunt Conicam sectionem in duobus istorum punctorum A et C, obtinebitur ut supra[437] ad ‖ Conicam sectionem æquatio hæc

‖[133]

$cfx=cxx+dy+exy+yy$. Deinde si supponatur ordinatam KL parallelam esse tangenti AT et concipiatur eam produci donec rursus occurrat Conicæ sectioni in M, et lineam illam LM accedere ad tangentem AT donec cum ea conveniat ad A: ultima ratio linearum KL et KM ad invicem erit ratio æqualitatis, ut contemplanti figuram constare potest.[438] Quamobrem in illo casu existentibus KL et KM sibi invicem æqualibus, hoc est duobus

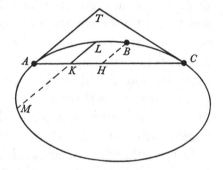

valoribus ipsius y (affrmativo scilicet KL, et negativo KM) æqualibus, debent æquationis $cfx=cxx+dy+exy+yy$ termini illi in quibus y est imparis dimensionis, hoc est termini $dy+exy$ respectu termini yy in quo y est paris dimensionis, evanescere. Aliter enim duo valores ipsius y, affirmativus et negativus, æquales

---

(435) Observe Newton's appeal to a limit ('ultimate') ratio of y to x at the origin A. Essentially, Newton proceeds to set $\lim\limits_{x,\,y\to0}(y/x)=p/g$ and then identify $-cf+dp/g=0$ as the 'ultimate' form (as x, y pass into zero) of $-cf+cx+dy/x+ey+y^2/x=0$.

$cfx = cx^2 + dy + exy + y^2$, $d = k - h - eg$ and $c = hk/(f-g)g$, imagine that the tangent AF meets the straight line EH in F and that the point L moves along the figure's perimeter CDE ... till finally it coincides with the point A: the last ratio of LK to AK will then be the ratio FH to AH, as may be agreed upon from inspection of the figure.[435] So call $FH = p$ and, in the present instance where LK is to AK in a last ratio, there will be $p:g = y:x$, that is, $x = gy/p$. Consequently, in place of x in the equation $cfx = cx^2 + dy + exy + y^2$ write gy/p and there will emerge $cfgy/p = cg^2y^2/p^2 + dy + egy^2/p + y^2$. Divide throughout by y and there will emerge $cfg/p = cg^2y/p^2 + d + egy/p + y$. Now, because the point L is supposed to coincide with point A, and hence KL or y to be infinitely small, that is, zero, delete terms multiplied by y and there will remain $cfg/p = d$. In consequence make $c = hk/(f-g)g$, then $d = cfg/p$ and finally $e = (k-h-d)/g$, and, with c, d and e ascertained, the equation $cfx = cx^2 + dy + exy + y^2$ will determine the conic.

If, finally,[436] three points A, B, C only are given, along with the position of two straight lines AT, CT touching the conic at two of those points, A and C, there will—as above[437]—be obtained this equation $cfx = cx^2 + dy + exy + y^2$ to the conic. Next, if the ordinate KL be supposed to be parallel to the tangent AT and conceived extended till it again meets the conic in M, while that line LM is imagined to approach the tangent AT till it finally coincides with it at A, then the last ratio of the lines KL and KM to each other will be one of equality, as inspection of the figure can establish.[438] In consequence, with KL and KM in that case proving equal to each other—in other words, with the two values of y (the positive one KL, namely, and the negative one KM) equal—, in the equation $cfx = cx^2 + dy + exy + y^2$ those terms in which y is of odd dimension (the terms $dy + exy$, that is, as against the term y^2 in which y is of even dimension) ought to vanish; for otherwise the two values of y, positive and negative, cannot

---

(436) Newton's Problem 57 on his pages 129–30 above.

(437) On supposing, as before, that the conic passes through $A(0, 0)$ and $C(f, 0)$. Newton will further restrict the conic to pass through $B(m, n)$ and be touched by the *y-axis* AT at A and by the line CT of slope $-g/f$ at C.

(438) Where N is the mid-point of LM (and hence AN is the diameter conjugate to LM), the ratio of AK to KN is given, while at A the ratio of both LK and KM to AK becomes infinite; in consequence, as LM comes to coincide with TA the 'last' ratio of both LK and KM to their difference $|LK - KM| = 2KN$ is infinite. Analytically, where $AK = x$, then $KL = -\frac{1}{2}ex + \frac{1}{2}\sqrt{[4cfx + (e^2 - 4c)x^2]}$ and $-KM = \frac{1}{2}ex + \frac{1}{2}\sqrt{[4cfx + (e^2 - 4c)x^2]}$, both of order $\sqrt{x}$ for small x. Newton's further *ad hoc* argument to show that $d = 0$ is the condition for the tangent at A to the conic to have infinite slope is excessively ponderous: the condition follows at once by requiring the defining equation $cfx/y = cx^2/y + d + ex + y$ to satisfy

$$\lim_{x,\,y \to 0} (x/y) = 0.$$

esse non possunt. Et in illo quidem casu AK infinite minor erit quam LK, hoc est x quam y, proinde et terminus exy quam terminus yy. Atcɜ adeo infinite minor existens, pro nihilo habendus erit. At terminus dy respectu termini yy, non evanescet ut oportet, sed eo major nisi d supponatur esse nihil. Delendus est itacɜ terminus dy, et sic restabit $cfx = cxx + exy + yy$ æquatio ad conicam sectionem. Concipiatur jam tangentes AT, CT sibi mutuò occurrere in T, et punctum L accedere ad punctum C donec in illud incidat. Et ultima ratio ipsius KL ad KC erit AT ad AC. KL erat y; AK, x; & AC, f; atcɜ adeo KC,

||[134] $f - x$. Dic $AT = g$ et ultima ratio y ad $f - x$ erit ea quæ est g || ad f. Æquatio $cfx = cxx + exy + yy$ subducto utrobicɜ cxx fit $cfx - cxx = exy + yy$, hoc est, $\overline{f - x}$ in $cx = y$ in $\overline{ex + y}$. Ergo est $y \cdot f - x :: cx \cdot ex + y$, adeocɜ $g \cdot f :: cx \cdot ex + y$. At puncto L incidente in C, fit y nihil. Ergo $g \cdot f :: cx \cdot ex$. Divide posteriorem rationem per x et evadet $g \cdot f :: c \cdot e$, et $\dfrac{cf}{g} = e$.[439] Quare si in æquatione $cfx = cxx + exy + yy$ scribas $\dfrac{cf}{g}$ pro e fiet $cfx = cxx + \dfrac{cf}{g}xy + yy$ æquatio ad conicam sectionem. Denicɜ ipsi KL seu AT a dato puncto B per quod Conica Sectio transire debet age parallelam BH occurrentem AC in H, et concipiendo LK accedere ad BH donec cum ea coincidat, in eo casu erit $AH = x$ et $BH = y$. Dic ergo datam $AH = m$ et datam $BH = n$ & perinde pro x et y in æquatione $cfx = cxx + \dfrac{cf}{g}xy + yy$ scribe m et n et orietur $cfm = cmm + \dfrac{cf}{g}mn + nn$.[440] Aufer utrobicɜ $cmm + \dfrac{cf}{g}mn$ et fiet $cfm - cmm - \dfrac{cf}{g}mn = nn$. Pone $f - m - \dfrac{fn}{g} = s$, et erit $csm = nn$. Divide utramcɜ partem æquationis per sm et orietur $c = \dfrac{nn}{sm}$. Invento autem c, determinata habetur æquatio ad Conicam sectionem $cfx = cxx + \dfrac{cf}{g}xy + yy$. Et inde per methodum Cartesij[441] Conica sectio datur et describi potest.

Octob: 1681.
Lect: 1.

Prob. 58.[442]

Dato globo A, positione parietis DE & centri globi B a pariete distantia BD: invenire molem globi B ea lege ut in spatijs liberis et vi gravitatis destitutis, si globus A, cujus

---

(439) In effect, Newton determines $T(0, g)$ to be in the tangent at $C(f, 0)$ to the conic $c(x - f)/y + e + y/x = 0$ by requiring that the line $(x - f)/y = -f/g$ joining T and C shall have a double meet with the conic at the latter point. Immediately $-cf/g + e = 0$.

(440) The condition, namely, for the conic to pass through $B(m, n)$.

(441) See note (433).

(442) This and the three final Problems 59–61 which follow adapt the basic results obtained earlier in Newton's arithmetical 'Prob 12' (on his pages 48–9 above) for the changes in speed attendant upon impact between perfectly elastic bodies—where, that is (see note (152))

be equal. But in that case AK will, in fact, be infinitely less than LK, that is, x infinitely less than y and accordingly the term exy infinitely less than the term y^2: hence, since it proves to be infinitely less, it will need to be considered as nothing. The term dy, however, will not, as required, vanish in comparison with the term y^2 but will indeed be greater than it unless d is taken to be zero. As a result the term dy must be deleted and there will thus remain $cfx = cx^2 + exy + y^2$ as the equation to the conic. Now conceive that the tangents AT, CT mutually intersect in T, and that the point L approaches the point C till it comes to coincide with it: the last ratio of KL to KC will then be AT to AC. Here KL was y, AK x and AC f, and hence KC is $f - x$. Call $AT = g$ and the last ratio of y to $f - x$ will be that of g to f. When cx^2 is subtracted from both sides of the equation $cfx = cx^2 + exy + y^2$, it becomes $cfx - cx^2 = exy + y^2$, that is, $(f - x)\,cx = y(ex + y)$. Therefore

$$y : (f - x) = cx : (ex + y) \quad \text{and so} \quad g : f = cx : (ex + y).$$

But as the point L coincides with C, y becomes zero, and therefore $g : f = cx : ex$. Divide the latter ratio by x and there will come to be $g : f = c : e$, and so $e = cf/g$.[439] Consequently, if in the equation $cfx = cx^2 + exy + y^2$ you write cf/g in place of e, the equation to the conic will become $cfx = cx^2 + (cf/g)\,xy + y^2$. Parallel to KL or AT, finally, from the given point B through which the conic ought to pass draw BH meeting AC in H, and on conceiving that LK approaches BH till it comes to coincide with it there will in that case be $AH = x$ and $HB = y$. So call the given length $AH = m$ and the given length $HB = n$, and in place of x and y correspondingly write m and n in the equation $cfx = cx^2 + (cf/g)\,xy + y^2$ and there will arise $cfm = cm^2 + (cf/g)\,mn + n^2$.[440] Take away $cm^2 + (cf/g)\,mn$ from either side and it will become $cfm - cm^2 - (cf/g)\,mn = n^2$. Put $f - m - fn/g = s$ and there will be $csm = n^2$. Divide each side of the equation by sm and there will ensue $c = n^2/sm$. Once c is found, however, the equation $cfx = cx^2 + (cf/g)\,xy + y^2$ to the conic is, in effect, determined. And from this by Descartes' method[441] the conic itself is given and can be described.

Problem 58[442]

Given a globe A, the position of a wall DE and the distance BD of the centre of a globe B from the wall, to find the mass of globe B satisfying the condition that, if the globe A—

conservation of total 'motion' (momentum) is assumed—to more complicated particular situations in which frictionless motion (either uniform or uniformly accelerated) occurs before and after impact. As an added twist Newton further introduces into Problems 58, 59 and 61, at right angles to the line of motion, an immovable, inelastic wall whose sole function is to reverse the direction of motion of any elastic body which collides with it. On any realistic viewpoint these problems are hopelessly idealized, but yet elegantly and ingeniously fulfil Newton's primary purpose of abstracting simple algebraic equations from a geometrical context.

‖[135] *centrum in linea BD, quæ ad parietem perpendi‖cularis est, ultra B producta consistit, uniformi cum motu versus D feratur donec is impingat in alterum quiescentem globum B; globus iste B postquam reflectitur a pariete, denuò occurrat globo A in dato puncto C.*

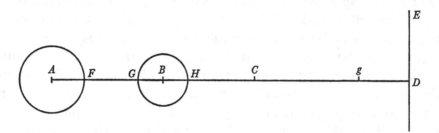

Sit globi A celeritas ante reflexionem a et erit per Prob 12 pag 48,[443] celeritas globi A post reflexionem $= \dfrac{aA - aB}{A+B}$, et celeritas globi B post reflexionem $= \dfrac{2aA}{A+B}$. Ergo celeritas globi A ad celeritatem globi B est ut $A-B$ ad $2A$. In GD cape $gD = GH$ diametro nempe globi B, et celeritates istæ erunt ut GC ad $Gg + gC$. Nam ubi Globus A impegit in globum B, punctum G quod in superficie globi B existens movetur in linea AD, perget per spatium Gg antequam globus ille B impinget in parietem, et per spatium gC postquam a pariete reflectitur;[444] hoc est per totū spatium $Gg + gC$, in eodem tempore quo globi A punctum F perget per spatium GC, eo ut globus uterꝗ rursus conveniant et in se mutuo impingant in puncto dato C. Quamobrem cum dentur intervalla BC & CD, dic $BC = m$, $BD + CD = n$, & $BG = x$, et erit $GC = m + x$ &

$$Gg + gC = GD + DC - 2gD = GB + BD + DC - 2GH = x + n - 4x$$

seu $= n - 3x$. Supra erat $A - B$ ad $2A$ ut celeritas globi A ad celeritatem globi B et celeritas globi A ad celeritatem globi B ut GC ad $Gg + gC$. Ergo cum sit

‖[136] $GC = m + x$ & $Gg + gC = n - 3x$ erit ‖ $A - B$ ad $2A$ sicut $m + x$ ad $n - 3x$. Porro globus A est ad globum B ut cubus radij ejus AF ad cubum radij alterius GB, hoc est si ponas radium AF esse s, ut s^3 ad x^3. Ergo

$$s^3 - x^3 . 2s^3 (:: A - B . 2A) :: m + x . n - 3x.$$

Et ductis extremis et medijs in se habebitur æquatio

$$s^3 n - 3s^3 x - nx^3 + 3x^4 = 2ms^3 + 2xs^3.$$

et per reductionem $3x^4 - nx^3 - 5s^3 x \genfrac{}{}{0pt}{}{+s^3 n}{-2s^3 m} = 0$. Cujus æquationis constructione dabitur globi B semidiameter x; quo dato datur etiam Globus ille. Q.E.F. Nota verò quod ubi punctum C jacet ad contrarias partes globi[445] B, debet signum quantitatis $2m$ mutari & scribi $3x^4 - nx^3 - 5s^3 x \genfrac{}{}{0pt}{}{+s^3 n}{+2s^3 m} = 0$.

whose centre is located in the line BD, perpendicular to the wall, in its extension beyond B— is borne along, in a space free (of resistance) and devoid of gravitational force, with a uniform motion in D's direction till it strikes against the other, stationary globe B, that globe B after rebounding from the wall meets the globe A a second time at the given point C.

Let the speed of globe A before impact be a, and then by Problem 12 on page 48[443] the speed of globe A after impact will be $= a(A-B)/(A+B)$ and the speed of globe B after impact $= 2aA/(A+B)$. Therefore the speed of globe A is to that of globe B as $A-B$ to $2A$. In GD take $gD = GH$, namely the diameter of globe B, and those speeds will be as GC to $Gg+gC$: for after the impact of globe A against globe B point G on the surface of globe B, moving in the line AD, proceeds through the space Gg before that globe B strikes the wall and through the space gC after it rebounds from the wall[444]—in other words, through the total space $Gg+gC$—in the same time as point F on globe A passes through the space GC, in order that both globes may again meet in mutual impact at the given point C. In consequence, since the intervals BC and CD are given, call $BC = m$, $BD+CD = n$ and $BG = x$, and there will then be $GC = m+x$, while

$$Gg+gC = (GD+CD-2gD \text{ or}) \ GB+BD+DC-2GH$$

$$= x+n-4x, \text{ that is, } n-3x.$$

Above, $A-B$ was to $2A$ as the speed of globe A to the speed of globe B, and the speed of globe A to that of globe B as GC to $Gg+gC$. Therefore, since $GC = m+x$ and $Gg+gC = n-3x$, there will be $A-B$ to $2A$ as $m+x$ to $n-3x$. Moreover, globe A is to globe B as the cube of the former's radius AF to the cube of the other's radius GB, that is, if you set the radius AF to be s, as s^3 to x^3. Therefore $s^3-x^3 : 2s^3 (= A-B : 2A) = m+x : n-3x$, and, when extremes and middles are multiplied into one another, the equation $s^3n-3s^3x-nx^3+3x^4 = 2ms^3+2xs^3$ will be obtained, and by reduction $3x^4-nx^3-5s^3x+(n-2m)\,s^3 = 0$. From the construction of this equation there will be given the radius x of the globe B; and when this is given the globe itself is given also. As was to be done. But note that when point C lies on the opposite side of globe[445] B, the sign of the quantity $2m$ should be changed and there then ought to be written

$$3x^4-nx^3-5s^3x+(n+2m)\,s^3 = 0.$$

---

(443) Compare note (152). Impact ('reflection') between perfectly elastic bodies is assumed; see previous note.

(444) This 'reflection' at the immobile wall DE serves merely to reverse the direction of motion; compare note (442).

(445) Newton first began to write 'corpor[is]' (body).

Si datus esset Globus B et quæreretur globus A ea lege ut globi duo post reflexionem convenirent in C, quæstio foret facilior. Nempe in inventa æquatione novissima supponendum esset x dari & s quæri. Qua ratione per debitam reductionem illius æquationis, translatis terminis $-5s^3x+s^3n-2s^3m$ ad æquationis partem contrariam ac divisa utraɋ parte per $5x-n+2m$, emerget $\frac{3x^4-nx^3}{5x-n+2m}=s^3$. Ubi per solam extractionem radicis cubicæ obtinebitur s.

Quod si dato Globo utroɋ quæreretur punctum C in quo post reflexionem ambo in se mutuo impingerent: eadem æquatio per debitam reductionem daret $m=\frac{1}{2}n-\frac{5}{2}x+\frac{x^4-x^3n}{2s^3}$, hoc est $BC=\frac{1}{2}Hg+\frac{1}{2}gC-\frac{B}{2A}\times\overline{HD+DC}$. Nam

‖[137] supra erat $n-3x=Gg+gC$. Unde si auferas $2x$ seu GH restabit ‖ $n-5x=Hg+gC$. Cujus dimidium est $\frac{1}{2}n-\frac{5}{2}x=\frac{1}{2}Hg+\frac{1}{2}gC$. Porrò de n seu $BD+CD$ aufer x seu BH et restabit $n-x$ seu $HD+CD$. Unde cum sit $\frac{x^3}{2s^3}=\frac{B}{2A}$ erit $\frac{x^3}{2s^3}\times\overline{n-x}$ seu $\frac{nx^3-x^4}{2s^3}=\frac{B}{2A}\times\overline{HD+CD}$. et signis mutatis $\frac{x^4-nx^3}{2s^3}=-\frac{B}{2A}\times\overline{HD+CD}$.

[1681]
Lect 2

Prob. 59.

Si globi duo A et B tenui jungantur filo PQ, et pendente globo B a globo A si dimittatur globus A, ita ut globus uterɋ simul sola gravitatis vi[446] *in eadem linea perpendiculari PQ cadere incipiat; dein globus inferior B, postquam a fundo seu plano horizontali FG sursum reflectitur, superiori decidenti globo A occurrat in puncto quodam D; ex data fili longitudine PQ et puncti illius D a fundo distantia DF, invenire altitudinem PF a qua globus superior A ad hunc effectū demitti debet.*

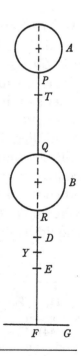

Sit fili PQ longitudo a. In perpendiculo $PQRF$ ab F sursum cape FE æqualem globi inferioris diametro QR, ita ut cum globi illius punctum infimum R incidit in fundum ad F punctum ejus supremum Q occupet locum E; sitɋ ED distantia per quam globus ille postquam a fundo reflectitur ascendendo transit antequā globo superiori decidenti occurrat in puncto D. Igitur

‖[138] ob datam ‖ puncti D a fundo distantiam DF globiɋ inferioris diametrum EF, dabitur eorum differentia DE. Sit ea $=b$. Sitɋ altitudo per quam globus ille inferior antequam impingit in fundum cadendo describit[447] RF vel $QE=x$, siquidem ea

---

(446) In other words, with a uniformly accelerated motion.

(447) A curious verb: in context we expect 'transit' and so we have translated its sense. In his 1722 revise Newton made the equivalent emendation of omitting the preposition 'per' before 'quam' in the preceding phrase.

If globe B had been given and globe A sought subject to the condition that the two globes should meet, after impact, in C, the question would become an easier one: specifically, in the equation most recently found you would need to suppose x given and s to be sought. On this account, by appropriate reduction of that equation (translation of the terms $-5s^3x + (n-2m)\,s^3$ to the opposite side of the equation, and division of each side by $5x - n + 2m$) there will emerge

$$s^3 = (3x^4 - nx^3)/(5x - n + 2m).$$

Here s will be obtained by extraction of a cube root alone.

But if, given both globes, the point C should be required in which, after rebound, each would hit against the other, the same equation would by appropriate reduction give $m = \tfrac{1}{2}n - \tfrac{5}{2}x + \tfrac{1}{2}(x^4 - nx^3)/s^3$, that is,

$$BC = \tfrac{1}{2}Hg + \tfrac{1}{2}gC - \tfrac{1}{2}(B/A)\,(HD + DC).$$

For above there was $n - 3x = Gg + gC$, and if you take $2x$ or GH from this there will remain $n - 5x = Hg + gC$, whose half is $\tfrac{1}{2}n - \tfrac{5}{2}x = \tfrac{1}{2}Hg + \tfrac{1}{2}gC$. Furthermore, from n or $BD + CD$ take away x or BH and there will remain $n - x = HD + CD$. Hence, since $\tfrac{1}{2}x^3/s^3 = \tfrac{1}{2}B/A$, there will be

$$(\tfrac{1}{2}(x^3/s^3)\,(n-x),\ \text{that is})\ \tfrac{1}{2}(nx^3 - x^4)/s^3 = \tfrac{1}{2}(B/A)\,(HD + CD),$$

and with signs changed $\tfrac{1}{2}(x^4 - nx^3)/s^3 = -\tfrac{1}{2}(B/A)\,(HD + CD)$.

Problem 59

If two globes A and B are joined by a slender thread PQ and if globe A, with globe B suspended from it, is released, so allowing each globe to begin to fall simultaneously under gravitational force alone[446] in the same vertical line PQ, while the lower globe B, after rebounding upwards from the base—the horizontal plane FG—, shall subsequently meet the upper globe A as it falls at a point D: given the length PQ of the thread and distance DF of that point D from the base, to find from these the height PF at which the upper globe A ought to be released to accomplish this.

Let the length of the thread PQ be a. In the vertical $PQRF$ upwards from F take FE equal to the diameter QR of the lower globe, so that when the lowest point R of that globe falls onto the base at the point F its uppermost point Q shall occupy the position E; then shall ED be the distance through which that globe passes in its ascent after rebounding from the base before it meets the upper globe as it falls at the point D. Consequently, because the distance DF of the point D from the base and the diameter EF of the lower globe are given, their difference DE will be given. Let it be $= b$. Let also the height RF or QE through which the lower globe passes in its fall before it strikes the base be $= x$, seeing

ignoretur. Et invento x si eidem addantur EF et PQ habebitur altitudo PF a qua globus superior ad effectum desideratum demitti debet.

Cum igitur sit $PQ=a$ & $QE=x$, erit $PE=a+x$. Aufer DE seu b et restabit $PD=a+x-b$. Est autem tempus descensus globi A ut radix spatij cadendo descripti seu $\sqrt{a+x-b}$, et tempus descensus globi alterius B ut radix spatij cadendo descripti, seu $\sqrt{x}$, et tempus ascensus ejusdem ut differentia radicis illius et radicis spatij quod cadendo tantum a Q ad D describeretur. Nam hæc differentia est ut tempus descensus a D ad E quod æquale est tempori ascensus ab E ad D. Est autem differentia illa $\sqrt{x}-\sqrt{x-b}$. Unde tempus descensus et ascensus conjunctim erit ut $2\sqrt{x}-\sqrt{x-b}$. Quamobrem cum hoc tempus æquetur tempori descensus globi superioris, erit $\sqrt{a+x-b}=2\sqrt{x}-\sqrt{x-b}$. Cujus æquationis partibus quadratis habebitur $a+x-b=5x-b-4\sqrt{xx-bx}$, seu $a=4x-4\sqrt{xx-bx}$, et ordinata æquatione $4x-a=4\sqrt{xx-bx}$. Cujus partes iterum quadrando oritur $16xx-8ax+aa=16xx-16bx$, seu $aa=8ax-16bx$. Et divisis omnibus per $8a-16b$, fiet $\dfrac{aa}{8a-16b}=x$. Fac igitur ut $8a-16b$ ad a ita a ad x, et habebitur x seu QE. Q.E.I.

|| [139]   Quod si ex dato QE quæreretur fili longitudo || PQ seu a: eadem æquatio $aa=8ax-16bx$ extrahendo affectam radicem quadraticam daret

$$a=4x-\sqrt{16xx-16bx}.^{(448)}$$

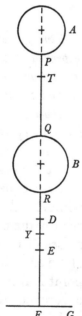

Id est si sumas QY mediam proportionalem inter QD et QE, erit $PQ=4EY$. Nam media illa proportionalis erit $\sqrt{x}\times\overline{x-b}$ seu $\sqrt{xx-bx}$ quod subductum de x seu QE relinquit EY, cujus quadruplum est $4x-4\sqrt{xx-bx}$.

Sin verò ex datis tum QE seu x tum fili longitudine PQ seu a, quæreretur punctum D in quo globus superior in inferiorem incidit; puncti illius a dato puncto E distantia DE seu b, e præcedente æquatione $aa=8ax-16bx$ eruetur transferendo aa et $16bx$ ad æquationis partes contrarias cum signis mutatis, & omnia dividendo per $16x$. Orietur enim $\dfrac{8ax-aa}{16x}=b$. Fac igitur ut $16x$, ad $8x-a$ ita a ad b et habebitur b seu DE.

Hactenus supposui globos tenui filo connexos simul dimitti. Quod si nullo connexi filo diversis temporibus dimittantur, ita ut globus superior A verbi gratia prius dimissus, descenderit per spatiũ PT antequam globus alter incipiat cadere; et ex datis distantijs PT, PQ ac DE quæratur altitudo PF a qua globus superior dimitti debet ea lege ut in inferiorem incidat ad punctũ D: sit $PQ=a$. $DE=b$. $PT=c$ & $QE=x$, et erit $PD=a+x-b$ ut

that it is not known. And, once x is found, if EF and PQ be added to it there will be obtained the height PF at which the upper globe ought to be released to achieve the result desired.

Since, then, $PQ = a$ and $QE = x$, there will be $PE = a+x$. Take away DE, that is, b, and there will be left $PD = a+x-b$. Now, however, the time of descent of globe A is as the root of the distance traversed in its fall, namely $\surd[a+x-b]$, and the time of descent of the other globe B is as the root of the distance traversed in its fall, namely $\surd x$, while the time of its ascent is as the difference of that root and the root of the distance it would traverse in falling merely from Q to D (for that difference is as the time of descent from D to E, which is equal to the time of ascent from E to D)—and that difference is $\surd x - \surd[x-b]$. Hence the combined time of descent and ascent will be as $2\surd x - \surd[x-b]$. Consequently, since this time is equal to the time of descent of the upper globe, there will be $\surd[a+x-b] = 2\surd x - \surd[x-b]$. When the sides of this equation are squared there will be had $a+x-b = 5x-b-4\surd[x^2-bx]$ or $a = 4x-4\surd[x^2-bx]$, and on ordering the equation $4x-a = 4\surd[x^2-bx]$. By again squaring the sides of this there arises $16x^2-8ax+a^2 = 16x^2-16bx$ or $a^2 = 8ax-16bx$, and after dividing through by $8a-16b$ there will come $x = a^2/(8a-16b)$. So make a to x as $8a-16b$ to a, and there will be had x, that is, QE. As was to be found.

But if, given QE, the length PQ or a of the thread should be sought, the same equation $a^2 = 8ax-16bx$ would, on extracting its 'affected' square root, yield $a = 4x-\surd[16x^2-16bx]$.[448] So that, if you take QY to be the mean proportional between QD and QE, there will be $PQ = 4EY$: for that mean proportional will be $\surd[x(x-b)]$ or $\surd[x^2-bx]$, and this, when subtracted from QE or x, will leave EY, four times which is $4x-4\surd[x^2-bx]$.

Or if, indeed, given both QE or x and the length PQ or a of the thread, there should be sought from these the point D at which the upper globe hits the lower one, the distance DE or b of that point from the given point E will be derived from the preceding equation $a^2 = 8ax-16bx$ by transposing a^2 and $16bx$ to opposite sides of the equation with their signs changed and dividing through by $16x$, when, to be sure, there will ensue $b = (8ax-a^2)/16x$. So make a to b as $16x$ to $8x-a$ and b, that is, DE, will be obtained.

Up to now I have supposed that globes linked together by a slender thread have been released simultaneously. But if no thread connects them and they are let fall at different times, so that, for example, the upper globe A (first to be released) has descended through the space PT before the other globe begins to fall, and if, given the distances PT, PQ and DE, there should be sought from them the height PF at which the upper globe ought to be released, stipulating that it shall strike the lower one at the point D: let $PQ = a$, $DE = b$, $PT = c$ and

---

(448) In line with the initial equation from which the preceding quadratic was derived, Newton correctly discards the unwanted positive sign of this square root.

supra. Et tempora quibus globus superior cadendo describat spatia PT ac TD, et globus inferior prius cadendo dein reascendendo describat summam spatiorum $QE+ED$ erunt ut $\sqrt{PT}$, $\sqrt{PD}-\sqrt{PT}$, & $2\sqrt{QE}-\sqrt{QD}$, hoc est ut $\sqrt{c}$, $\sqrt{a+x-b}-\sqrt{c}$ & $2\sqrt{x}-\sqrt{x-b}$. At ultima duo tempora, propterea quod spatia TD & $QE+ED$ simul describuntur, æqualia sunt. Ergo

$$\sqrt{a+x-b}-\sqrt{c}=2\sqrt{x}-\sqrt{x-b}.$$

‖ [140] Et ‖ partibus quadratis $a+c-2\sqrt{ca+cx-cb}=4x-4\sqrt{xx-bx}$. Pone $a+c=e$ et $a-b=f$ et erit per debitam reductionem

$$4x-e+2\sqrt{cf+cx}=4\sqrt{xx-bx}$$

et partibus quadratis

$$ee-8ex+16xx+4cf+4cx+\overline{16x-4e}\sqrt{cf+cx}=16xx-16bx.$$

Ac deletis utrobiꝗ $16xx$ et pro $ee+4cf$ scripto m nec non pro $8e-16b-4c$ scripto n, habebitur per debitam reductionem $\overline{16x-4e}\sqrt{cf+cx}=nx-m$. Et partibus quadratis

$$256cfxx+256cx^3-128cefx-128cexx+16ceex=nnxx-2mnx+mm.$$

Et ordinata æquatione $256cx^3 \begin{matrix} +256cf \\ -128ce \\ -nn \end{matrix}xx \begin{matrix} -128cef \\ +16cee \\ +2mn \end{matrix}x \begin{matrix} +6ceef \\ -mm \end{matrix}=0$. Cujus æquationis

constructione dabitur x seu QE: cui si addas datas distantias PQ et EF habebitur altitudo PF quam oportuit invenire.

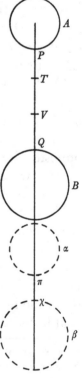

[1681]
Lect 3

Prob 60.

Si globi duo quiescentes superior A et inferior B diversis temporibus dimittantur, et globus inferior eo temporis momento cadere incipiat ubi superior cadendo jam descripsit spatium PT; invenire loca α, β quæ globi illi cadentes occupabunt ubi eorum intervallum $\pi\chi$ dato æquale est.

Cum dentur distantiæ PT, PQ & $\pi\chi$ dic primam a, secundam b, tertiam c, & pro $P\pi$ seu spatio quod globus superior antequam ‖ [141] pervenit ad locum quæsitum α cadendo describit ‖ ponatur x. Jam tempora quibus globus superior describit spatia PT, $P\pi$, $T\pi$ & inferior spatium $Q\chi$ sunt ut $\sqrt{PT}$, $\sqrt{P\pi}$, $\sqrt{P\pi}-\sqrt{PT}$, & $\sqrt{Q\chi}$. Quarum temporum posteriora duo, eo quod globi cadendo simul describant spatia $T\pi$ & $Q\chi$, sunt æqualia. Unde et $\sqrt{P\pi}-\sqrt{PT}$ æquale erit $\sqrt{Q\chi}$. Erat $P\pi=x$ & $PT=a$, et ad $P\pi$ addendo $\pi\chi$ seu c & a summa auferendo PQ seu b habebitur $Q\chi=x+c-b$. Quamobrem his substitutis fiet $\sqrt{x}-\sqrt{a}=\sqrt{x+c-b}$. Et æquationis partibus quadratis orietur

$QE = x$, and there will be $PD = a+x-b$, as above; also the times in which the upper globe in its fall shall describe the spaces PT and TD, and the lower globe first in its fall and then in its reascent describes the sum of the spaces QE and ED will be as $\sqrt{PT}$, $\sqrt{PD}-\sqrt{PT}$ and $2\sqrt{QE}-\sqrt{QD}$, that is, as $\sqrt{c}$, $\sqrt{[a+x-b]}-\sqrt{c}$, and $2\sqrt{x}-\sqrt{[x-b]}$. But the two latter times, for the reason that the spaces TD and $QE+ED$ are described simultaneously, are equal. Therefore

$$\sqrt{[a+x-b]}-\sqrt{c} = 2\sqrt{x}-\sqrt{[x-b]},$$

and with the sides squared $a+c-2\sqrt{[ca-cb+cx]} = 4x-4\sqrt{[x^2-bx]}$. Set $a+c = e$ and $a-b = f$, and by appropriate reduction there will be

$$4x-e+2\sqrt{[cf+cx]} = 4\sqrt{[x^2-bx]},$$

and with the sides squared

$$16x^2 - 8ex + e^2 + 4cf + 4cx + (16x-4e)\sqrt{[cf+cx]} = 16x^2 - 16bx.$$

There will then, with $16x^2$ deleted on each side and m written in place of e^2+4cf and also n in place of $8e-16b-4c$, be had by due reduction

$$(16x-4e)\sqrt{[cf+cx]} = nx-m,$$

and with the sides squared

$$256cfx^2 + 256cx^3 - 128cefx - 128cex^2 + 16ce^2x = n^2x^2 - 2mnx + m^2,$$

and when the equation is ordered

$$256cx^3 + (-128ce+256cf-n^2)\,x^2 + (16ce^2-128cef+2mn)\,x + 6ce^2f - m^2 = 0.$$

From the construction of this equation there will be given x, that is, QE; and if you add the given distances PQ and EF to this, there will be obtained the height PF which it was required to find.

Problem 60

If two stationary globes, the upper one A and the lower one B, are released at different times, and the lower one begins to fall at the exact instant when the upper one in its fall has already described the space PT, to find the positions α, β which those falling globes will occupy when their distance apart πχ is equal to a given magnitude.

Since the distances PT, PQ and $\pi\chi$ are given, call the first a, the second b, the third c, and for $P\pi$—the space, namely, which the upper globe describes in its fall before it reaches the required position α—let there be put x. Now the times in which the upper globe describes the spaces PT, $P\pi$, $T\pi$ and the lower one the space $Q\chi$ are as $\sqrt{PT}$, $\sqrt{P\pi}$, $\sqrt{P\pi}-\sqrt{PT}$ and $\sqrt{Q\chi}$. And of these times the two latter ones, inasmuch as the globes in their fall describe the spaces $T\pi$ and $Q\chi$ simultaneously, are equal: hence also $\sqrt{P\pi}-\sqrt{PT}$ will be equal to $\sqrt{Q\chi}$. Earlier $P\pi = x$ and $PT = a$, and by adding $\pi\chi$ or c to $P\pi$ and from the sum taking away PQ or b there will be had $Q\chi = x+c-b$. In consequence, when these values are substituted there will come to be $\sqrt{x}-\sqrt{a} = \sqrt{[x+c-b]}$. After the sides of the

$x+a-2\sqrt{ax}=x+c-b$. Ac deleto utrobiʒ x et ordinata æquatione habebitur $a+b-c=2\sqrt{ax}$. Et partibus quadratis erit quadratū de $a+b-c$ æquale $4ax$ et quadratum illud divisum per $4a$ æquale x, seu $4a$ ad $a+b-c$ sicut $a+b-c$ ad x. Ex invento autem x seu $P\pi$ datur globi superioris decidentis locus quæsitus α. Et per locorum distantiam simul datur etiam locus inferioris β.

Et hinc si punctum quæratur ubi globus superior cadendo tandem impinget in inferiorem; ponendo distantiam $\pi\chi$ nullam esse seu delendo c, dic $4a$ ad $a+b$ ut $a+b$ ad x seu $P\pi$, et punctum π erit quod quæris.

Et vicissim si detur punctum illud π vel χ in quo globus superior incidit in inferiorem et quæratur locus T quem superioris globi decidentis punctum P tunc ‖[142] occupabat cum globus inferior incipiebat cadere: ‖ quoniam est $4a$ ad $a+b$ ut $a+b$ ad x seu ductis extremis et medijs in se $4ax=aa+2ab+bb$, et per æquationis debitam ordinationem $aa=4ax-2ab-bb$; extrahe[449] radicem quadraticam et proveniet $a=2x-b-2\sqrt{xx-bx}$. Cape ergo $V\pi$ mediam proportionalem inter $P\pi$ et $Q\pi$ et versus V cape $VT=VQ$ et erit T punctum quod quæris. Nam $V\pi$ erit $=\sqrt{P\pi\times Q\pi}$ hoc est $=\sqrt{x\times\overline{x-b}}$ seu $=\sqrt{xx-bx}$: cujus duplum subductum de $2x-b$ seu de $2P\pi-PQ$ hoc est de $PQ+2Q\pi$ relinquit $PQ-2VQ$ seu $PV-VQ$ hoc est PT.

Si deniʒ globorum postquam superior incidit in inferiorem et impetu in se invicem facto inferior acceleratur, superior retardatur, desiderantur loci ubi inter cadendum distantiam datæ rectæ æqualem acquirent: Quærendus erit primo locus ubi superior impingit in inferiorem; dein ex cognitis tum magnitudinibus globorum tum eorum ubi in se[450] impingunt celeritatibus[,] inveniendæ sunt celeritates quas proxime post reflexionem habebunt, idʒ per modum Prob 12 pag 48.[451] Postea quærenda sunt loca summa ad quæ globi celeritatibus hisce si sursum ferrentur ascenderent & inde cognoscentur spatia quæ globi datis temporibus post reflexionem cadendo describent, ut et differentia spatiorum: et vicissim ex assumpta illa differentia per Analysin regredietur ad ipsa spatia cadendo descripta.

Ut si globus superior incidit in inferiorem ad punctum π et post reflexionem ‖[143] celeritas ‖ superioris deorsum tanta sit, ut si sursum esset ascendere faceret globum illum per spatium πN, et inferioris celeritas deorsum tanta esset ut, si

---

(449) Understand 'affectam' (affected): Newton intends that the (lesser) root of the preceding quadratic in a should be extracted. In the equivalent form $\sqrt{a}=\sqrt{x}-\sqrt{[x-b]}$ the ensuing equation results more straightforwardly, without need to make allowance for silent omission of the second root, by setting $c=0$ in the initial argument.

(450) Newton has here deleted 'invicem' (together), doubtless as being *de trop*.

(451) See note (152) and compare note (443).

equation are squared there will ensue $x+a-2\sqrt{[ax]} = x+c-b$, and then, with x deleted on each side and the equation ordered, there will be obtained

$$a+b-c = 2\sqrt{[ax]},$$

and, with the sides squared, the square of $a+b-c$ will be equal to $4ax$, and so that square divided by $4a$ equal to x: in other words, $4a$ is to $a+b-c$ as $a+b-c$ is to x. Once x, that is, $P\pi$, is found, however, from it is given the required position α of the upper descending globe; and by the distance of the globes there is immediately given also the position of the lower one β.

Hence, if the point be sought at which the upper globe in its fall shall at length strike the lower one, on setting the distance $\pi\chi$ to be zero—that is, deleting c—say: $4a$ is to $a+b$ as $a+b$ to (x or) $P\pi$, and the point π will be that required.

Conversely, if there be given that point π (or χ) at which the upper globe hits the lower one and the point T be sought which the point P on the upper descending globe occupied at the moment when the lower globe began to fall: since $4a$ is to $a+b$ as $a+b$ to x, and so, with extremes and middles multiplied into one another, $4ax = a^2+2ab+b^2$, and by appropriate ordering of the equation $a^2 = (4x-2b)\,a-b^2$, extract the[449] square root and there will be forthcoming $a = 2x-b-2\sqrt{[x^2-bx]}$. Take, therefore, $V\pi$ to be the mean proportional between $P\pi$ and $Q\pi$ and in the direction of V take $VT = VQ$, and then T will be the point you require: for $V\pi$ will be $= \sqrt{[P\pi \times Q\pi]}$, that is, $\sqrt{[x(x-b)]}$ or $\sqrt{[x^2-bx]}$, twice which when subtracted from $2x-b$, that is, from

$$(2P\pi-PQ \text{ or}) \quad PQ+2Q\pi \quad \text{leaves} \quad (PQ-2VQ \text{ or}) \; PV-VQ,$$

that is, PT.

Finally, if, after the impact of the upper of the globes on the lower one (when, with the transfer of momentum from one to the other, the lower one is accelerated, the upper one slowed), there are desired the positions at which, during their (subsequent) fall, they come to assume a distance apart equal to a given straight line, there will first need to be sought the place at which the upper one strikes the lower; then from the sizes of the globes and their speeds at impact[450] (both known) must be found their speeds immediately after rebound—this by the method of Problem 12 on page 48[451]; afterwards, there must be sought the highest points to which the globes would rise if carried upwards with these speeds and from this there will be known the spaces described by the globes in their fall at given times after their rebound, and so also the difference of those spaces: then, in turn, from the assumption of that difference regress will be made by analysis to the spaces themselves described in fall.

For instance, if the upper globe hits the lower one at the point π and, after rebound, the upper one's speed downwards is as much as would, if it were directed upwards, make that globe ascend through the space πN, while the lower one's speed downwards were as much as would, if directed upwards, make

sursum esset, ascendere faceret globum illum inferiorem per spatium πM: tum tempora quibus globus superior vicissim descenderet per spatia $N\pi$, NG et inferior per spatia $M\pi$, MH, forent ut $\sqrt{N\pi}$, $\sqrt{NG}$, $\sqrt{M\pi}$, $\sqrt{MH}$, adeoq̃ tempora quibus globus superior cōficeret spatium πG et inferior spatium πH forent ut $\sqrt{NG}-\sqrt{N\pi}$, ad $\sqrt{MH}-\sqrt{M\pi}$. Pone hæc tempora æqualia esse et erit $\sqrt{NG}-\sqrt{N\pi}=\sqrt{MH}-\sqrt{M\pi}$. Et insuper cum detur distantia GH pone $\pi G+GH=\pi H$. Et harum duarum æquationum reductione solvetur problema.[452] Ut si sit $M\pi=a$, $N\pi=b$, $GH=c$, $\pi G=x$: erit juxta posteriorem æquationem $x+c=\pi H$. Adde $M\pi$, fiet $MH=a+c+x$. Ad πG adde $N\pi$ et fiet $NG=b+x$. Quibus inventis, juxta priorem æquationem erit $\sqrt{b+x}-\sqrt{b}=\sqrt{a+c+x}-\sqrt{a}$. Scribatur e pro $a+c$ et $[-]\sqrt{f}$ pro $\sqrt{a}-\sqrt{b}$ et æquatio fiet $\sqrt{b+x}=\sqrt{e+x}+\sqrt{f}$. Et partibus quadratis $b+x=e+x+f+2\sqrt{ef+fx}$. seu $b-e-f=2\sqrt{ef+fx}$. Pro $b-e-f$ scribe g et fiet $g=2\sqrt{ef+fx}$ et partibus quadratis $gg=4ef+4fx$ et per reductionem $\dfrac{gg}{4f}-e=x$.

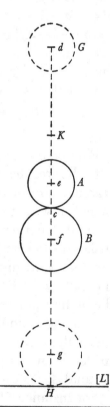

[1681]
Lect 4

Prob. 61.

‖[144] *Si duo sint globi A, B quorum superior ab altitudine G decidens, in alterum inferiorem* ‖ *B a fundo H versus superiora resilientem incidat, et hi globi ita per reflexionem ab invicem denuò recedant ut globus A vi reflexionis illius ad altitudinem priorem G redeat, idq̃ eodem tempore quo globus inferior B ad fundum H revertitur; dein globus A rursus decidat, et in globum B a fundo resilientem denuò incidat, idq̃ in eodem loco AB ubi prius in ipsum incidebat; et sic perpetuò globi ab invicem resiliant rursusq̃ ad eundem locum redeant:*[453] *ex datis globorum magnitudinibus, positione fundi et loco G a quo globus superior decidit, invenire locum ubi globi in se mutuò impingent.*

Sit e centrum Globi A, et f centrum globi B, d centrum loci G in quo globus superior in maxima est altitudine, g centrū loci globi inferioris ubi in fundum impingit, a semidiameter globi A, b semidiameter globi B, c punctum contactus globorum in se mutuo impingentium, & H punctum contactus globi inferioris et fundi. Et celeritas globi A ubi in globum B impingit ea erit quæ generatur casu globi ab altitudine de, adeoq̃ est ut $\sqrt{de}$. Hac eadem celeritate

‖[145] reflecti debet globus A versus superiora ut ad ‖ locum

---

(452) Newton's insistence on measuring the time fallen (under simple gravity) from rest by the square root of the geometrical distance fallen, rather than by that of its algebraic equivalent, makes his argument unnecessarily tedious to modern eyes.

that lower globe ascend through the space πM, then the times in which the upper globe would in turn descend through the spaces $N\pi$, NG and the lower one through the spaces $M\pi$, MH would be as $\sqrt{N\pi}$, $\sqrt{NG}$, $\sqrt{M\pi}$, $\sqrt{MH}$, and hence the times in which the upper globe would complete the space πG and the lower one the space πH would be as $\sqrt{NG} - \sqrt{N\pi}$ to $\sqrt{MH} - \sqrt{M\pi}$. Set these times equal and there will be $\sqrt{NG} - \sqrt{N\pi} = \sqrt{MH} - \sqrt{M\pi}$; in addition, since the distance GH is given, put $\pi G + GH = \pi H$. Then by the reduction of these two equations the problem will be solved.[452] So, if $M\pi = a$, $N\pi = b$, $GH = c$, $\pi G = x$, according to the latter equation there will be $\pi H = x + c$: add $M\pi$ and there will come $MH = a + c + x$. To πG add $N\pi$ and there will come to be $NG = b + x$. Once these are found, according to the former equation there will be

$$\sqrt{[b+x]} - \sqrt{b} = \sqrt{[a+c+x]} - \sqrt{a}.$$

Write e in place of $a+c$ and $\sqrt{f}$ in place of $\sqrt{b} - \sqrt{a}$, and the equation will become $\sqrt{[b+x]} = \sqrt{[e+x]} + \sqrt{f}$, and with the sides squared

$$b + x = e + x + f + 2\sqrt{[ef+fx]} \quad \text{or} \quad b - e - f = 2\sqrt{[ef+fx]}.$$

In place of $b - e - f$ write g and it will become $g = 2\sqrt{[ef+fx]}$ and, with its sides squared, $g^2 = 4ef + 4fx$, and then by reduction $x = \frac{1}{4}g^2/f - e$.

Problem 61

If there are two globes, A and B, the upper one of which as it falls down from the height G strikes the other, lower one B in its rebound aloft from the base H, and these globes so recede from each other in recoil that globe A by the force of that recoil returns to its former height G, and this in the same time as the lower globe B bounces back to the base H; and then the globe A again falls downward, once more hitting the globe B in its rebound from the base and this in the same place A/B where it hit it previously; and if the globes in this way perpetually rebound from one another and again return to the same place:[453] given the sizes of the globes, the position of the base and the place G down from which the upper globe falls, to find therefrom the place at which the globes mutually collide.

Let e be the centre of globe A and f that of globe B, d the centre of the place G in which the upper globe is at its greatest height, g the centre of the place of the lower globe when it strikes the base, a the radius of globe A, b the radius of globe B, c the point of contact of the globes as they mutually collide, and H the point of contact of the lower globe and the base. Then the speed of globe A when it strikes globe B will be that generated by the globe's fall from height de, and hence is as $\sqrt{de}$. With this same speed globe A ought to recoil upwards in

(453) A veritable perpetual motion machine! The remark is not without point, for Newton in his youth had spent a good deal of time contriving various types of 'perpetuall motion' under the attraction of gravity (by 'reflecting or refracting' its 'rays') or of an undefined 'Attraction magneticall'; see, in particular, pages 29 and 68 of his undergraduate 'Questiones quædam Philosoph[i]cæ' (ULC. Add. 3996: 102ʳ, 121ᵛ respectively).

priorem G redeat. Et globus B eadem celeritate deorsum reflecti debet qua ascenderat ut eodem tempore redeat ad fundum quo inde recesserat. Ut autem hæc duo eveniant, globorū motus inter reflectendum æquales esse debent. Motus autem ex globorum celeritatibus et magnitudinibus componuntur adeoꝗ quod fit ex globi unius mole et celeritate æquale erit ei quod fit ex globi alterius mole et celeritate.[454] Unde si factum ex unius globi mole et celeritate dividatur per molem alterius globi habebitur celeritas alterius globi proximè ante et post reflexionem seu sub fine ascensus et initio descensus. Erit igitur hæc celeritas ut $\dfrac{A\sqrt{de}}{B}$, seu cum globi sint ut cubi radiorum, ut $\dfrac{a^3\sqrt{de}}{b^3}$. Ut autem hujus celeritatis quadratum ad quadratum celeritatis globi A proximè ante reflexionem, ita altitudo ad quam globus B hac celeritate, si occursu globi A in eum decidentis non impediretur, ascenderet, ad altitudinem ed a qua globus $[A]$[455] descendit. Hoc est ut $\dfrac{A^q}{B^q}$ de ad de seu ut A^q ad B^q vel a^6 ad b^6 ita altitudo illa prior ad x si modo pro altitudine posteriore ed ponatur x.[456] Ergo hæc altitudo, ad quam nimirum B si non impediretur ascenderet, est $\dfrac{a^6}{b^6}x$. Sit ea fK. Ad fK adde fg seu $dH-de-ef-gH$, hoc est $p-x$ si modo pro dato $dH-ef-gH$ scribas p, et x pro incognito de;[457] et habebitur $Kg=\dfrac{a^6}{b^6}x+p-x$.

Unde celeritas globi B ubi decidit a K ad fundum, hoc est ubi decidit ‖[146] per spatium Kg quod centrum ‖ ejus inter decidendum describeret, erit ut $\sqrt{\dfrac{a^6}{b^6}x+p-x}$. At globus ille decidit a loco Bcf ad fundum eodem tempore quo globus superior A ascendit a loco Ace ad summam altitudinem d, aut vicissim descendit a d ad locum Ace et proinde cum gravium cadentium celeritates æqualibus temporibus æqualiter augeantur, celeritas globi B descendendo ad fundum tantum augebitur quanta est celeritas tota quam globus A eodem tempore cadendo a d ad e acquirat vel ascendendo ab e ad d amittat. Ad celeritatem itaꝗ quam globus B habet in loco Bcf adde celeritatem quam globus A habet in loco Ace et summa, quæ est ut $\sqrt{de}+\dfrac{a^3\sqrt{de}}{b^3}$ seu $\sqrt{x}+\dfrac{a^3}{b^3}\sqrt{x}$, erit celeritas globi B ubi is in fundum incidit. Proinde $\sqrt{x}+\dfrac{a^3}{b^3}\sqrt{x}$ æquabitur

---

(454) That is, the momentum of the two-body system is conserved.

(455) By a slip the manuscript has 'B', a reading passed by Whiston into his 1707 *editio princeps* but duly corrected by Newton in his 1722 revise.

(456) This last phrase is a late addition by Newton in the manuscript, made when he inverted an erroneous previous 'ita x ad hanc altitudinem' (...as x to this height).

order to return to its previous position G. And globe B ought to rebound downwards with the same speed as that with which it ascended in order to return to the base in the same time as it departed from it. For these two things to happen, however, the momenta of the globes at rebound ought to be equal. But the momenta are compounded of the speeds of the globes and their sizes, and hence the product of the mass of one globe and its speed will be equal to the mass of the other globe and its speed.[454] Hence, if the product of the mass of one globe and its speed be divided by the mass of the other globe, there will be had the speed of the second globe immediately before and after rebound, that is, at the end of ascent and start of descent. The latter speed will therefore be as $A\sqrt{de}/B$ or, since the globes are as the cubes of their radii, as $a^3\sqrt{de}/b^3$. But the square of this speed is to the square of globe A's speed immediately before rebound as the height to which globe B would, were it not prevented by its collision with globe A falling down onto it, ascend with this speed to the height ed from which globe A descends; in equivalent terms, $(A^2/B^2)\,de$ to de—or A^2 to B^2, that is, a^6 to b^6— is as the former height to x, if now for the latter height ed there be put x.[456] Consequently this height—that, I mean, to which B would ascend were it not prevented—is $(a^6/b^6)\,x$. Let it be Kf. To Kf add (fg or) $dH-de-ef-gH$, that is, $p-x$ if in place of the given magnitude $dH-ef-gH$ you write p (with x for the unknown de),[457] and there will be had $Kg = (a^6/b^6)\,x+p-x$. Whence the speed of globe B when it has fallen from K to the base, that is, fallen through the space Kg which its centre would describe during the fall, will be as

$$\sqrt{[(a^6/b^6-1)\,x+p]}.$$

But that globe falls from the position Bcf to the base in the same time as the upper globe A ascends from the position Ace to its greatest height d, or, conversely, descends from d to the position Ace, and accordingly, since the speeds of falling heavy bodies increase equally in equal times, the speed of body B in descending to the base will increase exactly as much as the total speed which globe A would in the same time acquire by falling from d to e, or lose by ascending from e to d. So to the speed which globe B has at the place Bcf add the speed which globe A has at the place Ace and the sum—which is as $(a^3/b^3+1)\sqrt{de}$, that is,

$$(a^3/b^3+1)\sqrt{x}$$

—will be the speed of the globe B when it hits the base. Consequently

$$(a^3/b^3+1)\sqrt{x} \quad \text{will equal} \quad \sqrt{[(a^6/b^6-1)\,x+p]}.$$

---

(457) This phrase, rendered superfluous by a preceding insertion (see note (456)), was nevertheless retained—unconsciously so?—by Newton in his 1722 revise and repeated in all later editions.

$\sqrt{\dfrac{a^6}{b^6}x+p-x}$. Pro $\dfrac{a^3+b^3}{b^3}$ scribe $\dfrac{r}{s}$ et pro $\dfrac{a^6-b^6}{b^6}$ $\dfrac{rt^{(458)}}{ss}$ et æquatio illa fiet

$\dfrac{r}{s}\sqrt{x}=\sqrt{\dfrac{rt}{ss}x+p}$ et partibus quadratis $\dfrac{rr}{ss}x=\dfrac{rt}{ss}x+p$. Aufer utrobiꝗ $\dfrac{rt}{ss}x$, duc

omnia in ss ac divide per $rr-rt$, et orietur $x=\dfrac{ssp}{rr-rt}$. Quæ quidem æquatio pro-

dijsset simplicior si modo assumpsissem $\dfrac{p}{s}$ pro $\dfrac{a^3+b^3}{b^3}$, prodijsset enim $\dfrac{ss}{p-t}=x$.

Unde faciendo ut sit $p-t$ ad s ut s ad x habebitur x seu $ed_{[,]}$ cui si addas ec habebitur dc et punctum c in quo globi in se mutuò impingent. Q.E.F.

Atꝗ hactenus varia evolvi Problemata. In scientijs enim addiscendis prosunt exempla magis quam præcepta. Qua de causa in his fusiùs expatiatus sum. Sed et aliqua quæ inter scribendum occurrebant immiscui sine Algebra soluta, ut insinuarem in problematîs quæ prima fronte difficilia videantur non semper ad ‖[147] Al‖gebram recurrendum esse. Sed tempus est jam æquationum resolutionem docere. Nam postquam Problema ad æquationem deductum est, radices illius æquationis quæ quantitates sunt Problemati satisfacientes extrahere oportebit.

[1681]
Lect 5

Quomodo æquationes resolvendæ sunt[459]

Postquam igitur in Quæstionis alicujus solutione ad æquationem perventum est, & æquatio illa debitè ordinata est et reducta; ubi quantitates quæ[460] pro datis habentur, revera dantur in numeris, pro ipsis substituendi sunt numeri illi in æquatione, & habebitur æquatio numeralis, cujus radix extracta tandem satisfaciet Quæstioni. Ut si in sectione anguli in quinꝗ partes æquales sumendo r pro radio circuli, q pro subtenso complementi anguli propositi ad duos rectos, et x pro subtensa complementi quintæ partis anguli illius, pervenissem ad hanc æquationem $x^5-5rrx^3+5r^4-r^4q=0$:[461] Uti in casu aliquo particulari dantur in numeris radius r & linea dati anguli complementum subtendens q; ut quod radius sit 10 & subtensa 3; substituo numeros illos in æquatione pro r et q &

---

(458) Where $t/s = a^3/b^3-1$, that is.

(459) In his library copy of Whiston's 1707 *editio princeps* Newton replaced this by the more accurate heading 'Inventio Æquationum numeralium' (The working out of numerical equations), but subsequently retained his earlier title in his 1722 revise. Over his next pages 147–210 he discusses the nature (real—positive or negative—and 'imaginary') of the roots of an algebraic equation, their delimitation in terms of the coefficients and, when possible, their accurate determination by factorizing the equation into linear, quadratic and higher-order components (both rational and 'surd') and, in the case of cubics and quartics, by sketching their full Cartesian solution. In the concluding appendix (on his pages 211–51) he elaborates, on the pattern of his earlier 'Problems for construing æquations' (ii: 450–516), several simple geometrical 'curvilinear' constructions of the roots of cubic and quartic

In place of a^3/b^3+1 write r/s and rt/s^2 in place of a^6/b^6-1,[458] and that equation will become $(r/s)\sqrt{x} = \sqrt{[(rt/s^2)\,x+p]}$, and with the sides squared

$$(r^2/s^2)\,x = (rt/s^2)\,x+p.$$

Take away $(rt/s^2)\,x$ from each side, multiply through by s^2 and divide by $r(r-t)$, and there will then ensue $x = ps^2/r(r-t)$. This equation would, of course, have proved simpler if only I had assumed p/s in place of a^3/b^3+1, for there would have resulted $x = s^2/(p-t)$. From this, by making s to x as $p-t$ to s there will be obtained x, that is, de: and if to this you add ec there will be had dc, and so the point c at which the globes mutually collide. As was to be done.

I have, thus far, developed a variety of problems. For in mastering a technical science examples are more helpful than precepts. And for this reason in discussing them I have wandered rather freely: indeed, certain things which came to me as I wrote I have also intermixed without algebraic solution in order to convey the point that in problems which at first glance seem difficult there is no need always to have recourse to algebra. But it is time now to explain the resolution of equations. For after a problem has been brought to an equation it will be necessary to extract the roots of that equation which are the quantities satisfying the problem.

HOW EQUATIONS ARE TO BE RESOLVED[459]

In the solution of some question, therefore, after an equation has been reached and it has been appropriately ordered and reduced, once the quantities which are[460] considered as given are in fact given in numerical terms, those numbers are to be substituted in their place in the equation and there will then be had a numerical equation whose root, when extracted, will at length satisfy the question. So if, in cutting an angle into five equal parts, I had, on taking r for the circle's radius, q for the complementary subtense of the angle proposed and x for the complementary subtense of a fifth of that angle, arrived at this equation $x^5-5r^2x^3+5r^4x-r^4q = 0$,[461] where the radius r and line q subtending the supplement of the given angle are, in some particular case, given in numbers— say that the radius is 10 and the subtense 3—, I substitute those numbers in the equation in place of r and q and there ensues the numerical equation

---

equations—this as an intended preliminary to an (unwritten) section on the approximate solution of general numerical equations (compare note (170) above).

(460) In his library copy of the 1707 edition Newton here inserted the phrase 'per species designantur &' (denoted by variables and) and subsequently incorporated it in his 1722 revise.

(461) See Newton's (geometrical) Problem 15 on his page 77 above, and compare I: 476. The five roots will be $x = 2r\cos\frac{1}{10}(\cos^{-1}[q/2r]+2k\pi)$, $k = 0, 1, 2, 3, 4$.

provenit æquatio numeralis $x^5 - 500x^3 + 50\,000x - 30\,000 = 0$, cujus radix tandem extracta erit x seu linea complementum quintæ partis anguli illius dati subtendens.

De natura radicum æquationis. ‖[148]

Radix vero numerus est qui si in æquatione pro litera vel specie radicem signifi‖cante substituatur, efficiet omnes terminos evanescere. Sic æquationis $x^4 - x^3 - 19xx + 49x - 30 = 0$, unitas est radix quoniam scripta pro x producit $1 - 1 - 19 + 49 - 30 = 0$ hoc est nihil. Sed æquationis ejusdem plures esse possunt radices. Nam si in hac eadem æquatione $x^4 - x^3 - 19xx + 49x - 30 = 0$, pro x scribas numerum 2 & pro potestatibus x similes potestates numeri 2, producetur $16 - 8 - 76 + 98 - 30$, hoc est nihil. Atcp ita si pro x scribas numerum 3 vel numerum negativum -5 utrocp casu producetur nihil, terminis affirmativis et negativis in hisce quatuor casibus se mutuo destruentibus.[462] Proinde cum numerorum 1, 2, 3 & -5 quilibet scriptus in æquatione pro x impleat conditionem ipsius x efficiendo ut termini omnes æquationis conjunctim æquentur nihilo, erit quilibet eorum radix æquationis.

Et ne mireris eandem æquationem habere posse plures radices, sciendum est plures esse posse solutiones ejusdem Problematis. Ut si circulorum duorum datorum quæreretur intersectio: duæ[463] sunt eorum intersectiones, atcp adeo quæstio admittit duo responsa; & perinde æquatio intersectionem determinans habebit duas radices quibus intersectionem utramcp determinet, si modo nihil in datis sit quo responsum ad unam intersectionem

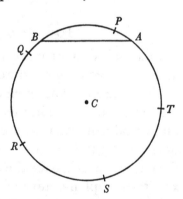

determinetur. Sic et si arcûs APB pars quinta AP invenienda esset, quamvis animum forte advertas tantum ad arcum APB, tamen æquatio qua quæstio solvetur determinabit quintam partem arcuum omnium qui terminantur ad puncta A et B: nempe quintam partem arcuum ASB, $APBSAPB$, $ASBPASB$ & $APBSAPBSAPB$, æque ac quintam

‖[149] ‖ partem arcus APB: quæ quintæ partes si dividas totam circumferentiam in æquales quincp partes PQ, QR, RS, ST, TP, erunt AT, AQ, ATS, AQR.[464] Quoniam igitur quærendo quintas partes arcuum quos recta AB subtendit, ad casus omnes determinandos circumferentia tota secari debet in quincp punctis P, Q, R, S, T, ideo æquatio ad omnes casus determinandos habebit radices quincp. Nam quintæ partes horum omnium arcuum pendent ab ijsdem datis et per ejusdem generis calculum inveniuntur ita ut in eandem semper æquationem incideris sive quæras quintam partem Arcus APB sive quintam

---

(462) In fact the quartic factorizes as $(x-1)(x-2)(x-3)(x+5) = 0$.

(463) Understand 'reales' (real). We might nowadays prefer also to include the pair of 'ideal' points $(\infty, \pm i\infty)$ in which they also intersect in the complex plane.

$x^5 - 500x^3 + 50\,000x - 30\,000 = 0$; and when its root x is at length extracted, that will be the length subtending the supplement of a fifth part of the angle given.

A root is, of course, a number which will, if it be substituted in the equation in place of the letter or variable denoting the root, make all its terms vanish. Thus in the equation $x^4 - x^3 - 19x^2 + 49x - 30 = 0$ unity is a root since, when it is written for x, there results $1 - 1 - 19 + 49 - 30 = 0$, that is, nothing. The same equation can, however, have several roots. For if in this same equation The nature of the roots of an equation

$$x^4 - x^3 - 19x^2 + 49x - 30 = 0$$

in place of x you write the number 2 (and in place of powers of x corresponding powers of the number 2), there will be produced $16 - 8 - 76 + 98 - 30$, that is, nothing. And likewise, if in place of x you write the number 3 or the negative one -5, in either case nothing will be produced, with positive and negative terms in these four cases mutually destroying one another.[462] Consequently, since any one of the numbers 1, 2, 3 and -5 when written in the equation in x's place fills x's stipulated rôle, making all the terms in the equation jointly equal to nothing, any one of these will be a root of the equation.

Lest you be astonished that the same equation can have several roots, you should be aware that there can be several solutions to the same problem. If, for instance, an intersection of two given circles be sought, they have two[463] intersections and hence the question admits of two answers; correspondingly, an equation determining the intersection will possess two roots, by which it shall determine the pair of intersections provided there be nothing in the given circumstances restricting the answer to a single intersection. So also, if the fifth part $\overset{\frown}{AP}$ of an arc APB were to be ascertained, even though you perhaps direct your attention only to the arc APB, the equation by which the question will be solved will nevertheless determine a fifth part of all arcs terminating at the points A and B: in other words, a fifth part of the arcs ASB, $APBSAPB$, $ASBPASB$ and $APBSAPBSAPB$ as well as a fifth of the arc APB. If you divide the total circumference into five equal parts $\overset{\frown}{PQ}$, $\overset{\frown}{QR}$, $\overset{\frown}{RS}$, $\overset{\frown}{ST}$, $\overset{\frown}{TP}$, these will be $\overset{\frown}{AT}$, $\overset{\frown}{AQ}$, $\overset{\frown}{ATS}$, $\overset{\frown}{AQR}$.[464] Seeing therefore that, in quest of the fifth parts of the arcs subtended by the line AB, the total circumference must, to determine all the cases, be cut at the five points P, Q, R, S, T, the equation to determine all the cases will in consequence have five roots. For the fifth parts of all these arcs are dependent on the same given conditions and are found by a computation of the same type, and as a result you will invariably come upon the same equation whether you seek a

---

(464) Where the circle's radius $AC = r$ and $AB = q$, the quinquisecting chords AP, AQ, AR, AS, AT are the roots $x = 2r\sin\frac{1}{10}(\sin^{-1}[q/2r] + 2k\pi)$, $k = 0, 1, 2, 3, 4$, of the quintic

$$x^5 - 5r^2x^3 + 5r^4x - r^4q = 0;$$

compare note (461) above, and also **I**: 478.

partem arcus *ASB* sive alterius cujusvis ex arcubus quintam partem. Unde si æquatio qua quinta pars arcus *APB* determinatur non haberet plures radices quam unam, dum quærendo quintam partem arcus *ASB* incidimus in eandem illam æquationem, sequeretur majorem hunc arcum habere eandem quintam partem cum priore qui minor est, eò quod subtensa ejus per eandem æquationis radicem exprimitur. In omni igitur Problemate necesse est æquationem qua respondetur tot habere radices, quot sunt quæsitæ quantitatis[465] casus diversi ab ijsdem datis pendentes et eadem argumentandi ratione determinandi.

    Potest[466] vero æquatio tot habere radices quot sunt dimensiones ejus, et non plures. Sic æquatio $x^4 - x^3 - 19xx + 49x - 30 = 0$ quatuor habet radices 1, 2, 3 & -5; non autem plures. Nam quilibet ex his numeris scriptus in æquatione pro ‖ [150] x efficiet terminos ‖ omnes se mutuò destruere ut dictum est; præter hos verò nullus est numerus cujus substitutione hoc eveniet. Cæterum numerus & natura radicum ex generatione æquationis optimè intelligetur. Ut si scire vellemus quomodo generetur æquatio cujus radices sint 1, 2, 3 & -5, supponendum erit x ambigue significare numeros illos, seu esse $x = 1$, $x = 2$, $x = 3$ & $x = -5$, vel quod perinde est, $x - 1 = 0$, $x - 2 = 0$, $x - 3 = 0$, & $x + 5 = 0$: et multiplicando hæc in se, prodibit multiplicatione $x - 1$ in $x - 2$ hæc æquatio $xx - 3x + 2 = 0$ quæ duarum est dimensionum ac duas habet radices 1 et 2. Et hujus multiplicatione in $x - 3$ prodibit $x^3 - 6xx + 11x - 6 = 0$ æquatio trium dimensionum totidemcp radicum, quæ iterum multiplicata per $x + 5$ fit $x^4 - x^3 - 19xx + 49x - 30 = 0$, ut supra. Cum igitur hæc æquatio generetur ex quatuor factoribus $x - 1$, $x - 2$, $x - 3$ & $x + 5$ in se continuo ductis, ubi factorum aliquis nihil est[,] quod sub omnibus fit nihil erit; ubi vero horum nullus nihil est, quod sub omnibus continetur nihil esse non potest. Hoc est, non potest $x^4 - x^3 - 19xx + 49x - 30$ esse nihilo æquale ut oportet, nisi his quatuor casibus ubi est $x - 1 = 0$ vel $x - 2 = 0$ vel $x - 3 = 0$ vel denicp $x + 5 = 0$, proinde soli numeri 1, 2, 3 & -5 valere possunt x seu radices esse æquationis. Et simile est ratiocinium de omnibus æquationibus. Nam tali multiplicatione imaginari possumus omnes generari, quamvis factores ‖ [151] ab invicem secernere solet ‖ esse difficillimum, & id ipsum est quod æquationem resolvere et radices extrahere. Habitis enim radicibus habentur factores.

    Radices vero sunt duplices, affirmativæ ut in allato exemplo 1, 2, & 3, &

---

    (465) Newton first repeated 'Problematis' (problem).

    (466) As a possibility, of course, when (as here) real roots are understood, and not of necessity. In this and following paragraphs Newton adheres closely to the sequence of the 'Caput Secundum. De natura æquationum, respectu radicum' of Kinckhuysen's *Algebra* (II: 320–3). The history of contemporary and succeeding attempts to prove this 'fundamental theorem of algebra'—which Newton is here content baldly to assert—is sketched by J. Tropfke in his *Geschichte der Elementar-Mathematik*, **3** (Berlin/Leipzig, ₂1922): 94–8. Gauss in his *Demonstratio nova theorematis omnem functionem algebraicam rationalem integram unius variabilis in factores reales primi vel secundi ordinis resolvi posse* (Helmstedt, 1799) [= *Werke*, **3** (Göttingen, 1876):

fifth of the arc *APB*, or a fifth of the arc *ASB*, or a fifth part of any other of the arcs. Hence, should the equation by which a fifth of the arc *APB* is determined not have more roots than one, while on seeking a fifth of the arc *ASB* we come upon that self-same equation, it would follow that this (larger) arc has the same fifth part as the previous (smaller) one simply because its subtense is expressed by the identical root of the equation. In every problem, therefore, it is necessary that the equation which leads to its answer shall have as many roots as the quantity sought[465] possesses distinct cases dependent on the same given conditions and requiring to be determined by the same manner of reasoning.

An equation can,[466] in truth, have as many roots as it possesses dimensions, and no more. Thus the equation $x^4 - x^3 - 19x^2 + 49x - 30 = 0$ has the four roots 1, 2, 3 and -5, but no more: for, as has been said, any one of these numbers when written in *x*'s place in the equation will make all the terms mutually destroy one another; while, with their exception, there is no number whose substitution will bring this about. For the rest, the number and nature of the roots will best be understood from the generation of the equation. Should we, for instance, wish to know how the equation whose roots are 1, 2, 3 and -5 is generated, we will need to suppose that *x* denotes those numbers ambiguously—that is, that $x = 1$, $x = 2$, $x = 3$ and $x = -5$, or, equivalently, that $x - 1 = 0$, $x - 2 = 0$, $x - 3 = 0$ and $x + 5 = 0$. Then, on multiplying these together, from the multiplication of $x - 1$ by $x - 2$ there will result this equation, $x^2 - 3x + 2 = 0$, which is of two dimensions and has the two roots 1 and 2; and from the multiplication of this by $x - 3$ there will result $x^3 - 6x^2 + 11x - 6 = 0$, an equation of three dimensions and having that many roots; and when this is once more multiplied by $x + 5$ there comes $x^4 - x^3 - 19x^2 + 49x - 30 = 0$, as above. Since, therefore, this equation is generated from the four factors $x - 1$, $x - 2$, $x - 3$ and $x + 5$ continually multiplied one into another, when any one of the factors is zero, the total product will be zero; while, indeed, when none of them is zero, the total content cannot be zero. In other words, $x^4 - x^3 - 19x^2 + 49x - 30$ cannot, as required, be equal to nothing except in the present four cases where $x - 1 = 0$ or $x - 2 = 0$ or $x - 3 = 0$, or, finally, $x + 5 = 0$, and consequently the numbers 1, 2, 3 and -5 can be values of *x*, that is, roots of the equation. A similar argument holds for all equations: for we may conceive that they are all generated by such a multiplication, though to separate factors one from another is usually very difficult, being the same thing as to resolve an equation and to extract its roots—of course, when the roots are had the factors are obtained.

Roots, however, are of two groups: positive (as 1, 2 and 3 in the example

---

1–30] was the first to give a fully adequate proof that every algebraic equation has one root in the complex plane. For Newton's youthful statement see 1: 520.

negativæ ut −5. Ex his vero aliquæ non raro evadunt impossibiles. Sic æquationis $xx - 2ax + bb = 0$ radices duæ quæ sunt $a + \sqrt{aa - bb}$ & $a - \sqrt{aa - bb}$ reales quidem sunt ubi aa majus est quam bb,[467] at ubi aa minus est quam bb evadunt impossibiles eò quod $aa - bb$ tunc evadet negativa quantitas, et negativæ quantitatis radix quadratica est impossibilis. Omnis enim radix possibilis sive affirmativa sit, sive negativa, si per seipsam multiplicetur, producet quadratum affirmativum: proinde impossibilis erit quæ quadratum negativum producere debet. Eodem argumento colligitur æquationem $x^3 - 4xx + [7]x - 6 = 0$[468] unam quidem realem radicem habere quæ est 2, duas vero impossibiles, $1 + \sqrt{-2}$ & $1 - \sqrt{-2}$. Nam quælibet ex his 2, $1 + \sqrt{-2}$, $1 - \sqrt{-2}$ scripta in æquatione pro x efficiet omnes ejus terminos se mutuo destruere: sunt vero $1 + \sqrt{-2}$ & $1 - \sqrt{-2}$ numeri impossibiles, eo quod extractionem radicis quadraticæ ex numero negativo −2 præsupponant.

Æquationum verò radices sæpe impossibiles esse æquum est ne casus problematum qui sæpe impossibiles sunt exhibeant possibiles. Ut si rectæ et circuli ‖[152] intersectio determinanda esset, et pro circuli radio et ‖ rectæ a centro ejus distantia ponantur literæ[469] duæ; ubi æquatio intersectionem definiens habetur si pro litera designante distantiam rectæ a centro ponatur numerus minor radio intersectio possibilis erit; sin major fiet impossibilis; et æquationis radices duæ quæ intersectiones duas determinant, debent esse perinde possibiles vel impossibiles ut rem ipsam verè exprimant. Atcg ita si circulus *CDEF* et Ellipsis *ACBF* se mutuo secent in punctis *C*, *D*, *E*, *F*, et ad rectam aliquam positione datam *AB*, demittantur perpendicula *CG*, *DH*, *EI*, *FK*, et quærendo longitudinem alicujus e perpendiculis, perveniatur tandem ad æquationem, æquatio illa ubi circulus secat Ellipsin in quatuor punctis habebit quatuor radices reales quæ erunt quatuor illa perpendicula. Quod si circuli radius manente centro ejus minuatur donec punctis *E* et *F* coalescentibus circulus tandem tangat Ellipsin, ex radicibus duæ illæ quæ perpendicula *EI* et *FK* jam coincidentia exprimunt evadent æquales. Et si circulus adhuc minuatur ut Ellipsin in puncto *EF* ne quidem tangat sed secet tantum in alteris duobus punctis *C*, *D*, tunc ex quatuor radicibus duæ illæ quæ perpendicula *EI FK* jam facta impossibilia exprimebant, fient una cum perpendiculis illis impossibiles. Et hoc modo in omnibus æquationibus augendo vel minuendo terminos earum, ex inæqualibus

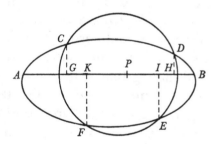

---

(467) Or, of course, when a^2 is equal to b^2.

(468) The manuscript reads ' $+4x$ ' for the third term, a slip passed by Whiston into his 1707 *editio princeps* but caught by Newton in his 1722 revise.

adduced) and negative (as -5 in it). Of these, indeed, several not infrequently prove to be impossible. Thus in the equation $x^2 - 2ax + b^2 = 0$ the two roots, namely $a + \sqrt{[a^2 - b^2]}$ and $a - \sqrt{[a^2 - b^2]}$, are assuredly real when a^2 is greater than b^2,[467] but when a^2 is less than b^2 they prove to be impossible ones, because $a^2 - b^2$ will then prove to be a negative quantity and the square root of a negative quantity is impossible. For every possible root, be it positive or negative, if it be multiplied by itself will produce a positive square; and consequently one which has to produce a negative square will be impossible. By the same reasoning it is gathered that the equation $x^3 - 4x^2 + 7x - 6 = 0$[468] has a single real root, namely 2, but two impossible ones, $1 + \sqrt{-2}$ and $1 - \sqrt{-2}$: for, when any of these, 2, $1 + \sqrt{-2}$ or $1 - \sqrt{-2}$, is written in the equation in x's place, it will make all its terms mutually destroy one another. (The numbers $1 + \sqrt{-2}$ and $1 - \sqrt{-2}$ are, of course, impossible inasmuch as they presuppose the extraction of the square root of the negative number 2.)

It is fair, however, that the roots of equations are often impossible; otherwise they would display as possible cases of problems which are often impossible. For instance, were the intersection of a straight line and a circle to be determined, if a pair of letters[469] are put for the circle's radius and the distance of the line from its centre, then, once an equation defining the intersection is obtained, the intersection will be possible if in place of the letter denoting the distance of the straight line from the centre be put a number less than the radius; but if it be greater, the intersection will become impossible: correspondingly, the two roots of the equation which determine the two intersections will need to be similarly possible or impossible in order to be a true expression of the situation. Likewise, if the circle *CDEF* and ellipse *ACBF* mutually cut one another in the points *C, D, E, F* and to some straight line *AB* given in position there be let fall the perpendiculars *CG, DH, EI, FK*, and if by seeking the length of any of the perpendiculars an equation be at length attained, then where the circle cuts the ellipse in four points that equation will have four real roots: namely, those four perpendiculars. But if, as its centre stays put, the circle's radius should diminish till at last, with points *E* and *F* coalescing, the circle comes to touch the ellipse, then those two of the roots which express the now coincident perpendiculars *EI* and *FK* will come to be equal. And if the circle diminishes still further so as no longer even to touch the ellipse in the point *E/F*, but shall now cut it merely in the two other points *C, D*, then those two of the four roots which expressed the perpendiculars *EI* and *FK* now made impossible will become impossible along with those perpendiculars. And in all equations, by increasing or diminishing their terms in this way, two of the unequal roots will usually come first to be

---

(469) That is, algebraic unknowns.

‖ [153] radicibus duæ primo æquales deinde impossibiles evadere solent. Et ‖ inde[470] fit quod radicum impossibilium numerus semper sit par.

[1681] Sunt tamen radices æquationum aliquando possibiles ubi schema impossibiles
Lect 6 exhibet. Sed hoc fit ob limitationem aliquam in schemate quod ad æquationem nil spectat.[471] Ut si in semicirculo *ADB* datis diametro *AB* et linea inscripta *AD* demissoq̃ perpendiculo *DC*, quærerem diametri seg- mentum *AC* foret $\frac{AD^q}{AB} = AC$. Et per hanc æquationem *AC* realis exhibetur quantitas ubi linea inscripta *AD* major est quam dia-

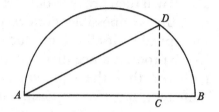

meter *AB*, per schema vero *AC* tunc evadit impossibilis. Nimirum in schemate linea *AD* supponitur inscribi in circulo atq̃ adeo diametro circuli major esse non potest, in æquatione vero nihil est quod a conditione illa pendeat. Ex hac sola linearum conditione colligitur æquatio, quod sint *AB*, *AD* et *AC* continuè proportionales.[472] Et quoniam æquatio non complectitur omnes conditiones schematis[,] non necesse est ut omnium conditionum teneatur limitibus. Quic- quid amplius est in schemate quam in æquatione potest illud limitibus arctare, hanc non item. Qua de causa ubi æquationes sunt imparium dimensionum adeoq̃ radices omnes impossibiles habere non possunt; schemata quantitatibus

‖ [154] a quibus ‖ radices omnes pendent sæpe limites imponunt quos transgredi servatis schematum conditionibus impossibile est.

Ex radicibus verò quæ reales sunt affirmativæ et negativæ ad plagas oppositas solent tendere. Sic in schemate penultimo[473] quærendo perpendiculum *CG* incidetur in æquationem cujus duæ erunt affirmativæ radices *CG* ac *DH* a punctis *C* et *D* tendentes versus unam plagam, et duæ negativæ *EI* et *FK* tendentes a punctis *E* et *F* versus plagam oppositam. Aut si in linea *AB* ad quam per- pendicula dimittuntur detur aliquod punctum *P* et pars ejus *PG* a puncto illo dato ad per-

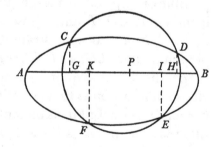

pendiculorum aliquod *CG* extendens quæratur[,] incidemus in æquationem quatuor radicum *PG*, *PH*, *PI*, *PK* quarum quæsita *PG* et quæ a puncto *P* ad

---

(470) We would now rather choose to say that this happens because complex roots of an algebraic equation (with real coefficients, it is understood) must occur in conjugate pairs.

(471) There is some confusion here between the real geometrical plane in which Newton's 'scheme' is drawn, and the related complex plane in which certain functions of the points may continue to have a real value. In the first volume (1: 266, note (65)) we picked up an error in Newton's discussion of extreme values of curvature in the parabola $rx = y(r-y)$, which

equal, and then impossible. In consequence of this[470] the number of impossible roots is always even.

On occasion, however, the roots of an equation are possible when the figure exhibits them as impossible. But this happens because of some limitation in the figure which does not affect the equation.[471] In the semicircle ADB, for instance, were I, given its diameter AB and the inscribed line AD, and with the perpendicular DC let fall, to seek the segment AC of the diameter, there would be $AC = AD^2/AB$; and by this equation AC is exhibited to be a real quantity when the inscribed line AD is greater than the diameter AB, whereas by the figure AC then proves to be impossible—to be sure, in the figure the line AD is supposed to be inscribed in a circle and hence cannot be greater than the circle's diameter, but in the equation there is nothing which may depend on that condition. The equation is gathered from the sole circumstance of the lines that AB, AD and AC are in continued proportion.[472] And because the equation does not embrace all the circumstances of the figure, it is not necessary that it be restrained by the confines of all those circumstances. Anything the figure possesses over and above what is in the equation can restrict the former by its confines, but not likewise the latter. For this reason, when equations are of odd dimension and hence cannot have all their roots impossible, figures often set limits to the quantities on which all the roots depend, and to transgress these while retaining the figures' circumstances is impossible.

Of roots which are real, however, positives and negatives customarily tend in opposite directions. Thus in the last figure but one[473] you will, on seeking the perpendicular CG, meet with an equation whose two positive roots will be CG and DH tending in one direction from the points C and D, and two negative ones EI and FK tending from the points E and F in the opposite direction. Or if there be given some point P in the line AB to which the perpendiculars are let fall and its section PG extending from that given point to some one, CG, of the perpendiculars be sought, we shall meet with an equation having four roots, and

---

derived from his failure to appreciate that complex points $(x > \frac{1}{4}r)$ on the parabola have real corresponding radii of curvature. In his following example, similarly, the distance

$$\sqrt{[x^2+y^2]} = \sqrt{[ax]}$$

from $A(0, 0)$ of any point $D(x, y)$ on the circle through A and $B(a, 0)$ continues to be real even when $x > a$ and the points $D(x, \pm i\sqrt{[x(x-a)]})$ are complex.

(472) Namely, in the terms of the preceding note, that

$$(AD^2 \text{ or}) \, x^2+y^2 = (AB \times AC \text{ or}) \, ax,$$

without restriction on the size of x.

(473) Here repeated for convenience.

easdem partes cum *PG* tendunt (ut *PK*) affirmativæ erunt, quæ vero tendunt ad partes contrarias (ut *PH, PI*) negativæ.

Ubi æquationis radices nullæ impossibiles sunt, numerus radicum affirmativarum & negativarum ex signis terminorum æquationis cognosci potest. Tot enim sunt radices affirmativæ quot signorum in continua serie mutationes de $+$ in $-$ & $-$ in $+$; cæteræ negativæ sunt.[474] Ut in æquatione

$$x^4 - x^3 - 19xx + 49x - 30 = 0,$$

ubi terminorum signa se sequuntur hoc ordine $+ - - + -_{[,]}$ variationes secundi $-$ a primo $+$, quarti $+$ a tertio $-$ & quinti $-$ a quarto $+$, indicant tres affirmativas esse radices adeoq quartam negativam esse. At ubi radices aliquæ impossibiles sunt regula non valet nisi quatenus ‖ impossibiles illæ quæ nec negativæ sunt nec affirmativæ pro ambiguis habeantur.[475] Sic in æquatione $x^3 + pxx + 3ppx - q = 0$ signa indicant unam esse affirmativam radicem & duas negativas. Finge $x = 2p$ seu $x - 2p = 0$, et multiplica æquationem priorem per hanc $x - 2p = 0$ et una adhuc radix affirmativa addatur prioribus et prodibit hæc æquatio $x^4 - px^3 + ppxx \genfrac{}{}{0pt}{}{-2p^3}{-q} x + 2pq = 0$, quæ habere deberet duas affirmativas ac duas negativas radices, habet tamen, si mutationem signorum spectes, affirmativas quatuor. Sunt ergo duæ impossibiles quæ pro ambiguitate sua priori casu negativæ, posteriori affirmativæ esse videntur.

Verum quot radices impossibiles sunt cognosci fere potest per hanc regulam. Constitue seriem fractionum quorum denominatores sunt numeri in hac progressione 1, 2, 3, 4, 5 &c pergendo ad numerum usq qui est dimensionum æquationis; numeratores verò eadem series numerorum in ordine contrario. Divide unamquamq fractionem posteriorem per priorem. Fractiones prodeuntes colloca super terminis medijs æquationis. Et sub quolibet mediorum terminorum, si quadratum ejus ductum in fractionem capiti imminentem sit majus quam rectangulum terminorum utrinq consistentium,[476] ‖ colloca signum $+$; sin minus signum $-$.[477] Sub primo vero et ultimo termino colloca signum $+$. et tot erunt radices impossibiles quot sunt in subscriptorum signorum serie mu-

‖ [155]

‖ [156]

---

(474) Newton takes over this (unproved) statement of Descartes' rule of signs from his youthful observation (1: 520) that 'If there bee none imaginary then the number of true and false rootes may bee knowne by ye signes of ye Equations termes: Namely there are so many true rootes as variations of signes & soe many false ones as successions of y^e same signes.'

(475) 'But if any roots bee imaginary, this rule...admitts of exception' (1: 520). Newton's following counter-instance is but slightly modified from his previous one, where the root $x = -2p$ is added to those of the cubic $x^3 - px^2 + 3p^2x - q^3 = 0$. We have seen (11: 322, note (54)) that Descartes himself was careful to say that his rule can set only upper bounds to the number of positive and negative real roots.

(476) Newton has here deleted 'vel si termini utrinq consistentes signis diversis afficiuntur'

of these the one required, *PG*, and those (here *PK*) tending the same way from point *P* as *PG* will be positive, but those (here *PH*, *PI*) tending the opposite way will be negative.

When none of the roots of an equation are impossible, the number of positive and of negative roots can be ascertained from the signs of the terms in the equation: for there are as many positive roots as changes of signs from + to − and from − to + in continued sequence, and the rest are negative ones.[474] So in the equation $x^4 - x^3 - 19x^2 + 49x - 30 = 0$, where the signs of the terms follow on in this order $+ - - + -$, the variations in the second (−) from the first (+), in the fourth (+) from the third (−), and in the fifth (−) from the fourth (+) indicate the presence of three positive roots and hence that the fourth is negative. But when certain roots are impossible, the rule is not valid except in the restricted form where the impossible ones—being neither negative nor positive—are treated as ambiguous.[475] Thus in the equation

$$x^3 + px^2 + 3p^2 x - q = 0$$

the signs indicate that one root is positive and that two are negative. Imagine $x = 2p$, that is, $x - 2p = 0$, and multiply the previous equation by this one, $x - 2p = 0$, so adding one further (positive) root to the previous ones: there will then result this equation $x^4 - px^3 + p^2 x^2 - (2p^3 + q)\,x + 2pq = 0$, which ought to have two positive roots and two negative ones, but yet, if you have regard for the change in signs, has four positive ones. Two, therefore, are impossible, these appearing—in conformity with their ambiguity—in the first case to be negative, but in the latter positive.

Yet the number of impossible roots present can almost always be found out by means of this rule. Set out a series of fractions whose denominators are numbers in this progression   1, 2, 3, 4, 5, …, proceeding as far as the number which is that of the equation's dimension, while the numerators are the same series of numbers in converse sequence. Divide each subsequent fraction by the previous one, place the resulting fractions above the middle terms of the equation, and, if the square of any middle term multiplied by the fraction stationed over its head be greater than the product of the terms set on its either side,[476] beneath it place the sign +, but otherwise the sign − :[477] beneath the first and last terms, however, place the sign +. There will then be as many impossible roots as changes from + to − and from − to + in the sequence of signs written

---

(or if the coefficients of the terms set on its either side are opposite in sign)—an evident particular case of the preceding condition.

   (477) In other words, Newton requires that over the term $a_i x^{n-i}$ of the given equation we set the fraction $k_i = \dfrac{n-i}{i+1} \Big/ \dfrac{n-i+1}{i} = \dfrac{i(n-i)}{(i+1)(n-i+1)}$, and then below it the sign + if $k_i a_i^2 > a_{i-1} \cdot a_{i+1}$, otherwise −.

tationes de $+$ in $-$ & $-$ in $+$.[478] Ut si habeatur æquatio $x^3 + pxx + 3ppx - q = 0$: divido seriei hujus $\frac{3}{1}. \frac{2}{2}. \frac{1}{3}$ fractionem secundam $\frac{2}{2}$ per primam $\frac{3}{1}$, et tertiam $\frac{1}{3}$ per secundam $\frac{2}{2}$, et fractiones prodeuntes $\frac{1}{3}$ & $\frac{1}{3}$ colloco super medijs terminis

$$\overset{\frac{1}{3}. \quad \frac{1}{3}.}{\underset{+ \quad - \quad + \quad +}{x^3 + pxx + 3ppx - q = 0.}}$$

æquationis ut sequitur. Dein quoniam quadratum

secundi termini pxx ductum in imminentem fractionem $\frac{1}{3}$, nimirum $\frac{ppx^4}{3}$ minus est quam primi termini x^3 et tertij $3ppx$ rectangulum $3ppx^4$, sub termino pxx colloco signum $-$. At quia tertij termini $3ppx$ quadratum $9p^4xx$ ductum in imminentem fractionem $\frac{1}{3}$, majus est quam nihil atcg adeo multo majus quam secundi termini pxx et quarti $-q$ rectangulum negativum, colloco sub tertio illo termino signum $+$. Dein sub primo termino x^3 et ultimo $-q$ colloco signa $+$. Et signorum subscriptorum quæ in hac sunt serie $+ - + +$ mutationes duæ,[479] una de $+$ in $-$, alia de $-$ in $+$ indicant duas esse radices impossibiles. Sic et æquatio

$$x^3 - 4xx + 4x - 6 = 0^{[480]}$$

$$\overset{\frac{1}{3}. \quad \frac{1}{3}.}{\underset{+ \quad + \quad - \quad +}{x^3 - 4xx + 4x - 6 = 0.}}$$

duas habet radices impossibiles. Æquatio item

$$x^4 - 6xx - 3x - 2 = 0$$

$$\overset{\frac{3}{8}. \quad \frac{4}{9}. \quad \frac{3}{8}.}{\underset{+ + \quad + \quad - \quad +}{x^4 \; * \; - 6xx - 3x - 2 = 0.}}$$

duas habet. Nam hæc fractionum series $[\frac{4}{1}. \frac{3}{2}. \frac{2}{3}. \frac{1}{4}]$

‖[157] dividendo secundam per ‖ primam, tertiam per secundam et quartam per tertiam, dat hanc seriem $\frac{3}{8}. \frac{4}{9}. \frac{3}{8}$ super medijs æquationis terminis collocandum. Dein secundi termini qui hic nihil est quadratum ductum in fractionem imminentem $\frac{3}{8}$ producit nihil, quod tamen majus est quàm rectangulum negativum $-6x^6$ sub terminis utrincg positis x^4 & $-6xx$ contentum. Quare sub termino illo deficiente scribo $+$. In cæteris pergo ut in exemplo superiori et signorum subscriptorum prodit hæc series $+ + + - +$ ubi duæ mutationes indicant duas radices impossibiles.

$$\overset{\frac{2}{5}. \quad \frac{1}{2}. \quad \frac{1}{2}. \quad \frac{2}{5}.}{\underset{+ \quad + \quad - \quad + \quad + +}{x^5 - 4x^4 + 4x^3 - 2xx - 5x - 4 = 0.}}$$

---

(478) Newton's (incomplete) rule for delimiting the number of 'impossible' roots, adapted from 1: [523–4, 526 →] 527–8: 'set downe a series of … fractions … whose numerators & denominators are a progression of units backward & forward. Divide each fraction by y^t p^{r}ceding it & set the quotes in order over all y^e middle termes of the Equation. Then observe of every middle terme whither its square multiplyed by y^e fraction over it bee greater equall or lesse y^n y^e rectangle of y^e two termes on either hand. If greater write $+$ underneath, if equall or lesse write $-$. Lastly set $+$ under y^e first & last terme & there shall bee soe many impossible roots as there are changes of signs.' It is easy to show (compare 1: 522, note (32))

beneath.[478] If, for instance, the equation $x^3 + px^2 + 3p^2x - q = 0$ be had, in this series, $\frac{3}{1}, \frac{2}{2}, \frac{1}{3}$ I divide the second fraction $\frac{2}{2}$ by the first one $\frac{3}{1}$, and the third $\frac{1}{3}$ by the second $\frac{2}{2}$, then place the resulting fractions $\frac{1}{3}$ and $\frac{1}{3}$ above the middle terms

in the equation as follows:
$$\overset{\quad\ \frac{1}{3}\quad\ \frac{1}{3}\ \ }{x^3 + px^2 + 3p^2x - q} = 0.$$
$$+\quad\ -\quad\ +\quad +$$
Then, seeing that the square of the second term px^2 multiplied by the fraction $\frac{1}{3}$ overhead—namely, $\frac{1}{3}p^2x^4$—is less than the product, $3p^2x^4$, of the first term x^3 and the third, $3p^2x$, beneath the term px^2 I place the sign $-$. But because the square $9p^4x^2$ of the third term $3p^2x$ multiplied by its overhead fraction $\frac{1}{3}$ is greater than nothing and hence much greater still than the (negative) product of the second term px^2 and the fourth $-q$, beneath the third term I place the sign $+$. Next, beneath the first term x^3 and the last one $-q$ I place $+$ signs. Then in the signs written beneath, which are in this sequence $+ - + +$, the two[479] changes (one from $+$ to $-$, the other from $-$ to $+$) indicate that there are two impossible roots. So, also, the equation $x^3 - 4x^2 + 4x - 6 = 0$[480] has two impossible roots.

Similarly, the equation $x^4 - 6x^2 - 3x - 2 = 0$ has two. For here this series of fractions $\frac{4}{1}, \frac{3}{2}, \frac{2}{3}, \frac{1}{4}$, on dividing the second by the first, the third by the second and the fourth by the third, yields this series $\frac{3}{8}, \frac{4}{9}, \frac{3}{8}$ to be placed above the middle terms in the equation. Next, the square of the second term—here zero—multiplied by its overhead fraction $\frac{3}{8}$ produces nothing, which is nevertheless greater than the (negative) product $-6x^6$ contained by the terms x^4 and $-6x^2$ set on its either side: beneath the missing term I therefore write $+$. In the case of the rest I proceed as in the preceding example, and for the signs written beneath there results this sequence $+ + + - +$, the two changes in which indicate two impossible

$$\overset{\quad\ \frac{1}{3}\quad\ \ \frac{1}{3}\ \ }{x^3 - 4x^2 + 4x - 6} = 0$$
$$+\quad\ +\quad\ -\quad\ +$$

$$\overset{\quad\ \frac{3}{8}\qquad\ \ \frac{4}{9}\quad\ \ \frac{3}{8}\ \ }{x^4 + 0[x^3] - 6x^2 - 3x - 2} = 0$$
$$+\quad +\quad\ +\quad\ -\quad\ +$$

$$\overset{\ \frac{2}{5}\quad\ \frac{1}{2}\quad\ \ \frac{1}{2}\quad\ \frac{2}{5}\ }{x^5 - 4x^4 + 4x^3 - 2x^2 - 5x - 4} = 0$$
$$+\quad +\quad\ -\quad\ +\quad\ +\quad +$$

---

that the rule validly determines the existence of one pair of complex roots, and we have previously suggested (1: 524, note (40)) that Newton was then led to make the logically unjustifiable inference that *each* occurrence of a double change of sign ...$+ - +$... marks a *separate* pair of complex roots. Rigorous justification of Newton's rule, first satisfactorily effected by J. J. Sylvester in 1865 (see *ibid.*), is both difficult and, because it involves the consideration of a number of special cases, not uncomplicated.

(479) Newton first wrote 'binæ' (pair of).

(480) A repeat of an equation which Newton earlier (see note (468)) erroneously computed to have the real root $x = 2$ and the complex pair $x = 1 \pm i\sqrt{2}$. The corrected cubic with these roots has the variant sign pattern $+ - - +$, the two variations in which correctly identify its pair of conjugate 'impossibles'.

Et ad eundem modum in æquatione $x^5 - 4x^4 + 4x^3 - 2xx - 5x - 4 = 0$ deteguntur impossibiles duæ.

[1681]
Lect 7
    Ubi termini duo vel plures simul desunt, sub primo terminorum deficientium collocandum est signum $-$, sub secundo signum $+$, sub tertio signum $-$ & sic deinceps, semper variando signa, nisi quod sub ultimo terminorum simul deficientium semper collocandum est signum $+$ ubi termini deficientibus utrincȝ proximi habent signa contraria.[481] Ut in æquationibus

$$x^5 + ax^4 * * * + a^5 = 0, \quad \text{et} \quad x^5 + ax^4 * * * - a^5 = 0.$$
$$+ \quad + - + - \quad + \qquad\qquad + \quad + - + + \quad +$$

quarum prior quatuor posterior duas habet impossibiles radices. Sic et æquatio

$$\overset{\frac{3}{7}\cdot}{}\ \overset{\frac{5}{9}\cdot}{}\ \overset{\frac{3}{5}\cdot}{}\ \overset{\frac{3}{5}\cdot}{}\ \overset{\frac{5}{9}\cdot}{}\overset{\frac{3}{7}\cdot}{}$$
$$x^7 - 2x^6 + 3x^5 - 2x^4 + x^3 \ * \ * - 3 = 0 \text{ sex habet impossibiles.}$$
$$+ \quad - \quad + \quad - \quad + - + +$$

|| [158]
    || Hinc etiam cognosci potest utrum radices impossibiles inter affirmativas radices latent an inter negativas. Nam signa terminorum signis subscriptis variantibus imminentium indicant tot affirmativas esse impossibiles quot sunt ipsorum variationes et tot negativas quot sunt ipsorum successiones sine variatione.[482] Sic in æquatione $x^5 - 4x^4 + 4x^3 - 2xx - 5x - 4 = 0$ quoniam signis
$$+ \quad + \quad - \quad + \quad + +$$
infra scriptis variantibus $+ - +$ quibus radices duæ impossibiles indicantur, imminentes termini $-4x^4 + 4x^3 - 2xx$ signa habent, quæ per duas variationes indicant duas affirmativas radices; ideo radices duæ impossibiles inter affirmativas latebunt. Cum itacȝ omnium æquationis terminorum signa $+ - + - - -$ per tres variationes indicant tres esse affirmativas radices et reliquas duas negativas esse, & inter affirmativas lateant duæ impossibiles, sequitur æquationis unam esse radicem verè affirmativam, duas negativas ac duas impossibiles. Quod si æquatio fuisset $x^5 - 4x^4 - 4x^3 - 2xx - 5x - 4 = 0$. tunc termini sub-
$$+ \quad + \quad - \quad - \quad + +$$
scriptis signis prioribus variantibus $+ -$ imminentes nimirum $-4x^4 - 4x^3$ per signa sua non variantia $-$ & $-$ indicant unam ex negativis radicibus impossibilem esse, et termini signis subscriptis posterioribus variantibus $- +$ imminentes, nimirum $-2xx - 5x$ per signa sua non variantia $-$ & $-$ indicant aliam ex negativis radicibus impossibilem esse. Quamobrem cum æquationis signa $+ - - - -$ per unam variationem indicent unam affirmativam radicem, cæteras quatuor negativas esse: sequitur || unam esse affirmativam

|| [159]

---

(481) Compare I: 530: 'If there bee two or more termes wanting set signes under them successively begining wth a negative, only end wth an affirmative if the terms on either hand have contrary signes.'

(482) Newton's complete rule for complex roots, distinguishing (to little real advantage) between positive and negative 'impossibles'.

roots. And in much the same way in the equation $x^5 - 4x^4 + 4x^3 - 2x^2 - 5x - 4 = 0$ two impossibles are disclosed.

When two or more terms together are simultaneously lacking, beneath the first of the deficient terms the sign $-$ must be placed, beneath the second the sign $+$, beneath the third the sign $-$, and so on in turn, always varying the signs except that beneath the last of the terms simultaneously lacking you must always place the sign $+$ when the terms next on either sides of the deficient ones have contrary signs.[481] So in the equations

$$x^5 + ax^4 + 0[x^3] + 0[x^2] + 0[x] + a^5 = 0 \quad \text{and} \quad x^5 + ax^4 + 0[x^3] + 0[x^2] + 0[x] - a^5 = 0,$$
$$+ \ \ + \ \ - \ \ \ \ + \ \ \ \ - \ \ \ \ + \qquad\qquad\quad + \ \ \ \ + \ \ \ \ - \ \ \ \ + \ \ \ \ + \ \ \ \ +$$

the first of which has four impossible roots, the latter two. So, again, the equation

$$\overset{\frac{3}{7}}{} \ \overset{\frac{5}{9}}{} \ \overset{\frac{3}{5}}{} \ \overset{\frac{3}{5}}{} \ \overset{\frac{5}{9}}{} \ \overset{\frac{3}{7}}{}$$
$$x^7 - 2x^6 + 3x^5 - 2x^4 + x^3 + 0[x^2] + 0[x] - 3 = 0 \text{ has six impossible roots.}$$
$$+ \ \ - \ \ + \ \ - \ \ + \ \ - \ \ + \ \ +$$

It may hence also be ascertained whether the impossible roots lie hidden among the positive roots or among the negative ones. For the signs of the terms over the heads of the varying subscript signs indicate that there are as many positive impossibles as variations in the signs of these, and as many negative ones as they have successions without variation.[482] So in the equation

$$x^5 - 4x^4 + 4x^3 - 2x^2 - 5x - 4 = 0,$$
$$+ \ \ + \ \ - \ \ + \ \ + \ \ +$$

seeing that the signs written below have the variations $+ - +$ by which two impossible roots are indicated, while the terms $-4x^4 + 4x^3 - 2x^2$ overhead have signs which by their two variations indicate two positive roots, in consequence the two impossibles will lie hidden among the positives. Accordingly, since the signs of all the terms in the equation, $+ - + - -$, by their three variations indicate that there are three positive roots and so the remaining two are negative, and because the two impossibles lie hidden among the positives, it follows that the equation possesses one really positive root, two negative ones and two impossibles. But had the equation been $x^5 - 4x^4 - 4x^3 - 2x^2 - 5x - 4 = 0$,
$$ + \ \ + \ \ - \ \ - \ \ + \ \ +$$
then the terms, namely $-4x^4 - 4x^3$, positioned over the first (pair of) varying subscript signs $+ -$ indicate, by their invariance in sign $- -$, that one of the negative roots is impossible, while the terms, namely $-2x^2 - 5x$, set over the latter (pair of) varying subscript signs $- +$ indicate, by their invariance in sign $- -$, that one or other of the negative roots is impossible. Consequently, since the equation's signs $+ - - - -$ indicate, by their single variation, that one root is positive and the remaining four negative, it follows that there is

duas negativas ac duas impossibiles. Atqȝ hæc ita se habent ubi non sunt plures impossibiles radices quam per regulam allatam deteguntur. Possunt enim plures esse, licet id perrarò eveniat.[483]

De trans-
mutationibus
æquationum.

Cæterum æquationis cujusvis radices omnes affirmativæ in negativas & negativæ in affirmativas mutari possunt, idqȝ mutando tantum signa terminorum alternorum.[484] Sic æquationis $x^5 - 4x^4 + 4x^3 - 2xx - 5x - 4 = 0$, radices tres affirmativæ[485] mutabuntur in negativas & duæ negativæ[485] in affirmativas mutando tantum signa secundi quarti et sexti termini ut hic fit

$$x^5 + 4x^4 + 4x^3 + 2xx - 5x + 4 = 0.$$

Easdem habet hæc æquatio radices cum priore nisi quod hic affirmativæ sunt quæ ibi erant negativæ et hic negativæ quæ ibi erant affirmativæ; et radices duæ impossibiles quæ ibi inter affirmativas latebant hic latent inter negativas, ita ut his deductis restet unica tantum radix vere negativa.

Sunt et aliæ æquationum transmutationes quæ diversis usibus inserviunt. Possumus enim supponere radicem æquationis ex cognita et incognita aliqua quantitate utcunqȝ componi et perinde pro ea substituere quod æquipollens esse fingitur.[486] Ut si supponamus radicem æqualem esse summæ vel differentiæ cognitæ alicujus et incognitæ quantitatis. Nam possumus hoc pacto radices

|| [160]   æquationis cognita illa quantita‖te augere vel diminuere vel de cognita quantitate subducere, atqȝ ita efficere ut earum aliquæ quæ prius erant negativæ jam fiant affirmativæ vel ut aliquæ ex affirmativis evadant negativæ; vel etiam ut omnes evadant affirmativæ aut omnes negativæ. Sic in æquatione

$$x^4 - x^3 - 19xx + 49x - 30 = 0,$$

si radices unitate augeri vellem, fingo $x + 1 = y$ seu $x = y - 1$ et perinde pro x scribo in æquatione $y - 1$ et pro quadrato, cubo, quadrato-quadrato de x similem potestatem de $y - 1$ ad hunc modum

$$
\begin{array}{r|l}
x^4. & y^4 - 4y^3 + 6yy - 4y + 1 \\
-x^3. & \quad\; -y^3 + 3yy - 3y + 1 \\
-19xx. & \qquad\quad -19yy + 38y - 19 \\
+49x. & \qquad\qquad\qquad +49y - 49 \\
-30 & \qquad\qquad\qquad\qquad -30 \\
\hline
\text{Summa} & y^4 - 5y^3 - 10yy + 80y - 96 = 0.
\end{array}
$$

---

(483) On his pages 160–1 below Newton will cite the instance of the cubic $x^3 - 3a^2x - 3a^3$, whose sign pattern $+\,+\,+\,+$ betrays no trace of its two complex roots

$$x/a = \omega . \sqrt[3]{[\tfrac{1}{2}(-3 + \sqrt{5})]} + \omega^{-1} . \sqrt[3]{[\tfrac{1}{2}(-3 - \sqrt{5})]}, \quad \omega = \tfrac{1}{2}(-1 \pm i\sqrt{3}).$$

(484) Compare the concluding section 'Quo pacto veræ radices æquationis mutentur falsis, et vicissim' of the second chapter—in Mercator's Latin version—of Kinckhuysen's *Algebra* (II: 322–3), itself narrowly based on Descartes' *Geometrie* [=*Discours de la Methode* (Leyden, 1637): 297–413]: 373–4.

one positive two negative and two impossible roots. This holds true, may I say, when there are no more impossible roots than the rule adduced discloses: for there may possibly be more, though that very rarely happens.[483]

For the rest, all the positive roots of any equation can be changed into negative ones and the negatives into positive ones, and this merely by changing the signs of alternate terms.[484] Thus in the equation

$$x^5 - 4x^4 + 4x^3 - 2x^2 - 5x - 4 = 0$$

Transmutations of equations.

the three positive roots[485] will be changed to negative ones and the two negatives[485] to positives merely by altering the signs of the second, fourth and sixth terms, thus: $x^5 + 4x^4 + 4x^3 + 2x^2 - 5x + 4 = 0$. This equation has the same roots as the previous one apart from those now being positive which were formerly negative and those now negative formerly being positive, while the two impossibles which formerly lay hidden among the positives are now concealed among the negatives, and as a result, when these are deducted, there remains but a single really negative root.

There are also other transmutations of equations which serve diverse purposes. We can, to be sure, suppose that an equation's root is in some manner composed of a known quantity and some unknown one, and correspondingly substitute in its place what is now conceived to be its equivalent.[486] In instance, we may suppose the root to be equal to the sum or difference of some known quantity and an unknown one: for by this approach we can increase or diminish the roots of an equation by that known quantity or subtract them from the known quantity, and so bring it about that some of these which were previously negative now become positive, or that some of the positives now come to be negatives, or even that all of them come to be positive or, conversely, all negative. Thus were I, in the equation $x^4 - x^3 - 19x^2 + 49x - 30 = 0$, to want the roots to be increased by unity, I conceive $x + 1 = y$, that is, $x = y - 1$, and correspondingly in x's place in the equation I write $y - 1$ and for the square, cube, square-square (fourth power) of x I enter the similar power of $y - 1$ to this effect:

x^4	$y^4 - 4y^3 + 6y^2 - 4y + 1$
$-x^3$	$-y^3 + 3y^2 - 3y + 1$
$-19x^2$	$-19y^2 + 38y - 19$
$+49x$	$+49y - 49$
-30	-30
Sum:	$y^4 - 5y^3 - 10y^2 + 80y - 96 = 0$

(485) Only one of the 'positive' roots ($x \approx 2\frac{1}{4}$) and neither 'negative' root of the given quintic is in fact real.

(486) Compare the section 'Quomodo radices alicujus æquationis... augeri vel diminui possint' (II: 323–4) of Kinckhuysen's 'Caput Tertium. De transformatione æquationum, ubi radices simul mutantur', itself based squarely on Descartes' *Geometrie* (note (484)): 374–6.

Et æquationis prodeuntis $y^4-5y^3-10yy+80y-96=0$[487] radices erunt 2, 3, 4, -4 quæ prius erant 1, 2, 3, -5, unitate jam factæ majores. Quod si pro x scripsissem $y+1\frac{1}{2}$ prodijsset æquatio $y^4+5y^3-10yy-\frac{5}{4}y+\frac{39}{16}=0$ cujus duæ fuissent radices affirmativæ $\frac{1}{2}$ & $1\frac{1}{2}$ ac duæ negativæ $-\frac{1}{2}$ & $-6\frac{1}{2}$. Pro x vero scribendo $y-6$ prodijsset æquatio[488] cujus radices fuissent 7, 8, 9, 1 omnes nimirum affirmativæ, et pro eodem scribendo $y+4$ radices jam numero quaternario diminutæ evasissent -3, -2, -1, -9 negativæ omnes.

Et hoc modo augendo vel diminuendo radices siquæ impossibiles sunt, hæ aliquando facilius detegentur quàm prius. Sic in æquatione $x^3-3aax-3a^3=0$

‖[161] radices nullæ per ‖ præcedentem regulam apparent impossibiles. At si augeas radices quantitate a scribendo $y-a$ pro x, in æquatione resultante

$$y^3-3ayy-a^3=0$$

radices duæ impossibiles jam per regulam illam detegi possunt.

[1681]   Eadem operatione possumus etiam secundos terminos æquationum tollere.[489]
Lect 8 Hoc enim fiet si cognitam quantitatem secundi termini æquationis propositæ per numerum dimensionū æquationis divisam, subducamus de quantitate quæ pro novæ æquationis radice significanda assumitur et residuum substituamus pro radice æquationis propositæ. Ut si proponatur æquatio $x^3-4xx+4x-6=0$, cognitam quantitatem secundi termini quæ est -4 divisam per numerum dimensionum æquationis 3 subduco de[490] specie quæ pro nova radice significanda assumitur, puta de y, & residuum $y+\frac{4}{3}$ substituo pro x et provenit

$$\begin{aligned}
y^3+4yy+\tfrac{16}{3}y &\;+\tfrac{64}{27} \\
-4yy-\tfrac{32}{3}y &\;-\tfrac{64}{9} \\
+4y &\;+\tfrac{16}{3} \\
&\;-6 \\
\hline
y^3 \quad * \quad -\tfrac{4}{3}y-\tfrac{146}{27} &=0.
\end{aligned}$$

Eadem methodo potest et tertius æquationis terminus tolli.[491] Proponatur æquatio $x^4-3x^3+3xx-5x-2=0$, et finge $x=y-e$ et substituendo $y-e$ pro x orietur hæc æquatio

$$y^4\begin{matrix}-4e\\-3\end{matrix}y^3\begin{matrix}+6ee\\+9e\\+3\end{matrix}yy\begin{matrix}-4e^3\\-9ee\\-6e\\-5\end{matrix}y\left.\begin{matrix}+e^4\\+[3]e^3\\+3ee\\+5e\\-2\end{matrix}\right\}=0.$$

---

(487) That is, $(y-2)(y-3)(y-4)(y+4)=0$, corresponding to the factorization
$$x^4-x^3-19x^2+49x-30=(x-1)(x-2)(x-3)(x+5)=0$$
on setting $x=y-1$.

(488) Specifically, $y^4-25y^3+215y^2-695y+504=0$.

The roots of the resulting equation $y^4 - 5y^3 - 10y^2 + 80y - 96 = 0$[487] will then be 2, 3, 4, −4, the previous ones 1, 2, 3, −5 now enlarged by unity. But if in place of x I had written $y + 1\frac{1}{2}$, there would have resulted the equation

$$y^4 + 5y^3 - 10y^2 - \tfrac{5}{4}y + \tfrac{39}{16} = 0,$$

possessing the two positive roots $\frac{1}{2}$ and $1\frac{1}{2}$ and the two negative ones $-\frac{1}{2}$ and $-6\frac{1}{2}$. On writing $y - 6$ in place of x, however, there would have resulted an equation[488] possessing the roots 7, 8, 9, 1, all positive, of course; while on entering $y + 4$ in its place the roots, now diminished by the number four, would have proved to be −3, −2, −1, −9, all negative.

And by increasing or diminishing the roots in this way those (if any) which are impossible may on occasion be more easily detected than before. Thus in the equation $x^3 - 3a^2x - 3a^3 = 0$ no roots appear, by the preceding rule, to be impossible. But if, by writing $y - a$ in place of x, you increase the roots by the quantity a, in the resulting equation $y^3 - 3ay^2 - a^3 = 0$ two impossible roots can now be detected by that rule.

By the same procedure we can also remove the second terms of equations.[489] This, to be sure, will happen if we subtract the known quantity of the second term in the equation proposed, after dividing it by the number of the equation's dimensions, from the quantity taken as the root of the new equation, and then substitute the residue in place of the root of the equation proposed. If, for instance, the equation $x^3 - 4x^2 + 4x - 6 = 0$ be proposed, I subtract the known quantity, here −4, of the second term, after dividing it by the number, 3, of the equation's dimensions, from the[490] character—y, say—taken to signify the new root, and then substitute the residue $y + \frac{4}{3}$ in x's place, when there ensues:

$$
\begin{array}{l}
y^3 \quad +4y^2 + \tfrac{16}{3}y \quad +\tfrac{64}{27} \\
\quad\quad\;\; -4y^2 - \tfrac{32}{3}y \quad -\tfrac{64}{9} \\
\quad\quad\quad\quad\quad\; +4y \quad +\tfrac{16}{3} \\
\quad\quad\quad\quad\quad\quad\quad\quad\;\; -6 \\
\hline
y^3 + 0[y^2] \quad -\tfrac{4}{3}y - \tfrac{146}{27} = 0.
\end{array}
$$

By the same method the third term in the equation can also be removed.[491] Let the equation $x^4 - 3x^3 + 3x^2 - 5x - 2 = 0$ be proposed and imagine $x = y - e$, and then on substituting $y - e$ in x's place this equation will arise:

$$y^4 - (4e + 3)\,y^3 + (6e^2 + 9e + 3)\,y^2 - (4e^3 + 9e^2 + 6e + 5)\,y$$
$$+ e^4 + 3e^3 + 3e^2 + 5e - 2 = 0.$$

---

(489) Compare Descartes' *Geometrie* (note (484)): 376 and especially Kinckhuysen's section 'Quomodo secundus terminus æquationis tolli possit' (II: 325).

(490) 'incognita' (unknown) is deleted.

(491) See Newton's additional section 'Quomodo tertius terminus tolli possit' in his 'Observationes' on Kinckhuysen's *Algebra* (II: 416). His present example is a little changed.

‖[162] Hujus æquationis tertius terminus est $6ee+9e+3$ ‖ ductum in yy. Ubi si $6ee+9e+3$ nullum esset eveniret id ipsum quod volumus. Fingamus itaqȝ nullum esse ut inde colligamus quinam numerus ad hunc effectum substitui debet pro e, & habebimus æquationem quadraticam $6ee+9e+3=0$ quæ divisa per 6 fiet $ee+\frac{3}{2}e+\frac{1}{2}=0$ seu $ee=-\frac{3}{2}e-\frac{1}{2}$, et extracta radice $e=-\frac{3}{4}\pm\sqrt{\frac{9}{16}-\frac{1}{2}}$, seu $=-\frac{3}{4}\pm\sqrt{\frac{1}{16}}$, hoc est $=-\frac{3}{4}\pm\frac{1}{4}$, atqȝ adeo vel $=-\frac{1}{2}$ vel $=-1$.[492] Unde $y-e$ erit vel $y+\frac{1}{2}$ vel $y+1$. Quamobrem cum $y-e$ scriptum fuit pro x, vice $y-e$ debet $y+\frac{1}{2}$ vel $y+1$ scribi pro x, ut tertius æquationis resultantis terminus nullus sit. Et in utroqȝ quidem casu id eveniet. Nam si pro x scribatur $y+\frac{1}{2}$ orietur hæc æquatio $y^4-y^3-\frac{15}{4}y-\frac{77}{16}=0$: sin scribatur $y+1$ orietur hæc $y^4+y^3-4y-12=0$.

Possunt et radices æquationis per datos numeros multiplicari vel dividi, et hoc pacto termini æquationum diminui, fractiones et radicales quantitates aliquando tolli.[493] Ut si æquatio sit $y^3-\frac{4}{3}y-\frac{146}{27}=0$ ad tollendas fractiones fingo esse $y=\frac{1}{3}z$ et perinde pro y substituendo $\frac{1}{3}z$ provenit æquatio nova

$$\frac{z^3}{27}-\frac{4z}{27}-\frac{146}{27}=0^{[494]}$$ et rejecto terminorum communi denominatore,

$$z^3-4z-146=0,$$

cujus æquationis radices sunt triplo majores quam ante. Et rursus ad diminuendos terminos æquationis hujus si scribatur $2v$ pro z prodibit $8v^3-8v-146=0$ et divisis omnibus per 8 fiet $v^3-v-18\frac{1}{4}=0$, cujus æquationis radices dimidia sunt
‖[163] radicum ‖ prioris. Et hic si tandem inveniatur v ponendum erit $2v=z$, $\frac{1}{3}z=y$ & $y+\frac{4}{3}=x$ et æquationis primò propositæ $x^3-4xx+4x-6=0$ habebitur radix x.

Sic et in æquatione $x^3-2x+\sqrt{3}=0$ ad tollendam quantitatem radicalem $\sqrt{3}$, pro x scribo $y\sqrt{3}$ et provenit æquatio $3y^3\sqrt{3}-2y\sqrt{3}+\sqrt{3}=0$ quæ divisis omnibus terminis per $\sqrt{3}$ fit $3y^3-2y+1=0$.[495]

Rursus æquationis radices in earum reciprocas transmutari possunt et hoc pacto æquatio aliquando ad formam commodiorē reduci. Sic æquatio novissima $3y^3-2y+1=0$. scribendo $\frac{1}{z}$ pro y evadit $\frac{3}{z^3}-\frac{2}{z}+1=0$ seu terminis omnibus multiplicatis per z^3 et ordine terminorum mutato $z^3-2zz+3=0$. Potest etiam æquationis terminus penultimus hoc pacto tolli si modo secundus priùs tollatur

---

(492) This derives more straightforwardly from the factorization

$$6e^2+9e+3 = 3(2e+1)(e+1) = 0.$$

Newton delays discussion of this technique, simpler if not always applicable, to the later section on reducing equations 'per divisores rationales' (on his pages 173–80 following).

(493) Compare Kinckhuysen's sections 'Quomodo radices multiplicari vel dividi possint...' and 'Quomodo fractiones æquationis mutentur in integra, et radicales in rationales, et alia nonnulla quæ usui esse possunt' (II: 329–30), based in turn on Descartes' *Geometrie*: 379–80.

In this equation the third term is $6e^2+9e+3$ multiplied by y^2, and if $6e^2+9e+3$ had been zero what we wanted would have occurred. Let us accordingly imagine that it is zero so as thereby to gather what number should be substituted for e to achieve this. We shall then have the quadratic equation $6e^2+9e+3=0$, which after division by 6 will become $e^2+\frac{3}{2}e+\frac{1}{2}=0$ or $e^2=-\frac{3}{2}e-\frac{1}{2}$, and when the root is extracted $e=-\frac{3}{4}\pm\sqrt{[\frac{9}{16}-\frac{1}{2}]}$ or $-\frac{3}{4}\pm\sqrt{\frac{1}{16}}$, that is, $-\frac{3}{4}\pm\frac{1}{4}$ and hence $-\frac{1}{2}$ or -1.[492] Whence $y-e$ will be $y+\frac{1}{2}$ or $y+1$, and therefore, since $y-e$ was written in x's place, instead of $y-e$ you must write $y+\frac{1}{2}$ or $y+1$ in x's place in order that the third term in the resulting equation shall be zero. That will, indeed, prove to happen in either case: for, if in x's place there be written $y+\frac{1}{2}$, this equation $y^4-y^3-\frac{15}{4}y-\frac{77}{16}=0$ will ensue; but if $y+1$ be written, there will arise this one $y^4+y^3-4y-12=0$.

The roots of an equation can also be multiplied or divided by given numbers, and by this means terms in equations can be diminished and on occasion fractions and radical quantities may be removed.[493] If, for instance, the equation be $y^3-\frac{4}{3}y-\frac{146}{27}=0$, to remove the fractions I imagine that $y=\frac{1}{3}z$ and, on making corresponding substitution of $\frac{1}{3}z$ in y's place, there ensues the new equation $\frac{1}{27}z^3-\frac{[12]}{27}z-\frac{146}{27}=0$ and so, when the common denominator of the terms is rejected, $z^3-[12]z-146=0$, an equation whose roots are three times larger than the previous ones. If now, further to diminish the terms in this equation, $2v$ be written in place of z, there will result $8v^3-[24]v-146=0$ and, after dividing through by 8, there will come to be $v^3-[3]v-18\frac{1}{4}=0$, an equation whose roots are half those of the former one. If, finally, v here be ascertained, you must set $2v=z$ and $\frac{1}{3}z=y$ and then $y+\frac{4}{3}=x$, and the root x of the equation $x^3-4x^2+4x-6=0$ first proposed will be had.

So also, to remove the radical quantity $\sqrt{3}$ in the equation $x^3-2x+\sqrt{3}=0$, in x's place I write $y\sqrt{3}$ and there comes out the equation $3y^3\sqrt{3}-2y\sqrt{3}+\sqrt{3}=0$, which becomes $3y^3-2y+1=0$ after dividing all the terms by $\sqrt{3}$.[495]

Again, an equation's roots can be transmuted into their reciprocals and by this means the equation may on occasion be reduced to a more convenient form. Thus, on writing $1/z$ in y's place the most recent equation comes to be

$$3/z^3-2/z+1=0;$$

that is, when all the terms are multiplied by z^3 and their order inverted, $z^3-2z^2+3=0$. The last term but one in an equation may also be removed in this manner provided the second one be first removed, as you see done in a

---

(494) As Whiston observed in his 1707 *editio princeps*, the second term should be $'-\frac{12}{27}z\,'$: the necessary emendations in the sequel are indicated in our English version.

(495) Compare Kinckhuysen's example $'x^3-xx\sqrt{3}+\frac{26}{27}x-\frac{8}{27\sqrt{3}}=0'$ on II: 330.

ut factum vides in exemplo præcedente. Aut si antepenultimum tolli cupias, id fiet si modo tertium prius tollas. Sed et radix minima hoc pacto in maximam convertitur et maxima in minimam: quod usum nonnullum habere potest in sequentibus. Sic in æquatione $x^4 - x^3 - 19xx + 49x - 30 = 0$, cujus radices sunt 3, 2, 1, -5, si scribatur $\frac{1}{y}$ pro x, resultabit æquatio $\frac{1}{y^4} - \frac{1}{y^3} - \frac{19}{yy} + \frac{49}{y} - 30 = 0$, quæ terminis omnibus multiplicatis per y^4 ac divisis per 30, signiscp mutatis fiet

‖[164] $y^4 - \frac{49}{30}y^3 + \frac{19}{30}yy + \frac{1}{30}y - \frac{1}{30} = 0$, cujus ‖ radices sunt $\frac{1}{3}$, $\frac{1}{2}$, 1, $-\frac{1}{5}$: radicum affirmativarum maxima 3 jam conversa in minimam $\frac{1}{3}$ et minima 1 jam facta maxima et radice negativa -5 quæ omnium maxime distabat a nihilo, jam omnium maxime accedente ad nihil.

Sunt et aliæ æquationum transmutationes[496] sed quæ omnes ad exemplum transmutationis illius ubi tertium æquationis terminum sustulimus confici possunt, ut non opus sit hac de re plura dicere. Addamus potius aliqua de limitibus æquationum.

[1681]
Lect 9
[Æquationum
generatio.][497]

Ex Æquationum generatione[498] constat quod cognita quantitas secundi termini æquationis, si signum ejus mutetur, æqualis sit aggregato omnium radicum sub signis proprijs; ea tertij æqualis aggregato rectangulorum sub singulis binis radicibus; ea quarti, si signum ejus mutetur, æqualis aggregato contentorum sub singulis ternis radicibus; ea quinti æqualis aggregato contentorum sub singulis quaternis; et sic in infinitum. Assumamus $x = a$, $x = b$, $x = -c$, $x = d$ &c seu $x - a = 0$, $x - b = 0$, $x + c = 0$, $x - d = 0$, et ex horum continua multiplicatione generemus æquationes, ut supra.[499] Jam multiplicando $x - a$ per $x - b$ producetur æquatio $xx \begin{smallmatrix} -a \\ -b \end{smallmatrix} x + ab = 0$: ubi cognita quantitas secundi termini si signa ejus mutentur, nimirum $a + b$, est summa duarum radicum a et b, et cognita tertij ab illud unicum quod sub utracp continetur rectangulum.

‖[165] Rursus multiplicando hanc ‖ æquationem per $x + c$ producetur æquatio cubica $x^3 \begin{smallmatrix} -a \\ -b \\ +c \end{smallmatrix} xx \begin{smallmatrix} +ab \\ -ac \\ -bc \end{smallmatrix} x + abc = 0$, ubi cognita quantitas secundi termini sub signis mutatis nimirum $a + b - c$ est summa radicum a, b et $-c$; cognita tertij $ab - ac - bc$ summa rectangulorum sub singulis binis a et b, a et $-c$, b et $-c$; et cognita quarti sub signo mutato $-abc$ illud unicum contentum est quod omnium

---

(496) Among other things Newton probably has in mind Kinckhuysen's section 'Quomodo penultimus terminus tolli possit' (II: 328–9) and the comparable removal of the antepenultimate term.

(497) We insert this explanatory marginal title for the purpose of distinguishing the present preliminary remarks from the narrower discussion 'De limitibus æquationum' which follows.

preceding example. Or should you desire to remove the one before that, that will come about provided you first remove the third one. Also, may I say, the least root is by this means converted into the greatest one, and the greatest into the least: this may possibly be of considerable use in the sequel. Thus, in the equation $x^4 - x^3 - 19x^2 + 49x - 30 = 0$ whose roots are 3, 2, 1 and -5, if $1/y$ be written in x's place there will result the equation

$$1/y^4 - 1/y^3 - 19/y^2 + 49/y - 30 = 0,$$

which, after all its terms are multiplied by y^4 and divided by 30 and its signs are changed, will become $y^4 - \frac{49}{30}y^3 + \frac{19}{30}y^2 + \frac{1}{30}y - \frac{1}{30}$, whose roots are $\frac{1}{3}$, $\frac{1}{2}$, 1 and $-\frac{1}{5}$: the greatest of the positive roots, 3, is now converted into their least, $\frac{1}{3}$, and the least one, 1, is now become the greatest, while the negative root -5, hitherto farthest distant of all from zero, is now the one approaching nearest of all to zero.

There exist other transmutations of equations also,[496] but these can all be accomplished on the pattern of the transmutation in which we took away the third term of an equation: as a result, there is no need to say more on this topic. Rather, let us add a few remarks regarding the limits of equations.

From the generation of equations[498] it is established that the 'known quantity' (coefficient) of the second term in an equation is, if its sign be changed, equal to the aggregate of all the roots under their proper signs; that of the third equal to the aggregate of the products of the separate roots two at a time; that of the fourth, if its sign be changed, equal to the aggregate of the (solid) products of the individual roots three at a time; that of the fifth equal to the aggregate of the products of the roots four at a time; and so on indefinitely. Let us assume $x = a$, $x = b$, $x = -c$, $x = d$, ... or $x - a = 0$, $x - b = 0$, $x + c = 0$, $x - d = 0$, ..., and from continued multiplication of these let us generate equations, as previously.[499] Now by multiplying $x - a$ by $x - b$ there will be produced the equation $x^2 - (a + b)x + ab = 0$, where the coefficient of the second term with its signs changed, namely $a + b$, is the sum of the two roots a and b, while the coefficient of the third, ab, is the unique product contained by both. Again, on multiplying this equation by $x + c$ there will be produced the cubic equation $x^3 - (a + b - c)x^2 + (ab - ac - bc)x + abc = 0$, where the coefficient of the second term with signs changed, namely $a + b - c$, is the sum of the roots a, b and $-c$; that of the third, $ab - ac - bc$, is the sum of their products two at a time, a with b, a with $-c$ and b with $-c$; that of the fourth with sign changed, $-abc$, the unique (solid) product generated by the continued multiplication of them all, a times b

[The generation of equations.][497]

(498) Understand by multiplying their linear factors (real or complex) together in the way pioneered at the beginning of the century by Thomas Harriot (in Walter Warner's posthumous edition of his *Artis Analyticæ Praxis*, London, 1631).

(499) See Newton's earlier section 'De forma Æquationis' on his page 30 above.

continua multiplicatione generatur, a in b in $-c$. Adhæc multiplicando cubicã illam æquationem per $x-d$ producetur hæcce quadrato-quadratica

$$x^4 \genfrac{}{}{0pt}{}{-a}{-b} x^3 \genfrac{}{}{0pt}{}{+ab}{-ac} xx \genfrac{}{}{0pt}{}{+abc}{-abd} x - abcd = 0:$$

$$\begin{array}{ccc} +c & -bc & +bcd \\ -d & +ad & +acd \\ & +bd & \\ & -cd & \end{array}$$

ubi cognita quantitas secundi termini sub signis mutatis $a+b-c+d$ est summa omnium radicum; ea tertij $ab-ac-bc+ad+bd-cd$ summa rectangulorum sub singulis binis; ea quarti sub signis mutatis $-abc+abd-bcd-acd$ summa contentorum sub singulis ternis; ea quinti $-abcd$ contentum unicum sub omnibus. Et hinc primò colligimus omnes æquationis cujuscunqȝ terminos nec fractos nec surdos habentis, radices non surdas et radicum binarum rectangula, ternarumqȝ aut pluriũ contenta esse aliquos ex divisoribus integris ultimi termini; atqȝ adeo ubi constiterit nullum ultimi termini divisorem esse aut radicem æquationis, aut **[166]** duarum radicum ‖ rectangulum pluriumve contentum, simul constabit nullam esse radicem[500] radicumve rectangulum aut contentum nisi quod sit surdum.

Ponamus jam cognitas quantitates terminorum æquationis sub signis mutatis esse p, q, r, s, t, v &c.[501] eam nempe secundi p, tertij q, quarti r, quinti s et sic deinceps. Et signis terminorum probe observatis fiat $p=a$. $pa+2q=b$.

$$pb+qa+3r=c. \quad pc+qb+ra+4s=d. \quad pd+qc+rb+sa+5t=e.$$

$pe+qd+rc+sb+ta+6v=f$, et sic in infinitum observata serie progressionis. Et erit a summa radicum, b summa quadratorum ex singulis radicibus, c summa cuborum, d summa quadrato-quadratorum, e summa quadrato-cuborum, f summa cubo-cuborum, & sic in reliquis.[502] Ut in æquatione

$$x^4 - x^3 - 19xx + 49x - 30 = 0,$$

ubi cognita quantitas secundi termini est -1, tertij -19, quarti $+49$, quinti -30; ponendum erit $1=p$. $19=q$. $-49=r$. $30=s$. et inde orientur $a=(p=)1$. $b=(pa+2q=1+38=)39$. $c=(pb+qa+3r=39+19-147=)-89$.

$$d=(pc+qb+ra+4s=-89+741-49+120=)723.$$

Quare summa radicum erit 1, summa quadratorum radicum 39, summa cuborum -89, & summa quadrato-quadratorum 723. Nimirum æquationis

---

(500) Understand 'realem' (real).

(501) Newton intends that the terms of the equation shall be alternatively plus and minus, so that it assumes the form $x^n - px^{n-1} + qx^{n-2} - rx^{n-3} + sx^{n-4} - tx^{n-5} + vx^{n-6} - \ldots = 0$.

(502) 'Newton's' rule for expressing the sum of the first, second, third, fourth, fifth, sixth, … powers of the roots, taken over with little change from 1: 519–20. The formulæ for the sum of the roots and those of their squares, cubes and fourth powers had already been published

times $-c$. Further, on multiplying that cubic equation by $x-d$ there will be produced this quartic one,

$$x^4 - (a+b-c+d)\,x^3 + (ab-ac-bc+ad+bd-cd)\,x^2$$
$$+ (abc-abd+bcd+acd)\,x - abcd = 0,$$

where the coefficient of the second term with its signs changed, $a+b-c+d$, is the sum of all the roots; that of the third, $ab-ac-bc+ad+bd-cd$, the sum of the products of the individual ones two at a time; that of the fourth with its signs changed, $-abc+abd-bcd-acd$, the sum of their (solid) products three at a time; that of the fifth, $-abcd$, the unique product of them all. Hence, first, we gather that in any equation having neither fractional nor surd terms all roots which are not surd—and their products two, three or more at a time—are one or other of the integral divisors of the final term; and in consequence, when it has been established that none of the divisors of the final term is either a root of the equation or the product of two or more roots, it will at once be affirmed that no[500] root or product of roots exists unless it be surd.

Let us now suppose that the coefficients, with signs [duly] changed of the terms of an equation be $p, q, r, s, t, v, \dots$:[501] precisely, let that of the second be p, that of the third be q, that of the fourth r, that of the fifth s, and so on. Then, with due observation of the signs of the terms, make $p = a$, $pa + 2q = b$,

$$pb + qa + 3r = c, \quad pc + qb + ra + 4s = d, \quad pd + qc + rb + sa + 5t = e,$$

$pe + qd + rc + sb + ta + 6v = f$, and so on indefinitely, observing the sequence of the progression: a will then be the sum of the roots, b the sum of the squares of the separate roots, c the sum of their cubes, d the sum of their fourth powers, e the sum of their fifth powers, f the sum of their sixth powers, and so on in the case of the rest.[502] In the equation $x^4 - x^3 - 19x^2 + 49x - 30 = 0$ as an example, where the coefficient of the second term is -1, that of the third -19, that of the fourth $+49$, and that of the fifth -30, you will have to set $p = 1$, $q = 19$, $r = -49$, $s = 30$ and there will arise therefrom $a = (p =) 1$,

$$b = (pa + 2q \text{ or } 1 + 38 =) 39, \quad c = (pb + qa + 3r \text{ or } 39 + 19 - 147 =) -89,$$

$d = (pc + qb + ra + 4s \text{ or } -89 + 741 - 49 + 120 =) 723, \dots$ Consequently the sum of the roots will be 1, the sum of the squares of the roots 39, the sum of their cubes -89, and the sum of their fourth powers 723. To be sure, the roots of that

---

by Albert Girard in his *Inuention Nouvelle en l'Algebre* (Amsterdam, 1629): fii^r, but it is unlikely that Newton was aware of the work of his predecessor. We have seen (1: 519, note (14)) that Newton had derived his rule as an unproved generalization of his computations on the coefficients (expressed as symmetric functions of the roots) of an equation of eighth degree. No rigorous demonstration of the rule was published till the content of Maclaurin's letter to Stanhope on 8 July 1743 systematically establishing it was printed by 'the Publisher' (Patrick Murdoch?) in Part II, Chapter XII of Maclaurin's posthumous *Treatise of Algebra* (London, 1748): 286–96.

illius radices sunt 1, 2, 3 & −5, et harum summa 1+2+3−5 est 1, summa quadratorum 1+4+9+25 est 39, summa cuborum 1+8+27−125 est −89 & summa quadrato-quadratorum 1+16+81+625 est 723.

De limitibus Et hinc colliguntur limites inter quos consistent radices æquationis ubi nulla
æquationum. earum impossibilis est. Nam cum radicum omnium quadrata sunt affirmativa, quadratorum summa affirmativa erit, ideoꝗ quadrato maximæ radicis major. ‖[167] Et eodem argumento, summa quadrato-quadratorum ‖ radicum omnium major erit quam quadrato-quadratum radicis maximæ, et summa cubo-cuborum major quam cubo-cubus radicis maximæ. Quamobrem si limitem desideres quem radices nullæ transgrediuntur, quære summam quadratorum radicum et extrahe ejus radicem quadraticam. Hæc enim radix major erit quam radix maxima æquationis. Sed ad radicem maximam propius accedes si quæras summam quadrato-quadratorum et extrahas ejus radicem quadrato-quadraticam, et adhuc magis si quæras summā cubo-cuborum et extrahas ejus radicem cubo-cubicam: et ita in infinitum.[503] Sic in æquatione præcedente radix quadratica summæ quadratorum radicum, seu $\sqrt{39}$ est $6\frac{1}{4}$ quam proximè, et $6\frac{1}{4}$ magis distat a nihilo quam ulla radicum 1, 2, 3, −5. At radix quadrato-quadratica summæ quadrato-quadratorum nempe $\sqrt[4]{723}$ quæ est $5\frac{1}{5}$ circiter propius accedit ad radicem a nihilo remotissimam −5.

[1681] Si inter summam quadratorum et summam quadrato-quadratorum radicum
Lect. 10. inveniatur media proportionalis, erit ea paulo major quam summa cuborum radicum sub signis affirmativis connexorum. Et inde hujus mediæ proportionalis & summæ cuborum sub proprijs signis, ut prius inventæ, semisumma erit major quam summa cuborum radicum affirmativarum, et semidifferentia major quam summa cuborum radicum negativarum. Atꝗ adeo maxima radicum ‖[168] affirmativarum minor erit quam ‖ radix cubica illius semisummæ, et maxima radicum negativarum minor quam radix cubica illius semidifferentiæ.[504] Sic in æquatione præcedente media proportionalis inter summam quadratorum radicum 39 & summam quadrato-quadratorum 723 est 168 circiter. Summa cuborum sub proprijs signis supra erat −89. Hujus et 168 semisumma est $39\frac{1}{2}$, semidifferentia $128\frac{1}{2}$. Prioris radix cubica, quæ est $3\frac{1}{2}$ circiter, major est quam maxima radicum affirmativarum 3. Posterioris radix cubica quæ est $5\frac{1}{21}$

---

(503) In general, if s_n is the sum of the n-th powers of the roots (all real), it is evident that $\sqrt[2n]{[s_{2n}]}$ is larger than the greatest root (positive or negative), but will approach it asymptotically as n increases. (Where all roots are not real, of course, the function will not necessarily tend to the greatest real root but may approximate the modulus of a pair of conjugate complex roots.) The remark is adumbrated on 1: 531–2.

(504) If the roots (supposed all real) of the equation are $\pm a_i$, a_i positive, and

$$s_k^+ = \Sigma(+a_i)^k, \quad s_k^- = \Sigma(|-a_i|)^k,$$

then the sum of the k-th powers of the roots and that of the k-th powers of the absolute values

equation are 1, 2, 3 and −5 and the sum of these (1 + 2 + 3 − 5) is indeed 1, the sum of their squares (1 + 4 + 9 + 25) is 39, the sum of their cubes (1 + 8 + 27 − 125) is −89, and the sum of their fourth powers (1 + 16 + 81 + 625) is 723.

And from this are gathered limits between which an equation's roots shall be positioned when none of them is impossible. For, since the squares of all roots are positive, the sum of their squares will be positive and in consequence greater than the square of the greatest root. And by the same argument the sum of the fourth powers of all the roots will be greater than the fourth power of the greatest root, and the sum of the sixth powers greater than the sixth power of the greatest root. Consequently, should you desire a limit which none of the roots shall surpass, ascertain the sum of the squares of the roots and extract its square root: for this root will be greater than the greatest root of the equation. But you will approach the greatest root more nearly if you seek the sum of the fourth powers and extract its fourth root; and still more so if you ascertain the sum of the sixth powers and extract its sixth root: and so indefinitely.[503] Thus in the preceding equation the square root of the sum of the squares of the roots, that is, $\sqrt{39}$, is $6\frac{1}{2}$ very nearly, and $6\frac{1}{2}$ is further distant from zero than any of the roots 1, 2, 3, −5. But the fourth root of the sum of their fourth powers, namely $\sqrt[4]{723}$—that is, $5\frac{1}{5}$ roughly—more nearly approaches the root, −5, farthest removed from zero.

If the mean proportional between the sum of the squares of the roots and the sum of their fourth power be found, it will be slightly greater than the sum of the cubes of the roots conjoined under positive signs. In consequence, the half-sum of this mean proportional and of the sum of the cubes under their proper signs (as before found) will be greater than the sum of the cubes of the positive roots, and their half-difference greater than the sum of the cubes of the negative roots; hence the greatest of the positive roots will be less than the cube root of that half-sum, and the greatest of the negative roots less than the cube root of that half-difference.[504] Thus in the preceding equation the mean proportional between the sum, 39, of the squares of the roots and the sum, 723, of their fourth powers is about 168, while, above, the sum of their cubes under their proper signs was −89. Of the latter and 168 the half-sum is $39\frac{1}{2}$, the half-difference $128\frac{1}{2}$. The cube root—about $3\frac{1}{2}$—of the former is greater than the greatest of the positive

of the roots are respectively $s_k = \sum_i (\pm a_i)^k = s_k^+ + (-1)^k \cdot s_k^-$ and $\sigma_k = \sum_i (\pm |a_i|)^k = s_k^+ + s_k^-$. From the Cauchy–Schwarz equality

$$\left(\sum_i a_i^{n-1}\right)\left(\sum_i a_i^{n+1}\right) - \left(\sum_i a_i^n\right)^2 = \sum_{i \neq j}(a_i a_j)^{n-1}(a_i - a_j)^2,$$

easily establishable for low values of i and n and then readily generalizable, it follows that $\sqrt{[s_{2p} \cdot s_{2p+2}]} \gtrless \sigma_{2p+1} = s_{2p+1}^+ + s_{2p+1}^-$. In particular $\sqrt{[s_2 s_4]} \gtrless s_3^+ + s_3^-$, whence, since $s_3 = s_3^+ - s_3^-$, at once $\frac{1}{2}(\sqrt{[s_2 s_4]} + s_3) \gtrless s_3^+$, $\frac{1}{2}(\sqrt{[s_2 s_4]} - s_3) \gtrless s_3^-$, with the greatest positive and negative roots bounded by $\sqrt[3]{s_3^+}$ and $\sqrt[3]{s_3^-}$ respectively.

proxime, transcendit[505] radicem negativam − 5. Quo exemplo videre est quam prope ad radicem hac methodo acceditur ubi unica tantum radix negativa est vel unica affirmativa. Et tamen propius adhuc accederetur, si inter summam quadrato-quadratorum et summam cubo-cuborum media proportionalis inveniretur atcჳ ex hujus et summæ quadrato-cuborum radicum semisumma et semidifferentia radices quadrato-cubicæ extraherentur.[506] Nam radix quadrato-cubica semisummæ transcenderet maximam radicem affirmativam, & radix quadrato-cubica semidifferentiæ maximam seu extimam negativam, sed excessu multo minore quam ante. Cum igitur radix quælibet, augendo vel diminuendo radices omnes fieri potest minima, dein minima in maximam

|| [169] converti[507] & postea omnes præter maximam fieri || negativæ, constat quomodo radix imperata quam proximè potest obtineri.

Si radices omnes præter duas negativæ sunt, possunt illæ duæ simul hoc modo erui. Inventa juxta methodum præcedentem summa cuborum duarum illarum radicum, ut et summa quadrato-cuborum et summa quadrato-quadrato-cuborum radicum omnium; inter posteriores duas summas quære mediam proportionalem, et ea erit differentia inter summam cubo-cuborum radicum affirmativarum et summam cubo-cuborum radicum negativarum quam proximè;[508] adeocჳ hujus mediæ proportionalis & summæ cubo-cuborum radicum omnium semisumma erit summa[509] cubo-cuborum radicum affirmativarum, et semidifferentia erit summa cubo-cuborum radicum negativarum. Habita igitur tum summa cuborum, tum summa cubo-cuborum radicum duarum affirmativarum, de duplo summæ posterioris aufer quadratum summæ prioris, & reliqui radix quadratica erit differentia cuborum duarum radicum. Habita vero tum summa tum differentia cuborum habentur cubi ipsi. Extrahe eorum radices cubicas et habebuntur æquationis radices duæ affirmativæ quam proxime.[510] Et si in altioribus potestatibus opus consimile institueretur magis

|| [170] adhuc accederetur ad radices. Sed hæ limitationes || ob difficilem calculum minus usui sunt et ad æquationes tantum extendunt quæ nullas habent radices imaginarias. Quapropter limites alia ratione invenire jam docebo quæ et facilior sit & ad omnes æquationes extendat.

---

(505) Understand in absolute value.

(506) In the terms of note (504), $\sqrt{[s_4 s_6]} \gtrless s_5^+ + s_5^-$ and $s_5 = s_5^+ - s_5^-$, whence

$$\tfrac{1}{2}(\sqrt{[s_4 s_6]} + s_5) \gtrless s_5^+ \quad \text{and} \quad \tfrac{1}{2}(\sqrt{[s_4 s_6]} - s_5) \gtrless s_5^-,$$

with $\sqrt[5]{s_5^+}$, $\sqrt[5]{s_5^-}$ narrow bounds to the greatest positive and negative roots.

(507) By the reciprocation $x \to 1/x$.

(508) In the terms of note (504), it follows from the general Schwarz equality that $\sqrt{[s_{2p-1} s_{2p+1}]} \approx s_{2p}^+ - s_{2p}^-$: Newton states the particular case for which $p = 3$.

(509) In a momentary slip Newton wrote 'semisumma' (half-sum). The lapse was passed by Whiston into his 1707 *editio princeps* but corrected by Newton in his 1722 revise.

(510) Since (see note (508)) $\sqrt{[s_5 s_7]} \approx s_6^+ - s_6^-$ and $s_6 = s_6^+ + s_6^-$, immediately

$$\tfrac{1}{2}(s_6 + \sqrt{[s_5 s_7]}) \approx s_6^+, \quad \tfrac{1}{2}(s_6 - \sqrt{[s_5 s_7]}) \approx s_6^-.$$

roots, 3; that of the latter—very nearly $5\frac{1}{21}$—surpasses[505] the negative root -5. From this example it may be seen that a near approach to a root is effected by this method when there is only a single negative root or but a single positive one. Yet a closer approach still would be made if the mean proportional between the sum of the fourth powers and the sum of the sixth ones were to be found, and then the fifth roots of the half-sum and half-difference of this and the sum of the fifth powers of the roots extracted.[506] For the fifth root of the half-sum would surpass the greatest positive root, and the fifth root of the half-difference the greatest (that is, most remote) negative one, but by an excess much less than before. Since therefore any root can, by increasing or diminishing all the roots, become the least one and this least one then converted into being the greatest,[507] while subsequently all the roots except the greatest one can be made negative, it is established how a very near approximation to any appointed root may be obtained.

If all roots except for two are negative, those two can simultaneously be derived in this manner. Having found, following the preceding method, the sum of the cubes of those two roots, along with the sums of the fifth and seventh powers of all the roots, ascertain the mean proportional between the two latter sums and it will be the difference between the sum of the sixth powers of the positive roots and the sum of the sixth powers of the negative ones to a very near approximation;[508] hence the half-sum of this mean proportional and the sum of the sixth powers of all the roots will be the sum of the sixth powers of the positive roots, and their half-difference will be the sum of the sixth powers of the negative ones. Having, then, both the sum of the cubes of the two positive roots and the sum of their sixth powers, from twice the latter sum take away the square of the former sum and the square root of the remainder will be the difference of the cubes of the two roots. Having both the sum and difference of the cubes, of course, you will have the cubes themselves. Extract their cube roots and very close approximations to the two positive roots of the equation will be had.[510] And if an exactly similar procedure were instituted in the case of higher powers, a still closer approach to the roots would be effected. But these limitings are of less service because of the arduousness of their computation and extend but to equations which have no imaginary roots. On this account I shall now explain how to find limits by another technique, one which at once is easier and extends to all equations.

Further, by the preceding it follows that $\frac{1}{2}(\sqrt{[s_2 s_4]} + s_3) \approx s_3^+$. Hence, since Newton assumes but two positive roots, a and b say, we have $\frac{1}{2}(s_6 + \sqrt{[s_5 s_7]}) \approx a^6 + b^6$ and $\frac{1}{2}(\sqrt{[s_2 s_4]} + s_3) \approx a^3 + b^3$, with $(a^3 + b^3)^2 + (a^3 - b^3)^2 = 2(a^6 + b^6)$ and so $\sqrt{[s_6 + \sqrt{[s_5 s_7]} - \frac{1}{2}(\sqrt{[s_2 s_4]} + s_3)]} \approx a^3 - b^3$; accordingly the roots a and b are very nearly the cube roots of the half-sum and half-difference of $\frac{1}{2}(\sqrt{[s_2 s_4]} + s_3)$ and $\sqrt{[s_6 + \sqrt{[s_5 s_7]} - \frac{1}{2}\sqrt{[s_2 s_4]} - \frac{1}{2}s_3]}$. All very ingenious but wholly impractical, as Newton proceeds to admit.

(511)Multiplicetur æquationis terminus unusquisꝗ per numerum dimensionum ejus et dividatur factum per radicem æquationis. Dein rursus multiplicetur unusquisꝗ terminorum prodeuntium per numerum unitate minorem quam prius, et factum dividatur per radicem æquationis. Et sic pergatur semper multiplicando per numeros unitate minores quam prius et factum dividendo per radicem donec tandem termini omnes destruantur quorum signa diversa sunt a signo primi seu altissimi termini præter ultimum. Et numerus ille erit omni affirmativa radice major qui in terminis prodeuntibus scriptus pro radice, efficit eorum qui singulis vicibus per multiplicationem producebantur aggregatum ejusdem semper esse signi cum primo seu altissimo termino æquationis.(512) Ut si proponatur æquatio $x^5 - 2x^4 - 10x^3 + 30xx + 63x - 120 = 0.$(513) Hanc

$$\overset{5 \quad\; 4 \qquad 3 \qquad\; 2 \quad\; 1 \quad\; 0}{}$$

primum sic multiplico $x^5 - 2x^4 - 10x^3 + 30xx + 63x - 120.$ Dein terminos pro-

$$\overset{4 \qquad 3 \qquad\; 2 \quad\; 1 \quad\; 0}{}$$

deuntes divisos per x rursum multiplico sic $5x^4 - 8x^3 - 30xx + 60x + 63.$ et terminos prodeuntes dividendo per x prodeunt $20x^3 - 24xx - 60x + 60$ quos ‖[171] minuendi gratia divido per maximum divisorem 4 et fiunt $5x^3 - ‖6xx - 15x + 15.$ Hi itidem multiplicati per progressionem 3. 2. 1. 0 et divisi per x fiunt $5xx - 4x - 5.$ Et hi multiplicati per progressionem 2. 1. 0 et divisi per $2x$ fiunt $5x - 2.$ Jam cum terminus æquationis altissimus x^5 affirmativus sit, tento quinam numerus scriptus in his productis pro x efficiet ea omnia affirmativa esse. Et quidem tentando 1, fit $5x - 2 = 3$ affirmativum sed $5xx - 4x - 5$ fit -4 negativum. Quare limes erit major quam 1. Tento itaꝗ numerum aliquem majorem puta 2. Et in singulis substituendo 2 pro x, evadunt

$$5x - 2 = 8.$$
$$5xx - 4x - 5 = 7.$$
$$5x^3 - 6xx - 15x + 15 = 1.$$
$$5x^4 - 8x^3 - 30xx + 60x + 63 = 79.^{(514)}$$
$$x^5 - 2x^4 - 10x^3 + 30xx + 63x - 120 = 46.^{(514)}$$

---

(511) This paragraph—and indeed the following section on the 'invention' of divisors—is taken over without essential variant from the corresponding section, entitled by us 'De limitibus æquationis', of the stray leaf reproduced as 1, §2 preceding.

(512) Compare 1, §2: note (31) above. If each of the derived functions

$$f^{[p]}(x) \equiv \sum_{0 \leqslant i \leqslant n-p} ((n-i)_p \, a_i x^{n-i-p})$$

of the given equation $f(x) \equiv \sum_{0 \leqslant i \leqslant n} (a_i x^{n-i})$ are positive, $p = 1, 2, 3, ..., n$, for a value $x = X$ for which $f(X) > 0$, then for all $h \geqslant 0$ it follows that

$$f(X+h) = f(X) + hf^{[1]}(X) + \tfrac{1}{2}h^2 f^{[2]}(X) + ... + h^n a_0$$

is positive, so that X is an upper bound to the (real) roots of $f(x) = 0$. As before, wherever possible Newton divides through each derivative function by the highest common factor of the coefficients (which for $f^{[p]}(x)$ will be some multiple of p).

[511]Let each individual term of an equation be multiplied by the number of its dimensions and the product divided by the equation's root; then again, let each of the resulting terms be multiplied by a number one less than before and the product divided by the equation's root; and so proceed, multiplying perpetually by numbers one less than before and dividing the product by the root, till at length all terms whose signs are different from that of the first (or highest) term are destroyed, the last one excepted. Then the number which, when entered in the resulting terms in place of the root, makes the aggregate of those produced by multiplication at each separate stage always of the same sign as the equation's first or highest term will be greater than every positive root of it.[512] If, for instance, the equation $x^5 - 2x^4 - 10x^3 + 30x^2 + 63x - 120 = 0$[513] be proposed I first multiply this as follows:

$$\begin{array}{ccccccc} 5 & 4 & 3 & 2 & 1 & 0 \\ x^5 - 2x^4 & - 10x^3 & + 30x^2 & + 63x & - 120; \end{array}$$

then the resulting terms divided by x I again multiply so:

$$\begin{array}{ccccc} 4 & 3 & 2 & 1 & 0 \\ 5x^4 - 8x^3 & - 30x^2 & + 60x & + 63, \end{array}$$

and on dividing the resulting terms by x there ensues $20x^3 - 24x^2 - 60x + 60$: in order to diminish these I divide through by their greatest divisor 4 and they become $5x^3 - 6x^2 - 15x + 15$. These likewise, after being multiplied by the progression 3, 2, 1, 0 and divided by x, come to be $5x^2 - 4x - 5$. And these, multiplied by the progression 2, 1, 0 and divided by $2x$, come to be $5x - 2$. Now since the highest term, x^5, of the equation is positive, I test for a number which, when written in these products in x's place, will render them all positive. On testing 1, indeed, there comes to be $5x - 2 = 3$, positive, but $5x^2 - 4x - 5$ comes to be -4, negative. Therefore the limit will be greater than 1. I accordingly test some larger number, 2 say. On substituting 2 in x's place in the individual products there then proves to be:

$$5x - 2 = 8,$$
$$5x^2 - 4x - 5 = 7,$$
$$5x^3 - 6x^2 - 15x + 15 = 1,$$
$$5x^4 - 8x^3 - 30x^2 + 60x + 63 = 79,\text{[514]}$$
$$x^5 - 2x^4 - 10x^3 + 30x^2 + 63x - 120 = 46.\text{[514]}$$

---

(513) The last but one term is changed from that in his previous example (compare 1, §2: note (33)), causing Newton some slight initial confusion; see next note.

(514) These values were originally entered as '81', '50' and '81. 50' respectively: evidently Newton momentarily forgot that in his present example the term '65x' in his original equation is (see note (513)) replaced by '63x', so necessitating a reduction of his previous values by 2 and 4 respectively.

Quare cum numeri prodeuntes 8. 7. 1. 79. 46[514] sint omnes affirmativi erit numerus 2 major quam radicum affirmativarum maxima. Similiter si limitem negativarum radicum invenire vellem, tento numeros negativos. Vel quod perinde est muto signa terminorum alternorum et tento affirmativos. Mutatis autem terminorum alternorum signis, quantitates in quibus numeri substituendi sunt fient

$$5x + 2$$
$$5xx + 4x - 5$$
$$5x^3 + 6xx - 15x - 15$$
$$\| 5x^4 + 8x^3 - 30xx - 60x + 63$$
$$x^5 + 2x^4 - 10x^3 - 30xx + 63x + 120.$$

‖ [172]

Ex his seligo quantitatem aliquam ubi termini negativi maximè prævalere videntur: puta $5x^4 + 8x^3 - 30xx - 60x + 63$, & hic substituendo pro x numeros 1 et 2 prodeunt numeri negativi -14 & -33. Unde limes erit major quam -2. Substituendo autem numerum 3 prodit numerus affirmativus 234. Et similiter in cæteris quantitatibus substituendo numerum 3 pro x prodit semper numerus affirmativus. Id quod ex inspectione sola colligere licet.[515] Quare numerus -3 transcendit omnes radices negativas. At<unk> ita habentur limites 2 & -3 inter quos radices omnes[516] consistunt.

Octob 1682
Lect 1.
[Inventio divisorum.]

[517]Horum vero limitum inventio usui est tum in reductione æquationum per radices rationales[,] tum in extractione radicum surdarum ex ipsis; ne forte radicem extra hos limites aliquando quæramus. Sic in æquatione novissima si radices rationales, siquas forte habeat, invenire vellem: ex superioribus certum est has non alias esse posse quam divisores ultimi termini æquationis qui hic est 120. Proin tentando omnes ejus divisores, si nullus eorum scriptus in æquatione pro radice x efficeret omnes[518] terminos evanescere: certum est æquationem non admittere radicem nisi quæ sit surda. At ultimi termini 120, divisores permulti sunt, nimirum 1. -1. 2. -2. 3. -3. 4. -4. 5. -5. 6. -6. 8. -8. 10.

(515) In the third line, for instance, since, when $x = 3$, at once $5x^3 > 15x$ and $6x^2 > 15$, immediately $5x^3 + 6x^2 - 15x - 15 > 0$; in the next line, similarly, $5x^4 > 30x^2$ and $8x^3 > 60x$.

(516) The given quintic has, in fact, a unique real root $x \approx 1\frac{1}{2}$, though its derivative is zero for two values in the interval $[-3, -1]$.

(517) The following section (on Newton's pages 173–87) is an extensively augmented revise of the concluding portion of the stray worksheet reproduced as 1, §2 above. It is clear that when Newton began to compose the present, lengthy manuscript he intended to add its content as a (considerably briefer) insert on his page 25 above (see note (71)), for in a set of 'Correctiones' added to the manuscript he wrote out a somewhat roughly penned first revise, titled 'De inventione Divisorum', where—to allow for the change of context—the present 'æquationes' are become mere general algebraic 'quantitates'. Failing to realize that the present text is an extended revision of the latter 'correction', Whiston religiously followed Newton's interim dictate that the preliminary revise be inserted on his earlier page 25 and then here reproduced the first paragraph only, continuing immediately after—without making

Therefore, since the resulting numbers 8, 7, 1, 79 and 46[514] are all positive, the number 2 will be greater than the greatest of the positive roots. Similarly, were I wishing to find a limit for the negative roots, I test negative numbers: or, what is exactly equivalent, I change the signs of alternate terms and text positive ones. When the signs of alternate terms are changed, however, the quantities in which numbers are to be substituted will become:

$$5x+2,$$
$$5x^2+4x-5,$$
$$5x^3+6x^2-15x-15,$$
$$5x^4+8x^3-30x^2-60x+63,$$
$$x^5+2x^4-10x^3-30x^2+63x+120.$$

From these I select some quantity in which negative terms appear to be most prevalent, $5x^4+8x^3-30x^2-60x+63$ say, and on substituting the numbers 1 and 2 in it in x's place there result the negative numbers -14 and -33: hence the limit will be greater than -2. On substituting the number 3, however, the positive number 234 results. And similarly, on substituting the number 3 in x's place in the remaining quantities, a positive number always results. Mere inspection alone permits that inference.[515] Consequently the number -3 surpasses all the negative roots. And in this way are had the limits 2 and -3 between which all the roots[516] are situated.

[517]The ascertaining of these limits is indeed of use both in the reduction of [The finding equations by means of rational roots and in the extraction from them of surd of divisors.] roots, saving us from on occasion searching for a root—as we well might— outside these limits. Thus were I to wish to find the rational roots in the most recent equation (if it should chance to have any), from the above it is certain that these can be no other than divisors of the equation's final term, here 120. Accordingly, if, on testing all its divisors, none of them when entered in the equation in x's place were to make all the[518] terms disappear, it is certain that the equation cannot admit of any root unless it be surd. However, the divisors of the final term 120 are very many, namely 1, -1, 2, -2, 3, -3, 4, -4, 5, -5,

---

any sign to his reader that more than fourteen pages of the manuscript were thereby silently omitted, or attempting to heal the break in continuity—with the now illogical opening sentence, 'Hactenus reductionem æquationum tradidi quæ rationales divisores admittunt....', of the next section 'Æquationum Reductio per divisores surdos' (on Newton's page 187 below). The text was not restored to its mature state by Newton in his 1722 revise—more out of oversight, we may suspect, than from any wish to sustain the illogical structure of Whiston's *editio princeps*—and the inferior version is reproduced in all printed editions. We here publish the extended revised version of the 'Inventio divisorum' in its logical place, repairing Whiston's inconsistencies and—we hope—adhering to Newton's own best intention. Variants from it in the interim 'corrected' text are recorded in the version reproduced in §2 below.

(518) 'ejus' (its) is cancelled.

$-10. 12. -12. 15. -15. 20. -20. 24. -24. 30. -30. 40. -40. 60. -60. 120.$ &

|| [173] -120, et || hos omnes divisores tentare, tædio esset. Cognito autem quod radices inter limites 2 et -3 consistunt, liberamur a tanto labore. Jam enim non opus erit divisores tentare nisi qui sunt inter hos limites, nimirum divisores $1, -1$ & -2. Nam si horum nullus radix est, certum est æquationem non habere radicem nisi quæ sit surda.

Attamen hoc non obstante inventio divisorum aliquando satis molesta esse potest. Ut si proponeretur æquatio $24x^5 - 50x^4 + 49x^3 - 40xx + 64x + 30 = 0$, multi quidem divisores intra limites radicum[519] hic prodirent quos omnes tentare fastidio esset. Quare methodum inveniendi divisores qua ab hujusmodi molestijs liberamur, hic tradere non pigebit.

Æquationum Reductio per divisores rationales.   Æquationem dispone secundum dimensiones radicis ejus, vel si literalis[520] est, secundum dimensiones literæ cujusvis quæ in ea reperitur: et pro litera illa sigillatim substitue tres vel plures terminos hujus progressionis $3. 2. 1. 0. -1. -2$, ac terminos totidem resultantes una cum omnibus eorum divisoribus statue e regione correspondentium terminorum progressionis, positis divisorum signis tam affirmativis quam negativis. Deinde e regione etiam statue progressiones arithmeticas quæ per omnium numerorum divisores percurrunt pergentes a majoribus terminis ad minores eodem ordine quo termini progressionis $3. 2. 1. 0. -1. -2$ pergunt et quarum termini differunt vel unitate vel numero aliquo qui dividit altissimum terminum æquationis. Siqua occurrit istiusmodi pro-

|| [174]   gressio, || iste terminus ejus qui stat e regione termini 0 progressionis primæ, divisus per differentiam terminorum et cum signo suo annexus litteræ præfatæ componet quantitatem per quam divisio tentanda est.[521]

Ut si æquatio sit $x^3 - xx - 10x + 6 = 0$, pro x substituendo sigillatim terminos progressionis $1, 0, -1$ orientur numeri $-4. +6. 14$ quos cum omnibus eorum divisoribus colloco e regione terminorum progressionis $1. 0. -1$ hoc modo

1	-4	1. 2. 4	$+4.$
0	6	1. 2. 3. 6	$+3.$
-1	14	1. 2. 7. 14	$+2.$

Dein quoniam altissimus terminus x^3 per nullum numerum præter unitatem divisibilis est, quæro in divisoribus progressionem arithmeticam cujus termini differunt unitate, & a superioribus ad inferiora pergendo decrescunt perinde ac

---

(519) The given equation has, in fact, only one real root $x \approx -\frac{1}{3}$.

(520) Having 'letters' (algebraic quantities) for the coefficients of its terms.

(521) As in the draft version (1, §2 above), understand that 'si per nullum hac methodo inventum divisorem res succedit, concludere potes æquationem reduci non posse per divisorem unius dimensionis' (if trial succeeds by no divisor found by this method, you can conclude that the equation is irreducible by a divisor of one dimension). Newton will again take the point up in the sequel.

6, −6, 8, −8, 10, −10, 12, −12, 15, −15, 20, −20, 24, −24, 30, −30, 40, −40, 60, −60, 120 and −120, and it would be wearisome to test all these divisors. But once it is known that the roots are situated between the limits 2 and −3 we are freed from so vast an effort, since now there will be no need to test any divisors but those between these limits, namely the divisors 1, −1 and −2. For if none of these is a root, it is settled that the equation has no root unless it be surd.

But this not withstanding, however, the finding of divisors can still on occasion be troublesome enough. If, for instance, the equation

$$24x^5 - 50x^4 + 49x^3 - 40x^2 + 64x + 30 = 0$$

were proposed, many divisors would indeed here prove to be within the limits of the roots[519] and it would be a loathsome matter to test them all. It will therefore not be an infliction to recount at this point a method of finding divisors by which we are freed from trouble of this sort.

Arrange the equation according to the dimensions of its root, or, if it is a literal one,[520] according to the dimensions of any one of the letters found in it, and in place of that letter separately substitute three or more terms of this progression [...] 3, 2, 1, 0, −1, −2, [...] and place the equal number of resulting terms together with all their divisors in line with corresponding terms of the progression, setting both positive and negative signs on the divisors. Next, also in line, place arithmetical progressions which run through the divisors of all the numbers, proceeding from the greater terms to the lesser ones in the same sequence as the terms of the progression 3, 2, 1, 0, −1, −2, ... go, and whose terms differ by unity or some number dividing the highest term of the equation. If any progression of this kind does occur, that term in it which is stationed in line with the term 0 of the first progression will, when divided by the difference of the terms and attached with its sign to the aforesaid letter, make up the quantity by which division is to be attempted.[521]

If the equation be, for instance, $x^3 - x^2 - 10x + 6 = 0$, on separately substituting for x the terms of the progression 1, 0, −1 the numbers −4, +6, +14 will ensue; these together with all their divisors I locate in line with the terms of the progression 1, 0, −1 in this way:

$$
\begin{array}{c|c|c|c}
1 & -4 & 1, 2, 4 & 4 \\
0 & 6 & 1, 2, 3, 6 & 3 \\
-1 & 14 & 1, 2, 7, 14 & 2.
\end{array}
$$

Next, since the highest term x^3 is divisible by no number except unity, I look among the divisors for an arithmetical progression whose terms differ by unity and in the sequence from upper to lower ones decrease in the same manner as

The reduction of equations by rational divisors.

termini progressionis lateralis 1. 0. -1. Et hujusmodi progressionem unicam tantum invenio nempe 4. 3. 2 cujus itaq3 terminum $+3$ seligo, qui stat e regione termini 0 progressionis primæ 1. 0. -1, tentoq3 divisionem per $x+3$ et res succedit prodeunte $xx-4x+2$.[522] Quamobrem concludo unam e tribus radicibus æquationis propositæ $x^3-xx-10x+6=0$, esse numerum negativum -3, alteras duas easdem esse cum radicibus hujus æquationis $xx-4x+2=0$. id est $2+\sqrt{2}$ & $2-\sqrt{2}$.

‖[175]   Haud secus si proponatur æquatio $x^4-6x^3-5xx+56x-30=0$, substituendo sigillatim pro x terminos hujus progressionis 1. 0. -1. -2. ‖ orientur numeri 16. -30. -84. -98. Quos cum eorum divisoribus colloco e regione substitutorum terminorum progressionis ut hic fit

1	16	1. 2. 4. 8. 16	$-4.\ 4.$
0	-30	1. 2. 3. 5. 6. 10. 15. 30	$-5.\ 3.$
-1	-84	1. 2. 3. 4. 6. 7. 12. 14. 21 &c	$-6.\ 2.$
-2	-98	1. 2. 7. 14. 49. 98	$-7.\ 1.$

Et in divisoribus animadverto duas progressiones esse deorsum decrescentes quarum termini differunt unitate, nempe -4. -5. -6. -7. & 4. 3. 2. 1. Quare terminos earum seligo qui stant e regione termini 0 progressionis primæ, nempe terminos -5 & $+3$, tentoq3 divisionem per divisores $x-5$ & $x+3$ & res per utrumq3 divisorem succedit prodeunte post geminam divisionem æquatione $xx-4x+2=0$, cujus itaq3 duæ radices $2+\sqrt{2}$ & $2-\sqrt{2}$ una cum -3 & $+5$ erunt radices quatuor æquationis propositæ.

Rursus si æquatio sit $6y^4-y^3-21yy+3y+20=0$, pro y substituo sigillatim 1. 0. -1 et numeros resultantes 7. 20. 1.[523] cum omnibus eorum divisoribus e regione colloco ut sequitur.

1	7	1. 7	7.
0	20	1. 2. 4. 5. 10. 20	4.
-1	1[523]	1	1.

Et in divisoribus hanc solam esse animadverto decrescentem progressionem arithmeticam $+7$. $+4$. $+1$. Hujus terminorum differentia 3 dividit altissimum æquationis terminum $6y^4$. Quare terminum $+4$ qui stat e regione termini 0

---

(522) The remainder of this paragraph and the whole of the next are not contained in the 'corrected' draft published by Whiston in 1707 and confirmed by Newton in his 1722 revise (see note (517)).

(523) As before (see 1, §2: note (40)) this should be '3', so—in context—permitting the two extra final columns '7. 5. 3' and '1. -1. -3': these may only be eliminated by computing further particular values of $f(y) \equiv 6y^4-y^3-21y^2+3y+20$, say, $f(2) = 30$ and $f(-2) = 34$. The slip was passed by Whiston in 1707, caught by Newton in his library copy of the *editio princeps* and duly corrected in his 1722 revise by appropriate amplification: '...pro y substituo

the terms of the side-progression 1, 0, -1. Of this sort I find but a single sequence, namely 4, 3, 2. Accordingly, I select its term $+3$ standing in line with the term 0 of the first progression 1, 0, -1 and attempt division by $x+3$: the approach succeeds, yielding $x^2-4x+2[=0]$.[522] Therefore I conclude that one of the three roots of the equation $x^3-x^2-10x+6 = 0$ proposed is the negative number -3, while the other two are identical with the roots of this latter equation $x^2-4x+2 = 0$; that is, $2+\sqrt{2}$ and $2-\sqrt{2}$.

No differently, if the equation $x^4-6x^3-5x^2+56x-30 = 0$ be proposed, on separately substituting in x's place the terms of this progression 1, 0, -1, -2 the numbers 16, -30, -84, -98 will ensue. These together with their divisors I locate in line with the substituted terms of the progression, as is here done:

1	16	1, 2, 4, 8, 16	$-4.\ 4.$
0	-30	1, 2, 3, 5, 6, 10, 15, 30	$-5.\ 3.$
-1	-84	1, 2, 3, 4, 6, 7, 12, 14, 21, ...	$-6.\ 2.$
-2	-98	1, 2, 7, 14, 49, 98	$-7.\ 1.$

I then notice that in the divisors there are two downwards decreasing progressions whose terms differ by unity, namely -4, -5, -6, -7 and 4, 3, 2, 1, and therefore select their terms stationed in line with the term 0 in the first progression, namely the terms -5 and $+3$, attempting division by the divisors $x-5$ and $x+3$. The approach succeeds by each divisor, yielding after a twin division the equation $x^2-4x+2 = 0$, whose two roots $2+\sqrt{2}$ and $2-\sqrt{2}$, accordingly, along with -3 and $+5$ will be the four roots of the equation proposed.

Again, should the equation be $6y^4-y^3-21y^2+3y+20 = 0$, in y's place I separately substitute 1, 0, -1 and locate the resulting numbers 7, 20, 1[523] together with all their divisors in line as follows:

1	7	1, 7	7.
0	20	1, 2, 4, 5, 10, 20	4.
-1	1	1	1.

I then notice that there is in the divisors this sole downwards decreasing arithmetical progression 7, 4, 1. The difference 3 of its terms divides the highest term $6y^4$ in the equation. Consequently, to the letter y I adjoin the term $+4$

---

sigillatim 2. 1. 0. -1. -2 & numeros resultantes 30. 7. 20. 3. 34 cum omnibus eorum divisoribus e regione colloco ut sequitur.

2	30	1. 2. 3. 5. 10. 15. 30	$+10.$
1	7	1. 7	$+7.$
0	20	1. 2. 4. 5. 10. 20	$+4.$
-1	3	1. 3	$+1.$
-2	34	1. 2. 17. 34	$-2.$

Et in divisoribus hanc solam...progressionem arithmeticam $+10$. $+7$. $+4$. $+1$. -2. ...'.

divisum per differentiam terminorum 3 adjungo literæ y tentoq divisionem
|| [176] || per $y + \frac{4}{3}$ vel quod perinde est per $3y + 4$ & res succedit prodeunte

$$2y^3 - 3yy - 3y + 5 = 0.$$

æquatione nova cujus tres radices una cum $-\frac{4}{3}$ sunt quatuor radices æquationis propositæ.

Atq ita si æquatio sit $24z^5 - 50z^4 + 49z^3 - 140zz + 64z + 30 = 0$ substituendo pro z terminos progressionis 2. 1. 0. -1 emergent numeri 42, 23, 30, 297[524] qui cum divisoribus suis e regione terminorum progressionis sic statuendi sunt

2	42	1. 2. 3. 6. 7. 14. 21. 42	$+3. +3. +7.$
1	23	1. 23	$+1. -1. +1.$
0	30	1. 2. 3. 5. 6. 10. 15. 30	$-1. -5. -5.$
-1	297	1. 3. 9. 11. [27.] 33. 99. 297	$-3. -9. -11.$

Tres occurrunt hic progressiones quorum termini $-1, -5, -5$ e regione termini 0 progressionis considerandi sunt. Hi autem divisi per differentias terminorum 2, 4, 6, dant tres divisores tentandos $z - \frac{1}{2}$, $z - \frac{5}{4}$ & $z - \frac{5}{6}$. Et divisio per ultimum divisorem $z - \frac{5}{6}$ seu $6z - 5$ succedit prodeunte æquatione nova

$$4z^4 - 5z^3 + 4zz - 20z - 6 = 0$$

cujus beneficio reliquæ radices æquationis propositæ investigandæ sunt.

[1682] Si nullus occurrit hac methodo divisor vel nullus qui dividit propositam
Lect. 2. æquationem, concludendum erit æquationem illam non admittere divisorem unius dimensionis. Potest tamen fortasse si plurium sit quam trium dimensionum divisorem admittere duarum. Quo casu divisor ille investigabitur hac methodo. In æquatione illa pro litera radicem significante substitue ut ante
|| [177] || quatuor vel plures terminos progressionis hujus 3. 2. 1. 0. $-1. -2. -3$. Divisores omnes numerorum resultantium sigillatim adde & subduc quadratis correspondentium terminorum progressionis illius ductis in divisorem aliquem numeralem altissimi termini æquationis propositæ, et summas differentiasq e regione progressionis colloca. Dein progressiones omnes collaterales nota quæ per istas summas differentiasq percurrunt. Sit $\mp C$ terminus istiusmodi progressionis qui stat e regione termini 0 progressionis primæ, $\mp B$ differentia quæ oritur subducendo $\mp C$ de termino proximè superiore[525] qui stat e regione termini 1 progressionis primæ, A prædictus termini altissimi divisor numeralis & l litera quæ in æquatione proposita radicem significat, et erit $All \pm Bl \pm C$ divisor tentandus.

Ut si æquatio reducenda sit $x^4 - x^3 - 5xx + 12x - 6 = 0$, pro x scribo successivè 3. 2. 1. 0. $-1. -2$ & resultantes numeros 39. 6. 1. $-6. -21. -26$, una cum

(524) Strictly '$-42. -23. +30. -297$'; see 1, §2: note (41). In the following table we have here inserted the trivially missing factor '27'.

which stands in line with the term 0, dividing it by the difference 3 of the terms, and so attempt division by $y + \frac{4}{3}$ or, what is equivalent, by $3y + 4$. The approach succeeds, yielding $2y^3 - 3y^2 - 3y + 5 = 0$, a new equation whose three roots, together with $-\frac{4}{3}$, are the four roots of the equation propounded.

And thus, if the equation be $24z^5 - 50z^4 + 49z^3 - 140z^2 + 64z + 30 = 0$, on substituting in place of z the terms of the progression 2, 1, 0, -1 there will emerge the numbers 42, 23, 30, 297 :[524] these together with their divisors are to be set out in line with the terms of the progression in this manner:

2	42	1, 2, 3, 6, 7, 14, 21, 42	+3. +3. +7.
1	23	1, 23	+1. −1. +1.
0	30	1, 2, 3, 5, 6, 10, 15, 30	−1, −5. −5.
−1	297	1, 3, 9, 11, 27, 33, 99, 297	−3. −9. −11.

Three progressions occur here, and their terms -1, -5, -5 in line with the term 0 of the progression have to be considered. These in fact, when divided by the differences 2, 4, 6 of the terms, yield three divisors to be tested: $z - \frac{1}{2}$, $z - \frac{5}{4}$ and $z - \frac{5}{6}$. Division by the last divisor $z - \frac{5}{6}$, that is, $6z - 5$, succeeds, producing the fresh equation $4z^4 - 5z^3 + 4z^2 - 20z - 6 = 0$ by means of which the remaining roots of the equation proposed are to be discovered.

If no divisor—or at least none which divides the equation proposed—offers itself by this method, you will be forced to conclude that that equation admits of no divisor of one dimension. Yet it can perhaps, if it is of more than three dimensions, admit of a divisor of two. In that case the divisor will be discovered by this method. In the equation in place of the letter denoting the root substitute, as before, four or more terms of this progression [...] 3, 2, 1, 0, -1, -2, -3 [...]. Add and subtract all divisors of the resulting numbers separately to and from the squares of corresponding terms of that progression multiplied by some numerical divisor of the highest term in the equation proposed, and place the sums and differences in line with the progression. Then note all progressions alongside which run through those sums and differences. Let $\mp C$ be the term in a progression of that type which stands in line with the term 0 in the primary progression, $\mp B$ the difference which arises on subtracting $\mp C$ from the term next above[525] (standing in line with the term 1 in the first progression), A the above-mentioned numerical divisor of the highest term and l the letter which signifies the root in the equation proposed: then $Al^2 \pm Bl \pm C$ will be the divisor to be attempted.

So, if the equation $x^4 - x^3 - 5x^2 + 12x - 6 = 0$ needs to be reduced, in x's place I write successively 3, 2, 1, 0, -1, -2 and the resulting numbers 39, 6, 1, -6,

---

(525) Namely $\mp B \mp C$, the value when $l = 1$ of $Al^2 - (Al^2 \pm Bl \pm C)$.

eorum divisoribus e regione dispono: addoqȝ et subduco divisores terminis progressionis illius quadratis ductisqȝ in divisorem numeralem altissimi æquationis termini x^4 qui unitas est, vizt terminis 9. 4. 1. 0. 1. 4, & summas differen

‖ [178] tiasqȝ e latere pariter dispono. Dein progressiones quæ in ijsdem obveniunt, ‖ e latere etiam scribo ut sequitur.

3	39	1. 3. 13. 39	9	−30. −4. 6. 8. 10. 12. 22. 48.	−4.	6.
2	6	1. 2. 3. 6	4	−2. 1. 2. 3. 5. 6. 7. 10	−2.	3.
1	1	1	1	0. 2	0.	0.
0	6	1. 2. 3. 6	0	−6. −3. −2. −1. 1. 2. 3. 6	2.	−3.
−1	21	1. 3. 7. 21	1	−20. −6. −2. 0. 2. 4. 8. 22.	4.	−6.
−2	26	1. 2. 13. 26	4	−22. −9. 2. 3. 5. 6. 17. 30	6.	−9.

Harum progressionum terminos 2 et −3 qui stant e regione termini 0 progressionis quæ in columna prima est usurpo successive pro $\mp C$. Differentias quæ oriuntur subducendo hos terminos de terminis superioribus 0 et 0, nempe −2 et +3 usurpo respectivè pro $\mp B$. Unitatem item pro A, et x pro l, et sic pro $All \pm Bl \pm C$ habeo divisores tentandos $xx+2x-2$ et $xx-3x+3$, per quorum utrumqȝ res succedit. Quamobrem æquationis propositæ quadrato-quadraticæ radices quatuor eædem erunt quæ harum duarum quadraticarum $xx+2x-2=0$ & $xx-3x+3=0$. adeoqȝ ad harum radicum inventionem jam nil amplius restat quam radicum quadraticarum extractio.

Proponatur jam æquatio $3y^5-6y^4+y^3-8yy-14y+14=0$, & operatio erit ut sequitur.[526]

3	170		27		−7.	17.
2	38	1. 2. 19. 38	12	−26. −7. 10. 11. 13. 14. 31. 50	−7.	11.
1	10	1. 2. 5. 10	3	−7. −2. 1. 2. 4. 5. 8. 13	−7.	5.
0	14	1. 2. 7. 14	0	−14. −7. −2. −1. 1. 2. 7. 14	−7.	−1.
−1	10	1. 2. 5. 10	3	−7. −2. 1. 2. 4. 5. 8. 13	−7.	−7.
−2	190		12		−7.	−13.

Primo rem tento addendo et subducendo divisores quadratis terminorum progressionis 4. 1. 0. 1, usurp[at]o 1 pro A sed res non succedit. Quare pro A

‖ [179] usurpo 3 alterum nempe termini altissimi $3y^5$ ‖ divisorem numeralem et quadratis istis multiplicatis per 3, hoc est numeris 12. 3. 0. 3 addo subducoqȝ divisores, et in terminis resultantibus progressiones hasce duas invenio −7. −7. −7. −7. & 11. 5. −1. −7. Expeditionis gratia neglexeram divisores extimorū numerorum 170 & 190. [Q]uare continuatis utrinqȝ progressionibus sumo proximos earum hinc inde terminos, vizt −7 & 17 superius & −7 ac −13

---

(526) Strictly, the second column should read '+170. −38. −10. +14. +10. −190', but only their absolute values are here relevant.

−21, −26 together with their divisors I arrange in line: I then add and subtract the divisors to and from the squared terms of that progression multiplied by a numerical divisor—here unity alone—of the highest term x^4 of the equation, that is, to and from the terms 9, 4, 1, 0, 1, 4, and likewise set the sums and differences out alongside. Next, the progressions which turn up in these I also write alongside as follows:

3	39	1, 3, 13, 39	9	−30, −4, 6, 8, 10, 12, 22, 48	−4.　6.
2	6	1, 2, 3, 6	4	−2, 1, 2, 3, 5, 6, 7, 10	−2.　3.
1	1	1	1	0, 2	0.　0.
0	6	1, 2, 3, 6	0	−6, −3, −2, −1, 1, 2, 3, 6	2. −3.
−1	21	1, 3, 7, 21	1	−20, −6, −2, 0, 2, 4, 8, 22	4. −6.
−2	26	1, 2, 13, 26	4	−22, −9, 2, 3, 5, 6, 17, 30	6. −9.

In these progressions the terms 2 and −3 standing in line with the term 0 of the progression in the first column I employ successively in place of $\mp C$. The differences which arise on subtracting these terms from the terms 0 and 0 above, namely −2 and +3, I use respectively in place of $\mp B$. I similarly put unity for A and x for l, and in this way in place of $Al^2 \pm Bl \pm C$ I have the divisors $x^2 + 2x − 2$ and $x^2 − 3x + 3$ to be tested. Through both of these the reduction succeeds. In consequence, the four roots of the quartic equation proposed will be identical with those possessed by these two quadratics, $x^2 + 2x − 2 = 0$ and $x^2 − 3x + 3 = 0$; and hence, to determine these roots, nothing more now remains except the extraction of quadratic roots.

Should now the equation $3y^5 − 6y^4 + y^3 − 8y^2 − 14y + 14 = 0$ be proposed, the procedure will be as follows:[526]

3	170		27		−7.　17.
2	38	1, 2, 19, 38	12	−26, −7, 10, 11, 13, 14, 31, 50	−7.　11.
1	10	1, 2, 5, 10	3	−7, −2, 1, 2, 4, 5, 8, 13	−7.　5.
0	14	1, 2, 7, 14	0	−14, −7, −2, −1, 1, 2, 7, 14	−7. −1.
−1	10	1, 2, 5, 10	3	−7, −2, 1, 2, 4, 5, 8, 13	−7. −7.
−2	190		12		−7. −13.

I first make a trial attempt by adding and subtracting the divisors to and from the squares 4, 1, 0, 1 of the terms of the progression, employing 1 in place of A, but the trial does not succeed. Consequently, in A's place I employ 3—the other numerical divisor, namely, of the highest term $3y^5$—and to and from those squares multiplied by 3, that is, to and from the numbers 12, 3, 0, 3, I add and subtract the divisors: in the terms then resulting I find these two progressions −7, −7, −7, −7 and 11, 5, −1, −7. Now for expediency I had before neglected the divisors of the outermost numbers 170 and 190. Therefore, continuing the progressions in each direction, I take their next terms on either side—to be precise, −7 and 17 above and −7 and −13 below—and test if,

inferius, ac tento si subductis his de numeris 27 ac 12 qui stant e regione in quarta columna differentiæ dividunt istos numero[s] 170 et 190 qui stant e regione in columna secunda. Et quidem differentia inter 27 & -7 id est 34 dividit 170 et differentia inter 12 et -7 id est 19 dividit 190. Item differentia inter 27 et 17 id est 10 dividit 170, sed differentia inter 12 et -13 id est 25 non dividit 190. Quare posteriorem progressionem rejicio. Juxta priorem $\mp C$ est -7 et $\mp B$ nihil, terminis progressionis nullam habentibus differentiam. Quare divisor tentandus $All \pm Bl \pm C$ erit $3yy+7$, et divisio succedit prodeunte Quoto $y^3-2yy-2y+2$.[527] Adeoꝗ proposita æquatio in has duas resolvitur $3yy+7=0$ & $y^3-2yy-2y+2=0$.

Proponatur deniꝗ æquatio $6x^7-10x^6-3x^5-6x^4-17x^3+55xx+19x-35=0$ et progressionis 3. 2. 1. 0. -1. -2 terminis pro x substitutis orientur numeri
‖[180] 4675. 23. 9. 35. 1. 1125[528] quos cum divisoribus suis dispono e regione ‖ terminorum progressionis ut hic fit

3	4675			54		
2	23	1. 23		24	1. 23. 25. 47	1.
1	9	1. 3. 9		6	-3. 3. 5. 7. 9. 15	3.
0	35	1. 5. 7. 35		0	-35. -7. -5. -1. 1. 5. 7. 35	5.
-1	1	1		6	5. 7	7.
-2	1125			24		

Imprimis autem rem tento addendo et subducendo divisores quadratis terminorum progressionis 9. 4. 1. 0. 4, sed frustra. Tento itaꝗ quod fiat addendo et subducendo divisores quadratis illis multiplicatis per altissimi æquationis termini divisorem 2, ut et per ejus divisorem 3, ac deniꝗ per ejus divisorem 6, quo casu res succedit. Etenim in terminis per additionem et subductionem resultantibus occurrit hæc progressio 1. 3. 5. 7 cujus terminum 5 qui stat e regione termini 0 ponendo $=\mp C$ & numerum -2 (qui subductione hujus 5 de superiore termino 3 relinquitur) $=\mp B$ et altissimi æquationis termini divisorem $6=A$, divisor $All \pm Bl \pm C$ fit $6xx+2x-5$. Et per hunc divisorem res succedit prodeunte Quoto $x^5-2x^4+x^3-3xx-x+7$. Quamobrem vice æquationis propositæ jam habemus duas simpliciores easdem radices involventes nimirum $6xx+2x-5=0$ & $x^5-2x^4+x^3-3xx-x+7=0$.[529]

[1682]
Lect. 3.
Si nullus inveniri potest hoc pacto divisor qui succedit, concludendum erit æquationem propositam non admittere divisorem duarum dimensionum. Posset eadem methodus extendi ad inventionem divisorum dimensionum plurium quærendo in prædictis summis differen‖tijsꝗ progressiones non arith-
‖[181]

---

(527) The preliminary version reproduced in 1, §2 above terminates (or was perhaps continued on a further sheet now lost?) at this point.

(528) More accurately, '$+4675$. $+23$. $+9$. -35. -1. -1125'.

(529) This example is not present in the 'correction' (§2 below) printed by Whiston in 1707 and sanctioned by Newton in 1722 (see note (517)).

after the numbers 27 and 12 standing in line in the fourth column are sub-tracted, the differences divide the numbers 170 and 190 standing in line in the second column. And, to be sure, the difference between 27 and -7 (that is, 34) divides 170 and the difference between 12 and -7 (that is, 19) divides 190; likewise, the difference between 27 and 17 (that is, 10) divides 170, but the difference between 12 and -13 (25, that is) does not divide 190. I consequently reject the latter progression. By the first one $\mp C$ is -7 and $\mp B$ zero, the terms of the progression having no difference. In consequence the divisor $Al^2 \pm Bl \pm C$ to be tested will be $3y^2 + 7$: the division succeeds, yielding the quotient $y^3 - 2y^2 - 2y + 2$,[527] and hence the proposed equation resolves itself into these two, $3y^2 + 7 = 0$ and $y^3 - 2y^2 - 2y + 2 = 0$.

Let there, finally, be proposed the equation

$$6x^7 - 10x^6 - 3x^5 - 6x^4 - 17x^3 + 55x^2 + 19x - 35 = 0,$$

and with the terms of the progression 3, 2, 1, 0, -1, -2 substituted in x's place there will ensue the numbers 4675, 23, 9, 35, 1, 1125;[528] these together with their divisors I set out in line with the terms of the progression, as is here done:

3	4675		54		
2	23	1, 23	24	1, 23, 25, 47	1
1	9	1, 3, 9	6	-3, 3, 5, 7, 9, 15	3
0	35	1, 5, 7, 35	0	-35, -7, -5, -1, 1, 5, 7, 35	5
-1	1	1	6	5, 7	7
-2	1125		24		

Now in the first instance I make a trial approach by adding and subtracting the divisors to and from the squares 9, 4, 1, 0, 4 of the terms in the progression, but to no avail. I accordingly test what may happen by adding and subtracting divisors to and from those squares multiplied by the divisor 2 of the highest term in the equation, and also by its divisor 3, and finally by its divisor 6—and in this case the trial succeeds. To be exact, in the terms resulting through addition and subtraction there occurs this progression 1, 3, 5, 7 and, on setting the term 5 in it standing in line with the term 0 equal to $\mp C$, the number -2 (left by subtracting this 5 from the term 3 above it) $= \mp B$, and the divisor 6 of the highest term in the equation $= A$, the divisor $Al^2 \pm Bl \pm C$ becomes $6x^2 + 2x - 5$. And by means of this divisor the approach succeeds, yielding the quotient $x^5 - 2x^4 + x^3 - 3x^2 - x + 7 = 0$. As a consequence, instead of the equation proposed we now have two simpler ones involving the same roots, namely $6x^2 + 2x - 5 = 0$ and $x^5 - 2x^4 + x^3 - 3x^2 - x + 7 = 0$.[529]

If no successful divisor can be found by this technique, you will have to conclude that the equation proposed can admit of no divisor of two dimensions. The same method could be extended to determining divisors of more dimensions still by looking in the above-mentioned sums and differences not for arithmetical

meticas quidem sed alias quasdem quarum terminorum differentiæ primæ secundæ tertiæ &c sunt in Arithmetica progressione:[530] at in his Tyro non est detinendus.

Ubi in proposita æquatione duæ sunt literæ, et omnes ejus termini ad dimensiones æque altas ascendunt, pro una istarum literarum pone unitatem, dein per regulas præcedentes quære divisorem, ac divisoris hujus comple deficientes dimensiones restituendo literam illam pro unitate. Ut si æquatio sit

$$6y^4 - cy^3 - 21ccyy + 3c^3y + 20c^4 = 0.$$

ubi termini omnes sunt quatuor dimensionū, pro c substituo 1 et æquatio evadit $6y^4 - y^3 - 21yy + 3y + 20 = 0$. Cujus divisor, ut supra est $3y + 4$ et completa deficiente dimensione posterioris termini 4 per dimensionem c, fit $3y + 4c$ divisor quæsitus. Ita si æquatio sit $x^4 - bx^3 - 5bbxx + 12b^3x - 6b^4 = 0$, posita 1 pro b et æquationis resultantis $x^4 - x^3 - 5xx + 12x - 6 = 0$ invento divisore $xx + 2x - 2$, compleo ejus deficientes dimensiones per dimensiones b. Et sic habeo divisorem quæsitum $xx + 2bx - 2bb$.

Ubi in æquatione proposita tres vel plures sunt literæ et ejus termini omnes ad easdem dimensiones ascendunt, potest divisor per præcedentes regulas inveniri, sed expeditius hoc modo. Quære omnes divisores terminorum omnium ‖[182] in quibus literarum ‖ aliqua non est, item terminorum omnium in quibus alia aliqua literarum non est, pariter et omnium in quibus tertia litera quartaცჳ et quinta non est si tot sunt literæ. Et sic percurre omnes literas, & e regione literarum colloca divisores respectivè. Dein vide si in serie aliqua divisorum per omnes literas pergente, partes omnes unicam tantum literam involventes tot vicibus reperiantur quot sunt literæ una dempta in æquatione proposita & partes literas duas involventes tot vicibus quot sunt literæ demptis duabus in eadem æquatione. Si ita est, partes istæ omnes sub signis suis semel sumptæ erunt divisor quæsitus.

Ut si proponatur æquatio

$$12x^3 - 14bxx + 9cxx - 12bbx - 6bcx + 8ccx + 8b^3 - 12bbc - 4bcc + 6c^3 = 0:$$

terminorum $8b^3 - 12bbc - 4bcc + 6c^3$ in quibus non est x, divisores unius dimensionis per præcedentes regulas inventi erunt $2b - 3c$ & $4b - 6c$: terminorum $12x^3 + 9cxx + 8ccx + 6c^3$ in quibus non est b, divisor unicus $4x + 3c$: ac terminorum

---

(530) These extended finite-differences tests for cubic and higher factors are exemplified on 1: 533–9. In footnote we there indicated (1: 535, note (84)) that an equivalent *Regula Generalis inveniendi divisores rationales compositos quantitatis Algebraïcæ* $a + bx + cxx + dx^3 + ex^4$, &c, provoked by the corresponding section 'De inventione Divisorum' in Whiston's *editio princeps* (*Arithmetica Universalis*, ₁1707: 43–7), was communicated by Niklaus I Bernoulli to his uncle Johann in spring 1708, and later published by him in (his?) *G. G. Leibnitii et Johan. Bernoulli Commercium*

progressions but for certain others, the first, second, third, ... differences of whose terms are in arithmetical progression:[530] but the beginner should not be tempted to linger over these side-issues.

When the equation proposed involves two letters and all its terms ascend to equally high dimensions, in place of one of those letters put unity, next by the preceding rules seek a divisor, and then fill out the missing dimensions of this divisor by restoring that dimension in unity's place. If the equation be, for instance, $6y^4 - cy^3 - 21c^2y^2 + 3c^3y + 20c^4 = 0$, all the terms in which are of four dimensions, in c's place I substitute 1 and the equation comes to be

$$6y^4 - y^3 - 21y^2 + 3y + 20 = 0.$$

As above, a divisor of this is $3y + 4$, and so, when the missing dimension of the latter term 4 is filled out by the dimension c, the required divisor becomes $3y + 4c$. So, if the equation be $x^4 - bx^3 - 5b^2x^2 + 12b^3x - 6b^4 = 0$, after 1 is set in b's place and of the equation $x^4 - x^3 - 5x^2 + 12x - 6 = 0$ resulting a divisor is found to be $x^2 + 2x - 2$, I fill out its missing dimensions by dimensions of b, and thus obtain the required divisor $x^2 + 2bx - 2b^2$.

When the equation proposed possesses three or more letters and all its terms ascend to the same dimensions, a divisor can be found by means of the preceding rules, but it is more speedily derived in this way. Ascertain all the divisors of all terms in which some one of the letters is not present, and those similarly of all terms in which any other of the letters is not present, and likewise those of all in which a third letter, a fourth one and a fifth is not present if there are that many letters. Run through all the letters in this manner, and in line with the letters place their respective divisors. Then see if, in some sequence of divisors proceeding through all the letters, all the parts involving merely a single letter shall be found as many times as there are letters, less one, in the equation proposed, and if the parts involving two letters are found as many times as there are letters, less two, in the same equation. If this is so, all those parts taken once under their proper signs will be the divisor required.

If, for instance, there be proposed the equation

$$12x^3 - 14bx^2 + 9cx^2 - 12b^2x - 6bcx + 8c^2x + 8b^3 - 12b^2c - 4bc^2 + 6c^3 = 0,$$

of the terms $8b^3 - 12b^2c - 4bc^2 + 6c^3$ in which x is not present the one-dimensional divisors found by the preceding rules will be $2b - 3c$ and $4b - 6c$, of the terms $12x^3 + 9cx^2 + 8c^2x + 6c^3$ in which b is not present there is a unique divisor $4x + 3c$,

Philosophicum et Mathematicum, **2** (Lausanne/Geneva): 189–209 in a much improved form. (Niklaus' original *Regula Generalis Inveniendi Æquationes, per quas alia quæpiam data, modo reducibilis sit, dividi potest*, as passed on to Leibniz in May 1708, is reproduced by C. I. Gerhardt in his edition of Leibniz' *Mathematische Schriften*, **3** (Halle, 1855): 827–35.)

$12x^3 - 14bxx - 12bbx + 8b^3$ in quibus non est c divisores $2x - b$ & $4x - 2b$.[531] Hos divisores e regione literarum x, b, c dispono ut hic vides

$$
\begin{array}{c|l}
x & 2b - 3c. \; 4b - 6c. \\
b & 4x + 3c. \\
c & 2x - b. \; 4x - 2b.
\end{array}
$$

Cum tres sint literæ et divisorum partes singulæ non nisi singulas literas[532] involvant, in serie divisorum debent partes illæ bis reperiri. Atqui divisorum

‖[183] $4b - 6c$ & $2x - b$ partes $4b$, ‖ $6c$, $2x$, b non nisi semel occurrunt. Extra divisorem illum cujus sunt partes non reperiuntur. Quare divisores illos negligo. Restant tantum tres divisores $2b - 3c$, $4x + 3c$ et $4x - 2b$. Hi in serie sunt per omnes literas x, b, c pergente, & eorum partes singulæ $2b$, $3c$, $4x$ bis reperiuntur in ipsis ut oportuit; idcg cum signis ijsdem si modò signa divisoris $2b - 3c$ mutentur et ejus loco scribatur $3c - 2b$. Nam signa omnia divisoris cujusvis mutare licet. Sumo itacg horum partes omnes $2b$, $3c$, $4x$ semel sub signis suis et aggregatum $-2b + 3c + 4x$ divisor erit quem invenire oportuit. Nam si per hunc dividas æquatiónem propositam prodibit $3xx - 2bx + 2cc - 4bb$. Adeocg æquationis illius tres radices eædem sunt quæ harum duarum $4x + 3c - 2b = 0$ &

$$3xx - 2bx + 2cc - 4bb = 0.$$

Rursus si æquatio sit[533]

$$12x^5 - 10ax^4 + 9bx^4 - 26aax^3 + 12abx^3 + 6bbx^3 + 24a^3xx - 8aabxx - 8abbxx$$

$$-24b^3xx - 4a^3bx + 6aabbx - 12ab^3x + 18b^4x + 12a^4b + 32aab^3 - 12b^5 = 0.$$

divisores terminorum in quibus x non est colloco e regione x; illos terminorum in quibus a non est, e regione a; et illos terminorum in quibus b non est, e regione b, ut hic vides.

$$
\begin{array}{c|l}
x & b. \; 2b. \; 4b. \; aa + 3bb. \; 2aa + 6bb. \; 4aa + 12bb. \; bb - 3aa. \; 2bb - 6aa. \; 4bb - 12aa. \\
a & 4xx - 3bx + 2bb. \; 12xx - 9bx + 6bb. \\
b & x. \; 2x. \; 3x - 4a. \; 6x - 8a. \; 3xx - 4ax. \; 6xx - 8ax. \; 2xx + ax - 3aa. \; 4xx + 2ax - 6aa.
\end{array}
$$

‖[184] ‖ Dein illos omnes qui sunt unius dimensionis rejiciendos sentio quia simplices b. $2b$. $4b$. x. $2x$ et partes compositorum $3x - 4a$. $6x - 8a$ non nisi semel in omnibus divisoribus reperiuntur, tres autem sunt literæ in æquatione proposita, &partes illæ unicam tantum involvunt, adeocg bis reperiri deberent. Similiter divisores

---

(531) Specifically $8b^3 - 12b^2c - 4bc^2 + 6c^3 = 2(2b - 3c)(2b^2 - 3c^2)$,

$$12x^3 + 9cx^2 + 8c^2x + 6c^3 = (4x + 3c)(3x^2 + 2c^2)$$

and, lastly, $12x^3 - 14bx^2 - 12b^2x + 8b^3 = 2(2x - b)(3x^2 - 2bx - 4b^2)$, whose quadratic factors are irreducible into rational linear ones.

(532) Newton first wrote 'unicam [literam]' (a single letter). The draft 'correction' (§2 below) likewise corrects this to the less misleading equivalent 'singulas literas'.

while of the terms $12x^3 - 14bx^2 - 12b^2x + 8b^3$ in which c is not present there are the divisors $2x - b$ and $4x - 2b$.[531] These divisors I arrange in line with the letters x, b, c as you see here:

$$\begin{array}{c|l} x & 2b - 3c,\ 4b - 6c. \\ b & 4x + 3c. \\ c & 2x - b,\ 4x - 2b. \end{array}$$

Since there are three letters and the separate components of the divisors involve none but single letters,[532] in the sequence of divisors those parts ought to be found twice each. However, the parts $4b$, $6c$, $2x$ and b of the divisors $4b - 6c$ and $2x - b$ occur but a single time: outside of the divisor whose components they are they are not found. Consequently, I neglect those divisors. There remain only the three divisors $2b - 3c$, $4x + 3c$ and $4x - 2b$. These are in a sequence proceeding through all the letters x, b, c and their individual parts $2b$, $3c$, $4x$ are found twice each in them, as was required, and then with the same signs provided the signs of the divisor $2b - 3c$ be changed and in its stead be written $3c - 2b$. (It is, of course, permissible to change all the signs together in any divisor.) I accordingly take all the parts $2b$, $3c$, $4x$ of these once each under their proper signs, and the aggregate $-2b + 3c + 4x$ will be the divisor it was required to find. For if you divide the propounded equation by this, there will result $3x^2 - 2bx + 2c^2 - 4b^2$, and hence the three roots of that equation will be the same as those of these two, $4x + 3c - 2b = 0$ and $3x^2 - 2bx + 2c^2 - 4b^2 = 0$.

Again, if the equation be

$$12x^5 - 10ax^4[-]9bx^4 - 26a^2x^3 + 12abx^3 + 6b^2x^3 + 24a^3x^2 - 8a^2bx^2 - 8ab^2x^2$$

$$-24b^3x^2 - 4a^3bx + 6a^2b^2x - 12ab^3x + 18b^4x + 12a^4b + 32a^2b^3 - 12b^5 = 0,$$

the divisors of the terms in which x is not present I place in line with x, those of the terms in which a is not present in line with a, and those of the terms in which b is not present in line with b, as you see here:

$$\begin{array}{c|l} x & b,\ 2b,\ 4b,\ a^2 + 3b^2,\ 2a^2 + 6b^2,\ 4a^2 + 12b^2,\ b^2 - 3a^2,\ 2b^2 - 6a^2,\ 4b^2 - 12a^2. \\ a & 4x^2 - 3bx + 2b^2,\ 12x^2 - 9bx + 6b^2. \\ b & x,\ 2x,\ 3x - 4a,\ 6x - 8a,\ 3x^2 - 4ax,\ 6x^2 - 8ax,\ 2x^2 + ax - 3a^2,\ 4x^2 + 2ax - 6a^2. \end{array}$$

Next, I perceive that all those which are of one dimension are to be rejected, because the simple ones b, $2b$, $4b$, x, $2x$ and the parts of the compound ones $3x - 4a$, $6x - 8a$ are found but once in all the divisors, whereas the equation proposed possesses three letters, and those parts involve but a single one and hence should be found twice. Similarly, the divisors of two dimensions $a^2 + 3b^2$,

---

(533) The third term following should read '$-9bx^4$', a slip passed by Whiston into his 1707 *editio princeps* but caught by Newton in his 1722 revise.

duarum dimensionum $aa+3bb$. $2aa+6bb$. $4aa+12bb$. $bb-3aa$. $4bb-12aa$ rejicio quia partes eorum aa. $2aa$. $4aa$. bb et $4bb$ unicam tantum literam involventes non nisi semel reperiuntur. Divisoris autem $2bb-6aa$, qui solus restat e regione x, partes $2bb$ & $6aa$, qui similiter unicam tantum literam involvunt, iterum reperiuntur: nempe pars $2bb$ in divisore $4xx-3bx+2bb$, & pars $6aa$ in divisore $4xx+2ax-6aa$. Quinetiam hi tres divisores in serie sunt stantes e regione trium literarum x, a, b. et omnes eorum partes $2bb$, $6aa$, $4xx$ quæ unicam tantum literam involvunt, bis reperiuntur in ipsis idɋ sub proprijs signis: partes vero $3bx$, $2ax$ quæ duas literas involvunt non nisi semel occurrunt in ipsis. Quare horum trium divisorum partes omnes diversæ $2bb$. $6aa$. $4xx$. $3bx$. $2ax$, sub signis suis connexæ divisorem desideratum $2bb-6aa+4xx-3bx+2ax$ conflabunt. Per hunc itaɋ divido æquationem propositam et oritur

$$3x^3-4axx-2aab-6b^3.$$

& concludo radices æquationis illius easdem esse atɋ harum duarum

||[185] $\quad 2bb-6aa+4xx-3bx+2ax=0,\quad ||\ \&\quad 3x^3-4axx-2aab-6b^3=0.$

[1682] Si quantitatis dividendæ seu æquationis reducendæ termini omnes non sunt
Lect. 4. æque alti, complendæ sunt dimensiones deficientes per dimensiones literæ cujusvis assumptæ, dein postquā per præcedentes regulas inventus est divisor litera assumpta delenda est. Ut si æquatio sit

$$12x^3-14bxx+9xx-12bbx-6bx+8x+8b^3-12bb-4b+6=0,$$

assume literā quamvis c et per dimensiones ejus comple deficientes dimensiones æquationis propositæ ad hunc modum

$$12x^3-14bxx+9cxx-12bbx-6bcx+8ccx+8b^3-12bbc-4bcc+6c^3=0.$$

Dein hujus invento divisore $4x-2b+3c$, dele c et habebitur divisor desideratus $4x-2b+3$.

Aliquando divisores facilius quam per has regulas inveniri possunt. Ut si litera aliqua in quantitate proposita unius sit tantùm dimensionis, quærendus erit maximus communis divisor terminorum in quibus litera illa reperitur et reliquorum terminorum in quibus non reperitur. Nam divisor ille totum dividet. Et si nullus est ejusmodi divisor communis, nullus erit divisor totius. Ut si proponatur æquatio

$$x^4-3ax^3-8aaxx+18a^3x-cx^3+acxx-8aacx+6a^3c-8a^4=0,$$

quæratur communis divisor terminorum $-cx^3+acxx-8aacx+6a^3c$, in quibus c unius est tantum dimensionis, et terminorum reliquorum

$$x^4-3ax^3-8aaxx+18a^3x-8a^4,$$

||[186] ac divisor ille, nempe || $xx+2ax-2aa$ dividet æquationem totam.

$2a^2 + 6b^2$, $4a^2 + 12b^2$, $b^2 - 3a^2$, $4b^2 - 12a^2$ I reject because their parts a^2, $2a^2$, $4a^2$, b^2 and $4b^2$ involving only a single letter are found but once. In the case of the divisor $2b^2 - 6a^2$ which alone remains in line with x, however, its parts $2b^2$ and $6a^2$ similarly involving only a single letter are found a second time: the part $2b^2$, namely, in the divisor $4x^2 - 3bx + 2b^2$, and the part $6a^2$ in the divisor

$$4x^2 + 2ax - 6a^2.$$

Moreover, indeed, these three divisors are in sequence as they stand in line with the three letters x, a, b and all their parts $2b^2$, $6a^2$, $4x^2$ involving only a single letter are twice found in them, and then under their proper signs, while, to be sure, their parts $3bx$, $2ax$ involving two letters occur but once in them. Consequently, all the different parts $2b^2$, $6a^2$, $4x^2$, $3bx$, $2ax$ of these three divisors will, when connected under their proper signs, constitute the desired divisor

$$2b^2 - 6a^2 + 4x^2 - 3bx + 2ax.$$

By this, accordingly, I divide the propounded equation and there arises $3x^3 - 4ax^2 - 2a^2b - 6b^3$. I then conclude that the roots of that equation are the same as those of these two, $2b^2 - 6a^2 + 4x^2 - 3bx + 2ax = 0$ and

$$3x^3 - 4ax^2 - 2a^2b - 6b^3 = 0.$$

If in the quantity to be divided or the equation to be reduced all terms are not equally high, the missing dimensions are to be filled out by the dimensions of any assumed letter and then, after a divisor has been found by means of the preceding rules, the letter assumed is to be deleted. So, if the equation be $12x^3 - 14bx^2 + 9x^2 - 12b^2x - 6bx + 8x + 8b^3 - 12b^2 - 4b + 6 = 0$, assume any letter c and by its dimensions fill out the missing dimensions of the propounded equation in this manner:

$$12x^3 - 14bx^2 + 9cx^2 - 12b^2x - 6bcx + 8c^2x + 8b^3 - 12b^2c - 4bc^2 + 6c^3 = 0.$$

Then, having found the divisor $4x - 2b + 3c$ of this, delete c and the desired divisor $4x - 2b + 3$ will be had.

On occasion divisors can be found more easily than by these rules. Thus, if some letter in the quantity proposed be only of one dimension, you will need to ascertain the greatest common divisor of the terms in which that letter is found and of the remaining terms in which it is not. For that divisor will divide the whole. And if there is no common divisor of this sort, the whole will have no divisor. If, for instance, there be proposed the equation

$$x^4 - 3ax^3 - 8a^2x^2 + 18a^3x - cx^3 + acx^2 - 8a^2cx + 6a^3c - 8a^4 = 0,$$

let there be sought the common divisor of the terms $-cx^3 + acx^2 - 8a^2cx + 6a^3c$, in which c is of one dimension only, and of the remaining terms

$$x^4 - 3ax^3 - 8a^2x^2 + 18a^3x - 8a^4,$$

and that divisor namely $x^2 + 2ax - 2a^2$, will divide the whole equation.

(534)Quod si in æquatione litera aliqua duarum sit et non plurium dimensionum, divisor ejus siquem habeat, aut erit communis divisor terminorum in quibus litera illa duarum est dimensionum et terminorum in quibus unius est dimensionis, ac terminorum in quibus non occurrit; aut invenietur extrahendo radicem illius literæ.(535) Ut si proponatur æquatio

$$ax^3 + bx^3 - aaxx - bbxx + a^3x + aabx - a^4 - aabb = 0,$$

in qua b ad duas tantum dimensiones ascendit, quæro communem divisorem terminorum $-bbxx - bbaa$ ubi b duarum est dimensionum et terminorum $bx^3 + aabx$ ubi b unius est dimensionis et terminorum $ax^3 - aaxx + a^3x - a^4$ ubi b non occurrit[,] & ejusmodi divisorem invenio $xx + aa$. Quare divido æquationem propositam per $xx + aa$ et prodit Quotiens $ax + bx - aa - bb$.

Atcɣ ita si æquatio esset $aax^3 - abx^3 + aabxx - ab^3x + b^4x - ab^4 = 0$, in qua litera a ad duas tantum dimensiones ascendit, tentarem primo siquis esset terminorum omnium $aax^3 + aabxx$ & $-abx^3 - ab^3x - ab^4$ et b^4x(536) divisor communis. Postea verò ubi ejusmodi divisorem nullum esse animadverter[a]m,(537) ordinarem æquationem secundum dimensiones literæ a ut hic fit

$$aa = \frac{bx^3 + b^3x + b^4}{x^3 + bxx}\, a - \frac{b^4}{xx + bx}.$$

‖[187] ‖Dein radicem quadraticam ejus extraherem et prodiret

$$a = \frac{bx^3 + b^3x + b^4 \pm \sqrt{bbx^6 - 2b^4x^4 - 2b^5x^3 + b^6xx + 2b^7x + b^8}}{2x^3 + 2bxx},$$

id est $a = \dfrac{bx^3 + b^3x + b^4 \pm bx^3 \mp b^3x \mp b^4}{2x^3 + 2bxx}$, seu $a = \dfrac{bx}{b + x}$ & $a = \dfrac{b^3}{xx}$. Unde concludo propositā æquationem dividi posse per $ax + ab - bx$, ut et per $axx - b^3$.

Æquationum   Hactenus reductionem æquationum tradidi quæ rationales divisores ad-
Reductio per   mittunt. Sed antequam æquationem quatuor sex aut plurium dimensionum
divisores   irreducibilem esse concludere possumus tentandum erit etiam annon per
surdos.(538)   surdum aliquem divisorem reduci queat; vel quod perinde est tentandum erit
                annon æquatio ita in duas æquales partes dividi possit ut ex utracɣ radix
                extrahatur. Id autem fiet per sequentem methodum.

---

(534) The two final paragraphs following are not present in the draft 'correction' (§ 2 below) and hence (see note (517)) were not printed by Whiston in his 1707 *editio princeps* or by Newton in his 1722 revise.

(535) And, of course, thereafter suitably ordering the result.

(536) Those involving square, unit and zero powers of a respectively.

(537) The manuscript reads 'animadverterem' by a slip.

(538) This brand new section (on Newton's pages 187–203 following) is evidently modelled on the factorization of $x^4 + a^4 = (x^2 + a^2)^2 - 2(ax)^2$ as $(x^2 + \sqrt{2}.ax + a^2)(x^2 - \sqrt{2}.ax + a^2)$ implied in the example of algebraic division inserted (at this time?) in the margin on his page 18 above;

(534)But if some letter in an equation be of two dimensions, and no more, its divisor—if it has any—will either be the common divisor of the terms in which that letter is of two dimensions, of the terms in which it is of one dimension and of the terms in which it does not occur; or it will be found by extracting that letter's root.(535) So, if there be proposed the equation

$$ax^3 + bx^3 - a^2x^2 - b^2x^2 + a^3x + a^2bx - a^4 - a^2b^2 = 0,$$

in which b ascends to two dimensions only, I seek the common divisor of the terms $-b^2x^2 - a^2b^2$ in which b is of two dimensions, of the terms $bx^3 + a^2bx$ in which b is of one dimension, and of the terms $ax^3 - a^2x^2 + a^3x - a^4$ where it does not occur, and I find $x^2 + a^2$ to be a divisor of this sort. Consequently, I divide the proposed equation by $x^2 + a^2$ and there results the quotient $ax + bx - a^2 - b^2$.

And thus, had the equation been $a^2x^3 - abx^3 + a^2bx^2 - ab^3x + b^4x - ab^4 = 0$, in which the letter a rises to two dimensions only, I should first have tested if there were any common divisor of all the terms $a^2x^3 + a^2bx^2$, $-abx^3 - ab^3x - ab^4$ and b^4x.(536) Subsequently, however, when I had noticed that there is no divisor of this sort, I would order the equation according to the dimensions of the letter a, as here: $a^2 = ((bx^3 + b^3x + b^4)/(x^3 + bx^2))\,a - b^4/(x^2 + bx)$. Then I would extract its square root and there would result

$$a = \frac{bx^3 + b^3x + b^4 \pm \sqrt{[b^2x^6 - 2b^4x^4 - 2b^5x^3 + b^6x^2 + 2b^7x + b^8]}}{2(x^3 + bx^2)},$$

that is, $a = \frac{1}{2}(bx^3 + b^3x + b^4 \pm [bx^3 - b^3x - b^4])/(x^3 + bx^2)$ or $a = bx/(b + x)$ and $a = b^3/x^2$. From which I conclude that the equation proposed can be divided by $ax - bx + ab$ and also by $ax^2 - b^3$.

Hitherto I have dealt with the reduction of equations which admit of rational divisors. But before we can conclude that an equation of four, six or more dimensions is irreducible, we shall need also to test whether or not it is possibly reducible by means of some surd divisor; or, what is equivalent, to test whether or not the equation can be divided into two equal portions in such a way that the root of each may be extracted. This will, however, be accomplished by the following method.

The reduction of equations by surd divisors.(538)

---

see note (53), and also compare IV: 205–8, especially 207. In modern terminology Newton here develops algorithms for splitting a given polynomial $f_{2m}(x) \equiv x^{2m} + px^{2m-1} + qx^{2m-2} + \dots$, irreducible over the rational field R, into factors $\phi_m(x) \times \psi_m(x)$ in the extended field $R(\sqrt{n})$ of quadratic surds. Subsequently explored by Maclaurin and Waring—who in his *Meditationes Algebraicæ* (Cambridge, 1770) gave some unimportant generalizations of the present Newtonian algorithms—, the topic was to lead to the definition and elaboration of the concept of an algebraic number by Kronecker, Dedekind and Hilbert; see I. G. Bachmakova, 'Quelques traits caractéristiques du développement de l'algèbre au XVIIIᵉ siècle', *Actes du XIᵉ Congrès International d'Histoire des Sciences*, **3** (Wroclaw, 1968): 216–20 [French version of the Russian original in *Istoriko-Mathematicheskie Issledovanya*, **17**, 1966: 317–23; see also *ibid.* **12**, 1959: 431–56].

Dispone æquationem secundum dimensiones literæ alicujus ita ut omnes ejus termini sub signis suis conjunctim æquales sint nihilo, & terminus altissimus affirmativo signo afficiatur. Deinde si æquatio quadratica sit (nam et hunc casum ob rei analogiam adjicere lubet) aufer utrobiꝗ terminum infimum & adde quartam partem quadrati cognitæ quantitatis termini medij. Ut si æquatio sit $xx - ax - b = 0$ aufer utrobiꝗ $-b$ et adde $\frac{1}{4}aa$, et emerget

$$xx - ax + \tfrac{1}{4}aa = b + \tfrac{1}{4}aa,$$

‖ [188] et extracta utrobiꝗ radice ‖ fiet $x - \frac{1}{2}a = \pm\sqrt{b + \frac{1}{4}aa}$ sive $x = \frac{1}{2}a \pm \sqrt{b + \frac{1}{4}aa}$.[539]

Quod si æquatio sit quatuor dimensionum, sit ea $x^4 + px^3 + qxx + rx + s = 0$, ubi p, q, r & s denotant cognitas quantitates terminorum æquationis signis proprijs adfectas. Fac $q - \frac{1}{4}pp = \alpha$. $r - \frac{1}{2}p\alpha^{[540]} = \beta$. $s - \frac{1}{4}\alpha\alpha = \zeta$. Dein pone pro n communem aliquem terminorum β et 2ζ divisorem integrum & non quadratum, qui et impar esse debet & per 4 divisus unitatem relinquere[541] si terminorum p et r alteruter sit impar. Pone etiam pro k divisorem aliquem quantitatis $\frac{\beta}{n}$ si p sit par vel imparis divisoris dimidium si p sit impar, vel nihil si dividuum β sit nihil. Aufer Quotum[542] de $\frac{1}{2}pk$ et reliqui dimidium dic l. Dein pro Q pone $\dfrac{\alpha + nkk}{2}$ & tenta si n dividat $QQ - s$, et Quoti radix[542] sit rationalis & æqualis l. Si hoc contigerit ad utramꝗ partem æquationis adde $nkkxx + 2nklx + nll$, & radicem extrahes utrobiꝗ, prodeunte $xx + \frac{1}{2}px + Q = n^{\frac{1}{2}}$ in $kx + l$.[543]

[1682]
Lect. 5.
Exempli gratia proponatur æquatio $x^4 + 12x - 17 = 0$. et quia p et q hic desunt et r est 12 & s est -17, substitutis hisce numeris fiet $\alpha = 0$, $\beta = 12$ & $\zeta = -17$, et ipsorum β et 2ζ seu 12 & -34 communis divisor unicus, nimirum 2, erit n. Porrò $\dfrac{\beta}{n}$ est 6 et ejus divisores 1, 2, 3 & 6 successivè tentandi sunt pro k & -3, -2,[544] -1, $-\frac{1}{2}$ pro l respective. Est autem $\dfrac{\alpha + nkk}{2}$ id est kk æquale Q. Est et

‖ [189] $\sqrt{\dfrac{QQ - s}{n}}$ id est $\sqrt{\dfrac{QQ + 17}{2}} = l$.[545] Ubi numeri ‖ pares 2 et 6 scribuntur pro k,

---

(539) Newton has already derived this general solution of the quadratic by 'completing the square' in 'Reg: 7' on his page 33.

(540) Misprinted '$r - \frac{1}{2}\alpha$' in Whiston's 1707 *editio princeps*, and corrected to '$r - \frac{1}{2}\alpha p$' in Newton's library copy, a variation followed in his 1722 revise.

(541) Newton first inaccurately wrote '…non quadratum, atꝗ etiam imparem' (non-square and also odd) simply. As will become evident the criterion is general for the quadratic field $R(\surd)$ of rationals and surds.

(542) Namely, β/nk and $\surd[(Q^2 - s)/n]$ respectively.

(543) Newton determines the conditions for the general quartic $x^4 + px^3 + qx^2 + rx + s = 0$ to be identical with

$$(x^2 + \tfrac{1}{2}px + \tfrac{1}{2}\alpha)^2 + \beta x + \zeta \equiv (x^2 + \tfrac{1}{2}px + Q)^2 - n(kx + l)^2 = 0,$$

Arrange the equation according to the dimensions of some letter so that all its terms joined together under their proper signs shall be equal to nothing and that its highest term has a positive sign affixed. Then if the equation be a quadratic (for it is agreeable, because of the analogy of its reduction, to adjoin this case too), take away the lowest term from each side and add a quarter of the square of the coefficient of the middle term. If, for instance, the equation be $x^2 - ax - b = 0$, take away $-b$ from each side and add $\frac{1}{4}a^2$, and there will emerge $x^2 - ax + \frac{1}{4}a^2 = b + \frac{1}{4}a^2$; then, when the root is extracted on each side, there will come to be $x - \frac{1}{2}a = \pm\sqrt{[b + \frac{1}{4}a^2]}$ or $x = \frac{1}{2}a \pm \sqrt{[b + \frac{1}{4}a^2]}$.[539]

But if the equation be of four dimensions, let it be $x^4 + px^3 + qx^2 + rx + s = 0$, where p, q, r and s denote the coefficients of the terms in the equation with their proper signs affixed. Make $q - \frac{1}{4}p^2 = \alpha$, $r - \frac{1}{2}p\alpha$[540] $= \beta$, $s - \frac{1}{4}\alpha^2 = \zeta$. Then for n put some integral, non-square common divisor of the terms β and 2ζ, which ought also to be odd, leaving unity when divided by 4,[541] if one or other of the terms p and r be odd. Put also for k some divisor of the quantity β/n if p be even, or half an odd divisor if p be odd, or nothing if the dividend β be nothing. Take the quotient[542] from $\frac{1}{2}pk$ and half the remainder call l. Then for Q put $\frac{1}{2}(\alpha + nk^2)$ and test if n shall divide $Q^2 - s$, and if the quotient's root[542] be rational and equal to l. Should this chance to be so, to each side of the equation add

$$nk^2x^2 + 2nklx + nl^2,$$

and then extract the root of each side, producing $x^2 + \frac{1}{2}px + Q = n^{\frac{1}{2}}(kx + l)$.[543]

For example, let the equation $x^4 + 12x - 17 = 0$ be proposed, and then, because p and q are here lacking, while r is 12 and s is -17, on substituting these numbers there will come $\alpha = 0$, $\beta = 12$ and $\zeta = -17$, and of β and 2ζ, that is, 12 and -34, the unique common divisor, namely 2, will be n. Furthermore β/n is 6, and so the divisors 1, 2, 3 and 6 of this are to be tested in succession in k's place, and -3, $-[\frac{3}{2}]$, -1, $-\frac{1}{2}$ respectively in l's place. However,

$$(\tfrac{1}{2}(\alpha + nk^2), \text{ that is}) \ k^2$$

is equal to Q, and so also $(\sqrt{[(Q^2 - s)/n]}, \text{ that is}) \ \sqrt{[\frac{1}{2}(Q^2 + 17)]} = l$.[545] When the even numbers 2 and 6 are written for k, Q becomes 4 and 36 and $Q^2 - s$ will be

---

n square-free. At once $\alpha = q - \frac{1}{4}p^2$, $\beta = r - \frac{1}{2}\alpha p$, $\zeta = s - \frac{1}{4}\alpha^2$, and also $\alpha = 2Q - nk^2$,

$$(\tfrac{1}{2}\alpha p + \beta \text{ or}) \ r = Qp - 2nkl, \quad (\tfrac{1}{4}\alpha^2 + \zeta \text{ or}) \ s = Q^2 - nl^2;$$

whence $Q = \frac{1}{2}(\alpha + nk^2)$, $l = \sqrt{[(Q^2 - s)/n]}$, $\beta/nk = \frac{1}{2}pk - 2l$ and $2\zeta/n = \alpha k^2 + \frac{1}{4}nk^4 - 2l^2$. Further, when p is odd—say $p = 2P + 1$—it follows that $2Q = q - P(P + 1) + \frac{1}{4}(4nk^2 - 1)$, so that $4nk^2$ is odd and so n and $2k$ are each odd, so that $\frac{1}{4}(nk^2 - 1)$ is integral and $n \equiv 1 \pmod 4$.

(544) Read '$-\frac{3}{2}$'. The error was passed by Whiston into the 1707 *editio princeps*, but caught by Newton in 1722.

(545) Newton first wrote '$\sqrt{\dfrac{QQ - s}{n}} = \dfrac{pQ - r}{2nk}$ id est $-\dfrac{3}{k} = l$'.

Q fit 4 et 36 et $QQ-s$ numerus erit impar adeoq dividi non potest per n seu 2. Quare numeri illi 2 & 6 rejiciendi sunt. Ubi vero 1 et 3 scribuntur pro k, Q fit 1 et 9, & $QQ-s$ fit 18 et 98, qui numeri dividi possunt per n & quotorū radices extrahi. Sunt enim ± 3 & ± 7: quarum tamen sola -3 congruit cum l. Pono itaq $k=1$, $l=-3$ & $Q=1$, & quantitatem $nkkxx+2nklx+nll$ id est $2xx-12x+18$ addo ad utramq partem æquationis et prodit $x^4+2xx+1=2xx-12[x]+18$ et extracta utrobiq radice, $xx+1=x\sqrt{2}-3\sqrt{2}$. Quod si radicis extractionem effugere malueris pone $xx+\frac{1}{2}px+Q=\sqrt{n}\times\overline{kx+l}$ et invenietur ut ante

$$xx+1=\pm\sqrt{2}\times\overline{x-3}.$$

Et ex hac æquatione si radices iterum extrahas proveniet

$$x=\pm\tfrac{1}{2}\sqrt{2}\,[\pm]\,\sqrt{-\tfrac{1}{2}\pm 3\sqrt{2}},^{(546)}$$

hoc est, secundum signorum variationes,

$$x=\tfrac{1}{2}\sqrt{2}+\sqrt{3\sqrt{2}-\tfrac{1}{2}} \quad \& \quad x=\tfrac{1}{2}\sqrt{2}-\sqrt{3\sqrt{2}-\tfrac{1}{2}}.$$

Item $x=-\tfrac{1}{2}\sqrt{2}+\sqrt{-3\sqrt{2}-\tfrac{1}{2}}$ & $x=-\tfrac{1}{2}\sqrt{2}-\sqrt{-3\sqrt{2}-\tfrac{1}{2}}$. Quæ quidem quatuor sunt radices æquationis sub initio propositæ $x^4+12x-17=0$. Sed earum ultimæ duæ sunt impossibiles.

Proponamus jam æquationem $x^4-6x^3-58xx-114x-11=0$, et scribendo -6, -58, -114, &-11 pro p, q, r, & s respectivè orietur $-67=\alpha$, $-315=\beta$ & $-1133\tfrac{1}{4}=\zeta$. Numerorum β & 2ζ seu -315 & $-\tfrac{4533}{2}$, com‖munis divisor est unicus 3 adeoq hic erit n, & ipsius $\dfrac{\beta}{n}$ seu -105 divisores sunt 3, 5, 7, 15, 21, 35 & 105, qui itaq tentandi sunt pro k. Quare tento primum 3 et quotum -35 qui prodit dividendo $\dfrac{\beta}{n}$ per k seu -105 per 3 subduco de $\tfrac{1}{2}pk$ seu -3×3 & restat 26, cujus dimidium 13 esse debet l. Sed $\dfrac{\alpha+nkk}{2}$ seu $\dfrac{-67+27}{2}$ id est -20 erit Q, & $QQ-s$ erit 411, qui dividi potest per n seu 3 sed quoti 137 radix non potest extrahi.$^{(547)}$ Quamobrem rejicio 3 et tento 5 pro k. Quotus qui jam prodit dividendo $\dfrac{\beta}{n}$ per k seu -105 per 5 est -21, & hunc subducendo de $\tfrac{1}{2}pk$ seu -3×5 restat 6 cujus dimidium 3 erit l. Est et Q seu $\dfrac{\alpha+nkk}{2}$ id est $\dfrac{-67+75}{2}$ numerus 4. Et $QQ-s$ seu $16+11$ dividi potest per n et Quoti qui est 9 radix extracta 3 congruit cum l. Quamobrem concludo esse $l=3$, $k=5$, $Q=4$ & $n=3$ et si $nkkxx+2nklx+nll$ id est $75xx+90x+27$ ad utramq partem æquationis

‖[190]

---

(546) Newton's ingenious rotated variant ' [±] ' (to distinguish it from the coupled ' ± ' before and after it) was standardized to ' ± ' in Whiston's 1707 *editio princeps* and not restored by Newton in 1722. We render it in our English version by '[±]'.

odd, and hence not divisible by n, that is, 2. Consequently, those numbers 2 and 6 must be rejected. But when 1 and 3 are written for k, Q becomes 1 and 9, and $Q^2 - s$ 18 and 98, numbers which can be divided by n and the roots of the quotients extracted. These are, to be sure, ± 3 and ± 7: however, of them -3 alone fits in with l. Accordingly I put $k = 1$, $l = -3$ and $Q = 1$, then add the quantity $(nk^2x^2 + 2nklx + nl^2$, that is) $2x^2 - 12x + 18$ to each side of the equation; there results $x^4 + 2x^2 + 1 = 2x^2 - 12x + 18$ and, with the root extracted,

$$x^2 + 1 = \sqrt{2} \cdot (x - 3).$$

But should you prefer to escape the root extraction, put

$$x^2 + \tfrac{1}{2}px + Q = \sqrt{n} \cdot (kx + l)$$

and there will be found as before $x^2 + 1 = \pm \sqrt{2} \cdot (x - 3)$. And if you again extract the roots of this equation, there will come out

$$x = \pm \tfrac{1}{2}\sqrt{2}[\pm]\sqrt{[-\tfrac{1}{2} \pm 3\sqrt{2}]};^{(546)}$$

that is, according to the variations in the signs, $x = \tfrac{1}{2}\sqrt{2} + \sqrt{[3\sqrt{2} - \tfrac{1}{2}]}$ and $x = \tfrac{1}{2}\sqrt{2} - \sqrt{[3\sqrt{2} - \tfrac{1}{2}]}$, and correspondingly $x = -\tfrac{1}{2}\sqrt{2} + \sqrt{[-3\sqrt{2} - \tfrac{1}{2}]}$ and $x = -\tfrac{1}{2}\sqrt{2} - \sqrt{[-3\sqrt{2} - \tfrac{1}{2}]}$. These are, to be sure, the four roots of the equation $x^4 + 12x - 17 = 0$ proposed at the beginning. But of these the two final ones are impossible.

Let us now propose the equation $x^4 - 6x^3 - 58x^2 - 114x - 11 = 0$: on writing -6, -58, -114 and -11 in place of p, q, r and s respectively there will then arise $\alpha = -67$, $\beta = -315$ and $\zeta = -1133\tfrac{1}{4}$. Of the numbers β and 2ζ, that is, -315 and $-\tfrac{4533}{2}$, there is a unique common divisor 3 and this will hence be n, while of β/n or -105 the divisors are 3, 5, 7, 15, 21, 35 and 105: these, accordingly, are to be tested in k's place. In consequence I first test 3 and the quotient -35 which results on dividing β/n by k, that is, -105 by 3, I subtract from $\tfrac{1}{2}pk$ or -3×3 and there remains 26, half which, 13, ought to be l. But $\tfrac{1}{2}(\alpha + nk^2)$ or $\tfrac{1}{2}(-67 + 27)$, that is, -20, will be Q and $Q^2 - s$ will be 411, which can be divided by n, or 3, but the root of the quotient 137 cannot be extracted.$^{(547)}$ Because of this I reject 3 and test 5 for k. The quotient which now results on dividing β/n by k, that is, -105 by 5, is -21 and on subtracting this from $\tfrac{1}{2}pk$ or -3×5 there remains 6, whose half, 3, will be l. Also Q or $\tfrac{1}{2}(\alpha + nk^2)$, that is, $\tfrac{1}{2}(-67 + 75)$, is the number 4 and $Q^2 - s$ or $16 + 11$ is divisible by n, while the root, 3, of the quotient—that is, 9—when extracted agrees with l. In consequence I conclude that $l = 3$, $k = 5$, $Q = 4$ and $n = 3$, and that, if

$$(nk^2x^2 + 2nklx + nl^2, \text{ that is}) \ 75x^2 + 90x + 27$$

---

(547) Newton's first hasty computation yielded '$QQ - s$ erit 409, qui dividi non potest per n seu 3' ($Q^2 - s$ will be 409, which cannot be divided by n, or 3)!

addatur, radicem utrobiꝗ extrahi posse et prodire $xx+\frac{1}{2}px+Q=\sqrt{n}\times\overline{kx+l}$, seu $xx-3x+4=\pm\sqrt{3}\times\overline{5x+3}$ et extracta iterum radice

$$x=\frac{3\pm5\sqrt{3}}{2}+\sqrt{17\pm\frac{21\sqrt{3}}{2}}.$$

Haud secus si proponatur æquatio hæcce $x^4-9x^3+15xx-27x+9=0$, scribendo -9, $+15$, -27, & $+9$ pro p, q, r, & s respectivè, emerget $-5\frac{1}{4}=\alpha$, $-50\frac{5}{8}=\beta$, & $2\frac{7}{64}=\zeta$. Ipsorum β & 2ζ seu $-\frac{405}{8}$ & $\frac{135}{32}$ communes divisores sunt ‖[191] 3, 5, 9, 15, 27, 45 & 135, sed 9 quadratus est, et 3, 15, 27, 135 divisi ‖ per numerum 4 non relinquunt unitatem ut ob imparem terminum p oporteret. His itaꝗ rejectis restant soli 5 & 45 tentandi pro n. Ponamus primo $n=5$ & ipsius $\frac{\beta}{n}$ seu $-\frac{81}{8}$ divisores impares dimidiati nempe $\frac{1}{2}$, $\frac{3}{2}$, $\frac{9}{2}$, $\frac{27}{2}$, $\frac{81}{2}$, tentandi erunt pro k.

Si k ponatur $\frac{1}{2}$, quotus $-\frac{81}{4}$ qui prodit dividendo $\frac{\beta}{n}$ per k, subductus de $\frac{1}{2}pk$ seu $-\frac{9}{4}$ relinquit 18 pro l, & $\frac{\alpha+nkk}{2}$ seu -2 est Q et $QQ-s$ seu -5 dividi quidem potest per n seu 5, sed Quoti negativi -1 radix impossibilis est quæ tamen deberet esse 18. Quare concludo k non esse $\frac{1}{2}$ et tento jam si sit $\frac{3}{2}$. Quotum qui oritur dividendo $\frac{\beta}{n}$ per k seu $-\frac{81}{8}$ per $\frac{3}{2}$ nempe Quotum $-\frac{27}{4}$ subduco de $\frac{1}{2}pk$ seu $-\frac{27}{4}$ & restat 0. Unde l jam nihil erit. Est autem $\frac{\alpha+nkk}{2}$ seu 3 æqualis Q, et $QQ-s$ nihil est, unde rursus l, qui hujus $QQ-s$ divisi per n radix est, invenitur nihil. Quamobrem his ita quadrantibus concludo esse $n=5$, $k=\frac{3}{2}$, $l=0$ & $Q=3$, adeoꝗ addendo ad utramꝗ partem æquationis propositæ terminos

$$nkkxx+2nklx+nll$$

id est $\frac{45}{4}xx$ et radicem quadraticam utrobiꝗ extrahendo prodire

$$xx+\tfrac{1}{2}px+Q=\sqrt{n}\times\overline{kx+l}, \quad \text{id est} \quad xx-4\tfrac{1}{2}x+3=[\pm]\sqrt{5}\times\tfrac{3}{2}x.$$

Eadem methodo reducuntur etiam æquationes literales. Ut si fuerit

$$x^4-2ax^3\genfrac{}{}{0pt}{}{+2aa}{-cc}xx-2a^3x+a^4=0,^{(548)}$$

‖[192] substituendo $-2a$, $2aa-cc$, ‖ $-2a^3$ & $+a^4$ pro p, q, r & s respectivè, obtinebuntur $aa-cc=\alpha$, $-acc-a^3=\beta$, & $\frac{3}{4}a^4+\frac{1}{2}aacc-\frac{1}{4}c^4=\zeta$. Quantitatum β & $2\zeta^{(549)}$ divisor communis est $aa+cc$ qui proinde erit n; & $\frac{\beta}{n}$ seu $-a$ divisores habet 1 et a. Sed quia n duarum est dimensionum & $k\sqrt{n}$ non nisi unius esse debet, ideo k nullius erit adeoꝗ non potest esse a. Sit ergo $k=1$, et diviso $\frac{\beta}{n}$ per k aufer quotum $-a$ de $\frac{1}{2}pk$ seu $-a$ et restabit nihil pro l. Porro $\frac{\alpha+nkk}{2}$ seu aa est Q et $QQ-s$ seu

be added to each side of the equation, the root on either part can be extracted, resulting in $(x^2 + \frac{1}{2}px + Q = \sqrt{n}.(kx+l)$ or) $x^2 - 3x + 4 = \pm\sqrt{3}.(5x+3)$, and with the root again extracted $x = \frac{1}{2}(3 \pm 5\sqrt{3})[\pm]\sqrt{[17 \pm \frac{21}{2}\sqrt{3}]}$.

No differently, if this equation $x^4 - 9x^3 + 15x^2 - 27x + 9 = 0$ be proposed, on writing -9, $+15$, -27 and $+9$ in place of p, q, r and s respectively, there will emerge $\alpha = -5\frac{1}{4}$, $\beta = -50\frac{5}{8}$ and $\zeta = 2\frac{7}{64}$. Of β and 2ζ, that is, $-\frac{405}{8}$ and $\frac{135}{32}$, the common divisors are 3, 5, 9, 15, 27, 45 and 135; but 9 is square, while 3, 15, 27 and 135 do not leave unity when divided by 4, as necessitated by reason of the odd term p. Accordingly, when these are rejected there remain 5 and 45 alone to be tested in n's place. Let us first set $n = 5$ and then the halves of the odd divisors of β/n or $-\frac{81}{8}$, namely $\frac{1}{2}$, $\frac{3}{2}$, $\frac{9}{2}$, $\frac{27}{2}$, $\frac{81}{2}$, will need to be tested for k. If k be set as $\frac{1}{2}$, the quotient $-\frac{81}{4}$ which results on dividing β/n by k leaves, when taken from $\frac{1}{2}pk$ or $-\frac{9}{4}$, the remainder 18 for l, and $\frac{1}{2}(\alpha + nk^2)$ or -2 is Q; $Q^2 - s$ or -5 is indeed divisible by n or 5, but the root of the negative quotient -1 is impossible though this ought to be 18. Consequently I conclude that k is not $\frac{1}{2}$, and now test if it be $\frac{3}{2}$. The quotient which arises on dividing β/n by k, that is, $-\frac{81}{8}$ by $\frac{3}{2}$, namely the quotient $-\frac{27}{4}$, I subtract from $\frac{1}{2}pk$ or $-\frac{27}{4}$ and there remains 0. Hence l will now be nothing. However, $\frac{1}{2}(\alpha + nk^2)$ or 3 is equal to Q and $Q^2 - s$ is nothing, whence l, the root of this $Q^2 - s$ divided by n, is again found to be nothing. In consequence, with these values squaring in this way, I conclude that $n = 5$, $k = \frac{3}{2}$, $l = 0$ and $Q = 3$, and therefore, on adding to each side of the equation proposed the terms $(nk^2x^2 + 2nklx + nl^2$, that is) $\frac{45}{4}x^2$ and extracting the square root on either side, that there results

$$(x^2 + \tfrac{1}{2}px + Q = \sqrt{n}.(kx+l), \text{ that is) } x^2 - 4\tfrac{1}{2}x + 3 = \pm\sqrt{5}.\tfrac{3}{2}x.$$

By the same method are reduced literal equations also. For instance, if there were $x^4 - 2ax^3 + (2a^2 - c^2)x^2 - 2a^3x + a^4 = 0$,[548] on substituting $-2a$, $2a^2 - c^2$, $-2a^3$ and $+a^4$ in place of p, q, r and s respectively there will be obtained $\alpha = a^2 - c^2$, $\beta = -a^3 - ac^2$ and $\zeta = \frac{3}{4}a^4 + \frac{1}{2}a^2c^2 - \frac{1}{4}c^4$. Of the quantities β and 2ζ[549] the common divisor is $a^2 + c^2$, and this as a result will be n; also β/n or $-a$ has divisors 1 and a. But, because n is of two dimensions and $k\sqrt{n}$ ought to be of but one, k will for that reason be of none and hence cannot be a. Let therefore $k = 1$ and, having divided β/n by k, take the quotient $-a$ from $\frac{1}{2}pk$ or $-a$ and there will remain nothing for l. Moreover, $\frac{1}{2}(\alpha + nk^2)$ or a^2 is Q and $Q^2 - s$ or $a^4 - a^4$ is

---

(548) The example is borrowed from the third book of Descartes' *Geometrie* (₁1637: 377 → *Geometria*, ₂1659: 77), where it derives from an analytical recasting of Heraclitus' solution of Apollonius, *On Inclinations* I, 8. Newton had met with it in his earliest researches into the theory and construction of equations (I: 509–11) and, again, at second hand in Kinckhuysen's *Algebra* (II: 325).

(549) That is, $-a(a^2 + c^2)$ and $\frac{1}{2}(a^2 + c^2)(3a^2 - c^2)$ respectively.

$a^4 - a^4$ nihil est; & inde rursus prodit nihil pro l. Quod arguit quantitates n, k, l & Q recte inventas esse et additis ad utramq̃ partem æquationis propositæ[550] terminis $nkkxx + 2nklx + nll$, id est $aaxx + ccxx$, radicem utrobiq̃ extrahi posse et extractione illa prodire $xx + \frac{1}{2}px + Q = \sqrt{n} \times \overline{kx+l}$, id est

$$xx - ax + aa = \pm x\sqrt{aa+cc}.$$

Et extracta iterum radice

$$x = \tfrac{1}{2}a \pm \tfrac{1}{2}\sqrt{aa+cc} + \text{vel} - {}^{(551)}\sqrt{\tfrac{1}{4}cc - \tfrac{1}{2}aa \pm \tfrac{1}{2}a\sqrt{aa+cc}}.$$

[1682]
Lect. 6.   Hactenus regulam applicui ad extractionem radicum surdarum: potest tamen eadem ad extractionem etiam rationalium applicari si modò pro quantitate n usurpetur unitas; eoq̃ pacto una vice examinare possumus utrum æquatio fractis et surdis terminis carens divisorem aliquem duarum dimensionum aut rationalem aut surdum admittat. Ut si æquatio $x^4 - x^3 - 5xx + 12x - 6 = 0$ proponatur, sub-

‖[193] stituendo -1, -5, $+12$, & -6 pro ‖ p, q, r & s respectivè, invenientur $-5\frac{1}{4} = \alpha$, $9\frac{3}{8} = \beta$ & $-10\frac{57}{64} = \zeta$. Terminorum β & 2ζ seu $\frac{75}{8}$ et $-\frac{697}{32}$ communis divisor est sola unitas. Quare pono $n = 1$. Qua[n]titatis $\frac{\beta}{n}$ seu $\frac{75}{8}$ divisores sunt 1, 3, 5, 15, 25, 75: quorum dimidia (siquidem p sit impar) tentanda sunt pro k. Et si pro k tentemus $\frac{5}{2}$, fiet $\frac{1}{2}pk - \frac{\beta}{nk} = -5$ et ejus dimidium $-\frac{5}{2} = l$. Item $\frac{\alpha + nkk}{2} = \frac{1}{2} = Q$ & $\frac{QQ - s}{n} = 6\frac{1}{4}$, cujus radix[552] congruit cum l. Concludo itaq̃ quantitates n, k, l, Q recte inventas esse et additis ad utramq̃ partem æquationis terminis

$$nkkxx + 2nklx + nll \quad \text{id est} \quad 6\tfrac{1}{4}xx - 12\tfrac{1}{2}x + 6\tfrac{1}{4},$$

radicem utrobiq̃ extrahi posse et extractione illa prodire

$$xx + \tfrac{1}{2}px + Q = \pm\sqrt{n} \times \overline{kx+l}.$$

id est $xx - \frac{1}{2}x + \frac{1}{2} = \pm 1 \times \overline{2\frac{1}{2}x - 2\frac{1}{2}}$, seu $xx - 3x + 3 = 0$ & $xx + 2x - 2 = 0$, adeoq̃ per hasce duas æquationes quadraticas, æquationem propositam quadrato-quadraticā dividi posse. Sed hujusmodi divisores rationales expeditius inveniuntur per aliam methodum supra[553] traditam.

Siquando quantitatis $\frac{\beta}{n}$ multi sunt divisores[,] ita ut omnes pro k tentare molestum fuerit, potest eorum numerus citò minui quærendo omnes divisores quantitatis $\alpha s - \frac{1}{4}rr$. Nam horum alicui aut imparis alicujus dimidio debet

‖[194] quantitas Q æqualis esse. Sic in exemplo novissimo ‖ $\alpha s - \frac{1}{4}rr$ est $-\frac{9}{2}$, e cujus divisoribus 1, 3, 9 aut ijsdem dimidiatis $\frac{1}{2}$, $\frac{3}{2}$, $\frac{9}{2}$ aliquis debet esse Q. Quare sigillatim tentando quantitatis $\frac{\beta}{n}$ divisores dimidiatos $\frac{1}{2}$, $\frac{3}{2}$, $\frac{5}{2}$, $\frac{15}{2}$, $\frac{25}{2}$ & $\frac{75}{2}$ pro k,

nothing; and from this there again results nothing for l. This confirms that the quantities n, k, l and Q have been correctly found and that, when to each side of the equation proposed[550] the terms $(nk^2x^2+2nklx+nl^2$, that is) $a^2x^2+c^2x^2$ are added, the root can be extracted on each side, the extraction yielding $(x^2+\frac{1}{2}px+Q = \sqrt{n}.(kx+l)$, that is) $x^2-ax+a^2 = \pm\sqrt{[a^2+c^2]}.x$. And when the root is extracted a second time

$$x = \tfrac{1}{2}a \pm \tfrac{1}{2}\sqrt{[a^2+c^2]} + \text{ or } -^{(551)} \sqrt{(\tfrac{1}{4}c^2 - \tfrac{1}{2}a^2 \pm \tfrac{1}{2}a\sqrt{[a^2+c^2]})}.$$

Thus far I have applied the rule to the extraction of surd roots. The same one can, however, be applied to the extraction of rational ones also, provided unity be employed in place of the quantity n; and in this manner we can at one go examine whether an equation lacking fractional and surd terms can admit of some divisor, either rational or surd, of two dimensions. For instance, if the equation $x^4-x^3-5x^2+12x-6 = 0$ be proposed, on substituting -1, -5, $+12$ and -6 in place of p, q, r and s respectively, there will be found $\alpha = -5\frac{1}{4}$, $\beta = 9\frac{3}{8}$ and $\zeta = -10\frac{57}{64}$. Of the terms β and 2ζ or $\frac{75}{8}$ and $-\frac{697}{32}$ the common divisor is unity alone. Consequently I put $n = 1$. The quantity β/n or $\frac{75}{8}$ has the divisors 1, 3, 5, 15, 25, 75: the halves of these (seeing that p is odd) are to be tested for k. Should we now test $\frac{5}{2}$ for k, there will come to be $\frac{1}{2}pk - \beta/nk = -5$ and its half $-\frac{5}{2} = l$. Likewise $Q = \frac{1}{2}(\alpha+nk^2) = \frac{1}{2}$ and $(Q^2-s)/n = 6\frac{1}{4}$, the[552] root of which agrees with l. Accordingly I conclude that the quantities n, k, l, Q have been correctly found and that, when to each side of the equation the terms $(nk^2x^2+2nklx+nl^2$, that is) $6\frac{1}{4}x^2-12\frac{1}{2}x+6\frac{1}{4}$ are added, the root can be extracted on each side, the extraction producing

$$(x^2+\tfrac{1}{2}px+Q = \pm\sqrt{n}.(kx+l), \text{ that is) } x^2-\tfrac{1}{2}x+\tfrac{1}{2} = \pm 1.(2\tfrac{1}{2}x-2\tfrac{1}{2}),$$

or $x^2-3x+3 = 0$ and $x^2+2x-2 = 0$; and hence that the quartic equation proposed can be divided by these two quadratic equations. But rational divisors of this kind are more speedily found by the other method delivered above.[553]

If at any time the quantity β/n proves to have many divisors, so that it would be troublesome to test them all in k's place, their number can be swiftly diminished by seeking out all divisors of the quantity $\alpha s - \frac{1}{4}r^2$. For to one or other of these, or its half when odd, the quantity Q ought to be equal. Thus in the most recent example $\alpha s - \frac{1}{4}r^2$ is $-\frac{9}{2}$, and one or other of its divisors 1, 3, 9 or their halves $\frac{1}{2}$, $\frac{3}{2}$, $\frac{9}{2}$ ought to be Q. Consequently, in separately testing the halved divisors $\frac{1}{2}$, $\frac{3}{2}$, $\frac{5}{2}$, $\frac{15}{2}$, $\frac{25}{2}$ and $\frac{75}{2}$ of the quantity β/n in k's place I reject all which

(550) This phrase replaces 'utrobiɋ' (to each side), doubtless to avoid a clash with the same word used, in a slightly different sense, in the next line.

(551) A variant on the notation ' +I ' above; see note (546).

(552) Understand 'negativa' (negative).

(553) On Newton's pages 176–80, especially 177–8.

rejicio omnes qui non efficiunt $\frac{1}{2}\alpha + \frac{1}{2}nkk$ seu $-\frac{21}{8} + \frac{1}{2}kk$ id est Q esse aliquem e numeris 1, 3, 9, $\frac{1}{2}$, $\frac{3}{2}$, $\frac{9}{2}$. Scribendo autem $\frac{1}{2}$, $\frac{3}{2}$, $\frac{5}{2}$, $\frac{15}{2}$ &c pro k, prodeunt respectivè $-\frac{5}{2}$, $-\frac{3}{2}$, $+\frac{1}{2}$, $+\frac{51}{2}$ &c pro Q, e quibus soli $-\frac{3}{2}$ & $\frac{1}{2}$ reperiuntur in prædictis numeris 1, 3, 9, $\frac{1}{2}$, $\frac{3}{2}$, $\frac{9}{2}$, adeoqʒ, cæteris rejectis, aut erit $k = \frac{3}{2}$ & $Q = -\frac{3}{2}$ aut $k = \frac{5}{2}$ & $Q = \frac{1}{2}$. Qui duo casus citò examinantur. Atqʒ hactenus de æquationibus quatuor dimensionum.

Si æquatio sex dimensionum reducenda est sit ea

$$x^6 + px^5 + qx^4 + rx^3 + sxx + tx + v = 0,$$

et fac $q - \frac{1}{4}pp = \alpha$. $r - \frac{1}{2}p\alpha = \beta$. $s - \frac{1}{2}p\beta = \gamma$. $\gamma - \frac{1}{4}\alpha\alpha = \zeta$. $t - \frac{1}{2}\alpha\beta = \eta$. $v - \frac{1}{4}\beta\beta = \theta$. $\zeta\theta - \frac{1}{4}\eta\eta = \lambda$. Dein sumatur pro n, communis aliquis terminorum 2ζ, η, 2θ divisor integer & non quadratus nec per numerum quadratum divisibilis, qui etiam per numerum 4 divisus relinquit unitatem si modo terminorum p, r, t aliquis sit impar. Pro k sumatur divisor aliquis integer quantitatis $\dfrac{\lambda}{2nn}$ si p sit par, vel divisoris imparis dimidium si p sit impar vel nihil si λ nihil sit. Pro Q, quantitas

‖[195] $\frac{1}{2}\alpha + \frac{1}{2}nkk$. Pro l, divisor aliquis ‖ quantitatis $\dfrac{Qr - QQp - t}{n}$ si Q sit integer vel divisoris imparis dimidium si Q sit fractus denominatorem habens numerum 2, vel nihil si dividuum[554] istud $\dfrac{Qr - QQp - t}{n}$ sit nihil. Et pro R quantitas

$$\tfrac{1}{2}r - \tfrac{1}{2}Qp + nkl.\text{[555]}$$

Dein tenta si $RR - v$ dividi possit per n et Quoti radix extrahi & præterea si radix ista æqualis sit tam quantitati $\dfrac{QR - \frac{1}{2}t}{nl}$ quam quantitati $\dfrac{QQ + pR - nll - s}{2nk}$. Si hæc omnia evenerint, dic radicem illam m et vice æquationis propositæ scribe hanc $x^3 + \frac{1}{2}pxx + Qx + [R] = \pm\sqrt{n} \times \overline{kxx + lx + m}$.[556] Etenim hæc æquatio quadrando partes et auferendo utrobiqʒ terminos ad dextram, producet æquationem propositam. Quod si ea omnia in nullo casu evenerint reductio erit impossibilis, si modò prius constet æquationem per divisorem rationalem reduci non posse.

---

(554) Namely, the numerator $Qr - Q^2p - t$.

(555) The manuscript reads '$+\frac{1}{2}nkl$' in error. Whiston failed to catch the slip in his 1707 *editio princeps* and it was left to Newton himself to delete the superfluous '$\frac{1}{2}$' in his 1722 revise. A few lines below we have replaced a careless 'r'.

(556) Newton's opening equalities result from the identity

$$x^6 + px^5 + qx^4 + rx^3 + sx^2 + tx + v \equiv (x^3 + \tfrac{1}{2}px^2 + \tfrac{1}{2}\alpha x + \tfrac{1}{2}\beta)^2 + \zeta x^2 + \eta x + \theta.$$

On further identifying with $(x^3 + \frac{1}{2}px^2 + Qx + R)^2 - n(kx^2 + lx + m)^2$ there ensues

$$2Q + \tfrac{1}{4}p^2 - nk^2 = q, \quad pQ + 2R - 2nkl = r, \quad Q^2 + pR - n(l^2 + 2km) = s, \quad 2QR - 2nlm = t$$

do not make Q, that is, $\frac{1}{2}(\alpha+nk^2)$ or $-\frac{21}{8}+\frac{1}{2}k^2$, one or other of the numbers 1, 3, 9, $\frac{1}{2}$, $\frac{3}{2}$, $\frac{9}{2}$. Now, on writing $\frac{1}{2}$, $\frac{3}{2}$, $\frac{5}{2}$, $\frac{15}{2}$, ... in k's place there result respectively $-\frac{5}{2}$, $-\frac{3}{2}$, $+\frac{1}{2}$, $+\frac{51}{2}$, ...: of these $-\frac{3}{2}$ and $\frac{1}{2}$ alone are found in the above-listed numbers 1, 3, 9, $\frac{1}{2}$, $\frac{3}{2}$, $\frac{9}{2}$ and hence, after the remainder are rejected, there will be either $k=\frac{3}{2}$ and $Q=-\frac{3}{2}$, or $k=\frac{5}{2}$ and $Q=\frac{1}{2}$. These two cases are swiftly examined. And so much for equations of four dimensions.

If an equation of six dimensions is to be reduced, let it be

$$x^6+px^5+qx^4+rx^3+sx^2+tx+v=0,$$

then make $q-\frac{1}{4}p^2=\alpha$, $r-\frac{1}{2}p\alpha=\beta$, $s-\frac{1}{2}p\beta=\gamma$, $\gamma-\frac{1}{4}\alpha^2=\zeta$, $t-\frac{1}{2}\alpha\beta=\eta$,

$$v-\frac{1}{4}\beta^2=\theta, \quad \zeta\theta-\frac{1}{4}\eta^2=\lambda.$$

Next, take for n some integral, non-square common divisor of the terms 2ζ, η, 2θ, which is not divisible by a square number and also leaves unity when divided by the number 4 should one or other of the terms p, r, t be odd. For k assume some integral divisor of the quantity $\frac{1}{2}\lambda/n^2$ if p be even, or half an odd divisor if p be odd, or nothing if λ be nothing; for Q put the quantity $\frac{1}{2}(\alpha+nk^2)$, and for l some divisor of the quantity $(Qr-Q^2p-t)/n$ if Q be an integer, or half an odd divisor if Q be a fraction having the number 2 for its denominator, or nothing if the dividend[554] of $(Qr-Q^2p-t)/n$ be nothing; and for R take the quantity $\frac{1}{2}r-\frac{1}{2}Qp+nkl$.[555] Then test if R^2-v may be divided by n and the root of the quotient extracted, and further if that root be equal both to the quantity $(QR-\frac{1}{2}t)/nl$ and to the quantity $\frac{1}{2}(Q^2+pR-nl^2-s)/nk$. If all these things happen, call that root m and instead of the equation proposed write this one, $x^3+\frac{1}{2}px^2+Qx+R=\pm\sqrt{n}\cdot(kx^2+lx+m)$:[556] for, to be sure, on squaring its sides and taking away the terms on the right from each side this equation will produce the one propounded. But if all these things happen in no case, then the reduction will be impossible provided it shall first be established that the equation cannot be reduced by a rational divisor.

---

and $R^2-nm^2=v$, whence $Q=\frac{1}{2}(\alpha+nk^2)$, $2nlm=Q(r-pQ+2nkl)-t$

or $\qquad\qquad (Qr-pQ^2-t)/n=2l(m-k)$, $\quad R=\frac{1}{2}(r-pQ)+nkl$

and $\qquad\qquad m=\sqrt{\dfrac{R^2-v}{n}}=\dfrac{2QR-t}{2nl}=\dfrac{Q^2+pR-nl^2-s}{2nk}$;

also $(\bmod n^2 k)$ $\zeta\equiv -nl^2+nk(lp-2m)$, $\frac{1}{2}\eta\equiv -nlm+nkl\beta$, $\theta\equiv -nm^2+nkl\beta$ and so

$$\lambda=\zeta\theta-\frac{1}{4}\eta^2\equiv 0.$$

Newton's further observations are immediate corollaries. So when p is even (and therefore α, β and γ are integral) k is integral and hence $\lambda\equiv 0\ (\bmod 2nk^2)$; but when p is odd (and so $2k$ is integral), say $p=2P+1$, it follows that

$$2Q=q-\frac{1}{4}p^2+nk^2=q-P(P+1)+\frac{1}{4}(4nk^2-1)$$

and accordingly $n\equiv 1\ (\bmod 4)$.

Exempli gratia proponatur æquatio

$$x^6 - 2ax^5 + 2bbx^4 + 2abbx^3 \begin{matrix} -2aabb \\ +2a^3b \\ -4ab^3 \end{matrix} xx \begin{matrix} +3aab^4 \\ -a^4bb \end{matrix} = 0,$$

et scribendo $-2a$, $+2bb$, $+2abb$, $-2aabb+2a^3b-4ab^3$, 0, & $3aab^4-a^4bb$ pro p, q, r, s, t et v respectivè, prodibunt $2bb-aa=\alpha$. $4abb-a^3=\beta$.

$$2a^3b + 2aabb - 4ab^3 - a^4 = \gamma. \quad -b^4 + 2a^3b + 3aabb - 4ab^3 - \tfrac{5}{4}a^4 = \zeta.$$

$\tfrac{1}{2}a^5 - a^3bb = \eta$. & $3aab^4 - a^4bb - \tfrac{1}{4}a^6 = \theta$. Et terminorum 2ζ, η, & 2θ communis

‖[196] divisor ‖ est $aa-2bb$, seu $2bb-aa$ perinde ut aa vel $2bb$ majus sit. Sed esto aa majus quam $2bb$ et $aa-2bb$ erit n. Debet enim n semper affirmativum esse.[557]

Porro $\dfrac{\zeta}{n}$ est $-\tfrac{5}{4}aa + 2ab + \tfrac{1}{2}bb$, $\dfrac{\eta}{n}$ est $\tfrac{1}{2}a^3$, & $\dfrac{\theta}{n}$ est $-\tfrac{1}{4}a^4 - \tfrac{3}{2}aabb$, adeoꝗ $\dfrac{\zeta}{2n} \times \dfrac{\theta}{n} - \dfrac{\eta\eta}{8nn}$,

seu $\dfrac{\lambda}{2nn}$ est $\tfrac{1}{8}a^6 - \tfrac{1}{4}a^5b + \tfrac{1}{4}a^4bb - \tfrac{1}{2}a^3b^3 - \tfrac{1}{8}aab^4$, cujus divisores sunt 1, a, aa, sed quia $\sqrt{n} \times k$ non nisi unius dimensionis esse potest, & $\sqrt{n}$ unius est, ideo k nullius erit, proinde non nisi numerus esse potest. Quare rejectis a et aa, restat solùm 1 pro k. Præterea $\tfrac{1}{2}\alpha + \tfrac{1}{2}nkk$ dat nihil pro Q, et $\dfrac{Qr - QQp - t}{n}$ etiam nihil est; adeoꝗ l qui ejus divisor esse debet, erit[558] nihil. Deniꝗ $\tfrac{1}{2}r - \tfrac{1}{2}pQ + nkl$[559] dat abb pro R. Et $RR - v$ est $-2aab^4 + a^4bb$, quod dividi potest per n seu $aa - 2bb$ et quoti $aabb$ radix extrahi, et radix illa negativè sumpta, nempe $-ab$, indefinitæ quantitati $\dfrac{QR - \tfrac{1}{2}t}{nl}$ seu $\dfrac{0}{0}$ non est inæqualis,[560] quantitati verò definitæ $\dfrac{QQ + pR - nll - s}{2nk}$ æqualis est. Quamobrem radix illa $-ab$ erit m, et loco æquationis propositæ scribi potest $x^3 - \tfrac{1}{2}pxx + Qx + R = \sqrt{n} \times \overline{kxx + lx + m}$, id est

$$x^3 - axx + abb = [\pm]\sqrt{aa - 2bb} \times \overline{xx - ab}.\text{[561]}$$

Cujus conclusionis veritatem probare potes quadrando partes æquationis inventæ et auferendo terminos ad dextram ex utraꝗ parte. Ea enim operatione producetur æquatio

$$x^6 - 2ax^5 + 2bbx^4 + 2abbx^3 - 2aabbxx + 2a^3bxx - 4ab^3xx + 3aab^4 - a^4bb = 0,$$

quæ reducenda proponebatur.

‖[197]   ‖ Si æquatio est octo dimensionum sit ea
[1682]
Lect. 7.

$$x^8 + px^7 + qx^6 + rx^5 + sx^4 + tx^3 + vxx + wx + z = 0,$$

---

(557) For it is necessary that $\pm \sqrt{n}$ be real.

(558) Presupposing that $k \neq m$!

(559) The manuscript again reads '$+\tfrac{1}{2}nkl$'; see note (555).

(560) A Wallisian touch; compare I: 124, note (9).

For example, let there be proposed the equation

$$x^6 - 2ax^5 + 2b^2x^4 + 2ab^2x^3 + (2a^3b - 2a^2b^2 - 4ab^3)\,x^2 - a^4b^2 + 3a^2b^4 = 0.$$

On writing $-2a$, $+2b^2$, $+2ab^2$, $2a^3b - 2a^2b^2 - 4ab^3$, 0 and $-a^4b^2 + 3a^2b^4$ in place of p, q, r, s, t and v respectively, there will result $\alpha = -a^2 + 2b^2$,

$$\beta = -a^3 + 4ab^2, \quad \gamma = -a^4 + 2a^3b + 2a^2b^2 - 4ab^3,$$

$$\zeta = -\tfrac{5}{4}a^4 + 2a^3b + 3a^2b^2 - 4ab^3 - b^4, \quad \eta = \tfrac{1}{2}a^5 - a^3b^2$$

and $\theta = -\tfrac{1}{4}a^6 - a^4b^2 + 3a^2b^4$. The terms 2ζ, η and 2θ have the common divisor $a^2 - 2b^2$ or $-a^2 + 2b^2$ according as a^2 or $2b^2$ is the greater. But let a^2 be greater than $2b^2$, and then $a^2 - 2b^2$ will be n: for n ought always to be positive.[557] Furthermore, ζ/n is $-\tfrac{5}{4}a^2 + 2ab + \tfrac{1}{2}b^2$, η/n is $\tfrac{1}{2}a^3$ and θ/n is $-\tfrac{1}{4}a^4 - \tfrac{3}{2}a^2b^2$, and hence $\tfrac{1}{2}(\zeta/n)\,(\theta/n) - \tfrac{1}{8}(\eta/n)^2$ or $\tfrac{1}{2}\lambda/n^2$ is $\tfrac{1}{8}a^6 - \tfrac{1}{4}a^5b + \tfrac{1}{4}a^4b^2 - \tfrac{1}{2}a^3b^3 - \tfrac{1}{8}a^2b^4$, the divisors of which are 1, a, a^2; but because $k\sqrt{n}$ can be of but one dimension and $\sqrt{n}$ is of one, k will in consequence be of zero dimension and as a result can be but a (pure) number. Therefore, on rejecting a and a^2, there remains only 1 for k. In addition, $\tfrac{1}{2}(\alpha + nk^2)$ yields nothing for Q and so $(Qr - Q^2p - t)/n$ is also nothing, and hence l, which must be a divisor of it, will[558] be nothing. Finally, $\tfrac{1}{2}r - \tfrac{1}{2}pQ + nkl$[559] yields ab^2 for R, and $R^2 - v$ is $a^4b^2 - 2a^2b^4$, which can be divided by n, that is, $a^2 - 2b^2$, and the root of the quotient a^2b^2 extracted; and when the negative value of that root, namely $-ab$, is taken, it is not unequal to the indefinite quantity $(QR - \tfrac{1}{2}t)/nl$ or $0/0$,[560] and is indeed equal to the definite quantity $\tfrac{1}{2}(Q^2 + pR - nl^2 - s)/nk$. In consequence, that root $-ab$ will be m, and in place of the equation proposed there can be written

$$(x^3 - \tfrac{1}{2}px^2 + Qx + R = \sqrt{n}\,.\,(kx^2 + lx + m),\ \text{that is})$$

$$x^3 - ax^2 + ab^2 = \pm\sqrt{[a^2 - 2b^2]}\,.\,(x^2 - ab).^{(561)}$$

The truth of this conclusion you can test by squaring the sides of the equation now found and taking the terms on the right from each side. For by that operation there will be produced the equation,

$$x^6 - 2ax^5 + 2b^2x^4 + 2ab^2x^3 + (2a^3b - 2a^2b^2 - 4ab^3)\,x - a^4b^2 + 3a^2b^4 = 0,$$

which was proposed for reduction.

If the equation is of eight dimensions, let it be

$$x^8 + px^7 + qx^6 + rx^5 + sx^4 + tx^3 + vx^2 + wx + z = 0$$

---

(561) Hence, on squaring, the given sextic is obtained in the form

$$(x^3 - ax^2 + ab^2)^2 = (a^2 - 2b^2)\,(x^2 - ab)^2.$$

Despite all Newton's hard work, we may well think it easier to test whether such reduction is possible by simple inspection.

et fiat $q-\frac{1}{4}pp=\alpha.\ r-\frac{1}{2}p\alpha=\beta.\ s-\frac{1}{2}p\beta-\frac{1}{4}\alpha\alpha=\gamma.\ t-\frac{1}{2}p\gamma-\frac{1}{2}\alpha\beta=\delta.\ v-\frac{1}{2}\alpha\gamma-\frac{1}{4}\beta\beta=\epsilon.$ $w-\frac{1}{2}\beta\gamma=\zeta.\ \&\ z-\frac{1}{4}\gamma\gamma=\eta.$[562] Et terminorum $2\delta,\ 2\epsilon,\ 2\zeta,\ 8\eta$ quære communem divisorem qui integer sit et non quadratus nec per quadratum divisibilis, quiꝗ etiam per 4 divisus relinquat unitatem si modo terminorum alternorum $p,\ r,\ t,$ w aliquis sit impar. Si nullus est ejusmodi divisor communis, certum est æquationem per extractionem surdæ radicis quadraticæ reduci non posse, et si non potest ea ita reduci, vix occurret illarum omnium quatuor quantitatum divisor communis.[563] Opusculum igitur hactenus institutum examinatio quædam est utrum æquatio reducibilis sit necne, adeoꝗ cum ejusmodi reductiones rarò possibiles sint, finem operi utplurimùm imponet.

Et simili ratione si æquatio sit decem duodecim vel plurium dimensionum impossibilitas reductionis cognosci potest. Ut si ea sit

$$x^{10}+px^9+qx^8+rx^7+sx^6+tx^5+vx^4+ax^3+bxx+cx+d=0,$$

faciendum erit $q-\frac{1}{4}pp=\alpha.\ r-\frac{1}{2}p\alpha=\beta.\ s-\frac{1}{2}p\beta-\frac{1}{4}\alpha\alpha=\gamma.\ t-\frac{1}{2}p\gamma-\frac{1}{2}\alpha\beta=\delta.$

$v=\frac{1}{2}p\delta-\frac{1}{2}\alpha\gamma-\frac{1}{4}\beta\beta=\epsilon.\quad a-\frac{1}{2}\alpha\delta-\frac{1}{2}\beta\gamma=\zeta.\quad b-\frac{1}{2}\beta\delta-\frac{1}{4}\gamma\gamma=\eta.\quad c-\frac{1}{2}\gamma\delta=\theta.$

$d-\frac{1}{4}\delta\delta=\kappa$, et quærendus communis divisor terminorum quinꝗ $2\epsilon,\ 2\zeta,\ 8\eta,\ 4\theta,$ 8κ qui integer sit & non quadratus, quiꝗ etiam per 4 divisus relinquat unitatem ‖[198] si modò terminorum ‖ alternorum $p,\ r,\ t,\ a,\ c$ aliquis sit impar.[564]

Sic si duodecim dimensionum æquatio sit

$$x^{12}+px^{11}+qx^{10}+rx^9+sx^8+tx^7+vx^6+ax^5+bx^4+cx^3+dxx+ex+f=0,$$

faciendum erit $q-\frac{1}{4}pp=\alpha.\ r-\frac{1}{2}p\alpha=\beta.\ s-\frac{1}{2}p\beta-\frac{1}{4}\alpha\alpha=\gamma.\ t-\frac{1}{2}p\gamma-\frac{1}{2}\alpha\beta=\delta.$

$v-\frac{1}{2}p\delta-\frac{1}{2}\alpha\gamma-\frac{1}{4}\beta\beta=\epsilon.\quad a-\frac{1}{2}p\epsilon-\frac{1}{2}\alpha\delta-\frac{1}{2}\beta\gamma=\zeta.\quad b-\frac{1}{2}\alpha\epsilon-\frac{1}{2}\beta\delta-\frac{1}{4}\gamma\gamma=\eta.$

$c-\frac{1}{2}\beta\epsilon-\frac{1}{2}\gamma\delta=\theta.\ d-\frac{1}{2}\gamma\epsilon-\frac{1}{4}\delta\delta=\kappa.\ e-\frac{1}{2}\delta\epsilon=\lambda.\ f-\frac{1}{4}\epsilon\epsilon=\mu$, & quærendus communis divisor integer et non quadratus terminorum sex $2\zeta,\ 8\eta,\ 4\theta,\ 8\kappa,\ 4\lambda,\ 8\mu$, qui per 4 divisus relinquat unitatem si modò terminorum alternorum $p,\ r,\ t,\ a,\ c,\ e$ aliquis sit impar.[565]

---

(562) These equalities are straightforward deductions from the identity

$$x^8+px^7+qx^6+rx^5+sx^4+tx^3+vx^2+wx+z$$

$$\equiv (x^4+\tfrac{1}{2}px^3+\tfrac{1}{2}\alpha x^2+\tfrac{1}{2}\beta x+\tfrac{1}{2}\gamma)^2+\delta x^3+\epsilon x^2+\zeta x+\eta.$$

Further identification with $(x^4+\frac{1}{2}px^3+Qx^2+Rx+S)^2-n(kx^3+lx^2+mx+h)^2$, where n is square-free (and $n\equiv 1\ (\mathrm{mod}\,4)$ if $p,\ r,\ t,\ w$ is odd), will determine n to be an integral divisor of $2\delta,\ 2\epsilon,$ 2ζ and $2(4\eta)$.

(563) Newton's caution is justified in the case of

$$x^8+x^4+2 = 0 \quad\text{or}\quad (x^4+\sqrt 2)^2 = (2\sqrt 2-1)\,x^4,$$

for which $2\delta = 2\epsilon = 2\zeta = 0$ and $8\eta = 16$.

and make $q-\frac{1}{4}p^2 = \alpha,\ r-\frac{1}{2}p\alpha = \beta,\ s-\frac{1}{2}p\beta-\frac{1}{4}\alpha^2 = \gamma,\ t-\frac{1}{2}p\gamma-\frac{1}{2}\alpha\beta = \delta,$

$$v-\frac{1}{2}\alpha\gamma-\frac{1}{4}\beta^2 = \epsilon,\quad w-\frac{1}{2}\beta\gamma = \zeta\ \text{ and }\ z-\frac{1}{4}\gamma^2 = \eta.^{(562)}$$

Of the terms 2δ, 2ϵ, 2ζ, 8η seek a common divisor which shall be integral and neither a square nor divisible by a square, and which shall also leave unity when divided by 4 should any one of the alternate terms p, r, t, w be odd. If there is no common divisor of this kind, it is certain that the equation cannot be reduced by the extraction of a surd square root, and if it cannot be so reduced a common divisor of all those four quantities is scarcely likely to occur.[563] The short procedure thus far instituted is therefore a kind of test whether an equation be reducible or not, and hence, since reductions of this type are rarely possible, will most commonly put an end to the work.

And if the equation be of ten, twelve or more dimensions, the impossibility of its reduction can be ascertained by a similar method. For instance, if it be

$$x^{10}+px^9+qx^8+rx^7+sx^6+tx^5+vx^4+ax^3+bx^2+cx+d = 0,$$

you will need to make $q-\frac{1}{4}p^2 = \alpha,\ r-\frac{1}{2}p\alpha = \beta,\ s-\frac{1}{2}p\beta-\frac{1}{4}\alpha^2 = \gamma,$

$$t-\frac{1}{2}p\gamma-\frac{1}{2}\alpha\beta = \delta,\quad v-\frac{1}{2}p\delta-\frac{1}{2}\alpha\gamma-\frac{1}{4}\beta^2 = \epsilon,\quad a-\frac{1}{2}\alpha\delta-\frac{1}{2}\beta\gamma = \zeta,$$

$b-\frac{1}{2}\beta\delta-\frac{1}{4}\gamma^2 = \eta,\ c-\frac{1}{2}\gamma\delta = \theta,\ d-\frac{1}{4}\delta^2 = \kappa$, and then look for a common divisor of the five terms 2ϵ, 2ζ, 8η, 4θ, 8κ which shall be integral and non-square, and which shall also leave unity when divided by 4 should any one of the alternate terms p, r, t, a, c be odd.[564]

And in this way, if the equation of twelve dimensions be

$$x^{12}+px^{11}+qx^{10}+rx^9+sx^8+tx^7+vx^6+ax^5+bx^4+cx^3+dx^2+ex+f = 0,$$

you will need to make $q-\frac{1}{4}p^2 = \alpha,\ r-\frac{1}{2}p\alpha = \beta,\ s-\frac{1}{2}p\beta-\frac{1}{4}\alpha^2 = \gamma,$

$$t-\frac{1}{2}p\gamma-\frac{1}{2}\alpha\beta = \delta,\quad v-\frac{1}{2}p\delta-\frac{1}{2}\alpha\gamma-\frac{1}{4}\beta^2 = \epsilon,\quad a-\frac{1}{2}p\epsilon-\frac{1}{2}\alpha\delta-\frac{1}{2}\beta\gamma = \zeta,$$

$b-\frac{1}{2}\alpha\epsilon-\frac{1}{2}\beta\delta-\frac{1}{4}\gamma^2 = \eta,\quad c-\frac{1}{2}\beta\epsilon-\frac{1}{2}\gamma\delta = \theta,\quad d-\frac{1}{2}\gamma\epsilon-\frac{1}{4}\delta^2 = \kappa,\quad e-\frac{1}{2}\delta\epsilon = \lambda,$

$f-\frac{1}{4}\epsilon^2 = \mu$, and then to look for an integral, non-square common divisor of the six terms 2ζ, 8η, 4θ, 8κ, 8μ, one which shall leave unity when divided by 4 should one or other of the alternate terms p, r, t, a, c, e be odd.[565]

---

(564) Analogous deductions from the identities

$$x^{10}+px^9+qx^8+rx^7+sx^6+tx^5+vx^4+ax^3+bx^2+cx+d$$

$$\equiv (x^5+\tfrac{1}{2}px^4+\tfrac{1}{2}\alpha x^3+\tfrac{1}{2}\beta x^2+\tfrac{1}{2}\gamma x+\delta)^2+\epsilon x^4+\zeta x^3+\eta x^2+\theta x+\kappa$$

$$\equiv (x^5+\tfrac{1}{2}px^4+Qx^3+Rx^2+Sx+T)^2-n(kx^4+lx^3+mx^2\ ...)^2.$$

(565) Newton here proceeds from the analogous identities (notice the double ϵ)

$$x^{12}+px^{11}+qx^{10}+rx^9+sx^8+tx^7+vx^6+ax^5+bx^4+cx^3+dx^2+ex+f$$

$$\equiv (x^6+\tfrac{1}{2}px^5+\tfrac{1}{2}\alpha x^4+\tfrac{1}{2}\beta x^3+\tfrac{1}{2}\gamma x^2+\tfrac{1}{2}\delta x+\tfrac{1}{2}\epsilon)^2+\epsilon x^6+\zeta x^5+\eta x^4+\theta x^3+\kappa x^2+\lambda x+\mu$$

$$\equiv (x^6+\tfrac{1}{2}px^5+Qx^4+Rx^3+Sx^2+Tx+V)^2-n(kx^5+lx^4+mx^3\ ...)^2.$$

Atcɜ ita in infinitum progredi licebit et æquatio proposita semper per extractionem surdæ radicis quadraticæ irreducibilis erit ubi ejusmodi divisor communis nullus est. Siquando vero ejusmodi divisor *n* inventus spem faciat futuræ reductionis potest ea institui insistendo vestigijs operis quod in æquatione octo dimensionum subjungimus.

Quære numerum quadratum cui per *n* multiplicato ultimus æquationis terminus *z* sub signo proprio adnexus quadratum numerum efficit. Id autem expeditè fiet si ad *z* ubi *n* est par vel ad 4*z* ubi *n* est impar successivè addantur *n*, 3*n*, 5*n*, 7*n*, 9*n*, 11*n*, & deinceps donec summa æqualis fiat numero alicui in tabula ‖ [199] numerorum quadratorum ‖ quam ad manus esse suppono. Et si nullus ejusmodi quadratus numerus priùs occurrit quàm summæ illius radix quadratica aucta radice quadratica excessus illius summæ supra ultimū æquationis terminum, quadruple major sit quam maximus terminorum æquationis propositæ *p, q, r, s, t, v* &c non opus erit rem ultra tentare. Æquatio enim reduci non potest. Sed si ejusmodi numerus quadratus prius occurrit, sit ejus radix *S* si *n* est par vel 2*S* si *n* est impar & $\sqrt{\dfrac{SS-z}{n}}$ dic *h*.[566] Debent autem *S* et *h* esse numeri integri si *n* est par, at si *n* impar est possunt esse fracti denominatorem habentes numerum binarium. Et si unus eorum fractus est, alter fractus esse debet. Quod idem de numeris *R* et *m*, *Q* & *l*, *P* et *k* post inveniendis observandum est.[567] Et omnes numeri *S* et *h* qui intra præfatum limitem inveniri possunt in catalogum referendi sunt.

Postea pro *k* tentandi sunt omnes numeri successive qui non efficiunt $nk\pm\frac{1}{2}p$[568] quadruplo majus quàm maximus terminus æquationis et ponendum est in omni casu $\dfrac{nkk+\alpha}{2}=Q$. Dein pro *l* tentandi sunt successive numeri omnes qui non efficiunt $nl\pm Q$[568] quadruplo majus quam maximus terminus æquationis et in omni tentamine ponendum $\dfrac{-npkk+2\beta}{4}+nkl=R$. Deniqɜ pro *m* tentandi sunt successivè omnes numeri qui non efficiunt $nm\pm R$[568] quadruplo ‖ [200] majus ‖ quàm maximus terminorum æquationis, et videndum an in casu quovis

---

(566) For in the octic case Newton identifies (see note (562))

$$x^8+px^7+qx^6+rx^5+sx^4+tx^3+vx^2+wx+z$$
$$\equiv (x^4+\tfrac{1}{2}px^3+Qx^2+Rx+S)^2-n(kx^3+lx^2+mx+h)^2,$$

whence $S^2=z+nh^2$ (or $\sum_{1\leqslant i\leqslant h}(2i-1)\,n$). Newton's criterion is that for reducibility

$$(|S|+\sqrt{[S^2-z]}\text{ or)}\ |S|+|h\sqrt{n}|$$

shall be not greater than four times the greatest coefficient in the equation. Given that $q=\tfrac{1}{4}p^2+2Q-nk^2$, $r=pQ-2nkl$, ..., $w=2RS-2nmh$, exact justification of this plausible inequality will be difficult, and we may be sure that Newton did not attempt to give one. In default of such a proof all of the many commentators on the published *Arithmetica*—Maclaurin,

You are free to go on indefinitely in this manner, and the equation proposed will ever be irreducible by the extraction of a surd square root when there is no common divisor of this kind. Whenever, indeed, a divisor n of this sort so found should encourage hope of a future reduction, that can be effected by following the steps in the working of the equation of eight dimensions which we now append.

Look for a square number such that, after it is multiplied by n, the final term z in the equation added to it under its proper sign shall make up a square number. (This will, I may interject, be speedily done if to z when n is even, or to $4z$ when n is odd, there be successively added n, $3n$, $5n$, $7n$, $9n$, $11n$ and so on till the sum becomes equal to some number in a table of squares which I suppose is to hand.) And if no square number of this kind occurs before the square root of that sum augmented by the square root of the excess of that sum over the final term in the equation shall be four times greater than the greatest of the terms p, q, r, s, t, v, ... in the equation proposed, there will be no need of further trial: for the equation cannot then be reduced. But if a square number of that kind does occur before that point, let its root be S if n is even, or $2S$ if n is odd, and $\sqrt{[(S^2-z)/n]}$ call h.[566] However, while S and h ought to be integers if n is even, if n is odd they can be fractions having the number 2 for denominator—and if one of them is a fraction, then the other must be a fraction. The same observation is to be made regarding the numbers R and m, Q and l, P and k to be found subsequently.[567] All numbers S and h which can be found within the stated bound should then be set out in a list.

After that, in k's place are successively to be tested all numbers which do not render $nk \pm \frac{1}{2}p$[568] four times greater than the greatest term in the equation, and in every case you must put $\frac{1}{2}(nk^2+\alpha) = Q$. Next, for l are to be successively tested all numbers which do not render $nl \pm Q$[568] four times greater than the greatest term in the equation, and at each trial you must put

$$\tfrac{1}{4}(-npk^2+2\beta)+nkl = R.$$

Finally, for m are successively to be tested all numbers which do not render $nm \pm R$[568] four times greater than the greatest of the terms in the equation, and

---

Castiglione, Lecchi, Euler, Beaudeux—are tactfully silent at this point. In Newton's following example ($n = 5$, $h = -1\frac{1}{2}$, $S = -2\frac{1}{2}$) $|S| + |h\sqrt{n}| \approx 5\cdot9$ is just over half the absolute value $(-r = -v = -w =)10$ of the equation's maximum coefficient, but—making use of the inequality $|S| + |h\sqrt{n}| < |S| + |nh|$?—he there introduces the modified condition that '$S+nh$' shall be not greater than four times the greatest coefficient. (See note (573).)

(567) These are straightforward consequences of the identity in the previous note.

(568) These should, perhaps, read more strictly '$k\sqrt{n} \pm \frac{1}{2}p$', '$l\sqrt{n} \pm Q$' and '$m\sqrt{n} \pm R$' on the previous pattern; compare note (566). They are doubtless to be derived from the inequalities $-q > nk^2 - \frac{1}{4}p^2$, $-s > nl^2 - Q^2$ and $-v > nm^2 - R^2$ in the same way that the preceding criterion regarding $|h\sqrt{n}| + |S|$ ensues from $-z = nh^2 - S^2$.

si fiat $s-QQ-pR+nll=2H$ et $H+nkm=S$, sit S aliquis numerorum[569] qui prius pro S in Catalogum relati erunt; et præterea si alter numerus ei S respondens, qui pro h in eundem Catalogum relatus erat sit his tribus $\dfrac{2RS-w}{2nm}$, $\dfrac{2QS+RR-v-nmm}{2nl}$ & $\dfrac{[p]S+2QR-t-2nlm}{2nk}$ æqualis.[570] Si hæc omnia in aliquo casu evenerint, vice æquationis propositæ scribenda erit hæcce

$$x^4+\tfrac{1}{2}px^3+Qxx+Rx+S=\sqrt{n}\times\overline{kx^3+lxx+mx+h}.$$

Exempli gratia proponatur æquatio

$$x^8+4x^7-x^6-10x^5+5x^4-5x^3-10xx-10x-5=0.$$

et erit $q-\tfrac{1}{4}pp=-1-4=-5=\alpha.$ $r-\tfrac{1}{2}p\alpha=-10+10=0=\beta.$

$s-\tfrac{1}{2}p\beta-\tfrac{1}{4}\alpha\alpha=5-\tfrac{25}{4}=-\tfrac{5}{4}=\gamma.$ $t-\tfrac{1}{2}p\gamma-\tfrac{1}{2}\alpha\beta=-5+\tfrac{5}{2}=-\tfrac{5}{2}=\delta.$

$v-\tfrac{1}{2}\alpha\gamma-\tfrac{1}{4}\beta\beta=-10-\tfrac{25}{8}=-\tfrac{105}{8}[=\epsilon].$ $w-\tfrac{1}{2}\beta\gamma=-10=\zeta.$

$z-\tfrac{1}{4}\gamma\gamma=-5-\tfrac{25}{64}=-\tfrac{345}{64}=\eta.$[571] Ergo 2δ, 2ϵ, 2ζ, 8η respectivè sunt -5, $-\tfrac{105}{4}$, -20, & $-\tfrac{345}{8}$ et earum divisor communis 5 qui per 4 divisus relinquit 1 perinde ut ob terminum imparem s oportuit. Cum itaᴄᴊ inventus sit divisor communis n seu 5 qui spem facit futuræ reductionis, quoniam iste impar est ad $4z$ seu -20 successivè addo n, $3n$, $5n$, $7n$, $9n$ &c seu 5, 15, 25, 35, 45 &c et prodeunt -15. 0. 25. 60. 105. 160. 225. 300. 385. 480. 585. 700. 825. 960. 1105. 1260. 1425. 1600. Ex quibus solum 0. 25. 225 & 1600 quadrati sunt. Quare horum radices

‖[201] dimidiatæ 0, $\tfrac{5}{2}$, $\tfrac{15}{2}$, 20, in catalogum referendæ sunt pro S,[572] ‖ et $\sqrt{\dfrac{SS-z}{n}}$, id est 1. $\tfrac{3}{2}$. $\tfrac{7}{2}$. 9 respective pro h. Sed quia $S+nh$[573] si scribatur 20 pro S et 9 pro h fit 65 numerus major quadruplo maximi terminorum æquationis, ideo rejicio 20 et 9[574] et reliquos solùm refero in tabulam ut sequitur

h	1. $\tfrac{3}{2}$. $\tfrac{7}{2}$.
S	0. $\tfrac{5}{2}$. $\tfrac{15}{2}$.

His ita dispositis, tento pro k numero[s][575] omnes qui non efficiunt

$$\tfrac{1}{2}[p]\pm nk^{[575]}\quad\text{seu}\quad 2\pm5k$$

majus quadruplo maximi termini æquationis 40, id est numeros -8. -7. -6.

---

(569) 'terminorum' (terms) was first written.

(570) These final conditions follow from the equalities $q=\tfrac{1}{4}p^2+2Q-nk^2$ and

$$s=pR+Q^2+2S-nl^2-2nkm,$$

whence $2Q=\alpha+nk^2$ and $s-Q^2-pR+nl^2=2(S-nkm)$; and from

$$t=pS+2QR-2nkh-2nlm,\quad v=2QS+R^2-nm^2-2nlh\quad\text{and}\quad w=2RS-2nmh,$$

whereby $h=(pS+2QR-t-2nlm)/2nk=(2QS+R^2-v-nm^2)/2nl=(2RS-w)/2nm.$

you must see whether, if there is made $s-Q^2-pR+nl^2=2H$ and $H+nkm=S$, this S is in any case one or other of the numbers[569] which were previously set out in the list for S; and, further, if the other number corresponding to that S, as set out in the same list for h, is equal to these three quantities $\frac{1}{2}(2RS-w)/nm$, $\frac{1}{2}(2QS+R^2-v-nm^2)/nl$ and $\frac{1}{2}(pS+2QR-t-2nlm)/nk$.[570] If all these conditions hold in some case, instead of the equation proposed you will be required to write this one, $x^4+\frac{1}{2}px^3+Qx^2+Rx+S=\sqrt{n}\,.\,(kx^3+lx^2+mx+h)$.

For example, let there be proposed the equation

$$x^8+4x^7-x^6-10x^5+5x^4-5x^3-10x^2-10x-5=0$$

and there will then be α (or $q-\frac{1}{4}p^2$) $=-1-4=-5$,

$$\beta\ (\text{or}\ r-\tfrac{1}{2}p\alpha)=-10+10=0,\quad \gamma\ (\text{or}\ s-\tfrac{1}{2}p\beta-\tfrac{1}{4}\alpha^2)=5-\tfrac{25}{4}=-\tfrac{5}{4},$$

$$\delta\ (\text{or}\ t-\tfrac{1}{2}p\gamma-\tfrac{1}{2}\alpha\beta)=-5+\tfrac{5}{2}=-\tfrac{5}{2},\quad \epsilon\ (\text{or}\ v-\tfrac{1}{2}\alpha\gamma-\tfrac{1}{8}\beta^2)=-10-\tfrac{25}{8}=-\tfrac{105}{8},$$

ζ (or $w-\frac{1}{2}\beta\gamma$) $=-10$, η (or $z-\frac{1}{4}\gamma^2$) $=-5-\frac{25}{64}=-\frac{345}{64}$.[571] Therefore 2δ, 2ϵ, 2ζ, 8η are respectively -5, $-\frac{105}{4}$, -20 and $-\frac{345}{8}$, with 5 their common divisor, leaving unity when divided by 4 exactly as required because of the term s being odd. So, since a common divisor n, that is, 5, has been found which raises hope of a future reduction, because it is odd to $4z$ or -20 I successively add n, $3n$, $5n$, $7n$, $9n$, ... or 5, 15, 25, 35, 45, ... and there ensues -15, 0, 25, 60, 105, 160, 225, 300, 385, 480, 585, 700, 825, 960, 1105, 1260, 1425 and 1600. Of these only 0, 25, 225 and 1600 are squares. Consequently their halved roots 0, $\frac{5}{2}$, $\frac{15}{2}$, 20 are to be entered in the list for S[572] and 1, $\frac{3}{2}$, $\frac{7}{2}$, 9 respectively for h, that is, $\sqrt{[(S^2-z)/n]}$. But because, if 20 be written for S and 9 for h, $S+nh$[573] comes to be 65, a number more than four times the greatest of the terms in the equation, for that reason I reject this 20 and 9,[574] and list the remaining ones only in a table as follows:

h	1, $\frac{3}{2}$, $\frac{7}{2}$
S	0, $\frac{5}{2}$, $\frac{15}{2}$

Having set these out in array, I test for k all numbers which do not make ($\frac{1}{2}p\pm nk$ or) $2\pm 5k$ greater than four times the greatest term, 40, in the equation—the numbers, -8, -7, -6, -5, -4, -3, -2, -1, 0, 1, 2, 3, 4, 5, 6, 7,

---

(571) In consequence (see note (562)) the given equation is reduced to the form

$$(x^4+2x^3-\tfrac{5}{2}x^2-\tfrac{5}{8})^2-\tfrac{5}{2}x^3-\tfrac{105}{8}x^2-10x-\tfrac{365}{64}=0.$$

(572) This replaces 'tentandi sunt pro S' (are to be tested...for S).

(573) See note (566).

(574) Understand for S and h respectively.

(575) The respective manuscript readings 'numero' and '$\frac{1}{2}\pm nk$' passed Whiston's none too critical eye into the 1707 *editio princeps*, but were correctly augmented by Newton himself in the 1722 revise.

$-5. -4. -3. -2. -1. 0. 1. 2. 3. 4. 5. 6. 7$, ponendo $\dfrac{nkk+\alpha}{2}$ seu $\dfrac{5kk-5}{2}$ id est

numeros $\frac{315}{2}. 120. \frac{175}{2}. 60. \frac{75}{2}. 20. \frac{15}{2}. 0. -\frac{5}{2}. 0. \frac{15}{2}. 20. \frac{75}{2}. 60. \frac{175}{2}. 120$ respective
pro Q. Imò vero cum $Q\pm nl$ & multo magis Q non debeat majus esse quam 40,
rejiciendos esse sentio $\frac{315}{2}. 120. \frac{175}{2}$ & 60 & qui his respondent $-8. -7. -6.$
$-5. 5. 6. 7$, adeoqʒ solos $-4. -3. -2. -1. 0. 1. 2. 3. 4$ pro k et $\frac{75}{2}. 20. \frac{15}{2}. 0.$
$-\frac{5}{2}. 0. \frac{15}{2}. 20. \frac{75}{2}$ pro Q respectivè tentandos. Tentemus autem -1 pro k et 0
pro Q et in hoc casu pro l tentandi deinceps erunt successivè omnes numeri qui
non efficiunt $Q\pm nl$ majus quam 40 id est omnes numeri inter 10 et -10, et
pro R respectivè numeri $\dfrac{2\beta-npkk}{4}+nkl$ seu $-5-5l$ id est $-55. -50. -45.$
$-40. -35. -30. -25. -20. -15. -10. -5. 0. 5. 10. 15. 20. 25. [30].$[576] $35.$
$40. 45$, quorum tamen tres priores et ultimum quia majores quam 40 negligere

|| [202]   licebit. Tentemus autem || -2 pro l et 5 pro R et in hoc casu pro m tentandi
præterea erunt omnes numeri qui non efficiunt $R\pm nm$ seu $5\pm[5m]$ majus quam
40 id est numeri omnes inter 7 et -9, et videndum an si ponendo

$$s-QQ-pR+nll \quad \text{id est} \quad 5-20+20 \quad \text{seu} \quad 5=2H,$$

sit $H+nkm$ seu $\frac{5}{2}-5m=S$, id est si ex his numeris $-\frac{65}{2}. -\frac{55}{2}. -\frac{45}{2}. -\frac{35}{2}.$
$-\frac{25}{2}. -\frac{15}{2}. -\frac{5}{2}. \frac{5}{2}. \frac{15}{2}. \frac{25}{2}. \frac{35}{2}. \frac{45}{2}. \frac{55}{2}. \frac{65}{2}. \frac{75}{2}. \frac{85}{2}$, aliquis æqualis sit alicui
numerorum $0. \pm\frac{5}{2}. \pm\frac{15}{2}$ qui prius in tabulam pro S relati erant. Et hujusmodi
quatuor occurrunt $-\frac{15}{2}. -\frac{5}{2}. \frac{5}{2}. \frac{15}{2}$ quibus respondent $\pm\frac{7}{2}. \pm\frac{3}{2}. \pm\frac{3}{2}. \pm\frac{7}{2}$ pro h
in eadem tabula scripti, ut et $2. 1. 0. -1$ pro m substituti. Verum tentemus $-\frac{5}{2}$
pro S, 1 pro m, et $\pm\frac{3}{2}$ pro h, et fiet $\dfrac{2RS-w}{2nm}=\dfrac{-25+10}{10}=-\dfrac{3}{2}$. &

$$\frac{2QS+RR-v-nmm}{2nl}=\frac{25+10-5}{-20}=-\frac{3}{2}.$$

& $\dfrac{pS+2QR-t-2nlm}{2nk}=\dfrac{-10+5+20}{-10}=-\dfrac{3}{2}$. Quare cum prodeat omni casu $-\frac{3}{2}$
seu h, concludo numeros omnes rectè inventos esse adeoqʒ vice æquationis pro-
positæ scribendum esse $x^4+\frac{1}{2}px^3+Qxx+Rx+S=\sqrt{n}\times\overline{kx^3+lxx+mx+h}$ id est
$x^4+2x^3+5x-2\frac{1}{2}=[\pm]\sqrt{5}\times\overline{-x^3-2xx+x-1\frac{1}{2}}$.[577] Etenim quadrando partes
hujus producetur æquatio illa octo dimensionum quæ sub initio proponebatur.

Quod si tentando casus omnes numerorum prædicti valores omnes ipsius h
nullo in casu inter se consensissent, argumento fuisset æquationem per extractionem surdæ radicis quadraticæ reduci non potuisse.

---

(576) This omitted multiple was overlooked by Whiston in 1707, but duly inserted by
Newton in his 1722 revise.

that is—, putting $(\frac{1}{2}(nk^2+\alpha)$ or) $\frac{1}{2}(5k^2-5)$, that is, the numbers $\frac{315}{2}$, 120, $\frac{175}{2}$, 60, $\frac{75}{2}$, 20, $\frac{15}{2}$, 0, $-\frac{5}{2}$, 0, $\frac{15}{2}$, 20, $\frac{75}{2}$, 60, $\frac{175}{2}$, 120 respectively, for Q. But here, to be sure, since $Q\pm nl$ and much more so Q ought not to be greater than 40, I perceive that $\frac{315}{2}$, 120, $\frac{175}{2}$ and 60, and hence $-8, -7, -6, -5, 5, 6, 7$ corresponding to them, are to be rejected and therefore only $-4, -3, -2, -1, 0, 1, 2, 3, 4$ and $\frac{75}{2}$, 20, $\frac{15}{2}$, 0, $-\frac{5}{2}$, 0, $\frac{15}{2}$, 20, $\frac{75}{2}$ are to be tested for k and Q respectively. Let us now test -1 for k and 0 for Q, and in this case for l there will need successively to be tested in turn all numbers which do not make $Q\pm nl$ greater than 40—all those between 10 and -10, that is—and in R's place the numbers $(\frac{1}{4}(2\beta-npk^2)+nkl$ or$)-5-5l$, that is, $-55, -50, -45, -40, -35, -30, -25, -20, -15, -10, -5, 0, 5, 10, 15, 20, 25, 30, 35, 40, 45$ respectively: because they are greater than 40, however, it will be permissible to neglect the first three and the last of these. Let us now test -2 for l and so 5 for R, and in this case there will further need to be tested for m all numbers which do not make $(R\pm nm$ or$)$ $5\pm 5m$ greater than 40—all numbers between 7 and -9, that is— and you must then see whether if, on putting

$$2H = (s-Q^2-pR+nl^2, \text{ that is) } 5-20+20 \text{ or } 5,$$

there is $S = (H+nkm$ or$)$ $\frac{5}{2}-5m$; in other words, if some one of these numbers $-\frac{65}{2}, -\frac{55}{2}, -\frac{45}{2}, -\frac{35}{2}, -\frac{25}{2}, -\frac{15}{2}, -\frac{5}{2}, \frac{5}{2}, \frac{15}{2}, \frac{25}{2}, \frac{35}{2}, \frac{45}{2}, \frac{55}{2}, \frac{65}{2}, \frac{75}{2}, \frac{85}{2}$ be equal to one or other of the numbers $0, \pm\frac{5}{2}, \pm\frac{15}{2}$ which were previously listed in the table for S. Of this kind four occur: $-\frac{15}{2}, -\frac{5}{2}, \frac{5}{2}, \frac{15}{2}$, and to these relate the entries $\pm\frac{7}{2}, \pm\frac{3}{2}, \pm\frac{3}{2}, \pm\frac{7}{2}$ for h in the same table, and also the numbers 2, 1, 0, -1 substituted in m's place. But let us test $-\frac{5}{2}$ for S, 1 for m and $\pm\frac{3}{2}$ for h: there will then come to be $(2RS-w)/2nm = (-25+10)/10 = -3/2$,

$$(2QS+R^2-v-nm^2)/2nl = (25+10-5)/-20 = -3/2$$

and $\qquad (pS+2QR-t-2nlm)/2nk = (-10+5+20)/-10 = -3/2.$

Consequently, since in each case there results $-\frac{3}{2}$, that is, h, I conclude that all the numbers are correctly found and hence that instead of the equation proposed there is to be written $x^4+\frac{1}{2}px^3+Qx^2+Rx+S = \sqrt{n}.(kx^3+lx^2+mx+h)$, that is, $x^4+2x^3+5x-2\frac{1}{2} = \pm\sqrt{5}.(-x^3-2x^2+x-1\frac{1}{2})$.[577] And certainly, on squaring the sides of this there will be produced the equation of eight dimensions which was initially proposed.

If however, on testing all the cases of the numbers, all the above-cited values of h had not agreed with one another in any case, that would have been proof that the equation could not be reduced by the extraction of a surd square root.

---

(577) Or, more accurately, its square $(x^4+2x^3+5x-2\frac{1}{2})^2 = 5(x^3+2x^2-x+1\frac{1}{2})^2$.

||[203] Deberent autem aliqua hic in operis || abbreviationem annotari, sed quæ brevitatis causa prætereo cum tantarum reductionum perexiguus sit usus et rei possibilitatem potius quam praxin commodissimam voluerim exponere. Sunt igitur hæ reductiones æquationum per extractionem surdæ radicis quadraticæ.

[1682] Adjungere jam liceret reductiones æquationum per extractionem surdæ
Lect. 8. radicis cubicæ,[578] sed et has, ut quæ perrarò utiles sint, brevitatis gratia præ-
[Reductiones tereo. Sunt tamen reductiones quædam cubicarum æquationum vulgò notæ,
quædam quas, si penitus præterirem, Lector[580] fortasse desideraret. Proponatur æquatio
notæ.][579] cubica $x^3 * + qx + r = 0$, cujus secundus terminus deest. Ad hanc enim formam æquationem omnem cubicam reduci posse constat ex præcedentibus.[581] Et supponatur x esse $= a + b$. Erit $a^3 + 3aab + 3abb + b^3$ (id est x^3) $+ qx + r = 0$. Sit $3aab + 3abb$ (id est $3abx$) $+ qx = 0$, et erit $a^3 + b^3 + r = 0$. Per priorem æquationem est $b = -\dfrac{q}{3a}$ et cubicè $b^3 = -\dfrac{q^3}{27a^3}$. Ergo per posteriorem est $a^3 - \dfrac{q^3}{27a^3} + r = 0$, seu

$a^6 + ra^3 = \dfrac{q^3}{27}$. et per extractionem affectæ radicis quadraticæ

$$a^3 = -\tfrac{1}{2}r \pm \sqrt{\tfrac{1}{4}rr + \frac{q^3}{27}}.$$

Extrahe radicem cubicam et habebitur a. Et supra erat $-\dfrac{q}{3a} = b$, et $a + b = x$.

Ergo $a - \dfrac{q}{3a}$ radix est æquationis propositæ.[582]

Exempli gratia proponatur æquatio $y^3 - 6yy + 6y + 12 = 0$. Ad tollendum secundū æquationis hujus terminum ponatur $x + 2 = y$ et orietur $x^3 * - 6x + 8 = 0$,

||[204] ubi est $q = -6$. || $r = 8$. $\tfrac{1}{4}rr = 16$. $\dfrac{q^3}{27} = -8$. $a^3 = -4 \pm \sqrt{8}$. $a - \dfrac{q}{3a} = x$ & $x + 2 = y$ id

est $2 + \sqrt[3]{-4 \pm \sqrt{8}} + \dfrac{2}{\sqrt[3]{-4 \pm \sqrt{8}}} = y$.[583]

---

(578) Evidently this reduction would proceed, in a computationally much more involved way, by identifying a given equation $x^{3n} + px^{3n-1} + qx^{3n-2} + rx^{3n-3} + \ldots + z = 0$ with the desired form $(x^n + \tfrac{1}{3}px^{n-1} + Qx^{n-2} \ldots + S)^3 - n(kx^{n-1} + lx^{n-2} \ldots + h)^3 = 0$, yielding a reduction to

$$x^n + (\tfrac{1}{3}p + k\sqrt[3]{n})\,x^{n-1} + (Q + l\sqrt[3]{n})\,x^{n-2} \ldots + S + h\sqrt[3]{n} = 0$$

and a corresponding equation $x^{2n} + (\tfrac{2}{3}p - k\sqrt[3]{n})\,x^{2n-1} \ldots + S^2 - Sh\sqrt[3]{n} + h^2\sqrt[3]{n^2} = 0$ of double degree. It is hard to imagine any occasion on which such a reduction would be useful, nor should we suppose that 'Newton a su résoudre le problème', as I. G. Bachmakova avers ('Quelques traits caractéristiques du développement de l'algèbre au XVIIIᵉ siècle' (note (538)): 218), or that he had any glimpse of the theory of ideals which is necessary to delimit its possibilities of solution.

(579) We add this marginal head as a convenient title descriptive of Newton's pages 203–7 following.

Certain remarks ought really to be passed here regarding contraction of the procedure, but these for brevity's sake I pass over since the usage of such involved reductions is exceedingly slight and I wanted to reveal the possibilities in the approach rather than develop the most serviceable technique for its practice. These, then, are the reductions of equations by the extraction of a surd square root.

It would be permissible for me now to add something on the reductions of equations by the extraction of a surd cube root,[578] but for brevity these also I pass by as being extremely rarely of use. There are, however, certain widely known reductions of cubic equations which, were I wholly to bypass them, the reader[580] would perhaps miss. Let there be proposed the cubic equation $x^3 + qx + r = 0$ whose second term is lacking: it is, of course, established by what has gone before[581] that every cubic equation can be reduced to this form. Suppose, now, that $x = a + b$ and there will be

[Certain well-known reductions.][579]

$$a^3 + 3a^2b + 3ab^2 + b^3 \text{ (that is, } x^3) + qx + r = 0.$$

Let $3a^2b + 3ab^2$ (that is, $3abx$) $+ qx = 0$, and there will be $a^3 + b^3 + r = 0$. By the first equation $b = -\frac{1}{3}q/a$ and, on cubing, $b^3 = -\frac{1}{27}q^3/a^3$ Therefore by the latter $a^3 - \frac{1}{27}q^3/a^3 + r = 0$ or $a^6 + ra^3 = \frac{1}{27}q^3$, and so by extracting the 'affected' square root $a^3 = -\frac{1}{2}r \pm \sqrt{[\frac{1}{4}r^2 + \frac{1}{27}q^3]}$. Extract the cube root and there will be had a. Above there was $b = -\frac{1}{3}q/a$ and $a + b = x$. Therefore $a - \frac{1}{3}q/a$ is the root of the equation proposed.[582]

For example, let the equation $y^3 - 6y^2 + 6y + 12 = 0$ be proposed. To remove the second term of this equation put $y = x + 2$ and there will ensue

$$x^3 - 6x + 8 = 0.$$

Here $q = -6$, $r = 8$ and so $\frac{1}{4}r^2 = 16$, $\frac{1}{27}q^3 = -8$, $a^3 = -4 \pm \sqrt{8}$ with $x = a - \frac{1}{3}q/a$ and $y = x + 2$, that is, $y = 2 + \sqrt[3]{(-4 \pm \sqrt{8})} + 2/\sqrt[3]{(-4 \pm 8)}$.[583]

---

(580) A clear indication of the 'audience' to which Newton addresses his present 'lectures'! There can be no real doubt that this manuscript was intended to be read at leisure; compare note (1). In the revised *Arithmetica* the 'reader' alone is addressed; see 2, [1]: note (120).

(581) See Newton's page 161 above.

(582) As we have seen (I: 119–20) Newton had noted this variant deduction of the Cardan solution $x = \sqrt[3]{[-\frac{1}{2}r + \sqrt{(\frac{1}{4}r^2 + \frac{1}{27}q^3)}]} + \sqrt[3]{[-\frac{1}{2}r - \sqrt{(\frac{1}{4}r^2 + \frac{1}{27}q^3)}]}$ in his youthful reading of John Wallis' *Adversus Marci Meibomii de Proportionibus Dialogum Tractatus Elencticus* (Oxford 1657): Dedicatio: 18–20; and had again met with it half a dozen years later in compiling his 'Observationes' on Kinckhuysen's *Algebra* (see II: 350–1). See also Descartes' *Geometrie* ($_1$1637: 298–300 → *Geometria*, $_2$1659: 93–4).

(583) Whence $y = 2 - \sqrt[3]{[4 + \sqrt{8}]} - \sqrt[3]{[4 - \sqrt{8}]} \approx -0.953$, the given cubic's unique real root.

Et hoc modo erui possunt radices omnium cubicarum æquationum ubi q affirmativum est vel etiam ubi q negativum est et $\dfrac{q^{3\,(584)}}{27}$ non majus quàm $\frac{1}{4}rr$, id est ubi duæ ex radicibus æquationis sunt impossibiles. At ubi q negativum est et $\dfrac{q^{3\,(584)}}{27}$ simul majus quam $\frac{1}{4}rr$ fit $\sqrt{\frac{1}{4}rr - \dfrac{q^{3\,(584)}}{27}}$ quantitas impossibilis, atqɜ adeo æquationis radix x vel y hoc casu impossibilis erit.[585] Scilicet hoc casu tres sunt radices possibiles quæ omnes eodem modo se habent ad æquationis terminos q et r et indifferenter designantur per literam x vel y, adeoqɜ omnes eadem deberent lege erui et exprimi qua una aliqua eruitur et exprimitur: sed omnes tres lege præfata exprimere impossibile est. Quantitas $a - \dfrac{q}{3a}$ qua x designatur multiplex esse non potest,[586] eaqɜ de causa Hypothesis quod x hoc in casu ubi triplex est, æqualis esse potest binomio $a - \dfrac{q}{3a}$ seu $a + b$ cujus nominum cubi $a^3 + b^3$ conjunctim æquentur r, & triplum rectangulum $3ab$ æquetur q, planè impossibilis est; et ex hypothesi impossibili conclusionem impossibilem colligi mirum esse non debet.[587]

Est et alius modus has radices exprimendi. Nimirum de $a^3 + b^3 + r$ id est de nihilo aufer $a^3 + r$ seu $\frac{1}{2}r \pm \sqrt{\frac{1}{4}rr + \dfrac{q^3}{27}}$ et restabit $b^3 = -\frac{1}{2}r \mp \sqrt{\frac{1}{4}rr + \dfrac{q^3}{27}}$. Est itaqɜ

‖[205] $a = \sqrt[3]{-\frac{1}{2}r + \sqrt{\frac{1}{4}rr + \dfrac{q^3}{27}}}$ & $b = \sqrt[3]{-\frac{1}{2}r - \sqrt{\frac{1}{4}rr + \dfrac{q^3}{27}}}$. ‖ vel $a = \sqrt[3]{-\frac{1}{2}r - \sqrt{\frac{1}{4}rr + \dfrac{q^3}{27}}}$

---

(584) Strictly, these should read ‘$-\dfrac{q^3}{27}$’ and ‘$\sqrt{\frac{1}{4}rr + \dfrac{q^3}{27}}$’ respectively.

(585) This strange phrase is explained by Newton in immediate sequel: though in this 'irreducible' case the cubic's roots are all real, it is impossible to derive them by Cardan's formula without an appeal to 'impossible' cube roots, and in consequence—in a context where implicit insistence is made that a real result should be forthcoming at each stage of the procedure—the real roots may not 'possibly' be derived in that way. A decade earlier, in his 'Observationes' on Kinckhuysen's *Algebra*, Newton had more realistically observed (II: 420) that 'radix per has regulas designata est revera impossibilis: et tamen per regulam extrahendi radices ex impossibilibus binomijs supra [II: 392–4] traditam reales radices exinde possunt obtineri'—specifically, in the case where $\frac{1}{27}q^3 > \frac{1}{4}r^2$, the triple root of the cubic $x^3 - qx + r = 0$ may be obtained in the (triply real) form

$$x = \omega . \sqrt[3]{[-\frac{1}{2}r + i\sqrt{(\frac{1}{27}q^3 - \frac{1}{4}r^2)}]} + \omega^{-1} . \sqrt[3]{[-\frac{1}{2}r - i\sqrt{(\frac{1}{27}q^3 - \frac{1}{4}r^2)}]},$$

in which $\omega = 1$ or $-\frac{1}{2}(1 \pm i\sqrt{3})$

are the cube roots of unity. In dubbing real roots so obtained as 'impossible' he now lays himself open to the paradox that, since these roots may also be had by angle-trisection (by constructing $\phi = \sin^{-1}(\frac{1}{4}r^2/\frac{1}{27}q^3)$ and then $x = 2\sqrt{(\frac{1}{3}q)} . \sin \frac{1}{3}(\phi + 2k\pi)$, $k = 0, 1, 2$) in an

And in this way may be derived the roots of all cubic equations in which q is positive or, alternatively, when q is negative and $[-]\frac{1}{27}q^3$ is not greater than $\frac{1}{4}r^2$; that is, when two of the equation's roots are impossible. But when q is negative and $[-]\frac{1}{27}q^3$ simultaneously greater than $\frac{1}{4}r^2$, the quantity

$$\sqrt{[\tfrac{1}{4}r^2[+]\tfrac{1}{27}q^3]}$$

becomes impossible, and hence the equation's root x or y will in this case be impossible.[585] Of course, in this case there are three 'possible' (real) roots, all of which bear the same relation to the terms q and r of the equation and are denoted without distinction by the letters x or y, and should therefore all be derived and expressed by the same rule that any one of them is. But it is impossible to express all three by the above-stated rule: the quantity $a - \frac{1}{3}q/a$ by which x is denoted cannot be multiple in value,[586] and for that reason the hypothesis that x in the present case, where it is triple in value, can be equal to the binomial $a - \frac{1}{3}q/a$ or $a + b$, the cubes $a^3 + b^3$ of whose components are together to be equal to r and three times their product $3ab$ equal to q, is quite evidently impossible—and it ought not to be astonishing that an impossible conclusion is gathered from an impossible hypothesis.[587]

There is also another method of expressing these roots. Namely, from $a^3 + b^3 + r$ (nothing, that is) take away $a^3 + r$ or $\frac{1}{2}r \pm \sqrt{[\frac{1}{4}r^2 + \frac{1}{27}q^3]}$ and there will remain $b^3 = -\frac{1}{2}r \mp \sqrt{[\frac{1}{4}r^2 + \frac{1}{27}q^3]}$. Accordingly $a = \sqrt[3]{(-\frac{1}{2}r + \sqrt{[\frac{1}{4}r^2 + \frac{1}{27}q^3]})}$ and $b = \sqrt[3]{(-\frac{1}{2}r - \sqrt{[\frac{1}{4}r^2 + \frac{1}{27}q^3]})}$ or, alternatively, $a = \sqrt[3]{(-\frac{1}{2}r - \sqrt{[\frac{1}{4}r^2 + \frac{1}{27}q^3]})}$ and

---

eminently real way, they are also 'possible'. Equally, the roots of $x^3 - qx - r = 0$, $\frac{1}{27}q^3 > \frac{1}{4}r^2$, are to be classified by Newton as 'impossible' when constructed by

$$x = \omega . \sqrt[3]{[\tfrac{1}{2}r + i\sqrt{(\tfrac{1}{27}q^3 - \tfrac{1}{4}r^2)}]} + \omega^{-1} . \sqrt[3]{[\tfrac{1}{2}r - i\sqrt{(\tfrac{1}{27}q^3 - \tfrac{1}{4}r^2)}]},$$

and as 'possible' when derived by $\phi = \cos^{-1}(\frac{1}{4}r^2/\frac{1}{27}q^3)$ and then

$$x = 2\sqrt{(\tfrac{1}{3}q)} . \cos \tfrac{1}{3}(\phi + 2k\pi), \quad k = 0, 1, 2.$$

(The latter trigonometrical technique was pioneered by François Viète in Caput VI, Theorema III of his *De Æquationum Recognitione et Emendatione* ([1591 →] Paris, 1615) [= *Opera Mathematica* (Leyden, 1646): 91], the former exemplified by Descartes in his *Geometrie* ($_1$1637: 398–400 → *Geometria*, $_2$1659: 93–5). Newton was familiar with both approaches—indeed he quotes the Cartesian version on his page 233 below, constructing the angle-trisection by a classical 'inclination' of a line segment through a point—and was at pains to clarify the distinction between such a 'trisection' and the finding of 'two meane proportionalls' (straightforwardly amenable to Cardan's algebraic formula) for Collins in the summer of 1670; see II: 418–20, note (101).)

(586) Again assuming (see previous note) that complex cube roots are impermissible.

(587) This is, of course, not a logical truth: a false premiss may very well lead deductively to a valid conclusion.

& $b = \sqrt[3]{-\tfrac{1}{2}r + \sqrt{\tfrac{1}{4}rr + \tfrac{q^3}{27}}}$, adeoq; horum summa

$$\sqrt[3]{-\tfrac{1}{2}r + \sqrt{\tfrac{1}{4}rr + \tfrac{q^3}{27}}} + \sqrt[3]{-\tfrac{1}{2}r - \sqrt{\tfrac{1}{4}rr + \tfrac{q^3}{27}}}$$

erit $=x$.[588]

Possunt etiam æquationum biquadraticarum radices mediantibus cubicis erui et exprimi.[589] Tollendus est autem primùm secundus æquationis terminus. Sit æquatio resultans $x^4 + qxx + rx + s = 0$. Pone hanc multiplicatione duarum $xx + ex + f = 0$ & $xx - ex + g = 0$ generari, id est eandem esse cum hac

$$x^4 * \begin{matrix} +f \\ +g \\ -ee \end{matrix} xx \begin{matrix} +eg \\ -ef \end{matrix} x + fg = 0.$$

et collatis terminis fiet $f + g - ee = q$, $eg - ef = r$, & $fg = s$. Quare $q + ee = f + g$.

$\dfrac{r}{e} = g - f$. $\dfrac{q + ee + \tfrac{r}{e}}{2} = g$. $\dfrac{q + ee - \tfrac{r}{e}}{2} = f$. $\dfrac{qq + 2eeq + e^4 - \tfrac{rr}{ee}}{4} (=fg) = s$, et per reduc-

tionem $e^6 + 2qe^4 \begin{matrix} +qq \\ -4s \end{matrix} ee - rr = 0$. Pro ee scribe y, et fiet $y^3 + 2qyy \begin{matrix} +qq \\ -4s \end{matrix} y - rr = 0$.

æquatio cubica cujus terminus secundus tolli potest et radix deinceps per regulam præcedentem vel secus extrahi. Dein habita illa radice regrediendum

erit ponendo $\sqrt{y} = e$, $\dfrac{q + ee - \tfrac{r}{e}}{2} = f$, $\dfrac{q + ee + \tfrac{r}{e}}{2} = g$, & æquationes duæ $xx + ex + f = 0$

& $xx - ex + g = 0$ extractis earum radicibus dabunt quatuor radices æquationis biquadraticæ $x^4 + qxx + rx + s = 0$, nimirum $x = -\tfrac{1}{2}e \pm \sqrt{\tfrac{1}{4}ee - f}$ et $x = \tfrac{1}{2}e \pm \sqrt{\tfrac{1}{4}ee - g}$.

‖ [206] Ubi notandum est quod si æquationis biquadraticæ radices quatuor ‖ possibiles

sunt,[590] æquationis cubicæ $y^3 + 2qyy \begin{matrix} +qq \\ -4s \end{matrix} y - rr = 0$ radices tres possibiles erunt,

atq; adeo per regulam præcedentem extrahi nequeunt. Sic et si æquationis quinq; vel plurium dimensionum radices affectæ in radices non affectas medijs æquationis terminis quoquo pacto sublatis convertantur,[591] illa radicum expressio semper erit impossibilis ubi plures quam una radix in æquatione imparium dimensionum possibiles sunt, aut plures quàm duæ in æquatione parium dimensionum quæ per extractionem surdæ radicis quadraticæ methodo supra exposita reduci nequeunt.

---

(588) See note (582).

(589) As Newton acknowledges in the sequel, this present construction of the auxiliary cubic—not essentially different from Scipione del Ferro's of *c.* 1515 (see II: 421, note (107))—is taken over from Descartes' *Geometrie* (₁1637: 383–7 = *Geometria*, ₂1659: 79–82), perhaps by way of Kinckhuysen's analogous section 'De solutione æquationum quatuor dimensionum, quando problema est planum' (II: 333–4).

$b = \sqrt[3]{(-\frac{1}{2}r + \sqrt{[\frac{1}{4}r^2 + \frac{1}{27}q^3]})}$, and hence their sum

$$\sqrt[3]{(-\frac{1}{2}r + \sqrt{[\frac{1}{4}r^2 + \frac{1}{27}q^3]})} + \sqrt[3]{(-\frac{1}{2}r - \sqrt{[\frac{1}{4}r^2 + \frac{1}{27}q^3]})}$$

will be equal to x.[588]

The roots of quartic equations, too, may be derived and expressed by means of intermediary cubics.[589] The equation's second term must, however, first be removed. Let the resulting equation be $x^4 + qx^2 + rx + s = 0$. Suppose this to be generated by the multiplication of the two quadratics $x^2 + ex + f = 0$ and $x^2 - ex + g = 0$, that is, to be identical with

$$x^4 + (f + g - e^2)\, x^2 + e(g - f)\, x + fg = 0,$$

and when terms are compared there will come to be $q = f + g - e^2$, $r = e(g - f)$ and $s = fg$. Consequently $q + e^2 = f + g$, $r/e = g - f$ and so $g = \frac{1}{2}(q + e^2 + r/e)$, $f = \frac{1}{2}(q + e^2 - r/e)$, while $\frac{1}{4}(q^2 + 2e^2q + e^4 - r^2/e^2) = (fg \text{ or})\, s$ and by reduction $e^6 + 2qe^4 + (q^2 - 4s)\, e^2 - r^2 = 0$. In place of e^2 write y and this will become $y^3 + 2qy^2 + (q^2 - 4s)\, y - r^2 = 0$, a cubic equation whose second term can be removed and the root subsequently, by the preceding rule or otherwise, be extracted. Then, once that root is obtained, the procedure will be reversed by setting $\sqrt{y} = e$, $\frac{1}{2}(q + e^2 - r/e) = f$, $\frac{1}{2}(q + e^2 + r/e) = g$, and when their roots are extracted the two equations $x^2 + ex + f = 0$ and $x^2 - ex + g = 0$ will yield the four roots of the quartic equation $x^4 + qx^2 + rx + s = 0$: specifically,

$$x = -\tfrac{1}{2}e \pm \sqrt{[\tfrac{1}{4}e^2 - f]} \quad \text{and} \quad x = \tfrac{1}{2}e \pm \sqrt{[\tfrac{1}{4}e^2 - g]}.$$

Here you should note that, if the four roots of the quartic are real,[590] then the three roots of the cubic equation $y^3 + 2qy^2 + (q^2 - 4s)\, y - r^2 = 0$ will be real and hence cannot be extracted by the preceding rule. So also, if the affected roots of an equation of five or more dimensions be changed into non-affected (pure) ones by somehow or other removing the middle terms of the equation,[591] that way of expressing the roots will always be impossible when in an equation of odd dimensions more than one root is real, or when in an equation of even dimensions, irreducible by the method of extracting a surd square root expounded above, more than two are.

---

(590) And hence the quartic may be split into component quadratics in three distinct ways.

(591) Like his recently deceased Scot compatriot James Gregory (see H. W. Turnbull's *James Gregory Tercentenary Memorial Volume* (London, 1939): 382–9) and his German contemporary Walther von Tschirnhaus (in his 'Methodus auferendi omnes terminos intermedios ex data æquatione', *Acta Eruditorum* (May 1683): 204–7), Newton here seems to be toying with the idea that a given equation of n-th degree, say $x^n + ax^{n-1} + bx^{n-2} \ldots + k = 0$, may, by elimination with an auxiliary equation $x^m + (ry + s)\, x^{m-1} + \ldots + vy^m = 0$, $m \leqslant n - 1$, be reduced to an equivalent one $y^p + \alpha y^{p-1} + \ldots + \epsilon = 0$, $p \leqslant n - 1$, of lower degree. (For the quintic and higher equations we now, of course, know this to be impossible.) See also note (90).

Docuit Cartesius[592] æquationem biquadraticam per regulas ultimò traditas reducere. E.g. proponatur æquatio a nobis supra[593] reducta

$$x^4 - x^3 - 5xx + 12x - 6 = 0.$$

Tolle secundum terminum scribendo $v + \tfrac{1}{4}$ pro x et orietur

$$v^4 - \tfrac{43}{8}vv + \tfrac{75}{8}v - \tfrac{851}{256} = 0.$$

Ad tollendas fractiones scribe $\tfrac{1}{4}z$ pro v et orietur $z^4 - 86zz + 600z - 851 = 0$. Hic est $-86 = q$. $600 = r$ et $-851 = s$, adeoꝗ $y^3 + 2qyy \genfrac{}{}{0pt}{}{+qq}{-4s}y - rr = 0$ substitutis æquipollentibus fiet $y^3 - 172yy + 10800y - 360000 = 0$. Ubi tentando omnes ultimi termini divisores $1, -1, 2, -2, 3, -3, 4, -4, 5, -5$ & deinceps usꝗ ad 100 invenietur tandem $y = 100$.[594] Quod idem multò expeditius per methodum a nobis supra expositam inveniri potuit.[595] Dein habito y, radix ejus 10 erit e,

et $\dfrac{q + ee - \dfrac{r}{e}}{2}$ id est $\dfrac{-86 + 100 - 60}{2}$ seu -23 erit f et $\dfrac{q + ee + \dfrac{r}{e}}{2}$ seu 37 erit g, adeoꝗ

‖ [207]   æquationes ‖ $xx + ex + f = 0$ et $xx - ex + g = 0$ scripto z pro x et substitutis æquipollentibus evadent $zz + 10z - 23 = 0$ et $zz - 10z + 37 = 0$. Restitue v pro $\tfrac{1}{4}z$ et orientur $vv + 2\tfrac{1}{2}v - \tfrac{23}{16} = 0$ et $vv - 2\tfrac{1}{2}v + \tfrac{37}{16} = 0$. Restitue insuper $x - \tfrac{1}{4}$ pro v et emergent $xx + 2x - 2 = 0$ et $xx - 3x + 3 = 0$, æquationes duæ quarum radices quatuor $x = -1 \pm \sqrt{3}$ et $x = 1\tfrac{1}{2} \pm \sqrt{-\tfrac{3}{4}}$, eædem sunt cum radicibus quatuor æquationis biquadraticæ sub initio propositæ $x^4 - x^3 - 5xx + 12x - 6 = 0$. Sed hæ facilius per methodum inveniendi divisores a nobis supra explicatam inveniri poterunt.

[1682]
Lect. 9.
Extractio
Radicum
ex binomijs.   [596]Hactenus æquationum reductiones modis ni fallor facilioribus et magis generalibus quam ab alijs factum est tradidisse suffecerit. Sed quoniam in hujusmodi operationibus sæpe devenimus ad radicales complexas quæ ad simpliciores reduci possunt, convenit etiam harum reductiones exponere. Eæ

---

(592) See note (589) above.

(593) This example has twice been adduced, in fact: once (on Newton's pages 177–8 above) as an illustration of his finite-difference method of splitting an equation into rational components; again (on his following pages 192–3) as an instance of how his method of determining 'surd' factors of an equation may be applied to rational cases also.

(594) The remaining pair of roots $y = 36 \pm 48i$ is complex, and hence the given quartic can be split into component (rational) quadratics in only one way, so that not all its roots are real; see note (590).

(595) Namely (see his pages 173–80 above) by computing low values

$$\ldots, \quad f_1 = -349371 = -3^2 . 11 . 3529, \quad f_0 = -360000 = -2^4 . 3^2 . 5^2,$$

$f_{-1} = -370973 = -101 \cdot 3673, \ldots$ of $f_y \equiv y^3 - 172y^2 + 10800y - 360000$, and then determining the unique arithmetical progression $\ldots, 99, 100, 101, \ldots$, which runs through the table of their factors.

Descartes[592] has explained how to reduce a quartic equation by means of the rules lately presented. For example, let there be proposed the equation $x^4 - x^3 - 5x^2 + 12x - 6 = 0$ reduced by us above.[593] Remove the second term by writing $v + \frac{1}{4}$ in x's place and there will arise $v^4 - \frac{43}{8}v^2 + \frac{75}{8}v - \frac{851}{256} = 0$. To remove fractions write $\frac{1}{4}z$ for v and there will arise $z^4 - 86z^2 + 600z - 851 = 0$. Here $q = -86$, $r = 600$ and $s = -851$, and hence $y^3 + 2qy^2 + (q^2 - 4s)y - r^2 = 0$ will, when equivalents are substituted, become

$$y^3 - 172y^2 + 10\,800y - 360\,000 = 0.$$

On testing in this all divisors of the last term, $1, -1, 2, -2, 3, -3, 4, -4, 5, -5$ and so on in turn up to 100, there will at length be found $y = 100$:[594] the same result could have been much more speedily established by the method we expounded previously.[595] Next, when y is had, its root 10 will be e and

$$(\tfrac{1}{2}(q + e^2 - r/e), \text{ that is)} \ \tfrac{1}{2}(-86 + 100 - 60) \text{ or } -23$$

will be f and $(\frac{1}{2}(q + e^2 + r/e)$ or$)$ 37 will be g, and hence, with z written for x and equivalents substituted, the equations $x^2 + ex + f = 0$ and $x^2 - ex + g = 0$ will prove to be $z^2 + 10z - 23 = 0$ and $z^2 - 10z + 37 = 0$. Restore v for $\frac{1}{4}z$ and there will ensue $v^2 + 2\frac{1}{2}v - \frac{23}{16} = 0$ and $v^2 - 2\frac{1}{2}v + \frac{37}{16} = 0$. Restore, moreover, $x - \frac{1}{4}$ in place of v and there will emerge $x^2 + 2x - 2 = 0$ and $x^2 - 3x + 3 = 0$, two equations whose four roots $x = -1 \pm \sqrt{3}$ and $x = 1\frac{1}{2} \pm \sqrt{-\frac{3}{4}}$ are the same as the four roots of the quartic equation initially proposed, $x^4 - x^3 - 5x^2 + 12x - 6 = 0$. But these could more easily have been found by the method of ascertaining divisors earlier explained by us.

[596]Up to now it has sufficed to present reductions of equations by techniques which are, unless I am deceived, easier and more general than what has been accomplished by others. But, seeing that in operations of this sort we often come up against compound radicals which can be reduced to simpler ones, it is convenient to display their reductions also. These are achieved by extracting

Extraction of the roots of binomials.

---

(596) Having in would-be 'correction' previously, in immediate continuation of Newton's page 29, inserted from his terminal pages 266–8 the much reworked draft 'De Reductione Radicaliū ad simpliciores radicales per extractionem radicum' (see note (85) above), Whiston in his 1707 *editio princeps* printed only the first, introductory paragraph which follows, parenthetically justifying his omission of the remainder of this section with the editorial remark 'Verum cum hoc jamdudum præstitum sit..., pluribus impræsentiarum [*sc.* extractionum] supersedemus'. Whiston's amputation of the remaining text (inessentially variant from his printed draft, to be sure) was confirmed by Newton in his 1722 revise, where the first paragraph too is omitted as *de trop*, leaving behind it no trace of its prior existence at the present place and destroying the smooth transition to the geometrical construction of cubics and quartics which was Newton's original intention. As we have previously announced (see note (85)) we here reproduce the page-sequence of the manuscript, freed from such later tamperings (sanctioned by Newton or no).

fiunt per extractiones radicum ex binomijs, aut ex quantita[ti]bus magis compositis quæ ut binomia considerari possunt.

Radices quantitatum quæ ex integris et radicalibus quadraticis[(597)] componuntur, sic extrahe. Designet A quantitatis alicujus partem majorem, B partem minorem et erit $\dfrac{A+\sqrt{AA-BB}}{2}$ quadratum majoris partis radicis et $\dfrac{A-\sqrt{AA-BB}}{2}$ quadratum partis minoris, quæ quidem parti majori adnectenda est cum signo ipsius B.[(598)] Ut si quantitas sit $3+\sqrt{8}$ cujus radicem quadraticam

||[208] extrahere || oportet, scribendo 3 pro A et $\sqrt{8}$ pro B fiet $\sqrt{AA-BB}=1$, indeqȝ quadratum majoris partis radicis $\dfrac{3+1}{2}$ id est 2 et quadratum minoris partis $\dfrac{3-1}{2}$ id est 1. Ergo radix quæsita est $1+\sqrt{2}$. Rursus si ex $\sqrt{32}-\sqrt{24}$ radix extrahenda sit, ponendo $\sqrt{32}$ pro A et $\sqrt{24}$ pro B erit $\sqrt{AA-BB}=\sqrt{8}$ et inde $\dfrac{\sqrt{32}+\sqrt{8}}{2}$ et $\dfrac{\sqrt{32}-\sqrt{8}}{2}$ hoc est $3\sqrt{2}$ et $\sqrt{2}$ quadrata partium radicis. Radix itaqȝ est $\sqrt[4]{18}-\sqrt[4]{2}$.

Eodem modo si de $aa+2x\sqrt{aa-xx}$ radix quadratica extrahi debet, pro A scribe aa et pro B $2x\sqrt{aa-xx}$ et erit $AA-BB=a^4-4aaxx+4x^4$ cujus radix est $aa-2xx$. Unde quadratum unius partis radicis erit $aa-xx$, illud alterius xx, adeoqȝ radix $x+\sqrt{aa-xx}$. Rursus si habeatur $aa+5ax-2a\sqrt{ax+4xx}$, scribendo $aa+5ax$ pro A et $2a\sqrt{ax+4xx}$[(599)] pro B, fiet $AA-BB=a^4+6a^3x+9aaxx$ cujus radix est $aa+3ax$. Unde quadratum majoris partis radicis erit $aa+4ax$, illud minoris ax, et radix $\sqrt{aa+4ax}-\sqrt{ax}$.

Deniqȝ si habeatur $6+\sqrt{8}-\sqrt{12}-\sqrt{24}$ ponendo $6+\sqrt{8}=A$ et $-\sqrt{12}-\sqrt{24}=B$, fiet $AA-BB$[(600)]$=8$. Unde radicis pars major $\sqrt{3+\sqrt{8}}$, hoc est (ut supra) $1+\sqrt{2}$ et pars minor $\sqrt{3}$ atqȝ adeo radix ipsa $1+\sqrt{2}-\sqrt{3}$. Cæterum ubi plures sunt hujusmodi termini radicales, possunt partes radicis citiùs inveniri dividendo factum quarumvis duarum radicalium per tertiam aliquam radicalem quæ producit Quotum rationalem et integrum. Nam Quoti illius radix erit duplum

||[209] partis radicis || quæsitæ.[(601)] Ut in exemplo novissimo $\dfrac{\sqrt{8}\times\sqrt{12}}{\sqrt{24}}=2$. $\dfrac{\sqrt{8}\times\sqrt{24}}{\sqrt{12}}=4$. $\dfrac{\sqrt{12}\times\sqrt{24}}{\sqrt{8}}=6$. Ergo partes radicis sunt 1, $\sqrt{2}$, $\sqrt{3}$, ut supra.

---

(597) Quantities involving square roots but no higher surds.

(598) This classical Euclidean rule—whose validity is attested by straightforwardly computing $(\sqrt{a}+\sqrt{b})^2$ and identifying the resulting $(a+b)+\sqrt{(4ab)}$ with $A+B$ term for term— is doubtless a borrowing from Kinckhuysen's *Algebra* (see II: 312–13) by way of his appended 'Observatio' theorem (II: 372–6); see also III: 370, note (44).

(599) Newton first wrote more accurately '$-2a\sqrt{aa+4xx}$'. In cancelling the minus sign he probably assumes that the preceding rule is to be understood in the general form

$$\sqrt{[A\pm B]} = \sqrt{[\tfrac{1}{2}(A+\sqrt{(A^2-B^2)})]} \pm \sqrt{[\tfrac{1}{2}(A-\sqrt{(A^2-B^2)})]}.$$

the roots of binomials or of more complicated quantities which may be considered as binomials.

The roots of quantities which are made up of integers and quadratic radicals[597] extract in this way. Let A denote the greater part of some quantity, B its lesser part, and then $\frac{1}{2}(A+\sqrt{[A^2-B^2]})$ will be the square of the greater part of its root and $\frac{1}{2}(A-\sqrt{[A^2-B^2]})$ the square of the lesser part, this, of course, to be attached to the greater part with B's sign.[598] For instance, if $3+\sqrt{8}$ be the quantity whose square root it is necessary to extract, on writing 3 for A and $\sqrt{8}$ for B there will come to be $\sqrt{[A^2-B^2]}=1$ and in consequence the square of the greater part of the root is $\frac{1}{2}(3+1)$, that is, 2, and the square of the lesser part is $\frac{1}{2}(3-1)$, that is, 1. Therefore the required root is $1+\sqrt{2}$. Again, if the root of $\sqrt{32}-\sqrt{24}$ is to be extracted, on putting $\sqrt{32}$ for A and $\sqrt{24}$ for B there will be $\sqrt{[A^2-B^2]}=\sqrt{8}$ and thence $\frac{1}{2}(\sqrt{32}+\sqrt{8})$ and $\frac{1}{2}(\sqrt{32}-\sqrt{8})$, that is, $3\sqrt{2}$ and $\sqrt{2}$ are the squares of the root's parts. The root is accordingly $\sqrt[4]{18}-\sqrt[4]{2}$.

In the same way, if the square root of $a^2+2x\sqrt{[a^2-x^2]}$ has to be extracted, in A's place write a^2 and in B's $2x\sqrt{[a^2-x^2]}$, and then $A^2-B^2=a^4-4a^2x^2+4x^4$, whose root is a^2-2x^2. From which the square of one part of the root will be a^2-x^2, that of the other x^2, and hence the root is $x+\sqrt{[a^2-x^2]}$. Again, if there be had $a^2+5ax-2a\sqrt{[ax+4x^2]}$, on writing a^2+5ax for A and $2a\sqrt{[ax+4x^2]}$[599] for B there will come $A^2-B^2=a^4+6a^3x+9a^2x^2$, whose root is a^2+3ax. Whence the square of the root's greater part will be a^2+4ax, that of its lesser one ax, and the root itself $\sqrt{[a^2+4ax]}-\sqrt{[ax]}$.

Finally, if $6+\sqrt{8}-\sqrt{12}-\sqrt{24}$ be had, on setting

$$6+\sqrt{8}=A \text{ and } -\sqrt{12}-\sqrt{24}=B$$

there will come A^2-B^2[600]$=8$. From this the greater part of the root is $\sqrt{[3+\sqrt{8}]}$, that is, (as above) $1+\sqrt{2}$, and its lesser part $\sqrt{3}$, and hence the root itself is $1+\sqrt{2}-\sqrt{3}$. However, when there are several radical terms of this sort, the parts of the roots can more swiftly be found by dividing the product of any two radicals by some third radical which produces a rational, integral quotient: for the root of that quotient will be twice the part of the root sought.[601] In the most recent example, for instance, $\sqrt{8}\times\sqrt{12}/\sqrt{24}=2$, $\sqrt{8}\times\sqrt{24}/\sqrt{12}=4$ and $\sqrt{12}\times\sqrt{24}/\sqrt{8}=6$: therefore the parts of the root are 1, $\sqrt{2}$ and $\sqrt{3}$, as above.

---

(600) That is, $(6+2\sqrt{2})^2-12(1+\sqrt{2})^2$.

(601) Since $(\sqrt{a}+\sqrt{b}+\sqrt{c})^2=(a+b+c)+2\sqrt{(ab)}+2\sqrt{(bc)}+2\sqrt{(ca)}$, identification of this with $A+2\sqrt{B}+2\sqrt{C}+2\sqrt{D}$ determines not only Newton's conditions $\sqrt{B}\times\sqrt{D}/\sqrt{C}=a$, $\sqrt{B}\times\sqrt{C}/\sqrt{D}=b$ and $\sqrt{C}\times\sqrt{D}/\sqrt{B}=c$, but also that $A=a+b+c$. In Newton's following example ($A=6$, $B=2$, $C=3$, $D=6$ and so) $a=2$, $b=1$, $c=3$, the latter requirement is evidently also satisfied.

Est et regula[602] extrahendi altiores radices ex quantitatibus numeralibus duarum potentia commensurabilium partium.[603] Sit quantitas $A \pm B$. Ejus pars major A. Index radicis extrahendæ e. Quære minimum numerum N cujus potestas N^e dividi potest per $AA - BB$ sine residuo et sit Quotus Q.[604] Computa $\sqrt[e]{\overline{A + B} \times \sqrt{Q}}$ in numeris integris proximis. Sit illud r. Divide $A\sqrt{Q}$ per maximum divisorem rationalem. Sit Quotus s, sitcg $\dfrac{r + \dfrac{N}{r}}{2s}$ in numeris integris proximis t, et erit $\dfrac{ts \pm \sqrt{ttss - N}}{\sqrt[2e]{Q}}$ radix quæsita, si modo radix extrahi potest.[605]

Ut si radix cubica extrahenda sit ex $\sqrt{968} + 25$[606] erit $AA - BB = 343$. Ejus divisores 7, 7, 7. Ergo $N = 7$ & $Q = 1$. Porrò $\overline{A + B} \times \sqrt{Q}$ seu $\sqrt{968} + 25$, extracta prioris partis radice fit paulo major quam 56. Ejus radix cubica in numeris proximis est 4. Ergo $r = 4$. Insuper $A\sqrt{Q}$ seu $\sqrt{968}$ extrahendo quicquid rationale est fit $22\sqrt{2}$. Ergo $\sqrt{2}$, ejus pars radicalis, est s et $\dfrac{r + \dfrac{N}{r}}{2s}$ seu $\dfrac{5\frac{3}{4}}{2\sqrt{2}}$ in numeris integris proximis est 2. Ergo $2\sqrt{2} + 1$ est radix quæsita, si modò radix extrahi potest.

---

(602) Namely the Waessenaer–Descartes rule, first published by the latter anonymously in Waessenaer's *Den On-wissen Wis-konstenaer*... (Leyden, 1640) as a contribution to his attack on Stampioen (see note (342) above), and subsequently refashioned by Frans van Schooten in the concluding section of his *Additamentum* to his Latin editions of Descartes' *Geometrie* (*Geometria*, ₁1649: 329–36 = ₂1659: 394–400; compare II: 315, note (39)), whence Kinckhuysen took the somewhat garbled version he presented in his 1661 *Algebra* (see II: 314–17). Pierre Costabel in his 'Descartes et la racine cubique des nombres binômes' (*Revue d'Histoire des Sciences et de leurs Applications*, **22**, 1969: 97–116) has recently given a critical account, based in part on newly unearthed documents, of the historical circumstances in which Descartes was led to formulate his rule; see also II: 315, note (39) and 316, note (41). It will be evident (see note (605) below) that Newton's present rule slightly refines Descartes'.

(603) At this point in his later revise of the present manuscript (as the unfinished 'Arithmeticæ Universalis Liber Primus' reproduced in Part 2) Newton added the schoolmasterish aside: 'Hæc rarò usu veniet & a Tyrone prætermitti potest' (This will rarely prove to be of use and can be passed over by the novice); see **2**, [1]: note (39). On the rule's strictly mathematical limitations consult note (605).

(604) In other words, set $N^e/(A^2 - B^2) = Q$. The effect of this factor $\sqrt{Q}$ is to convert $\sqrt[e]{(A \pm B)}$ into $\sqrt[e]{[(A \pm B)\sqrt{Q}]} \equiv \sqrt[e]{[a \pm b]}$, in which $\sqrt[e]{[a^2 - b^2]}$ is integral.

(605) In justification of this refinement of Descartes' rule we may (see the previous note) consider the reduced binomial $\sqrt[e]{[a \pm b]} = x \pm \sqrt{y}$, in which

$$x(= st),\ y \text{ and } N = \sqrt[e]{[a^2 - b^2]} = x^2 - y$$

are integral. Descartes then instructs (compare II: 314, note (39)) that we determine the integer $2m$ satisfying $m - \frac{1}{2} < \sqrt[e]{[a + b]} < m$, so yielding $x = [\frac{1}{2}(m + N/m)]$; for, on setting $\alpha_m = m + N/m - 2x = ((m - x)^2 - y)/m$, we may remake the inequality as $\sqrt{y} < m - x < \sqrt{y} + \frac{1}{2}$, whence $0 < (m - x)^2 - y < \frac{1}{4} + \sqrt{y} < x + \sqrt{y}$ and therefore $0 < \alpha_m < (x + \sqrt{y})/m < 1$, so that $(\frac{1}{2}\alpha_m \text{ or}) \frac{1}{2}(m + N/m) - x < \frac{1}{2}$. Newton here effectively adds the amendment that, since also $\sqrt{y} > m - \frac{1}{2} - x > \sqrt{y} - \frac{1}{2}$, therefore $0 > (m - \frac{1}{2} - x)^2 - y > -(\sqrt{y} - \frac{1}{4}) > -(\sqrt{y} + x - \frac{1}{2})$ and so

There is also a rule[602] for extracting higher roots of numerical quantities compounded of two parts whose (square) powers are commensurable.[603] Let the quantity be $A \pm B$ with A its greater part and e the index of the root to be extracted. Look for the least number N whose power N^e can be divided by $A^2 - B^2$ without remainder, and let the quotient be Q.[604] Compute

$$\sqrt[e]{[(A+B)\sqrt{Q}]}$$

to the nearest integer and let it be r; divide $A\sqrt{Q}$ by its greatest rational divisor, let the quotient be s and $\frac{1}{2}(r+N/r)/s$ to the nearest integer be t: then

$$(st \pm \sqrt{[s^2 t^2 - N]})/\sqrt[2e]{Q}$$

will be the required root, provided the root can be extracted.[605]

For instance, if the cube root of $\sqrt{968}+25$[606] is to be extracted, there will be $A^2 - B^2 = 343$ and its divisors $7 \times 7 \times 7$: therefore $N = 7$ and $Q = 1$. Moreover, when the root of the first part of $((A+B)\sqrt{Q}$ or) $\sqrt{968}+25$ is extracted it proves to be a little greater than 56, and the cube root of this to the nearest integer is 4: therefore $r = 4$. Further, by extracting from $(A\sqrt{Q}$ or) $\sqrt{968}$ all that is rational, it becomes $22\sqrt{2}$. Therefore $\sqrt{2}$, its radical part, is s and

$$(\tfrac{1}{2}(r+N/r)/s \text{ or}) \ \tfrac{1}{2}(5\tfrac{3}{4})/\sqrt{2}$$

to the nearest integer is 2. Therefore $2\sqrt{2}+1$ is the root sought, provided the

---

$0 > \alpha_{m-\frac{1}{2}} > -(x+\sqrt{y}-\frac{1}{2})/(m-\frac{1}{2}) > -1$, whence $(-\frac{1}{2}\alpha_{m-\frac{1}{2}}$ or) $x-\frac{1}{2}(m-\frac{1}{2}+N/(m-\frac{1}{2})) < \frac{1}{2}$. Accordingly, it is enough to determine $x = [\frac{1}{2}(m+N/m)]$ in which either $2m$ or $2m-1$ is the nearest integer (greater or less) to $2\sqrt[e]{[a+b]}$. The requirement here that x (and hence $y = x^2 - N$) be integral may always be met by choosing a large enough multiplier Q, but the least Q which renders $\sqrt[e]{[(A^2-B^2)Q]} = \sqrt[e]{[a^2-b^2]}$ integral will not always define corresponding integral values of t and so x. No instances will be found among Newton's following examples, but we may cite the Eulerian illustrations

$$\sqrt[3]{[\sqrt{5}+2]} = \tfrac{1}{2}\sqrt{5}+\tfrac{1}{2} \quad \text{and} \quad \sqrt[3]{[2\sqrt{7}+3\sqrt{3}]} = \tfrac{1}{2}\sqrt{7}+\tfrac{1}{2}\sqrt{3}:$$

in the former $(Q = 1, s = \sqrt{5})$ Newton's test yields successively $N = 1$, $r = 1$ and $t = 0$; while in the latter $(Q = 1, s = \sqrt{7})$ it produces $N = 1$, $r = 2$ and again $t = 0$. These—and such instances of higher roots as $\sqrt[5]{[10\sqrt{10}+22\sqrt{2}]} = \tfrac{1}{2}\sqrt{10}+\tfrac{1}{2}\sqrt{2}$, for example—may be subsumed by determining $2t$ to be the nearest integer to $(r+N/r)/s$. With this modification the test correctly yields for the last example $(Q = 1, s = \sqrt{10})$ the values $N = 2$, $r = 2$ and so (since $2t = 1$ is the nearest integer to $3/\sqrt{10}$) $t = \frac{1}{2}$. The deficiency in Newton's manuscript rule was first pointed out by Euler in 1741 in an extended discussion 'De extractione radicum ex quantitatibus irrationalibus' (*Commentarii Academiæ Scientiarum Petropolitanæ*, 13, 1741–3 [1751]: 16–60 [= *Leonhardi Euleri Opera Omnia* (1) 6, 1921: 31–77]) where (compare Edward Waring, *Meditationes Algebraicæ* (Cambridge, 1770): Caput v, Problema xxxix, Exemplum 2: 174–6) a second modification is proposed: in essence, since $x \pm \sqrt{y} = \sqrt[e]{[(A \pm B)\sqrt{Q}]}$ where $x^2 - y = N$, at once $x^2 + y = \frac{1}{2}(\sqrt[e]{[(A+B)^2 Q]}+\sqrt[e]{[(A-B)^2 Q]})$, from which x^2 and y are straightaway computed—a rule which, in Euler's words, 'minori opera quam NEUTONIANA ad calculum revocatur [ac] etiam tutius operatur' ('De extractione...': 25 [= *Opera* (1) 6: 40]).

(606) That is, $22\sqrt{2}+25$. This is one of Descartes' worked examples in Schooten's *Additamentum* (*Geometria*, ₁1649: 326–7 =₂1659: 392–3).

Tento itaq per multiplicationem si cubus ipsius $2\sqrt{2}+1$ fiat $\sqrt{968}+25$ et res succedit.

‖ [210]   ‖ Rursus si radix cubica extrahenda sit ex $68-\sqrt{4374}$[607] erit $AA-BB=250$, cujus divisores sunt 5, 5, 5, 2. Ergo $N=5\times2=10$, et $Q=4$, et $\sqrt[3e]{\overline{A+B}\times\sqrt{Q}}$ seu $\sqrt[3]{68+\sqrt{4374}\times2}$ in numeris proximis integris est $7=r$.[608] Insuper $A\sqrt{Q}$ seu

$68\sqrt{4}$ extrahendo quicquid rationale est fit $136\sqrt{1}$. ergo $s=1$ et $\dfrac{r+\dfrac{N}{r}}{2s}$ in numeris integris proximis est $4=t$. Ergo $ts=4$ et $\sqrt{ttss-N}=\sqrt{6}$ et $\sqrt[2e]{Q}=\sqrt[6]{4}$ seu $\sqrt[3]{2}$, atq adeo radix tentanda $\dfrac{4-\sqrt{6}}{\sqrt[3]{2}}$.[609]

Porrò si radix quadrato-cubica extrahenda sit ex $29\sqrt{6}+41\sqrt{3}$ erit $AA-BB=3$, adeoq $N=3$. $Q=81$. $r=5$.[610] $s=\sqrt{6}$. t[611]$=1$. $ts=\sqrt{6}$. $\sqrt{ttss-N}=\sqrt{3}$ & $\sqrt[2e]{Q}=\sqrt[10]{81}$ seu $\sqrt[5]{9}$. atq adeo radix tentanda $\dfrac{\sqrt{6}+\sqrt{3}}{\sqrt[5]{9}}$.[612]

Cæterum in hujusmodi operationibus si quantitas cujus radicem extrahere oportet fractio sit, vel partes ejus communem habent divisorem, radices denominatoris et factorum seorsim extrahendæ erunt. Ut si ex $\sqrt{242}-12\frac{1}{2}$[613] radix cubica extrahenda sit, hoc reductis partibus ad communem denominatorem fiet $\dfrac{\sqrt{968}-25}{2}$. Dein extracta seorsim numeratoris ac denominatoris radice cubica orietur $\dfrac{2\sqrt{2}-1}{\sqrt[3]{2}}$. Rursus si ex $\sqrt[3]{3993}+\sqrt[6]{17578125}$[613] radix aliqua extrahenda sit, divide partes per maximum communem divisorem $\sqrt[3]{3}$ et emerget $11+\sqrt{125}$. Unde quantitas proposita valet $\sqrt[3]{3}\times\overline{11+\sqrt{125}}$, cujus radix invenietur extrahendo seorsim radicem factoris utriusq $\sqrt[3]{3}$ & $11+\sqrt{125}$.[614]

‖ [211]
Æquationum
constructio
linearis.[615]

‖ Hactenus æquationum proprietates, transmutationes, limites et omnis generis reductiones docui. Demonstrationes non semper adjunxi quoniam satis

---

(607) That is, $68-27\sqrt{6}$.

(608) Accurately, $r=4+\sqrt{6}=7_-$.

(609) Here 'res succedit' since $\frac{1}{2}(4-\sqrt{6})^3=68-27\sqrt{6}$.

(610) Accurately, $r=\sqrt{6}+\sqrt{3}=4\cdot18_+$ and hence to the nearest integer '$r=4$'; the lapse does not, however, affect the sequel—happily so, since it is not corrected in any printed edition.

(611) Doubtless computed as $[5\frac{3}{5}/2\sqrt{6}]$ rather than $[4\frac{3}{4}/2\sqrt{6}]$; compare the previous note.

(612) The trial succeeds since $\frac{1}{9}(\sqrt{6}+\sqrt{3})^5=29\sqrt{6}+41\sqrt{3}$.

(613) Minimal variants on Descartes' equivalent examples

$$\sqrt[3]{[\sqrt{242}+12\tfrac{1}{2}]}=\sqrt[3]{\tfrac{1}{2}}+\sqrt[6]{128}\quad\text{and}\quad\sqrt[5]{[\sqrt[3]{3993}+\sqrt[6]{17578125}]}=(1+\sqrt{5})\cdot\sqrt[15]{3}\sqrt[5]{16}$$

in Schooten's *Additamentum* (*Geometria,* ₁1649: 323–4/329 =₂1659: 389–90/394).

root can be extracted. I accordingly test by multiplication if the cube of $2\sqrt{2}+1$ does make $\sqrt{968}+25$, and the trial succeeds.

Again, if the cube root to be extracted be that of $68-\sqrt{4374}$,[607] there will be $A^2-B^2 = 250$, the divisors of which are $5\times5\times5\times2$: therefore $N = 5\times2 = 10$ and $Q = 4$, while $(\sqrt[e]{[(A+B)\sqrt{Q}]}$ or) $\sqrt[3]{[(68+\sqrt{4374})\times2]}$ to the nearest integer is $7 = r$.[608] Furthermore, on extracting all that is rational from it, $(A\sqrt{Q}$ or) $68\sqrt{4}$ becomes $136\sqrt{1}$: therefore $s = 1$ and $\frac{1}{2}(r+N/r)/s$ to the nearest integer is $4 = t$: therefore $st = 4$ and $\sqrt{[s^2t^2-N]} = \sqrt{6}$, while $\sqrt[2e]{Q} = (\sqrt[6]{4}$ or$)$ $\sqrt[3]{2}$, and hence the root to be tested is $(4-\sqrt{6})/\sqrt[3]{2}$.[609]

Moreover, if the fifth root is to be extracted from $29\sqrt{6}+41\sqrt{3}$, there will be $A^2-B^2 = 3$ and hence $N = 3$, $Q = 81$, $r = 5$,[610] $s = \sqrt{6}$, t[611] $= 1$ and so $st = \sqrt{6}$, and $\sqrt{[s^2t^2-N]} = \sqrt{3}$ and $\sqrt[2e]{Q} = (\sqrt[10]{81}$ or$)$ $\sqrt[5]{9}$; hence the root to be tested is $(\sqrt{6}+\sqrt{3})/\sqrt[5]{9}$.[612]

For the rest, if in operations of this type the quantity whose root it is necessary to extract be a fraction or its parts have a common divisor, the roots of the denominator and factors will need to be separately extracted. If, for instance, the cube root of $\sqrt{242}-12\frac{1}{2}$[613] is to be extracted, when its parts are reduced to a common denominator this will become $(\sqrt{968}-25)/2$; then, after the cube root of numerator and denominator has been separately extracted, there will ensue $(2\sqrt{2}-1)/\sqrt[3]{2}$. Again, if some root of $\sqrt[3]{3993}+\sqrt[6]{17\,578\,125}$[613] has to be extracted, divide the parts by the greatest common divisor $\sqrt[3]{3}$ and there will emerge $11+\sqrt{125}$: whence the value of the quantity proposed is

$$\sqrt[3]{3}\,.\,(11+\sqrt{125}),$$

the root of which will be found by separately extracting the root of each factor $\sqrt[3]{3}$ and $11+\sqrt{125}$.[614]

So far, I have developed properties of equations, explaining their transmutations, limits and reductions of every kind. Proofs I have not always attached

The linear construction of equations.[615]

(614) Observe that the ensuing chapter 'De Reductione fractionum et Radicalium ad series convergentes' in Newton's draft (see §2: note (36) following) has now vanished without trace. This unimplemented section-title was not printed by Whiston in his 1707 *editio princeps*, nor did Newton make any attempt to revive or augment it in his 1722 revise.

(615) Understand by curves, both 'geometrical' (algebraic) and 'mechanical' (non-algebraic). Perhaps to emphasize his near-total omission of the preceding section 'Extractio Radicum ex binomiis' (see note (596) above) Whiston in his 1707 *editio princeps* left a whole page blank and then entered the present marginal head as a main title on his page 279—in larger, blacker type than that allowed to the book's half title on page 1!—, so leaving the distinctly misleading impression that the following 'linear' construction of equations was an appendix to the preceding algebraic text. Though in his 1722 revise Newton lazily consented to this restyling of his original intention (and indeed confirmed it by excising the one paragraph of the preceding section which Whiston retained in 1707), in his library copy of the *editio princeps* he had earlier recorded his thorough disapproval of Whiston's editorial inter-

[1682]
Lect 10
faciles mihi visæ sunt,[616] et nonnunquam absq nimijs ambagibus tradi non possent. Restat jam tantum ut æquationum postquam ad formam commodissimam reductæ sunt radices in numeris extrahere doceam. Et hic præcipua difficultas est in figuris duabus vel tribus prioribus obtinendis. Id quod commodissimè per æquationis constructionem aliquam seu Geometricam sive Mechanicam confit. Qua de causa non pigebit hujusmodi constructiones aliquas subjungere.

Veteres ut ex Pappo discimus[617] trisectionem anguli & inventionem duarum mediè proportionalium, sub initio per rectam lineam et circulum frustra aggressi sunt. Postea considerare cœperunt alias permultas lineas ut Conchoidem Cissoidem et Conicas sectiones et per harum aliquas solverunt Problemata. Tandem re penitius examinata et Conicis Sectionibus in Geometriam receptis, Problemata distinxerunt in tria[618] [genera; Plana quæ] per lineas a pla[no originem derivantes,] Rectam nempe [et Circulum solvi possunt,] solida quæ per [lineas ortum a solidi id est Coni] consideratione d[erivantes solvebantur, &] linearia ad quo[rum solutionem] requirebantur [lineæ magis compositæ. Et juxta] hanc distinctione[m problemata solida per alias] ‖ lineas quam Conicas sectiones solvere a Geometria alienum est,[619] præsertim si nullæ aliæ lineæ præter rectam circulum et Conicas sectiones in

‖[212]

---

pretation of the status of the ‘Æquationum constructio linearis’ in his manuscript scheme. Not only did he there cancel Whiston's main title, replacing it with a new marginal head ‘Inventio primarum figurarum Radicis’ (The finding of the first figures in the root) immediately following the previous text (on Whiston's page 277) and closing up the intervening blank, but he iconoclastically pared this final section to the bone, leaving only the first paragraph following and the (now unproved) geometrical constructions of the cubics

$$x^3 + qx + r = 0, \quad x^3 + px^2 + r = 0 \quad \text{and} \quad x^3 + px^2 + qx + r = 0$$

enunciated on his pages 217–18/223, 219–20/226–7 and 220–1/228–9 respectively below, together with the stray observation on his page 232 regarding the indeterminacy of n. (So much for those who single out the several remarks on the primacy of synthetic geometry—all marked for deletion in his library copy—as clear evidence of Newton's mature dislike of algebraic analysis.) It was doubtless his intention in this revised section to conclude with an outline of method for extracting the roots of equations *in numeris* to any desired accuracy (II: 218–20, III: 42–6) to give weight to his earlier reference (on his page 52 above; see note (170)) to ‘regulæ post docendæ’ for so resolving numerical equations. In the outcome, however, though in his 1722 revise he struck out Edmond Halley's surrogate ‘Methodus Nova Accurata & facilis inveniendi Radices Æquationum quarumcunque generaliter, sine prævia Reductione’ (which Whiston, *faute de mieux*, had reprinted on pages 327–43 of his *editio princeps* from *Philosophical Transactions* 18, No. 210 [for May 1694]: 136–48), Newton there contented himself (see note (170)) with a brief reference in footnote to finding ‘figuras primas radicis per constructionem quamvis mechanicam et reliquas per methodum Vietæ’.

(616) When we look back at Newton's rule for delimiting the complex roots of an equation— first given rigorous proof by Sylvester in 1865 (see note (478))—, we may be allowed a disbelieving smile at this remark.

since they have seemed easy enough to me[616] and they could frequently not possibly be presented without excessive digression. It now remains merely to explain how to extract the roots of equations numerically once they have been reduced to their most convenient form. Here, the especial difficulty lies in obtaining their first two or three figures. That is most conveniently accomplished by some construction or other of the equation, be it a geometrical or a mechanical one. And for this reason it will not be an infliction to append certain constructions of the kind.

The Ancients, as we learn from Pappus,[617] initially attacked the trisection of an angle and the finding of two mean proportionals by way of the straight line and circle, but to no effect. Subsequently, they began to take numerous other lines—such as the conchoid, cissoid and the conic sections—into consideration and by means of certain of these they solved those problems. At length, having pondered the matter more deeply and accepted conic sections into geometry, they distinguished problems into three types: plane ones, solvable by lines— namely, the straight line and circle—deriving their origin from the plane; solid ones, solved by lines deriving their source from consideration of a solid—a cone, to be exact—; and linear ones, for whose solution more complicated lines were required. Following this distinction it is[619] alien to geometry to solve solid problems by any lines other than conics, especially if no other lines except the straight line, circle and conics are to be accepted into geometry. But mathe-

---

(617) Newton evidently refers to the preamble (§36) to Pappus, *Mathematical Collection* IV, 31 ff.: 'When the ancient geometers sought to divide a given rectilineal angle into three equal parts they were at a loss.... We say that there are three kinds of problems in geometry, some being called *plane*, some *solid*, some *linear*. Those which can be solved by means of a straight line and a circumference of a circle are properly called plane; for the lines by which such problems are solved have their origin in a plane. Such problems, however, as are solved by using for their discovery one or more of the sections of the cone are called solid; for in the construction it is necessary to use surfaces of solid figures, I mean the conic surfaces. There remains a third kind of problem called linear; for other lines...are used for their construction, having a more complicated...origin as they are generated from more irregular surfaces and intricate movements. ...lines of this kind are spirals and quadratrices [quadratrixes] and cochloids [conchoids] and cissoids. It appears to be no small error for geometers when a plane problem is solved by conics or other curved lines and in general when any problem is solved by an inappropriate kind' (Ivor Thomas, *Selections illustrating the history of Greek Mathematics. I: From Thales to Euclid* (London, 1939): 347–51). The various classical solutions of the problems of angle-trisection and cube-duplication by means of conics, the cissoid and various conchoidal 'vergings' are conveniently listed by T. L. Heath in his *History of Greek Mathematics*, 1 (Oxford, 1921): 235–68. A number of these are introduced by Newton into his following text.

(618) The outside bottom corner of the present manuscript page (and its verso, page 212) has, at some time since its text was printed by Whiston, been torn off and is now lost. As needed, we have in the sequel restored certain fragmented phrases from the corresponding pages 279–81 of the 1707 *editio princeps*.

(619) The less incisive 'foret' (would be) was first written.

Geometriam recipiantur. At Recentiores[620] longius progressi receperunt lineas omnes in Geometriam quæ per æquationes exprimi possunt, et pro dimensionibus æquationum distinxerunt lineas illas in genera, legemɋ tulerunt non licere Problema per lineam superioris generis construere quod construi potest per lineam inferioris. In lineis contemplandis et eruendis earum proprietatibus distinctionem earum in genera juxta dimensiones æquationum per quas definiuntur laudo. At æquatio non est sed descriptio quæ curvam Geometricam efficit. Circulus linea Geometrica est non quod per æquationem exprimi potest sed quod descriptio ejus postulatur. Æquationis simplicitas non est sed descriptionis facilitas quæ lineam ad constructiones Problematum prius admittendam esse indicat. Nam æquatio ad Parabolam simplicior est quam æquatio ad circulum et tamen circulus ob simpliciorem descriptionem prius admittitur. Circulus et Coni sectiones si æquationum dimensiones spectentur [ejusdem sunt or]dinis, et tamen circulus in [constructione problematum] non connumeratur [cum his sed ob simpliciorem] descriptionem deprimitur [ad ordinem inferiorem lineæ rectæ,] ita ut per circulum [construere quod per rectas] construi potest, non [sit illicitum; per Conicas v]erò sectiones con[struere quod per circulum] construi potest, [vitio vertatur. Aut igitur] legem a ‖ dimensionibus æquationum in circulo observandam esse statue et sic distinctionem inter problemata plana et solida ut vitiosam tolle, aut concede legem illam in lineis superiorum generum non ita observandam esse quin aliquæ ob simpliciorem descriptionem præferantur alijs ejusdem ordinis et in constructione Problematum cum lineis inferiorum ordinum connumerentur. In constructionibus quæ sunt æque Geometricæ præferendæ semper sunt simpliciores. Hæc lex omni exceptione major est. Ad simplicitatem verò constructionis expressiones Algebraicæ nil conferunt.[621] Solæ descriptiones linearū hic in censum veniunt.

‖ [213]

---

(620) Above all Descartes, who in the opening of the second book of his *Geometrie* (₁1637: 315) likewise echoed Pappus: 'Les anciens ont fort bien remarqué, qu'entre les Problesmes de Geometrie, les vns sont plans, les autres solides, & les autres lineaires, c'est a dire, que les vns peuuent estre construits, en ne traçant que des lignes droites, & des cercles; au lieu que les autres ne le peuuent estre, qu'on n'y employe pour le moins quelque section conique; ni enfin les autres, qu'on n'y employe quelque autre ligne plus composée. Mais ie m'estonne de ce qu'ils n'ont point outre cela distingué diuers degrés entre ces lignes plus composées.' (He goes on to suggest that possibly the 'Ancients' were discouraged from so doing by meeting with such curves as 'la Spirale, la Quadratrice, & semblables, qui n'appartienent veritable qu'aux Mechaniques' before encountering the more tractable higher algebraic curves, such as the cissoid and conchoid.) We have already noticed (IV: 341, note (23)) that by 'degré' Descartes meant not the modern concept of algebraic degree—first used systematically to classify curves by Newton himself!—but a Cartesian 'grade' whose n-th order comprises the pair of the $(2n-1)$-th and $2n$-th algebraic degrees.

(621) Here, of course, 'simplicity' is being used in a multiple sense whose strands Newton would have done better to distinguish. Thus a classification of curves by degree (in its Newtonian significance) is 'simple' in that it is well-defined (being that of its defining equation

maticians of more recent times[620] have, in their further progress, welcomed into geometry all lines which can be expressed by means of equations, and have distinguished those lines into classes according to the dimensions of those lines, laying down the formal rule that it is not permissible to construct a problem by means of a line of a superior class when it can be constructed by one of a lower. In contemplating curves and deriving their properties I commend their distinction into classes in line with the dimensions of the equations by which they are defined. Yet it is not its equation but its description which produces a geometrical curve. A circle is a geometrical line not because it is expressible by means of an equation but because its description (as such) is postulated. It is not the simplicity of its equation but the ease of its description which primarily indicates that a line is to be admitted into the construction of problems. To be sure, the equation to a parabola is simpler than that to a circle, and yet because of its simpler construction the circle is given prior admission. A circle and the conics are, if regard be paid to the dimensions of their equations, of the same order, and yet in the construction of problems a circle is not numbered with these latter curves but, because of its simpler description, is reduced to the lower order of the straight line; as a result it is not impermissible to construct by means of a circle what can be constructed by straight lines, but to construct by means of conics what can be constructed by a circle is to be reckoned a fault. Either, therefore, decree that the law (differentiating) by the dimensions of the equations is to be observed in the case of the circle and thus elide the distinction between plane and solid problems as being faulty, or allow that that law is not to be observed in lines of higher class in a way which precludes some, because of their simpler description, being given preference over others of the same order and, in the construction of problems, being ranked with lines of lower order. In constructions which are of equal geometrical rating the simpler ones are always to be preferred. This law overrides all exception. On the simplicity, indeed, of a construction the algebraic representation has no bearing.[621] Here the descrip-

---

in a standard Cartesian system of coordinates) and invariant under a number of simple geometrical transformations (notably that of optical projection from a point onto a second plane); whereas constructive 'simplicity' is a loose and often inconsistent blend of such *desiderata* as elegance, economy and accuracy of solution. His present preference of the circle over the parabola as 'simpler' curve (though its Cartesian equation is the more complicated) does not, we may add, accord well with his contemporary attempt (IV: 344) to disprove Descartes' assertion (*Geometrie*, ₁1637: 309) that 'la premiere, & la plus simple [ligne] de toutes aprés les sections coniques' is the Cartesian trident $(x-a)(x-b)^2 = cxy$, constructible (*ibid.*: 322, 343–4; compare II: 363, note (123)) as the intersection of a straight line, rotating round a fixed pole, with a parabola of parameter c moving along a line, distant a away, such that the rotating line passes always through a point in its axis distant b^2/c from its vertex: neither of Newton's 'simpler' counter-instances—the Wallis cubic $x(x-a)(x-b) = cy$ and the hyperbolic hyperbolism $x(x-a)y = b$—are easily to be drawn except by marking a

Has solas considerabant Geometræ qui circulum conjungebant cum recta. Prout hæ sunt faciles vel difficiles constructio facilis vel difficilis redditur. Adeoqȝ a rei natura alienum est leges constructionibus aliunde præscribere. Aut igitur lineas omnes præter rectam et circulum et forte Conicas sectiones e Geometria cum Veteribus excludamus aut admittamus omnes secundū descriptionis simplicitatem. Si Trochoides in Geometriam reciperetur, liceret ejus beneficio angulum in data ratione secare.[622] Numquid ergo reprehenderes siquis hac linea ad dividendū angulum in ratione numeri ad numerum uteretur & contenderes hanc lineam per æquationem[623] non definiri, lineas verò quæ per æquationes definiuntur adhibendas esse. Igitur si angulus e.g. in 10001 partes dividendus esset, teneremur curvam lineam æquatione plusquam

‖ [214]   centum ‖ dimensionum[624] definitam in medium afferre, quam tamen nemo mortalium describere nedum intelligere valeret, et hanc anteponere Trochoidi quæ linea notissima est et per motum rotæ vel circuli facillimè describitur.[625] Quod quam absurdum sit quis non videt. Aut igitur Trochoides in Geometriam non est admittenda aut in constructione Problematum curvis omnibus difficilioris descriptionis anteferenda. Et eadem est ratio de reliquis curvis. Quo nomine trisectiones anguli per Conchoidem quas Archimedes in Lemmatis et Pappus in collectionibus[626] posuere præ aliorum hac de re inventis omnibus

---

sufficiently dense set of their points, though as we have seen (II: 498–500) in his earlier 'Problems for construing æquations' he had given an organic construction of the instance $x^3 = cy$ of the former from a describend hyperbola.

(622) The cycloid is readily seen (compare III: 162, note (303)) to be defined as the locus of points C in CBB', perpendicular to the diameter AD of the semicircle ABD and intersecting it in B, such that instantaneously $CB = \widehat{AB}$: we may evidently construct any defined portion $\widehat{AOb}$ of $\widehat{AOB}$ (where O is the semicircle's centre) by determining the parallel cb to CB which is

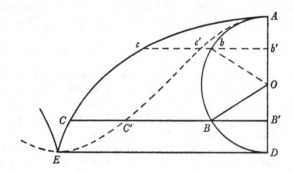

to it in the same ratio. The latter construction affords no difficulty if we also construct the sinusoidal 'compagne de la roulette' (see III: 204, note (434)), defined as the locus of points C' in CB' such that $C'B' = CB$: for at once cb meets this latter curve such that

$$c'b' \text{ (or } cb) : C'B' \text{ (or } CB) = \widehat{AOb} : \widehat{AOB}.$$

tions of curves alone come into the reckoning. This consideration alone swayed the geometers who joined the circle with the straight line. According as these descriptions are easy or difficult the construction is rendered easy or difficult. It is hence alien to the nature of the subject to prescribe laws for constructions on any other basis. Either, then, we are, with the Ancients, to exclude from geometry all lines except the straight line and circle and maybe the conics, or we are to admit them all according to the simplicity of their description. Were the cycloid to be accepted into geometry, it would be allowable by its aid to cut up an angle in a given ratio.[622] Could you then, if someone were to use this line to divide an angle in an integral ratio, see anything reprehensible in this and contend that this line is not defined by an[623] equation, but that lines defined by equations need to be employed? We would in consequence, were the angle to be divided into (for instance) 10001 parts, be compelled to bring into play a curve defined by an equation of more than a hundred dimensions:[624] this, however, no mortal would be capable of describing, let alone comprehending and valuing above the cycloid—a curve which is exceedingly well known and very easily described through the motion of a wheel or a circle.[625] How absurd this is, any one may see. Either, then, the cycloid is not to be admitted into geometry; or in the construction of problems it is to be preferred to all curves having a more difficult description. And the same reasoning goes for the rest of curves. On that head we commend the trisections of an angle by a conchoid (which Archimedes in his *Lemmas* and Pappus in his *Collection*[626] placed above

---

(623) Understand 'algebraicam' (algebraic).

(624) The largest prime factor of 10001 is, in fact, 137: Newton will also require a subsidiary angle-section into 73 parts. Both sections will require the solution of complicated equations of equivalent degree, derived on the style of those of 'Prob. 15' on his page 77 above.

(625) By 'trochoid' understand, of course, not only the primary branch of the cycloid but also its infinity of conjugates, for as David Gregory noted down in May 1694 after a visit to Newton 'Nulla...est suppositio quæ rotam describentem sistat' (Royal Society. Gregory MS C44, reproduced in *The Correspondence of Isaac Newton*, **3**, 1961: 335). See also Descartes' letter to Mersenne on 23 August 1638 (N.S.) (*Œuvres de Descartes*, **2**, 1898: 307–13, especially 309–10), first published as 'verba authoris' in Schooten's Latin commentary on Book 2 of Descartes' *Geometrie* (*Geometria*, ₁1649: 226–9 = ₂1659: 267–70).

(626) Newton here presumably refers to the elegant νεῦσις construction (Archimedes, *Lemmas* 8 = Pappus, *Mathematical Collection* IV, 32) which he discusses, in generalized form, on his pages 234–5 below; see note (675). Pappus' *Collection* also contains (IV, 34; compare Heath, *Greek Mathematics*, **1**: 243) a second trisection which constructs the meets of a circle arc subtending the given angle with the hyperbolic locus of the vertex of a triangle, one of whose base angles is twice the other. Since the latter locus—independently derived by Newton in his preceding 'Prob. 36' (on his pages 98–9) without reference to Pappus and without noticing, as his predecessor had done, that point *B* in his figure is a focus (see note (320))—is, in present context, less simple than the Archimedean νεῦσις involving a straight line and a circle (or, in Pappus' variant, a second line), this second trisection is manifestly not Newton's present 'pre-eminent' one.

laudamus: siquidem lineas omnes præter rectam et circulum e Geometria excludere debeamus aut secundum descriptionis simplicitatem admittere, & Conchoides simplicitate descriptionis nulli curvarum præter circulum cedit. Æquationes sunt expressiones computi arithmetici, et in Geometria locum propriè non habent nisi quatenus quantitates verè Geometricæ (id est lineæ, superficies, solida & proportiones) aliquæ alijs æquales enunciantur. Multiplicationes Divisiones et ejusmodi computa in Geometriam recens introducta sunt idcg inconsultò et contra primum institutum scientiæ hujus.[627] Nam qui constructiones Problematum per rectam et circulum a primis Geometris[628] adinventas considerabit facile sentiet Geometriam excogitatam esse ut expedito linearum ductu effugeremus computandi tædium. Proinde hæ duæ scientiæ confu[n]di non debent. Veteres tam sedulo distinguebant eas ab invicem ut in ‖ Geometriam terminos Arithmeticos nunquam introduxerint. Et recentes utramcg confundendo amiserunt simplicitatem in qua Geometriæ elegantia omnis consistit. Est itacg Arithmeticè quidem simplicius quod per simpliciores æquationes determinatur, at Geometricè simplicius est quod per simpliciorem ductum linearum colligitur,[629] et in Geometria prius et præstantius esse debet quod est ratione Geometrica simplicius. Mihi igitur vitio vertendum non erit si cum Mathematicorum Principe Archimede alijscg Veteribus Conchoidem ad solidorum problematum constructionem adhibeam. Attamen siquis aliter senserit, sciat me hic de constructione non Geometrica sed qualicuncg sollicitum esse qua radices æquationum in numeris proximè assequar. Cujus rei gratia præmitto hoc problema Lemmaticum.

‖ [215]

Inter datas duas lineas[630] *AB, AC rectam datæ longitudinis BC*
ponere quæ producta transeat per datum punctū P.

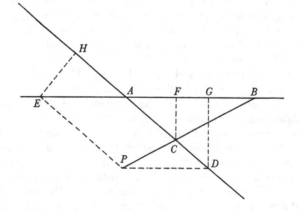

Si circa polum *P* gyret linea *BC* et simul termino ejus *C* incedat super recta *AC*, ejus alter terminus *B* describet Conchoidem Veterum. Secet hæc lineam *AB*

all the findings of others on this topic) inasmuch as one ought either to exclude from geometry all lines except the straight line and circle or to admit them according to the simplicity of their description, while the conchoid in the simplicity of its description yields to no curve except the circle. Equations are expressions belonging to arithmetical computation and in geometry properly have no place except in so far as certain truly geometrical quantities (lines, surfaces and solids, that is, and their ratios) are stated to be equal to others. Multiplications, divisions and computations of that sort have recently been introduced into geometry, but the step is ill-considered and contrary to the original intentions of this science:[627] for anyone who examines the constructions of problems by the straight line and circle devised by the first geometers[628] will readily perceive that geometry was contrived as a means of escaping the tediousness of calculation by the ready drawing of lines. Consequently these two sciences ought not to be confused. The Ancients so assiduously distinguished them one from the other that they never introduced arithmetical terms into geometry; while recent people, by confusing both, have lost the simplicity in which all elegance in geometry consists. Accordingly, the arithmetically simpler is indeed that which is determined by simpler equations, while the geometrically simpler is that which is gathered[629] by a simpler drawing of lines—and in geometry what is simpler on geometrical grounds ought to be first and foremost. It will not therefore be interpreted as a fault in me if with the prince of mathematicians, Archimedes, and others of the Ancients I should employ a conchoid in the construction of solid problems. Nonetheless, if anyone does feel differently, I want him to know that my immediate concern is not for a construction which is geometrical, but for one of any sort whereby I may attain a numerical approximation to the roots of equations. With this motive I premise this lemmatical problem:

Between two given lines[630] *AB, AC to place a straight line BC of given length which, when produced, shall pass through the given point P.*

If the line *BC* should revolve round the pole *P* while simultaneously at its end-point *C* it moves along on the straight line *AC*, its other end-point *B* will describe the Ancients' conchoid. Let this intersect the line *AB* in the point *B*,

---

(627) Yet one more dig at Descartes (to whom Newton owed too much not to resent his debt?). The same criticism is elaborated in Newton's contemporary preface (IV: 420–2) to his 'Geometria Curvilinea', and again, with particular reference to the classical Greek 3/4-line locus, in his 'Veterum Loca solida restituta' (IV: 274–82, especially 276).

(628) 'Veteribus' (Ancients) is cancelled. Newton, of course, means the early Greek geometers, up to (and probably including) Euclid.

(629) 'innotescit' (is made known) was first written.

(630) Understand 'rectas' (straight).

in puncto *B*. Junge *PB*, et ejus pars *BC* erit recta quam ducere oportuit. Et eadem lege linea *BC* duci potest ubi vice rectæ *AC* linea aliqua curva adhibetur.[631]

‖[216]
Octob. 1683.
Lect 1

‖ Sicui constructio hæcce per Conchoidem minus placeat, potest alia per conicam Sectionem ejus vice substitui. A puncto *P* ad rectas *AD*, *AE* age *PD*, *PE* constituentes parallelogrammum *EADP*, et a punctis *C* ac *D* ad rectam *AB* demitte perpendicula *CF*, *DG*, ut et a puncto *E* ad rectam *AC* versus *A* productam perpendiculum *EH*, et dictis $AD = a$. $PD = b$. $BC = c$. $AG = d$. $AB = x$ et $AC = y$ erit $AD.AG::AC.AF$ adeoq́ $AF = \dfrac{dy}{a}$. Erit et $AB.AC::CD.PD$ seu $x.y::b.a-y_{[;]}$ Ergo $by = ax - yx$ quæ æquatio est ad Hyperbolam.[632] Rursus per 13. II Elem erit $BC^q = AC^q + AB^q - 2FAB$ id est $cc = yy + xx - \dfrac{2dxy}{a}$.[633]

Prioris æquationis partes ductas in $\dfrac{2d}{a}$ aufer de partibus hujus et restabit $cc - \dfrac{2bdy}{a} = yy + xx - 2dx$, æquatio ad circulum ubi x et y ad rectos sunt angulos.[634] Quare si hasce duas lineas Hyperbolam et circulum ope harum æquationum componas_{[;]} earum intersectione habebis x et y seu *AB* et *AC* quæ positionem rectæ *CB* determinant. Componentur autem lineæ illæ ad hunc modum.

Duc rectas duas quasvis *KL* æqualem *AD* et *KM* æqualem *PD* continentes angulum rectum *MKL*. Comple parallelogrammum *KLMN* et asymptotis *LN*, *MN* per punctū *K* describe Hyperbolam *IKX*.

---

(631) The line *BC* may, of course, lie on either side of *AD*: in Newton's configuration as drawn the second solution *B'C'* is possible; while if *AB* crosses the node (or cusp) of the

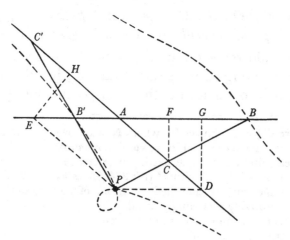

join PB and its section BC will be the straight line it was required to draw. And by the same rule the line BC can be drawn when instead of the straight line AC some curved line is employed.[631]

Should this construction by means of a conchoid be not quite to anyone's liking, another by the aid of a conic can be substituted in its stead. From the point P to the straight lines AD, AE draw PD, PE forming the parallelogram $EADP$, then from the points C and D to the straight line AB let fall the perpendiculars CF, DG and also from the point E to the straight line AC extended on A's side the perpendicular EH. On calling $AD = a$, $PD = b$, $BC = c$, $AG = d$, $AB = x$ and $AC = y$, there will be $AD:AG = AC:AF$ and hence $AF = (d/a)\,y$. Also, there will be $AB:AC = CD:PD$ or $x:y = b:(a-y)$ and therefore

$$by = ax - yx,$$

the equation to a hyperbola.[632] Again, by *Elements*, ii, 13, there will be

$$BC^2 = AC^2 + AB^2 - 2AF \times AB,$$

that is, $c^2 = y^2 + x^2 - 2(d/a)\,xy$.[633] Take the sides of the previous equation multiplied by $2d/a$ away from the sides of the present one and there will remain $c^2 - 2(bd/a)\,y = y^2 + x^2 - 2dx$, the equation to a circle when x and y are at right angles.[634] Consequently, if you build up these two lines, hyperbola and circle, with the help of these equations, from their intersection you will have x and y, that is, AB and AC, so determining the position of the straight line CB. Those lines, however, will be constructed in this manner.

Draw any two straight lines, KL equal to AD and KM equal to PD, containing the right angle $\widehat{MKL}$. Complete the rectangle $KLMN$ and with asymptotes LN, MN through the point K describe the hyperbola IKX. Again, in KM

---

conchoid's lower branch (through P), the two remaining solutions will also be real. The (ten-month?) annual break in Newton's 'lectures' indicated by the marginal chronology at this point is thoroughly unbelievable.

(632) In a standard Cartesian system of coordinates, that is. In his sequel Newton will for convenience take the ordinate angle (between $KY = x$ and $YX = y$) to be right, constructing this hyperbola $(x+b)\,(a-y) = ab$ through the origin $K(0, 0)$ and with asymptotes

$$LM\ (x = -b), \quad LN\ (y = a).$$

(633) '...quæ æquatio est ad Ellipsin' (the equation to an ellipse). Newton proceeds to derive the circle which is through the intersections of this ellipse with the preceding hyperbola.

(634) In his following scheme Newton will contruct this circle

$$(x-d)^2 + (y + bd/a)^2 = c^2 + d^2(1 + b^2/a^2)$$

through $S(0, \sqrt{[c^2 + (bd/a)^2]} - bd/a)$ and on centre $T(d, -bd/a)$, where $KP = d$, $KQ = c$, $RK = bd/a$ and $RS = RQ$.

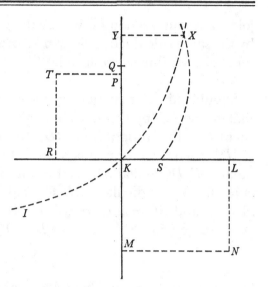

In *KM* versus *K* producta cape *KP* æqualem *AG* et *KQ* æqualem *BC*. Et ‖[217] ‖ in *KL* producta versus *K* cape *KR* æqualem *AH* et *RS* æqualem *RQ*. Comple parallelogrammum *PKRT* et centro *T* intervallo *TS* describe circulum. Secet hic Hyperbolā in puncto *X*. Ad *KP* demitte perpendiculum *XY* et erit *XY* æqualis *AC* & *KY* æqualis *AB*. Quæ duæ lineæ *AC* et *AB* vel una earum cum puncto *P* determinant positionem quæsitā rectæ *BC*.[635] Cui constructioni demonstrandæ et ejus casibus secundum casus Problematis determinandis non immoror.

Hac, inquam, constructione solvi potest Problema sicui ita visum sit. Sed hæc solutio magis composita est quam ut usibus ullis inservire possit. Nuda speculatio est, et speculationes Geometricæ tantum habent elegantiæ quantum simplicitatis tantumq̃ laudis merentur quantum utilitatis secum afferrunt. Ea de causa constructionem per Conchoidem præfero ut multo simpliciorem et non minus Geometricam & quæ resolutioni æquationum a nobis propositæ optimè conducit.[636] Præmisso igitur præcedente Lemmate construimus Problemata cubica et quadrato-quadratica[637] ut sequitur.

Proponatur æquatio cubica $x^3 * + qx + r = 0$ *cujus terminus secundus deest*, tertius vero sub signo suo designatur per $+q$ et quartus per $+r$. Duc quamlibet[638] *KA* quam

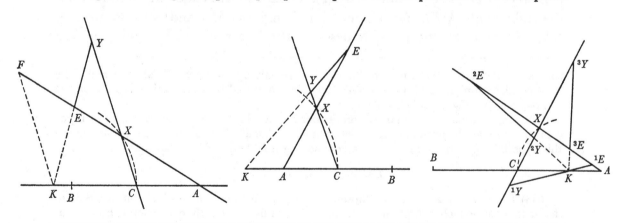

(635) The second real solution in Newton's preceding lemmatical problem (see note (631)) corresponds to the second real intersection—not indicated by Newton—between the hyperbola *IKX* and the circle *SX*. Should two further real solutions of the problem exist, the circle *SX* would also intersect the second branch (not here drawn) of the hyperbola.

extended on K's side take KP equal to AG and KQ equal to BC, and in KL extended on K's side take KR equal to AH and RS equal to RQ. Complete the rectangle $PKRT$ and with centre T, radius TS describe a circle. Let this cut the hyperbola in the point X. To KP let fall the perpendicular XY and then XY will be equal to AC and KY equal to AB. These two lines AC and AB—or one of them with the point P—determine the required position of the straight line BC.[635] I do not dwell upon the demonstration of this construction and the determination of its cases in line with the cases of the problem.

By this construction, let me repeat, the problem can be solved if anyone is of such a mind. But this solution is too complicated to be able to serve any practical use: it is a barren speculation. Geometrical speculations possess elegance only in so far as they have simplicity, and merit praise only inasmuch as utility is their concomitant. For that reason I prefer the construction by the conchoid as being simpler and no less geometrical and an excellent contribution to the resolution of equations proposed by us.[636] With the preceding lemma premised, therefore, we construct cubic and square-square (quartic)[637] problems as follows.

Let there be proposed the cubic equation $x^3 + qx + r = 0$, *whose second term is lacking,* but (the coefficient of) the third is denoted under its proper sign by $+q$ and the fourth by $+r$. Draw any[638] KA you please and call it n. In KA extended in

---

(636) In fact, if A is taken as origin and $B(X, Y)$ a general point in the perpendicular Cartesian coordinate system in which AD is the X-axis, then the preceding problem may be constructed through the meets of the line AE defined by $Y = (a\beta/d)X$ and the conchoid

$$(X-\alpha)^2 Y^2 = (Y+\beta)^2 (c^2 - Y^2),$$

where $\alpha = a - bd/a$ and $\beta = b\sqrt{[a^2 - d^2]}/a$. On eliminating Y and then transforming to $x = (a/d)X$, there results the same quartic equation

$$x^4 + 2(b-d)x^3 + (a^2 + b^2 - 2bd - c^2)x^2 - 2bc^2x - b^2c^2 = 0$$

as would ensue by eliminating y between the above defining equations for the hyperbola and circle by means of which Newton previously constructed the problem.

(637) No such constructions of quartics by a conchoidal νεῦσις are contained in the present, manifestly incomplete section, but it will be clear from notes (635) and (636) preceding that such conchoidal constructions of general fourth-degree equations having one pair (at least) of roots real and one root opposite in sign to the rest are possible. In his following constructions of cubics—which repeat without essential variant corresponding resolutions in his earlier 'Problems for construing æquations' (see II: 470–6, 508–12)—Newton ensures that these conditions are met in each case by attaching a known root of appropriate sign. In the *editio princeps* at this point Whiston in 1707 added the thoroughly misleading remark—retained by Newton in his 1722 revise, doubtless to avoid the lacuna which Whiston so neatly bridged— 'utpote quæ ad cubica reduci possunt' (seeing that they can be reduced to cubic ones), namely, as Newton outlined on his pages 205–6 above, by employing an auxiliary cubic to factorize the quartic equation posed by a general 'solid' problem into its quadratic components.

(638) Understand 'rectam' (straight line).

dic n. In KA utrinꝗ producta cape $KB = \dfrac{q}{n}$ ad easdem partes cum KA si habeatur

|| [218] $+q$, aliter ad || contrarias. Biseca BA in C et centro K radio KC fac circulū CX,

cui inscribe rectam CX æqualem $\dfrac{r}{nn}$ et produc eam utrinꝗ. Dein junge AX et

produc eam utrinꝗ. Deniꝗ inter has lineas CX et AX inscribe EY ejusdem longitudinis cum CA quæꝗ producta transeat per punctum K, et XY erit radix æquationis. Et ex his radicibus affirmativæ erunt quæ cadunt ad partes X versus C et negativæ quæ cadunt ad partes contrarias si habeatur $+r$ et contra si habeatur $-r$.[639]

Demonstratio.

Ad demonstrationem præmittimus Lemmata sequentia.

Lē. 1. Est YX ad AK ut CX ad KE: Etenim age KF parallelam CX et ob similia triangula ACX, AKF, et EYX EKF erit AC ad AK ut CX ad KF et YX ad YE seu AC ut KF ad $KE_{[,]}$ adeoꝗ YX ad AK ut CX ad KE. Q.E.D.

Lem. 2. Est YX ad AK ut CY ad AK+KE. Nam componendo est YX ad AK ut $YX + CX$ id est CY ad $AK + KE$. Q.E.D.

|| [219] || *Lem. 3. Est KE−BK*[640] *ad YX ut YX ad AK.* Nam per 12. II Elem est

$$YK^q - CK^q = CY^q - CY \times CX = CY \times YX.$$

hoc est si theorema resolvatur in proportionem CY ad $YK - CK$ ut $YK + CK$ ad YX. Sed est $YK - CK = YK - YE + CA - CK = KE - BK$. et

$$YK + CK = YK - YE + CA + CK = KE + AK.$$

Adeoꝗ est CY ad $KE - BK$ ut $KE + AK$ ad YX. Sed per Lemma secundum erat CY ad $KE + AK$ ut YK ad AK. Ergo ex æquo est YX ad $KE - BK$ ut AK ad YX seu $KE - BK$ ad YX ut YX ad AK. Q.E.D.

His præmissis Demonstrabitur Theorema ut sequitur. In primo Lemmate erat YX ad AK ut CX ad KE seu $KE \times YX = AK \times CX$. In tertio erat $KE - BK$ ad YX ut YX ad AK. Unde si prioris rationis termini ducantur in YX fiet

$$KE \times YX - BK \times YX \text{ ad } YX^q \text{ ut } YX \text{ ad } AK,$$

id est $AK \times CX - BK \times YX$ ad YX^q ut YX ad $AK_{[,]}$ et ductis extremis et medijs in

(639) A but slightly augmented repeat of *Modus* 1 of Problem 2 on II: 470. The following demonstration (wholly cancelled in Newton's library copy; see note (615)) clarifies that on II: 470–2 by distinguishing it, as sketched in the 'Aut sic melius' on II: 472, into three component lemmas. As before (see II: 471, note (62)) we may restore Newton's preliminary analysis by setting $KX = a$, $KC = b$, $CX = c$, $CA = YE = d$ and the unknowns $XY = x$, $EK = y$, and then deriving from the geometrical relationship $KA \times CX \times YE = CA \times XY \times EK$ the algebraic equivalent $(b+d)c = xy$: whence, since

$$2b \cos \hat{C} = (b^2 + c^2 - a^2)/c = (b^2 + (x+c)^2 - (y+d)^2)/(x+c),$$

either direction take $KB = q/n$: the same way as KA if $+q$ be had, but otherwise the opposite way. Bisect BA in C and with centre K, radius KC construct the circle CX and in it inscribe the straight line CX equal to r/n^2, producing it each way. Next, join AX and extend it each way. Finally, between these lines CX and AX inscribe EY of the same length as CA such that, when produced, it shall pass through the point K, and XY will be a root of the equation. Of these roots the ones falling on the side of X towards C will be positive and those falling on the opposite side negative if $+r$ be had, but the converse if $-r$ be supposed.[639]

Demonstration

In proof we premise the following lemmas:

Lemma 1. XY is to KA as CX to KE. For, to be sure, draw KF parallel to CX and then, because of the similar triangles ACX, AKF and EYX, EKF, there will be CA to KA as CX to KF and XY to EY (or CA) as KF to KE, and therefore XY to KA as CX to KE. As was to be proved.

Lemma 2. XY is to KA as CY to KA + KE. For, by compounding, XY is to KA as $XY + CX$ (that is, CY) to $KA + KE$. As was to be proved.

Lemma 3. KE − KB[640] *is to XY as XY to KA.* For by *Elements*, II, 12, there is $KY^2 - KC^2 = CY^2 - CY \times CX = CY \times XY$; that is, if the proposition be resolved into a proportion, CY is to $KY - KC$ as $KY + KC$ to XY. But

$$KY - KC = KY - EY + CA - KC = KE - KB$$

and $$KY + KC = KY - EY + CA + KC = KE + KA.$$

Hence CY is to $KE - KB$ as $KE + KA$ to XY. But by the second lemma CY was to $KE + KA$ as XY to KA. Therefore *ex æquo* XY is to $KE - KB$ as KA to XY, or $KE - KB$ to XY as XY to KA. As was to be proved.

With these premises the theorem will be demonstrated as follows. In the first lemma there was XY to KA as CX to KE, that is, $KE \times XY = KA \times CX$. In the third, $KE - KB$ was to XY as XY to KA: hence if the members of its first ratio be multiplied by XY, there will come to be $KE \times XY - KB \times XY$ to XY^2 as XY to KA, that is, $KA \times CX - KB \times XY$ to XY^2 as XY to KA; and when extremes and

---

on eliminating y and discarding the unwanted root $x = -c$ (corresponding to YE coincident with CA) there ensues $x^3 + (a^2 - b^2)c^{-1}x^2 + (b^2 - d^2)x - c(b+d)^2 = 0$. On equating this with $x^3 + qx + r = 0$ term by term there results $a^2 - b^2 = 0$, $b^2 - d^2 = q$ and $c(b+d)^2 = -r$, so that $KX = KC$ and, on making $KA = b + d = n$ of arbitrary length, $KB = b - d = q/n$ and $CX = -r/n^2$ also. Newton's third figure illustrates the case where the equation has three real roots, and so three positions of EY through K are possible.

(640) Newton thoughtlessly first copied this as '$BK + KE$' from his earlier 'Demonstratio' (II: 472), in whose accompanying figure $BK = -KB$ is drawn in the opposite sense to that of his first two present figures (here implied).

se $AK^q \times CX - AK \times BK \times YX = YX^{cub}$. Deniæ pro YX, AK, BK et CX restitutis x, n, $\frac{q}{n}$, & $\frac{r}{nn}$ orietur $r - qx = x^3$. Q.E.D. Quod verò ad signorum variationes attinet, istis secundum casus Problematum determinandis non immoror.

[1683]
Lect 2

Proponatur jam æquatio cujus tertius terminus deest $x^3 + pxx + r = 0$. *Et ad ejus* constructionem assumpto quolibet n, cape in recta aliqua longitudines duas

$KA = \frac{r}{nn}$ & $KB = p$ idæ ad easdem partes

‖ [220]

si r et p habeant eadem signa, aliter ad contrarias. Biseca BA in C et ‖ centro K radio KC describe circulum cui inscribe CX æqualem n, et produc eam utrinæ. Item junge AX et produc eam utrinæ. Deniæ inter has lineas CX et AX inscribe EY ejusdem longitudinis cum CA ita ut ea si producatur transeat per K, et KE erit radix æquationis. Radices autem affirmativæ sunt ubi punctum Y cadit a parte puncti X versus C & negativæ ubi

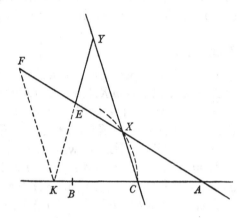

punctum Y cadit ad alteras partes puncti X si modo habeatur $+r$ & contra si habeatur $-r$.[641]

Ad hujus Propositionis demonstrationem schemata[642] et Lemmata de priori propositione mutuò sumantur, et Demonstratio erit ut sequitur.

Per Lemma 1 erat YX ad AK ut CX ad KE seu $YX \times KE = AK \times CX$, & per Lemma 3 $KE - KB$ ad YX ut YX ad AK, aut (sumpto KB ad contrarias partes) $KE + KB$ ad YX ut YX ad AK adeoæ $KE + KB$ in KE ad $YX \times KE$ seu $AK \times CX$ ut YX ad AK seu CX ad KE. Quare ductis extremis et medijs in se est

$$KE^{cub} + KB \times KE^q = AK \times CX^q,^{[643]}$$

et ipsarum KE, KB, AK et CX restitutis valoribus supra assignatis, $x^3 + pxx = r$.[644]

Proponimus jam æquationem trium dimensionum $x^3 + pxx + qx + r = 0$[645] *nullo termino carentem et cujus tres radices non sunt omnes affirmativæ neæ omnes negativæ.*[646]

---

(641) Much as before (see note (639)) on setting $x = (b+d)c/y$ in the previous cubic $(a = b)$: $x^3 + (b^2 - d^2)x - c(b+d)^2 = 0$ there results (after reordering)

$$y^3 - (b-d)y^2 - (b+d)c^2 = 0.$$

When y is replaced by x and the cubic then identified with $x^3 + px^2 + r = 0$, it follows that $-KB = -(b-d) = p$ and, where $c = n$ is arbitrary, $-KA = r/n^2$. Newton's following equivalent geometrical demonstration is, as we have remarked (note (615)), deleted in his library copy of the *editio princeps*. Both construction and demonstration are repeated from II: 474–6.

middles are multiplied together $KA^2 \times CX - KA \times KB \times XY = XY^3$. Finally, after x, n, q/n and r/n^2 are restored in place of XY, KA, KB and CX, there will arise $r - qx = x^3$. As was to be demonstrated. In regard to the variations in the signs, however, I do not linger over determining these in accordance with the various cases of the problem.

Let there now be proposed an equation lacking its third term, $x^3 + px^2 + r = 0$. Having, to effect its construction, assumed an arbitrary n, in some straight line take two lengths, $KA = r/n^2$ and $KB = p$: these in the same direction if r and p have the same sign, otherwise in opposite ones. Bisect BA in C and with centre K, radius KC describe a circle and in it inscribe CX equal to n, extending it each way. Likewise, join AX and extend it each way. Finally, between these lines CX and AX inscribe EY of the same length as CA so that, if produced, it shall pass through K, and KE will be a root of the equation. The roots, however, are positive when the point Y falls on the side of point X towards C, and negative when the point Y falls on the opposite side of point X, provided $+r$ be had; and the converse if $-r$ be supposed.[641]

In proof of this proposition let the diagrams[642] and lemmas be borrowed from the previous proposition, and the demonstration will then be as follows.

By Lemma 1, XY was to KA as CX to KE, that is, $XY \times KE = KA \times CX$, while by Lemma 3 $KE - KB$ was to XY as XY to KA, or alternatively (when KB is taken the opposite way) $KE + BK$ to XY as XY to KA; hence $(KE + BK) \times KE$ is to $(XY \times KE$ or) $KA \times CX$ as $(XY$ to KA or) CX to KE. Consequently, when extremes and middles are multiplied together, there is

$$KE^3 + BK \times KE^2 = KA \times CX^2,^{[643]}$$

and so, after the values of KE, BK, KA and CX assigned above are restored, $x^3 + px^2 = r$.[644]

We now propose a third-degree equation, $x^3 + px^2 + qx + r = 0$, not deficient in any term and with its three roots neither all positive nor all negative.[646] Then, first, if the

---

(642) For ease of reference we have repeated the first of these in the Latin text.

(643) This follows at once by substituting $YX = AK \times CX/KE$ in the preceding cubic (compare note (641)).

(644) Or more accurately '$-r$' since r is chosen above to have the same sign as p when (as here) KA and KB are drawn the same way from K.

(645) Originally '$x^3 + pxx + rx + s = 0$', it would appear (see note (651) below). The change from r, s to q, r respectively has, where not made by Newton himself, been silently adjusted in the sequel.

(646) It will be evident from the following figure that when, correspondingly, three positions of EY are possible—namely, when K lies between A and C, or above A but close enough to it—these will lie one to the left of X and the other two on its right in all cases. When a pair of roots of the cubic are not real, the two latter related positions of EY will no longer be

Et primò si terminus q negativus est, in recta aliqua KB capiantur longitudines duæ $KA = \frac{r}{q}$ et $KB = p$, idcg ad easdem partes puncti K si p et $\frac{r}{q}$ habent signa diversa; aliter

‖[221] ad contrarias. ‖ Biseca AB in C, et ad punctum illud C erige perpendiculum CX æquale radici quadraticæ termini q: Et inter lineas rectas AX et CX utrincg productas in infinitum inscribatur recta EY quæ æqualis sit rectæ AC et producta transeat per punctum

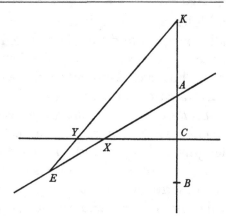

K, atcg KE erit radix æquationis, quæ quidem affirmativa erit si punctum X cadat inter puncta A et E, negativa verò si punctum E cadat ad partes puncti X versus A.[647]

Quod si terminus q affirmativus est[,] in recta KB capiantur longitudines illæ duæ $KA = \sqrt{\frac{-r}{p}}$ & $KB = \frac{q}{KA}$, idcg ad easdem partes puncti K si $\sqrt{\frac{-r}{p}}$ et $\frac{q}{KA}$ habent signa diversa; aliter ad contrarias. Biseca AB in C et ad punctum illud C erige perpendiculum CX æquale termino p: et inter lineas rectas AX et CX utrincg productas in infinitum inscribatur recta EY quæ æqualis sit rectæ AC et producta transeat per punctum K atcg XY erit radix æquationis[,] quæ quidem negativa erit si punctum X cadat inter puncta A et E, affirmativa verò si punctum Y cadat ad partes puncti X versus punctum C.[648]

---

possible, with K lying too far above A to allow them: in either case following, the analytical condition for this will be that

$$(p^2 - 3q)^3 < (p^3 - \tfrac{9}{2}pq + \tfrac{27}{2}r)^2.$$

(This was first explicitly derived by Edmond Halley in 'De Numero Radicum in Æquationibus Solidis ac Biquadraticis, sive tertiæ ac quartæ potestatis, earumcg limitibus, tractatulus', *Philosophical Transactions* **16**, No. 190 [for November 1687]: 387–402, especially 390–1. As Halley went on to demonstrate the more primitive inequalities

$$p^2 < 3q \quad \text{and} \quad p^3 > 27r$$

derived by Thomas Harriot (*Artis Analyticæ Praxis* (London, 1631): Sectio v, Propositio 5: 83–6) are necessary but not sufficient.) In his illustration Newton wisely contents himself with the position of EY which is always real.

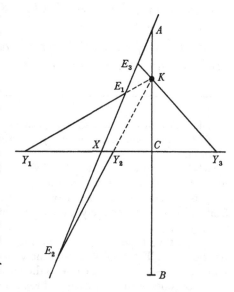

coefficient q is negative, in some straight line KB let two lengths be taken, $KA = r/q$ and $KB = p$: this on the same side of point K if p and r/q have different signs, otherwise on contrary ones. Bisect AB in C, and at that point C erect the perpendicular CX equal to the square root of the coefficient q. Then between the lines AX and CX extended indefinitely either way inscribe the straight line EY which shall be equal to the line AC and, when produced, shall pass through the point K, and KE will be a root of the equation—positive, to be sure, should the point X fall between the points A and E, but negative if the point E fall on the side of X towards A.[647]

But if the coefficient q is positive, let there be taken in the straight line KB the two lengths $KA = \sqrt{[-r/p]}$ and $KB = q/KA$: this on the same side of point K if $\sqrt{[-r/p]}$ and q/KA have different signs, otherwise on contrary ones. Bisect AB in C and at that point C erect the perpendicular CX equal to the coefficient p: then between the lines AX and CX indefinitely extended each way inscribe the straight line EY which shall be equal to the straight line AC and, when produced, shall pass through the point K, and XY will be a root of the equation. This, to be sure, will be negative if point X should fall between the points A and E, but positive if the point Y falls on the side of point X towards the point C.[648]

---

(647) Newton earlier gave an analytical discussion of this conchoidal construction in his 'Problems for construing æquations' (II: 508–10). Essentially (compare II: 510, note (147)), the line $EY = AC$ cut off from EK by two arbitrary lines XA, XC is determined (see note (639)) by $xy = (b+d)c$ and $(b^2+c^2-a^2)(x+c) = c(b^2+(x+c)^2-(y+d)^2)$ on setting $KX = a$, $KC = b$, $CX = c$, $CA = YE = d$ and also $XY = x$, $EK = y$; whence, on eliminating x and discarding the unwanted root $y = b+d$ (corresponding to YE coincident with CA) there results

$$y^3 - (b-d)y^2 - (a^2-b^2)y - c^2(b+d) = 0.$$

When y is replaced by x and the cubic then equated term by term with $x^3+px^2+qx+r = 0$ (q negative) it follows that $KB = -(b-d) = p$, $a^2-b^2 = -q$ and $KA = -(b+d) = r/c^2$. For simplicity Newton makes $a^2-b^2 = c^2$, so that $\widehat{ACX}$ is right. His further observations are immediate corollaries.

(648) When, in the previous note, y is eliminated and the unwanted root $x = -c$ analogously discarded, there results (with again $a^2-b^2 = c^2$) the cubic

$$x^3 + cx^2 + (b^2-d^2)x - c(b+d)^2 = 0.$$

Identification with $x^3+px^2+qx+r = 0$ (q positive) yields $c = p$ and in consequence

$$KA = b+d = \sqrt{[-r/p]} \quad \text{and} \quad KB = (b-d) = q/KA.$$

Newton's further remarks all follow from this in an immediate way.

Demonstratio casus prioris.[649]

|| [222] Per Lemma primum erat KE ad CX ut AK ad YX et ita (componendo) est $KE + AK$ id est $KY + KC$ ad $CX + YX$ id est CY. Sed in triangulo || rectangulo KCY est YC^q æquale $YK^q - KC^q$ id est æquale $KY + KC$ in $KY - KC$ et resolvendo terminos æquales in proportionales, $KY + KC$ ad CY ut CY ad $KY - KC$, seu $KE + AK$ ad CY ut CY ad $EK - KB$. Quare cum in hac proportione fuerit KE ad $CX_{[,]}$ duplicetur proportio et erit KE^q ad CX^q ut $KE + AK$ ad $KE - KB$, et ductis extremis et medijs in se $KE^{cub} - KB \times KE^q = CX^q \times KE + CX^q \times AK$. Et restitutis valoribus supra assignatis $x^3 - pxx = qx + r$.

Demonstratio casus secundi.[649]

Per Lemma primum est KE ad CX ut AK ad YX, ductisꝗ extremis et medijs in se fit $KE \times YX = CX \times AK$. Scribe ergo in superioribus $KE \times YX$ pro $CX \times AK$ et fiet $KE^{cub} - KB \times KE^q = CX^q \times KE + CX \times KE \times YX$. Et applicatis omnibus ad KE erit $KE^q - KB \times KE = CX^q + CX \times YX$, ductisꝗ omnibus in AK habebitur

$$AK \times KE^q - AK \times KB \times KE = AK \times CX^q + AK \times CX \times YX:$$

Ac rursus scripto $KE \times YX$ pro $CX \times AK$ fiet

$$AK \times KE^q - AK \times KB \times KE = KE \times YX \times CX + KE \times YX^q:$$

et applicatis omnibus ad KE orietur $AK \times KE - AK \times KB = YX \times CX + YX^q$, ductisꝗ omnibus in YX emerget

$$AK \times KE \times YX - AK \times KB \times YX = YX^q \times CX + YX^{cub.}$$

et pro $KE \times YX$ scriptis in primo termino $CX \times AK$ fiet

$$CX \times AK^q - AK \times KB \times YX = CX \times YX^q + YX^{cub:}$$

seu quod perinde est $YX^{cub.} + CX \times YX^q + AK \times KB \times YX - CX \times AK^q = 0.$[650]

atꝗ pro YX, CX, AK et KB substitutis valoribus supra assignatis x, p, $\sqrt{\dfrac{-r}{p}}$,

$\dfrac{q}{\sqrt{\dfrac{-r}{p}}}$ [651] emerget tandem $x^3 + pxx + qx + r = 0,$[651] æquatio construenda.

---

(649) As we have observed (see note (615)) these newly contrived synthetic proofs were later to be deleted by Newton in his library copy of Whiston's 1707 *editio princeps*.

(650) This results at once from the cubic derived in the first case on there substituting $KE = AK \times CX/YX$ and reordering.

(651) Through an oversight Newton has here omitted to alter his original expressions ‘ $\sqrt{\dfrac{-s}{p}}$, $\dfrac{r}{\sqrt{\dfrac{-s}{p}}}$ ’ and the equation ‘$x^3 + pxx + rx + s = 0$’ to accord with his previous text (see

Demonstration of the former case[649]

By the first lemma KE was to CX as KA to XY, and so *componendo* is

$$(KE+KA, \text{ that is}) \ KY+KC \text{ to } (CX+XY, \text{ that is}) \ CY.$$

But in the right-angled triangle KCY there is CY^2 equal to KY^2-KC^2, that is, $(KY+KC)(KY-KC)$, and by resolving the equality into a proportion $KY+KC$ is to CY as CY to $KY-KC$, or $KE+KA$ to CY as CY to $KE-KB$. Consequently, since KE was to CX in this proportion, 'double' (square) the proportion and there will be KE^2 to CX^2 as $KE+KA$ to $KE-KB$, and so, once extremes and middles are multiplied together, $KE^3-KB\times KE^2 = CX^2\times KE+CX^2\times KA$. And on restoring the values assigned above $x^3-px^2 = qx+r$.

Demonstration of the second case[649]

By the first lemma KE is to CX as KA to XY, and so, when extremes and middles are multiplied together, there comes to be $KE\times XY = CX\times KA$. In the preceding result, therefore, write $KE\times XY$ in place of $CX\times KA$ and there will come $KE^3-KB\times KE^2 = CX^2\times KE+CX\times KE\times XY$. Then, on dividing through by KE, there will be $KE^2-KB\times KE = CX^2+CX\times XY$ and, on multiplying through by KA, there will be had

$$KA\times KE^2-KA\times KB\times KE = KA\times CX^2+KA\times CX\times XY.$$

Again, when $KE\times XY$ is written in place of $KA\times CX$, there will come to be $KA\times KE^2-KA\times KB\times KE = KE\times XY\times CX+KE\times XY^2$; and, on dividing through by KE, there will arise $KA\times KE-KA\times KB = XY\times CX+XY^2$ and then, on multiplying through by XY, there will emerge

$$KA\times KE\times XY-KA\times KB\times XY = CX\times XY^2+XY^3,$$

and with $CX\times KA$ written in place of $KE\times XY$ in the first term there will come to be $CX\times KA^2-KA\times KB\times XY = CX\times XY^2+XY^3$ or, what is exactly the same, $XY^3+CX\times XY^2+KA\times KB\times XY-CX\times KA^2 = 0$.[650] And when in place of XY, CX, KA and KB their values x, p, $\sqrt{[-r/p]}$ and $q/\sqrt{[-r/p]}$ assigned above are substituted, there will at length emerge $x^3+px^2+qx+r = 0$,[651] the equation to be constructed.

---

note (645)), but we have made the trivial substitution of q and r for r and s required for consistency. Equivalent changes (here and above) are marked in the margin of the manuscript in Whiston's hand. By confusing the '$\frac{q}{-r}$' set in correction of the second expression, the printer of the 1707 *editio princeps* made it out to be '$\sqrt{\dfrac{q-r}{p}}$', an error corrected by Newton in his 1722 revise into the accurate equivalent '$q\sqrt{\dfrac{p}{-r}}$' which appears in all subsequent editions.

‖ Solvuntur etiam hæ æquationes ducendo rectam lineam datæ longitudinis inter circulum et aliam rectam lineam positione datos, ea lege ut recta illa ducta convergat ad punctum datum.[652]

Proponatur enim *æquatio cubica* $x^3 * + qx + r = 0$, *cujus terminus secundus deest.* Duc rectam *KA* ad arbitrium. Eam dic *n*.

In *KA* utrinꝗ producta cape $KB = \dfrac{q}{n}$

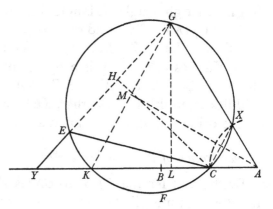

idꝗ ad easdem partes puncti *K* cum linea *KA* si modo habeatur $-q$,[653] aliter ad diversas. Biseca *BA* in *C* et centro *A* intervallo *AC* describe circulum *CX*. Ad hunc apta lineam rectam $CX = \dfrac{r}{nn}$ et per puncta *K*, *C* et *X* describe circulum *KCXG*. Junge *AX* et junctam produc donec ea iterum secet circulum ultimò descriptum *KCXG* in puncto *G*. Deniꝗ inter hunc ultimò descriptum circulum et rectam *KC* utrinꝗ productam inscribe rectam *EY* ejusdem longitudinis cum recta *AC* ita ut ea convergat ad punctum *G*. Et acta recta *EC* erit una ex radicibus æquationis. Radices autem affirmativæ sunt quæ cadunt in majori circuli segmento *KGC* et negativæ quæ in minori *KFC* si habeatur $-r$, et contra si habeatur $+r$ affirmativæ in minori segmento *KFC* negativæ in majori *KGC* reperientur.[654]

Ad hujus verò constructionis demonstrationem præmittimus Lemmata sequentia.

---

(652) This construction—much as the preceding one (see note (636))—will in fact yield a quartic relationship in the first of the two cases following, where in effect the intersections of a conchoid $x^2 y^2 = (x+a)^2(b^2-x^2)$ with a circle $(x+a)(x-c) + y(y-d) = 0$ through its pole $(-a, 0)$ are determined: for, on eliminating y, there results

$$d^2 x^2 (x+a)^2 (b^2-x^2) = (x^2(x+a)(x-c) + (x+a)^2(b^2-x^2))^2,$$

that is, (since $x \neq -a$!) $((a+c)x^2 - b^2(x+a))^2 + d^2 x^2(x^2-b^2) = 0$, a quartic which cannot have an odd number of negative roots. As before (compare note (637)) Newton reduces this to a general cubic by supposing one of its roots is known. The second case, where the conchoid's intersections with a circle $(x-e)^2 + y^2 = s^2$ having its centre $(e, 0)$ on the former's axis of symmetry $(y = 0)$ yields straightforwardly the cubic condition

$$(x+a)^2(b^2-x^2) = x^2(s^2-(x-e)^2),$$

that is, $2(a+e)x^3 + (a^2-b^2+s^2-e^2)x^2 - 2ab^2x - a^2b^2 = 0$.

(653) Momentarily forgetting that he has here altered his original canonical equation $x^3 = qx+r$, Newton first copied '$+q$' from II: 472 (compare the next note).

These equations are also solved by drawing a straight line of given length between a circle and another straight line, (both) given in position, with the stipulation that the drawn line shall be directed through a given point.[652]

For *let the cubic equation* $x^3 + qx + r = 0$, *lacking its second term, be proposed*. Draw the straight line KA arbitrarily, and call it n. In KA extended each way take $KB = q/n$: this on the same side of the point K as the line KA provided $-q$[653] be had, otherwise on the opposing one. Bisect BA in C and with centre A, radius AC describe the circle-arc CX. Into this fit the straight line $CX = r/n^2$ and through the points K, C and X describe the circle $KCXG$. Join AX and produce the joining line till it again meets the circle last described in the point G. Finally, between this last-described circle and the line KC extended each way inscribe the straight line EY of the same length as the line AC such that it is directed through the point G. Then, when the straight line CE is drawn, this will be one of the roots of the equation. Positive roots, however, are those which fall in the circle's greater segment KGC and negatives ones in the lesser segment KFC if $-r$ be had; and conversely so, if $+r$ be had, the positives will be located in the lesser segment KFC and the negatives in the greater one KGC.[654]

But in proof of this construction we premise the following lemmas.

---

(654) These observations relate, of course, to Newton's figure as drawn. The construction—along with the synthetic demonstration (wholly cancelled in his library copy of the *editio princeps*; see note (615))—which follows in sequel—is a trivially revised and augmented repeat of *Modus* 2 of Problem 2 of his earlier 'Problems for construing æquations' (II: 472–4). We may restore the analysis much as before (see II: 473, note (66)) by setting $CK = a$, $AC = b$, $AX = YE = c$, $GK = d$, $CE = x$ and $KY = y$. For it then follows that $AG = b(a+b)/c$ and also, because the triangles CEY, GKY are similar (or $CE:EY = GK:KY$) and therefore $xy = cd$, that $GY = y(y+a)/c = d(ax+cd)/x^2$. Further, since

$$2GC \cdot \cos \widehat{GCK} = (a^2 + GC^2 - d^2)/a = -(b^2 + GC^2 - b^2(a+b)^2/c^2)/b,$$

at once $GC^2 = ab^2(a+b)/c^2 - ab + bd^2/(a+b)$. Hence, since similarly

$$2d \cdot \cos \widehat{GKY} = (y^2 + d^2 - y^2(y+a)^2/c^2)/y = -(a^2 + d^2 - GC^2)/a,$$

there results, after discarding the factor $y + a + b = 0$ (corresponding to YE coincident with AX), the cubic $y^3 + (a-b)y^2 + (b^2-c^2)y - c^2d^2/(a+b) = 0$; and from this, on substituting $y = cd/x$, the required condition $x^3 - [(a+b)(b^2-c^2)/cd]x^2 - (a^2-b^2)x - (a+b)cd = 0$. Identification of this with $x^3 + qx + r = 0$ yields $b = c$ (or $AC = AX$), $a^2 - b^2 = -q$ and $(a+b)cd = -r$, so that on taking $KA = -(a+b) = n$, free, there ensues $KB = -(a-b) = -q/n$ and (since now CX is parallel to KG) $CX = cd/(a+b) = -r/n^2$. Since negative values of x correspond to negative ones of y—when, that is, Y is to the right of K in Newton's figure (drawn for negative r)—these will manifestly be constructed (where possible) by points E in the lesser arc $\widehat{CK}$. (The condition for the reality of this latter pair will clearly, on both geometrical and algebraic grounds, be $\frac{1}{4}q^2 \geqslant -\frac{1}{27}r^3$.)

|| [224]   *Lemma 1. Positis quæ in constructione supe*||*riore*₍,₎ *est CE ad KA ut CE+CX ad AY, & CX ad KY.*

Nam rectâ *KG* ductâ est *AC* ad *AK* ut *CX* ad *KG* idcʒ ob similia triangula *ACX, AKG.* Sunt etiam triangula *YEC YKG* similia: quippe quæ communem habent angulum ad *Y* et angulos ad *G* et *C* in eodem circuli *KCG* segmento *EGCK* atcʒ adeo æquales. Inde fit *CE* ad *EY* ut *KG* ad *KY*₍,₎ id est *CE* ad *AC* ut *KG* ad *KY* eo quod *EY* et *AC* juxta Hypothesin æquantur. Collata autem hacce cum superiore proportionalitate colligitur ex æquo perturbatè quod sit *CE* ad *KA* ut *CX* ad *KY*, et vicissim *CE* ad *CX* ut *KA* ad *KY*. Unde componendo fit *CE+CX* ad *CX* ut *KA+KY* ad *KY* id est ut *AY* ad *KY*, et vicissim *CE+CX* ad *AY* ut *CX* ad *KY* hoc est ut *CE* ad *KA*. Q.E.D.

Lemma 2. Demisso ad lineam GY perpendiculo CH, fiet rectangulum 2HEY æquale rectangulo CE×CX.

Nam demisso etiam ad lineam *AY* perpendiculo *GL*, triangula *KGL ECH* rectos habentia angulos ad *L* et *H* et angulos ad *K* et *E* in eodem circuli *CGK* segmento *CKEG*, adeocʒ æquales, æquiangula sunt et proinde similia. Est ergo *KG* ad *KL* ut *EC* ad *EH*. Porro a puncto *A* ad lineam *KG* demisso perpendiculo *AM*, ob æquales *AK AG* bisecabitur *KG* in *M* et triangula *KAM KGL* ob angulum ad *K* communem et angulos ad *M* et *L* rectos fient similia, et inde est *AK* ad *KM* ut *KG* ad *KL*. Sed ut est *AK* ad *KM* ita est 2*AK* ad 2*KM* seu *KG* et ita (ob similia triangula *AKG ACX*) est 2*AC* ad *CX* et (ob æquales *AC* et *EY*) ita est 2*EY* ad *CX*. Ergo est 2*EY* ad *CX* ut *KG* ad *KL*. Sed erat *KG* ad *KL* ut *EC* ad *EH*, ergo est 2*EY* ad *CX* ut *EC* ad *EH*, atcʒ adeo rectangulum 2*HEY*

|| [225]   (ductis nimirum extremis et medijs in se) æquale || est rectangulo *EC×CX*. Q.E.D.

Assumpsimus hic lineas *AK AG* æquales esse. Nimirum rectangula *CAK, XAG* (per Corol. Prop. 36. lib. [3]⁽⁶⁵⁵⁾ Elem.) æqualia sunt, atcʒ adeo ut *CA* est ad *XA* ita *AG* est ad *AK*. Sed *CA, XA* æquales sunt per Hypothesin, ergo et *AG, AK*.

Lemma 3. Constructis omnibus ut supra, tres lineæ BY, CE, KA, sunt continuè proportionales.

Nam (per 12 Prop. lib. 2, Elem.) est $CY^q = EY^q + CE^q + 2EY \times EH$. Et ablato utrincʒ EY^q fit $CY^q - EY^q = CE^q + 2EY \times EH$. Sed $2EY \times EH$ (per Lem 2) æquale est rectangulo $CE \times CX$, et addito utrincʒ CE^q fit

$$CE^q + 2EY \times EH = CE^q + CE \times CX.$$

Ergo $CY^q - EY^q$ æquale est $CE^q + CE \times CX$ id est $CY + EY$ in $CY - EY$ æquale

---

(655) By a venial confusion Newton has written '6' in the manuscript: Whiston has corrected the slip to '3' in the margin of the manuscript, and his *editio princeps* reads 'III'.

Lemma 1. With the suppositions of the above construction, CE is to AK as CE+CX to AY and as CX to KY.

For, when the straight line KG is drawn, AC is to AK as CX to KG: this because of the similar triangles ACX, AKG. The triangles YEC, YKG also are similar, seeing that they have a common angle at Y and their angles at G and C in the same circle segment KCG and hence equal. As a result CE comes to be to EY as KG to KY, that is, CE to AC as KG to KY inasmuch as EY and AC are, by hypothesis, equal. On comparing this proportion with the previous one, however, it is gathered *ex æquo* by cross-compounding that CE is to AK as CX to KY, and, by permuting, CE to CX as AK to KY. Whence, on compounding, there comes to be $CE+CX$ to CX as $AK+KY$ to KY, that is, as AY to KY, and, by permuting, $CE+CX$ to AY as CX to KY, that is, as CE to AK. As was to be proved.

Lemma 2. When the perpendicular CH is let fall to the line GY, the product $2HE \times EY$ will prove to be equal to the product $CE \times CX$.

For, when there is also let fall the perpendicular GL to the line AY, the triangles KGL, ECH, having right angles at L and H and their angles at K and E in the same segment $CKEG$ of the circle CGK and hence equal, are equiangular and consequently similar: therefore GK is to LK as CE to HE. Moreover, when the perpendicular AM is dropped from the point A to the line KG, because AK, AG are equal KG will be bisected at M and the triangles KAM, KGL, because of their common angle at K and their angles at M and L being right, will prove to be similar: whence AK is to MK as GK to LK. But AK is to MK as $2AK$ to $(2MK$ or$)$ GK, and so (because of the similar triangles AKG, ACX) is $2AC$ to CX, that is, (because AC and EY are equal) $2EY$ to CX: therefore $2EY$ to CX is as GK to LK. But GK was to LK as CE to HE, and therefore $2EY$ is to CX as CE to HE; hence (on multiplying extremes and middles together, of course) the product $2HE \times EY$ is equal to the product $CE \times CX$. As was to be proved.

We have here assumed that the lines AK, AG are equal. To be exact, the products $AC \times AK$ and $AX \times AG$ are (by *Elements*, III, 36, Corollary) equal, and hence AC is to AX as AG is to AK. But AC, AX are equal by hypothesis, and therefore so too are AG, AK.

Lemma 3. With everything constructed as above, the three lines BY, CE and AK are in continued proportion.

For (by *Elements*, Book II, Proposition 12) there is

$$CY^2 = EY^2 + CE^2 + 2HE \times EY;$$

and, when EY^2 is taken off on each side, there comes

$$CY^2 - EY^2 = CE^2 + 2HE \times EY.$$

But (by Lemma 2) $2HE \times EY$ is equal to the product $CE \times CX$, and when CE^2 is added on each side there comes $CE^2 + 2HE \times EY = CE^2 + CE \times CX$. Therefore $CY^2 - EY^2$ is equal to $CE^2 + CE \times CX$, that is, $(CY+EY)(CY-EY)$ is equal to

est $CE^q + CE \times CX$. et resolutis æqualibus rectangulis in latera proportionalia fit $CE + CX$ ad $CY + EY$ ut $CY - EY$ ad CE. Sunt autem tres lineæ EY, CA, CB æquales et inde $CY + EY = CY + CA = AY$ et $CY - EY = CY - CB = BY$. Scribantur itaꝗ AY pro $CY + EY$ et BY pro $CY - EY$ et fiet $CE + CX$ ad AY ut BY ad CE. Sed (per Lem. 1) est CE ad KA ut $CE + CX$ ad AY. Ergo est CE ad KA ut BY ad $CE_{[,]}$ hoc est lineæ tres BY, CE, KA sunt continuè proportionales. Q.E.D.

Tandem ope horum Lemmatum constructio superioris Problematis sic demonstratur.

Per Lemma 1 est CE ad KA ut CX ad KY, adeoꝗ $KA \times CX = KY \times CE$ et applicatis his æqualibus extremorum et mediorum rectangulis ad CE fit $\dfrac{KA \times CX}{CE} = KY$. His lateribus æqualibus adde BK et æqualia erunt

$$BK + \frac{KA \times CX}{CE} \quad \& \quad BY.$$

‖[226] Unde per Lemma 3um ‖ est $BK + \dfrac{KA \times CX}{CE}$ ad CE ut CE ad KA, et inde, ductis extremis et medijs in se provenit CE^q æquale $BK \times KA + \dfrac{KA^q \times CX}{CE}$, et omnibus præterea ductis in CE fit CE^{cub} æquale $BK \times KA \times CE + KA^q \times CX$.[656] CE erat radix æquationis dicta x, KA erat n, KB $\dfrac{q}{n}$ et CX $\dfrac{r}{nn}$. His pro CE, KA, KB et CX substitutis oritur $x^3 = qx + r$ seu $x^3 - qx - r = 0$, æquatio construenda; ubi q et r negativæ prodeunt sumptis KA et KB ad easdem partes puncti K et radice affirmativa in majori segmento CGK existente. Hic unus casus est Constructionis demonstrandæ. Ducatur KB ad partes contrarias id est mutetur signum ejus seu signum ipsius $\dfrac{q}{n}$ vel quod perinde est signum termini q, et habebitur constructio æquationis $x^3 + qx - r = 0$: Qui casus est alter. In his casibus CX et radix affirmativa CE cadunt ad easdem partes lineæ AK. Cadant CX et radix negativa ad easdem mutato signo ipsius CX seu $\dfrac{r}{nn}$ vel (quod perinde est) signo ipsius r et habebitur casus tertius $x^3 + qx + r = 0$ ubi radices omnes sunt negativæ. Et mutato rursus signo ipsius KB seu $\dfrac{q}{n}$ vel solius q, incidetur in casum quartum $x^3 - qx + r = 0$. Quorum omnium casuum constructiones percurrere licebit et sigillatim demonstrare ad modum casus primi. Nos uno casu demonstrato cæteros leviter attingere satis esse putavimus. Hi verbis ijsdem mutato solùm linearum situ demonstrantur.

---

(656) The general cubic relationship determining the unknown length CE, which Newton proceeds to identify with $x^3 = \pm qx \pm r$.

$CE^2 + CE \times CX$; and, when the equal products are resolved into the 'sides' of a proportion, there comes to be $CE + CX$ to $CY + EY$ as $CY - EY$ to CE. However, the three lines EY, AC, CB are equal, and in consequence

$$CY + EY = CY + AC = AY \quad \text{and} \quad CY - EY = CY - CB = BY.$$

Accordingly, let AY be written in place of $CY + EY$ and BY in place of $CY - EY$, and there will prove to be $CE + CX$ to AY as BY to CE. But (by Lemma 1) $CE + CX$ to AY is as CE to AK. Therefore CE is to AK as BY to CE; in other words, the three lines BY, CE and AK are in continued proportion. As was to be proved.

With the help of these lemmas the construction of the above problem is at length demonstrated in this fashion.

By Lemma 1 CE is to AK as CX to KY, and hence $AK \times CX = KY \times CE$; and when these equal products of the extremes and middles are divided by CE there comes $AK \times CX/CE = KY$. To these equal sides add BK, and then

$$BK + AK \times CX/CE \quad \text{and} \quad BY$$

will be equal. Hence, by Lemma 3, $BK + AK \times CX/CE$ is to CE as CE to AK, and thence, when extremes and middles are multiplied together, there ensues CE^2 equal to $AK \times BK + AK^2 \times CX/CE$, and, further, on multiplying through by CE there comes to be CE^3 equal to $AK \times BK \times CE + AK^2 \times CX$.[656] Now CE was the equation's root called x, AK was n, BK q/n and CX r/n^2. When these values are substituted in place of CE, AK, BK and CX, there arises

$$x^3 = qx + r \quad \text{or} \quad x^3 - qx - r = 0,$$

the equation to be constructed: here negative q and r result when AK and BK are taken to be on the same side of point K, the positive root being located in the greater segment CGK. This is one case of the construction to be demonstrated. Let BK be drawn the opposite way, that is, let its sign—or the sign of q/n or, what is the equivalent, the sign of q—be changed, and there will be had the construction of the equation $x^3 + qx - r = 0$: this is a second case. In these cases CX and the positive root CE fall on the same side of the line AK. Let CX and the negative root fall on the same side and, with the sign of CX or r/n^2 (or, equivalently, the sign of r) changed, there will be had the third case $x^3 + qx + r = 0$ in which all the roots are negative. On changing, again, the sign of BK or q/n— or of q alone—you will arrive at the fourth case $x^3 - qx + r = 0$. You are at liberty to run through the constructions of all these cases, demonstrating them separately after the manner of the first case. We have considered it enough, having demonstrated one case, to touch lightly on the rest—which, indeed, are proved in the same words with only the position of the lines changed.

[1683]
Lect. 4.

Construenda jam sit æquatio cubica $x^3 + pxx * + r = 0$, cujus tertius terminus deest.

In figura superiore[657] assumpta longitudine quavis n, capias in recta quavis infinita AY KA et KB quarum KA

valeat $\frac{r}{nn}$ & KB valeat p. Has cape

ad easdem partes puncti K si modò

‖ [227] signa terminorum p et r ‖ sint eadem, secus ad contrarias. Biseca BA in C et centro $[A]$ intervallo $[A]C$ describe circulum CX.[658] In eo aptes rectam CX æqualem longitudini assumptæ n. Junge AX et produc junctam ad G ita ut fiat AG æqualis AK et per puncta K, C, X, G, describe circulum. Deniꝗ

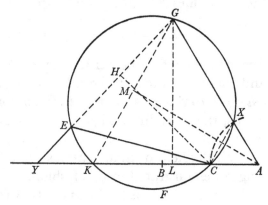

inter hunc circulum et rectam KC utrinꝗ productam inscribe rectam EY ejusdem longitudinis cum recta AC ea lege ut hæc inscripta recta transeat per punctum G, si modò ipsa producatur: et acta recta KY erit una ex radicibus æquationis. Sunt autem radices affirmativæ quæ cadunt ad partes puncti K versus punctum A si modò habeatur $+r$, sin habeatur $-r$, affirmativæ sunt quæ cadunt ad partes contrarias. Et si affirmativæ radices jacent ex una parte puncti A, negativæ sunt quæ jacent ex altera.[659]

Demonstratur autem hæc constructio ope Lemmatum trium novissimorum in hunc modum.

Per Lemma tertium sunt BY, CE, KA continuè proportionales, et per Lemma primum ut est CE ad KA ita est CX ad KY. Ergo BY est ad CE ut CX ad KY. BY idem est quod $KY - KB$. Ergo $KY - KB$ est ad CE ut CX ad KY. Sed ut est $KY - KB$ ad CE ita est $KY - KB$ in KY ad CE in KY, idꝗ per prop 1 Lib. 6 Elementorum, et ob proportionales CE ad KA ut CX ad KY est CE in KY æquale KA in CX. Ergo $KY - KB$ in KY est ad KA in CX (ut $KY - KB$ ad CE hoc est) ut CX ad KY. Et ductis extremis et medijs in se invicem fit $KY - KB$

‖ [228] in KY^q ‖ æquale KA in CX^q: id est $KY^{\text{cub}} - KB \times KY^{\text{quad.}}$ æquale $KA \times CX^{\text{quad.}}$.[660]
Erat autem in constructione, KY radix æquationis dicta x, KB æqualis p, KA

æqualis $\frac{r}{nn}$, et CX æqualis n. Scribantur igitur $x, p, \frac{r}{nn}$ et n pro KY, KB, KA et CX

respectivè et fiet $x^3 - pxx = r$ seu $x^3 - pxx - r = 0$.

---

(657) Here repeated for convenience of consultation.

(658) The nonsensical manuscript phrase 'centro K intervallo KC describe circulum CXG', duly passed by Whiston into his 1707 *editio princeps* and missed by Newton's usually sharp eye in his 1722 revise, was subsequently emended by Castiglione in his commented edition (**2**: 246–7) of the published *Arithmetica* in 1761—but the illogicality was retained by Horsley in his *Newtoni Opera Omnia*, **1** (London, 1779): 211.

Let it now be required to construct the cubic equation $x^3 + px^2 + r = 0$ lacking its third term.

Assuming in the above figure[657] any length for n, you should take AK and BK in any unbounded straight line AY such that the value of AK is r/n^2, and that of KB is p. Take these on the same side of the point K provided the signs of the terms p and r are the same, otherwise on opposite ones. Bisect BA in C and with centre A, radius AC describe the circle-arc CX.[658] In it you are to fit the straight line CX equal to the length assumed for n. Join AX and extend the join to G so as to make AG equal to AK, and through the points K, C, X, G describe a circle. Finally, between this circle and the straight line CK extended each way inscribe the straight line EY of the same length as the line AC, with the restriction that this inscribed line shall, if only it be produced, pass through the point G: then, after the straight line KY is drawn, it will be one of the roots of the equation. Positive roots, however, are those which fall on the side of point K towards A provided $+r$ be had; but if $-r$ be had, positives are those falling on the opposite side. And if positive roots lie on one side of the point A, the negatives are those lying on the other.[659]

This construction is, however, demonstrated with the aid of the three most recent lemmas in this manner.

By the third lemma BY, CE, AK are in continued proportion, while by the first CE is to AK as CX to KY. Therefore BY is to CE as CX to KY. Now BY is the same as $KY - KB$, and therefore $KY - KB$ is to CE as CX to KY. But $KY - KB$ is to CE as $(KY - KB) \times KY$ is to $CE \times KY$—this by *Elements*, Book VI, Proposition 1—and, because of the proportionals CE to AK as CX to KY, $CE \times KY$ is equal to $AK \times CX$. Therefore $(KY - KB) KY$ is to $AK \times CX$ as $(KY - KB$ to CE, that is) CX to KY. And on then multiplying extremes and middles one into the other there comes to be $(KY - KB) KY^2$ equal to $AK \times CX^2$; that is,

$$KY^3 - KB \times KY^2 \quad \text{equal to} \quad AK \times CX^2.\text{[660]}$$

In the construction, however, KY was the equation's root called x, KB equal to p, AK equal to r/n^2, and CX equal to n. Let, then, x, p, r/n^2 and n be written in place of KY, KB, AK and CX respectively, and there will come to be $x^3 - px^2 = r$ or $x^3 - px - r = 0$.

---

(659) On interchanging x and y in note (654), the segment $KY = x$ is, in general, determined by the cubic $x^3 + (a - b) x^2 + (b^2 - c^2) x - c^2 d^2/(a + b) = 0$. When identification with $x^3 + px^2 + r = 0$ is made, again $b = c$ (or $AC = AX$) but now $-KB = a - b = p$ and, where $CX = cd/(a + b) = n$ is taken without restriction, also $-KA = a + b = r/n^2$. Newton's further observations follow in an immediate way. As before, the present construction and its following synthetic demonstration (wholly cancelled in Newton's library copy of the *editio princeps*; see note (615)) repeat an equivalent passage—the *Idem aliter* to Problem 3 (II: 476)—in his earlier 'Problems for construing æquations'.

(660) Once more (compare note (650)) the derivation of this variant cubic follows immediately from the previous one by the substitution $CE = KA \times CX/KY$ and reordering.

Resolvi potest constructio demonst[r]anda in hosce quatuor æquationum casus $x^3-pxx-r=0$, $x^3-pxx+r=0$, $x^3+pxx-r=0$ et $x^3+pxx+r=0$. Casum primum jam demonstratum dedi, cæteri tres ijsdem verbis mutato tantum linearū situ demonstrantur. Nimirum uti sumendo KA et KB ad easdem partes puncti K et radicem affirmativam KY ad contrarias partes jam prodijt

$$KY^{\text{cub}}-KB\times KY^q=KA\times CX^q \quad \text{et inde} \quad x^3-pxx-[r]=0:^{(661)}$$

sic sumendo KB ad contrarias partes puncti K prodibit simili argumentationis progressu $KY^{\text{cub}}+KB\times KY^q=KA\times CX^q$ et inde $x^3+pxx-r=0$. Et in hisce duobus casibus si mutetur situs radicis affirmativæ KY sumendo eam ad alteram partem puncti K, per similem argumentationis seriem devenietur ad alteros duos casus $KY^{\text{cub.}}+KB\times KY^q=-KA\times CX^q$ seu $x^3+pxx+r=0$ et $KY^{\text{cub}}-KB\times KY^q=-KA\times CX^q$ seu $x^3-pxx+r=0$. Qui omnes casus erant demonstrandi.

[1683]
Lect 5
Proponatur jam æquatio cubica $x^3+px^2+qx+r=0$, nullo (nisi fortè tertio) termino carens. Ea construetur ad hunc modum.

Cape ad arbitrium longitudinem n. Ejus dimidio æqualem duc rectam

‖[229] quamvis GC, et ad punctum G erige perpendiculum GD æquale ‖ $\sqrt{\dfrac{r}{p}}$. Deinde

si termini p et r habent contraria signa, centro C intervallo CD describe circulum PBE. Sin eadem sunt eorum signa, centro D intervallo GC describe circulum occul-

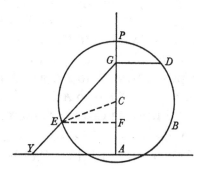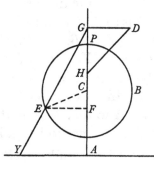

tum$^{(662)}$ secantem rectam GA in $H_{[,]}$ dein centro C intervallo GH describe circulum PBE. Tum fac $GA=-\dfrac{q}{n}-\dfrac{r}{np}$, eamc̦ duc in linea GC ad partes puncti

G versus C si modò quantitas $-\dfrac{q}{n}-\dfrac{r}{np}$ (signis terminorum p, q, r in æquatione construenda probè observatis) affirmativa obvenerit: secus age GA ad alteras partes puncti G, et ad punctum A erecto perpendiculo AY, inter hoc et circulum PBE superius descriptum inscribe lineam EY æqualem termino p ea lege ut hæc inscripta convergat ad punctum G. Quo facto et producta illa EY ad G, erit linea EG una ex radicibus æquationis construendæ. Quæ quidem radices affirmativæ sunt ubi punctum E cadit inter puncta G et Y, & negativæ ubi E cadit extra, si modò habeatur $+p$; et contra si $-p$.$^{(663)}$

The construction to be demonstrated can be resolved into these four cases of equations: $x^3 - px^2 - r = 0$, $x^3 - px^2 + r = 0$, $x^3 + px^2 - r = 0$ and $x^3 + px^2 + r = 0$. Of the first case I have just given the proof, and the other three are demonstrated in the same words with but the position of the lines changed. To be exact, in the same way that, on taking AK and BK on the same side of point K and positive root KY to be on the opposite one, there just now resulted

$$KY^3 - KB \times KY^2 = AK \times CX^2$$

and thereby $x^3 - px^2 - r = 0$;[661] so, on taking BK to be on the opposite side of point K, there will result by a similar sequence of argument that

$$KY^3 + KB \times KY^2 = AK \times CX^2 \quad \text{and thereby} \quad x^3 + px^2 + r = 0.$$

And if in these two cases the location of the positive root KY be changed by taking it on the other side of the point K, by a similar progression of argument you will arrive at the other two cases $KY^3 + KB \times KY^2 = -AK \times CX^2$ or $x^3 + px^2 + r = 0$ and $KY^3 - KB \times KY^2 = -AK \times CX^2$ or $x^3 - px^2 + r = 0$. These were all the cases needing to be proved.

Let there now be proposed the cubic equation $x^3 + px^2 + qx + r = 0$ *lacking no term* (*unless maybe its third one*). It will be constructed in this manner.

Take the length n at will. Draw any straight line GC equal to its half, and at the point G erect the perpendicular GD equal to $\sqrt{[r/p]}$. Next, if the coefficients p and r have contrary signs, with centre C and radius CD describe the circle PBE; but if their signs are the same, with centre D and radius GC describe a concealed circle[662] cutting the straight line GA in H, and then with centre C and radius GH describe the circle PBE. Subsequently, make $GA = -q/n - r/np$, drawing it in the line GC in C's direction provided the quantity $-q/n - r/np$ (with due regard paid to the signs of the coefficients p, q, r in the equation to be constructed) comes out to be positive; otherwise, draw GA on the further side of point G and, having erected the perpendicular AY at the point A, between this and the circle PBE earlier drawn inscribe a line EY equal to the coefficient p, with the stipulation that this inscribed line shall be directed through the point G. When this is done and EY produced to G, the line GE will be one of the roots of the equation to be constructed. These roots are, to be sure, positive when the point E falls between the points G and Y, and negative when E falls outside, provided $+p$ be had; and conversely so if it be $-p$.[663]

---

(661) The manuscript reads '$x^3 - pxx - q = 0$' by a trivial slip.

(662) 'Hidden' from the reader's view because it is not shown in Newton's figures.

(663) Newton has given the analysis of this construction in outline in his earlier 'Problems for construing æquations' (II: 512): he now adds a number of details and a newly contrived, somewhat ponderous geometrical synthesis (once again wholly cancelled in Newton's library copy of the 1707 *editio princeps*; see note (615)). To repeat, since, on setting $GC = a$, $GA = b$,

Demonstrationi hujus constructionis præmittimus Lemmata sequentia.

Lem 1. Demisso ad AG perpendiculo EF et acta recta EC: est

$$EG^q + GC^q = EC^q + 2CGF.$$

Nam per prop. 12 lib. 2 Elem. est $EG^q = EC^q + GC^q + 2GCF$. Addatur utrinꝗ GC^q et fiet ‖ $EG^q + GC^q = EC^q + 2GC^q + 2GCF$. Sed $2GC^q + 2G[C]F$ est $2GC$ in $GC + CF$ id est $2CGF$. Ergo $EG^q + GC^q = EC^q + 2CGF$. Q.E.D.

‖ [230]

Lem 2. In Constructionis casu primo ubi circulus PBE transit per punctum D est $EG^q - GD^q = 2CGF$. Nam per Lemma primum est $EG^q + GC^q = EC^q + 2CGF$ et ablato utrinꝗ GC^q fit $EG^q = EC^q - GC^q + 2CGF$. Sed $EC^q - GC^q$ idem est quod $CD^q - GC^q$ hoc est idem quod GD^q. Ergo $EG^q = GD^q + 2CGF$ et subducto utrobiꝗ GD^q fit $EG^q - GD^q = 2CGF$. Q.E.D.

Lem 3. In constructionis casu secundo ubi circulus P[BE] non transit per punctum D est $EG^q + GD^q = 2CGF$. Namꝗ in Lemmate primo erat $EG^q + GC^q = EC^q + 2CGF$. Aufer utrinꝗ EC^q et fiet $EG^q + GC^q - EC^q = 2CGF$. Sed $GC = DH$ et $EC = CP = GH$. Ergo $GC^q - EC^q = DH^q - GH^q = GD^q$, atꝗ adeo

$$EG^q + GD^q = 2CGF. \quad \text{Q.E.D.}$$

Lem 4. Est 2CGF in GY = 2CG in AGE. Namꝗ ob similia triangula GEF, GYA est GF ad GE ut AG ad GY hoc est (per prop 1 lib 6 Elementorū) ut $2CG \times AG$ ad $2CG \times GY$. Ducantur extrema et media in se et fiet

$$2CG \times GY \times GF = 2CG \times AG \times GE. \quad \text{Q.E.D.}$$

Tandem ope horum Lemmatum Constructio Problematis sic demonstratur.

In casu primo est (per Lem 2) $EG^q - GD^q = 2CGF$ et ductis omnibus in GY fit $EG^q \times GY - GD^q \times GY = 2CGF \times GY$ (hoc est per Lemma 4) $= 2CG \times AGE$. Pro GY scribe $EG + EY$ et fiet

$$EG^{\text{cub.}} + EY \times EG^q - GD^q \times EG - GD^q \times EY = 2CGA \times EG$$

seu $EG^{\text{cub.}} + EY \times EG^q \genfrac{}{}{0pt}{}{-GD^q}{-2CGA} \times EG - GD^q \times EY = 0.$

$CP(= CE = CD$ or $HG) = c$, $EY = d$ and $EG = x$, the triangles CGE, AGY share a common angle at G, immediately $(2\cos\hat{G} =) (x^2 + a^2 - c^2)/ax = 2b/(x+d)$ and therefore

$$x^3 + dx^2 + (a^2 - 2ab - c^2)x + (a^2 - c^2)d = 0:$$

whence, by identifying this with $x^3 + px^2 + qx + r = 0$, there ensues $EY = (d =) p$ and also $a^2 - 2ab - c^2 = q$, $(a^2 - c^2)d = r$, so that $a^2 - c^2 = r/p$ and $2ab = r/p - q$. Taking $GC = a = \frac{1}{2}n$ arbitrarily, Newton makes $GA = (b =) r/pn - q/n$ and then in the two cases where $a^2 > r/p$ and $a^2 < r/p$ constructs the circle (B) through P, where $CP = (c =) \sqrt{[a^2 - r/p]}$.

For the demonstration of this construction we premise the following lemmas.

Lemma 1. On letting fall the perpendicular EF to AG and drawing the straight line EC, there is $GE^2 + GC^2 = EC^2 + 2GC \times GF$. For by *Elements*, Book II, Proposition 12, $GE^2 = EC^2 + GC^2 + 2GC \times CF$. Let GC^2 be added to each side and there will come to be $GE^2 + GC^2 = EC^2 + 2GC^2 + 2GC \times CF$. But $2GC^2 + 2GC \times CF$ is $2GC(GC + CF)$, that is, $2GC \times GF$. Therefore $GE^2 + GC^2 = EC^2 + 2GC \times GF$. As was to be proved.

Lemma 2. In the construction's first case, where the circle PBE passes through the point D, there is $GE^2 - GD^2 = 2GC \times GF$. For by the first lemma

$$GE^2 + GC^2 = EC^2 + 2GC \times GF,$$

and when GC^2 is taken away from each side there comes

$$GE^2 = EC^2 - GC^2 + 2GC \times GF.$$

But $EC^2 - GC^2$ is the same as $CD^2 - GC^2$, that is, as GD^2. Therefore

$$GE^2 = GD^2 + 2GC \times GF$$

and, after GD^2 is subtracted on either side, there comes $GE^2 - GD^2 = 2GC \times GF$. As was to be proved

Lemma 3. In the construction's second case, where the circle PBE does not pass through the point D, $GE^2 + GD^2 = 2GC \times GF$. For in the first lemma there was

$$GE^2 + GC^2 = EC^2 + 2GC \times GF.$$

Take away EC^2 on each side and there will be $GE^2 + GC^2 - EC^2 = 2GC \times GF$.

But $GC = DH$ and $EC = CP = GH$. Therefore $GC^2 - EC^2 = DH^2 - GH^2 = GD^2$, and hence $GE^2 + GD^2 = 2GC \times GF$. As was to be proved.

Lemma 4. There is $2GC \times GF \times GY = 2CG \times GA \times GE$. For because of the similar triangles GEF, GYA there is GF to GE as GA to GY, that is, (by *Elements*, Book VI, Proposition 1) as $2GC \times AG$ to $2GC \times GY$. Let extremes and middles be multiplied together and there will come to be $2GC \times GY \times GF = 2GC \times GA \times GE$. As was to be proved.

With the aid of these lemmas the construction of the problem is at length demonstrated in this manner.

In the first case there is (by Lemma 2) $GE^2 - GD^2 = 2GC \times GF$ and on multiplying throught by GY there comes

$$GE^2 \times GY - GD^2 \times GY = (2GC \times GF \times GY, \text{ that is, by Lemma 4) } 2GC \times GA \times GE.$$

In place of GY write $GE + EY$ and there will come to be

$$GE^3 + EY \times GE^2 - GD^2 \times GE - GD^2 \times EY = 2GC \times GA \times GE$$

or $GE^3 + EY \times GE^2 + (-GD^2 - 2GC \times GA) GE - GD^2 \times EY = 0$.

In casu secundo est (per Lem 3) $EG^q + GD^q = 2CGF$ et ductis omnibus in GY fit $EG^q \times GY + GD^q \times GY = 2CGF \times GY$ (hoc est per Lemma 4) $= 2CG \times AGE$. Pro GY [scribe] $EG + EY$ et fiet

$$EG^{\text{cub.}} + EY \times EG^q + GD^q \times EG + GD^q \times EY = 2CGA \times EG$$

seu $EG^{\text{cub.}} + EY \times EG^q \begin{array}{c} +GD^q \\ -2CGA \end{array} \times EG + GD^q \times EY = 0.$

‖[231]   ‖Jam verò erat EG radix æquationis constructæ dicta x: item $GD = \sqrt{\dfrac{r}{p}}$,

$EY = p$, $2CG = n$ et $GA = -\dfrac{q}{n} - \dfrac{r}{np}$, id est in casu primo ubi terminorum p et r diversa sunt signa: at in casu secundo ubi alterutrius p vel r mutatur signum fiet $-\dfrac{q}{n} + \dfrac{r}{np} = GA$. Scribantur igitur pro EG, GD, EY, $2CG$ et GA quantitates

x, $\sqrt{\dfrac{r}{p}}$, p, n et $-\dfrac{q}{n} \mp \dfrac{r}{np}$, et casu primo fiet $x^3 + px^2 \begin{array}{c} +\dfrac{r}{p} \\ +q - \dfrac{r}{p} \end{array} x - r = 0$ id est

$x^3 + pxx + qx - r = 0$, casu autem secundo $x^3 + px^2 \begin{array}{c} +\dfrac{r}{p} \\ +q - \dfrac{r}{p} \end{array} x + r = 0$ id est

$x^3 + pxx + qx + r = 0$. Est igitur in utroȝ casu EG vera longitudo radicis x. Q.E.D.

Subdistinguitur autem casus uterȝ in casus plures particulares:[664] Nimirum prior in hosce $x^3 + px^2 + qx - r = 0$, $x^3 + pxx - qx - r = 0$, $x^3 - pxx + qx + r = 0$, $x^3 - pxx - qx + r = 0$, $x^3 + pxx - r = 0$ et $x^3 - pxx + r = 0$; posterior in hosce

$$x^3 + pxx + qx + r = 0, \quad x^3 + pxx - qx + r = 0, \quad x^3 - pxx + qx - r = 0,$$

$x^3 - pxx - qx - r = 0$, $x^3 + pxx + r = 0$ et $x^3 - pxx - r = 0$. Quorum omnium demonstrationes verbis ijsdem ac duorum jam demonstratorū, mutato tantum linearum situ, compinguntur.

[1683]   Hæ sunt Problematum constructiones præcipuæ per inscriptionem rectæ
Lect. 6.   longitudine datæ inter circulum et rectam lineam positione datam ea lege ut inscripta ad datum punctum convergat. Inscribitur autem talis recta ducendo Conchoidem veterum, cujus Polus sit punctum illud ad quod recta inscribenda debet convergere, Regula seu Asymptotos recta altera positione data et intervallum longitudo rectæ inscribendæ. Secabit enim hæc Conchoides circulum præfatum in puncto E per quod recta inscribenda duci debet.[665] Suffecerit

---

(664) The two groups following are differentiated by having p and r of opposite or the same signs respectively.

In the second case there is (by Lemma 3) $GE^2 + GD^2 = 2GC \times GF$ and on multiplying throughout by GY there comes to be

$GE^2 \times GY + GD^2 \times GY = (2GC \times GF \times GY,$ that is, by Lemma 4) $2GC \times GA \times GE.$

In place of GY write $GE + EY$ and there will come to be

$$GE^3 + EY \times GE^2 + GD^2 \times GE + GD^2 \times EY = 2GC \times GA \times GE$$

or $GE^3 + EY \times GE^2 + (GD^2 - 2GC \times GA)\,GE - GD^2 \times EY = 0.$

Now indeed GE was the constructed equation's root called x; likewise

$$GD = \sqrt{[r/p]}, \quad EY = p, \quad 2GC = n \quad \text{and} \quad GA = -q/n - r/np:$$

in the first case, that is, when the coefficients p and r have opposite signs, but in the second case, where the sign of one or other of p or r changes, there will come to be $GA = -q/n + r/np$. Let, then, the quantities x, $\sqrt{[r/p]}$, p, n and $-q/n \mp r/np$ be written in place of GE, GD, EY, $2GC$ and GA, and in the first case there will come $x^3 + px^2 + (r/p + q - r/p)\,x - r = 0$, that is, $x^3 + px^2 + qx - r = 0$; in the second case, however, $x^3 + px^2 + (r/p + q - r/p)\,x + r = 0$, that is,

$$x^3 + px^2 + qx + r = 0.$$

In either case, therefore, GE is the true magnitude of the root x. As was to be proved.

Each case is further distinguished, however, into several particular cases:[664] to be precise, the first one into these, $x^3 + px^2 + qx - r = 0$, $x^3 + px^2 - qx - r = 0$, $x^3 - px^2 + qx + r = 0$, $x^3 - px^2 - qx + r = 0$, $x^3 + px^2 - r = 0$ and $x^3 - px^2 + r = 0$; and the latter one into these, $x^3 + px^2 + qx + r = 0$, $x^3 + px^2 - qx + r = 0$,

$$x^3 - px^2 + qx - r = 0, \quad x^3 - px^2 - qx - r = 0, \quad x^3 + px^2 + r = 0$$

and $x^3 - px^2 - r = 0$. The proofs of all these may be framed in the same words as those of the two cases now demonstrated, with merely the position of the lines changed.

These are the principal constructions of problems by the inscription of a straight line given in length between a circle and a straight line given in position, with the stipulation that the inscribed line shall be directed through a given point. Such a straight line is inscribed, however, by drawing a classical conchoid whose pole shall be the point through which the line to be inscribed ought to be directed, its 'rule' or asymptote the other straight line given in position, and its interval the length of the line to be inscribed. This conchoid will, of course, cut the previously mentioned circle in the point(s) E through which the straight line to be inscribed ought to be drawn.[665] But it will be

---

(665) Compare notes (636) and (652).

verò in rebus practicis rectam illam inter circulum et alteram positione datam rectam ratione quacunꝗ mechanica[(666)] interponere.

‖[232]   ‖In hisce autem constructionibus notandum est quod quantitas *n* ubiꝗ indeterminata est et ad arbitrium assumenda relinquitur, id adeò ut singulis problematis constructiones commodius aptentur.[(667)] Hujus rei exempla in inventione duarum mediè proportionalium et anguli trisectione dabimus.

    Inveniendæ sint inter a et b duæ mediè proportionales x et y. Quoniam sunt *a. x. y. b* continuè proportionales erit a^2 ad x^2 ut *x* ad *b* adeoꝗ $x^3 = aab$ seu $x^3 - a^2b = 0$. Hic desunt æquationis termini *p* et *q* et loco termini *r* habetur $-a^2b$. Igitur in constructionum formula prima, ubi recta *EY* ad datum punctum *K* convergens inseritur inter alias duas positione datas rectas *EX* et *YC* et recta *CX* ponitur æqualis $\dfrac{r}{nn}$ id est æqualis $-\dfrac{aab}{nn}$, assumo *n* æqualem *a* et sic fit *CX* æqualis $-b$.

Unde talis emergit constructio.[(668)]

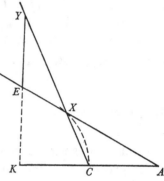

    Duco quamvis *KA* æqualem *a*, eamꝗ biseco in *C* centroꝗ *K* intervallo *KC* describo circulum *CX* ad quem apto rectam *CX* æqualem *b* et inter rectas *AX CX* infinitè productas pono *EY* æqualem *CA* et convergentem ad punctum *K*. Sic erunt *KA*, *XY*, *KE*, *CX* continuè proportionales id est *XY* et *KE* duæ mediè proportionales inter *a* et *b*. Constructio nota est.[(669)]

---

(666) Not least conveniently so, of course, by 'eye' alone.

(667) Retaining this sentence alone, Newton in his library copy of Whiston's *editio princeps* has cancelled the preceding paragraph and the whole of the remainder (pages 232–51) of his manuscript text (see note (615)).

(668) This is considerably recast in detail—but without essential structural alteration—from the *Idem aliter* of the introductory Problem 15 (see II: 460) of Newton's earlier 'Problems for construing æquations'. For its particular analysis see II: 461, note (34).

(669) It is, of course, Nicomedes'; see T. L. Heath's *Greek Mathematics*, **1**, 1921: 260–2. As presented by Eutocius (in his comments upon the second book of Archimedes' *Sphere and Cylinder* [ed. Heiberg, **3**: 106]) and by Pappus (*Mathematical Collection* III, 5 ≈ IV, 24). Nicomedes' proof is long and circuitous, in effect constructing $C\alpha = a$ at right angles to $CX = b$, bisecting $C\alpha$, CX in ζ and ϵ, and then, with the parallelogram $\beta\alpha CX$ completed, drawing the cross-line *XE* parallel to δK (where δ is the meet of *CX* and $\zeta\beta$). On making $KC = KX = \frac{1}{2}C\alpha$ and then inserting the equal line-segment *EY* in the angle $\widehat{YXE}$, the construction is shown to yield the required proportion $C\alpha : XY = XY : \alpha\gamma = \alpha\gamma : CX$ (where γ is the meet of $Y\beta$ with $C\alpha$) by proving that,

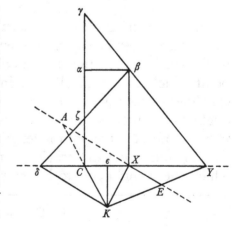

enough in practice to set that line between the circle and the other straight line given in position by any 'mechanical' method.[666]

In these constructions you should, however, note that the quantity n is everywhere undetermined and left free to be assumed at will: this so that the constructions may the more fitly be adapted to individual problems.[667] We shall give instances of this point in the finding of two mean proportionals and the trisection of an angle.

Let there be required to find two mean proportionals x and y between a and b. Seeing that a, x, y, b are in continued proportion, there will be a^2 to x^2 as x to b and hence $x^3 = a^2b$ or $x^3 - a^2b = 0$. Here the terms $p[x^2]$ and $q[x]$ are lacking, while in place of the term r there is had $-a^2b$. Therefore in the first general form of construction, where the straight line EY directed through the given point K is inserted between two other straight lines XE and CY given in position, and the straight line CX is set equal to r/n^2, that is, $-a^2b/n^2$, I assume n equal to a and thus CX comes to be equal to $-b$. The construction[668] which hence emerges is this.

I draw any KA equal to a and bisect it in C, then with centre K, radius KC I describe the circle-arc CX and in it fit the straight line CX equal to b, and between the indefinitely extended straight lines AX and CX I place EY equal to CA and directed through the point K. In this way KA, XY, KE, CX will be in continued proportion, that is, XY and KE are the two mean proportionals between a and b. The construction is well known.[669]

---

since $\alpha\zeta = EY$ and also $\gamma\alpha/\alpha\zeta = (2\gamma\alpha/\alpha C$ or$)$ $2CX[= \delta X]/XY = KE/EY$, or $\gamma\zeta = KY$, therefore $\gamma\alpha \times \gamma C = \gamma\zeta^2 - \alpha\zeta^2 = KY^2 - KX^2 = CY \times XY$ and in consequence

$$XY : \alpha\gamma = C\gamma : CY = (\beta X \text{ or}) C\alpha : XY = \alpha\gamma : (\alpha\beta \text{ or}) CX.$$

The present considerable simplification in which the line EX is constructed through point A (in KC such that $KC = CA$) was, in effect, introduced by Johann Molther in his *Problema Deliacum, de Cubi Duplicatione...expedite & Geometrice solutum: Ubi...simul nonnulla de Anguli trisectione, Heptagoni fabrica, circulique Quadratura...inseruntur* (Frankfurt, 1619): Caput II. 'Mesolabii secundi expositio': 51–8 [summarized by Marin Mersenne in his *Harmonie Universelle, contenant la Theorie et la Pratique de la Musique*, **1** (Paris, 1636): Livre II, Proposition VII: 65–70], but it is unlikely that Newton was aware of the researches of his German predecessor. Lacking Nicomedes' own derivation of the construction (in his lost work on conchoids) we cannot now say with certainty whether his insight into the possibilities of conchoidal constructions in solving problems was more profound than is revealed by the cumbrous, far from obvious proof preserved by Pappus. The reader will find it interesting to contrast Newton's analysis—which here, in effect, identifies $x^3 - n^2m = 0$ with the canonical cubic

$$x^3 + (b^2 - d^2)\,x - c(b+d)^2 = 0$$

by setting $KB = b - d = 0$ (that is, B coincident with K), $KA = b + d = n$ and $CX = c = m$ (see note (639))—with that given recently (in ignorance of Newton's method) in A. Seidenberg's 'Remarks on Nicomedes' Duplication' (*Archive for History of Exact Sciences*, **3**, 1966: 97–101).

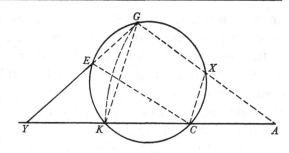

In altera autem constructionum formula ubi recta EY ad datum punctum G convergens ponitur inter circulum $GECX$ et rectam AK, estᴄ̧ $CX = \dfrac{r}{nn}$ id est (in hoc Proble-mate) $= \dfrac{-aab}{nn}$[,] pono ut prius $n = a$

et sic fit $CX = -b$, cæteraᴄ̧ peraguntur ut sequitur.[(670)]

‖ [233]   ‖ Duco rectam quamvis KA æqualem a, eamᴄ̧ biseco in C & centro A intervallo AK describo circulum KG ad quem apto rectam KG æqualem $2b$ et constituendo triangulum æquicrurum AKG. Dein per puncta C, K, G circulum describo et inter hujus perimetrum et rectam productam AK inscribo rectam EY æqualem KC et convergentem ad punctum G. Quo facto continuè proportionales erunt $AK, EC, KY, \frac{1}{2}KG$, id est EC et KY duæ mediè proportionales erunt inter datas a et b.

Secandus jam sit angulus in partes tres æquales. Sitᴄ̧ angulus secandus ACB, partes ejus inveniendæ ACD, DCE, ECB. Centro C intervallo CA describatur circulus $ADEB$ secans rectas $CA, CD, CE,$ CB in A, D, E, B. Jungantur AD, DE, EB ut et AB secans rectas CD, CE in F et H et ipsi CE parallela agatur DG occurrens AB in G. Ob similia[(671)] triangula $CAD, ADF,$ DFG, continuè proportionales sunt CA, AD, DF, FG. Ergo si dicatur $AC = a$, et $AD = x$, fiet $DF = \dfrac{xx}{a}$ et $FG = \dfrac{x^3}{aa}$. Est

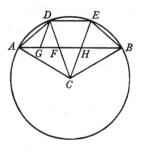

autem $AB = BH + HG + FA - GF = 3AD - GF = 3x - \dfrac{x^3}{aa}$.

Dic $AB = b$ et fiet $b = 3x - \dfrac{x^3}{aa}$ seu $x^3 - 3aax + aab = 0$.[(672)] Hic deest æquationis terminus secundus p et loco q et r habentur $-3aa$ et aab. Ergo in constructionum

‖ [234]   formula prima ubi erat $p = 0$, $KA = n$, $KB = \dfrac{q}{n}$ et $CX = \dfrac{r}{nn}$ id est in proble‖mate

---

(670) This effectively repeats the second *Idem aliter* to Newton's earlier 'Prob 15' on II: 462. A direct demonstration of the construction is readily obtained, on setting

$$AC = CK = AX = XG = EY = \tfrac{1}{2}a,$$

$CX (= \frac{1}{2}GK) = b$ and $KY = y$, by determining $YG = KY \times CY/EY = y(y + \frac{1}{2}a)/\frac{1}{2}a$ and then equating $(y^2 + 4b^2 - y^2(y + \frac{1}{2}a)^2/\frac{1}{4}a^2)/4by = (-\cos \widehat{GKC}$ or$) -\frac{1}{2}b/\frac{1}{2}a$, whence

$$(y + a)(y^3 - ab^2) = 0.$$

Since $y = -a$ corresponds to EY coincident with XA, the approach constructs the required equation $y^3 = ab^2$: in consequence, on putting $CE = x = (EY \times CK/KY$ or$) \, ab/y$, at once

In the other form of construction, however, where the straight line EY directed through the point G is set between the circle $GECX$ and the straight line AK, while $CX = r/n^2$, that is, (in the present problem) $-a^2b/n^2$, I put $n = a$ as before and thus there comes to be $CX = -b$, and then the rest will be accomplished as follows.[670]

I draw any straight line KA equal to a and bisect it in C, then with centre A, radius AK I describe the circle-arc KG, and into it fit the straight line KG equal to $2b$, so forming the isosceles triangle AKG. Then through the points C, K, G I describe a circle, and between the circumference of this and the extension of the straight line AK I inscribe the straight line EY equal to KC and directed through the point G. When this is done, AK, EC, KY and $\frac{1}{2}KG$ will be in continued proportion, that is, EC and KY will be the two mean proportionals between the given quantities a and b.

Let it now be required to cut an angle into three equal parts. Let the angle to be cut be $\widehat{ACB}$, and $\widehat{ACD}$, $\widehat{DCE}$, $\widehat{ECB}$ its sections to be found. With centre C, radius CA describe the circle $ADEB$ cutting the straight lines CA, CD, CE, CB in A, D, E, B. Join AD, DE, EB and also AB cutting the straight lines CD, CE in F and H, and parallel to CE draw DG meeting AB in G. Because of the similar[671] triangles CAD, ADF, DFG the lines CA, AD, DF, FG are in continued proportion. Therefore if AC be called $= a$ and $AD = x$, there will come to be $DF = x^2/a$ and $FG = x^3/a^2$. However, $AB = (AF + GH + HB - GF$ or$) \; 3AD - GF = 3x - x^3/a^2$. Call $AB = b$ and there will come to be $b = 3x - x^3/a^2$ or $x^3 - 3a^2x + a^2b = 0$.[672] Here the second term $p[x^2]$ in the equation is lacking and in place of q and r there are had $-3a^2$ and a^2b. Therefore in the first general form of construction, where $p = 0$, $KA = n$, $KB = q/n$ and $CX = r/n^2$—that is, in the problem now to

---

$a:x = x:y = y:b$. François Viète had used essentially this method in his Propositio VII of *Supplementum Geometriæ* (Tours, 1593) [= *Opera Mathematica* (Leyden, 1646): 244–5] to construct the particular case of the classical cube-duplication $(a = 2b)$: here the triangles ACX, AKG are equilateral, with GC perpendicular to AK and it is readily shown that

$$(GK =) \; 2CX:DY = DY:KY = KY:CX,$$

where D is the meet with GY of DK perpendicular to AK. (Viète does not make the final reduction to Newton's form by drawing CE and then demonstrating that it is equal to DY.)

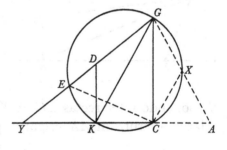

(671) And, indeed, 'isoscelia' (isosceles) since $CA = CD = CE$ and hence

$$\widehat{CDE} \text{ (or } \widehat{GFD}) = \widehat{CED} \text{ (or } \widehat{FGD}).$$

(672) This elegant derivation of the basic trisection cubic is taken over from Descartes (*Geometrie*, $_1$1637: 396–7 = *Geometria*, $_2$1659: 91–2), perhaps by way of pages 92–3 of Kinckhuysen's *Algebra ofte Stelkonst* (see ii: 351–2).

jam construendo $KB = -\dfrac{3aa}{n}$ et $CX = \dfrac{aab}{nn}$, ut hæ quantitates evadant quàm simplicissimæ pono $n = a$, et sic fit $KB = -3a$ et $CX = b$. Unde talis emergit Problematis constructio.[673]

Ago quamvis $KA = a$ et ad contrarias partes $KB = 3a$. Biseco BA in C centroq K intervallo KC describo circulum cui inscribo rectam $CX = b$. et acta recta AX inter ipsam infinitè productam et rectam CX pono rectam EY æqualem AC et convergentem ad punctum K. Sic fit $XY = x$. Quinetiam ob æquales circulos $ADEB$, CXA et æquales subtensas AB, CX nec non æquales subtensarum partes BH, XY, æquales erunt anguli ACB, CKX ut et anguli BCH, XKY atqɜ adeo anguli CKX tertia pars erit angulus XKY. Dati igitur cujusvis anguli CKX pars tertia

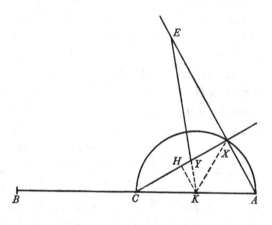

XKY invenietur ponendo inter chordas CX, AX infinitè productas rectam EY æqualem diametro AC et convergentem ad circuli centrum K.[674]

Hinc si a circuli centro K ad subtensam CX demittas perpendiculum KH erit angulus HKY tertia pars anguli HKX, adeo ut si detur quilibet angulus HKX inveniri possit ejus pars tertia HKY demittendo a quolibet lateris utriusvis KX puncto X ad latus alterum KH perpendiculum XH et lateri KH ducendo parallelam XE, dein rectam YE duplam ipsius KX et convergentem ad punctum

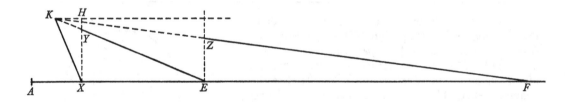

K ponendo inter rectas XH et XE. Vel sic. Detur angulus quilibet AXK. Ad

‖[235] ‖latus alterutrum AX erigatur perpendiculum XH et a lateris alterius XK

---

(673) As Newton adds in a late note at the end of his next paragraph but one, the construction is found classically in Pappus, *Mathematical Collection* IV, 32, though it is clear that he himself has come to it by way of his undergraduate reading of Viète's *Supplementum Geometriæ* (see 1: 73—4), whose discussion of the problem of trisection devolves from the equivalent construction expounded in Archimedes' *Lemmas*; see note (626) above and note (675) following.

be constructed $KB = -3a^2/n$ and $CX = a^2b/n^2$—, to make these quantities turn out as simple as possible I set $n = a$ and thus there comes to be $KB = -3a$ and $CX = b$. From this there emerges the following construction of the problem.[673]

I draw any $KA = a$ and, in the opposite direction, $KB = 3a$. Then I bisect BA in C and with centre K, radius KC I describe a circle in which I inscribe the straight line $CX = b$. And, having drawn the straight line AX, between its indefinite extension and the straight line CX I set the straight line EY equal to AC and directed through the point K. In this way there comes into being $XY = x$. In fact, because the circles $ADEB$, CXA, the subtenses AC, CX and also the sections $\overset{\frown}{BH}$, $\overset{\frown}{XY}$ are each equal, the angles $\widehat{ACB}$, $\widehat{CKX}$ and also the angles $\widehat{BCH}$, $\widehat{XKY}$ will be equal, and hence the angle $\widehat{XKY}$ will be a third of the angle $\widehat{CKX}$. Of any arbitrary given angle $\widehat{CKX}$, therefore, its third part $\widehat{XKY}$ will be found by setting between the indefinitely extended chords CX, AX the straight line EY equal to the diameter AC and directed through the circle's centre K.[674]

Hence if from the circle's centre K you let fall the perpendicular KH to the subtense CX, the angle $\widehat{HKY}$ will be a third of the angle $\widehat{HKX}$; consequently, if any angle $\widehat{HKX}$ be given, you can find its third part $\widehat{HKY}$ by letting fall from a point X in either side KX the perpendicular KH to the other, drawing the parallel XE to the side KH, and then setting between the straight lines XH and XE the straight line YE, twice (the size) of KX and directed through the point K. Or (do it) this way. Let there be given any angle $\widehat{AXK}$. To one or other side AX erect the perpendicular XH and from any point K in the other side draw the

---

(674) As Newton well knew (compare note (585)) the trisection cubic $x^3 - 3a^2x + a^2b = 0$ has the three real roots $x = 2a\sin\frac{1}{3}(\sin^{-1}[b/2a] + 2k\pi)$, $k = 0, 1, 2$: in his present figure, correspondingly, there will be three possible positions $E_kD_kY_k$, $k = 0, 1, 2$, of EY constructing the trisections $\widehat{XKD_k} = \frac{1}{3}(\widehat{CKX} + 2k\pi)$ of $\widehat{CKX} = 2\sin^{-1}[b/2a]$.

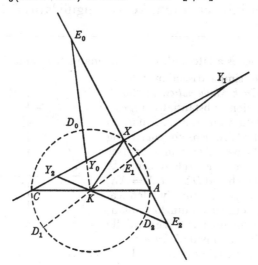

puncto quovis K agatur recta KE cujus pars YE interjacens lateri AX producto et ejus perpendiculo XH sit dupla lateris XK et erit angulus KEA tertia pars anguli dati AXK. Tum rursus erecto perpendiculo EZ et acta KF cujus pars ZF inter EF et EZ sit dupla ipsius KE fiet angulus KFA tertia pars anguli KEA. et sic pergitur per continuam anguli trisectionem in infinitum. Extat autem hæc trisectio apud Pappum, lib. 4 Prop 32.[675]

[1683]   Quod si angulum per alteram constructionum formulam ubi recta inter
Lect. 7   aliam rectam et circulum ponenda est, trifariam dividere malueris: hic etiam erunt $KB = \dfrac{q}{n}$ & $CX = \dfrac{r}{nn}$, id est in problemate de quo nunc agimus $KB = -\dfrac{3aa}{n}$ et $CX = \dfrac{aab}{nn}$ adeoꝗ ponendo $n = a$[676] fiet $KB = -3a$ et $CX = b$. Et inde talis emerget constructio.

A puncto quovis K ducantur ad easdem partes rectæ duæ $KA = a$ & $KB = 3a$. Biseca AB in C, centroꝗ A intervallo AC describe circulum. In eo pone rectam $CX = b$. Junge AX et junctam produc donec ea iterum secet circulum jam descriptum in G. Tum inter hunc circulum et rectam KC infinitè productam pone rectam EY æqualem rectæ AC et convergentem ad punctum G et acta recta EC erit longitudo quæsita x qua tertia pars anguli dati subtenditur.[677]

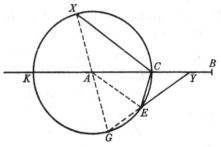

Talis constructio consequitur formulam superiùs allatam: quæ tamen sic evadet concinnior. Ob æquales circulos KXG et $ADEB$ et æquales subtensas
‖[236] CX et AB æquales sunt anguli CAX sive KAG et ‖ ACB, adeoꝗ CE subtensa est tertiæ partis anguli KAG. Quare dato quovis angulo KAG, ut ejus inveniatur pars tertia CAE, pone inter circulum KCG et anguli latus KA infinitè productum

---

(675) This last sentence is a late addition in the manuscript and should be understood, of course, to refer only to the simple trisection of $A\widehat{X}K$; compare T. L. Heath, *Greek Mathematics*, 1: 235–6. Newton's present extension substantially repeats the earlier 'Prob 17. Datum angulum...trifariam secare idꝗ continuò' in his 'Problems for construing æquations' (II: 464). As we have remarked (see notes (626) and (673)), since the circle of centre X through K meets EY in D such that $DE = DX = XK$, the present construction is structurally identical with the one 'per alteram constructionum formulam' following, which (as Newton observes) is essentially the eighth proposition in Archimedes' *Lemmas*.

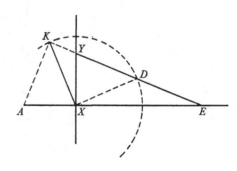

(676) Whence KA is the circle's radius and A its centre.

straight line KE whose section YE, lying between the side AX produced and its perpendicular XH, shall be twice the side XK: the angle $\widehat{KEA}$ will then be a third of the given angle $\widehat{AXK}$. Again, when the perpendicular EZ is erected and there is drawn KF, the section ZF of which between EF and EZ shall be twice KE, then the angle $\widehat{KFA}$ will prove to be a third of the angle $\widehat{KEA}$. And you may proceed in this way by continual angle-trisection indefinitely. This trisection, however, exists in Pappus' [*Collection*,] Book IV, Proposition 32.[675]

But should you prefer to divide an angle into three by the other form of construction in which a straight line has to be set between another straight line and a circle, here too there will be $KB = q/n$ and $CX = r/n^2$, that is, in the problem which is our present concern, $KB = -3a^2/n$ and $CX = a^2b/n^2$; hence on setting $n = a$[676] there will come to be $KB = -3a$ and $CX = b$. From this will emerge the following construction.

From any point K draw in the same direction two straight lines $KA = a$ and $KB = 3a$, bisect AB in C and with centre A, radius AC describe a circle. In it place the straight line $CX = b$, join AX and produce the joining line till it once more intersects the circle just now described in G. Then between this circle and the straight line KC, indefinitely extended, place the straight line EY equal to the line AC and directed through the point G: the straight line EC will then, when drawn, be the required length x which subtends a third part of the given angle.[677]

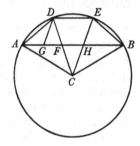

Such a construction narrowly follows the form previously laid down: but it will prove to be neater, however, when done in this way. Because the circles KXG, $ADEB$ and the subtenses CX, AB are each equal, the angles $\widehat{CAX}$ (or $\widehat{KAG}$) and $\widehat{ACB}$ are equal, and hence CE is the subtense of a third part of the angle $\widehat{KAG}$. Consequently, given any angle $\widehat{KAG}$, to find its third part $\widehat{CAE}$ place between the circle KCG and the angle's

---

(677) There will again (compare note (674)) be three possible positions $E_k Y_k$, $k = 0, 1, 2$, of EY corresponding to the three trisections $\widehat{CAE_k} = \frac{1}{3}(\widehat{CAX} + 2k\pi)$ of $\widehat{CAX} = 2\sin^{-1}[b/2a]$.

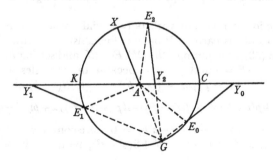

rectam *EY* æqualem circuli semidiametro *AG* et convergentem ad punctum *G*. Sic docuit Archimedes angulum trifariam secare.[678] Eædem constructiones facilius explicari possint quàm hic factum est, sed in his volui ostendere quomodo ex generalibus Problematum constructionibus superius expositis, constructiones simplicissimas particularium problematum derivare liceat.

Præter constructiones hic expositas adjungere liceret alias quamplurimas. Ut *si inter a et b inveniendæ essent duæ mediè proportionales*, Age quamvis *AK=b* et huic perpendiculare *AB=a*. Biseca *AK* in *I* et in eadem*AK*, subtensæ *BI* æqualem pone *AH* ut et in linea *AB* producta subtensæ *BH* æqualem *AC*. Tum in linea *AK* ad alteras partes puncti *A* cape *AD* cujusvis longitudinis, et huic æqualem

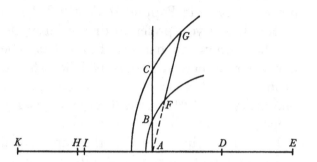

DE, centrisꝗ *D* et *E* intervallis *DB, EC* describe circulos duos *BF CG* et inter eos pone rectam *FG* æqualem rectæ *AI* et convergentem ad punctum *A* et erit *AF* prima duarum mediè proportionalium quas invenire oportuit.[679]

Docuerunt Veteres[680] inventionem duarum medie proportionalium per Cissoidem sed lineæ hujus descriptionem commodam manualem nemo quod scio apposuit. Sit *AP*[681] diameter et *F* centrum circuli ad quem Cissois pertinet.

---

(678) In the margin of the manuscript Whiston has inserted the correct reference to 'Lemmat Archim: 8', that is, Lemma 8 of the Archimedean *Lemmata* or *Liber Assumptorum*; see T. L. Heath, *Greek Mathematics*, 1: 240–1. The source of both Newton's and Whiston's knowledge of this νεῦσις—equivalent, as we have seen (note (675) above), to the preceding trisection, that of Pappus' *Mathematical Collection* IV, 32—was probably Isaac Barrow's recent 'editio nova' of these '*Lemmat[a]* Archimedis, quæ vocuntur' in his *Archimedis Opera...Methodo Nova Illustrata, & Succinctè Demonstrata* (London, 1675): 265–76, rather than the independent *editiones principes* of Samuel Foster (London, 1659) and G. A. Borelli (Florence, 1661), to both of which Barrow acknowledges a prefatory debt. In the latter edition Borelli notes of the present Lemma 8—which is surely Archimedes' own?—that 'Hæc...propositio elegantissima est' (*Archimedis Liber Assumptorum Interprete Thebit Ben-Kora Exponente Al Machtasso Ex Codice Arabico manuscripto* [set in appendix to his *princeps* edition of *Apollonii Pergæi Conicorum Lib. V. VI. VII*: 379–413]: 400).

(679) This construction is repeated with only trivial variants from the opening solution (II: 460) of Problem 15 of his earlier 'Problems for construing æquations'. Much as before (compare II: 461, note (33)), on joining *DB, DF, EC, EG* and setting *FG = p, DB = DF = q, EC = EG = r, AD = s, AE = t* and *AF = x*, because the triangles *ADF, AEG* share a common angle at *A* there is $(x^2+s^2-q^2)/sx = ((x+p)^2+t^2-r^2)/t(x+p)$ or

$$(t-s)x^3+p(t-2s)x^2+(t(s^2-q^2)-s(p^2+t^2-r^2))x+pt(s^2-q^2) = 0;$$

and when this is identified with $(t-s)(x^3-a^2b) = 0$ there comes to be $t = 2s$ (or $AE = 2AD$), $t(s^2-q^2) = s(p^2+t^2-r^2)$ and so $r^2-t^2 = p^2+2(q^2-s^2)$, with finally $pt(s^2-q^2)/(t-s) = -a^2b$

side *KA*, infinitely extended, the straight line *EY* equal to the circle's radius *AG* and directed through the point *G*. Archimedes has taught how to cut an angle into three in this manner.[678] The same constructions might more easily be explained than I have done so here, but in these I have wanted to reveal how, from the general constructions of problems previously exhibited, one is free to derive the simplest constructions of particular problems.

Apart from the constructions here expounded it would have been permissible for me to introduce a considerable number of others. For instance, *were two mean proportionals to be found between a and b*, draw any *AK = b* and, perpendicular to this, *AB = a*. Bisect *AK* in *I* and, in the same line *AK*, place *AH* equal to the subtense *BI*, also, in the line *AB* produced, *AC* equal to the subtense *BH*. Then in the line *AK* on the other side of point *A* take *AD* of any length, and *DE* equal to this, then with centres *D* and *E* and radii *DB*, *EC* describe two circles *BF*, *CG* and between them set the straight line *FG* equal to the line *AI* and directed through the point *A*: *AF* will then be the first of the two mean proportionals it was required to find.[679]

The Ancients[680] have explained the finding of two mean proportionals by help of the cissoid, but no one, so far as I know, has set out a convenient manual description of this curve. Let *A*[*G*][681] be the diameter and *F* the centre

---

or $2p(q^2-s^2) = a^2b$. For simplicity Newton then makes $\widehat{BAD} = \frac{1}{2}\pi$ and sets $FG = p = \frac{1}{2}b$: in this circumstance $AB = \sqrt{[q^2-s^2]} = a$ and $AC = \sqrt{[r^2-t^2]} = \sqrt{[\frac{1}{4}b^2+2a^2]}$, which is constructed from $AI = FG = \frac{1}{2}b$ as equal to

$$BH = \sqrt{[AB^2 + (AH^2 \text{ or}) BI^2]} = \sqrt{[AI^2 + 2AB^2]}.$$

(680) Following the lead of Diocles in his treatise περὶ πυρείων (on burning-mirrors), extant only in an unpublished Arabic version (Dublin, Chester Beatty Library. Arabic MS 5255: 1ʳ–26ʳ) recently located by G. J. Toomer. Diocles' construction by the cubic curve which is traditionally—since the mid-seventeenth century, and not least by Newton (see II: 60; III: 158, 462)—identified with Proclus' 'ivy leaf' has hitherto been known only in a fragment preserved by Eutocius in his commentary on the second book of Archimedes' *Sphere and Cylinder* (ed. Heiberg, **3**: 66–70, given in English translation in Ivor Thomas, *Selections illustrating...Greek Mathematics* (note (617)), **1**: 271–9; compare Heath's *Greek Mathematics*, **1**: 264–6). We may add that Newton's attribution of 'Cissois Dioclëa' reveals no profound classical knowledge on his part or awareness of the (then) unpublished writings of Fermat, Roberval and Beaugrand in the late 1630's in which, it would appear, the identification was first made, but merely echoes Barrow's 'Cissois vulgaris, seu Dioclea' (*Lectiones Geometricæ: In quibus (præsertim) Generalia Curvarum Linearum Symptomata declarantur* (London, 1670): Lectio IX, §XVI: 75).

(681) Newton's manuscript figure, here accurately reproduced in the Latin text, is badly wrong—as will be immediately obvious—in supposing that the locus (*C*) of the midpoint of *DE* can pass through the pole *P* of the *norma PED*, whose side *PE* can never be closer to *C* (at any given moment) than the distance *EC*. In his 1707 *editio princeps* Whiston made the accurate emendation, adopted without comment in Newton's 1722 revise, of setting *P* in *AF* such that *FP = 2AF* and then renaming '*G*' the now unmarked lowest point of the circle: the redrawn figure so necessitated is given in the English version, reproduced from Newton's 1722 revise.

Ad punctum *F* erigatur normalis *FD* eaꝗ producatur in infinitum.[682] Moveatur
|| [237]   norma[683] rectangula || *PED* ea lege ut crus ejus *EP* perpetuo transeat per
punctum *P*, et crus alterū *ED* circuli Diametro *AP* æquale, termino suo *D*

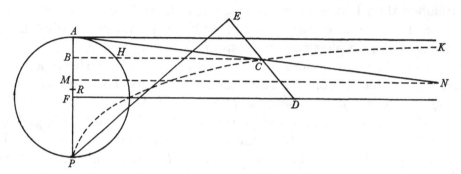

tangat semper lineam *FD*, et cruris hujus medium punctum *C* describet
Cissoidem desideratam *PCK* ut supra exposui.[684] Quare si inter duas quasvis *a*
et *b* inveniendæ sint duæ mediæ proportionales: cape *AM*=*a*. erige perpendiculum *MN*=*b*. Junge *AN* et lege præfata moveatur norma *PED* usꝗ dum
punctum ejus *C* incidat in rectam *AN*. Tum demisso ad *AP* perpendiculo *CB*
cape *t* ad *BH* et *v* ad *BP* ut est *MN* ad *BC* et ob continuè proportionales *AB*,
BH, *BP*, *BC* erunt etiam continuè proportionales *a*, *t*, *v*, *b*.[685]

Simili normæ applicatione construi possunt etiam alia Problemata solida.
Verbi gratia *proponatur æquatio cubica* $x^3 . pxx - qx + r = 0$: *ubi q semper negativum sit,*

r affirmativum et p signi utriusvis. Fac $AP = \dfrac{r}{q}$ eamꝗ biseca in *F*, et cape $FR = \frac{1}{2}p$ idꝗ

versus *A* si habeatur $+p_{[,]}$ aliter versus *P*. Fac insuper $AB = \sqrt{q}$ et erige normales
FD, *BC*. In normæ autem crure *ED* cape *ED* & *EC* ipsis *AP* et *AR* æquales
respectivè et applicetur deinceps norma ad schema sic ut punctum ejus *D*
tangat rectam *FD* et punctum *C* rectam *BC*, et erit *BC* æquationis radix quæsita
x.[686] Sed in his nimius sum.

(The further point *L* and the line *CQ* there marked relate to Newton's revise in 1722 of the
next paragraph; see note (686).) In the Latin text following we faithfully reproduce the text
of the deposited manuscript, but correct its confusions of '*P*' and '*G*' appropriately in our
English version.

(682) In line with his corrected diagram (see previous note) Whiston here added in his 1707
editio princeps the explanatory phrase: 'Et producatur *FG* ad *P*, ut *FP* æqualis sit circuli
Diametro' (And produce *FG* to *P* so that *FP* shall be equal to the circle's diameter).

(683) For the terminology see II: 118, 156; also compare III: 269, note (603).

(684) In 'Prob. 20' on his pages 81–2; see note (262).

(685) Whence the mean proportionals are $t = (MN/BC)\,BH$ and $v = (MN/BC)\,BP$. This
elegant variant on Diocles' classical construction (see note (680)) obviates the need to construct a separate cissoid for each new pair of quantities *a*, *b*.

(686) As Newton realized at a very late stage in the printing off in 1722 of his revise
of Whiston's *editio princeps*, this construction is irretrievably blemished. Where now the

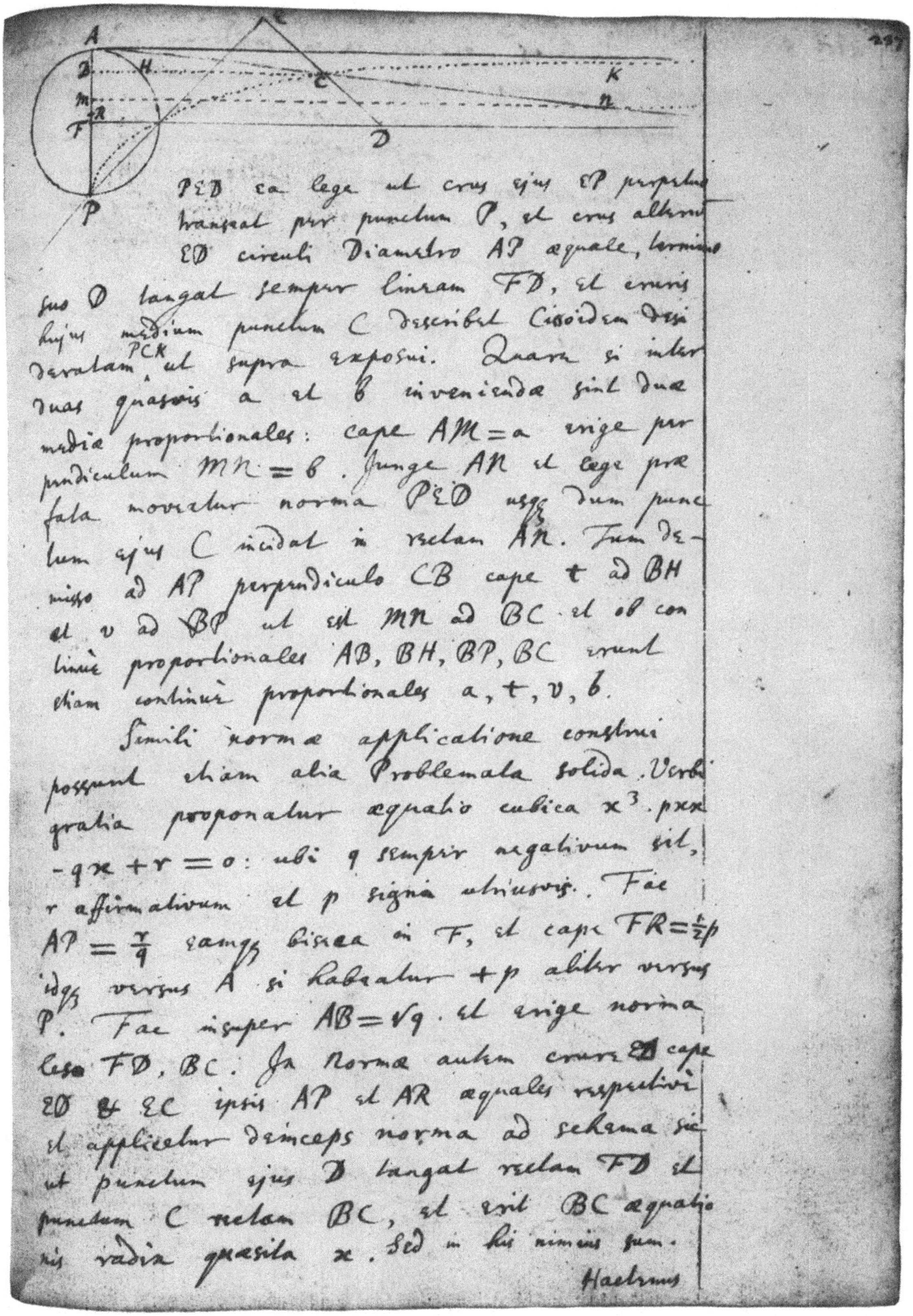

$P \varepsilon D$ ea lege ut crus ejus εP perpetuo transeat per punctum P, et crus alterum εD circuli Diametro AP aequale, termino suo D tangat semper lineam FD, et cruris hujus medium punctum C describet Cissoiden desideratam, ut supra exposui. Quare si inter duas quasvis a et b invenienda sint duae mediae proportionales: cape $AM = a$ erige perpendiculum $MN = b$. Junge AN et lege praefata moveatur norma $P\varepsilon D$ usque dum punctum ejus C incidat in rectam AN. Tum demisso ad AP perpendiculo CB cape t ad BH et v ad BP ut est MN ad BC. et ob continue proportionales AB, BH, BP, BC erunt etiam continue proportionales a, t, v, b.

Simili normae applicatione construi possunt etiam alia Problemata solida. Verbi gratia proponatur aequatio cubica $x^3 . p x x - q x + r = 0$: ubi q semper negativum sit, r affirmativum et p signi utriusvis. Fac $AP = \frac{r}{q}$ eamque biseca in F, et cape $FR = \frac{1}{2} p$ idque versus A si habeatur $+ p$ aliter versus P. Fac insuper $AB = \sqrt{q}$. et erige normae latus FD, BC. Ja Norma autem cruri εD cape εD & εC ipsis AP et AR aequales respective et applicetur deinceps norma ad schema sic ut punctum ejus D tangat rectam FD et punctum C rectam BC, et erit BC aequationis radix quaesita x. Sed in his nimius sum.

Hactenus

Plate III. Faulty construction of the cissoid of Diocles by a moving angle ($1, 2, \S 1$).

of the circle to which the cissoid relates. At the point F erect the normal FD and extend it indefinitely.[682] Move the right-angled 'sector'[683] PED, restricting its leg EP perpetually to pass through the point P, while its other leg ED, equal to

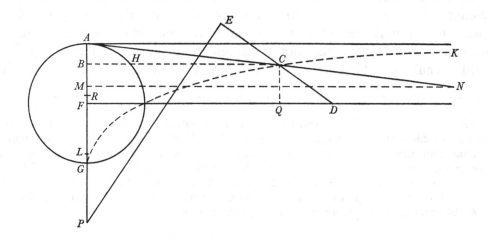

the circle's diameter $A[G]$, shall at its end-point D ever touch the line FD: the mid-point C of the latter leg will then describe the desired cissoid PCK, as I have shown above.[684] Consequently, if it be required to find two mean proportionals between any two quantities a and b, take $AM = a$, erect the perpendicular $MN = b$, join AN and let the sector PED move with the prescribed restraint until its point C falls onto the straight line AN. Then, letting fall the perpendicular CB to AP, take t to BH and v to $B[G]$ as MN is to BC, and then, because AB, BH, $B[G]$ and BC are in continued proportion, so too will a, t, v, b be in continued proportion.[685]

By a similar application of the sector other 'solid' problems also can be solved. For example, *let the cubic equation* $x^3 \pm px^2 - qx + r = 0$ *be proposed, in which q shall always be negative and r positive, while p may be of either sign.* Make $A[G] = r/q$ and bisect it in F, and take $FR = \frac{1}{2}p$: this in A's direction if $+p$ be had, otherwise towards P. In addition, make $AB = \sqrt{q}$ and erect the normals FD, BC. Now in the leg ED of the sector take ED and EC equal to $A[G]$ and AR respectively, and thereafter apply the sector to the diagram in such a way that its point D shall touch the line FD and its point C the line BC: then will BC be the required root x of the equation.[686] But I am overlong on this topic.

---

point C is fixed in general position in the side ED of the *norma PED*, say, by $CD = ka$ on taking

$$AF = FG = GP = \tfrac{1}{2}ED = a,$$

we may readily determine the curve (C) traced by C as the end-point D of the *norma* travels

‖ [238]
[1683]
Lect 8 ‖ Hactenus constructionem solidorum Problematum per operationes quarum praxis manualis maximè simplex est & expedita exponere visum fuit. Sic Veteres postquam confectionem horum problematum per compositionem locorum solidorum assecuti fuerant, sentientes ejusmodi constructiones ob difficilem Conicarum sectionum descriptionem inutiles esse, quærebant constructiones faciliores per Conchoidem, Cissoidem, extensionem filorum et figurarum adaptiones quascunᶐ mechanicas: prælata mechanica utilitate inutili speculationi Geometricæ ut ex Pappo discimus.[687] Sic magnus ille

---

along FD while its other side PE continues to pass through the pole P by taking (much as in Newton's Problem 20 on his pages 81–2 above) $FB = X$, $BC = y$ to be the (perpendicular) Cartesian coordinates of $C(X, y)$: for, on taking FD and PE to meet instantaneously in I and drawing CQ normal to FD, at once $FQ + QD = FI + (ID$ or$) IP = (QC + CD) PF/QD$ and consequently $y + \sqrt{[k^2a^2 - X^2]} = (X + ka) 2a/\sqrt{[k^2a^2 - X^2]}$; whence, after discarding the factor $X + ka = 0$ (corresponding to the singular locus furnished by the straight line through $S(-ka, 0)$ parallel to FD), there will, upon squaring, ensue

$$y^2(X - ka) + (X + ka)(X + (2 - k) a)^2 = 0,$$

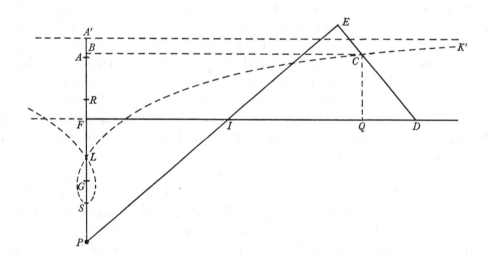

a nodal cubic with a double point at $L(-(2-k)a, 0)$ and having for asymptote the line through $A'(ka, 0)$ parallel to FD. Hence, when the locus is referred to the coordinates $LB = x = X + (2 - k) a$, $BC = y$ its Cartesian equation assumes the form

$$y^2(x - 2a) + x^2(x + 2(k - 1) a) = 0.$$

(In particular, when $k = 1$, the defining equation of the Dioclean cissoid in the preceding paragraph is seen to be $y^2(x - 2a) + x^3 = 0$; compare II: 61, note (58).) And when this is re-ordered as $x^3 + 2(k - 1) ax^2 + y^2x - 2ay^2 = 0$, identification with the given equation

$$x^3 + px^2 + qx + r = 0$$

Up to now it has seemed appropriate to expound the construction of solid problems by means of operations whose manual technique is the simplest and speediest. So, after the Ancients had achieved the accomplishment of these problems by compounding 'solid' loci, feeling that constructions of this sort are, because of the difficulty of describing conics, of no practical use, they looked for easier constructions by means of the conchoid and cissoid, the extending of threads and any kind of mechanical application of figures: as we learn from Pappus,[687] mechanical usefulness was preferred to useless geometrical specula-

---

requires that $GL = (k-1)a = p$, $FQ = (BC \text{ or}) y = \sqrt{q}$ (where q must be positive) and $AG = -2a = (r/y^2 \text{ or}) r/q$. Realizing this late in the printing of his 1722 edition, Newton replaced its (original) pages 315–16 with a guarded sheet containing the diagram reproduced above in our English text and the following newly framed argument: '... Verbi gratia proponatur æquatio cubica $x^3 \pm pxx + qx - r = 0$: ubi q semper affirmativum sit, r negativum, & p signi utriusvis. Fac $AG = \dfrac{r}{q}$, eamque biseca in F, & cape FR & $GL = \frac{1}{2}p$, idque versus A si habeatur $+p_{[,]}$ aliter versus P. Erige insuper normalem FD, inque ea cape $FQ = \sqrt{q}$ [et] huic etiam erige normalem QC. In normæ autem crure ED cape ED & EC ipsis AG & AR æquales respective, & applicetur deinceps norma ad schema sic ut punctum ejus D tangat rectam FD, & punctum C rectam QC, tum si compleatur parallelogrammum BQ erit LB æquationis radix quæsita x' (For example, let the cubic equation $x^3 \pm px^2 + qx - r = 0$ be proposed, in which q shall always be positive, r negative, while p may be of either sign. Make $AG = r/q$ and bisect it in F, and take $FR = GL = \frac{1}{2}p$: this in A's direction if $+p$ be had, otherwise towards P. In addition erect the normal FD, taking in it $FQ = \sqrt{q}$, and normal to this erect also QC. Now in the leg ED of the sector take ED and EC equal to AG and AR respectively, and thereafter apply the sector to the diagram in such a way that its point D shall touch the line FD and its point C the line BC: if, then, the rectangle $[C]B[F]Q$ be completed, LB will be the required root x of the equation). It is a comment on the algebraic competence of his British and Continental contemporaries that no one seems to have brought this *faux pas* to Newton's attention during the decade and a half following its publication in 1707: we may well picture the overblown disgust which Johann Bernoulli would have evinced, however, and be glad that it lay undetected by him (even though he read through Whiston's *editio princeps* soon after it appeared, doubtless on look-out for such major blemishes). In his 1722 revise Newton, realizing in hindsight how callow it now appeared when he had, in fact, spent too little time on the present construction, silently omitted the final remark which follows.

(687) See the preamble to Pappus' *Mathematical Collection*, IV, 31, quoted in our note (617) above. In an earlier, equivalent passage leading into III, 5, Pappus had written (in our translation) of 'the third [linear] type of problem... admitting other lines in their construction and of a more varied and complicated origin, such as spirals, quadratrixes, conchoids and cissoids, which possess numerous, astonishing properties.... The ancient geometers were unable to construct the problem [of finding two mean proportionals], 'solid' by its nature, in conformity with this geometrical scheme [of constructing 'solid' problems by 'solid' conic-loci], because it is not easy to trace out sections of a cone in the plane; however, they succeeded admirably in making use of instruments suitable for achieving the construction easily and manually, as will readily be agreed in the case of Eratosthenes' *Mesolabe* [whose machine is described in III, 5 following; see Heath, *Greek Mathematics*, 1: 258–60]..., while Nicomedes resolved the problem by employing the conchoid, by means of which he also accomplished the trisection of an angle [described by Pappus in his IV, 32; see note (673)].'

Archimedes trisectionem anguli per coni sectiones a superioribus Geometris expositam neglexit et in Lemmatis suis angulum modo a nobis superius⁽⁶⁸⁸⁾ exposito trifariam secare docuit. Si veteres problemata per figuras ea tempestate in Geometriam non receptas construere maluerint, quanto magis præferendæ nunc sunt illæ figuræ in Geometriam æque ac ipsæ coni sectiones a pleriscq receptæ.

Verum tamen novo huic Geometrarum generi⁽⁶⁸⁹⁾ haud assentior qui figuras hasce omnes in Geometriam recipiunt. Eorum regula admittendi lineas omnes ad constructionem Problematum eo ordine quo æquationes quibus lineæ illæ definiuntur numero dimensionum ascendunt,⁽⁶⁹⁰⁾ arbitraria est et in Geometria fundamentum non habet. Imò falsa est propterea quod circulus hac lege cum Coni sectionibus conjungendus esset, quem tamen Geometræ omnes cum linea recta conjungunt.⁽⁶⁹¹⁾ Vacillante autem hac regula tollitur fundamentum admittendi certo ordine lineas omnes Analyticas in Geometriam. In Geometriam ‖[239] planam meo quidem judicio lineæ nullæ præter rectam et ‖ circulum admitti debent, nisi forte linearum distinctio aliqua priùs excogitetur qua linea circularis conjungatur cum recta et a reliquis omnibus segregetur.⁽⁶⁹²⁾ Quinimò ne tum quidem augenda est Geometria plana numero linearum. Nam figuræ omnes sunt planæ quæ admittuntur in Geometriam planam₍,₎ id est quas Geometræ postulant in plano describere. Et problema omne planum est quod per figuras

(688) On his pages 235–6; see note (678). Since the only known classical trisections making direct use of conics in their construction are those given by Pappus in his *Mathematical Collection* iv, 34 (see Heath, *Greek Mathematics*, **1**: 240–3; compare note (626) above), Newton's suggestion that Archimedes 'neglected' these is an anachronism.

(689) Personified by Descartes, of course. To continue the passage from his *Geometrie* quoted in note (620), Descartes went on: '...ie m'estonne de ce que [les Anciens] n'ont point... distingué diuers degrés entre ces lignes plus composées, & ie ne sçaurois comprendre pourquoy ils les ont nommées mechaniques, plutost que Geometriques....Il est vray qu'ils n'ont pas aussy entierement receu les sections coniques en leur Geometrie, & ie ne veux pas entreprendre de changer les noms qui ont esté approuués par l'vsage; mais il est, ce me semble, tres clair, que prenant comme on fait pour Geometrique ce qui est precis & exact, & pour Mechanique qui ne l'est pas; & considerant la Geometrie comme vne science, qui enseigne generalement a connoistre les mesures de tous les cors, on n'en doit pas plutost exclure les lignes les plus composées que les plus simples, pourvû qu'on les puisse imaginer estre descrites par vn mouuement continu, ou par plusieurs qui s'entresuiuent & dont les derniers soient entierement reglés par ceux qui les precedent. car par ce moyen on peut tousiours auoir vne connoissance exacte de leur mesure' (*Geometrie*, ₁1637: 315–16). Descartes does not thereby restrict his basic system of coordinates to be standard Cartesian—indeed, in sequel he employs bipolars in defining his 'ovals', though he is careful to reduce the latter system to a perpendicular Cartesian one—but we may gather from this passage that it was his intention to define as 'geometrical' those (and only those) curves which are 'exactly' and 'precisely' defined in an allowable coordinate system, and which may accordingly, when reduced to standard Cartesian coordinates, be defined by an algebraic equation. (In iii: 141, note (236) we pointed to the possible inconsistency implicit in Descartes' refusal to allow curvilinear lengths, even when 'geo-

tion. Thus the mighty Archimedes ignored the trisection of an angle by means of conics expounded by his predecessors in geometry, and in his *Lemmas* taught how to cut an angle into three by the method we just now exhibited.[688] If the Ancients preferred to construct problems by means of figures not at that period received into geometry, how much greater should our preference now be for those figures when by most they are received into geometry on an equal footing with conics themselves?

But yet I do not at all approve of the new generation of geometers[689] who welcome all these curves into geometry. Their rule of admitting all (curved) lines into the construction of problems in the ascending sequence of the number of dimensions in the equations by which those lines are defined[690] is arbitrary and has no foundation in geometry. Indeed, it is false since on that principle the circle would need to be combined with the conics, and yet all geometers place it jointly with the straight line.[691] The wavering in this rule destroys it as the basis for admitting all analytical curves into geometry in a precise order. Into plane geometry—in my personal opinion—no curves except the straight line and circle ought to be admitted, unless maybe some preliminary distinction between lines should be devised by which a circular line is to be conjoined with a straight one and so segregated from all others.[692] But, to be sure, not even then must plane geometry be increased in the number of its lines. For all figures which are admitted into plane geometry—those, that is, which geometers postulate to be describable in a plane—are plane. And every problem which can be con-

---

metrically' rectifiable, into his *Geometrie*: he would, in consequence, not allow as 'geometrical' a curve defined by an algebraic relationship between the arc-length

$$s = \tfrac{8}{27}[(1 + \tfrac{9}{4}z)^{\tfrac{3}{2}} - 1]$$

of a semi-cubical parabola $z = x^{\tfrac{2}{3}}$ and corresponding ordinate $t = y - z$, even though this is reducible in an immediate way to an equivalent algebraical relationship between the perpendicular Cartesian coordinates x and y.)

(690) We have noticed (see note (620)) that this 'Cartesian' classification by algebraic degree of the defining Cartesian equation is not found in Descartes' *Geometrie* but was first systematically pursued by Newton himself in his early mathematical researches.

(691) As members of the class of classical 'plane' loci. Not all of Newton's contemporaries did, in fact, so conjoin them—in particular, Frans van Schooten had devoted the whole of the second book of his *Exercitationum Mathematicarum Libri Quinque* (Leyden, 1657): 113–90 to the 'Constructio Problematum Simplicium Geometricorum, seu quæ solvi possunt, ducendo tantùm rectas lineas'—but this does not fault Newton's argument that the circle, by reason of its relative ease of construction, ought not to be classified with the conics in a single class of curves determined simply as having a second-degree Cartesian defining equation. He here, however, ignores the very real rewards to be gained by classifying the circle and conics together and then generalizing their common properties to higher curves: a method used with profit in his own earlier 'Conick propertys to bee examined in other curves' (II: 90).

(692) Hence removing the need for defining such a distinction in the context of plane geometry.

planas construi potest. Sic igitur admissis in Geometriam planam conicis sectionibus alijscʒ magis compositis figuris, problemata omnia solida et plus quam solida[693] quæ per has figuras construi possunt evadent plana. Sunt autem problemata omnia plana ejusdem ordinis. Linea recta Analyticè simplicior est quam circulus: hoc non obstante Problemata ejusdem sunt ordinis quæ per rectas solas et quæ per circulos construuntur. Solis postulatis reducitur circulus ad eundem ordinem cum recta. Siquis speculando Ellipsin incideret in problema aliquod solidum et ipsum beneficio ejusdem Ellipseos et circuli[694] construeret: hoc problema jam pro plano habendum esset, eò, quòd Ellipsis jam ante in plano descripta haberi supponitur et constructio omnis quæ superest absolvitur per circuli solius descriptionem. Eadem de causa problemata quævis plana per datam Ellipsin[695] construere licitum est. Verbi gratia si

|| [240] ||datæ Ellipseos *ADFG* requireretur centrum $O_{[,]}$ ducerem parallelas duas *AB*, *CD* Ellipsi occurrentes in *A*, *B*, *C*, *D* aliascʒ duas *EF*, *GH* Ellipsi occurrentes in *E*, *F*, *G*, *H*. Has bisecarem in *I*, *K*, *L*, *M* & junctas, *IK*, *LM* producerem uscʒ ad concursum suum in *O*. Legitima est hæc constructio plani problematis per Ellipsin. Nil refert quod Ellipsis Analyticè definiatur per æquationem duarum dimensionum. Nil quod Ellipsis Geometricè generetur sectione figuræ solidæ.[696]

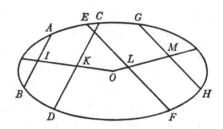

Hypothesis sola, quòd Ellipsis jam descripta habetur in plano, problemata omnia solida per ipsam constructa reducit ad ordinem planorum efficitcʒ ut plana omnia per ipsam legitimè construantur. Et eadem est ratio Postulati. Quod vi postulatorum fieri potest ut jam factum & datum assumere concessum est. Postuletur igitur Ellipsin in plano describere et ad ordinem planorum problematum reducentur ea omnia quæ per Ellipsin construi possunt, planacʒ omnia per Ellipsin licebit construere.[697]

[698]Necesse est igitur aut Problemata plana et solida inter se confundi aut lineas omnes rejici a Geometria plana præter rectam et circulum et siqua forsan alia detur aliquando in statu construendi alicujus Problematis. Verum genera problematum confundi nemo certè permiserit. Rejiciantur igitur e Geometria plana sectiones Conicæ aliæcʒ figuræ omnes præter rectam et circulum et quas

---

(693) That is, 'linear' in the classical Greek classification; see Newton's page 21 above, and also note (617).

(694) 'rectǽve' (or straight line) is deleted.

(695) Understand 'in plano constructam' (constructed *in plano*)—whether by points, by a trammel or organic construction or by any other convenient method, 'geometrical' or 'mechanical'.

structed by means of plane figures is plane. So then, once conics and other more complicated figures are admitted into geometry, all 'solid' problems and ones more than solid[693] which can be constructed by these figures will turn out to be plane. All plane problems are, however, of the same order. A straight line is analytically simpler than a circle: this notwithstanding, problems constructible by straight lines alone and those constructible by circles are of the same order— it is by postulate alone that a circle is reduced to the same order as a straight line. Were one, while considering an ellipse, to meet with some solid problem and construct it with the aid of that same ellipse and a circle,[694] this problem would need now to be classed as a plane one on the understanding that the ellipse is supposed already to be had, as being previously described *in plano*, and all the outstanding construction is accomplished by the description of a circle alone. For the same reason you are at liberty to construct any plane problems you wish by means of a given ellipse[695]. For example, if in the given ellipse *ADFG* the centre *O* were required, I should draw two parallels *AB*, *CD* meeting the ellipse in *A*, *B*, *C*, *D* and two others *EF*, *GH* meeting the ellipse in *E*, *F*, *G*, *H*; then bisect these in *I*, *K*, *L*, *M* and extend the joins *IK*, *LM* to their meet in *O*. This is a legitimate construction of a plane problem by an ellipse. It is not relevant that an ellipse is analytically to be defined by an equation of two dimensions; nor that an ellipse is geometrically to be generated by the section of a solid figure.[696] The sole hypothesis that the ellipse is already had, described *in plano*, reduces all solid problems constructed by it to the order of plane ones and determines that all plane ones may legitimately be constructed by it. And the reason for its postulation is the same. What can come to be by force of postulation may permissibly be assumed as already done and given. Let it therefore be postulated that an ellipse is described *in plano*, and then to the order of plane problems will be reduced all those which can be constructed by the ellipse, while it will be allowable to construct[697] all plane ones by the ellipse.

[698]It is hence necessary either that (the classes of) plane and solid problems shall be confounded one with the other, or that all lines be rejected from plane geometry except for the straight line and circle and any other which may, on occasion, chance to be given in the circumstances of some problem to be constructed. But no one, certainly, would permit classes of problems to be confused. So let there be rejected from plane geometry the conics and all other figures except for the straight line and circle and those which happen to be

(696) A right cone or cylinder, or just possibly the hyperboloid of revolution whose plane section Newton demonstrated to be a conic in 'Prob: 19' on his pages 80–1 above.

(697) Newton first wrote 'legitimè construentur' (will be legitimate to construct).

(698) In the margin of the manuscript at this place Newton first wrote and then cancelled 'Lect. 8', evidently deciding in afterthought that its revised position at the head of his page 238 signalled a more natural break between 'lectures'.

||[241]

contigerit in statu problematum dari. Alienæ sunt igitur a Geometria[699] descriptiones illæ omnes conicarum sectionum in plano quibus hodierni Geometræ[700] tantoperè indulgent. Nec || tamen ideò Coni sectiones e Geometria rejiciendæ erunt. Hæ in plano non describuntur Geometricè, generantur verò in solidi Geometrici superficie plana.[701] Conus constituitur Geometricè et plano Geometrico secatur. Tale Coni segmentum figura Geometrica est eundemꝗ habet locum in Geometria solida ac segmentum circuli in plana, et hac ratione basis ejus, quam Coni sectionem vocant, figura Geometrica est. Locum igitur habet Coni sectio in Geometria quatenus ea superficies est solidi Geometrici. Alia autem nulla ratione Geometrica quam solidi sectione generatur, et ideò non nisi in Geometriam solidam antiquitus admissa fuit. Talis autem Conicarum sectionum generatio difficilis est & in rebus practicis, quibus Geometria potissimùm inservire debet, prorsus inutilis. Ideò veteres se ad varias figurarum in plano descriptiones mechanicas receperunt, et nos ad eorum exemplar constructiones præcedentes concinnavimus. Sunto constructiones illæ Mechanicæ: sic et constructiones per Coni sectiones in plano (ut jam moris est) descriptas mechanicæ sunt. Sunto constructiones per datas Coni sectiones Geometricæ: sic et constructiones per alias quascunꝗ figuras datas Geometricæ sunt et ejusdem ordinis cum constructionibus planorum Problematum. Nulla ratione præferendæ sunt in Geometria Sectiones conicæ figuris alijs nisi quatenus hæ a sectione Coni, praxi ad solutionem problematum prorsus inutili, derivantur. Verum tamen ne constructiones per conicas sectiones omninò præteream, visum fuit aliqua de his subjungere in quibus etiam praxi manuali non incommodæ consulatur.

[702]Conicarum sectionum simplicissima est Ellipsis. Hæc notior est et circulo

---

(699) Continue to understand 'plana' (plane).

(700) Doubtless, Newton is thinking especially of René–François de Sluse and Philippe de La Hire who, in their *Mesolabum, seu Duæ mediæ Proportionales inter Extremas datas per Circulum et Ellipsim vel Hyperbolam infinitis modis exhibitæ. Accedit Problematum quorumlibet solidorum Effectio per easdem Curvas* (Liège, ₁1659 → ₂1668) and *Nouveaux Elemens des Sections Coniques, Les Lieux Geometriques, La Construction ou Affectation des Equations* (Paris, 1679) respectively, developed a variety of ways of constructing cubic and quartic equations by intersecting conics. (Newton's library copies of these—the former in its much less rare second edition—are now Trinity College, Cambridge. NQ.9.40 and NQ.9.164.) As Collins wrote in review of Sluse's book (*Philosophical Transactions*, 4, No. 45 [for 25 March 1669]: 903–9, especially 904–5), 'The Author observes, that amongst those, that solve this Probleme [of finding two Means] by the *Conick Sections*, they seem to have afforded fewer Effections thereof, than there have been Ages, since it was first proposed. Very few by ayd of a *Circle* and an *Hyperbola* or *Parabola*: by a *Circle* and *Ellipsis* none, that he could observe to have been published. The which the Author considering, and studying how to supply, he found out not onely one, but infinite such Effections, and that not in one Method, but many; following the guidance of which Methods, by the like felicity he hath constructed all solid Problems infinite ways, by a *Circle* and an *Ellipsis* or *Hyperbola*..., without reduction of the Æquations proposed; and sheweth a *general* Construction for all *Cubick* and *Bi-quadratick* Æquations by ayd of a *Circle* and a *Parabola*'. The

given in the circumstances of problems. Alien to [699]geometry, therefore, are all those descriptions of conics *in plano* in which present-day geometers[700] so grossly indulge. But conics will not, however, need to be thrown out of geometry on that account. While these are not described geometrically *in plano* they are indeed generated in the plane surface[701] of a geometrical solid, while the cone is geometrically constituted and cut by a geometrical plane. Such a segment of a cone is a geometrical figure, occupying the same place in solid geometry as the segment of a circle in a plane one, and for this reason its base—which people call a conic section—is a geometrical figure. The conic section, therefore, has a place in geometry insofar as it is the (plane) surface of a geometrical solid. It is, however, generated by no other geometrical method than the sectioning of a solid, and for that reason was in antiquity admitted into none but solid geometry. Now such a generation of conics is difficult and for practical purposes—in which above all geometry ought to be of service—wholly useless. Because of that the ancients had recourse to various mechanical descriptions of figures *in plano* and we ourselves fashioned in their pattern the constructions which precede. Decree that those constructions are mechanical, then so too are constructions by conic sections described (as is now customary) *in plano*. But grant that constructions by given conics are geometrical, then so also are constructions by any other given figures geometrical and of the same order as the constructions of plane problems. There is in geometry no ground for preferring conics to other curves unless it be that they are—by a technique wholly without useful application to the resolution of problems—derived from the sectioning of a cone. But so as not completely to pass over constructions by means of conics, however, I have thought fit to add some remarks on this topic and, in so doing, give consideration also to their not unhandy manual performance.

[702]Of conics the ellipse is the simplest. This is better known, more closely

---

simple method of constructing two means x, y between a and b by constructing the meets of the parabola defined, in standard Cartesian coordinates, by $x^2 = ay$ with the parabola or hyperbola defined by $y^2 = bx$ and $xy = ab$ respectively is, in effect, that proposed by Menæchmus (whose solutions are preserved by Eutocius in his commentary on the second book of Archimedes' *Sphere and Cylinder* [ed. Heiberg, **3**: 78–80]; see Heath's *Greek Mathematics*, **1**: 251–5). Descartes' improvement on Menæchmus of constructing these intersections as the meets of the parabola $x^2 = ay$ with the circle $x^2 + y^2 = bx + ay$ (of centre $(\frac{1}{2}b, \frac{1}{2}a)$ and through the origin) was first published in 1636 anonymously—as a 'solution...inuentée depuis peu par vn homme de condition & de merite, qui pour son rare esprit est l'vn des plus grands ornemens de notre France'—and without proof by Mersenne in his *Harmonie Universelle* (note (669)): Livre VI, Proposition XLV: 408–11, and reprinted the next year in Descartes' own *Geometrie* ($_1$1637: 395–6).

(701) Understand 'sectionis' (of section).

(702) Newton subsequently entered and then cancelled in the margin alongside 'Lect. 9'. Its repositioning on his page 244 below scarcely marks a more plausible break between 'lectures', while stretching the already considerable length of 'Lect. 8' by three further pages.

‖[242] magis affinis ‖ et praxi manuali faciliùs describitur in plano. Parabolam præferunt pleriꜹ ob simplicitatem æquationis per quam ea exprimitur. Verum hac ratione Parabola ipso etiam circulo præferenda esset contra quam fit. Falsa est igitur argumentatio a simplicitate æquationum. Æquationum speculationi nimium indulgent hodierni Geometræ. Harum simplicitas est considerationis Analyticæ. Nos in compositione versamur et Compositioni leges dandæ non sunt ex Analysi. Manuducit Analysis ad Compositionem: sed Compositio non priùs vera confit quam liberatur ab omni Analysi. Insit Compositioni vel minimum Analyseos et Compositionem veram nondum assecutus es. Compositio in se perfecta est et a mixtura[703] speculationum Analyticarum abhorret. Pendet Figurarum simplicitas a simplicitate geneseos et Idearū, et æquatio non est sed descriptio (sive Geometrica sive Mechanica) qua figura generatur et redditur conceptu facilis. Ellipsi igitur primum locum tribuentes docebimus jam quomodo æquationes per ipsam construere licet.[704]

Proponatur æquatio quævis cubica $x^3 = pxx + qx + r$, ubi p, q et r datas coefficientes cum signis suis $+$ et $-$ significant et alteruter terminorum p et q vel etiam uterꜹ deesse potest. Sic enim æquationum omnium cubicarum constructiones una illa operatione quæ sequitur exhibebimus.

A puncto quovis B in recta quavis data cape duas quascunꜹ rectas BC BE ad easdem partes ut et inter ipsas mediam proportionalem BD. Et BC dictâ n, cape etiam in eadem recta $BA = \dfrac{q}{n}$, idꜹ versus punctum C si habeatur $-q$, aliter ad partes contrarias. Ad punc-

‖[243] tum A ‖ erige perpendiculum $A[I]$, inꜹ eo cape AF æqualem p, FG æqualem AF, FI æqualem $\dfrac{r}{nn}$, et FH in ratione ad FI ut est BC ad BE. FH verò et FI capiendæ sunt ad partes puncti F versus G si termini p et r habent eadem signa, aliter ad partes versus A. Compleantur

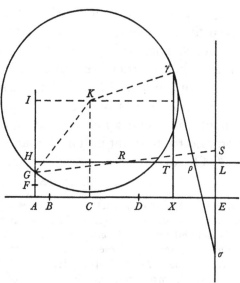

parallelogramma $IACK$ & $HAEL$, centroꜹ K et intervallo KG describatur circulus. Tum in linea HL capiatur ad utramvis partem puncti H longitudo HR

---

(703) Observe the (al)chemical metaphor. As his contemporary papers and notebook records (ULC. Add. 3973 and 3975 especially) show, Newton was at this time (spring 1684?) busy in his Trinity elaboratory boiling metals and sublimating a wide variety of compounds, so that a chemical analogy would easily suggest itself.

related to the circle and more easily described *in plano* in manual practice. A considerable number prefer the parabola because of the simplicity of the equation by which it is expressed. But on this basis the parabola would need also to be preferred to the circle itself, contrary to what occurs. Reasoning from the simplicity of their equations is therefore false. Present-day geometers indulge too much in speculation from equations. The simplicity of these is a consideration belonging to analysis: we are here occupied with composition, and laws are not to be laid down for composition from an analytical standpoint. Analysis guides us to the composition, but true composition is not achieved before it is freed from all its analysis. Let even the slightest trace of the analysis be present in the composition and you will not yet have attained the true composition. Composition is perfect in itself and shrinks from any admixture[703] with analytical speculations. Simplicity in figures is dependent on the simplicity of their genesis and conception, and it is not its equation but its description (whether geometrical or mechanical) by which a figure is generated and rendered easy to conceive. Assigning this primary place to the ellipse, therefore, we shall now explain how we are at liberty to construct equations by its means.[704]

Let there be proposed any cubic equation $x^3 = px^2 + qx + r$, where p, q and r denote given coefficients with their signs $+$ and $-$, and one or other (or indeed both) of p and q may be lacking. In this way, to be sure, we shall exhibit the construction of all cubic equations in the single procedure which follows.

From an arbitrary point B in any arbitrary given straight line take any two straight lines BC, BE in the same direction, and also the mean proportional BD between these. Then, calling BC n, take also $BA = q/n$ in the same straight line: this towards the point C if $-q$ be had, otherwise the opposite way. At the point A erect the perpendicular AI and in it take $AF = p$, $FG = AF$, $FI = r/n^2$ and FH in ratio to FI as BC is to BE: FH, however, and FI are to be taken on the side of point F towards G if the coefficients p and r have the same sign, otherwise towards A. Complete the rectangles $IACK$ and $HAEL$, and with centre K, radius KG describe a circle. Subsequently, in the line HL on either side of point H you wish take a length HR which shall be to HL as BD to BE, draw GR cutting

---

(704) With this somewhat rambling and not entirely coherent exposition of his views on the relative rôles of (algebraic) analysis and pure (synthetic) composition in geometry, and his rather heavy-footed insistence on ease of description rather than simplicity of the defining Cartesian equation thus—in a loud echo of his earlier remarks on his pages 211–15 preceding—set down as justification for now daring, in defiance of his previously stated position, to employ an ellipse (either constructed by a trammel or merely 'given' without specification of its description, as we shall see) in the geometrical construction of equations, Newton now proceeds to repeat all but *verbatim* a further slab (see II: 478–90) of his previous 'Problems for construing æquations'.

quæ sit ad *HL* ut *BD* ad *BE*; Agatur *GR* secans *EL* in *S* et moveatur linea *GRS* puncto ejus *R* super linea *HL* et puncto *S* super linea *EL* incedente, donec tertium ejus punctum *G* describendo Ellipsin, occurrat circulo, quemadmodum videre est in positione $\gamma\rho\sigma$. Nam dimidium perpendiculi γX ab occursus illius puncto γ in rectam *AE* demissi erit radix æquationis. Potest autem Regulæ *GRS* vel $\gamma\rho\sigma$ terminus *G* vel γ circulo in tot punctis occurrere quot sunt possibiles radices. Et e radicibus hæ sunt affirmativæ quæ cadunt ad eas partes rectæ *AE* ad quas recta *FI* ducitur a puncto *F* et illæ negativæ quæ cadunt ad contrarias partes lineæ *AE*, si modò habeatur $+r$: et contra si habeatur $-r$.[705]

‖ [244]
[1683]
Lect 9

‖ Demonstratur autem hæc constructio subsidio Lemmatum sequentium.[706]

Lemma 1. Positis quæ in superiore constructione, est

$$2CAX - AX^q = \gamma X^q - 2AI \times \gamma X + 2AG \times FI.$$

Namcɜ ex natura circuli est $K\gamma^q - CX^q$ æquale quadrato ex $\gamma X - AI$. Sed est $K\gamma^q$ æquale $GI^q + AC^q$ et CX^q æquale quadrato ex $AX - AC$ hoc est æquale $AX^q - 2CAX + AC^q$, atcɜ adeo horum differentia $GI^q + 2CAX - AX^q$ æquatur quadrato ex $\gamma X - AI$ id est ipsi $\gamma X^q - 2AI \times \gamma X + AI^q$. Auferatur utrincɜ GI^q et manebunt æqualia $2CAX - AX^q$ et $\gamma X^q - 2AI \times \gamma X + AI^q - GI^q$. Verum AI^q (per Prop. 4 lib. 2 Elem) æquale est $AG^q + 2AGI + GI^q$ atcɜ adeo $AI^q - GI^q$ æquale est $AG^q + 2AGI$ hoc est æquale $2AG$ in $\frac{1}{2}AG + GI$ seu æquale $2AG \times FI$, et proinde $2CAX - AX^q$ æquale est $\gamma X^q - 2AI \times \gamma X + 2AG \times FI$. Q.E.D.

Lemma 2. Positis quæ in superiore constructione est

$$2EAX - AX^q \ \text{æquale} \ \frac{FI}{FH} X\gamma^q - \frac{2FI}{FH} AH \times X\gamma + 2AG \times FI.$$

---

(705) The construction of 'Prob 4' on II: 478 reproduced virtually word for word. Much as before (see II: 478, note (77)), on adding the root $x - p = 0$ to the given cubic $x^3 = px^2 + qx + r$, the resulting equation $x^2(x-p)^2 = (qx+r)(x-p)$ may be expressed as the eliminant of $x(x-p) = ny$ and $n^2y^2 = qny + r(x-p)$. On substituting $x = \frac{1}{2}X$, $y = \frac{1}{2}Y$ (to avoid fractions in the sequel) there ensue $X^2 = 2pX + 2nY$ and $n^2Y^2 = 2qnY + 2rX - 4pr$, the defining equations in a standard, perpendicular Cartesian coordinate system, of a pair of parabolas, one meet of which is evidently the point $(2p, 0)$. More generally, through their intersections passes the ellipse $X^2 + (n^2/k^2) Y^2 = 2(p + r/k^2) X + 2n(1 + q/k^2) Y - 4pr/k^2$ or

$$(X - (p + r/k^2))^2 + (n^2/k^2)(Y - (k^2 + q)/n)^2 = (-p + r/k^2)^2 + (k + q/k)^2,$$

of centre $(p + r/k^2, (k^2 + q)/n)$ and with its *latus rectum* in proportion to its main axis (*latus transversum*) $2(k/n) \sqrt{[(-p + r/k^2)^2 + (k + q/k)^2]}$ in the ratio n to k; that is, in modern terms, of eccentricity $\sqrt{[1 - n^2/k^2]}$. In particular, when $n = k$ this is the defining equation of a circle $(X - (p + r/n^2))^2 + (Y - (n + q/n))^2 = (-p + r/n^2)^2 + (n + q/n)^2$ of centre $(p + r/n^2, n + q/n)$. Having chosen *A* to be the origin of the Cartesian coordinates $AX = Y$, $X\gamma = X$ and set $AB = q/n$, $BC = n$, $BD = k$ and so $BE = k^2/n$, $AF = FG = p$, $FI = r/n^2$ and so $FH = r/k^2$, Newton constructs the ellipse with axes $HL(X = p + r/k^2)$ and $LS(Y = (k^2 + q)/n)$ by means of the trammel $\gamma\rho\sigma$ (where ρ, σ move freely in HL, LS respectively) in which

$$\gamma\rho : \gamma\sigma = GR : GS = HR : HL = (k : k^2/n \ \text{or}) \ n : k.$$

EL in *S* and move the line *GRS*, with its point *R* travelling on the line *HL* and point *S* on the line *EL*, till its third point *G*—describing an ellipse—meets the circle, as may be seen in the position $\gamma\rho\sigma$. Then, to be sure, half the perpendicular γX let fall to the straight line *AE* from the meeting point γ will be a root of the equation. The end-point *G* (or γ) of the rule *GRS* (or $\gamma\rho\sigma$) can, however, meet the circle in as many points as there are roots possible. And of these roots the positives are those which fall on the side of the line *AE* on which the line *FI* is drawn from point *F*, the negatives those falling on the opposite side of the line *AE*, provided there be had $+r$; the converse holds if $-r$ be had.[705]

This construction is demonstrated, however, with the support of the following lemmas.[706]

Lemma 1. With the suppositions of the previous construction there is

$$2AC \times AX - AX^2 = X\gamma^2 - 2AI \times X\gamma + 2AG \times FI.$$

For it is a property of a circle that $K\gamma^2 - CX^2$ is equal to $(X\gamma - AI)^2$. But $K\gamma^2$ is equal to $GI^2 + AC^2$ and CX^2 equal to $(AX - AC)^2$, that is,

$$AX^2 - 2AC \times AX + AC^2,$$

and hence their difference $GI^2 + 2AC \times AX - AX^2$ is equal to $(X\gamma - AI)^2$, that is, to $X\gamma^2 - 2AI \times X\gamma + AI^2$. Take away GI^2 from each side and there will remain the equals $2AC \times AX - AX^2$ and $X\gamma^2 - 2AI \times X\gamma + AI^2 - GI^2$. But (by *Elements*, Book II, Proposition 4) AI^2 is equal to $AG^2 + 2AG \times GI + GI^2$ and hence $AI^2 - GI^2$ is equal to $AG^2 + 2AG \times GI$, that is, $2AG(\frac{1}{2}AG + GI)$ or $2AG \times FI$, and consequently $2AC \times AX - AX^2$ is equal to $X\gamma^2 - 2AI \times X\gamma + 2AG \times FI$. As was to be proved.

Lemma 2. With the suppositions of the previous construction there is

$$2AE \times AX - AX^2 = (FI/FH) X\gamma^2 - 2(FI/FH) AH \times X\gamma + 2AG \times FI.$$

---

and $\gamma\sigma = GS = \sqrt{[(GH \times GS/GR)^2 + HL^2]} = \sqrt{[(-p + r/k^2)^2 k^2/n^2 + (k^2 + q)^2/n^2]}$, equal to half the main axis, whence $\gamma\rho = GR$ is half the minor axis; likewise the circle is constructed with centre *K*, the meet of *IK* ($X = p + r/n^2$) and *CK* ($Y = n + q/n$), and to pass through $G(2p, 0)$. Since *n* and *k* are arbitrary, the construction has two degrees of freedom of choice.

(706) These four lemmas (amplified versions of *Conclusiones* 2, 1, 3 and 4 on II: 478–80) develop from the construction, reversing the analysis of the preceding note, the respective defining equations $2(n + q/n) Y - Y^2 = X^2 - 2(p + r/n^2) X + 4p(r/n^2)$,

$$2(k^2/n + q/n) Y - Y^2 = (k^2/n^2) (X^2 - 2(p + r/k^2) X) + 4p(r/n^2),$$

$Y/(X - 2p) = X/2n$ and $2(r/n^2)/(Y - 2q/n) = Y/(X - 2p)$ of the circle, ellipse and pair of component parabolas through their intersections: from the latter, on implicitly discarding the factor $X - 2p = 0$, Newton deduces the required eliminant $(\frac{1}{2}X)^3 = p(\frac{1}{2}X)^2 + q(\frac{1}{2}X) + r$, where $\frac{1}{2}X = x$, exactly as in *Conclusio* 5 on II: 482.

Notum est enim quod punctum γ motu regulæ $\gamma\rho\sigma$ superius assignato describit Ellipsin cujus centrum est L et axes duo cum rectis LE et LH coincidunt, quorum qui in LE æquatur $2\gamma\rho$ sive $2GR$ et alter in LH æquatur $2\gamma\sigma$ sive $2GS$.[707] Et horum ratio ad invicem ea est quæ lineæ HR ad lineam HL sive lineæ BD ad lineam BE. Unde latus transversum est ad latus rectum principale ut BE ad BC sive ut FI ad FH. Quare cum γT ordinatim applicetur ad HL, erit ex natura Ellipseos $GS^q - LT^q$ æquale $\frac{FI}{FH} T\gamma^q$. Est autem LT æquale $AE - AX$ et $T\gamma$ æquale $X\gamma - AH$. Scribantur horum quadrata pro LT^q et $T\gamma^q$ et fiet

$$GS^q - AE^q + 2EAX - AX^q = \frac{FI}{FH} \text{ in } X\gamma^q - 2AH \times X\gamma + AH^q.$$

Est autem $GS^q - AE^q$ æquale quadrato ex $GH + LS$ propterea quod GS hypo-
||[245] tenusa est ‖ trianguli rectanguli cujus latera sunt ipsis AE et $GH + LS$ æqualia. Est et (ob similia triangula RGH RSL) LS ad GH ut LR ad HR et componendo $GH + LS$ ad GH ut HL ad $HR_{[,]}$ et duplicando rationes quadratum ex $GH + LS$ est ad GH^q ut HL^q ad HR^q hoc est (per constructionem) ut BE^q ad BD^q id est ut BE ad BC seu FI ad $FH_{[,]}$ adeoꝗ quadratum ex $GH + LS$ æquale est $\frac{FI}{FH} GH^q$. Est itaꝗ $GS^q - AE^q$ æquale $\frac{FI}{FH} GH^q$, atꝗ adeo

$$\frac{FI}{FH} GH^q + 2EAX - AX^q = \frac{FI}{FH} \text{ in } X\gamma^q - 2AH \times X\gamma + AH^q.$$

Auferatur utrinꝗ $\frac{FI}{FH} GH^q$ et restabit

$$2EAX - AX^q = \frac{FI}{FH} \text{ in } X\gamma^q - 2AH \times X\gamma + AH^q - GH^q.$$

Est $AH = AG + GH$ adeoꝗ $AH^q = AG^q + 2AGH + GH^q$ et subducto utrinꝗ GH^q restat $AH^q - GH^q = AG^q + 2AGH$ hoc est $= 2AG$ in $\frac{1}{2}AG + GH$ seu $= 2AG \times FH$, atꝗ adeo est $2EAX - AX^q = \frac{FI}{FH} \text{ in } X\gamma^q - 2AH \times X\gamma + 2AG \times FH$

$$\text{id est } = \frac{FI}{FH} X\gamma^q - \frac{2FI}{FH} AH \times X\gamma + 2AG \times FI. \text{ Q.E.D.}$$

Lemma 3. Ijsdem positis est AX ad $X\gamma - AG$ ut $X\gamma$ ad $2BC$.
Nam si de æqualibus in Lemmate secundo subducantur æqualia in Lemmate primo, restabunt æqualia $2CE \times AX$ & $\frac{HI}{FH} X\gamma^q - \frac{2FI}{FH} AH \times X\gamma + 2AI \times X\gamma$. Ducatur pars utraꝗ in FH et fiet $2FH \times CE \times AX$ æquale

$$HI \times X\gamma^q - 2FI \times AH \times X\gamma + 2AI \times FH \times X\gamma.$$

It is, of course, known that the point γ by the motion of the rule $\gamma\rho\sigma$ above assigned describes an ellipse whose centre is L and having its two axes coincident with the lines LE and LH: of these that in LE is equal to $(2\gamma\rho$ or$)$ $2GR$, and the other in LH is equal to $(2\gamma\sigma$ or$)$ $2GS$;[707] and their ratio one to the other is that of the line HR to the line HL, that is, of the line BD to the line BE. Hence the major axis is to the principal *latus rectum* as BE to BC, or as FI to FH. Consequently, since $T\gamma$ is applied as an ordinate to HL, by the nature of an ellipse there will be $GS^2 - TL^2 = (FI/FH)\,T\gamma^2$. However, TL is equal to $AE - AX$ and $T\gamma$ equal to $X\gamma - AH$. Let the squares of these latter be written in place of TL^2 and $T\gamma^2$ and there will come to be

$$GS^2 - AE^2 + 2AE \times AX - AX^2 = (FI/FH)\,(X\gamma^2 - 2AH \times X\gamma + AH^2).$$

But $GS^2 - AE^2$ is equal to $(GH + LS)^2$, seeing that GS is the hypotenuse of a right-angled triangle whose sides are equal to AE and $GH + LS$. Also (because the triangles RGH, RSL are similar) LS is to GH as LR to HR, and so, by compounding, $GH + LS$ is to GH as HL to HR, and on doubling the ratios $(GH + LS)^2$ is to GH^2 as HL^2 to HR^2, that is, (by construction) as BE^2 to BD^2, which is as BE to BC or FI to FH: hence $(GH + LS)^2 = (FI/FH)\,GH^2$. Accordingly

$$GS^2 - AE^2 = (FI/FH)\,GH^2$$

and in consequence

$$(FI/FH)\,GH^2 + 2AE \times AX - AX^2 = (FI/FH)\,(X\gamma^2 - 2AH \times X\gamma + AH^2).$$

Take $(FI/FH)\,GH^2$ from each side and there will remain

$$2AE \times AX - AX^2 = (FI/FH)\,(X\gamma^2 - 2AH \times X\gamma + AH^2 - GH^2).$$

Now $AH = AG + GH$ and therefore $AH^2 = AG^2 + 2AG \times GH + GH^2$, and when GH^2 is subtracted on each side there remains $AH^2 - GH^2 = AG^2 + 2AG \times GH$, that is, $2AG(\frac{1}{2}AG + GH)$ or $2AG \times FH$. Hence there is

$$2AE \times AX - AX^2 = (FI/FH)\,(X\gamma^2 - 2AH \times X\gamma + 2AG \times FH),$$

that is, $(FI/FH)\,X\gamma^2 - 2(FI/FH)\,AH \times X\gamma + 2AG \times FI$. As was to be proved.

Lemma 3. With the same suppositions, AX is to $X\gamma - AG$ as $X\gamma$ to $2BC$.

For if from the equality in Lemma 2 there be taken away the equality in Lemma 1 there will remain the equality

$$2CE \times AX = (HI/FH)\,X\gamma^2 - 2(FI/FH)\,AH \times X\gamma + 2AI \times X\gamma.$$

Multiply each side by FH and there will come

$$2FH \times CE \times AX = HI \times X\gamma^2 - 2FI \times AH \times X\gamma + 2AI \times FH \times X\gamma.$$

---

(707) Newton had learnt this trammel construction from Schooten's *Exercitationum Mathematicarum Libri Quinque* (Leyden, 1657): 314–17/323–5 as an undergratuate; see I: 31[f].

Est autem $AI = AH + HI$ adeoꝗ

$$2FI \times AH - 2FH \times AI = 2FI \times AH - 2FHA - 2FHI.$$

Sed $2FI \times AH - 2FHA = 2AHI$ et $2AHI - 2FHI = 2HI \times AF$. ergo

$$2FI \times AH - 2FH \times AI = 2HI \times AF$$

adeoꝗ $2FH \times CE \times AX = HI \times X\gamma^q - 2HI \times AF \times X\gamma$ et inde HI ad FH ut $2CE \times AX$ ad $X\gamma^q - 2AF \times X\gamma$. Sed per constructionem HI est ad FH ut CE ad BC atꝗ adeo ut $2CE \times AX$ ad $2BC \times AX_{[,]}$ et proinde

$$2BC \times AX \quad \text{et} \quad X\gamma^q - 2AF \times X\gamma$$

|| [246] (per prop 9 lib 5 || Elem) erunt æqualia. Æqualium verò rectangulorum proportionalia sunt latera, AX ad $X\gamma - 2AF$ (id est ad $X\gamma - AG$) ut $X\gamma$ ad $2BC$. Q.E.D.

Lemma 4. Ijsdem positis, est $2FI$ ad $AX - 2AB$ ut $X\gamma$ ad $2BC$.

Nam de æqualibus in Lemmate tertio, nimirum

$$2BC \times AX = X\gamma^q - 2AF \times X\gamma$$

subducantur æqualia in Lemmate primo, et restabunt æqualia

$$-2AB \times AX + AX^q = 2FI \times X\gamma - 2AG \times FI,$$

hoc est AX in $AX - 2AB = 2FI$ in $X\gamma - AG$. Æqualium verò rectangulorum proportionalia sunt latera, $2FI$ ad $AX - 2AB$ ut AX ad $X\gamma - AG$, hoc est (per Lem 3) ut $X\gamma$ ad $2BC$. Q.E.D.

Præstratis his Lemmatibus Constructio Problematis sic tandem demonstratur.

Per Lemma 4^{tum} est $X\gamma$ ad $2BC$ ut $2FI$ ad $AX - 2AB$, hoc est (per Prop 1 lib 6 Elem) ut $2BC \times FI$ ad $2BC \times \overline{AX - 2AB}$, seu ad $2BC \times AX - 2BC \times 2AB$. Sed per Lemma 3 est AX ad $X\gamma - 2AF$ ut $X\gamma$ ad $2BC$ seu $2BC \times AX = X\gamma^q - 2AF \times X\gamma$ adeoꝗ $X\gamma$ est ad $2BC$ ut $2BC \times FI$ ad $X\gamma^q - 2AF \times X\gamma - 2BC \times 2AB$, et ductis extremis et medijs in se fit $X\gamma^{\text{cub.}} - 2AF \times X\gamma^q - 4BC \times AB \times X\gamma = 8BC^q \times FI$. Addantur utrinꝗ $2AF \times X\gamma^q + 4BC \times AB \times X\gamma$ et fiet

$$X\gamma^{\text{cub}} = 2AF \times X\gamma^q + 4BC \times AB \times X\gamma + 8BC^q \times FI.$$

Erat autem in constructione demonstranda, $\frac{1}{2}X\gamma$ radix æquationis dicta x, nec non $AF = p$. $BC = n$. $AB = \frac{q}{n}$ & $FI = \frac{r}{nn}$, adeoꝗ $BC \times AB = q$. et $BC^q \times FI = r$. Quibus substitutis fiet $x^3 = px^2 + qx + r$. Q.E.D.

Now $AI = AH + HI$ and hence

$$2FI \times AH - 2FH \times AI = 2FI \times AH - 2FH \times AH - 2FH \times HI.$$

But $2FI \times AH - 2FH \times AH = 2AH \times HI$ and

$$2AH \times HI - 2FH \times HI = 2HI \times AF.$$

Therefore $2FI \times AH - 2FH \times AI = 2HI \times AF$ and hence

$$2FH \times CE \times AX = HI \times X\gamma^2 - 2HI \times AF \times X\gamma,$$

and in consequence HI is to FH as $2CE \times AX$ to $X\gamma^2 - 2AF \times X\gamma$. But, by construction, HI is to FH as CE to BC and so as $2CE \times AX$ to $2BC \times AX$, and therefore (by *Elements*, Book v, Proposition 9) $2BC \times AX$ and $X\gamma^2 - 2AF \times X\gamma$ will be equal. Of equal products, however, the members are in proportion: AX to ($X\gamma - 2AF$, that is) $X\gamma - AG$ as $X\gamma$ to $2BC$. As was to be proved.

Lemma 4. With the same suppositions, $2FI$ is to $AX - 2AB$ as $X\gamma$ to $2BC$.

For from the equality in Lemma 3, specifically $2BC \times AX = X\gamma^2 - 2AF \times X\gamma$, take away the equality in Lemma 1 and there will remain the equality $-2AB \times AX + AX^2 = 2FI \times X\gamma - 2AG \times FI$, that is,

$$AX(AX - 2AB) = 2FI(X\gamma - AG).$$

But of equal products the members are in proportion:

$$2FI \text{ to } AX - 2AB \text{ as } AX \text{ to } X\gamma - AG,$$

that is, (by Lemma 3) as $X\gamma$ to $2BC$. As was to be proved.

Once these preliminary lemmas are laid down, the construction of the problem is at length demonstrated in this way.

By Lemma 4 there is $X\gamma$ to $2BC$ as $2FI$ to $AX - 2AB$, that is, (by *Elements*, Book vi, Proposition 1) as $2BC \times FI$ to $2BC(AX - 2AB)$ or

$$2BC \times AX - 2BC \times 2AB.$$

But, by Lemma 3, AX is to $X\gamma - 2AF$ as $X\gamma$ to $2BC$, that is,

$$2BC \times AX = X\gamma^2 - 2AF \times X\gamma:$$

hence $X\gamma$ is to $2BC$ as $2BC \times FI$ to $X\gamma^2 - 2AF \times X\gamma - 2BC \times 2AB$, and when extremes and middles are multiplied together there comes

$$X\gamma^3 - 2AF \times X\gamma^2 - 4BC \times AB \times X\gamma = 8BC^2 \times FI.$$

Add $2AF \times X\gamma^2 + 4BC \times AB \times X\gamma$ to each side and it will become

$$X\gamma^3 = 2AF \times X\gamma^2 + 4BC \times AB \times X\gamma + 8BC^2 \times FI.$$

In the construction to be demonstrated, however, $\frac{1}{2}X\gamma$ was the equation's root called x, along with $AF = p$, $BC = n$, $AB = q/n$ and $FI = r/n^2$, so that $BC \times AB = q$ and $BC^2 \times FI = r$. And when these are substituted there will come to be $x^3 = px^2 + qx + r$. As was to be proved.

Cor. Hinc si *AF* et *AB* ponantur nulla per Lemma[708] tertium et quartum fiet 2*FI*[709] ad *AX* ut *AX* ad *Xγ* et *Xγ* ad 2*BC*.[709] Unde constat inventio duarum mediè proportionalium inter datas ‖ quaslibet *FI* et *BC*.[710]

‖ [247]

[1683]
Lect 10

Scholium. Hactenus æquationis cubicæ constructionem per Ellipsin solummodo exposui: sed regula sua natura generalior est, sese ad omnes coni sectiones indifferenter extendens. Nam si loco Ellipseos velis Hyperbolam adhiberi, cape lineas *BC, BE* ad contrarias partes puncti *B,* dein puncta *A, F, G, I, H, K, L* et *R* determinentur ut ante, excepto tantum quod *FH* debet sumi ad partes ipsius *F* contra *I* et quod *HR* non in linea *HL* sed in linea *AI* ad utramꝗ partem puncti *H* capi debet et vice rectæ *GRS* duæ aliæ rectæ a puncto *L* ad puncta duo *R* & *R* hinc inde duci pro asymptotis Hyperbolæ. Cum istis itaꝗ Asymptotis *LR, LR* describe Hyperbolam per punctum *G,* ut et circulum centro *K* intervallo *KG:* et dimidia perpendiculorum ab eorum intersectionibus ad rectam *AE* demissorum erunt radices æquationis propositæ. Quæ omnia, signis + et − probè mutatis, demonstrantur ut prius.[711]

Quod si Parabolam velis adhiberi, abibit punctum *E* in infinitum atꝗ adeo nullibi capiendum erit et punctum *H* cum puncto *F* coincidet, eritꝗ Parabola circa axem *HL* cum latere recto principali *BC*[712] per puncta *G* et *A* describenda, sito vertice ad partes puncti *F* ad quas punctum *B* situm est respectu puncti *C.*[713]

Sic sunt constructiones per Parabolam, si simplicitatem analyticam spectes, simplicissimæ omnium: Eæ per Hyperbolam proximum locum obtinent &

---

(708) Newton first thoughtlessly began to copy 'per Conclusi[onem 4]' (by Conclusion 4) from the equivalent corollary on II: 482.

(709) That is, '2*AI*' and '2*AC*' respectively, since *B* and *F* are now coincident with *A.*

(710) Or *AI* and *AC* (see previous note). The explicit construction is given at length in Newton's earlier 'Problems for construing æquations' (see II: 462). In this particular case the equation $x^3 - r = 0$ is constructed as the eliminant ($x = \frac{1}{2}X \neq 0$) of the equations $X^2 = 2nY$ and $Y^2 = 2(r/n^2) X$, *n* free, and hence as the intersections (other than the origin *A/B/F/G*) of the circle

$$X^2 + Y^2 = 2(r/n^2) X + 2nY$$

and the family of ellipses (*k* free) $X^2 + (n^2/k^2) Y^2 = 2(r/k^2) X + 2nY$; compare II: 463, note (42).

(711) In the notation of note (704) Newton in this variant replaces the family of ellipses by the equivalent family of hyperbolas (all through *G*(2*p*, 0) and the three other intersections with the circle of centre *K*) defined by

$$X^2 - (n^2/k^2) Y^2 = 2(p - r/k^2) X$$
$$+ 2n(1 - q/k^2) Y + 4pr/k^2$$

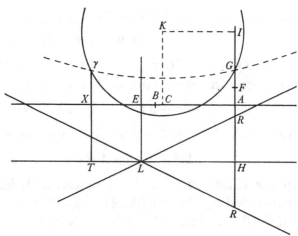

Corollary. Hence if AF and AB be set as nothing, by Lemmas[708] 3 and 4 it will come that $2FI$[709] is to AX as AX to $X\gamma$ and as $X\gamma$ to $2BC$.[709] From this the finding of two mean proportionals between any two given lengths $[A]I$ and $[A]C$ is manifest.[710]

Scholium. So far I have expounded the construction of a cubic equation by an ellipse only: but by its very nature the rule is more general, extending to all conics indifferently. Should you, for instance, in place of an ellipse wish to employ a hyperbola, take the lines BC, BE on opposite sides of the point B, and then determine the points A, F, G, H, I, K, L and R as before, except only that FH ought to be laid off on the side of F away from I, that HR ought to be taken not in the line HL but in the line AI on each side of point H, and that instead of introducing the straight line GRS two other straight lines ought to be drawn either way from the point L through the two points R to serve as the hyperbola's asymptotes. With that pair of asymptotes LR, accordingly, describe the hyperbola through point G, and also a circle with centre K, radius KG: the halves of the perpendiculars let fall from their intersections to the straight line AE will then be the roots of the equation proposed. All these things are, with appropriate interchange of the signs $+$ and $-$, demonstrated as in the above.[711]

But should you wish to employ a parabola, the point E will go off to infinity and hence will not need to be taken anywhere, while point H will coincide with point F, and the parabola will have to be described about axis HL through the point G and A and with principal *latus rectum* $[2]BC$, its vertex lying on the same side of F on which B is located with respect to the point C.[713]

Thus, constructions by a parabola are, if you regard their analytical simplicity, simplest of all; those by a hyperbola hold next place; and the last place is retained by those accomplished by an ellipse. But if regard be had for the

---

or $$(X-(p-r/k^2))^2 - (n^2/k^2)\,(Y-(q-k^2)/n)^2 = (p+r/k^2)^2 + (k-q/k)^2,$$

whose asymptotes LR have the equation

$$X-(p-r/k^2) = \pm\,(n/k)\,(Y-(q-k^2)/n)$$

and hence pass through the centre $L(p-r/k^2,\ (q-k^2)/n)$ at a slope $HR/HL = \pm n/k$. Compare II: 484, note (90).

(712) A trivial slip for '$2BC$', copied without correction from the equivalent text in Newton's 'Problems...' (II: 484; compare 485, note (92)).

(713) In this limiting case ($k = \infty$) Newton, exactly as earlier (compare II: 485, note (93)), constructs the meets of the circle with the parabola defined by $X^2 = 2pX + 2nY$, through $A(0, 0)$ and $G(2p, 0)$ and with axis $X = p$ (through F) meeting it in the vertex $(p,\ -\tfrac{1}{2}p^2/n)$ in the direction of C from AI ($Y = 0$).

ultimum locum tenent quæ per Ellipsin absolvuntur. Quod si praxeos manualis in describendis figuris spectetur simplicitas, mutandus est ordo.

In hisce autem constructionibus observandum venit quod proportione lateris ∥[248] recti principalis ad latus transversum determinatur species ∥ Ellipseos et Hyperbolæ et proportio illa[714] eadem est quæ linearum *BC* et *BE* atcɜ adeo assumi potest. Parabolæ verò species est unica quam artifex ponendo *BE* infinitè longam[715] assequitur. Sic igitur penes Artificem est æquationem quamcunɢɜ cubicam per Conicam sectionem imperatæ speciei construere. A figuris autem specie datis ad figuras magnitudine datas devenietur augendo vel diminuendo in ratione data lineas omnes quibus figuræ specie dabuntur, atcɜ ita æquationes omnes cubicas per datam quamvis Conicam sectionem construere licebit. Id quod sic plenius explico.[716]

Proponatur æquationem quamcunɢɜ cubicam $x^3 = pxx . qx . r$, ope datæ cujuscunɢɜ sectionis conicæ construere.

A puncto quovis *B* in recta quavis infinita *BCE* cape duas quascunɢɜ longitudines *BC BE* ad easdem partes si data Coni sectio sit Ellipsis, ad contrarias si ea sit Hyperbola. Sit autem *BC* ad *BE* ut datæ sectionis latus rectum principale ad latus trans-

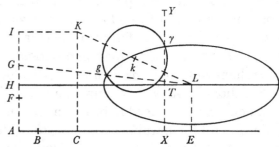

versum, et *BC* nominata *n*, cape $BA = \dfrac{q}{n}$ idcɜ versus *C* si habeatur $-q$, aliter ad partes contrarias. Ad punctum *A* erige perpendiculum *AI* incɜ eo cape *AF* æqualem *p* et *FG* æqualem *AF*; item *FI* æqualem $\dfrac{r}{nn}$. Capiatur vero *FI* versus *G* si termini *p* et *r* habent eadem signa, aliter versus *A*. Dein fac ut sit ∥[249] *FH* ad *FI* ut *BC* ad *BE* et hanc *FH* cape a puncto *F* ∥ versus *I* si sectio sit Ellipsis, aut ad partes contrarias si ea sit Hyperbola. Porrò compleantur parallelogramma *IACK* et *HAEL* et hæ omnes jam descriptæ lineæ transferantur ad datam sectionem Conicam, aut quod perinde est his superponatur curva ita ut axis ejus sive transversa diameter principalis conveniat cum recta *LH* et

---

(714) Namely n^2/k^2; see note (705).

(715) That is, $k = \infty$. A second parabola defined (see note (705)) by

$$n^2Y^2 = 2qnY + 2rX - 4pr$$

will, in fact, ensue on setting $k = 0$, when points *D* and *E* come to coincide with *B*, and point *H* passes to infinity in *AG*.

(716) The following construction is all but a *verbatim* repeat of 'Prob. 5' of his earlier

simplicity of their manual procedure in describing figures, that order must be inverted.

In these constructions, however, it occurs to me to observe that by the ratio of the principal *latus rectum* to the major axis the species of ellipse and hyperbola is determined, and that that ratio[714] is the proportion of the lines *BC* and *BE*, and hence we can assume it. In the case of the parabola, however, there is a unique species: this the practised geometer attains by setting *BE* infinite in length.[715] It is thus within the skilled geometer's power, therefore, to construct any cubic equation by a conic of appointed species. Figures given (also) in magnitude are arrived at from ones given in species, however, by increasing or diminishing in given proportion all lines which determine the figures in species, and in this way we are at liberty to construct all cubic equations by means of any given conic. This I explain more fully thus:[716]

Let it be proposed to construct any cubic equation $x^3 = px^2 + qx + r$ whatsoever with the help of any given conic.

From any arbitrary point *B* in any unbounded straight line *BCE* lay off any two lengths *BC*, *BE*— on the same side if the conic be an ellipse, on opposite ones if it be a hyperbola: let, however, *BC* be to *BE* as the principal *latus rectum* of the given conic to its major axis. Then, on naming *BC* n, take $BA = q/n$: this towards *C* if there be had $-q$, otherwise in the opposite direction. At the point *A* erect the perpendicular *AI*, and in it take $AF = p$, $FG = AF$ and likewise $FI = r/n^2$: take *FI* towards *G*, however, if the coefficients p and r have the same sign, otherwise take it towards *A*.

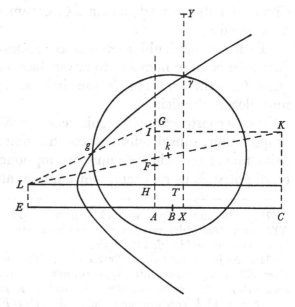

otherwise take it towards *A*. Next, make *FH* to *FI* as *BC* to *BE*, taking this *FH* from point *F* towards *I* if the conic be an ellipse, but, alternatively, the opposite way if it be a hyperbola. Furthermore, complete the rectangles *IACK* and *HAEL*, and transfer all these lines now described to the given conic—or, what is exactly the same, superimpose that curve on these so that its (major) axis or principal transverse diameter coincides with the straight line *LH* and its centre

'Problems for construing æquations' (see II: **484**–8). For convenience we have set Newton's pair of diagrams one with his Latin text and the other with our English version.

centrum cum puncto *L*. His ita constitutis agatur recta *KL* ut et recta *GL* secans conicam sectionem in *g*. In *LK* cape *Lk* quæ sit ad *LK* ut *Lg* ad *LG*, centroq; *k* et intervallo *kg* describe circulum. A punctis ubi hic secuerit curvam impositam demitte perpendicula ad lineā *LH*, cujusmodi sit γ*T*. Deniq; versus γ cape *TY* quæ sit ad *T*γ ut *LG* ad *Lg* et hæc *TY* producta secet rectam *AB* in *X*, eritq; recta [½]*XY*[(717)] una ex radicibus æquationis. Sunt autem radices affirmativæ quæ jacent ad partes rectæ *AB* ad quas recta *FI* jacet a puncto *F*, & negativæ quæ jacent ad contrarias partes si modò habeatur +*r*, et contra si −*r* obvenerit.[(718)]

Hoc modo construuntur æquationes cubicæ per Ellipses et Hyperbolas datas. [Q]uod si detur Parabola, capienda est *BC*[(719)] æqualis lateri recto ipsius, Dein punctis *A*, *F*, *G*, *I* et *K* inventis ut ante, centro *K* intervallo *KG* describendus est circulus, et Parabola ita applicanda ad schema jam descriptum (aut schema ad Parabolam) ut ipsa transeat per puncta *A* et *G*, & axis ejus ipsi *AC* parallelus per punctum *F*, cadente vertice ad partes puncti illius *F* ad quas punctum *B* cadit a puncto *C*. His ita constitutis, si perpendicula ab ejus occursibus cum circulo demittantur ad lineam *BC*, eorum dimidia erunt radices æquationis construendæ.

[‖ [250]] ‖ Et notes quod ubi secundus æquationis terminus deest et latus rectum Parabolæ ponitur numerus binarius, hæc constructio evadet eadem cum illa quam Cartesius attulit in Geometria[(720)] sua, præterquam quod lineamenta hic sunt illorum duplicia.

Hæc est constructionum regula generalis. Verum ubi Problemata particularia proponuntur, consulendum est constructionum formulis simplicissimis. Libera enim manet quantitas *n* cujus assumptione constructio plerumq; simplicior reddi potest. Ejus rei exemplum unum et alterum[(721)] subjungo.

---

(717) The manuscript reads '*XY*' simply, a slip in transcribing the earlier 'dimidium ipsius *XY*' (see II: 488) which was passed by Whiston into the 1707 *editio princeps*, to be later corrected by Newton himself in his 1722 revise.

(718) As in his preceding 'Problems...' Newton adapts the previous construction to a given ellipse of given eccentricity ϵ—or, more accurately, one in which the ratio $(1-\epsilon^2)$ of its *latus rectum* to its main axis (*latus transversum*) is known—by equating the latter ratio to $BC/BE = n^2/k^2$, *k* free, whence *n* and so the points *F*, *G*, *H*, *I*, *K*, *L* are fixed. On setting the given ellipse with its centre at *L* and main axis along *LH*, and then drawing *LG* to intersect the ellipse in *g*, the preceding construction may be carried out 'in miniature' on the given ellipse if we determine the meets γ of a circle through *g* and of centre *k*, where

$$Lk/LK = BC/BE = n^2/k^2 (= 1-\epsilon^2)$$

and then, on reversing the homethetic transformation, 'expand' the scale to derive the intersections γ (in the previous construction) of the circle through *G* of centre *K* with the similarly expanded ellipse (of same centre *L*) through *G*. (As will be clear from the figures on II: 487, where the latter intersections are denoted by Γ, these will in fact lie at the meet of *L*γ with the parallel to *LH* through *Y*, but of course here only their distances $T\gamma \times LG/Lg = TY$ are of interest.) The following, newly added paragraph explicitly lays out the analogous extension of the parabolic case of the previous construction.

with the point L. With this set-up, draw the straight line KL and also the straight line GL cutting the conic at g; in LK take Lk, which shall be to LK as Lg to LG, and with centre k, radius kg describe a circle; then from the points in which this cuts the superimposed curve let fall perpendiculars—such as γT— to the line LH. Finally, towards γ take TY, which shall be to $T\gamma$ as LG to Lg, and let this TY, when produced, intersect the straight line AB in X: the line $\frac{1}{2}XY$[717] will then be one of the roots of the equation. Positive roots, however, are those which lie on the same side of the line AB as that on which the line FI lies away from the point F, the negatives those lying on the opposite one, provided there be had $+r$; the converse holds if $-r$ occurs.[718]

In this manner cubic equations are constructed by means of given ellipses and hyperbolas. But if a parabola be given, you should take $[2]BC$[719] equal to its *latus rectum* and then, having ascertained the points A, F, G, I and K as before, you must describe a circle with centre K and radius KG; the parabola is now to be applied to the configuration already described (or the configuration to the parabola) so that it passes through the points A and G, and its axis—parallel to AC—through the point F, its vertex lying on the same side of F as that on which point B falls from point C. With this set-up, if perpendiculars are let fall from its meets with the circle to the line BC, their halves will be the roots of the equation to be constructed.

And note that when the second term in the equation is lacking and the parabola's *latus rectum* is set to be the number two, this construction will prove to be identical with that imparted by Descartes in his *Geometry*,[720] except for the individual lines here being double those there.

This is a general rule for constructions. But when particular problems are proposed, you must look out for the simplest forms of construction: for the quantity n remains free and its careful choice can very often make the construction simpler. I append an illustration or two[721] of this point.

---

(719) Read '$2BC$' since the defining equation of the present parabola is (compare note (713)) $(X-p)^2 = 2n(Y+\frac{1}{2}p^2/n)$, of parameter $2n$. The slip stands uncorrected in the printed editions.

(720) *Geometrie*, $_1$1637: 388–95 [= *Geometria*, $_2$1659: 85–90]; compare II: 491, note (101). In this case, $p = 0$, the cubic $x^3 = qx+r$ is (compare note (705)) constructed by adding the root $x = 0$ and then expressing $x^4 = qx^2+rx$ as the eliminant of $x^2 = ny$ and $n^2y^2 = qny+rx$, and so of $x^2+y^2 = (r/n^2)x+(n+q/n)y$. On putting $n = 1$ and defining Cartesian coordinates $X = 2x$, $Y = 2y-(q+1)$, the Cartesian construction by the parabola $X^2 = 2(Y+q+1)$ and circle $(X-r)^2+Y^2 = (q+1)^2+r^2$ results except for a doubling in its 'lineaments'.

(721) In a clear attempt to disguise the essential incompleteness of the present manuscript, which contains only one further example in its existing state, Whiston altered this to read 'exemplum unum' simply, changing the sense to 'a single illustration'. The emendation was, consciously or no, accepted by Newton in his 1722 revise. In line with his pages 233–6 above it would have been entirely appropriate for Newton to have constructed, as his second example, the trisection cubic $x^3 - 3a^2x = \pm a^2b$.

Detur Ellipsis et inter datas lineas a et b inveniendæ sint duæ mediæ proportionales. Sit earum prima x et a. x. $\dfrac{xx}{a}$. b erunt continuè proportionales adeoqʒ $ab = \dfrac{x^3}{a}$ seu $x^3 = aab$ æquatio est quam construere oportet. Hic desunt termini p, et q et terminus r est aab adeoqʒ BA et AF nullæ sunt et FI est $\dfrac{aab}{nn}$. Ut terminus novissimus evadat simplicior assumatur $n = a$ et fiet $FI = b$. Deinde constructio ita se habebit.

A puncto quovis A in recta quavis infinita AE cape $AC = a$ et ad easdem partes puncti A cape AE ad AC ut est Ellipseos latus transversum ad latus rectum principale.[(722)] Tum in perpendiculo AI cape $AI = b$ et AH ad AI ut est AC ad AE. Compleantur parallelogramma $IACK$, $HAEL$. Jungantur LA LK. Huic schemati imponatur Ellipsis data.[(723)] Secet ea rectam AL in puncto g. Fiat Lk ad LK ut Lg ad LA. Centro k intervallo kg describatur circulus secans Ellipsin in γ. Ad AE demittatur perpendiculum γX secans HL

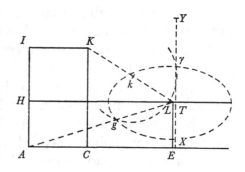

in T et producatur id ad Y ut sit TY ad $T\gamma$ sicut $[L]A$ ad $[L]g$.[(724)] Sic fiet ‖[251] $[\tfrac{1}{2}]XY$[(725)] prima ‖ duarum mediè proportionalium x.[(726)]

---

(722) By a thoughtless slip—reproduced by Whiston in his *editio princeps* but corrected by Newton himself in his 1722 revise—the manuscript reads invertedly '...ut est Ellipseos latus rectum principale ad latus transversum'.

(723) Understand so that its centre is at L and main axis along LH.

(724) The manuscript carelessly reads 'TA ad Tg', a slip reproduced by Whiston in his *editio princeps* but corrected by Newton in 1722.

(725) Again (compare note (717)) the manuscript reads 'XY', to be reproduced by Whiston in 1707 and corrected by Newton in 1722.

(726) The corollary to the previous theorem on his pages 246–7 which determined the means $\tfrac{1}{2}X\gamma = x$ and $\tfrac{1}{2}AX = y$ between $AC = n$ and $AI = r/n^2$ by constructing the roots of the cubic $x^3 - r = 0$ as the intersections ($x = \tfrac{1}{2}X$, $y = \tfrac{1}{2}Y$) of the circle $X^2 + Y^2 = 2(r/n^2)X + 2nY$ and the ellipse $X^2 + (n^2/k^2)Y^2 = 2(r/k^2)X + 2nY$ is now reproduced in minuscule as the ellipse $g\gamma$ of equal eccentricity $\sqrt{[1 - n^2/k^2]}$.

In sequel, though a little before (see note (721)) he had clearly stated his intention of adding a second example in illustration of this elegant homothetic transformation of the preceding construction, Newton breaks off abruptly without peroration of any kind (and leaving a number of blank pages following). Freed from his statutory obligation to gather together enough—and no more!—material for deposit as the permanent record of the preceding eleven years, 1673–83, of his lectures from the Lucasian chair, he might well have gone on to repeat from his earlier 'Problems for construing æquations' the ensuing conic constructions of the

Given an ellipse, let it be required to find two mean proportionals between given lines a and b.

Let the first of these be x and then a, x, x^2/a, b will be in continued proportion, so that $ab = x^3/a$ or $x^3 = a^2b$ is the equation which it is necessary to construct. Here the terms in p and q are lacking, while the term r is a^2b; hence BA and AF are zero and FI is a^2b/n^2. To render the most recent quantity simpler, assume $n = a$ and there will come to be $FI = b$. In sequel, the construction will then be conducted as follows.

From an arbitrary point A in any unbounded straight line AE take $AC = a$, and on the same side of point A take AE to AC as the ellipse's major axis to its principal *latus rectum*. Then in the perpendicular AI take $AI = b$ and then AH to AI as AC is to AE. Complete the rectangles $IACK$, $HAEL$ and join LA, LK. On this configuration superimpose the given ellipse.[723] Let it cut the straight line AL in the point g, and then make Lk to LK as Lg to LA, and with centre k, radius kg describe a circle intersecting the ellipse in γ. To AE let fall the perpendicular γX cutting HL in T and produce it to Y so that TY shall be to $T\gamma$ as LA to Lg.[724] In this way $\frac{1}{2}XY$[725] will prove to be the first, x, of the two mean proportionals.[726]

---

general quartic $x^4 + px^3 + qx^2 + rx + s = 0$ 'modis dupliciter infinitis' (II: 490–8) and perhaps also its final constructions of higher equations (up to ninth degree) as the meet of a given Wallisian cubic with conics or other cubics (II: 498–508). Perhaps here, too, he intended to elaborate the conchoidal constructions of quartic equations to which he referred *en passant* on his page 217 above (compare note (637)). We have already suggested (see note (615)) that it was probably originally his plan to conclude with a number of *regulæ* for approximately extracting the roots of cubic, quartic (and higher?) equations *in numeris* on the lines of the rules in his earlier *De Analysi* (see II: 218–20; and compare note (170) above). But this—and all else—must remain conjectural in default of any documented knowledge of what exactly Newton further intended to include in his deposited manuscript.

§2. NEWTON'S 'CORRECTIONS' TO THE MAIN TEXT.[1]

CORRECTIONES.

Ad pag 2. l 1.[2]

DE VOCUM QUARUNDAM ET NOTARUM SIGNIFICATIONE.

1.[3] Per *Numerum* non tam multitudinem unitatum quam abstractam quantitatis cujusvis ad aliam ejusdem generis quantitatem quæ pro unitate habetur rationem intelligimus. Estɋ triplex; integer, fractus et surdus: Integer quem unitas metitur, fractus quem unitatis pars submultiplex[4] metitur, et surdus cui unitas est incommensurabilis.

Integrorum numerorum notas 0, 1, 2, 3, 4, 5, 6, 7, 8, 9, et notarum —— millesimas &c. Hos autem dicimus fractos *decimales* quòd in ratione decimali perpetuò decrescant. Et ad distinguendum integros a decimalibus interjici solet comma vel punctum vel etiam lineola —— millesimas partes. Surdorum et aliorum fractorum notæ in sequentibus habentur.

Cum rei alicujus ——

Quantitates vel affirmativæ sunt ——

In aggregato quantitatum nota + significat quantitatem suffixam esse cæteris addendam et nota − esse subducendam. Et has notas —— valet 8. Item $a+b$ valet summam quantitatum a et b et $a-b$ valet differentiam quæ

---

(1) These draft revisions of portions of the preceding main text of Newton's algebraic '*lectiones*', set in the manuscript on fourteen unnumbered pages in immediate sequel (ULC. Dd. 9.68: 253–6/259–68), should be treated with some caution. The opening 'correction', marked to replace the earlier opening section 'De Notatione' (on his preceding pages 2–5), is penned in a careful hand and, as we have noticed (§1: note (8)), Newton made a slight corresponding change in his main text to allow for its insertion; however, in his subsequent unfinished revise, the 'Arithmeticæ Universalis Liber Primus' (ULC. Add. 3993, reproduced in Part 2 below), it is significant that he preferred to repeat his original section 'De Notatione' without taking heed of the additional passages in the present augmented version. The two remaining sets of 'corrections', much more roughly scrawled with extensive cancellations and insertions, have a yet more dubious status: the section 'De inventione Divisorum', here intended to be included at the foot of his preceding page 23, was in fact subsequently entered in considerably revised and augmented form on his later pages 173–87 in a somewhat different context; while the concluding paragraphs 'De Reductione Radcaliū ad simpliciores radicales per extractionem radicum', scheduled here to be set at the end of his previous page 29 'in calce capitis *de Reductione radicalium*', was afterwards inserted on his following pages 207–10 in a more finished version later repeated, with the trivial addition of but a single phrase (see §1: note (603)), in the corresponding section on the 'Extractio Radicum' in his subsequent 'Arithmeticæ Universalis Liber Primus'. Given our total ignorance of the historical circumstances in which Newton made deposit of his algebraic Lucasian *lectiones* in the middle 1680's, we may readily accept the opening 'correction' as a genuine improvement upon the related

Translation

CORRECTIONS

At page 2, line 1.[2]

THE MEANING OF CERTAIN APPELLATIONS AND SYMBOLS

1.[3] By a 'number' we understand not so much a multitude of units as the abstract ratio of any quantity to another quantity of the same kind which is considered to be unity. It is threefold: integral, fractional and surd. An integer is measured by unity, a fraction by a submultiple part of unity,[4] while a surd is incommensurable with unity.

The symbols (0, 1, 2, 3, 4, ...) for integers and [the values] of [those] symbols ... thousandths and so forth. These, however, we call 'decimal' fractions because they perpetually decrease in a decimal ratio. And to distinguish integers from decimals a comma, a point or even a short line is usually interposed. ... thousandths. Symbols for surds and other fractions are treated in the sequel.

When [the quantity] of some object ....

Quantities are either positive....

In a collection of quantities the symbol $+$ signifies that the quantity appended to it must be added to the rest, the symbol $-$ that it must be taken away from them. And these symbols ... the value [of $6-1+3$] is 8. Likewise the value of $a+b$ is the sum of the quantities a and b, and that of $a-b$ is the difference which

section in the main text preceding, but we cannot but be mystified over the reason why—if this was more than a pure accident—the two latter inferior draft revises came to be included with it. In his 1707 *editio princeps* of the text (see §1: note (2)) William Whiston uncritically accepted these 'corrections' at their face-value, incorporating them at their indicated places in the pages of his printed *Arithmetica* and, in the case of the two latter sections, omitting the corresponding revised versions with little thought for the hiatus left by their deletion (compare §1: notes (517) and (596)). Other than making some small attempt to smooth over the two gaps thus opened up by Whiston's editing of his manuscript, Newton in his 1722 revise accepted this restructuring as a *fait accompli* and it is accordingly preserved in all subsequent editions.

In our reproduction of the text of these 'corrections' we have, following Newton's precedent in the opening paragraphs, extensively omitted sections of text where these are identical with the equivalent portions of the main manuscript (reproduced in §1 preceding). To forestall any confusion, we may add that the use of a long hyphen '——' to indicate an omitted passage is Newton's convention (here accurately reproduced without editorial tampering), while the equivalent employment of sets of suspension points '...' is our modern editorial insertion. In either case the missing portions are readily filled in by referring back to the main text preceding, while the complete texts are given in all printed editions of the *Arithmetica*.

(2) Understand that the following revision is to be inserted in place of the section 'De Notatione' on Newton's pages 2–5 in the preceding main text; compare §1: note (9).

(3) The manuscript has no other related section-numbers '2', '3', ... In his 1707 *editio princeps* (see §1: note (2)) Whiston silently omitted this stray, purposeless unit.

(4) A 'proper' part of unity, that is—one of which an integral multiple is unity.

oritur subducendo b ab a. Et $a-b+c$ valet summam istius differentiæ et quantitatis c. Puta si a sit 5, b 2, et c 8: tum $a+b$ valebit 7 et $a-b$ 3, et $a-b+c$ 11.[5] Hæ autem notæ $+$ et $-$ dicuntur *signa*. Et ubi neutrum initiali quantitati præfigitur, signum $+$ subintelligi debet.

Multiplicatio propriè dicitur quæ fit per numeros integros, utpote quærendo novam quantitatem tot vicibus[6] majorem quantitate multiplicanda quot numerus multiplicans sit major unitate. Sed aptioris vocabuli defectu Multiplicatio etiam dici solet quæ fit per fractos aut surdos numeros quærendo novam quantitatem in ea quacunꝗ ratione[7] ad quantitatem multiplicandam quam habet multiplicator ad unitatem. Necꝗ tantùm fit per abstractos numeros sed etiam per concretas quantitates, ut per lineas, superficies, motum localem, pondera &c, quatenus hæ ad aliquam sui generis notam quantitatem tanquam unitatem relatæ, rationes numerorum exprimere possunt & vices supplere. Quemadmodum si quantitas A multiplicanda sit per lineam duodecim pedum, posito quod linea bipedalis sit unitas, producentur per istam multiplicationem $6A$, sive sexies A, perinde ac si A multiplicaretur per abstractum numerum 6, siquidem $6A$ sit in ea ratione ad A quam habet linea duodecim pedum ad unitatem bipedalem.[8] Atꝗ ita si duas quasvis lineas AC et AD per se multiplicare oportet, capiatur AB unitas, et agatur BC eiꝗ parallela DE, et AE productum erit hujus multiplicationis, eo quod sit ad AD ut AC ad unitatem AB. Quinetiam mos obtinuit ut genesis seu descriptio superficiei per lineam super alia linea ad rectos angulos moventem dicatur multiplicatio istarum

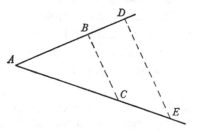

linearum.[9] Nam quamvis linea utcunꝗ multiplicata non possit evadere superficies, adeoꝗ hæc superficiei e lineis generatio longè alia sit a multiplicatione, in hoc tamen conveniunt quod numerus unitatum in alterutra linea multiplicatus per numerum unitatum in altera producat abstractum numerum unitatum in superficie lineis istis comprehensa, si modò unitas superficialis definiatur, ut solet, Quadratum cujus latera sunt unitates lineares.[10] Quemadmodum si recta AB constet quatuor unitatibus et AC tribus, tum

---

(5) Without any manuscript authority Whiston here in his 1707 *editio princeps* reinserted from Newton's main text (on his page 3) the omitted sentences 'Item $2a+3a$ valet $5a$. Et $3b-2a-b+3a$ valet $2b+a$; nam $3b-b$ valet $2b$ et $-2a+3a$ valet a, quorum aggregatum est $2b+a$. Et sic in alijs' (Likewise the value of $2a+3a$ is $5a$. And the value of $3b-2a-b+3a$ is $2b+a$; for that of $3b-b$ is $2b$ and that of $-2a+3a$ is a, the aggregate of which is $2b+a$. And so in other cases). Since Newton subsequently retained this interpolated passage in his 1722 revise, its present omission may well have been a mere oversight.

(6) Quite inexplicably, Whiston in his 1707 *editio* replaced this by the illogical and ungrammatical phrase 'toties quoties': in an elegant correction first made in his library copy of that edition and subsequently introduced into his 1722 revise, Newton retained 'toties' in the present position and converted the following 'quot' into a corresponding 'quoties'.

ensues on taking b away from a. And the value of $a-b+c$ is the sum of that difference and of the quantity c. If, say, a be 5, b 2 and c 8, then the value of $a+b$ will be 7, that of $a-b$ 3, and that of $a-b+c$ 11.[5] These symbols + and −, however, are called 'signs'. And when neither is prefixed to an initial quantity, the sign + should be understood.

Multiplication, properly so called, is effected by means of integers—to wit, by ascertaining a new quantity the same number of times greater than the quantity to be multiplied as the multiplying number is greater than unity. But for want of a more suitable word the name 'multiplication' is also usually given to the procedure of ascertaining, by means of fractions or surds, a new quantity in proportion to the quantity to be multiplied as the ratio, whatsoever it is,[7] which the multiplier has to unity. Nor is it effected merely by means of abstract numbers, but also through the agency of concrete quantities—as by lines, surfaces, local motion, weights and so on—in so far as these, when referred to some known quantity of their own kind as unity, are able to express the ratios of numbers and fill their rôle. Thus, for instance, if the quantity A has to be multiplied by a line of twelve feet, supposing that a two-foot line is the unit, there will be produced by that multiplication $6A$, or six times A, exactly as if A were to be multiplied by the abstract number 6, seeing that $6A$ is in proportion to A in the ratio of the line of twelve feet to the two-foot unit.[8] And so, if it is necessary to multiply any two lines AC and AD together, take AB to be the unit, draw BC and parallel to it DE, and then AE will be the product of this multiplication, for the reason that it is to AD as AC to the unit AB. In fact, the custom obtains of calling the genesis or description of a surface (plane area) by a line moving at right angles upon another line the multiplication of those lines.[9] For, although a line, however it be multiplied, cannot come to be a surface, and hence this generation of a surface by lines is far different from multiplication, yet they concur inasmuch as the number of units in one or other of the lines when multiplied by the number of units in the second produces the abstract number of units in the surface comprised by those lines, provided that the surface unit be (as usually it is) a square whose sides are linear units.[10] If, for instance, the

---

(7) Namely, integral, fractional or surd (general algebraic).

(8) Newton first concluded '...siquidem duodecim illi pedes valebunt sex unitates' (seeing that those twelve feet will have the value of six units).

(9) Compare the scholium (III: 344) to Theorem 5 of Newton's earlier fluxional addendum to his 1671 tract. The present adaptation of Euclid, *Elements*, VI, 2, as a geometrical model for constructing the product $AE = a \times b$ of the magnitudes $AC = a$ and $AD = b$ in terms of the linear unit $AB = 1$ is taken over from the first book of Descartes' *Geometrie* (*Discours de la Methode* (Leyden, 1637): 298 = *Geometria* (Amsterdam, $_2$1659): 2). We have seen (IV: 490–2, 520) that Newton in his contemporary 'Geometria Curvilinea' makes good use of the model in his alternative derivation of the rule for expressing the fluxion of the product of two variable quantities.

(10) Or, more generally, by a rectangle of unit area.

rectangulum *AD* constabit quater tribus seu duodecim unitatibus quadratis
ut inspicienti schema patebit. Estqǫ similis analogia
solidi et ejus quod continua trium quantitatum
multiplicatione producitur. Et hinc vicissim evenit
quod vocabula *ducere, contentum, rectangulum, quadratum, cubus, dimensio, latus,* et similia quæ ad Geometriam spectant, Arithmeticis tribuantur operationibus. Nam per *quadratum,* vel *rectangulum,* vel
quantitatem duarum dimensionum non semper intelligimus

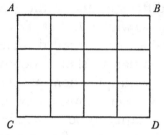

superficiem, sed utplurimùm quantitatem alterius cujuscunqǫ generis quæ
multiplicatione aliarum duarum quantitatum producitur, et sæpissimè lineam
quæ producitur multiplicatione aliarum duarum linearum. Atqǫ ita dicimus
cubum vel *Parallelepipedum,*[11] vel *quantitatem trium dimensionum* pro eo quod binis
multiplicationibus producitur, *latus* pro radice, *ducere* pro multiplicare, & sic
in alijs.[12]

Numerus speciei alicui immediatè præfixus denotat speciem illam toties
sumendam sive per numerum istum multiplicandam esse. Sic 2*a* denotat duo
a, 3*b* tria *b,* 15*x* quindecim *x.*

Duæ vel plures species ——

Inter quantitates se multiplicantes nota × vel vocabulum *in* ——

Divisio propriè est quæ fit per numeros integros quærendo novam quantitatem tot vicibus minorem quantitate dividenda quot unitas sit minor Divisore.
Sed ob analogiam vox etiam usurpari solet
cum nova quantitas in ratione quacunqǫ ad
quantitatem dividendam quæritur quam habet
unitas ad divisorem: sive divisor ille sit fractus
aut surdus numerus aut alia cujusvis generis
quantitas. Sic ad dividendum lineam *AE* (fig
1)[13] per lineam *AC,* existente *AB* unitate:
agenda est *ED* parallela *CB,* et erit *AD* Quotiens.
Imò et Divisio propter similitudinem quandam

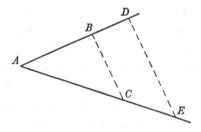

dicitur cum rectangulum ad datam lineam tanquā Basem applicatur ut inde
noscatur altitudo.

Quantitas infra quantitatem ——
Aliqando divisor ——
Etsi multiplicatio ——
Si quantitas seipsam multiplicet ——[14]

---

(11) The context requires that this 'parallelepiped' shall have rectangular faces.

(12) This Cartesian divorce of arithmetical (algebraic) from spatial (geometrical) dimension was already made in late Greek geometry—Pappus, for instance, did not hesitate to
generalize the Euclidean *locus ad tres et quatuor lineas* to comprehend the comparable loci defined

straight line AB consists of 4 units and AC of 3, then the rectangle (product) AD will consist of four times three or twelve square units, as will be evident by an inspection of the diagram. There is a like analogy between a solid and the product of the continued multiplication of three quantities. And hence, conversely, it comes about that the vocabulary of 'drawing [into]', 'content', 'rectangle', 'square', 'cube', 'dimension', 'side' and similar words with geometrical reference are bestowed on arithmetical operations. So, for example, by a 'square' or 'rectangle' or 'quantity of two dimensions' we do not always understand a plane area, but very frequently a quantity of any other kind which is produced by the multiplication of two other quantities, and most often a line produced by the multiplication of two other lines. And thus we say 'cube' or 'solid rectangle'[11] or 'quantity of three dimensions' to denote the product of two multiplications at a time, 'side' for root, 'drawing [into]' for multiplying, and the like in other instances.[12]

A number prefixed immediately before some variable signifies that that variable is to be taken an equal number of times, or multiplied by that number. Thus $2a$ denotes two a's, $3b$ three b's, $15x$ fifteen times x.

Two or more variables....

Between quantities multiplying each other the symbol '$\times$' or the word 'into'....

Division is properly what is effected by means of integers on ascertaining a new quantity which is as many times less than the quantity to be divided as unity is less than the divisor. But by analogy the appellation is usually also employed when a new quantity is sought in proportion to the quantity to be divided in the ratio, whatever it be, of unity to the divisor—and whether that divisor be a fraction or a surd or another quantity of any kind. Thus, to divide the line AE (first figure)[13] by the line AC, with AB again the unit, ED must be drawn parallel to CB and then AD will be the quotient. To be sure, because of a certain similarity it is even called division when a rectangle is applied to a given line as base so as to find out its height therefrom.

A quantity [set] below a quantity....

Sometimes the divisor....

Even though multiplication....

If a quantity should multiply itself....[14]

---

by the products of four, five and more lines, though he was careful to add that no geometrical 'rectangle' of more than three dimensions was possible (see his remarks on the third book of Apollonius' *Conics* in the opening preamble to Book VII of his *Mathematical Collection*)—but the distinction still led to much philosophical dispute in Newton's own day. His present discussion of the point is refreshingly brief and uncluttered.

(13) Here repeated for convenience.

(14) Observe that this paragraph is now inserted after the three preceding ones (which follow it in Newton's earlier main version; see §1: note (18)).

Cùm autem radix per seipsam multiplicata producat quadratum, et quadratum illud iterum per radicem multiplicatum producat cubum &c: erit (ex definitione Multiplicationis) ut unitas ad radicem ita radix ad quadratum et quadratum ad cubum &c. Adeoꝗ quantitatis cujuscunꝗ radix quadratica erit medium proportionale inter unitatem et quantitatem illam, et radix cubica primum e duobus mediè proportionalibus, et radix quadrato-quadratica primum e tribus, et sic præterea. Duplici igitur affectione radices innotescunt, tum quod seipsas multiplicando producant superiores potestates, tum quod sint e medijs proportionalibus inter istas potestates et unitatem. Sic numeri 64 radicem quadraticam esse 8 et cubicam 4, vel ex eo patet quod 8×8 et $4 \times 4 \times 4$ valeant 64, vel quod sit 1 ad 8 ut 8 ad 64, et 1 ad 4 ut 4 ad 16 et 16 ad 64. Et hinc si lineæ alicujus *AB* radix quadratica extrahenda est, produc ad *C* ut sit *BC* unitas, dein super *AC* describe semicirculum et ad *B* erige perpendiculum huic circulo occurrens in *D*, eritꝗ *BD* radix quia media proportionalis est inter *AB* et unitatem *BC*.[15]

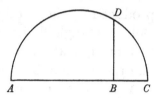

Ad designandam radicem alicujus quantitatis præfigi solet nota $\sqrt{}$ si radix sit quadratica, et $\sqrt{3}$: vel $\sqrt{③}$ si sit cubica, et $\sqrt{4}$: vel $\sqrt{④}$ si quadrato-quadratica &c: Sic $\sqrt{64}$ denotat 8; et $\sqrt{3}$: 64, vel $\sqrt{③}$ 64 denotat 4. Et $\sqrt{aa}$ denotat *a*; et $\sqrt{ax}$ denotat radicem quadraticam ex *ax*; et $\sqrt{3}$:4*axx* radicem cubicam ex 4*axx*. Ut si *a* sit 3, et *x* 12; tum $\sqrt{ax}$ erit $\sqrt{36}$, seu 6; et $\sqrt{3}$:4*axx* erit $\sqrt{3}$:1728, seu 12. Et hæ radices ubi non licet extrahere dicuntur surdæ quantitates ut $\sqrt{ax}$ vel surdi numeri ut $\sqrt{12}$.

Nonnulli pro designanda ——

Deniꝗ notarum quæ ex his componuntur interpretatio per Analogiam facilè innotescet. Sic enim $\frac{3}{4}a^3bb$ denotat tres quartas partes ipsius a^3bb, et $3\frac{a}{c}$ ter $\frac{a}{c}$, et $7\sqrt{ax}$ septies $\sqrt{ax}$. Item $\frac{a}{b}x$ denotat id quod fit multiplicando *x* per $\frac{a}{b}$, et $\frac{5ee}{4a+9e}z^3$ id quod fit multiplicando z^3 per $\frac{5ee}{4a+9e}$, hoc est per Quotum exortum divisione $5ee$ per $4a+9e$; et $\frac{2a^3}{9c}\sqrt{ax}$ id quod fit multiplicando $\sqrt{ax}$ per $\frac{2a^3}{9c}$; et $\frac{7\sqrt{ax}}{c}$ quotum exortum divisione $7\sqrt{ax}$ per *c*; et $\frac{8a\sqrt{cx}}{2a+\sqrt{cx}}$ quotum exortum divisione $8a\sqrt{cx}$ per $2a+\sqrt{cx}$. Et sic $\frac{3axx-x^3}{a+x}$ denotat quotum exortum divisione differentiæ $3axx-x^3$

---

(15) As with the preceding geometrical model for multiplication (note (9) above), this adaptation of Euclid, *Elements*, ɪɪ, 14, is taken over from Descartes' *Geometrie* (₁1637: 298 = *Geometria*, ₂1659: 2).

Since, however, a root multiplied by itself produces its square, and that square multiplied again by the root produces its cube, and so on, there will (from the definition of multiplication) be unity to the root as the root to its square, and as its square to its cube, and so on. In consequence, the square root of any quantity whatsoever will be the mean proportional between unity and that quantity, its cube root the first of two mean proportionals, the square-square (fourth) root the first of three, and so forth. Roots are accordingly characterized by a twofold property, both as producing higher powers on being multiplied, and as being among the mean proportionals between those powers and unity. Thus, that the square root of 64 is 8 and its cube root 4 is clear either from the value of 8×8 and $4 \times 4 \times 4$ being 64, or because 1 to 8 is as 8 to 64 and 1 to 4 as 4 to 16 and as 16 to 64. Hence also, if the square root of some line AB has to be extracted, produce it to C so that BC is unity, then on AC describe a semicircle and at B erect a perpendicular meeting this circle in D: BD will then be the root since it is the mean proportional between AB and the unit BC.[15]

To designate the root of some quantity the symbol $\sqrt{}$ is usually prefixed if the root be square, $\sqrt{3}$: or $\sqrt{③}$ if it be cube, and $\sqrt{4}$: or $\sqrt{④}$ if it be a fourth one, and so on. Thus $\sqrt{64}$ denotes 8, and $\sqrt{3}$: 64 or $\sqrt{③}$ 64 denotes 4. And so $\sqrt{aa}$ denotes a, and $\sqrt{ax}$ the square root of ax; while $\sqrt{3}$: $4axx$ signifies the cube root of $4ax^2$. So if a be 3 and x 12, $\sqrt{ax}$ will then be $\sqrt{36}$ or 6; and $\sqrt{3}$: $4axx$ will be $\sqrt[3]{1728}$ or 12. And when it is impermissible to extract them, these roots are called surd quantities (as, for instance, $\sqrt{ax}$) or surd numbers (such as $\sqrt{12}$).

Some people [use q] to designate....

Lastly, the interpretation of symbolic expressions which are compounded from these will readily come to be known by analogy. Thus, for example, $\frac{3}{4}a^3bb$ designates three quarters of a^3b^2, $3\frac{a}{c}$ three times a/c and $7\sqrt{ax}$ seven times $\sqrt{ax}$. Likewise $\frac{a}{b}x$ designates the effect of multiplying x by a/b, and $\frac{5ee}{4a+9e}z^3$ that of multiplying z^3 by $\frac{5ee}{4a+9e}$, that is, by the quotient arising from the division of $5e^2$ by $4a+9e$; while $\frac{2a^3}{9c}\sqrt{ax}$ is the result of multiplying $\sqrt{ax}$ by $2a^3/9c$, $\frac{7\sqrt{ax}}{c}$ the quotient from the division of $7\sqrt{ax}$ by c, and $\frac{8a\sqrt{cx}}{2a+\sqrt{cx}}$ the quotient arising from the division of $8a\sqrt{cx}$ by $2a+\sqrt{cx}$. And thus $\frac{3axx-x^3}{a+x}$ denotes the quotient arising from the division of the difference $3ax^2-x^3$ by the

per summam $a+x$, et $\sqrt{\dfrac{3axx-x^3}{a+x}}$ radicem ejus Quoti, et $\overline{2a+3c}\sqrt{\dfrac{3axx-x^3}{a+x}}$ id quod fit multiplicando radicem illam per summam $2a+3c$. Sic etiam $\sqrt{\tfrac{1}{4}aa+bb}$ denotat radicem summæ quantitatem $\tfrac{1}{4}aa$ & bb, et $\sqrt{\tfrac{1}{2}a+\sqrt{\tfrac{1}{4}aa+bb}}$ radicem summæ quantitatum $\tfrac{1}{2}a$ et $\sqrt{\tfrac{1}{4}aa+bb}_{[,]}$ et $\dfrac{2a^3}{aa-zz}\sqrt{\tfrac{1}{2}a+\sqrt{\tfrac{1}{4}aa+bb}}$ radicem illam multiplicatam per $\dfrac{2a^3}{aa-zz}$. Et sic in alijs.

Cæterùm nota quod in hujusmodi complexis quantitatibus non opus erit ad significationem singularum literarum semper attendere sed sufficiet in genere tantum imaginari, e.g. quod $\sqrt{\tfrac{1}{2}a+\sqrt{\tfrac{1}{4}aa+bb}}$ significat radicem aggregati $\tfrac{1}{2}a+\sqrt{\tfrac{1}{4}aa+bb}$; quodcuncꝫ tandem prodeat illud aggregatum cum numeri vel lineæ pro literis substituuntur. Atcꝫ ita quod $\dfrac{\sqrt{\tfrac{1}{2}a+\sqrt{\tfrac{1}{4}aa+bb}}}{a-\sqrt{ab}}$ significat quotum exortum divisione quantitatis $\sqrt{\tfrac{1}{2}a+\sqrt{\tfrac{1}{4}aa+bb}}$ per quantitatem $a-\sqrt{ab}$, perinde ac si quantitates illæ simplices essent et cognitæ, etsi quænam sint impræsentia prorsus ignoretur, et ad singularium partium constitutionem aut significationem neutiquam attendatur. Id quod monendum esse duxi ne complexione terminorum Tyrones quasi conterriti in limine hæreant.[16]

[*Ad pag 25. l. 28.*][17]

DE INVENTIONE DIVISORUM.

Huc spectat inventio divisorum per quos quantitas aliqua dividi possit. Si quantitas simplex est[18] divide eam per minimum ejus divisorem, et quotum per minimum divisorem ejus, donec quotus restet indivisibilis, et omnes quantitatis divisores primos habebis. Dein horum divisorum singulos[,] binos, ternos, quaternos, &c duc in se, & habebis etiam omnes divisores compositos. Ut si numeri 60 divisores omnes desiderentur, divide eum per 2, et quotum 30 per 2, et quotum 15 per 3 et restabit quotus indivisibilis 5. Ergo divisores primi sunt 1, 2, 2, 3, 5: ex binis compositi 4, 6, 10, 15: ex ternis 12, 20, 30, ex omnibus[19] 60.

---

(16) It is assumed that we now continue with the section 'De Additione' in the main text (on Newton's pages 5 ff.).

(17) See §1: note (71). It will be evident that the present intermediate 'correction' differs from the corresponding portion of Newton's initial draft (reproduced in 1, §2 preceding) and the much augmented revise which he subsequently inserted in his main text (see §1: note (517)) essentially in finding the 'divisors' (factors) of algebraic 'quantitates' rather than of the 'æquationes' which ensue on equating these to zero. The gaps in the much abbreviated text we here give will readily be filled from unchanged equivalent passages on Newton's pages 173–86 of the preceding main text.

sum $a+x$, $\sqrt{\dfrac{3axx-x^3}{a+x}}$ the root of that quotient, and $\overline{2a+3c}\sqrt{\dfrac{3axx-x^3}{a+x}}$ the result of multiplying that root by the sum $2a+3c$. So also $\sqrt{\tfrac{1}{4}aa+bb}$ denotes the root of the sum of the quantities $\tfrac{1}{4}a^2$ and b^2, $\sqrt{\tfrac{1}{2}a+\sqrt{\tfrac{1}{4}aa+bb}}$ the root of the sum of the quantities $\tfrac{1}{2}a$ and $\sqrt{[\tfrac{1}{4}a^2+b^2]}$, and $\dfrac{2a^3}{aa-zz}\sqrt{\tfrac{1}{2}a+\sqrt{\tfrac{1}{4}aa+bb}}$ that root multiplied by $2a^3/(a^2-z^2)$. And the like in other instances.

But note, however, that in complex quantities of this sort there will be no need always to pay heed to the meaning of individual letters. Rather, it will in general be sufficient merely to imagine, for instance, that $\sqrt{\tfrac{1}{2}a+\sqrt{\tfrac{1}{4}aa+bb}}$ signifies the root of the aggregate $\tfrac{1}{2}a+\sqrt{[\tfrac{1}{4}a^2+b^2]}$, whatever that aggregate proves at length to be when numbers or lines are substituted in place of the letters; and, similarly, that $\dfrac{\sqrt{\tfrac{1}{2}a+\sqrt{\tfrac{1}{4}aa+bb}}}{a-\sqrt{ab}}$ signifies the quotient arising from the division of the quantity $\sqrt{\tfrac{1}{2}a+\sqrt{\tfrac{1}{4}aa+bb}}$ by the quantity $a-\sqrt{ab}$ exactly as if those quantities were simple and known, even though what they may conceivably be is for the present utterly beyond our ken and it be pointless to give attention to the constitution or meaning of their individual parts. I have thought fit to give this warning lest beginners, scared stiff, as it were, by the complexity of terms, should stick fast on the very threshold.[16]

[*At page 25, line 28.*][17]

THE FINDING OF DIVISORS

Related to this is the finding of divisors by which some quantity may be divisible. If the quantity is simple,[18] divide it by its least divisor, and the quotient by its least divisor, until there remains an indivisible quotient, and you will have all prime divisors of the quantity. Then multiply these divisors together one, two, three, four, ... at a time and you will have also all the composite divisors. Should, for instance, all divisors of the number 60 be desired, divide it by 2, and the quotient 30 by 2, and the quotient 15 by 3, and there will remain the indivisible quotient 5. Therefore the prime divisors are 1, 2, 2, 3, 5; the compounds of these two at a time 4, 6, 10, 15; those of these three at a time 12, 20, 30; that of them all[19] 60. Again, should all divisors of the quantity $21ab^2$

---

(18) After this slightly altered opening Newton in sequel repeats the words of his previous draft (1, §2).

(19) This replaces 'quaternis' (four at a time).

Rursus si quantitatis 21*abb* divisores omnes desiderentur, divide eam per 3, et quotum 7*abb* per 7, et quotum *abb* per *a*, et quotum *bb* per *b*, et restabit quotus primus *b*. Ergo divisores primi sunt 1, 3, 7, *a*, *b*, *b*; ex binis compositi 21, 3*a*, 3*b*, 7*a*, 7*b*, *ab*, *bb*; ex ternis 21*a*, 21*b*, 3*ab*, 3*bb*, 7*ab*, 7*bb*, *abb*; ex quaternis 21*ab*, 21*bb*, 3*abb*, 7*abb*; ex quinis 21*abb*. Eodem modo ipsius 2*abb* − 6*aac* divisores omnes sunt 1, 2, *a*, *bb* − 3*ac*, 2*a*, 2*bb* − 6*ac*, *abb* − 3*aac*, 2*abb* − 6*aac*.[20]

Ubi quantitas, postquam divisa est per omnes simplices divisores manet composita et suspicio est eam compositum aliquem divisorem habere: dispone eam secundum dimensiones literæ alicujus quæ in ea est,[21] substitue sigillatim terminos hujus progressionis arithmeticæ 2, 1, 0, −1, et collige in totidem summas terminos resultantes & in ordine dispone summas juxta seriem dictæ progressionis. Tunc summæ cujusꝗ quære divisores omnes, eosꝗ e regione correspondentium terminorum progressionis dispositos, ordine percurrendo, nota in ijs progressiones arithmeticas quarum terminorum differentia dividit altissimum terminum propositæ quantitatis.[22] Siqua obvenerit ejusmodi progressio, terminum ejus qui stat e regione 0 conjunge cum litera illa per signum + si series a majoribus terminis ad minores eodem[23]

Si quantitas postquam divisa est per omnes simplices divisores manet composita et suspicio est eam compositum aliquem divisorem habere, dispone eam secundum dimensiones literæ alicujus quæ in ea est, et pro litera illa substitue sigillatim tres vel plures terminos hujus progressionis arithmeticæ, 3, 2, 1, 0 −1, −2, ac … per quam divisio tentanda est.

Ut si quantitas sit $x^3 − xx − 10x + 6$, pro x substituendo sigillatim … … res succedit, prodeunte $xx − 4x + 2$.

Rursus si quantitas sit $6y^4 − y^3 − 21yy + 3y + 20$. pro y substituo sigillatim …… res succedit prodeunte $2y^3 − 3yy − 3y + 5$.

Atꝗ ita si quantitas sit $24a^5 − 50a^4 + 49a^3 − 140aa + 64a + 30$: operatio erit ut sequitur.

2	42	1. 2. 3. 6. 7. 14. 21. 42.	+3. +3. +7
1	23	1. 23.	+1. −1. +1
0	30	1. 2. 3. 5. 6. 10. 15. 30.	−1. −5. −5
−1	297	1. 3. 9. 11. 33. 99. 297.	−3. −9. −11

Tres occurrunt hic progressiones quarum termini −1. −5. −5[24] divisi per differentias terminorum 2. 4. 6, dant tres divisores tentandos $a − \frac{1}{2}$, $a − \frac{5}{4}$ &

---

(20) Since this opening paragraph effectively repeats the last paragraph on Newton's page 27 in the preceding main text, the latter was silently omitted by Whiston in his 1707 *editio princeps* (compare §1: note (71)).

(21) Newton first wrote simply 'pro litera aliqua quæ in illa est' (in place of some letter in it).

(22) This replaces the more concise 'terminum istum' (that term).

(23) Newton leaves the sentence unfinished, but we may understand 'ordine progredit.' as its completion.

be desired, divide it by 3, and the quotient $7ab^2$ by 7, and the quotient ab^2 by a, and the quotient b^2 by b, and there will remain the prime quotient b. Therefore the prime divisors are 1, 3, 7, a, b, b; the compounds of these two at a time 21, $3a$, $3b$, $7a$, $7b$, ab, b^2; those of these three at a time $21a$, $21b$, $3ab$, $3b^2$, $7ab$, $7b^2$, ab^2; those of these four at a time $21ab$, $21b^2$, $3ab^2$, $7ab^2$; that of these five at a time $21ab^2$. In the same way all the divisors of $2ab^2 - 6a^2c$ are 1, 2, a, $b^2 - 3ac$, $2a$, $2b^2 - 6ac$, $ab^2 - 3a^2c$, $2ab^2 - 6a^2c$.[20]

When a quantity, after it has been divided by all its simple divisors, remains composite and there is a suspicion that it has some compound divisor, arrange it according to the dimensions of some letter in it,[21] separately substitute the terms of this arithmetic progression [...] 2, 1, 0, -1, [...], then collect the resulting terms into their several sums and arrange the sums in order according to the sequence of the stated progression. Thereafter, ascertain all the divisors of each sum, set these out in line with corresponding terms in the progression and, running through them in sequence, note the arithmetic progressions in them, the interval between whose terms divides the highest term in the quantity proposed.[22] If any progression of this sort does occur, join its term standing in line with 0 to the letter by $+$ if the series [proceeds] from greater terms to lesser ones in the same [sequence].[23]

If a quantity, after it has been divided by all its simple divisors, remains composite and there is a suspicion that it has some compound divisor, arrange it according to the dimensions of some letter in it, and in place of that letter separately substitute three or more terms of this arithmetic progression [...] 3, 2, 1, 0, -1, -2, [...] and ... by which division is to be attempted.

If the quantity be, for instance, $x^3 - 10x^2 - 10x + 6$, on separately substituting for x ... ... the approach succeeds, yielding $x^2 - 4x + 2$.

Again, should the quantity be $6y^4 - y^3 - 21y^2 + 3y + 20$, in y's place I separately substitute ... ... the approach succeeds, yielding $2y^3 - 3y^2 - 3y + 5$.

And thus, if the quantity be $24a^5 - 50a^4 + 49a^3 - 140a^2 + 64a + 30$, the procedure will be as follows:

2	42	1, 2, 3, 6, 7, 14, 21, 42	$+3. +3. +7.$
1	23	1, 23	$+1. -1. +1.$
0	30	1, 2, 3, 5, 6, 10, 15, 30	$-1. -5. -5.$
-1	297	1, 3, 9, 11, 33, 99, 297	$-3. -9. -11.$

Three progressions occur here, and their terms -1, -5, -5,[24] when divided by the differences 2, 4, 6 of the terms, yield three divisors to be tested: $a - \frac{1}{2}$,

---

(24) Understand 'e regione termini 0 progressionis' (in line with the term 0 in the progression) as on Newton's page 176 in the revised main text.

$a - \frac{5}{6}$. Et divisio per ultimum divisorem $a - \frac{5}{6}$ seu $6a - 5$ succedit prodeunte $4a^4 - 5a^3 + 4a^2 - 20a - 6$.

Si nullus occurrit hac methodo divisor ... . In quantitate illa pro litera substitue, ut ante, ... erit $All \pm Bl \pm C$ divisor tentandus.

Ut si quantitas proposita sit $x^4 - x^3 - 5xx + 12x - 6$, pro x ... ductiscp in divisorem numeralem termini x^4 qui ... ... per quorum utrumcp res succedit.

Rursus si proponatur quantitas $3y^5 - 6y^4 + y^3 - 8yy - 14y + 14$, operatio erit ut sequitur. ... ... divisio succedit prodeunte $y^3 - 2yy - 2y + 2$.

Si nullus inveniri potest hoc pacto divisor ... ...in his Tyro non est detinendus.

Ubi in quantitate proposita duæ sunt literæ ... ... divisorem quæsitum $xx + 2bx - 2bb$.

Ubi in quantitate proposita tres vel plures sunt literæ, ... erunt divisor quæsitus.

Ut si proponatur quantitas $12x^3 - 14bxx + 9cxx$ ... ... prodibit

$$3xx - 2bx + 2cc - 4bb.$$

Rursus si quantitas sit $12x^5 - 10ax^4 + 9bx^4$ ... ... oritur $3x^3 - 4axx - 2aab - 6b^3$.

Si quantitatis alicujus termini omnes non sunt æque alti, ... habebitur divisor desideratus $4x - 2b + 3$.

Aliquando divisores ... . Ut si proponatur quantitas $x^4 - 3ax^3 ... - 8a^4$: quæratur ... dividet totam quantitatem.[25]

Cæterum maximus duorum numerorum divisor communis, si prima fronte non innotescit, invenitur perpetua ablatione minoris de majori et reliqui de ablato. Nam quæsitus erit divisor qui tandem nihil relinquit. [——]

Hic perge ut in pag 26 & 27 mutatis[26] mutandis uscp ad verba: [*abscp residuo.*]

---

(25) Newton has cancelled a following sentence: 'Ubi verò nullus est divisor communis, nullus erit divisor totius' (When, in fact, there is no common divisor, there will be no divisor of the whole).

(26) An unnecessary 'paucis' (little) is here deleted. Newton's last paragraph repeats the opening (on pages 25–6 of his main text) to the passage he now orders to follow.

$a-\frac{5}{4}$ and $a-\frac{5}{6}$. Division by the last divisor $a-\frac{5}{6}$, that is, $6a-5$, succeeds, producing $4a^4-5a^3+4a^2-20a-6$.

If no divisor offers itself by this method …. In that quantity in place of the letter substitute, as before, … … then $Al^2 \pm Bl \pm C$ will be the divisor to be attempted.

So, if the quantity proposed be $x^4-x^3-5x^2+12x-6$, in x's place … multiplied by a numerical divisor of the term x^4 of … …. Through both of these the division succeeds.

Again, should the quantity $3y^5-6y^4+y^3-8y^2-14y+14$ be proposed, the procedure will be as follows: … … the division succeeds, yielding y^3-2y^2-2y+2.

If no divisor can be found by this technique … …but the beginner should not linger over these side-issues.

When the quantity proposed involves two letters … …the required divisor $x^2+2bx-2b^2$.

When the quantity proposed involves three or more letters, … … will be the divisor required.

If, for instance, there be proposed the quantity $12x^3-14bx^2+9cx^2$ … … there will result $3x^2-2bx+2c^2-4b^2$.

Again, if the quantity be $12x^5-10ax^4+9bx^4$ … … there arises

$$3x^3-4ax^2-2a^2b-6b^3.$$

If in some quantity all terms are not equally high, … … the desired divisor $4x-2b+3$ will be had.

On occasion divisors … …. If, for instance, there be proposed the quantity $x^4-3ax^3 … -8a^4$, let there be sought … will divide the whole quantity.[25]

For the rest, if the greatest common divisor of two numbers does not come to be known at first inspection, it is found by continually taking away the lesser from the greater and then the remainder from the subtrahend. For the required divisor will be the number which at length leaves nothing. …

Here proceed as in pages 26 and 27, changing what[26] needs to be, as far as the words 'absᵱ residuo' (… without remainder).

[*Ad pag 29 l. 37.*]

In calce capitis *de Reductione radicalium* hæc adde.[27]

De Reductione Radicaliū ad simpliciores radicales per extractionem radicum.

Radices quantitatum quæ ex integris et radicalibus quadraticis componuntur,

six extrahe. ... ... et inde $\dfrac{\sqrt{32}+\sqrt{8}}{2}$ & $\dfrac{\sqrt{32}-\sqrt{8}}{2}$ hoc est[28] $3\sqrt{2}$ & $\sqrt{2}$ quadrata

partium radicis. Radix itaꝗ est $\sqrt[4]{18}-\sqrt[4]{2}$. Eodem modo si ... et radix

$\sqrt{aa+4ax}-\sqrt{ax}$. Deniꝗ si habeatur ... . Nam Quoti istius radix erit duplum[29]

... sunt 1, $\sqrt{2}$, $\sqrt{3}$ ut supra.

Est et regula extrahendi altiores radices ... ... cujus potestas N^c dividitur per

$AA-BB$ sine residuo, et sit quotus Q. Computa $\sqrt[c]{A+B}\times\sqrt{Q}$ in numeris integris

proximis. Sit illud r.[30] Divide $A\sqrt{Q}$ ... erit $\dfrac{ts\pm\sqrt{ttss-N}}{2c\sqrt{Q}}$ radix quæsita, si modo

radix extrahi potest.

Ut si radix cubica extrahenda sit ex $\sqrt{968}+25$ ... et res succedit.

Rursus si radix cubica extrahenda sit ex $\sqrt{2312}-\sqrt{2187}$: erit $AA-BB=25$;

ejus divisores 5, 5.[31] Ergo $N=5$ & $Q=5$. Porro $A\pm B\times Q$[32] seu

$$\overline{\sqrt{2312}-\sqrt{2187}}\times 5$$

in numeris est $6{\lfloor}1$ proxime, & ejus radix cubica in numeris integris proximis

est 2. Ergo $r=2$. Insuper $A\sqrt{Q}$ seu $\sqrt{2312}\times\sqrt{5}$ extrahendo quicquid rationale

est fit $34\sqrt{10}$. Ergo $s=10$[33]

Rursus si radix cubica extrahenda sit ex $68-\sqrt{4374}$ ... et $\sqrt[c]{A+B}\times\sqrt{Q}$ seu ...

& $^{2c}\!\sqrt{Q}=\sqrt[6]{4}$ seu $\sqrt[3]{2}$ atꝗ adeo radix tentanda $\dfrac{4-\sqrt{6}}{\sqrt[3]{2}}$.

---

(27) See §1: notes (85) and (596). Newton's revised version of this draft 'correction' was subsequently inserted by him on pages 207–10 of the main text under the more exact marginal subhead 'Extractio Radicum ex binomijs' (The extraction of roots from binomials). As before we here concern ourselves only with divergences from the latter text.

(28) Newton has cancelled a parenthetic 'ut e reductione radicalium mox patebit' (as will soon be evident from reduction of the radicals) and a following intermediate '$\dfrac{4\sqrt{2}+2\sqrt{2}}{2}$ &

$\dfrac{4\sqrt{2}-2\sqrt{2}}{2}$ seu '.

(29) 'Nam Quotus iste erit duplum quadratum' (For that quotient will be twice the square) was first written in error.

(30) Newton first continued: 'Sit etiam ipsius $A\sqrt{Q}$ pars radicalis simplicissima s sitꝗ...' (Let also the simplest radical part of $A\sqrt{Q}$ be s and let...).

[*At page 29, line 37.*]

At the foot of the chapter, 'The Reduction of Radicals', add the following:[27]

THE REDUCTION OF RADICALS TO SIMPLER RADICALS
BY THE EXTRACTION OF ROOTS

The roots of quantities which are made up of integers and quadratic radicals extract in this way. ... ... and thence $\frac{1}{2}(\sqrt{32}+\sqrt{8})$ and $\frac{1}{2}(\sqrt{32}-\sqrt{8})$, that is,[28] $3\sqrt{2}$ and $\sqrt{2}$, are the squares of the root's parts. The root is accordingly $\sqrt[4]{18}-\sqrt[4]{2}$. In the same way, if ... and the root $\sqrt{[a^2+4ax]}-\sqrt{[ax]}$. Finally if ... be had ... . For the root of that quotient will be twice[29] ... are 1, $\sqrt{2}$ and $\sqrt{3}$, as above.

There is also a rule for extracting higher roots ... ... whose power N^c is divided by A^2-B^2 without remainder, and let the quotient be Q. Compute

$$\sqrt[c]{[(A+B)\sqrt{Q}]}$$

to the nearest integer. Let that be r.[30] Divide $A\sqrt{Q}$ ...: then

$$(st\pm\sqrt{[s^2t^2-N]})/2\sqrt[c]{Q}$$

will be the required root, provided the root can be extracted.

For instance, if the cube root of $\sqrt{968}+25$ is to be extracted ... and the trial succeeds.

Again, if the cube root to be extracted be that of $\sqrt{2312}-\sqrt{2187}$, there will be $A^2-B^2=25$ and its divisors 5×5:[31] therefore $N=5$ and $Q=5$. Furthermore $((A\pm B)\times Q^{(32)}$ or$)$ $(\sqrt{2317}-\sqrt{2187})\times5^{(32)}$ is numerically $6\cdot1$ very nearly, and the cube root of this to the nearest integer is 2: therefore $r=2$. In addition, by extracting from $(A\sqrt{Q}$ or$)$ $\sqrt{2312}\times\sqrt{5}$ all that is rational, it becomes $34\sqrt{10}$. Therefore $s=10$.[33]

Again, if the cube root to be extracted be that of $68-\sqrt{4374}...$, while $(\sqrt[c]{[(A+B)\sqrt{Q}]}$ or$)$ ... and $2\sqrt[c]{Q}=(\sqrt[6]{4}$ or$)$ $\sqrt[3]{2}$, and hence the root to be tested is $(4-\sqrt{6})/\sqrt[3]{2}$.

---

(31) Since $A^2=2312$ and $B^2=2187$, this should read 'erit $AA-BB=125$; ejus divisores $5, 5, 5$' (there will be $A^2-B^2=125$ and its divisors $5\times5\times5$). This, together with the slip in the next line, completely vitiates the sequel.

(32) Read '$\sqrt{Q}$' and '$\sqrt{5}$' respectively (on allowing the preceding erroneous deduction that $Q=5$).

(33) Newton cancels this example without bothering to correct it. Accurately (see note (31)) $A^2-B^2=5^3$, whence $N=5$ and $Q=1$, and also $\sqrt[3]{[A-B]}=\sqrt[3]{[34\sqrt{2}-27\sqrt{3}]}(=2\sqrt{2}-\sqrt{3})$ is *in proximis numeris* $r=1$; further, s (the radical part of $2\sqrt{2}$) is $\sqrt{2}$, and therefore t (the nearest integer to $(1+5)/2\sqrt{2}=\frac{3}{2}\sqrt{2})$ is 2. At once the correct cube root follows as

$$st-\sqrt{[(st)^2-N]}=2\sqrt{2}-\sqrt{3}.$$

Iterum si radix quadrato-cubica extrahenda sit ex $29\sqrt{6}+41\sqrt{3}:\ldots$ radix tentanda $\dfrac{\sqrt{6}+\sqrt{3}}{\sqrt[5]{9}}$.

Cæterum in hujusmodi operationibus si quantitas[34] fractio sit vel partes ejus communem habent divisorem: radices denominatoris & factorum[35] seorsim extrahe. Ut si ex $\sqrt{242}-12\frac{1}{2}$ radix cubica extrahenda sit: … … radicem factoris utriusᴄᵦ $\sqrt[3]{3}$ & $11+\sqrt{125}$.

De Reductione fractionum et Radicalium
ad series convergentes.[36]

---

§3. THE 'SYNTHETIC' RESOLUTION OF PROBLEMS IN NEWTON'S LECTURES.[1]

De Problematum resolutione Synthetica.

Problemata in Lectionibus meis[2] sic resolvo syntheticè.

Prob. 1.[3] Datis $\angle B$, $\angle C$, latere BC invenire AB, BD. Datur ratio AD ad BD et AD ad DC[,] ergo datur ratio AD ad $BD+[D]C$. et BD ad $[D]C$.[4]

Prob 3.[5] Dat[is] perim[etro] $AB+AC+BC$ et area $\frac{1}{2}AB\times AC$ invenire tri[angulum]. Datur $Q:\overline{AB+AC+BC}$ hoc est $BC^q+2BC\times\overline{AB+AC}+\overline{AB+AC}\,|^q$. sive

$$2BC^q+2BC\times\overline{AB+AC}+2AB\times AC.$$

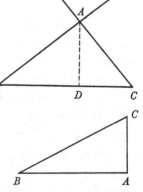

---

(34) Understand 'cujus radicem extrahere oportet' (whose root it is required to extract) as in Newton's revise.

(35) Newton first wrote 'divisoris et reliquæ quantitatis' (divisor and remaining quantity).

(36) Whiston in his 1707 *editio princeps* silently omitted this unimplemented section-title (compare §1: note (614)). It was evidently Newton's intention at this point to insert a brief discussion of the reduction of binomial unit-fractions and roots to convergent infinite series by the primitive methods *dividendo* and *radicem extrahendo* which he had earlier exemplified in his 'De Analysi' (II: 212–16) and equivalently in his 1671 tract (III: 34–42). Since neither of these texts were available to Whiston at the time he prepared his edition of Newton's manuscript, we may forgive him for not attempting any editorial interpolation comprehending and completing Newton's intended section.

(1) Nothing is known of the background to this fragmentary piece, roughly written on two sides (ULC. Add. 3963.1: 1ʳ/1ᵛ) of a stray folded folio sheet. By its handwriting it was drafted by Newton soon after he entered in the main text (§1 preceding) the geometrical Problems 1, 3,

Further, if the fifth root is to be extracted from $29\sqrt{6}+41\sqrt{3}$, ... root to be tested is $(\sqrt{6}+\sqrt{3})/\sqrt[5]{9}$.

For the rest, if in operations of this type the quantity[34] be a fraction or its parts have a common divisor, separately extract the roots of the denominator and factors.[35] If, for instance, the cube root of $\sqrt{242}-12\frac{1}{2}$ is to be extracted, ... the root of each factor $\sqrt[3]{3}$ and $11+\sqrt{125}$.

The reduction of fractions and radicals to convergent series[36]

Translation

The synthetic resolution of problems

Problems in my lectures[2] I resolve synthetically in this manner.

Problem 1.[3] Given $\hat{B}$, $\hat{C}$ and the side BC, to find AB and BD. There is given the ratio of AD to BD and that of AD to DC, and hence there is given that of AD to $BD+DC$ and that of BD to DC.[4]

Problem 3.[5] Given the perimeter $AB+AC+BC$ and the area $\frac{1}{2}AB\times AC$, to find the triangle. There is given $(AB+AC+BC)$,[2] that is,

$$BC^2+2BC(AB+AC)+(AB+AC)^2 \quad \text{or} \quad 2BC^2+2BC(AB+AC)+2AB\times AC.$$

4, 6, 9, 10, 13 and 32 of which variant 'synthetic' analyses are now given. (Despite Newton's designation of their mode as a 'resolutio synthetica', these are not synthetically composed in the style of Euclid's *Elements* but closely follow the analytical approach of his *Data*.) If we are correct in our identification (note (25) below) of the following unnumbered 'Prob. De inventione distantiæ Cometæ in Systemate Copernicæa' with Problem 52 in Newton's main text, there given a synthetic solution exactly of the form here required, it may well be that this represents his first enunciation of the cometary problem before he wrote out its elegant geometrical analysis in finished form. Such a growing desire on his part, even as he formulated their intendedly algebraic solution, to resolve his exemplifying geometrical problems by such classical means rather than by reducing them to algebraic equations would explain his unexpected decision to present geometrical analyses of his following Problems 55–7, adding equivalent algebraic solutions only in a subdued scholium following (compare §1: notes (416), (418) and (420)). The final problem 'De inventione Apheliorum...Planetarum' seems distinctly out of place in the present context, for its equant hypothesis of planetary motion—true only to a near approximation—cannot be substantiated without appeal to a series expansion or some sophisticated equivalent geometrical argument from bounding limits.

(2) Understand the exemplifying geometrical problems set down by Newton on pages 67–146 of the deposited text of his Lucasian lectures on algebra (reproduced in §1 preceding).

(3) See §1: Newton's pages 67–8.

(4) 'Unde datur AD, BD et DC.'

(5) See §1: Newton's pages 68–9.

Aufer datum $2AB \times AC$ et restabit datum $2BC^q + 2BC \times \overline{AB+AC}$. hoc est $2BC \times \overline{AB+AC+BC}$. Sed $AB+AC+BC$ datur[,] ergo $2BC$ datur.

Prob. 4.[6] Dat[is] $AB+AC+BC=M$. area $\frac{1}{2}AB \times CD$ et Ang[ulo] A invenire tri[angulum]. Datur rat[io] AC et AD ad CD. Ergo $AC \times BA$ & $AD \times BA$ datur. Datur

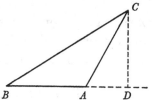

$$M^q = BC^q + 2BC \times \overline{AB+AC} + AB^q + AC^q + 2AB \times AC.$$

Aufer datum $2AB \times AC$[,] adde datum $2AB \times AD$ et dabitur

$$2BC^q + 2BC \times \overline{AB+AC}.$$

hoc est $2BC \times M$. Ergo datur BC.

Prob 6.[7] Datis AB, $AC+BC$ et ang C invenire tri. Datur ratio AC ad DC et BC ad EC adeoq $AC+BC$ ad $DC+EC$[,] ergo datur $DC+EC$. Aufer de dato $AC+BC$ et dabitur $B[D]-AE$. Datur ratio AC ad AD et BC ad BE & $AC+BC$ ad $AD+BE$[,] ergo datur $AD+BE$. Sit $AD+BE=N$, et $BD-AE=M$ & erit $AD=N-BE$ & $BD=M+AE$. Datur $AB^q=AD^q+BD^q$[,] ergo datur N^q-2BE, $N+BE^q+M^q+2M \times AE+AE^q$. Aufer datum N^q+M^q ut et datum BE^q+AE^q[8] [et] restabit datum $2M \times AE - 2N \times BE$. Simili argumento dabitur

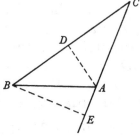

$2M \times BD - 2N \times AD$.[9] Datur ergo $\frac{M}{N}AE-BE$ et $\frac{M}{N}BD-AD$. adde datū

$AD+BE$ et dabitur $\frac{M}{N}AE+AD$. adde datum $\frac{M}{N}BD-AD$ [et] dabitur

$\frac{M}{N}AE+\frac{M}{N}BD$. datur ergo $AE+BD$ adeoq[10] datur BD: [ut et] ang ABD, cujus sinus est ad AC ut sinus C ad AB.

Prob 9. De Perambulatorio.[11] Est [triangulum] $AEI=AEK$. Ergo $ADQEK$ ad $ADQI$ ut EQ ad AD. Ergo area tota data ad quatuor parallelogramata AQ, AL, BN, CP est ut $EQ+FK+GM+HO$ ad datum $AB+BC+CD+AD$. Et area tota data applicata ad datum $AB+BC+CD+AD$ est ad latitudinem parallelogrammorum AI ut

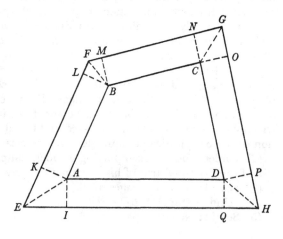

Take away the given magnitude $2AB \times AC$ and there will remain the given one $2BC^2 + 2BC(AB + AC)$, that is, $2BC(AB + AC + BC)$. But $AB + AC + BC$ is given, and therefore $2BC$ is given.

Problem 4.[6] Given that $AB + AC + BC = M$, the area $\frac{1}{2}AB \times CD$ and the angle $\hat{A}$, to find the triangle. The ratios of AC and AD to CD are given; therefore $AC \times BA$ and $AD \times BA$ are given. There is given

$$M^2 = BC^2 + 2BC(AB + AC) + AB^2 + AC^2 + 2AB \times AC.$$

Take away the given $2AB \times AC$, add the given $2AB \times AD$, and there will then be given $2BC^2 + 2BC(AB + AC)$, that is, $2BC \times M$. Therefore BC is given.

Problem 6.[7] Given AB, $AC + BC$ and the angle $\hat{C}$, to find the triangle. There is given the ratio of AC to DC and that of BC to EC, and hence that of $AC + BC$ to $DC + EC$; therefore $DC + EC$ is given. Take it from $AC + BC$ given and there will be given $BD - AE$. There is given the ratio of AC to AD and that of BC to BE, and so of $AC + BC$ to $AD + BE$; therefore $AD + BE$ is given. Let $AD + BE = N$ and $BD - AE = M$, and there will be $AD = N - BE$, $BD = M + AE$. Now $AD^2 + BD^2 = AB^2$, given, and therefore there is given

$$N^2 - 2BE \times N + BE^2 + M^2 + 2M \times AE + AE^2.$$

Take away the given $N^2 + M^2$ and also $BE^2 + AE^2$[8] given, and there will remain the given magnitude $2M \times AE - 2N \times BE$. By a similar reasoning there will be given $2M \times BD - 2N \times AD$.[9] There is therefore given $(M/N)AE - BE$ and $(M/N)BD - AD$. Add $AD + BE$ given and there will be given $(M/N)AE + AD$; add $(M/N)BD - AD$ given and there will be given $(M/N)(AE + BD)$. There is therefore given $AE + BD$ and hence[10] BD is given, along with angle $\widehat{ABD}$, whose sine is to AC as the sine of $\hat{C}$ to AB.

Problem 9. The promenade.[11] The triangles AEI and AEK are equal; therefore $(ADQEK)$ is to $(ADQI)$ as EQ to AD. Consequently the total given area is to the four rectangles (AQ), (AL), (BN), (CP) as $EQ + FK + GM + HO$ to the given sum $AB + BC + CD + DA$; and so the total given area, divided by the given $AB + BC + CD + DA$, is to the rectangles' width AI as $EQ + FK + GM + HO$ to

---

(6) See §1: Newton's page 69.

(7) See §1: Newton's page 70. It is understood that AD, BE are the perpendiculars from A, B onto BC, AC respectively.

(8) That is, AB^2.

(9) This should be '$2M \times BD + 2N \times AD$', that is,

$$M^2 + N^2 + AD^2 + BD^2 - (N - AD)^2 - (BD - M)^2.$$

So corrected, Newton's further argument will readily be seen to be circular: for on now subtracting the given quantity $AE + (N/M)AD$ from the given $BD + (N/M)AD$ there will again ensue $BD - AE$ given.

(10) Since $BD - AE$ was earlier found to be given.

(11) See §1: Newton's pages 71–2.

$EQ+FK+GM+HO$ ad datum $AB+BC+CD+AD$. Ergo datur rectangulum $AI \times EQ+F[K]+GM+HO$. Sed $EQ+F[K]+GM+HO$ componitur ex datis $AB+BC+CD+AD$ (quorum summa sit N) et ex non datis EI, FL, NG, PH, Quæ[12] sunt ad AI ut summa tangentium angulorum EAI FBL GCN HDP, ad radium, atcʒ adeo in ratione data. Sit ratio ista M ad N et erit

$$EQ+F[K]+GM+HO = N+\frac{N}{M}AI.$$ Datur ergo $N \times AI + \frac{N}{M}AI^q$. atcʒ adeo

datur $M \times AI + AI^q$. Ut et $\frac{1}{4}MM + M \times AI + AI^q$ et radix ejus $AI \pm \frac{1}{2}M$.[13] Datur ergo AI.

Prob. 10.[14] Datur $CE \times ED$.[15] Est et $CD . CE :: BD . BF$. Ergo[16] datur $CE \times BD$. Quare datur ratio ED ad BD atcʒ adeo ratio sinus anguli EBD ad sinum anguli dati BED. et proinde datur angulus EBD.

Prob 13.[17] Produc CD et erige $\perp$[18] FG ac demitte $\perp$[18] GH et junge EG. Et erit

$$DG^q = EG^q - ED^q$$

$$= EF^q + FG^q - ED^q = EF^q + GH^q.$$

Ergo datur DG. Super diametro CG describe circ[ulum] secantē BH in F et dabitur F.[19]

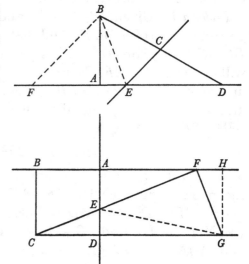

(12) Newton first continued 'singulæ sunt ad AI in ratione data' (are individually to AI in given ratio).

(13) Read either '$AI + \frac{1}{2}M$' or '$\pm AI \pm \frac{1}{2}M$'.

(14) See §1: Newton's page 72. It is again understood (though the present figure is re-lettered) that the pole B and lines EC, ED are given in position and the area of the triangle CED, cut off by the transversal BCD, is given in magnitude.

(15) That is, $2\triangle CED/\sin \hat{E}$.

(16) There is a manifest lacuna here, for Newton has not shown that the conditions of the problem fix CD or, indeed, that its length is uniquely defined. In fact, on setting $BF = a$, $FE = b$ and $CE \times ED = c^2$, immediately $CE/ED = a/(ED+b)$ and so $b \times CE - a \times ED + c^2 = 0$, yielding $CE = \frac{1}{2}c(-c \pm \sqrt{[4ab+c^2]})/b$ and so

$$CE+ED = \frac{1}{2}(a+b)c(-c \mp \sqrt{[4ab+c^2]})/ab - c^2/b,$$

whence finally $CD^2 = (CE+ED)^2 - 2c^2(1+\cos \hat{E})$ will yield two values for CD (corresponding to the two solutions possible; see §1: note (230)). But this seems a very circuitous approach, since as soon as the dual value of CE is known the magnitude of $\widehat{EBC}$ ($= \widehat{EBD}$) is at once determined.

(17) See §1: Newton's pages 74–6.

(18) 'perpendiculum'.

the given $AB+BC+CD+DA$. Therefore the product $AI(EQ+FK+GM+HO)$ is given. But $EQ+FK+GM+HO$ is composed of the given lengths

$$AB+BC+CD+DA$$

(whose sum take to be N) and of EI, FL, NG, PH not given. The latter[12] are to AI as the sum of the tangents of the angles $\widehat{EAI}$, $\widehat{FBL}$, $\widehat{GCN}$, $\widehat{HDP}$ to the radius—in a given ratio, accordingly. Let that ratio be M to N and there will then be $EQ+FK+GM+HO=N+(N/M)AI$. There is therefore given

$$N\times AI+(N/M)\,AI^2,$$

and hence $M\times AI+AI^2$ is given, and so also $\frac{1}{4}M^2+M\times AI+AI^2$ and its root $\frac{1}{2}M+AI$. Therefore AI is given.

Problem 10.[14] $CE\times ED$[15] is given. Also $CD:CE=BD:BF$; therefore[16] $CE\times BD$ is given. Consequently there is given the ratio of ED to BD, and hence of the sine of angle $\widehat{EBD}$ to the sine of the given angle $\widehat{BED}$. As a result the angle $\widehat{EBD}$ is given.

Problem 13.[17] Produce CD and erect the perpendicular FG, let fall the perpendicular GH and join EG. There will then be

$$DG^2 = (EG^2-ED^2 \text{ or}) EF^2+FG^2-ED^2 = EF^2+GH^2.$$

Therefore DG is given. On CG as diameter describe a circle intersecting BH in F and F will be given.[19]

---

(19) As we have seen (IV: 264) Newton had inserted a slightly earlier form of this construction of a classical Apollonian verging problem (see 1: 509, note (9)) a few years before among his solutions of a miscellaneous group of problems in his Waste Book. He was doubtless aware that it was a minimal improvement of the construction attributed to Heraclitus by Pappus (*Mathematical Collection*, IV, 72). A stray, unlettered figure in Newton's text at this point

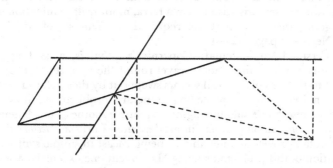

suggests that he tentatively sought to adapt his construction of Apollonius' problem of 'inclinations' to the more general case where $ABCD$ is no longer a square but a rhombus (compare IV: 260, note (33)): a later successful solution along these lines will be reproduced in the seventh volume.

Prob 32.[20] Datur rat[io] $CE.BD$ et $HE.CE_{[,]}$ ergo rat. $HE.BD$. Sit $AH.MB$ in eadem rat. et dabitur rat. $AE.MD$.[21] Comple pgrū[22] $MAEN$ et age ND et dabitur rat. MN ad MD. Ergo datur ang MND. Hunc aufer de dato MNE, et dabitur ang DNE. in triangulo NED dantur latera NE,[23] ED, et ang DNE. ergo datur ang DEN, cujus complementum est ADF. In tri. BDF dantur ang. B, D et latus $DF_{[,]}$ ergo dantur et later[a] BD, BF.

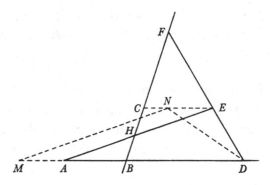

Prob.   *De inventione distantiæ Cometæ in Systemate Copernicæa.*[24]

Prob.   *De inventione Apheliorum & verorum motuum Planetarum.*[25]

Sit S Sol, T Planeta.[26] A punctum aliquod datum inter solem et Perihelion Planetæ. [G]SB medius motus Planetæ. SB linea data.[27] Angulus BST ad angulum SBA in data ratione puta ut 8 ad 5. [G]ST vera longitudo Planetæ ab Aphelio.[28]

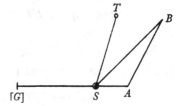

In hac Hypothesi dentur tres observati loci Achronici.[29] quæritur locus

---

(20) See §1: Newton's pages 92–3. As before, the problem is to insert the transversal DEF between the lines, AD, AE and BF such that the intercepted segments DE, EF are of given length: it is again understood that EC is drawn parallel to DB.

(21) That is, '$AH+HE. MB+BD$'.

(22) 'parallelogrammum'.

(23) Equal to the given length MA.

(24) This would appear to be a variant preliminary enunciation of what Newton subsequently set in his main text as 'Prob. 52. E Cometæ motu uniformi rectilineo per cœlum trajicientis locis quatuor observatis, distantiam a terra, motusq determinationem, in Hypothesi Copernicæa colligere', there giving it the required synthetic proof which is here wholly omitted; see §1: Newton's pages 121–4.

(25) Again understand 'in Systemate Copernicæa', that is, according to the modified Keplerian hypothesis in which the planets travel round the Sun, set at a focus, in ellipses in times proportional to the area of the focal sectors swept out by their solar *radii vectores* (compare §1: note (400)). In line with the preceding cometary problem, we may conjecture that Newton intended to insert this highly original equant hypothesis of planetary motion (see note (28) below) as a companion problem in his deposited Lucasian '*lectiones*', but abandoned the idea when its detail proved unamenable to being recast in simple synthetic form. A first, cancelled enunciation of this problem as being 'De inventione excentricitatis planetæ' would appear to have required the construction of the planet's elliptical orbit given the three focal radii ST, SV, SW (see note (32)).

(26) Paradigmatically, 'T[erra]' (the Earth). In sequel Newton has cancelled the unnecessary phrase 'ST distantia Planetæ a Sole' (ST the distance of the planet from the Sun).

(27) Understand 'magnitudine' (but not 'positione').

Problem 32.[20] The ratios of *CE* to *BD* and of *HE* to *CE* are given, and therefore that of *HE* to *BD*. Let *AH* to *MB* be in the same ratio, and the ratio of *AE* to *MD*[21] will then be given. Complete the parallelogram *MAEN* and draw *ND*, and the ratio of *MN* to *MD* will be given; therefore the angle $M\widehat{N}D$ is given. Take this from the given $M\widehat{N}E$ and the angle $D\widehat{N}E$ will be given. In the triangle *NED* the sides *NE*,[23] *ED* and angle $D\widehat{N}E$ are given; therefore the angle $D\widehat{E}N$ is given, and its supplement $A\widehat{D}F$. In the triangle *BDF* the angles $\hat{B}$, $\hat{D}$ and the side *DF* are given; so too, therefore, are the sides *BD* and *BF*.

   Problem. Finding a comet's distance in the 'Copernican' system.[24]

   Problem. Finding the aphelia and true motions of the planets.[25]

   Let *S* be the Sun, *T* a planet,[26] *A* some given point between the Sun and the planet's perihelion, $G\widehat{S}B$ the planet's mean motion, *SB* a line given [in magnitude]. Take angle $B\widehat{S}T$ to angle $S\widehat{B}A$ in some given ratio—put it as 8 to 5—and $G\widehat{S}T$ will be the true longitude of the planet (measured) from aphelion.[28]

   In this hypothesis let three observed acronychal positions[29] be given: the

---

(28) So following the usual seventeenth-century convention in which, contrary to modern practice, planetary mean and true motion is measured from the aphelion line. (In a first version, interestingly, Newton followed current habit, writing 'Sit…*ASB* medius motus… *AST* longitudo Planetæ a Perihelio'.) The ratio of 8 to 5 is not arbitrary, as Newton seems to suggest, but is theoretically the 'best' value to set on the ratio of the angles $B\widehat{S}T$ and $S\widehat{B}A$ when —as Newton here clearly intends—the radius $SB = SC = SD$ of the sun-centred equant circle *BCD* is taken equal to the mean solar distance $\frac{1}{2}GH$. For, on taking $ST = r$, $G\widehat{S}T = \phi$, $G\widehat{S}B = T$ and the solar eccentricity $OS = ea$ where $GH = 2a$ is the main axis of the elliptical orbit *TVW* of centre *O*, then (by Kepler's areal law) $r^2 . d\phi/dT = a^2 \sqrt{[1-e^2]}$ where, since $T(r, \phi)$ is a general point on the ellipse referred to $S(0, 0)$ as origin of polar coordinates, $r = a(1-e^2)/(1-e\cos\phi)$, so that

$$T = (1-e^2)^{\frac{3}{2}} \int_0^\phi (1-e\cos\phi)^{-2} . d\phi = \phi + 2e\sin\phi + \tfrac{3}{4}e^2\sin 2\phi \ldots.$$

Further, if the equant point *A* (in *SH*) be fixed by $SA = \lambda ea$ and we set $S\widehat{B}A/B\widehat{S}T = k$, since $SB . \sin S\widehat{B}A = SA . \sin S\widehat{A}B$ there results $\sin k(T-\phi) = \lambda e \sin(T - k(T-\phi))$ and in consequence $2ke(\sin\phi + \tfrac{3}{2}e\sin 2\phi \ldots) = \lambda e(\sin\phi + (1-k)e\sin 2\phi \ldots)$, so that to $O(e^3)$ $k = \frac{1}{2}\lambda = \frac{5}{8}$. It follows that $SA = \frac{5}{4} \times OS$ and also (as Newton directs) $B\widehat{S}T : S\widehat{B}A = 8:5$. This elegant equant hypothesis, much like equivalent approximations of Boulliau, Nicolaus Mercator and Vincent Wing which Newton knew well, relates the true and mean motions of the three 'upper' planets Mars, Jupiter and Saturn accurately to within the limits of observational error (say 2′ of arc) obtaining in his day; see D. T. Whiteside, 'Newton's Early Thoughts on Planetary Motion: A Fresh Look', *British Journal for the History of Science*, **2** (1964): 117–37, especially 123–8.

(29) The 'twilight' positions when the Sun and planet, as seen from the Earth, are in momentary opposition. Once the period of the planet's orbit and the corresponding distances of the Earth from the Sun are known, it is a simple matter to compute these positions from the related angular sightings made from the Earth when the planet has, after a period of time, returned exactly to its acronychal position. The technique was pioneered by Kepler in his long series of researches into the orbit of Mars during 1601–5 which culminated in his recognition of its 'true' (elliptical) shape and were subsequently condensed in the central portions of

puncti *A* respectu solis. Sit *TSV* differentia[30] loci primi et secundi. *VSW* diff[30] sc̄dī et q^{ti}.[31] *BSC* diff medij motus in priori casu. *CSD* diff medij motus in sc̄dō. Ergo

$$TSB . SBA :: 8 . 5 :: VSC . SCA ::$$

$$WSD . SDA.^{[32]}$$

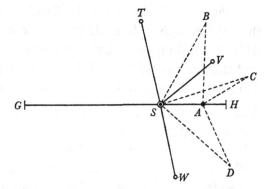

his *Astronomia Nova* (Prague, 1609); compare A. Pannekoek, *A History of Astronomy* (London, 1961): 236–41. For obvious reasons of practicality these ideal 'twilight' positions were usually computed from accurately observed positions near to dusk (or dawn) or even from the exactly timable ones measured at midnight on the same or previous day. A century before Copernicus had written of 'opposition[es] solar[es] antiqu[æ], ...quas acronychias ipsarum fulsiones Græci [*sc.* Ptolemy and his followers] appellant' (*De Revolutionibus Orbium Cœlestium, Libri VI* (Nuremberg, 1543): 143^r).

(30) Understand 'angularis' (angular).

(31) 'secundi et quarti'—but the latter should read 'tertij'!

(32) It is difficult to see how this approach can be made to yield the position of *A*: doubtless Newton saw as much when he abandoned it. On the other hand, given three focal *radii vectores* *ST*, *SV*, *SW* it is relatively simple to fix the main axis *GH* of the ellipse *TVW* both in magnitude and position, and we need then only construct $SA = \frac{5}{4}OS$, where *O* is the ellipse's centre. Thomas Harriot, in still unpublished researches of about 1610, had reduced the former problem to the construction of the ellipse's second focus, initially by a somewhat complicated algebraic construction, but afterwards by elegantly transforming it to the particular case of the classical Apollonian 3-circles problem where these are all through a unique point. Sixty years later James Gregory and Edmond Halley gave further solutions of the problem; see Gregory's

position of point A in regard to the Sun is required. Let $\widehat{TSV}$ be the[30] difference between the first and second positions, $\widehat{VSW}$ the[30] difference between the second and [third]; and $\widehat{BSC}$ the difference in mean motion in the first case, $\widehat{CSD}$ the difference in mean motion in the second. Accordingly

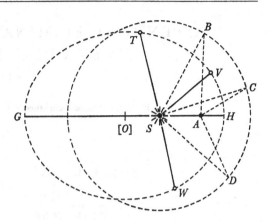

$$\widehat{TSB}:\widehat{SBA} = 8:5 = \widehat{VSC}:\widehat{SCA}$$
$$= \widehat{WSD}:\widehat{SDA}.^{(32)}$$

letter to Colin Campbell on 30 April 1674 (first published in *Archæologia Scotica*, **3**, 1831: §xxv: 280–2 [=ed. H. W. Turnbull, *James Gregory Tercentenary Memorial Volume* (London, 1939): 280–1]), and Halley's to Oldenburg in early July 1676 (S. P. Rigaud, *Correspondence of Scientific Men of the Seventeenth Century*, **1** (Oxford, 1841): 226–36, especially 230–4) and his subsequent 'A direct Geometrical Process to find the Aphelion, Eccentricities, and Proportions of the Orbs of the Primary Planets, without the Supposition...of the Equality of the Motion at the other Focus of the Ellipsis' (*ibid.*: 237–41, subsequently published in a slightly abridged Latin version, 'Methodus directa & Geometrica...', in *Philosophical Transactions* 11, No. 128 [for 25 September 1676]: 683–6). Gregory's method, making elegant use of the focus-directrix property of an ellipse, was rediscovered by Philippe de La Hire a few years later and set by him as Propositio xxv of Liber viii of his *Sectiones Conicæ* (Paris, 1685): 191–2. The trivially variant version repeated by Newton in his *Philosophiæ Naturalis Principia Mathematica* (London, ₁1687): Liber i, Propositio xxi, Scholium: 68–9 is, in manifest ignorance of Gregory's priority, attributed to 'Clarissimus Geometra *De la Hire*'.

APPENDIX 1. ELIMINATION BETWEEN TWO GIVEN CUBIC EQUATIONS.[1]

[*c.* 1720]

From the original worksheet[2] in the Bodleian Library, Oxford

[1]

$$\underset{①}{} \qquad \qquad \qquad \underset{②}{}$$

$$\begin{array}{c} a+bx+cxx+dx^3 \\ +\overline{e+fx+gxx}\,|\text{ in }y \\ +\overline{h+ix}\,|\text{ in }yy \end{array} = y^3 = \begin{array}{c} k+lx+mxx+nx^3 \\ +\overline{p+qx+rxx}\,|\text{ in }y \\ +\overline{s+tx}\,|\text{ in }yy \end{array}.$$

$$\underset{③}{}$$

Whence[3] $\dfrac{\begin{array}{c}A+Bx+Cxx+Dx^3\\+\overline{E+Fx+Gxx}\,|\,y\end{array}}{H+Ix}=yy=\dfrac{\begin{array}{c}a+bx+cxx+dx^3\\+\overline{e+fx+gxx}\,|\,y\end{array}}{y-h-ix}$. Which by reduction[4]

becomes of this form

$$\underset{④}{}$$

$$\dfrac{\begin{array}{c}K+Lx+Mxx+Nx^3+Px^4\\+\overline{Q+Rx+Sxx+Tx^3}\,|\,y\end{array}}{E+Fx+Gxx}=yy=\dfrac{\begin{array}{c}A+Bx+Cxx+Dx^3\\+\overline{E+Fx+Gxx}\,|\,y\end{array}}{H+Ix}.$$

And by reduction $\qquad \underset{⑤}{} \dfrac{\alpha+\beta x+\gamma xx+\delta x^3+\epsilon x^4+\zeta x^5}{\theta+\eta x+\lambda x^2+\mu x^3+\nu x^4}=y.$

[Similarly $\qquad\underset{⑥}{}$ ----- $=yy=$ -----.[5] And by reduction] $\underset{⑦}{}\dfrac{\xi\text{------}+\pi x^5}{\rho\text{------}+\sigma x^4}=y.$[6]

As ① & ② give ③ & ③ & ① give ④ & ④ & ③ give ⑤ so ③ & ② give ⑥, & ⑥ & ③ give ⑦.[7]

---

(1) Nothing is known of the circumstances in which Newton came to prepare this schematic outline of eliminating a component variable between two given cubic equations involving it. Though its handwriting clearly shows the piece to be very late—it may well have been drafted at the time (1721–2) Newton was revising Whiston's *editio princeps* of the *Arithmetica* for its second edition—, we here reproduce it for its very clear exposition of his manner of deriving the corresponding *Regula 4* in his deposited copy of his Lucasian 'lectures' on algebra (see §1: note (117)).

(2) MS New College 361.3: 2ᵛ, a stray page otherwise containing some late, partially cancelled sentences on Biblical chronology.

(3) On reordering the quadratic in y which results from eliminating y^3 between ① and ②, and then equating its highest term to a value of y^2 derived from ①.

(4) By cross-multiplying the numerators and denominators in ③ and then dividing through by $E+Fx+Gx^2$. Newton then again equates the resulting value of y^2 to that obtained in ③.

[2]

$$\frac{a-k+\overline{b-l}\,|\,x+\overline{c-m}\,|\,xx+\overline{d-n}\,|\,x^3 \atop +\overline{e-p}\,|\,y+\overline{f-q}\,|\,xy+\overline{g-r}\,|\,xxy}{s-h+\overline{t-i}\,|\,x}=yy=\frac{A+Bx+Cxx+Dx^3 \atop +Ey+Fxy+Gxxy}{H+Ix}.\quad{}^{(8)}$$

$$a+bx+cxx+dx^3 \atop +ey+fxy+gxxy}=\overline{y-h-ix}\times\frac{A+Bx+Cxx+Dx^3 \atop [+]Ey+Fxy+Gxxy}{H+Ix}.\quad{}^{(9)}$$

[Whence]

$$\begin{aligned}&+aH+bHx+cHxx+dHx^3\\&\qquad+aI\quad+bI\quad+cI\quad+dIx^4=Eyy+Fxyy+Gxxyy\\&+Ah+Bh\quad+Ch\quad+Dh\\&\qquad+Ai\quad+Bi\quad+Ci\quad+Di\end{aligned}$$

$$\begin{aligned}&eHy+fHxy+gHxxy\\&\quad+eI\quad+fI\quad+gIx^3y\\&-Ay\quad-Bxy\quad-Cxxy-Dx^3y\\&+hE\quad+hFxy+hGx^2y\\&\quad+iE\quad+iF\quad+iG\end{aligned}=\frac{\begin{aligned}&EA+EBx+ECx^2+EDx^3\\&+EEy+2EFxy+2EGxxy\\&+FAx+FBxx+FCx^3+FDx^4\\&+FFxxy+2FGx^3y+GGx^4y\\&+GAxx+GBx^3+GCx^4+GDx^5\end{aligned}}{H+Ix}.\quad{}^{(10)}$$

[And by reduction]

$$\frac{aHH-EA+AHh\cdots+dIIx^5+DIIx^5}{-eHH+AH-EhH+EE\cdots-gIIx^4+DI[x^4]-GiIx^4+GGx^4}=y.$$

[That is] $ass-2ash+ahh-ae+ap+ke-kp-ksh+ash-ahh+khh^{(11)}$

---

(5) By a similar elimination of y^2 between ② and ③.

(6) The final eliminant is straightaway obtained by equating the values for y obtained in ⑤ and ⑦.

(7) Or, on restating Newton's scheme of reduction as a flow diagram,

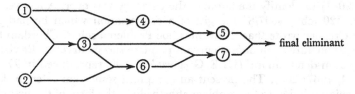

(8) Newton now explicitly identifies the coefficients in ③ with those of ①–②.

(9) That is, $(y-h-ix)\,y^2$. He proceeds at once to multiply through by $H+Ix$ and reorder the resulting quadratic in y^2.

(10) That is, $(E+Fx+Gx^2)\,y^2$ where the value for y^2 is that obtained in ③.

(11) On inserting $H^2=(s-h)^2$, $EA=(e-p)\,(a-k)$, $AH=(a-k)\,(s-h)$, ... in the preceding numerator. Newton breaks off his computation when its detail becomes unmanageably complex.

APPENDIX 2. THE THEORY OF THE
GENERALLY LOADED CHAIN.[1]

[*c.* 1720]

From the autograph[2] in the University Library, Cambridge

[1]

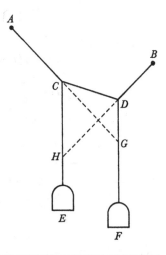

Ex ijsdem Principijs facile deducitur quod si Pondera duo *E* et *F* a filo *ACDB* gravitate destituto ad ejus pun[c]ta *C* et *D* mediantibus filis *CE* et *DF* suspendantur; et producantur *AC* et *BD* donec filis *DF* et *CE* in punctis *G* et *H* occurrant: tensiones filorum *AC*, *CD* et *CE* erunt ad invicem ut *CG*, *CD* et *DG*, et tensiones filorum *CD*, *DB* et *DF* erunt ad invicem ut *CD DH* et *CH*[,] et propterea tensiones filorum *CE* et *DF* (seu pondera *E* et *F*) erunt ad invicem ut *DG* et *CH*.[3]

Et propterea si filum gravitate destitutum *ABCDEFGH* ponderibus plu[r]ibus æqualibus

---

(1) This 'obvious' extension of Problem 44 of Newton's deposited Lucasian 'lectures' to embrace the theory of the generally loaded chain (see §1: note (368) above) was not in fact, as the handwriting of the present text proves, made by him till long after Huygens, Pardies, Jakob and Johann Bernoulli, Leibniz and subsequently others had resolved the basic particular cases of the uniformly (horizontally) loaded suspension bridge and uniformly thick hanging chain. (See C. Truesdell, *The Rational Mechanics of Flexible or Elastic Bodies, 1638–78* [=*Leonhardi Euleri Opera Omnia* (2) **11**.2 (Zurich, 1960): 1–435]: 44–6, 50–3, 64–87; also J. E. Hofmann, 'Vom öffentlichen Bekanntwerden der Leibnizschen Infinitesimalmathematik' (*Sitzungsberichte der Österreichischen Akademie der Wissenschaften, Mathem.-naturw. Klasse* (II), **175** (Vienna, 1966): 209–54: 225–39.) Though we may see in hindsight that Newton already had the solution of the general problem of the arbitrarily loaded string in his grasp by about 1680, there is no indication that he was aware of this till another decade had passed by and the race (during 1690–1) to identify the curve of the catenary was over. A few years afterwards, noting down on '20 feb[rii] 1697/8' the gist of a conversation which he had just had with Newton, David Gregory wrote that 'Newtonus hoc Problema solvit. Chordam flexilem ita in diversis sui punctis onerare ut in datam curvam pondere suo incurvetur. Et vicissim Fili dato modo onusti figuram determinare' (ULE. Gregory MS A90, reproduced in *The Correspondence of Isaac Newton*, **4**, 1967: 265). The present autograph, if somewhat later in date, expounds Newton's Huygenian solution to the problem effectively in the form of the first-order differential equation $dy/dx = k.w_s$; see note (3) below. Earlier still, we may remark, Gregory had noted in a 'Charta confuse de Neutoni cogitatis' (of May 1694?) that his mentor had attained the insight that 'Catenariam erectam esse arcum fortissimum' (ULE. Gregory MS C57, reproduced in *The Correspondence of Isaac Newton*, **3**, 1961: 345), adding as its 'Newtoni Demonstratio' a not very clear, wholly unexplained figure of an erect catenary. (Hooke had attained this result in the 1670's, passing it on to Wren—for incorporation in the brick cone supporting the lantern of St. Paul's?—and 'registering' his discovery in anagram form in a postscript to

appensis *I, K, L, M, N, O*[4] disten-
datur, et fila *BI, CK* &c quorum ope
pondera appenduntur æqualibus
ab invicem intervallis distant: filum
distentum angulis suis *C, D, E, F, G*
&c Parabolam tanget.[5]

Et augendo vel minuendo lo[n]gi-
tudines partium rectilinearum fili
BC, CD, DE &c cæteris manentibus,
fieri potest ut ejus anguli *B, C, D*
&c tangant figuram quamcunqɜ
datam.[6]

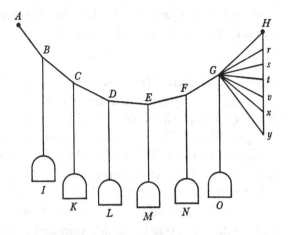

his *A Description of Helioscopes. And some other Instruments* (London, 1676): 31, later decoded by
Richard Waller in his edition of Hooke's *Posthumous Works* (London, 1705): xxi; compare
M. 'Espinasse, *Robert Hooke* (London, 1956): 71–2, 173, note 20.) In default of any knowledge
of the background to the present manuscript we may surmise somewhat hesitantly that it was
intended to replace the last paragraph of his earlier Problem 44 in his 1722 revision of Whiston's
editio princeps of his Lucasian lectures. But, if so, why Newton came to retain the latter unchanged (as Problem 49; see §1: note (216)) in the printed edition requires to be explained.
We would guess that—here as elsewhere—his reluctance in old age to undergo the mental
strain of radically restructuring an existing printed text won through, rather than any qualms
that its technical subtlety would be out of place in a (largely) elementary introduction to
algebra.

(2) Add. 3965.13: 375ᵛ. The squatness of the handwriting of this much cancelled and
overwritten piece, one characteristic of Newton's old age, supports the internal evidence of its
text (see note (1)) that it was composed in the last decade of his life.

(3) This manifestly replaces the last lines of Problem 44 of the main text (§1: Newton's
page 115), beginning 'Ex præcedentis Problematis solutione satis facilè colligetur hæcce
solutio hujus....': indeed, the former unproved conclusion that 'erit pondus *E* ad pondus *F*
ut *DG* ad *CH*' (see §1: note (373)) is now given explicit demonstration on the Hookean
principle 'ut tensio, sic vis [ponderis]'. Newton's figure remains essentially unchanged from
that drawn by him forty years before. In sequel he will suppose that *CD* is a vanishing increment, say *ds* (of weight $d(w_s)$ equal to that of *F*), of the arc *AC* = *s* whose instantaneous slope
dy/dx at *C*(*x, y*) is such that its increment $d(dy/dx) = DG - CH$ is proportional to the difference
in tension at the points *C* and *D*, and hence to the weight $d(w_s)$ of *CD*. Accordingly the basic
defining differential equation of the loaded string ensues in the form $dy/dx = k.w_s$. In yet one
more insight paralleling Newton's (and, like it, unpublished till the present century), Christiaan
Huygens came early in April 1691 (see his *Œuvres complètes*, **19**: 67–8) to abstract an identical
'Fundamentum omnium eorum quæ de Curva Catenæ reperimus [*sc.* in September 1690;
ibid.: 66–7]'.

(4) 'mediantibus filis parallelis' is cancelled.

(5) The Huygenian case of the suspension bridge, where (see note (3)) $w_s \propto x$ and hence
$dy/dx = x/a$, so yielding the parabola $2ay = x^2 + b^2$ as the Cartesian equation of its curve.
Huygens had established this result, by using elementary properties of the parabola and a
principle of statical balance derived from Stevin, in his 'De Catena pendente' (*Œuvres
complètes*, **11**: 37–44) written, when still only seventeen years old, in October 1646 but unprinted
in his lifetime. A first adequate published proof was given by Pardies in §LXXVI of his *La*

Et si partes illæ *BC*, *CD*, *DE* &c sint æquales, anguli *B*, *C*, *D*, *E* &c tangent lineam quæ Catenaria dici solet.[7]

[2][8]   Ex ijsdem Principijs deducitur quod si de filo *DACBF* circa paxillos duos *A*, *B* labile suspendantur tria pondera *D*, *E*, *F*; *D* et *F* de extremitatibus fili, et *E* de medio ejus puncto *C*[9] per filum *CE*[,] et pondera stant in æquilibrio et producatur linea *BC* donec filo *AD* occurrat in *H*:[10] pondera *D*, *E*, *F*, atcp adeo tensiones filorum *AC*, *CB* et *CE* erunt inter se, ut sunt lineæ *AC*, *CH* et *AH*. Nam si tensio fili *AC* ponderi *D* proportionalis exponatur per lineam *AC*, tensio fili *BC* ponderi *F* proportionalis exponetur per lineam *CH*, et hæ tensiones co[m]-ponunt tensionem *HA*, cui contraria et æqualis est tensio fili *CE* ponderi *E* [proportionalis].[11]

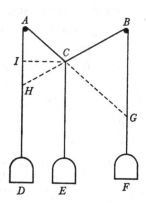

Statique, ou La Science des Forces Mouvantes (Paris, 1673): 132–4, making use of the more sophisti-cated principle of balancing tangential tensions now known (correctly for once) by his name. See Truesdell's *Rational Mechanics...1638–78* (note (1) above): 44–6, 81–2.

(6) See note (3).

(7) Here (in the terms of note (3)) $w_s \propto s$, whence the defining equation of the catenary is $dy/dx = s/a$; on taking the arc-length s as parameter we derive the equations

$$dx/ds = a/\sqrt{[s^2+a^2]} \quad \text{and} \quad dy/ds = s/\sqrt{[s^2+a^2]},$$

and so after integration $x/b = \log((s+\sqrt{[s^2+a^2]})/a)$ and $y = \sqrt{[s^2+a^2]}$, yielding therefore the explicit Cartesian equation $x/b = \log((y+\sqrt{[y^2-a^2]})/a)$—or, in more familiar modern form, $y = a\cosh(x/b)$—for the catenary. This is effectively the solution presented (independently) by Johann Bernoulli, Leibniz and Huygens in the *Acta Eruditorum* (June 1691): 274–6, 277–81 and 281–2 respectively in answer to Jakob Bernoulli's challenge of May 1690; see §1: note (368), and compare Truesdell's *Rational Mechanics...1638–1788* (note (1) above): 64–75. In sequel to the memorandum of '20 feb^{rii} 1697/8' quoted above (in note (1)) David Gregory added that 'Credit [Newtonus] rationem Catenariæ [s] ad suam abscissam [$y = \sqrt{(s^2+a^2)}$] esse geometrice et facile exprimibilem, Catenariæ ad suum applicatum [$x = b\cosh^{-1}\sqrt{(s^2/a^2+1)}$] aliquanto difficilius'; from which it would appear that Newton had by then derived the basic defining equation $dy/dx = s/a$. Gregory himself, starting from an erroneous deduction of this same equation, had just a few months earlier published an elaborate 'demonstration' of the solutions of Huygens, Leibniz and Johann Bernoulli in his 'Catenaria' (*Philosophical Trans-actions*, **19**, No. 231 [for August 1697]: 637–52 = *Acta Eruditorum* (July 1698): 305–21). Leibniz was not slow to point out Gregory's basic deficiency in his anonymous 'Animadversio ad *Davidis Gregorii* schediasma de catenaria' (*Acta Eruditorum* (February 1699): 87–91, unjustly attributing it to a like defect in Newton's fluxional method; subsequently, in what Truesdell justly calls a 'pitiful attempt to salvage his proof' (*Rational Mechanics...*: 85, note 4), Gregory unwisely attempted a naïve counterblast in his 'Responsio ad [Leibnitii] Animadversionem...' (*Philosophical Transactions*, **21**, No. 259 [December 1699]: 419–26 = *Acta Eruditorum* (July 1700): 301–6). Throughout, Newton maintained a diplomatic silence.

(8) Newton recasts his first draft into a slightly more finished form.

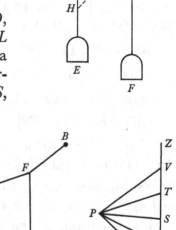

Et si filum *ACDB* extremitatibus suis ad paxillos duos *A*, *B* positione datos alligetur & ab ejus punctis duobus *C* et *D* mediantibus filis *CE* et *DF*, suspendantur pondera duo *E* et *F*, & producantur rectæ *AC*, *BD* donec filis *DF* et *CE* in *G* et *H* occurrant: tensiones filorum *AC*, *CD*, *DB*, *CE* ac *DF* erunt inter se ut sunt rectæ *CG*, *CD*, *DH*, *DG* et *CH*.

Et si de fili *ACDEFB*[12] punctis quotcunꝗ *C*, *D*, *E*, *F* mediantibus filis [*C*]*H*, [*D*]*I*, [*E*]*K*, [*F*]*L* suspendantur pondera totidem *H*, *I*, *K*, *L*; et a puncto aliquo dato *P* ad rectam horizonti perpendicularem *QZ* ducantur rectæ *PQ*, *PR*, *PS*, *PT*, *PV* fili *AB* partibus *AC*, *CD*, *DE*, *EF*, *FB* parallelæ, tensiones partium illarum erunt ad invicem ut sunt rectæ illæ ductæ, et ad tensiones filorum [*C*]*H*, [*D*]*I*, [*E*]*K*, [*F*]*L*, ut sunt eædem rectæ ductæ ad rectæ *QZ* partes *QR*, *RS*, *ST*, *TV*. Adeoꝗ si partes illæ *QR*, *RS*, *ST*, *TV* sint inter se æquales,

pondera *H*, *I*, *K*, *L* erunt inter se æqualia, & contra. Et si præterea intervalla[13] filorum [*C*]*H*, [*D*]*I*, [*E*]*K*, [*F*]*L* sint inter se æqualia, filum *AB* angulis suis *C*, *D*, *E*, *F* tanget Parabolam Conicam;[14] sin fili *AB* partes *AC*, *CD*, *DE*, *EF*, *FB* sint inter se æquales, filum *AB* angulis *C*, *D*, *E*, *F* tanget Lineam Catenariam.[15] Et augendo vel minuendo longitudines partium fili *AB*, fieri potest ut filum illud angulis suis *C*, *D*, *E*, *F* tangat curvam quancunꝗ datam, et partibus infinite brevibus et infinite multis cum Curva illa coincidat.

---

(9) 'inter paxillos posito' is deleted.

(10) Newton first wrote '...producatur linea *AC* donec filo *BF* occurrat in *G*'; correspondingly, in his figure he first extended *AC* to meet *BF* in *G*. The line *CI*, which does not appear in his following demonstration, is evidently the perpendicular from *C* onto *AH*.

(11) The manuscript reads 'æqualis' in error.

(12) Again understand 'gravitate destituti'. The manuscript lacks a corresponding figure: that reproduced is our restoration.

(13) That is, the horizontal distances.

(14) See note (5).   (15) See note (7).

APPENDIX 3. COMPUTATIONS OF RECTILINEAR COMETARY PATHS.[1]

[*c.* October 1685?]

From the originals in the University Library, Cambridge

[1][2] Dec 21 [6^h. 36′. 59″] A. [Dec] 30 [8^h. 10′. 26″] B. Jan 13 [7^h. 8′. 55″] C. Mar 2 [8^h. 9′. 44″] D.[3]

$ASB = 9^{[gr]}$. $14^{[′]}$. $10^{[″]} = 9^{[gr]} {}_{\llcorner} 23611$.

$ASD = 23$. 25. $56 = 23 {}_{\llcorner} 43222$.

$ASE = 71$. 45. $29 = 71 {}_{\llcorner} 7575$.[4]

$AS[= BS = DS = ES] = 10000000$.

[prodit] $AB = 1610260$. $AD = 4061253$. $AE = 11721437$. [ut et] $SAB = 85 {}_{\llcorner} 381945$. $SAD = 78 {}_{\llcorner} 28389$. $SAE = 54 {}_{\llcorner} 12125$. [Rursus] $AfB = 42$. 29. $27 = 42 {}_{\llcorner} 49083$.

$AgD = $ comp 80. 51. 56

     $= $ compl $80 {}_{\llcorner} 86555 = 99 {}_{\llcorner} 13445$.

$AhE = $ comp 113. 4. 51

     $= $ comp $113 {}_{\llcorner} 08083 = 66 {}_{\llcorner} 91917$.

$SAa = 23$. 59. $28 = 23 {}_{\llcorner} 99111$.

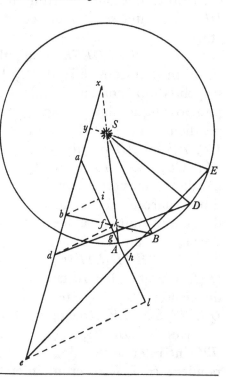

(1) The innocent reader of Newton's algebraic *lectiones* will scarcely gain the impression that his Problem 52 (§1: 121–4)—determining the path of a solar comet in the Wrennian hypothesis (see §1: note (400)) that it is (at least to sufficient approximation over a given time-interval) a uniformly traversed straight line—ever held more than a passing, theoretical interest for him. In fact, spurred on by the appearance in November 1680 of a bright new comet clearly visible in England, during the following winter and early spring of 1681 Newton conducted—initially through a common Cambridge acquaintance, James Crompton—a brisk correspondence with John Flamsteed (Astronomer Royal at Greenwich) regarding its observed apparent path and the nature of its true orbit (see *The Correspondence of Isaac Newton*, **2**, 1960: 340–7, 358–67): in this he was highly critical of Flamsteed's suggestion to him, repeated in a letter to Halley on 17 February 1680/1 (*ibid.*: 336–8), that a comet is 'made to fetch a compass about the sun', perhaps by a continual 'magnetic' attraction towards it. Only after autumn 1684, when he came to accept the universality of inverse-square gravitation within the solar system (and to reject all Cartesian solar, terrestrial, planetary and lunar deferent vortices and their interactions), did Newton slowly come also to agree that a comet's path is, to a near approximation anyway, a highly elongated 'planetary' orbit and hence curved round the Sun (set at its focus) as Flamsteed had earlier urged. As late as 19 September 1685 he could write to Flamsteed that 'I have not yet computed ye orbit of a comet but am now going about it: & taking that of 1680 into fresh consideration, it seems very probable that those of November &

December are yᵉ same comet.... My calculation of yᵉ [parabolic] orbit will depend only on three observations & if I can get three at convenient distances exact to a minute or less yᵉ orbit will answer exactly enough not only to yᵉ observations of December January February & March but also to those of November before yᵉ Comet was conjoyned wᵗʰ yᵉ sun' (*Correspondence of Isaac Newton*, **2**, 1960: 419–20). (The cometary method here referred to is presumably the clumsy *ad hoc* graphical-cum-arithmetical technique sketched at the end of his subsequently rejected 'De Motu Corporum Liber Secundus' (ULC. Add. 3990: 51–6 [= *De Mundi Systemate Liber* (London, 1728): 99–106]) rather than the more sophisticated Proposition 41 of the third book of his published *Principia*, ₁1687: 487–90.) Even for a while afterwards, it would appear, Newton clung tenaciously to Wren's rectilinear hypothesis, hoping that it would, despite its manifest imperfections, prove to be computationally viable as a first rough approximation to the truth. We here reproduce two (of several) of his extant calculations of the orbit of the 1680/1 comet made on this basis, and a preliminary algebraic analysis of the Wrennian construction: the former to illustrate numerically the relative inefficacy and practical short-comings of the rectilinear hypothesis; the latter to be compared with the equivalent synthetic presentation of the hypothesis given in Problem 52 of the preceding *Arithmetica*. The date of composition assigned in sequel, strongly consistent with Newton's handwriting style in the manuscript, is further supported by his observation to Halley on 20 June 1686 regarding 'yᵉ Theory of Comets' that 'In Autumn last [1685] I spent two months in calculations to no purpose for want of a good method' (*Correspondence*, **2**: 437) and the related surmise (see note (3)) that the cometary sightings here used were those, corrected for atmospheric refraction, which Flamsteed passed on to Newton in late September of that year.

(2) Add. 3965.14: 589ʳ. For clarity we have slightly compressed Newton's computations.

(3) These 'true times' of sightings of the 1680/1 comet (along *Aa*, *Bb*, *Dd*, *Ee*) from the Earth at *A*, *B*, *D*, *E* as it circles the Sun *S* at a constant distance $AS = BS = DS = ES$ of 10^7 units (the 'radius' of the trigonometrical tables Newton here uses) are restored on the basis of the table on page 490 of the first edition of Newton's *Principia* (London, ₁1687). Observe that the differences between solar and cometary longitudes there listed for observations made on 21 and 30 December 1680 and 13 January 1680/1 are exactly the angles $\widehat{SAa} = 23° \, 59' \, 28''$, $\widehat{SBb} = 57° \, 14' \, 45''$ and $\widehat{SDd} = 81° \, 25' \, 28''$ now employed while the differences in times of observation at *B* and *D* from that at *A* are likewise the 'Temp[us] *a* [→] *b*' and 'Temp[us] *a*[→]*d*' here recorded. The time of observation on 2 March 1680/1 is computed by adding the 'Temp[us] *a*[→]*e*' to the time of observation on the previous 21 December: it may represent a corrected *tempus verum* of Newton's own observation of the comet at Cambridge at 'hora 8' on that day (see the table 'Ex observationibus proprijs Cometæ anni 1680/1' reproduced from Newton's Waste Book (ULC. Add. 4004: 97ʳ) in his *Correspondence*, **2**: 357). It will be clear that the values for the sightings at *A*, *B* and *D* differ significantly in value from those, uncorrected for refraction, which Flamsteed passed on to Newton on 7 March 1680/1 (*ibid.*: 354). J. A. Ruffner believes that Flamsteed's corrected observations were sent to Newton as part of a (now lost) 'Included tablet' accompanying his letter of 26 September 1685, 'in which you will not wonder...to find a difference of some few minutes from yᵉ former I sent you' (*ibid.*: 422). If this plausible conjecture is correct, the present manuscript was composed not long after Newton's receipt of the latter letter—more than a year, that is, after he had proclaimed his belief, in the Scholium to Problem 4 of his tract *De Motu Corporum* (compare J. W. Herivel, *The Background to Newton's Principia* (Oxford, 1966): 267–8), in an elliptical cometary path. We may perhaps surmise that Newton is here giving the simple Wrennian hypothesis a final chance to prove its computational inadequacy before finally, in default, rejecting its effectiveness even as an auxiliary numerical technique.

(4) This third difference in solar longitude may have been computed by Newton himself or alternatively, like the previous pair (compare note (3)), taken from the (now lost) corrected table sent by Flamsteed (in September 1685?) and later printed in *Principia*: 490.

[adeoqȝ] $fAB = 109\_373056$. $gAD = $ compl $102\_27500 = 77\_72500$. [ut et]

$hAE = $ compl 78_11236. $SBb = 57.\ 14.\ 45 = 57\_24583$.

[hoc est] $ABf = 28\_136112$. $SDd = 81.\ 25.\ 28 = 81\_42444$. [sive] $ADg = 3\_14055$.
[Deniqȝ] $SEe = 65.\ 18.\ 52 = 65\_314444$. [ut et] $AE[h] = 11\_19319$.

[log] sin AfB	42_49083.	9.8296075		[log] sin AgD	99_13445.	9.9944573
sin ABf	28_136112.	9.6735442		sin ADg	3_14055.	8.7386655
AB	1610260.	4.2068960		AD	4061253.	4.6086599
[fit] Af	1124172.	4.0508327.		[fit] Ag	225355_75.	3.3528681.

[log] sin AhE	66_91917.	9.9637655
sin AEh	11_19319.	9.2880653
AE	11721437.	5.0689808
[fit] Ah	2473322.	4.3932806.

[Prodit][5] $1349528 = fg$. $3597494 = fh$. $2247966 = gh$.

Temp[us] $ab = 9^{\mathrm{d}}.\ 1^{\mathrm{h}}.\ 33'.\ 27'' = 217^{\mathrm{h}}\_5575$. $bd = 334\_974722$.
Temp $ad = 23.\ 0.\ 31.\ 56 = 552\_53222$. $be = 1487\_98833$ [6]
Temp $ae = 71.\ 1.\ 32.\ 45 = 1705\_54583$.

$fi\,.\,bi[::\mathrm{rad} = 10000000\,.\,\mathrm{Tang}\,AfB = 9162765] + bi\,.\,dk[::ab\,.\,ad]$
$+ dk\,.\,gk[::\mathrm{Tang}\,AgD = 62192668\,.\,\mathrm{rad}] = fi\,.\,gk$.

$fi\,.\,bi[+]bi\,.\,el[::ab\,.\,ae+]el\,.\,hl[::fi\,.\,hl.]$

[Unde] $fi\,.\,gk\,.\,hl::10000000\,.\,3740811\_3\,.\,30602395$.

[adeoqȝ] $fi\,.\,fi - gk = ik - fg\,.\,fi + hl = il - fh::10000000\,.\,6259188\_7\,.\,40602395$.
[Jam posito quod] $ik\,.\,il[::bd\,.\,be]::fg\,.\,N::ik - fg\,.\,il - N$. [erit] $N = 5994713$.
[Iterum quia] $il - N\,.\,ik - fg\ +\ ik - fg\,.\,il - fh[= il - N\,.\,il - fh.\ \mathrm{erit}]$

$il - N\,.\,il - fh$. (divisim) $N - fh::6847844\,.\,10000000\,.\,3152156$.

[adeoqȝ] $N - fh = 2397219$. Ergo $il - fh = 7605014$, nec non $fi = 1873045\_7$.
[adeoqȝ] $fi - gk = ik - fg = 1172374\_65$. [hoc est] $gk = 700671\_05$. $ik = 2521902\frac{2}{3}$.
[&] $il = 11202508$. [Rursus] Tang $AfB = 9162765$. [&] Tang $AgD = 62192668$.

---

(5) On computing $fg = Af + Ag, fh = Af + Ah$ and $gh = Ag - Ah$. Since

$$\widehat{SDA}(= \widehat{SAD}) < \widehat{SDd},$$

Dd must meet *Aa* between *A* and *h* (and not between *f* and *A* as shown).

(6) By hypothesis the comet traverses the straight line *ae* uniformly, hence covering distances *ab*, *ad*, *ae* proportional to the times between the observations at *A* and *B*, *D*, *E* respectively.

[ergo] $bi^{(7)} = 1716227{_|}76.$ $dk^{(7)} = 4357660{_|}2.$ [ut et]

$dk - bi(= 2641432{_|}44) \cdot ik(= 2521902\frac{2}{3}) :: bi \cdot ai = 1638565{_|}3 :: r \cdot ct : bai.$

[adeoqʒ] Ang $bai = 46^{[gr]}{_|}32613.$ [&] $Aa^{(8)} = 4635783.$

s:	$Axa,^{(9)}$	$22{_|}33502.$	9.5798079
	Aa	$4635783.$	4.6661231
s:	Aab	$46{_|}32613.$	9.8592634
[fit]	Ax	$= 8823134{_|}46.^{(10)}$	4.9456229.
s[:]	xAa	$23{_|}99111.$	9.6091430
[fit]	ax	$= 4959948.$	4.6954771.
s:	Axa	$[22{_|}33502.]$	9.5798079
cs:	Axa		9.9661318
	$Sx^{(11)}$	$1176866.$	4.0707270
[fit]	Sy	$[=]447234{_|}12.$	3.6505349
[et]	xy	$[=]1088575{_|}6.$	4.0368588

[adeoqʒ] $ay = 3871372.^{(12)}$

---

(7) That is, $fi \times \tan \widehat{AfB}$ and $gk \times \tan \widehat{AgD}$ respectively. Observe that Newton somewhat inconsistently takes the radius of his natural trigonometrical functions to be 10^7, that of the logarithmic ones to be 10^{10}.

(8) Or $Af + fi + ia$. On Newton's contraction 'ct:' for 'cotangens' see I: 467, (note 11).

(9) Namely, $\widehat{bai} - \widehat{SAa}$.

(10) Whence $Ax < AS$ and so (contrary to Newton's figure) x lies between A and S, that is, the cometary path as constructed lies on the wrong side of the Sun S!

(11) Namely, $AS - Ax$. On the contractions 's:' and 'cs:' see I: 466, note (6).

(12) These final computations determine that Sy is the perpendicular from S onto ae. Newton's method of defining the rectilinear orbit $abde$, not essentially different from that given in his Problem 52 above (compare §1: note (406)), can be summarized as follows. From the given circle radii $AS = BS = DS = ES$ and the given angles $\widehat{ASB}$, $\widehat{ASD}$, $\widehat{ASE}$ (subtended by AB, AD, AE at S) and sightings $\widehat{SAa}$, $\widehat{SBb}$, $\widehat{SDd}$, $\widehat{SEe}$, Newton computes by simple trigonometry the lengths of fg and fh and the values of $\widehat{AfB}$, $\widehat{AgD}$ and $\widehat{AhE}$. Further the ratios $ab : bd : de$ are readily determined from the differences in the times of sighting at A, B, D and E (see note (6)). Then by compounding the ratios

$$fi/bi \ (= \cot \widehat{AfB}), \quad bi/dk \ (= ba/da) \quad \text{and} \quad dk/gk \ (= \tan \widehat{AgD}),$$

and again fi/bi, $bi/el \ (= ba/ea)$ and $el/hl \ (= \tan \widehat{AhE})$, the value of the proportion $fi : gk : hl$ is readily established. If now fg/N is set equal to $ik/il \ (= bd/be$, known), then N is known and also the equal ratio $(ik - fg)/(il - N)$; accordingly, since the ratio of $ik - fg = fi - gk$ to $il - fh = fi + hl$ is also readily determined from the preceding proportion, the ratio

$$(il - fh)/(il - N)$$

is known and hence too the ratio of $il - fh$ to the known quantity $N - fh$, so that $il - fh = fi + hl$ is known and in consequence also fi, gk and hl. The remaining elements of the figure are then

[2]$^{(13)}$

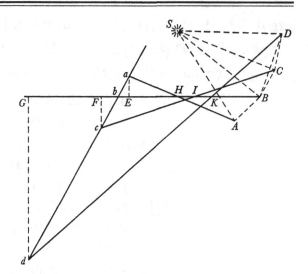

Dec 26. [5^h. 20'. 44".   *A*]
Dec 30. [8^h. 10'. 26".   *B*]
Jan 5. [6^h. 1'. 38".   *C*]
Jan 13. [7^h. 8'. 55".   *D*]$^{(14)}$

$ASB = 4^g. 11'. 42'' = 4{\llcorner}195.$ $ABS = 87{\llcorner}9025.$ $ABH = 30{\llcorner}65667.$
$CSB = 6. 0. 59 = 6{\llcorner}0163889.$ $CBS[=]86{\llcorner}9918.$ $CBI = 144{\llcorner}23764.$
$DSB = 14. 11. 46 = 14{\llcorner}1961111.$ $DBS[=]82{\llcorner}90194.$ $DBK = 140{\llcorner}14777.$
$SBH[=]57. 14. 45 = 57{\llcorner}24583.$
$SAH = 42. 13. 28 = 42{\llcorner}22444.$ $AHB = 19{\llcorner}2163888.$ $BAH = 130{\llcorner}12694.$
$SCI = 72. 25. 51 = 72{\llcorner}43083.$ $CIB = 21{\llcorner}201388.$ $BCI = 14{\llcorner}56097.$
$SDK = 81. 25. 28 = 81{\llcorner}42444.$ $DKB = 38{\llcorner}374722.$ $BDK = 1{\llcorner}47750.$
$AS[=BS=CS=DS] = 10000000.$
$BA = 732001{\llcorner}38.$ Log BA 3.8645113.
$BC = 1049574{\llcorner}1.$ Log BC 4.0210132.
$BD = 2471356.$ Log BD 4.3929352.

---

straightforwardly determinable by successive trigonometrical solution of triangles. Since the ratios $fg/fh = 1\cdot349528/2\cdot698678$ and $bd/be = 334\cdot97472/1487\cdot98833$ are significantly different—even when allowance is made for the fact that the late sighting *Ehe* from *E* (on 2 March 1680/1) is not good, while the postulate of uniform motion along *ad* considerably underestimates the true value of *bd/be*—, there will (compare §1: note (405)) be no danger of the solution being indeterminate.

(13) Add. 3965.14: 587^r. Newton repeats his previous computation of the rectilinear path of the 1680/1 comet, now taking four sightings over a much narrower time-interval of eighteen (as compared to his previous ninety-one) days in order to attain greater accuracy—and not least (compare note (10)) to construct an orbit which passes the Sun on its correct side.

(14) As before (see notes (3) and (4)) these *tempora vera* of the sightings at *A*, *B*, *C* and *D* are, like the following differences $\widehat{ASB}$, $\widehat{BSC}$, $\widehat{BSD}$ in solar longitude and the differences $\widehat{SAa}$, $\widehat{SBb}$, $\widehat{SCc}$, $\widehat{SDd}$ between these and the corresponding longitudes of the cometary sightings *Aa*, *Bb*, *Cc*, *Dd*, taken from the corrected table of Flamsteed's observations (sent in September 1685?) which is reproduced in Newton's *Principia*, $_1$1687: 490.

Log sin *AHB*. 9.5173763.   Log *BH*=4.23057967.   *BH*=1700512.

    BIC. 9.5582750.   Log *BI*=3.8631209.   *BI*=729660⌊68.

    BKD. 9.7929530.   Log *BK*=3.01133705.   *BK*=102644⌊8.

    BAH. 9.88344467.   Log tan *EHa* 9.5422795.

    BCI. 9.4003827.   Log tan *FIc*   9.5887233.

    BDK. 8.41135485.   Log tan *GKd* 9.8986549.

$$L: \frac{G[d]}{F[c]} = \qquad\qquad L: \frac{G[d]}{aE} =$$

Ex rationibˢ *KG* ad *Gd*, *Gd* ad *Fc*, *Fc* ad *FI* fit ratio *KG* ad *FI* ut 10000000 ad 8644884. Ex rationibus *KG* ad *Gd*, *Gd* ad *aE*, *aE* ad *EH* fit ratio *KG* ad *EH* ut 10000000 ad 6702623⁴⁄₉. Nam tempora *ab*, *bc*, *bd* sunt 98ʰ⌊8283333, 141ʰ⌊853333, 334ʰ⌊97472. Eorum Logⁱ 1.9948814, 2.1518395, 2.5250120. Et inde

$$\text{Log}\,\frac{FI}{KG}^{-1} = \ \ 0.9367591 \quad \text{et} \quad \log\frac{EH}{KG}^{-1} = \ \ 0.8262448.$$

Sunt ergo   *KG*.*FI*.*EH*::10000000.8644884.6702623. Et divisim [fit] *KG*.*KG*−*FI*(*IK*+*FG*).*KG*−*EH*(*HK*+*EG*)::10000000.1355116.3297377. Fiat *N* ad *HK* ut *FG* ad *EG* seu *cd* ad *ac* id est ut 193ʰ⌊12139 ad 240ʰ⌊681666 (logarithmice ut 4.2858304 ad 4.3814430) et cum *HK* sit 1597867⌊2 (logarithmice 4.2035406) erit *N*=1282118 et *N*+*FG* ad *HK*+*EG* in eadem ratione[15] adeoꝗ ad *IK*+*FG* in ratione composita ex ratione 4.2858304 ad 4.3814430 logarithmice, et ratione 3297377 ad 1355116[16] (logarithmicè 4.5181687 ad 4.1319764) hoc est in ratione (logarithmicè) 4.2905797 ad 4.0000000 seu numericè 1952440⌊9 ad 1000000. Et divisim *N*−*IK* ad *IK*+*FG* ut 952440⌊9 ad 1000000. Est *IK*[17]=627015⌊9   et   *N*−*IK*=655102⌊1. Ergo *IK*+*FG*=687813⌊8 et *FG*=60797⌊9. et ex ratione superiore *KG*=7956311. *KF*=7895513⌊1.

    IF=7268497⌊2. Log *KG*=4.9007117. Log *Gd*=3.7993666.

Gd=630037⌊94. Log *IF*=4.8614445⅖. Log *Fc*=3.4501678. *Fc*=281947⌊27. *Gd*−*Fc*=348080⌊67 . *GF*=60797⌊9::*dG*.*Gb*::rad.cotang.*G*[*bd*].

    Log *Gd*−*Fc*=3.5416798. Log *GF*=2.7838821.

Ergo Log *Gb*=30415683[18] & Log tang *Gbd*=10.7577977.[19]

---

(15) Namely '*N* ad *HK*'.

(16) That is, '(*HK*+*EG* =) *KG*−*EH* ad (*IK*+*FG* =) *KG*−*FI*'.

(17) Or *BI*−*BK*.

(18) The last two figures should be '93'.

(19) It follows that *Gb* = 110044·5 and $\widehat{Gbd}$ = 80° 5′ 33″, whence

      BG = (*BK*+*KG* or) 8058956   and so   *Bb* = 7948912;

accordingly the cometary path *ad* meets *BS* in *x* (beyond *S*) such that *Sx* = 1555779, and the

[3][20]

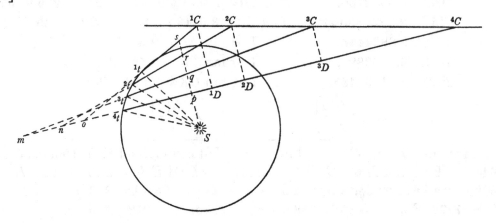

[Pone] $^1C^4C = ax.$ $^2C^4C = bx.$ $^3C^4C = cx.$ $pq = h.$ $pr = k.$ $ps = l.$
$^1C^1D = ay.$ $^2C^2D = by.$ $^3C^3D = cy.$[21]

[ut et] $r = $ radio. $e = $ tang $pso.$ $f = $ tang $prn.$ $g = $ tang $pqm.$

[Erit] $r \cdot e :: l \cdot po = \dfrac{el}{r}.$ [similiter] $np = \dfrac{fk}{r}.$ $pm = \dfrac{gh}{r}.$

$r \cdot e :: ay \cdot \dfrac{eay}{r} = {}^1Do.$ $\dfrac{fby}{r} = {}^2Dn.$ $\dfrac{gcy}{r} = {}^3Dm.$

[Porro] $\begin{aligned} -{}^1Do + po + {}^2Dn - pn &= {}^1D{}^2D. \\ {}^3Dm - pm + po - {}^1Do &= {}^1D{}^3D. \end{aligned}$ $\begin{aligned} a - b &= t. \\ a - c &= v. \end{aligned}$ [hoc est]

$$\frac{tgcy}{r} - \frac{teay}{r} + \frac{veay}{r} - \frac{vfby}{r} = tpm - tpo + vpo - vpn = \frac{tgh - tel + vel - vfk}{r}.$$

---

perpendicular distance Sy of S from ad is $1\,054\,347$. (Newton must have been relieved to find that this time the computed path lies on the correct side of the Sun!) Since the ratios

$$ac/cd \; (= 240{\cdot}68166/193{\cdot}12139) \quad \text{and} \quad HI/IK \; (= 0{\cdot}970851/0{\cdot}627016)$$

differ, there is again (compare §1: note (405)) no worry about possible indeterminacy in the solution. The method of solution is, *mutatis mutandis*, the same as before: the proportion $KG:FI:EH$ is determined by compounding the known ratios KG/Gd, Gd/FC, FC/FI and KG/Gd, Gd/aE, aE/EH respectively, while by setting $N/HK = FG/EG$ ($= cd/ac$) the ratios $(N+FG)/(HK+EG)$ and hence $(N+FG)/(IK+FG)$ and so $(N-IK)/(IK+FG)$ are determined in consequence; hence, since HI and IK are known, so also is FH and therefore so too KG, FI and EH, and the remaining elements of the configuration are easily computed.

(20) Add. 3965.11: 155[r]. Newton sets out an algebraic analysis of the general Wrennian construction of a comet, supposing that it move uniformly in a straight line. Some unfinished computations on the same sheet (and on f. 154[r] preceding) seek to apply the method to observations of the comet of 1664/5 made between 'Dec. 3[d]. 17[h]. $14\frac{1}{2}'$ [1664]' and 'Mart 1. 7. 30 [1664/5]'.

(21) Evidently $^1C^4C/^1C^1D = {}^2C^4C/^2C^2D = {}^3C^4C/^3C^3D$.

[seu] $\dfrac{tgc}{e}-ta+va-\dfrac{vfb}{e}\Big\}y=\dfrac{tgh}{e}-tl+vl-\dfrac{vfk}{e}$. [vel]

$$\dfrac{g}{e}c-a+\dfrac{v}{t}a-\dfrac{f}{e}\times\dfrac{v}{t}b \text{ in } y=\dfrac{g}{e}h-l+\dfrac{v}{t}l-\dfrac{f}{e}\times\dfrac{v}{t}k.$$

[Cape] $\dfrac{f}{e}=p.\ \dfrac{g}{e}=q.\ \dfrac{v}{t}=w.$ [Fit] $qc-a+w\times\overline{a-pb}$ in $y=qh-l+w\times\overline{l-pk}.$

[adeoqʒ] $y=\dfrac{qh-l+w\times\overline{l-pk}}{qc-a+w\times\overline{a-pb}}.$ (22)

---

(22) Compare this deduction with that given in §1: note (406). The present solution is evidently indeterminate when $qh-l+w(l-pk) = qc-a+w(a-pb) = 0$, that is, when

$$(qh-l)/(pk-l) = w = (a-c)/(a-b)\quad\text{and also}\quad(qc-a)/(pb-a) = (a-c)/(a-b):$$

geometrically, these are (compare §1: note (405)) the conditions $mo/no = {}^{1}C^{3}C/{}^{1}C^{2}C$ and $o\alpha/o\beta = {}^{3}C^{4}C/{}^{2}C^{4}C$, where α, β are the meets of $o{}^{1}C$ with $m{}^{3}C$ and $n{}^{2}C$ respectively.

APPENDIX 4. COMPUTATIONS FOR AN UNIDENTIFIED GEOMETRICAL PROBLEM.[1]

[Late 1670's]

From a worksheet[2] in the University Library, Cambridge

$BD + DN (BG). B[E :: BF. BD]. CE = DG - DF.$
$CO = CF = CD - DF. CE - CO = DG - DC.$
$BD \times BG = BF \times BE.$ Et

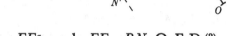

$$2BD \times BG + BN^q = 2BD^q + 2BD \times DG + BN^q$$
$$= BD^q + 2BD \times DG + DG^q = BG^q = BE^q + EG^q$$
$$= BE^q + BF^q = 2BF^q + 2BF \times FE + FE^q$$
$$= 2BF \times BE + FE^q.$$

Aufer utrincʒ æqualia et restabunt $BN^q = FE^q$. unde $FE = BN$. Q.E.D.[3]

---

(1) These unidentified calculations, now located among drafts of Newton's optical correspondence during the late 1670's though this is probably merely fortuitous, are not related to any known context in an immediately obvious way. The near-trivial mathematical result derived seems to have no direct application and the terminal 'Q.E.D.' suggests that it may have been intended as a geometrical exercise. We reproduce it here largely for want of a better place to do so, though it may just possibly relate to some problem intended to go into one or other of Newton's Lucasian lectures of the period.

(2) Add. 3970.3: 570^v, the final verso of a double sheet which also contains the draft of a letter (to Aubrey?) on his earlier optical correspondence with Lucas and Francis Hall (reproduced under the date of '?June 1678' in *The Correspondence of Isaac Newton*, **2**, 1960: 266–8), a second draft letter (also to Aubrey or perhaps Wallis?) regarding 'what papers I found in yor D. Wilkins's Real Character...' (*ibid.*: 296, tentatively—and inconsistently?— dated '1679'), and the miscellaneous notes on the factorization of algebraic binomials reproduced in the present edition on IV: 205–8. The handwriting fits well with the date of 1677–8 thereby suggested. We omit a cancelled couple of lines in the middle of the text, and insert an appropriate figure to incorporate the gist of Newton's computation.

(3) It would appear that Newton's theorem here proved is: given a right triangle *BEG* with $BF = EG$ set off in the base *BE* and *FD* let fall perpendicularly from *F* to the hypotenuse *BG*, to show that the circle on centre *D* through *G* meets the perpendicular at *B* to *BG* in points *N* such that $BN = FE$. (Or is some other geometrical figure consistent with the scheme of computation possible?) The line *CE* used in construction is, in this interpretation, the perpendicular from *E* onto *FD*. Clearly, if *EA* is drawn perpendicular to *BG*, then

$$CE = (DA \text{ or}) DG - AG, \quad \text{that is,} \quad DG - DF$$

since the triangles *EAG*, *BDF* are congruent.

PART 2

THE
'ARITHMETICÆ UNIVERSALIS
LIBER PRIMUS'
(1684)

INTRODUCTION

As with so many of his other minor mathematical tracts and revises, Newton's 'First Book of Universal Arithmetic' presents us with the essentially irresolvable problem of restoring from the internal evidence of its text the background of a manuscript whose history is otherwise unknown. Its nineteenth-century cataloguer—here probably J. C. Adams—was content to list it loosely as 'apparently an early copy...of the *Arithmetica Universalis*',[1] but this is misleading and inaccurate. For, though its central portion is written out largely in the hand of Humphrey Newton[2] and its early paragraphs are indeed virtually unchanged transcriptions of equivalent sections of the deposited Lucasian 'lectures' which subsequently appeared in Whiston's *editio princeps* as Newton's *Arithmetica*,[3] both its opening and closing pages as well as a few middle paragraphs are penned in Newton's own handwriting and he has himself also corrected much of what his secretary has copied out for him, while the central sections of the manuscript are radically restructured from their parent text, containing much in the way of general observation and particular illustration which is new. Here, evidently, is the vestige of an intendedly elaborate revision by Newton of his earlier algebraic *lectiones* as they were deposited by him about the spring of 1684—one now, to be sure, addressed specifically to a private 'reader' and not collectively to a public academic gathering.[4] But because he never found opportunity to complete even the first book of this grand revision, we cannot be sure how extended and comprehensive the project was meant to be.

In the much foreshortened manuscript as we have it the preliminary sections on 'Notation', 'Addition', 'Subtraction', 'Division' and 'Root extraction'—now comprehending Newton's modified Cartesian rule for extracting cubic and higher roots of 'asymmetric' (surd) binomials[5]—reveal (as we have remarked) few changes from their equivalents in the parent *Arithmetica*, while that dealing with 'The finding of divisors' is a but lightly remoulded abridgment of his earlier initial revise destined for insertion 'Ad pag 25' in the deposited text.[6]

---

(1) *A Catalogue of the Portsmouth Collection of Books and Papers written by or belonging to Sir Isaac Newton* (Cambridge, 1888): 47: '[vii.] 5. A MS. copy of a portion of the *Arithmetica Universalis*, apparently an early copy.'

(2) Newton's amanuensis at Trinity College in the middle 1680's; see iv: 169, note (2) and [1]: notes (1) and (2) below.

(3) See **1, 2**, §1: note (2).

(4) See [1]: note (120).

(5) Compare [1]: note (38).

(6) See **1, 2**, §2: note (17). The relatively sophisticated finite-difference techniques for determining the linear, quadratic and higher factors of given equations in a single variable were, it is clear (compare [1]: note (32)), intended to be expounded in a later (unwritten) section of the present revise.

The following 'Reduction of fractions and surds' is likewise, except for some compression, only trivially different from the corresponding portion of the parent text which it subsumes.[7] In strong contrast the manuscript's central sections on 'The finding [and] reduction of equations' introduced in afterthought[8] by preliminary observations on 'An equation and its roots', present a widely rewritten and much augmented exposition of the application of this basic concept of classical algebra to the solving of problems by 'translating' their defining conditions into appropriate algebraic equivalents. To fill out his general remarks, moreover, Newton works elaborately through twelve typical arithmetical and geometrical problems, emphasizing the unity of his method of approach, but complementing these minutely detailed reductions by a score more borrowed without essential change (except for the final 'Prob [21]')[9] from those exhibited, more concisely and elegantly if in less systematic manner, in his earlier *Arithmetica*. At this point, and without any prior warning, the text abruptly terminates.[10]

On Newton's intentions in the projected sequel certain forward references in the completed portion shed a little light. It is, in particular, clear that (in the style of pages 173–203 of his parent *Arithmetica*) he had it in mind to incorporate further sections treating of the rational and surd factorization of given algebraic equations;[11] correspondingly, no doubt, he would have planned to complete these exact determinations of roots with a summary (on the lines of his earlier pages 203–7) of the standard contemporary solution of the general cubic[12] and quartic equations. But what, if any, of his previous appendix on the geometrical construction of equations (pages 211–51 of the original *Arithmetica*) he had here decided to retain is nowhere indicated in the completed part of the revise and must remain highly uncertain.[13] Yet more tenuous still must be any conjecture regarding the content of the complementary 'Second Book'. We may not unreasonably guess that it would, among other things, perhaps have contained a summary—not unlike that he was to present just a few weeks later in his 'De

---

(7) Newton's subsequent intention of adding a further section at this point 'De Reductione fractionum et Radicalium ad series convergentes' (see **1**, 2, §1: note (614) and §2: note (36)) is still, as yet, firmly in the future.

(8) See [1]: note (46).

(9) Compare [1]: note (134).

(10) We may guess that Newton's attention was suddenly diverted by external circumstance, perhaps the arrival out of the blue in mid-June 1684 of David Gregory's presentation copy of his *Exercitatio Geometrica de Dimensione Figurarum* which led him at once to draft his own 'Matheseos Universalis Specimina' in hurried riposte (compare IV: 413–18).

(11) See [1]: notes (32) and (99).

(12) Compare [1]: note (119).

(13) As we have mentioned (**1**, 2, §1: note (615)), Newton came later to cancel much of this appendix in his library copy of Whiston's *editio princeps* of the deposited Lucasian 'lectures', probably soon after its publication in 1707.

computo serierum'[14]—of his several techniques for expanding given algebraic functions into equivalent infinite series and of their applications in extracting the roots of equations approximately *in numeris* (in cases intractable by an exact algebraic approach) and, maybe, also in intercalating required quantities by finite-difference interpolation.[15] But this is essentially undemonstrable.

Here we reproduce (in [1]) only those parts of Newton's revised *Arithmetica Universalis* which differ significantly from the corresponding portions of the parent deposited text,[16] implicitly referring back to the unaltered original equivalents wherever omission is now made of passages in the manuscript. As a bonus we have (in [2]) added a stray Newtonian excursus on 'The rule of brothers', whose autograph dates it to the same period as the preceding revised *Arithmetica* and whose text elaborates the simplification in algebraic reduction attendant on a prior identification of pairs of 'twin' solutions to a problem[17] which he had previously expounded more succinctly (and with the single example of 'Prob [21]' in illustration) in the concluding section[18] of that revise. The appended, tentatively presented thesis that problems are 'usually' to be graded according to the dimensions of the equations to which their conditions (in algebraic equivalent) reduce is significant only in that it contradicts Newton's firm standpoint in the parent *Arithmetica*[19] that the dimension of its algebraic equivalent has effectively nothing to do with the 'simplicity' of the construction by which a problem is geometrically resolved.

---

(14) See IV: 590–616. Compare also the preceding 'Matheseos Universalis Specimina' (IV: 526–88, especially 540–64).

(15) See IV, 1, 1: *passim*.

(16) Reproduced as 1, 2, §§1/2 above.

(17) See [2]: note (147); compare also 1, 2, §1: note (245).

(18) Itself a lightly remoulded version of (geometrical) Problem 13 on pages 74–6 of the earlier *Arithmetica*; see [1]: note (134).

(19) 1, 2, §1: Newton's pages 211–14 above.

THE 'FIRST BOOK OF
UNIVERSAL ARITHMETIC'

[Spring 1684?][1]

From the originals in the University Library, Cambridge

[1][2]   ARITHMETICÆ UNIVERSALIS
             LIBER PRIMUS.[3]

Arithmetica vulgaris et Algebra ijsdem computandi fundamentis innituntur et ad eandem metam collimant. At Arithmetica[4] definitè et particulariter Quæstiones tractat, Algebra autem indefinitè et universaliter, sic ut enunciata ferè omnia quæ in ejus computo habentur, et præsertim conclusiones, Theoremata dici possint. Verùm Algebra maximè præcellit quod cum in Arithmetica Quæstiones solvantur tantum[4] progrediendo a datis quantitatibus ad quæsitas,[4] hæc a quæsitis tanquam datis ad datas tanquam quæsitas plerumqʒ regreditur ut ad conclusionem aliquam, seu æquationem, quocunqʒ demum modo perveniatur, ex qua quantitatem quæsitam elicere liceat. Eoqʒ pacto conficiuntur difficillima problemata quorum resolutiones ex Arithmetica sola frustra peterentur. Arithmetica tamen Algebræ in omnibus ejus operationibus ita inservit,[4] ut non nisi unicam perfectam computandi scientiam constituere videantur; et utramqʒ propterea[5] conjunctim explicabo.

Quisquis hanc scientiam aggreditur imprimis vocum et notarum significationes intelligat et fundamentales addiscat operationes, Additionem, Subductionem, Multiplicationem, Divisionem, Extractionem radicum, Reduc-

---

(1) Strictly, this is a *terminus ante quem non* for the main text [1], which is in large part (see note (2)) written out in the hand of Humphrey Newton, Newton's amanuensis at Trinity from some time in 'yᵉ last year of K. Charles 2ᵈ' (see IV: 169, note (2)), and hence some time during the 4–5 years from the late winter of 1683–4. Our narrower surmise that it was composed in the spring (or perhaps early summer) of 1684 accords well with Newton's handwriting in the introductory and closing portions of the manuscript and with the plausible inference that it succeeded its parent text (1, 2, §§1/2) closely in time. The allied autograph fragment reproduced in [2] is manifestly, on both contextual and stylistic grounds, of the same period as [1].

(2) Add. 3993: 1–93 (followed by some 150 blank pages on which Newton doubtless intended to set the remainder of this unfinished text). Of these pages 1–9, 44 [bottom]–53 [top] and 92–3 are autograph, while the remainder is transcribed by Humphrey Newton from an original now lost (where at least it differs from the parent *Arithmetica*); pages 10–44 and 53–68 of the latter, however, are corrected in Newton's hand. In addition, a folded folio sheet, now inserted between pages 92 and 93, contains (ff. 93a/93c) an amplification of the original opening paragraph of the 'Inventio Æquationum' on pages 42–3, (f. 93d) an insertion on 'p. 54. l. 2' after 'Reg. 2' and a new 'Reg. 13' on page 60, and (ff. 93b/93c) the little variant preliminary draft for pages 92–3: all these are introduced at appropriate places in our edited

Translation

Universal Arithmetic
Book One[3]

Common arithmetic and algebra rest on the same computational foundations and are directed to the same end. But whereas arithmetic treats questions in a definite, particular way, algebra does so in an indefinite, universal manner, with the result that almost all pronouncements which are made in this style of computation—and its conclusions especially—may be called theorems. However, algebra most excels, in contrast with arithmetic where questions are solved merely by progressing from given quantities to those sought, in that for the most part it regresses from the sought quantities, treated as given, to those given, as though they were the ones sought, so as at length and in any manner to attain some conclusion—that is, equation—from which it is permissible to derive the quantity sought. In this fashion the most difficult problems are accomplished, ones whose solution it would be useless to seek of arithmetic alone. Yet arithmetic is so instrumental to algebra in all its operations that they seem jointly to constitute but a unique, complete computing science, and for that reason I shall explain both together.[5]

Whoever enters upon this science should in the first place understand the meanings of the expressions and symbols and also learn the fundamental operations: addition, subtraction, multiplication, division, the extraction of

text. The inside front cover of the manuscript's contemporary binding not unexpectedly bears the legend 'Sep 25 1727 Not fit to be printed. Tho Pellet' (compare I: xix, note (10)).

(3) No corresponding 'Liber secundus' (Book Two) would appear ever to have been drafted, since even the extant text of the present first book is palpably unfinished; see notes (32) and (99) below. We have sketched in the preceding introduction what little can be gathered of the projected content (rational and surd factorization of algebraic quantities, the particular construction—geometrical and numerical as well as algebraic?—of equations) of the later portion of Book 1, adding our conjecture that Book 2 may, in part at least, have comprehended the expansion of algebraic quantities into equivalent infinite series, and the analogous expression of the (real) roots of given equations in terms of the coefficients, by the methods developed in his earlier 'De Analysi' and 1671 tract (II: 212–30/III: 36–66) and refined in his contemporary 'Matheseos Universalis Specimina' and 'De Computo Serierum' (IV: 540–64/590–616). Observe that Newton here for the first time uses the title of *Arithmetica Universalis* which Whiston was to set on the *editio princeps* of his parent Lucasian 'lectures' in 1707 (see 1, 2, §1: note (3)), and which we too have used in our reproduction of the deposited manuscript. Whether Whiston was primed by Newton or whether this now classical short title was his own independent précis of the introductory paragraph of his published text is not known.

(4) In a closer imitation of the phrasing of the first paragraph in the parent *Arithmetica* (1, 2, §1) Newton here originally wrote 'Arithmetica quidem', 'tantum resolvantur', 'a datis ad quæsitas quantitates' and 'subservit' respectively.

(5) Newton here first introduced and then cancelled the repetitive phrase 'tanquam unam universalem Arithmeticam' (as if they were a single universal arithmetic).

tiones fractionum & radicalium quantitatum, et modos ordinandi terminos Æquationum ac incognitas quantitates, ubi plures sunt, exterminandi. Deinde has operationes reducendo Problemata ad æquationes exerceat, et ultimò naturam et resolutionem æquationum contempletur.

DE NOTATIONE.[6]

Integrorum numerorum notas 0, 1, 2, 3, 4, &c et notarum ubi plures inter se nectuntur, significationes nemo non intelligit. Quemadmodum verò numeri in primo loco ante unitatem, sive ad sinistram, scripti denotant denas unitates, in secundo loco centenas, in tertio millenas: sic numeri in primo loco post unitatem scripti denotant decimas partes unitatis, in secundo centesimas, in tertio millesimas & sic in infinitum. Dicuntur verò fracti decimales qui sic post unitatem scribuntur. Et ad distinguendum integros a decimalibus, interjici solet comma vel punctum vel lineola. Sic numerus 732'569 denotat septingentas triginta duas unitates, una cum quinqȝ decimis sex centesimis et novem millesimis partibus unitatis. Qui et sic 732.569, vel etiam sic 732ₗ569[7] nonnunquam scribitur. Atqȝ ita numerus 57104'2083 denotat quinquaginta septem mille centum et quatuor unitates una cum duabus decimis, octo millesimis et tribus decimis millesimis partibus unitatis. Et numerus 0'064 denotat sex centesimas & quatuor millesimas partes.

Cum rei alicujus quantitas ignota est vel indeterminate spectatur ita ut per numerū exprimi nequeat,[8] solemus per speciem aliquam seu literam designare. Et siquando cognitas quantitates tanquam indeterminatas spe[c]tamus, discriminis causa designamus has initialibus Alphabetæ literis *a, b, c, d*, et incognitas finalibus *z, y, x, v*, &c.[9]

Quantitates vel affirmativæ sunt seu majores nihilo vel negativæ seu nihilo minores. Sic in rebus humanis possessiones dici possunt bona affirmativa, debita verò bona negativa. Inqȝ motu locali progressus dici potest motus affirmativus & regressus motus negativus, quia prior auget, posterior diminuit iter confectum. Unde et in Geometria si linea motu progressivo versus plagam aliquam ducta pro affirmativa habeatur, negativa erit quæ facto regressu versus plagam oppositam ducitur. Veluti si *AB* dextrorsum ducatur et *BC* sinistrorsum, atqȝ *AB* statuatur affirmativa tunc *BC* pro negativa habebitur eo quod inter ducendum

A   C   B

---

(6) Except for two transpositions of paragraphs (see notes (9) and (14)) in line with the initial revise 'De vocum quarundam et notarum significatione' (1, 2, §2), this section narrowly follows the section 'De Notatione' in the parent deposited manuscript (1, 2, §1: Newton's pages 2–5).

(7) Notice that Newton's preferred private notation with a comma separatrix on the line, **732,569** (see **1, 2**, §1: note (11)), is now ignored by him. He will rapidly lapse back into it; compare note (21).

roots, the reductions of fractions and radical quantities, and the ways of ordering the terms of equations and of eliminating unknown quantities therein (when they have several). Then let him practise these operations by reducing problems to equations, and, finally, contemplate the nature and solution of equations.

NOTATION[6]

The symbols 0, 1, 2, 3, 4, ... for integers, and the significance of those symbols when several are joined together, everyone understands. To be sure, numbers written, for instance, in the first place immediately before a unit (on its left, that is) denote tens of units, those in the second place hundreds, in the third thousands, ...; likewise, numbers written in the first place after a unit denote tenths of a unit, those in the second place hundredths, in the third thousandths, ... and so on indefinitely. Those written in this way after a unit are, however, called decimal fractions. And to distinguish integers from decimals a comma, point or short line is usually interposed. Thus the number 732'569 denotes seven hundred and thirty-two units, together with five tenths, six hundredths and nine thousandths of a unit. This is also sometimes written 732.569 or thus 732ı569 even.[7] And similarly the number 57104'2083 denotes fifty-seven thousand, one hundred and four units, together with two tenths, eight thousandths and three ten-thousandths of a unit. And the number 0'064 denotes six hundredths and four thousandths.

When the quantity of some object is unknown or indeterminately regarded (so that it cannot be expressed by a number[8]), we usually designate it by some species or letter. And should we ever regard known quantities as indeterminate, for distinction's sake we designate these by initial letters of the alphabet a, b, c, d, ... and the unknowns by final ones z, y, x, ...[9]

Quantities are either positive, that is, greater than zero, or negative, that is, less than zero. So in human affairs possessions can be called positive goods, but debts negative ones. And in local motion an advance can be called a positive motion and a retreat a negative one, since the former increases the path completed and the latter diminishes it. Hence also in geometry, if a line drawn with advancing motion in some direction be considered as positive, then its negative is one drawn retreating in the opposite direction. For instance, if AB be drawn rightwards and BC leftwards, and AB is decreed to be positive, then BC will be

---

(8) Newton first wrote 'ita ut per numeros non liceat exprimere' (so that it is not permitted to express it by numbers), as in the parent text (**1**, 2, §1).

(9) In sequel Newton first entered the paragraph 'Numerus speciei alicui immediatè præfixus...quindecim x' from the parent *Arithmetica* and then cancelled it, delaying it, as in his initial revise (**1**, 2, §2), till after the next paragraph but one. See note (12).

diminuit AB redigitɋ ad breviorem AC. Et universaliter id omne negativum dicere solemus quo positivum diminuitur. Negativæ autem quantitati designandæ signum $-$, affirmativæ signum $+$ præfigi solet. Et ubi neutrum præfigitur, signum $+$ subintelligendum est. Signum incertum denotamus nota $\pm$ et huic contrarium signum nota $\mp$.

In aggregato quantitatum nota $+$ significat quantitatem suffixam esse cæteris addendam & nota $-$ esse subducendam:[10] Sic $2+3$ valet summam numerorum 2 et 3 hoc est 5 et $5-3$ valet differentiam quæ oritur subducendo 3 de 5, hoc est 2. Et $-5+3$ valet differentiam quæ oritur subducendo 5 de 3 hoc est -2, et $6-1+3$ valet 8.[11] Item $a+b$ valet summam quantitatum a et b. et $a-b$ differentiam quæ oritur subducendo b de a, et $a-b+c$ summam istius differentiæ et quantitatis c. Puta si a sit 5, b 2 et c 8, tum $a+b$ valebit 7, $a-b$ 3, et $a-b+c$ 11.

Numerus speciei alicui immediatè præfixus denotat speciem illam toties sumendam esse. Sic $2a$ denotat duo a, $3b$ tria b, $15x$ quindecim x.[12]

Duæ vel plures species immediatè connexæ designant factum seu quantitatem quæ fit per multiplicationem ambarum vel omnium in se invicem. Sic ab vel ba (nam perinde est quo ordine literæ scribantur) denotat quantitatem quæ fit multiplicando b per a vel a per b & abx denotat quantitatem quæ fit multiplicando factum illum ab per x. Ut si a sit 2, b 3, et x 5, tum ab erit 6 et abx 30.

Inter quantitates se multiplicantes[13] nota $\times$ vel vocabulum *in* ad factum designandum nonnunquam scribitur. Sic 3×5 vel 3 in 5 denotat 15. Sed usus harum notarum præcipuus est ubi compositæ quantitates sese multiplicant. Veluti si $y+b$ multiplicet $y-2b$ terminos utriusɋ multiplicatoris lineola superimposita connectimus et scribimus $\overline{y+b} \times \overline{y-2b}$.[14]

Quantitas infra quantitatem cum lineola interjecta denotat quotum, seu quantitatem quæ oritur divisione superioris quantitatis per inferiorem. Si[c] $\frac{6}{2}$ denotat quantitatem quæ oritur dividendo 6 per 2, hoc est 3 et $\frac{5}{8}$ quantitatem quæ oritur dividendo 5 per 8 hoc est octavam partem numeri 5. et $\frac{a}{b}$ quantitatem quæ prodit dividendo a per b. puta si a sit 15 et b 3, tum $\frac{a}{b}$ denotat 5. Et sic $\frac{ab-bb}{a+x}$

---

(10) A following sentence 'Et has notas vocabulis plus et minus exprimere solemus' (And these symbols we usually express by the words 'plus' and 'minus') was first copied from the parent manuscript and then deleted. In the next sentence, correspondingly, Newton transcribed and then deleted the phrases 'sive 2 plus 3' and 'sive 5 minus 3'.

(11) Observe that the sequel is somewhat changed from the text of the parent *Arithmetica*.

(12) See note (9).

(13) In sequel Newton began to copy 'c[omma vel]' (comma or) from the parent manuscript and then decided to omit this reference to a variant notation much used by him in his private papers; compare **1**, **2**, §1: note (17).

considered as negative because, as it is being drawn, it diminishes AB, reducing it to the shorter length AC. And, universally, we usually call negative everything by which a positive is diminished. To denote a negative quantity, however, the sign $-$ is usually prefixed, to a positive one the sign $+$. And when neither is prefixed, the sign $+$ must be understood. An uncertain sign we denote by the symbol $\pm$, and the sign opposite to it by the symbol $\mp$.

In a collection of quantities the symbol $+$ signifies that the quantity appended must be added to the rest, the symbol $-$ that it must be taken away.[10] So $2+3$ is in value the sum of the numbers 2 and 3, that is, 5, and the value of $5-3$ is the difference which arises on taking 3 from 5, that is, 2. And the value of $-5+3$ is the difference which arises on taking 5 from 3, that is, -2, while that of $6-1+3$ is 8.[11] The value of $a+b$, likewise, is the sum of the quantities a and b, and that of $a-b$ the difference which arises on taking b from a, while $a-b+c$ is the sum of that difference and the quantity c. If, say, a be 5, b 2 and c 8, then the value of $a+b$ will be 7, that of $a-b$ 3, and that of $a-b+c$ 11.

A number set immediately in front of some variable denotes that the variable is to be taken an equivalent number of times. Thus $2a$ denotes two a, $3b$ three b, $15x$ fifteen x.[12]

Two or more variables immediately connected denote their product, or the quantity which results by multiplying both or all in turn together. Thus ab or ba (for it is just the same, whatever order the letters be written in) denotes the quantity which comes from multiplying b by a or a by b, and abx denotes the quantity which comes from multiplying that product ab by x. So, if a be 2, b 3 and x 5, then ab will be 6 and abx 30.

Between quantities multiplying one another[13] the symbol $\times$ or the word 'into' is not infrequently written to designate the product. Thus 3×5 or 3 into 5 denotes 15. But the most important use of these symbols is when compound quantities multiply one another. If, for instance, $y+b$ should multiply $y-2b$, we connect the terms in each multiplier by a short, overhead line and write '$\overline{y-b}\times\overline{y-2b}$'.[14]

A quantity set below a quantity with a short intervening rule denotes their quotient, that is, the quantity which arises from the division of the upper quantity by the lower one. Thus $\frac{6}{2}$ denotes the quantity which arises on dividing 6 by 2, that is 3; and $\frac{5}{8}$ the quantity which ensues on dividing 5 by 8, that is, an eighth of the number 5; and $\frac{a}{b}$ the quantity which results on dividing a by b—if, say, a be 15 and b 3, then $\frac{a}{b}$ denotes 5. And so

---

(14) In line with his initial revise (**1**, 2, §2) Newton has here deferred for three paragraphs the one, beginning 'Si quantitas seipsam multiplicet...', which follows in immediate sequel in the parent *Arithmetica*.

denotat quantitatem quæ oritur dividendo $ab - bb$ per $a + x$. Atcɜ ita in alijs. Hujusmodi autem quantitates dicuntur Fractiones, parscɜ superior Numerator & inferior Denominator.

Aliquando divisor quantitati divisæ, interjecto arcu præfigitur. Sic ad designandam quantitatem quæ oritur divisione $\dfrac{axx}{a + x}$ per $a - b$, scribi potest $\overline{a - b} \Big) \dfrac{axx}{a + x}$.

Etsi multiplicatio per immediatam quantitatum conjunctionem denotari solet, tamen numerus integer ante numerum fractum denotat summam utriuscɜ. Sic $3\frac{1}{2}$ denotat tria cum semisse.

Si quantitas seipsam multiplicet, numerus factorum compendij gratia, suffigi solet. Sic pro aaa scribimus $a^3{}_{[,]}$ pro $aaaa$ scribimus a^4 et pro $aaabb$ scribimus a^3bb vel a^3b^2. Ut si a sit 5 et b 2, tum a^3 erit $5 \times 5 \times 5$ seu 125 et a^4 erit $5 \times 5 \times 5 \times 5$ seu 625, atcɜ a^3b^2 erit $5 \times 5 \times 5 \times 2 \times 2$ seu 500: Ubi nota quod numerus inter duas species immediatè scriptus, ad priorem semper pertinet. Sic 3 in quantitate a^3bb non denotat bb ter capiendum esse sed a in se bis ducendum. Nota etiam quod hæ quantitates tot dimensionum potestatum vel dignitatum esse dicuntur quot factoribus seu quantitatibus se multiplicantibus constant. Et numerus suffixus vocatur index potestatum vel dimensionum. Sic aa est duarum potestatum & a^3 trium ut indicat numerus suffixus 3. Dicitur etiam aa quadratum, a^3 cubus, a^4 quadrato-quadratum, a^5 quadrato-cubus, a^6 cubo-cubus, a^7 quadrato-quadrato-cubus &c. Et quantitas a cujus in se multiplicatione hæ potestates generantur dicitur earum radix, nempe radix quadratica quadrati a, cubica cubi a^3 &c.

Ad designandam radicem quantitatis alicujus præfigi solet nota $\sqrt{}$ si radix sit quadratica et $\sqrt[3]{}$ vel $\sqrt{}3$: si sit cubica, et $\sqrt[4]{}$ seu $\sqrt{}4$: si quadrato-quadratica &c.[15] Sic $\sqrt{}64$ denotat 8; et $\sqrt[3]{}64$ denotat 4. Item $\sqrt{}aa$ denotat a et $\sqrt{}ax$ radicem quadraticam ex ax et $\sqrt[3]{}4axx$ radicem cubicam ex $4axx$. Ut si a sit 3 et x 12, tum $\sqrt{}ax$ erit $\sqrt{}36$ seu 6, & $\sqrt[3]{}4axx$ erit $\sqrt[3]{}1728$ seu 12. Et hæ radices, ubi non licet extrahere, dicuntur surdæ quantitates ut $\sqrt{}ax$, vel surdi numeri ut $\sqrt{}12$.

Nonnulli pro designanda quadratica potestate usurpant q, pro cubica c, pro quadrato-quadratica qq, pro quadrato-cubica qc &c. Et ad hunc modum pro quadrato, cubo, et quadrato-quadrato ipsius A scribitur A^q, $A^c{}_{[,]}$ A^{qq} &c. Et pro radice cubica ex $abb - x^3$ scribitur $\sqrt{}c: \overline{abb - x^3}$. Alij alias notas adhibent sed quæ jam fere exoleverunt.

[16]Cæterum quantitates in geometrica progressione ut a^3. $aa\sqrt{}a$. a^2. $a\sqrt{}a$. a. $\sqrt{}a$. 1. $\dfrac{1}{\sqrt{}a} \cdot \dfrac{1}{a} \cdot \dfrac{1}{a\sqrt{}a} \cdot \dfrac{1}{aa}$ &c per indices in arithmetica progressione speciei suffixas hoc

---

(15) Notice that Newton now omits mention of Stevin's variant 'ring' notations, $\sqrt{}③$ and $\sqrt{}④$ (see **1**, **2**, §1: note (25)), for the cube and fourth root, inserting in their place Girard's more familiar radicals $\sqrt[3]{}$ and $\sqrt[4]{}$; compare note (44).

(16) The following paragraph is new in the present revise.

$(ab-b^2)/(a+x)$ denotes the quantity which arises on dividing $ab-b^2$ by $a+x$. And the like in other instances. Quantities of this sort, however, are called fractions, the upper part the numerator and the lower one the denominator.

Sometimes the divisor is prefixed to the quantity divided, with an arc inserted between. Thus, to designate the quantity which arises from the division of $\frac{ax^2}{a+x}$ by $a-b$, there can be written '$\overline{a-b})\frac{axx}{a+x}$'.

Even though multiplication is usually denoted by an immediate joining of quantities, an integer before a fraction none the less denotes the sum of each. So $3\frac{1}{2}$ denotes three together with a half.

If a quantity should multiply itself, for shortness the number of factors is usually appended. Thus in place of aaa we write a^3, in place of $aaaa$ we write a^4, and in place of $aaabb$ we write a^3bb or a^3b^2. If, say, a be 5 and b 2, then a^3 will be $5\times5\times5$ or 125, a^4 will be $5\times5\times5\times5$ or 625, and a^3b^2 will be $5\times5\times5\times2\times2$ or 500. Here note that a number written immediately between two variables relates always to the preceding one. Thus 3 in the quantity a^3b^2 does not signify that b^2 is to be taken three times, but that a has twice to be multiplied into itself. Note also that these quantities are said to be of as many dimensions, powers or 'dignities' as they comprise factors or quantities multiplying one another, and that the number appended is called the index of their power or dimension. Thus a^2 is of two dimensions, while a^3, as its appended number 3 indicates, is of three. Also, a^2 is said to be the square, a^3 the cube, a^4 the square-square (fourth power), a^5 the square-cube (fifth power), a^6 the cube-cube (sixth power), a^7 the square-square-cube (seventh power) and so forth. And the quantity a from whose self-multiplication these powers are generated is said to be their root—precisely, the square root of the square a^2, the cube one of the cube a^3 and so on.

To designate the root of some quantity the symbol $\sqrt{}$ is usually prefixed if the root be square, $\sqrt[3]{}$ or $\sqrt{}3$: if it be cube, and $\sqrt[4]{}$ or $\sqrt{}4$: if it be a fourth one, and so on.[15] Thus $\sqrt{64}$ denotes 8, and $\sqrt[3]{64}$ denotes 4. Likewise $\sqrt{aa}$ denotes a, $\sqrt{ax}$ the square root of ax and $\sqrt[3]{4axx}$ the cube root of $4ax^2$. So if a be 3 and x 12, then $\sqrt{ax}$ will be $\sqrt{36}$ or 6, and $\sqrt[3]{4axx}$ will be $\sqrt[3]{1728}$ or 12. And when it is impermissible to extract them, these roots are called surd quantities (as, for instance, $\sqrt{ax}$) or surd numbers (such as $\sqrt{12}$).

Some people use q to designate a square power, c for a cube, qq for a square-square (fourth) power, qc for a square-cube (fifth) one, and so on. And in this manner for the square, cube and square-square of a is written A^q, A^c, A^{qq}, and so forth. And for the cube root of ab^2-x^3 is written '$\sqrt{c}:\overline{abb-x^3}$'. Others employ still other symbols, but these have now practically gone out of use.

[16]For the rest, quantities in geometrical progression, such as a^3, $a^2\sqrt{a}$, a^2, $a\sqrt{a}$, a, $\sqrt{a}$, 1, $1/\sqrt{a}$, $1/a$, $1/a\sqrt{a}$, $1/a^2$, ..., we designate[18] by means of indices in

modo a^3. $a^{\frac{5}{2}}$. a^2. $a^{\frac{3}{2}}$. a^1 (seu a). $a^{\frac{1}{2}}$. a^0.[17] $a^{-\frac{1}{2}}$. a^{-1}. $a^{-\frac{3}{2}}$. a^{-2} &c respectivè designamus.[18] Qua ratione $a^{\frac{1}{3}}$ denotat radicem cubicam ipsius a, et $a^{-\frac{1}{3}}$ idem quod $\dfrac{1}{a^{\frac{1}{3}}}$,[19] et $3a^{-\frac{1}{3}}b^2$ idem quod $\dfrac{3bb}{a^{\frac{1}{3}}}$,[19] et $\overline{a^3+ab^2}^3$ cubum ipsius a^3+ab^2, et $\overline{a^3+ab^2}^{\frac{2}{3}}$ radicem quadrato-cubicam illius cubi, et $\overline{a^3+ab^2}^{-\frac{2}{3}}$ reciprocum illius radicis, seu $\dfrac{1}{\overline{a^3+ab^2}^{\frac{2}{3}}}$. Quod Notationis genus longè commodissimum judico: quippe quo Analogia inter Multiplicationem Divisionem et Extractionem radicum exprimitur.

Nota :: significat quantitates hinc inde proportionales esse. Sic $a \cdot b :: d \cdot e$ significat esse a ad b ut d ad e vel a ad d ut b ad e. Et $a \cdot b \cdot c :: d \cdot e \cdot f$ significat esse a, b et c inter se ut sunt d e et f inter se respectivè vel esse a ad d, b ad e et c ad f in eadem ratione. Deniq3 nota $=$ designat quantitates hinc inde æquales esse. ut $x = b$.[20]

ADDITIO.	...	...	...[21]
SUBDUCTIO.	...	...	...[22]
MULTIPLICATIO.	...	...	...[23]
DIVISIO.	...	...	...[24]

INVENTIO DIVISORUM.[25]

Divisioni affinis est inventio divisorum. Si dividuum sit quantitas simplex, divide eam per minimum ejus divisorem, et quotum per minimum divisorem

---

(17) Newton has cancelled a superfluous 'seu 1' (that is, 1).

(18) This replaces the more florid 'designare perquam naturale est' (it is completely natural to designate).

(19) Read '$\dfrac{1}{a^{\frac{1}{3}}}$' and '$\dfrac{3bb}{a^{\frac{1}{3}}}$' respectively.

(20) Understand 'designat x æqualem esse b' (denotes x to be equal to b), as in the parent *Arithmetica*.

(21) This section, written out in Newton's own hand in the manuscript, is only trivially different from the equivalent one 'De Additione' in the parent *Arithmetica* (**1**, 2, §1: Newton's pages 5–8) and we need therefore only list the few inessential changes introduced in amelioration of its original. In the first line of its opening paragraph Newton first faithfully copied 'Numerorum ubi non sunt admodum compositi...' and then smoothed its verbal flow by replacing 'ubi' with 'qui' (which [are not...]). Three sentences later an ugly 'Quemadmodum' is likewise first transcribed and then deleted in favour of the equivalent 'Ut'. More interestingly, in the example of addition *in numeris decimalibus* on the next page Newton, forgetful of his earlier preference for a superscript comma (see note (7)), lowers the decimal separatrix onto the line, writing

$$
\begin{array}{r}
630{,}953 \\
51{,}0807 \\
305{,}27 \\
\hline
987{,}3037.
\end{array}
$$

arithmetical progression appended to the variable in this manner: a^3, $a^{\frac{5}{2}}$, a^2, $a^{\frac{3}{2}}$, a, $a^{\frac{1}{2}}$, a^0,[17] $a^{-\frac{1}{2}}$, a^{-1}, $a^{-\frac{3}{2}}$, a^{-2}, ... respectively. By this principle $a^{\frac{1}{3}}$ denotes the cube root of a, $a^{-\frac{1}{3}}$ the same as $1/a^{[\frac{1}{3}]}$ and $3a^{-\frac{1}{3}}b^2$ the same as $3b^2/a^{[\frac{1}{3}]}$, while $(a^3+ab^2)^3$ denotes the cube of a^3+ab^2, $(a+ab^2)^{\frac{3}{5}}$ the fifth root of that cube, and $(a^3+ab^2)^{-\frac{3}{5}}$ the reciprocal of this last root, that is, $1/(a^3+ab^2)^{\frac{3}{5}}$. This type of notation I judge to be by far the most convenient, seeing that by it is expressed the analogy between multiplication, division and the extraction of roots.

The symbol $::$ signifies that the quantities on its either side are proportional. Thus $a\,.\,b::d\,.\,e$ signifies that a is to b as d to e, or a to d as b to e; and

$$a\,.\,b\,.\,c::d\,.\,e\,.f$$

signifies that a, b and c are to one another respectively as d, e and f are to each other, or that a to d, b to e and c to f are in the same ratio. The symbol $=$, finally, denotes that the quantities on its either side are equal: as $x=b$.[20]

ADDITION   ...   .[21]

SUBTRACTION   ...   .[22]

MULTIPLICATION   ...   .[23]

DIVISION   ...   .[24]

THE FINDING OF DIVISORS[25]

Akin to division is the finding of divisors. If the dividend be a simple quantity, divide it by its least divisor, and the quotient by its least divisor, until there

---

Two paragraphs later, finally, Newton—making here (as elsewhere in this and the next paragraph) the trivial substitution of 'fiunt' for his original 'faciunt'—transcribed the example '$6\sqrt{ab-xx}+7\sqrt{ab-xx}$ fiunt $13\sqrt{ab-xx}$' and then for no good reason altered the notation, writing in its place '$6\overline{ab-xx}^{\frac{1}{2}}+7\overline{ab-xx}^{\frac{1}{2}}$ fiunt $13\overline{ab-xx}^{\frac{1}{2}}$'.

(22) This section, written out wholly in Humphrey Newton's hand, is unchanged from its parent 'De Subductione' (**1**, 2, §1: Newton's pages 8–10) except that Newton has corrected Humphrey's copied title into its present nominative form and his mistranscription 'minus' (of 'nimis') in its opening line.

(23) Apart from the trivial change of its title, this section has been transcribed word for word by Humphrey Newton from its parent 'De Multiplicatione' (**1**, 2, §1: Newton's pages 10–12).

(24) This section, too, has minimal deviations from its parent text 'De Divisione' (**1**, 2, §1: Newton's pages 13–18). Newton himself has transcribed the opening lines (down to '...dic quoties 7 continetur'), substituting 'fit' for 'instituitur' and adding an emphatic 'idɋ' to make the first sentence now read 'Divisio fit in numeris quærendo...et scribendo totidem idɋ unitates in Quoto'. The whole remaining manuscript section (from 'in 37. Resp: 5....') is accurately copied in Humphrey Newton's hand, but in its last sentence but one Newton has himself subsequently changed '...a prioribus' to the equivalent 'a primis'.

(25) This revised section adapts the opening and closing portions of Newton's earlier initial revise 'De inventione Divisorum' (reproduced in **1**, 2, §2), omitting its exposition of the cumbrous finite-difference technique for discovering linear and higher factors of given

ejus, donec quotus restat indivisibilis, et omnes quantitatis divisores primos habebis. Dein horum divisorū singulos, binos ternos, quaternos, &c duc in se, et habebis etiam omnes divisores compositos. Ut si numeri 60 divisores omnes desiderentur, divide eum per 2, et quotum 30 per 2, et quotum 15 per 3 et restabit quotus indivisibilis 5. Ergo divisores primi sunt 1, 2, 2, 3, 5; ex binis compositi 4, 6, 10, 15; ex ternis 12, 20, 30; ex omnibus 60. Rursus si quantitatis $21abb$ divisores omnes desiderentur, divide eam per 3, et quotum $7abb$ per 7, et quotum abb per a, et quotum bb per b, et restabit quotus primus b. Ergo divisores primi sunt 1, 3, 7, a, b, b; ex binis compositi 21, $3a$, $3b$, $7a$, $7b$, ab, bb; ex ternis $21a$, $21b$, $3ab$, $3bb$, $7ab$, $7bb$, abb; ex quaternis $21ab$, $21bb$, $3abb$, $7abb$; ex quinis $21abb$. Eodem modo ipsius $2abb-6aac$ divisores omnes sunt 1, 2, a, $bb-3ac$, $2a$, $2bb-6ac$, $abb-3aac$, $2abb-6aac$.

Si quantitas postquam divisa est per omnes suos simplices divisores manet composita, et suspicio est eam compositum aliquem divisorem admittere, dispone eam secundum dimensiones literæ alicujus quæ in ea est, et[26] tenta si divisor aliquis primi termini + vel − divisore aliquo termini ultimi dividat totum. Tentare autem sufficit divisore[s] quorum dimensio non major est dimidia dimensione dividui. Ut si dividuum sit $x^3*-11x+14$ divisor primi termini x^3 est x, ij ultimi 14 sunt 1, 2, 7, 14. Tentandi sunt itacg divisores $x+1$, $x-1$, $x+2$, $x-2$, $x+7$, $x-7$, $x+14$, $x-14$ et res per divisorem $x-2$ succedit prodeunte quoto $xx+2x-7$.

Si dividuum sit $2a^4-4a^3+7aa-6a+6$ tento primi termini divisorem aliquem a, $2a$, aa, $2aa$, + vel − aliquo ultimi termini divisore 1, 2, 3, 6 et res succedit per $2aa+3$.[27]

Porro si dividuum sit $a^3 {+2aac \atop -aab} -3abc+bbc$ tento primi termini divisorem a + vel − aliquo ultimi termini divisore b, vel c et res succedit per divisorem compositum $a-b$.[28]

Quod si dividuum sit $y^6 {[+]aa \atop -2c[c]} y^4 {-a^{[4]} \atop +c^4} yy {-a^6 \atop -2a^{[4]}cc \atop [-aac^4]}$ considero divisores termini primi y, yy, y^3, eoscg ultimi a, aa, $aa+cc$, a^3+acc et ex ijs compono divisores tentandos $y+a$, $y-a$, $yy+aa$, $yy-aa$, $yy+aa+cc$, $yy-aa-cc$, y^3+a^3+acc, y^3-a^3-acc et succedit divisio per $yy-aa-cc$.[29] Termini autem ultimi divisores omnes a, aa, $aa+cc$ &c methodo jamjam exposita inveniuntur.

algebraic quantities, but elaborating the simpler test by trial divisors and incorporating the technique for determining the highest common factor of two algebraic quantities expounded in the parent *Arithmetica* (**1**, 2, §1: Newton's pages 25–7). The opening eleven words of the manuscript text are in Newton's hand, the remainder being Humphrey Newton's transcription.

(26) The sequel diverges considerably from the text (see note (25)) which has till now been faithfully followed.

remains an indivisible quotient, and you will have all prime divisors of the quantity. Then multiply these divisors together one, two, three, four, ... at a time and you will have also all the composite divisors. Should, for instance, all divisors of the number 60 be desired, divide it by 2, and the quotient 30 by 2, and the quotient 15 by 3, and there will remain the indivisible quotient 5. Therefore the prime divisors are 1, 2, 2, 3, 5; the compounds of these two at a time 4, 6, 10, 15; those of these three at a time 12, 20, 30; that of them all 60. Again, should all divisors of the quantity $21ab^2$ be desired, divide it by 3, and the quotient $7ab^2$ by 7, and the quotient ab^2 by a, and the quotient b^2 by b, and there will remain the prime quotient b. Therefore the prime divisors are 1, 3, 7, a, b, b; the compounds of these two at a time 21, $3a$, $3b$, $7a$, $7b$, ab, b^2; those of these three at a time $21a$, $21b$, $3ab$, $3b^2$, $7ab$, $7b^2$, ab^2; those of these four at a time $21ab$, $21b^2$, $3ab^2$, $7ab^2$; that of these five at a time $21ab^2$. In the same way all the divisors of $2ab^2 - 6a^2c$ are 1, 2, a, $b^2 - 3ac$, $2a$, $2b^2 - 6ac$, $ab^2 - 3a^2c$, $2ab^2 - 6a^2c$.

If a quantity after it has been divided by all its simple divisors, remains composite and there is a suspicion that it admits some composite divisor, arrange it according to the dimensions of some letter in it and[26] test if some divisor of the first term $+$ or $-$ some divisor of the final term shall divide the whole. It is sufficient, however, to test divisors whose dimension is not greater than half the dimension of the dividend. For instance, if the dividend be $x^3 - 11x + 14$, the divisor of the first term is x, those of the final one 14 are 1, 2, 7, 14. Accordingly the divisors $x+1$, $x-1$, $x+2$, $x-2$, $x+7$, $x-7$, $x+14$, $x-14$ are to be tested, and trial succeeds by the divisor $x-2$, yielding the quotient $x^2 + 2x - 7$.

If the dividend be $2a^4 - 4a^3 + 7a^2 - 6a + 6$, I test some divisor a, $2a$, a^2, $2a^2$ of the first term $+$ or $-$ some divisor 1, 2, 3, 6 of the final term, and trial by $2a^2 + 3$ succeeds.[27]

If, moreover, the dividend be $a^3 + a^2(-b + 2c) - 3abc + b^2c$, I test the divisor a of the first term $+$ or $-$ some divisor, b or c, of the final term, and trial by the composite divisor $a-b$ succeeds.[28]

But if the dividend be $y^6 + (a^2 - 2c^2)y^4 + (-a^4 + c^4)y^2 - (a^6 + 2a^4c^2 + a^2c^4)$, I consider the divisors y, y^2, y^3 of the first term and those, a, a^2, $a^2 + c^2$, $a^3 + ac^2$, of the final one, and from them compose the divisors $y + a$, $y - a$, $y^2 + a^2$, $y^2 - a^2$, $y^2 + a^2 + c^2$, $y^2 - (a^2 + c^2)$, $y^3 + a^3 + ac^2$, $y^3 - (a^3 + ac^2)$ to be tested. Division by $y^2 - (a^2 + c^2)$ succeeds.[29] All divisors a, a^2, $a^2 + c^2$, ... of the final term are, however, found by the method now already expounded.

---

(27) Producing the second divisor $a^2 - 2a + 2$.

(28) So yielding the irreducible quadratic factor $a^2 + 2ac - bc$.

(29) The division has just been performed in the first of the terminal worked examples in the preceding section 'Divisio'; compare **1**, **2**, §1: note (52) for the history of this Cartesian sextic.

Ad inveniendos divisores magis compositos quam qui per allatam regulam prodire solent, excogitatæ sunt ab Analystis multo cum sudore regulæ variæ particulares[30] quibus forsan addiscendis Tyro magis impediretur quam juvaretur utendis. Quare cum regula generalis hactenus desiderata fuit, incidi verò in quandam qua dividuorum omnium tres vel plures literas involventium divisores omnes prodeunt, lubet eam hic exponere. Detur imprimis dividuum quodvis cujus omnes termini ascendunt ad eundem numerum dimensionum. [31]Quære omnes divisores terminorum omnium in quibus literarum aliqua non est, item... divisor quæsitus.

Ut si proponatur quantitas

$$12x^3 - 14bxx + 9cxx - 12bbx - 6bcx + 8ccx + 8b^3 - 12bbc - 4bcc + 6c^3 : \ldots$$

invenire oportuit. Nam si per hunc dividas quantitatem propositam prodibit $3xx - 2bx + 2cc - 4bb$.

Rursus si quantitas sit $12x^5 - 10ax^4 [-]9bx^4 - 26aax^3 \ldots + 32aab^3 - 12b^5$: divisores... divisorem desideratum $2cc - 6aa \ldots + 2ax$ conflabunt. Per hunc itaɋ divido quantitatem propositam et oritur $3x^3 - 4axx - 2aab - 6b^3$.

Si quantitatis alicujus termini omnes non sunt æque alti, complendæ sunt dimensiones deficientes per dimensiones literæ cujusvis assumptæ, dein per præcedentes regulas invento divisore, litera assumpta delenda est. Ut si quantitas sit $12x^3 - 14bxx \ldots - 4b + 6$: assume... et habebitur divisor desideratus $4x - 2b + 3$.

Hoc pacto divisores omnium quantitatum inveniuntur præterquam earum quæ unicam tantum literam involvunt quæve ad unicam reduci possunt. Nam si æquatio involvit duas literas et termini omnes non sunt æque alti, reducetur ea ad tres supplendo dimensiones deficientes ut supra: sin termini omnes æque alti sunt reducetur ea ad unicam scribendo unitatem pro literarum alterutra. Reliquum est igitur ut methodo generali sciamus invenire divisores quantitatum quæ unicam tantum literam involvunt. Ea de re fusiùs scripturi sumus in sequentibus.[32]

Siquando maximus duarum quantitatum divisor desideratur prodit is quærendo divisores unius et tentando siquis eorum dividat alterum. Sed

---

(30) In Newton's day these rules were effectively restricted to the simplest cases in which an immediate reduction to the difference of two squares, $\alpha^2 - \beta^2 = (\alpha + \beta)(\alpha - \beta)$, was possible; otherwise—if at all—factors of algebraic quantities were determined by straightforward trial and error. Failure to notice that particular algebraic expressions were reducible spoilt many a seventeenth-century mathematical argument, as we have noticed in the case of Huygens (III: 506, note (53)) and Newton himself (III: 123, note (192) and 134, note (218)). A few years earlier Newton had come to considerable grief in attempting the general factorization of the simple binomial $x^n \pm 1$, n integral; see IV: 205–8.

(31) In sequel the manuscript returns to transcribing the parent text (compare 1, 2, §1: Newton's pages 181–6) and we have seen no need to make complete reproduction of it where it is unchanged.

To determine divisors more composite than those which are usually forth-coming by the rule adduced, algebraists have with a deal of sweat contrived various particular rules,[30] but learning these would perhaps hinder the beginner more than their practice would help him. Till now a general rule has been wanting. Therefore, since I have come upon one, indeed, which yields all divisors of all dividends involving three or more letters, it is agreeable to expound it here. In the first place let there be given any dividend in which all the terms ascend to the same number of dimensions. [31]Ascertain all the divisors of all terms in which some one of the letters is not present, and those similarly ... divisor required.

If, for instance, there be proposed the quantity

$$12x^3 - 14bx^2 + 9cx^2 - 12b^2x - 6bcx + 8c^2x + 8b^3 - 12b^2c - 4bc^2 + 6c^3 : \ldots$$

it was required to find. For if you divide the propounded quantity by this, there will result $3x^2 - 2bx + 2c^2 - 4b^2$.

Again, if the quantity be $12x^5 - 10ax^4 - 9bx^4 - 26a^2x^3 \ldots + 32a^2b^3 - 12b^5$, the divisors ... will ... constitute the desired divisor $2b^2 - 6a^2 \ldots + 2ax$. By this, accordingly, I divide the propounded quantity and there arises

$$3x^3 - 4ax^2 - 2a^2b - 6b^3.$$

If in some quantity all terms are not equally high, the missing dimensions are to be filled out by the dimensions of any assumed letter and then, after a divisor has been found by means of the preceding rules, the letter assumed is to be deleted. So, if the quantity be $12x^3 - 14bx^2 \ldots - 4b + 6$, assume ... and the desired divisor $4x - 2b + 3$ will be had.

By this technique divisors of all quantities are found, with the exception of those involving but a single letter or of those reducible to a single one. For if an equation involves two letters and all its terms are not equally high, it will be reduced to (one involving) three by supplying the missing dimensions as above; but if all terms are equally high, it will be reduced to a single one by writing unity in place of one or other of its letters. It is therefore left for us to know how, by a general method, we may find divisors of quantities involving but a single letter. We will treat that topic more copiously in the sequel.[32]

Should at any time the greatest divisor of two quantities be desired, this results from ascertaining the divisors of one and testing if any of them shall divide

---

(32) Since the unfinished existing manuscript contains no such future elaboration of Newton's finite-difference technique for determining the linear, quadratic and higher factors of given quantities involving but one algebraic variable—whether as such or in the context of equations (see **1**, 2, §1: Newton's pages 174–81; compare I: 532–9) where their zeros are all-important and the linear factors give their roots—, we infer that this omitted discussion was scheduled to form part of the unwritten concluding portion of the present 'Liber primus'.

methodus generalior et plerumcꝫ expeditior est auferre minorem quantitatem aut ejus multiplicem de majore et reliquum aut ejus multiplicem de ablato: Nam quæsitus erit divisor qui tandem nihil relinquit. Sic ad inveniendum maximum communem divisorem numerorum 203 et 667, aufer ter 203 de 667, et reliquum 58 ter de 203, et reliquum 29 bis de 58 restabitcꝫ nihil: quod indicat 29 esse divisorem quæsitum.[33]

Eadem obtinet methodus in speciebus, si modò et quantitates illæ et residua juxta literæ alicujus dimensiones ut in Divisione ostensum est, ordinentur, et qualibet vice concinnentur dividendo illas per suos omnes divisores qui aut simplices sunt, aut singulos terminos instar simplicium dividunt. Sic ad inveniendum communem divisorem quantitatum

$$x^4 - 3ax^3 - 8aaxx + 18a^3x - 8a^4 \quad \text{et} \quad x^3 - axx - 8aax + 6a^3$$

multiplica posteriorem per x ut primus ejus terminus evadat idem cum primo termino prioris. Dein aufer, et restabit $-2ax^3 + 12a^3x - 8a^4_{[,]}$ quod concinnatum dividendo ipsum per $-2a$ evadit $x^3 - 6aax + 4a^3$. Hoc aufer de posteriore et restabit $-axx - 2aax + 2a^3$, quod itidem per $-a$ divisum fit $xx + 2ax - 2aa$. Hoc autem per x multiplica ut ejus primus terminus evadat idem cum primo termino novissimi ablati $x^3 - 6aax + 4a^3$, de quo auferendum est. Multiplicatū igitur aufer et restabit $-2axx - 4aax + 4a^3$, quod per $-2a$ divisum fit etiam

$$xx + 2ax - 2aa.$$

Et hoc residuum cum idem sit ac superius proindecꝫ ablatum relinquat nihil, maximus erit quantitatum divisor communis quem invenire oportuit.[34]

Porro si quantitates sunt $6a^5 + 15a^4b - 4a^3cc - 10aabcc$ et

$$9a^3b - 27aabc - 6abcc + 18bc^3$$

eas concinno dividendo priorem per $aa_{[,]}$ posteriorem per $3b$. Sic[35] restat $\genfrac{}{}{0pt}{}{15b}{+18c} aa \genfrac{}{}{0pt}{}{-10bcc}{-12c^3}$, quod concinnatum dividendo utrumcꝫ[36] per $5b + 6c$ proinde ac si $5b + 6c$ simplex esset quantitas, evadit $3aa - 2cc$. Hoc multiplicatum per a aufero de $3a^3 - 9aac - 2a[c]c + 6c^3$ et secunda vice restat $-9aac + 6c^3$ quod itidem concinnatum dividendo terminos ejus per $-3c$, evadit etiam $3aa - 2cc$ ut ante. Quare $3aa - 2cc$ quæsitus est propositarum quantitatum divisor.[37]

---

(33) This and the remaining paragraphs of the present revised section are lightly adapted from the latter half of the parent *Arithmetica*'s 'De Reductione Fractionum ad minimos terminos' (**1**, 2, §1: Newton's pages 25–7), the reduction there, of course, being made by eliminating the greatest common divisor of numerator and denominator of the given fraction.

(34) As before (**1**, 2, §1: Newton's page 26), when the greatest common factor is divided out, the two given quantities reduce to

$$x^2 + 2ax(-2a^2)(x - 3a) \text{ and } (x^2 + 2ax - 2a^2)(x^2 - 5ax + 4a^2).$$

the other. But a more general and for the most part readier method is to take away the lesser quantity, or a multiple of it, from the greater one and then the remainder, or a multiple of it, from the magnitude subtracted: for, to be sure, the required divisor will be the subtrahend which at length leaves nothing. Thus, to find the greatest common divisor of the numbers 203 and 667, take three times 203 from 667, and three times the remainder 58 from 203, and twice the remainder 29 from 58, and nothing will be left: this reveals that 29 is the required divisor.[33]

The same method obtains in the case of algebraic quantities, provided that those quantities and the residues are, as was shown in 'DIVISION', ordered according to the dimensions of some letter and, at each and any stage, refined by dividing them through by all their divisors which either are simple or, in the fashion of simple ones, divide their individual terms. Thus, to find the common divisor of the quantities $x^4 - 3ax^3 - 8a^2x^2 + 18a^3x - 8a^4$ and $x^3 - ax^2 - 8a^2x + 6a^3$, multiply the latter by x so that its first term comes to be the same as the first term of the former; then take it away and there will remain $-2ax^3 + 12a^3x - 8a^4$, which, when refined by dividing through by $-2a$, comes to be $x^3 - 6a^2x + 4a^3$. Take this from the latter and there will remain $-ax^2 - 2a^2x + 2a^3$, which, after a like division by $-a$, becomes $x^2 + 2ax - 2a^2$. Now multiply this by x so that its first term comes to be identical with the first term of the quantity most recently subtracted, $x^3 - 6a^2x + 4a^3$, from which it is to be taken. Thereafter, take the multiplied quantity away and there will remain $-2ax^2 - 4a^2x + 4a^3$, which, after division by $-2a$, also becomes $x^2 + 2ax - 2a^2$. Since this is the same as the preceding residue and hence, when taken away, will leave nothing, it will be the greatest common divisor of the quantities: as was required to be found.[34]

Moreover, if the quantities are $6a^5 + 15a^4b - 4a^3c^2 - 10a^2bc^2$ and

$$9a^3b - 27a^2bc - 6abc^2 + 18bc^3,$$

I refine these by dividing the former through by a^2, the latter through by $3b$. In this way[35] there remains $(15b + 18c)\, a^2 - (10bc^2 + 12c^3)$, and when this is refined by dividing each (term) by $5b + 6c$, just as though $5b + 6c$ were a simple quantity, it comes to be $3a^2 - 2c^2$. This, multiplied by a, I take away from $3a^3 - 9a^2c - 2ac^2 + 6c^3$ and at the second stage there remains $-9a^2c + 6c^3$: that, when likewise refined by dividing its terms by $-3c$, comes also to be $3a^2 - 2c^2$, as before. Consequently $3a^2 - 2c^2$ is the required divisor of the propounded quantities.[37]

---

(35) Understand, as in the parent text (**1**, 2, §1: Newton's page 26), 'ablato bis $3a^3 - 9aac - 2acc + 6c^3$ de $6a^3 + 15aab - 4aac - 10bcc$' (on taking away twice $3a^3 - 9a^2c - 2ac^2 + 6c^3$ from $6a^3 + 15a^2b - 4a^2c - 10bc^2$).

(36) A necessary 'terminum' is accidentally omitted.

(37) As before (**1**, 2, §1: Newton's page 27), on dividing out their common divisor the given quantities prove to be $(3a^2 - 2c^2)\, a^2(2a + 5b)$ and $(3a^2 - 2c^2)\, 3b(a - 3c)$ respectively.

Prodit aliquando divisor communis e terminis per quos Quantitates et residua concinnantur. Ut si proponantur quantitates $aadd - ccdd - aacc + c^4$ et $4aad - 4acd - 2acc + 2c^3$ seu, terminis earum juxta dimensiones literæ d dispositis, $\begin{smallmatrix} aa \\ -cc \end{smallmatrix} dd \begin{smallmatrix} -aacc \\ +c^4 \end{smallmatrix}$ & $\begin{smallmatrix} 4aa \\ -4ac \end{smallmatrix} d \begin{smallmatrix} -2acc \\ +[2]c^3 \end{smallmatrix}$. Has oportet concinnare dividendo utrumcᵹ prioris terminum per $aa - cc$ et utrumcᵹ posterioris per $2a - 2c$, perinde ac si $aa - cc$ & $2a - 2c$ essent simplices quantitates. Atcᵹ ita vice prioris emergit $dd - cc$, et vice posterioris $2ad - cc$, ex quibus sic præparatis nullus communis divisor obtineri potest. Sed e terminis $aa - cc$ et $2a - 2c$ per quos quantitates abbreviatæ sunt, prodit ejusmod[i] divisor $a - c$. Quod si necᵹ termini $aa - cc$ & $2a - 2c$ communem divisorem habuissent, conclusissem quod quantitatum propositarū nullus fuisset divisor communis.

EXTRACTIO RADICUM.[38]

Cum numeri alicujus radix quadratica ... ... prodit $zz + 2z - 4$. Atcᵹ ita in altioribus radicibus.

Radices quantitatum quæ ex partibus asymmetris componuntur, sic extrahe. ... ... sunt 1, $\sqrt{2}$, $\sqrt{3}$, ut supra.

Est et regula extrahendi altiores radices ex quantitatibus numeralibus duarum potentia commensurabilium partium. Hæc rarò usu veniet & a Tyrone prætermitti potest,[39] describam tamen nequid hac in parte desideretur. Sit quantitas $A \pm B$. ... ... radicem factoris utriuscᵹ $\sqrt[3]{3}$ & $11 + \sqrt{125}$.

REDUCTIO FRACTARUM ET RADICALIUM.[40]

Præcedentibus operationibus inservit reductio quantitatum fractarum et radicalium, idcᵹ vel ad minimos terminos vel ad eandem denominationem.

Quantitas fracta reducitur ad minimos terminos dividendo numeratorem et denominatorem per divisorem maximum communem. Sic $\dfrac{aac}{bc}$ reducitur ad simpliciorem $\dfrac{aa}{b}$ dividendo utrumcᵹ aac et bc per c; et $\frac{203}{667}$ reducitur ad simpli-

---

(38) In line with the initial revise (**1**, 2, §2, where the latter portion is directed to be added 'in calce capitis de Reductione Radicalium'), this section comprehends both the parent *Arithmetica*'s 'De extractione Radicum' (**1**, 2, §1: Newton's pages 18–24) and its later 'Extractio Radicum ex binomijs' (*ibid.*: Newton's pages 207–10). Since with one exception (see next note) it exactly repeats these preliminary texts, we have been content merely to reproduce its key opening and terminating phrases.

(39) In addition to its impracticality Newton's modified Cartesian rule is, as we have mentioned (**1**, 2, §1: note (605)), generally true only when A^2 and B^2 are integers—so much for Newton's following hope that the rule will complete his exposition of techniques for exact root-extraction.

Sometimes a common divisor results from the terms by means of which the quantities and residues are refined. So, if there be propounded the quantities $a^2d^2 - c^2d^2 - a^2c^2 + c^4$ and $4a^2d - 4acd - 2ac^2 + 2c^3$, that is, when their terms are arranged according to the dimensions of the letter d, $(a^2 - c^2)\,d^2 - (a^2c^2 - c^4)$ and $(4a^2 - 4c^2)\,d - (2ac^2 - 2c^3)$, the requirement is that we refine these by dividing each term of the former by $a^2 - c^2$ and each of the latter by $2a - 2c$, just as if $a^2 - c^2$ and $2a - 2c$ were simple quantities. In this way in the former's stead there emerges $d^2 - c^2$, and in the latter's $2ad - c^2$; from these, when thus prepared, no common divisor can be obtained. But from the terms $a^2 - c^2$ and $2a - 2c$ by which the quantities have been shortened there does result a divisor of this type, $a - c$. However, if the terms $a^2 - c^2$ and $2a - 2c$ had not had a common divisor, I should have concluded that there was no common divisor of the quantities proposed.

EXTRACTION OF ROOTS.[38]

When the square root of some number … … … … proves to be $z^2 + 2z - 4$. And similarly in the case of higher roots.

The roots of quantities which are made up of irrational parts extract in this way. … … are 1, $\sqrt{2}$ and $\sqrt{3}$, as above.

There is also a rule for extracting higher roots of numerical quantities of two parts whose (square) powers are commensurable. This will rarely prove to be of use and can be passed over by the novice,[39] yet I will describe it so that nothing will be felt to be wanting in this section. Let the quantity be $A \pm B$ … … the root of each factor $\sqrt[3]{3}$ and $11 + \sqrt{125}$.

THE REDUCTION OF FRACTIONS AND RADICALS[40]

Of service in the preceding operations is the reduction of fractional and radical quantities: this either to least terms or to a common denominator.

A fractional quantity is reduced to least terms by dividing numerator and denominator by their greatest common divisor. Thus a^2c/bc is reduced to the simpler equivalent a^2/b by dividing each of a^2c and bc by c; $\frac{203}{667}$ is reduced to the

---

(40) We should understand 'QUANTITATUM' in the Latin title. This section compresses together, without introducing any essential novelties, the parent *Arithmetica*'s 'De Reductione Fractionum ad minimos terminos'—retaining only one instance (see note (41)) of a reduction worked by the generalized Euclidean algorithm for finding common divisors which here, in revise, is abstracted to form the latter half of the preceding section on 'Inventio Divisorum' (see note (33))—and its following 'De Reductione Fractionum ad communem Denominationem' (**1**, 2, §1: Newton's pages 25–7/27–8), along with the allied 'De reductione Radicalium ad minimos terminos' and 'De Reductione Radicalium ad eandem denominationem' (*ibid.*: Newton's pages 28–9).

ciorem $\frac{7}{23}$ dividendo utrumcg 203 et 667 per 29; et $\frac{203aac}{667bc}$ reducitur ad $\frac{7aa}{23b}$

dividendo per 29c. Atcg ita $\frac{6a^3-9acc}{6aa+3ac}$ evadit $\frac{2aa-3cc}{2a+c}$ dividendo per 3a. Et

$\frac{a^3-aab+abb-b^3}{aa-bb}$ evadit $\frac{aa+bb}{a[+b]}$ dividendo per $a-b$. Et

$$\frac{x^4-3ax^3-8aaxx+18a^3x-8a^4}{x^3-axx-8aax+6a^3} \quad \text{fit} \quad \frac{xx-5ax+4aa}{x-3a}$$

mediante divisore $xx+2ax-2aa$ per methodum superius expositam[41] invento. Utilis est autem hæc reductio ad abbreviandas quantitates quæ per multiplicationem vel divisionem prodeunt. Ut si multiplicare oportet $\frac{2ab^3}{3ccd}$ per $\frac{9acc}{bdd}$ vel id

dividere per $\frac{bdd}{9acc}$ prodibit $\frac{18aab^3cc}{3bccd^3}$ et per reductionem $\frac{6aabb}{d^3}$. Sed in hujusmodi casibus præstat ante operationem concinnare terminos, dividendo per maximum communem divisorem quos postea dividere oporteret. Sic in allato exemplo si dividam 2ab^3 et bdd per communem divisorem b, et 3ccd ac 9acc per communem

divisorem 3cc, emerget fractio $\frac{2abb}{d}$ multiplicanda per $\frac{3a}{dd}$ vel dividenda per $\frac{dd}{3a}$,

prodeunte tandem $\frac{6aabb}{d^3}$ ut supra. Atcg ita $\frac{aa}{c}$ in $\frac{c}{b}$ evadit $\frac{aa}{1}$ in $\frac{1}{b}$ seu $\frac{aa}{b}$. Et $\frac{aa}{c}$

divis: per $\frac{b}{c}$ evadit aa div: per b seu $\frac{aa}{b}$. Et $\frac{a^3-axx}{x}$ in $\frac{cx}{aa+ax}$ evadit $\frac{a-x}{x}$ in $\frac{c}{1}$ seu

$\frac{ac}{x}-c$. Et 28 div: per $\frac{7}{3}$ evadit 4 div: per $\frac{1}{3}$ seu 12.

Quantitates fractæ reducuntur ad communem denominatorem multiplicatione mutua terminorum utriuscg per denominatorem alterius. Sic habitis $\frac{a}{b}$ et $\frac{c}{d}$

duc terminos unius $\frac{a}{b}$ in d, et vicissim terminos alterius $\frac{c}{d}$ in b, et emergent $\frac{ad}{bd}$ et

$\frac{bc}{bd}$, quarum communis est denominator bd. Ubi verò denominatores communem habent divisorem sufficit multiplicatio alterna per divisores reliquos.[42] Sic

fractiones $\frac{a^3}{bc}$ et $\frac{a^3}{bd}$ quarum denominatores communem habent divisorem b

reducuntur ad hasce $\frac{a^3d}{bcd}$ et $\frac{a^3c}{bcd}$ multiplicando eas alternè per divisores reliquos c ac d. Hæc autem reductio præcipuè usui est in Additione et Subductione fractionum, quæ si diversos habent denominatores, ad eundem reducendæ sunt

antequam uniri possunt. Sic $\frac{a}{b}+\frac{c}{d}$ per reductionem evadit $\frac{ad}{bd}+\frac{bc}{bd}$, sive $\frac{ad+bc}{bd}$.

simpler $\frac{7}{23}$ by dividing each of 203 and 667 by 29; and $\frac{203}{667}a^2c/bc$ is reduced to $\frac{7}{23}a^2/b$ by dividing through by $29c$. And similarly $(6a^3-9ac^2)/(6a^2+3ac)$ comes to be $(2a^2-3c^2)/(2a+c)$ on dividing through by $3a$; and

$$(a^3-a^2b+ab^2-b^3)/(a^2-b^2)$$

comes to be $(a^2+b^2)/(a+b)$ on dividing through by $a-b$; while

$$\frac{x^4-3ax^3-8a^2x^2+18a^3x-8a^4}{x^3-ax^2-8a^2x+6a^3} \quad \text{becomes} \quad \frac{x^2-5ax+4a^2}{x-3a}$$

by means of the divisor $x^2+2ax-2a^2$ found by the method previously expounded.[41] This reduction is, may I say, useful for abbreviating quantities resulting by multiplication or division. If, for instance, it is necessary to multiply $2ab^3/3c^2d$ by $9ac^2/bd^2$—or to divide it by $bd^2/9ac^2$—, there will result

$$18a^2b^3c^2/3bc^2d^3$$

and by reduction $6a^2b^2/d^3$. But in cases of this sort it is better to refine the terms before the operation, dividing through by the greatest common divisor, which it would afterwards be necessary to divide out. Thus, in the example presented, should I divide $2ab^3$ and bd^2 by their common divisor b, and $3c^2d$ and $9ac^2$ by their common divisor $3c^2$, it will emerge that the fraction $2ab^2/d$ has to be multiplied by $3a/d^2$—or divided by $d^2/3a$—, at length yielding $6a^2b^2/d^3$ as above. And similarly a^2/c times c/b comes to be $a^2/1$ times $1/b$, or a^2/b; and a^2/c divided by b/c comes to be a^2 divided by b, or a^2/b. Again $(a^3-ax^2)/x$ times $cx/(a^2+ax)$ comes to be $(a-x)/x$ times $c/1$, or $ac/x-c$. And 28 divided by $\frac{7}{3}$ comes to be 4 divided by $\frac{1}{3}$, or 12.

Fractional quantities are reduced to a common denominator by a mutual multiplication of the terms of each by the denominator of the other. Thus, when a/b and c/d are had, multiply the terms of one, a/b, into d and in sequel the terms of the other, c/d, into b, and there will emerge ad/bd and bc/bd, having the common denominator bd. When, in fact, denominators have a common divisor, alternate multiplication by the remaining divisors is enough.[42] Thus the fractions a^3/bc and a^3/bd, whose denominators have the common divisor b, are reduced to these, a^3d/bcd and a^3c/bcd, on multiplying them alternately by the remaining divisors c and d. This reduction is, however, of especial use in the addition and subtraction of fractions; for, if they have differing denominators, they must be reduced to the same one before they can be united. Thus $a/b+c/d$ comes by reduction to be $ad/bd+bc/bd$, that is, $(ad+bc)/bd$. And $a+ab/c$ becomes

---

(41) See pages 552 above; and compare note (34).

(42) Newton's correction of 'sufficit multiplicare alternè per Quotientes' (it is sufficient to multiply alternately by the quotients), faithfully transcribed by Humphrey Newton from the parent text (**1**, 2, §1: Newton's page 28).

Et $a+\dfrac{ab}{c}$ fit $\dfrac{ac+ab}{c}$. Et $\dfrac{a^3}{bc}-\dfrac{a^3}{bd}$ fit $\dfrac{a^3d-a^3c}{bcd}$ vel $\dfrac{d-c}{bcd}a^3$. Et $\dfrac{c^4+x^4}{cc-xx}-cc-xx$ evadit

$\dfrac{2x^4}{cc-xx}$. Atcp ita $\frac{2}{3}+\frac{5}{7}$ fit $\frac{14}{21}+\frac{15}{21}$ sive $\dfrac{14+15}{21}$ hoc est $\frac{29}{21}$. Et $\frac{11}{6}-\frac{3}{4}$ fit $\frac{22}{12}-\frac{9}{12}$ sive

$\frac{13}{12}$. Et $\frac{3}{4}-\frac{5}{12}$ fit $\frac{9}{12}-\frac{5}{12}$ sive $\frac{4}{12}$ hoc est $\frac{1}{3}$. Et $3\frac{4}{7}$ sive $\frac{3}{1}+\frac{4}{7}$ evadit $\frac{21}{7}+\frac{4}{7}$ sive $\frac{25}{7}$.

Et $25\frac{1}{2}$ evadit $\frac{51}{2}$. Sic etiam $\dfrac{aa}{x}-a+\dfrac{2x[x]}{3a}-\dfrac{a[x]}{a-x}$ auferendo a de $\dfrac{aa}{x}$ et ad resi-

duum $\dfrac{aa-ax}{x}$ addendo $\dfrac{2xx}{3a}$ atcp de summa $\dfrac{3a^3-3aax+2x^3}{3ax}$ auferendo $\dfrac{ax}{a-x}$ fit

$\dfrac{3a^4-6a^3x+2ax^3-2x^4}{3aax-3axx}$. Nec secus $3\frac{4}{7}-\frac{2}{3}$ per successivas operationes[43] fit $\frac{61}{21}$.

Quantitas radicalis ubi radix totius nequit extrahi plerumcp reducitur ad simpliciorem extrahendo radicem divisoris alicujus. Sic $\sqrt{aabc}$ extrahendo radicem divisoris aa fit $a\sqrt{bc}$. Et $\sqrt{48}$ extrahendo radicem divisoris 16 fit $4\sqrt{3}$. Et

$\sqrt{48}\,aabc$ extrahendo radicem divisoris $16aa$ fit $4a\sqrt{3bc}$. Et $\sqrt{\dfrac{a^3b-4aabb+4ab^3}{cc}}$

extrahendo radicem divisoris $\dfrac{aa-4ab+4bb}{cc}$ fit $\dfrac{a-2b}{c}\sqrt{ab}$. Et $\sqrt{\dfrac{aaoomm}{ppzz}+\dfrac{4aam^3}{pzz}}$

extrahendo radicem divisoris $\dfrac{aamm}{ppzz}$ fit $\dfrac{am}{pz}\sqrt{oo+4mp}$. Et $6\sqrt{\frac{75}{98}}$ extrahendo

radicem divisoris $\frac{25}{49}$ fit $\frac{30}{7}\sqrt{\frac{3}{2}}$, sive $\frac{30}{7}\sqrt{\frac{6}{4}}$ radicemcp denominatoris adhuc extra-

hendo, fit $\frac{15}{7}\sqrt{6}$. Et sic $a\sqrt{\dfrac{b}{a}}$ sive $a\sqrt{\dfrac{ab}{aa}}$ extrahendo radicem denominatoris fit

$\sqrt{ab}$. Et $\sqrt[3]{8a^3b+16a^4}$ extrahendo radicem cubicam divisoris $8a^3$ fit $2a\sqrt[3]{b+2a}$.

Haud secus $\sqrt[4]{a^3x}$ extrahendo radicem quadraticam divisoris aa fit $\sqrt{a}$ in $\sqrt[4]{ax}$

vel extrahendo radicem quadrato-quadraticam divisoris a^4 fit $a\sqrt[4]{\dfrac{x}{a}}$. Atcp ita

$\sqrt{6}: a^7x^5$ convertitur in $a\sqrt{6}:ax^5$ vel in $ax\sqrt{6}[:]\dfrac{a}{x}$ vel etiam in $\sqrt{ax}\times\sqrt{3}:aax$.

Inservit autem hæc reductio Additioni et Subductioni radicalium: quippe quæ ad formam simplicissimam reductæ conveniant ex parte radicali si modò uniri possunt. Sic $\sqrt{48}+\sqrt{75}$ per reductionem evadit $4\sqrt{3}+5\sqrt{3}$ hoc est $9\sqrt{3}$. Et $\sqrt{48}-\sqrt{\frac{16}{27}}$ per reductionem evadit $4\sqrt{3}-\frac{4}{9}\sqrt{3}$ hoc est $\frac{32}{9}\sqrt{3}$. Et sic

$$\sqrt{\dfrac{4ab^3}{cc}}+\sqrt{\dfrac{a^3b-4aabb+4ab^3}{cc}}$$

extrahendo quicquid est rationale, evadit $\dfrac{2b}{c}\sqrt{ab}+\dfrac{a-2b}{c}\sqrt{ab}$ hoc est $\dfrac{a}{c}\sqrt{ab}$. Et

$\sqrt[3]{8a^3b+16a^4}-\sqrt[3]{b^4+2ab^3}$ evadit $2a\sqrt{3}:\overline{b+2a}-b\sqrt{3}:\overline{b+2a}$ hoc est

$$\overline{2a-b}\sqrt{3}:\overline{b+2a}.\text{[44]}$$

---

(43) These are elaborated in the parent *Arithmetica* (**1**, 2, §1: Newton's page 28).

$(ac+ab)/c$. While $a^3/bc-a^3/bd$ becomes $(a^3d-a^3c)/bcd$ or $((d-c)/bcd)\,a^3$. And $(c^4+x^4)/(c^2-x^2)-(c^2+x^2)$ comes to be $2x^4/(c^2-x^2)$. Similarly $\frac{2}{3}+\frac{5}{7}$ becomes $\frac{14}{21}+\frac{15}{21}$ or $\dfrac{14+15}{21}$, that is, $\frac{29}{21}$. And $\frac{11}{6}-\frac{3}{4}$ becomes $\frac{22}{12}-\frac{9}{12}$ or $\frac{13}{12}$. And $\frac{3}{4}-\frac{5}{12}$ becomes $\frac{9}{12}-\frac{5}{12}$ or $\frac{4}{12}$, that is, $\frac{1}{3}$. While $3\frac{4}{7}$ or $\frac{3}{1}+\frac{4}{7}$ comes to be $\frac{21}{7}+\frac{4}{7}$ or $\frac{25}{7}$. And $25\frac{1}{2}$ comes to be $\frac{51}{2}$. So also $a^2/x-a+2x^2/3a-ax/(a-x)$, by taking a from a^2/x, and to the residue $(a^2-ax)/x$ adding $2x^2/3a$, and then from the sum

$$(3a^3-3a^2x+2x^3)/3ax$$

taking away $ax/(a-x)$, becomes $(3a^4-6a^3x+2ax^3-2x^4)/(3a^2x-3ax^2)$. No differently, $3\frac{4}{7}-\frac{2}{3}$ comes by successive operations[43] to be $\frac{61}{21}$.

When the root of its whole cannot be extracted, a radical quantity is commonly reduced to a simpler one by extracting the root of some divisor of it. Thus, by extracting the root of its divisor a^2, $\sqrt{[a^2bc]}$ becomes $a\sqrt{[bc]}$. And $\sqrt{48}$, on extracting the root of the divisor 16, becomes $4\sqrt{3}$. And $\sqrt{[48a^2bc]}$, on extracting the root of the divisor $16a^2$, becomes $4a\sqrt{[3bc]}$. While

$$\sqrt{[(a^3b-4a^2b^2+4ab^3)/c^2]},$$

on extracting the root of the divisor $(a^2-4ab+4b^2)/c^2$, becomes $((a-2b)/c)\sqrt{[ab]}$. And $\sqrt{[a^2o^2m^2/p^2z^2+4a^2m^3/pz^2]}$, on extracting the root of the divisor a^2m^2/p^2z^2, becomes $(am/pz)\sqrt{[o^2+4mp]}$. Again $6\sqrt{\frac{75}{98}}$, on extracting the root of the divisor $\frac{25}{49}$, becomes $\frac{30}{7}\sqrt{\frac{3}{2}}$, that is, $\frac{30}{7}\sqrt{\frac{6}{4}}$, and by further extracting the root of the denominator it becomes $\frac{15}{7}\sqrt{6}$. And thus $a\sqrt{[b/a]}$ or $a\sqrt{[ab/a^2]}$ on extracting the root of the denominator becomes $\sqrt{[ab]}$. And $\sqrt[3]{[8a^3b+16a^4]}$ on extracting the cube root of the divisor $8a^3$ comes to be $2a\sqrt[3]{[b+2a]}$. No differently, $\sqrt[4]{[a^3x]}$ on extracting the square root of the divisor a^2 becomes $\sqrt{a}\times\sqrt[4]{[ax]}$; alternatively, on extracting the fourth root of the divisor a^4 it becomes $a\sqrt[4]{[x/a]}$. And so $\sqrt[6]{[a^7x^5]}$ is converted to $a\sqrt[6]{[ax^5]}$, or to $ax\sqrt[6]{[a/x]}$, or even to $\sqrt{[ax]}\times\sqrt[3]{[a^2x]}$. This reduction, further, is of service in the addition and subtraction of radicals, seeing that these, when reduced to simplest form, have to agree in their radical part if they can be united. Thus $\sqrt{48}+\sqrt{75}$ comes by reduction to be $4\sqrt{3}+5\sqrt{3}$, that is, $9\sqrt{3}$. And $\sqrt{48}-\sqrt{\frac{16}{27}}$ by reduction comes to be $4\sqrt{3}-\frac{4}{9}\sqrt{3}$, that is, $\frac{32}{9}\sqrt{3}$. And thus $\sqrt{[4ab^3/c^2]}+\sqrt{[(a^3b-4a^2b^2+4ab^3)/c^2]}$ on extracting whatever is rational proves to be $(2b/c)\sqrt{[ab]}+((a-2b)/c)\sqrt{[ab]}$, that is, $(a/c)\sqrt{[ab]}$. And $\sqrt[3]{[8a^3b+16a^4]}-\sqrt[3]{[b^4+2ab^3]}$ comes to be $2a\sqrt[3]{[b+2a]}-b\sqrt[3]{[b+2a]}$, that is, $(2a-b)\sqrt[3]{[b+2a]}$.[44]

---

(44) Observe that Newton—or at least his amanuensis Humphrey Newton acting under his instruction—has partially replaced his notations $\sqrt{③}$ (or $\sqrt{3:}$) and $\sqrt{④}$ for cube and fourth roots in his original text (**1**, **2**, §1: Newton's page 29) by their now standard equivalents $\sqrt[3]{}$ and $\sqrt[4]{}$ (first suggested by Albert Girard in his *Inuention Nouuelle en l'Algebre* (Amsterdam,

Cum in radicalibus diversæ denominationis instituenda est multiplicatio vel divisio reducendæ sunt hæ ad eandem denominationem, id verò præfigendo utriꝗ signum radicale cujus index est minimus numerus multiplex indice utriusꝗ prioris, et perinde augendo dimensiones suffixarum quantitatum. Sic enim...sive $\sqrt{2ab}$.[45]

ÆQUATIO ET EJUS RADICES.[46]

Expositis computandi modis superest eorum usus in resolutione problematum. Ea consistit in æquationum Inventione Reductione et Resolutione. Proposito aliquo problemate imprimis ex conditionibus ejus eliciendæ sunt æquationes seu quantitatum æqualitates, dein hæ (si plures sunt) in unam transformandæ, et hæc reducenda ad aliquam ex sequentibus formulis

$$x - a = 0.$$

$$xx - ax + b = 0.$$

$$x^3 - ax^2 + bx - c = 0.$$

$$x^4 - ax^3 + bxx - cx + d = 0. \quad [\&c.]$$

Ubi x designat quantitatem quæsitam & a, b, c, d &c quantitates quascunꝗ datas, signa autem $+$ et $-$ modis omnibus variari possunt et ex intermedijs terminis unus vel plures deesse.[47] Dicitur autem æquationū prima simplex, secunda quadratica[,] tertia cubica & sic deinceps pro dimensione maxima quantitatis quæsitæ. Labor ultimus est ut ex hujusmodi æquatione eruatur quæsita illa quantitas, id est ut inveniatur[48] quantitas quæ si in æquatione scribatur pro x, efficiet ut ejus termini omnes conjunctim sumpti evad[a]nt nihilo æquales. Ut si incideretur in æquationem $x^3 - xx[-]14x[+]24 = 0$[49] quoniam scribendo 2 pro x, termini æquationis fiunt $8 - 4[-]28[+]24$ & hi conjunctim sunt nihilo æquales, erit x incognita quantitas [æqualis 2].

1629); see F. Cajori, *A History of Mathematical Notations*, **1** (Chicago, 1929): 159). The expression $\sqrt[3]{8a^3b + 16a^4} - \sqrt[3]{b^4 + 2ab^3}$ was initially copied by Humphrey Newton as

$$\sqrt{3} : \overline{8a^3b + 16a^4} - \sqrt{3} : \overline{b^4 + 2ab^3},$$

Newton himself making subsequent substitution of $\sqrt[3]{}$ for $\sqrt{3}$:.

(45) Except for the trivial omission of a terminal 'Et sic in alijs' (And so in other instances) and the universal replacement of the notations $\sqrt{3}$:, $\sqrt{4}$: and $\sqrt{6}$: by their more familiar equivalents $\sqrt[3]{}$, $\sqrt[4]{}$ and $\sqrt[6]{}$ (compare note (44)), this omitted portion of text exactly repeats that of the parent *Arithmetica* (**1**, 2, §1: Newton's page 29).

(46) A much reworked and augmented version of Newton's previous section 'De forma Æquationis' (**1**, 2, §1: Newton's pages 30–1). The manuscript title was originally 'INVENTIO ÆQUATIONUM' (THE FINDING OF EQUATIONS) but is replaced as shown on Newton's inserted correction sheet (f. 93c; see note (2)). From the same correction sheet (ff. 93a/93c) we have

When multiplication or division must be instituted in radicals of differing denomination, they must be reduced to the same denomination—this, of course, by prefixing to each a radical sign whose index is the least number which is a multiple of the indices of each of the previous ones, and by correspondingly increasing the dimensions of the attached quantities. For thus ... or $\sqrt{[2ab]}$.[45]

AN EQUATION AND ITS ROOTS[46]

Having exhibited the modes of computation, it remains for us to display their use in the solution of problems. This consists in the finding of equations, their reduction and their resolution. When some problem is proposed, in the first instance from its conditions equations—or equalities between quantities—must be contrived; then these (if there are several) must be transformed into a single one and this reduced to some one of the following general forms:

$$x - a = 0, \qquad\qquad x^2 - ax + b = 0,$$
$$x^3 - ax^2 + bx - c = 0, \qquad x^4 - ax^3 + bx^2 - cx + d = 0, \dots.$$

Here x denotes the quantity sought and a, b, c, d, ... any given quantities whatsoever, while the signs $+$ and $-$ can vary in all possible ways, and one or more of the intervening terms can be lacking.[47] The first of the equations, however, is called simple, the second quadratic, the third cubic, and so on in conformance with the greatest dimension of the quantity sought. The final task is to root out the quantity sought from an equation of this type, that is, to find[48] a quantity which, if it be written in the equation in x's place, will make all its terms taken jointly together come to be equal to nothing. For instance, were you to come upon the equation $x^3 - x^2 - 4x + 24 = 0$,[49] since on writing 2 in x's place the terms in the equation become $8 - 4 - 28 + 24$ and these are jointly equal to nothing, the unknown quantity x will then be equal to 2.

---

introduced extensive additions which expand the single paragraph of the initial version into the revised text now reproduced.

(47) The initial version, complemented by a first addition on ff. 93a/93c of the manuscript insert, reads in sequel: 'Est ultimò ex hujusmodi æquatione eruenda quæsita quantitas, hoc est quantitas quæ si in æquatione scribatur pro x, termini æquationis evadent nihilo æquales. Dicitur autem quæsita quantitas æquationis radix et in ejus extractione labor ultimus consistit' (From an equation of this type, finally, there needs to be hunted out the required quantity, that is, one such that, if it be written in x's place in the equation, the equation's terms will prove to be (together) equal to nothing. The required quantity, however, is called the root of the equation and the final task consists in its extraction).

(48) Newton first repeated 'eruatur' (root out).

(49) By a slip Newton has here (and correspondingly in the next paragraph) written '$x^3 - xx + 14x - 24 = 0$' as the explicit form of his intended cubic $(x-2)(x-3)(x+4) = 0$. In immediate sequel this produces the interesting corollary that, when $x = 2$, there ensues '$8 - 4 + 28 - 24$' = 0.

Dicitur autem incognita quantitas radix æquat[ionis] suntɋ tot æquationis cujusɋ radices quot dimensiones incognitæ quantitatis. Sic

$$x^3 - xx[-]14x[+]24 = 0$$

quoniam ad tres dimensiones ascendit, tres habebit radices. Eæ sunt 2, 3 et −4. Nam harum quælibet scripta in æquatione pro *x* efficit ut termini omnes conjunctim æquentur nihilo. Cujus rei ratio est quod æquatio ex radicibus continue multiplicatis confit. Ponentur $x = 2$, $x = 3$ et $x = -4$ seu (quod perinde est) $x - 2 = 0$, $x - 3 = 0$ et $x + 4 = 0$ et harum continua multiplicatione fiet $x^3 - xx[-]14x[+]24 = 0$. Duæ autem vel quatuor ex his radicibus vel etiam plures si æquatio est plurium dimensionum, possibilitatem quandoɋ amittunt & tunc imaginariæ vel impossibiles dici solent. Sic æquatio $x^3 - 3xx - 9x + b = 0$ si *b* non major sit quam $3^{(50)}$ neɋ minor quam $-1^{(50)}$ tres habebit radices reales. At harum duæ si *b* fiat 3 vel $-1^{(51)}$ evadent æquales, deinde si *b* hos limites transcendat, possibilitatem amittent. Nam radices binæ possibiles prius evadunt æquales q̄$^{(52)}$ possibilitatem amittunt. Tot sunt Problematis cujusɋ solutiones quot sunt æquationis ad quam deducitur reales radices. Earum affirmativæ pro solutionibus veris, negativæ pro falsis$^{(53)}$ habentur, et solutiones illæ omnes simul investigantur deducendo problema ad æquationem et hujus radices extrahendo.

Æquationis radix quandoɋ prodit addendo vel auferendo quantitates quasdem utrobiɋ & radicē utrāɋ extrahendo. Et hoc pacto radices omnis quadraticæ invenire licet. Ut si $- -^{(54)}$

---

(50) These values should be '27' and '−5' respectively. When, however, *b* takes these values, the given cubic will have the double roots $x = 3$ and −1 respectively. Newton doubtless derived the latter from the derivative condition $3x^2 - 6x - 9 = 0$ and then, in haste, forgot to compute the corresponding values of *b*. On setting the cubic in the equivalent form $(x-1)^3 = 12(x-1) - (b-11)$, the Cardan formula determines the roots to be

$$x = 1 + \sqrt[3]{(-\tfrac{1}{2}(b-11) + \tfrac{1}{2}\sqrt{[(b-11)^2 - 16^2]})} + \sqrt[3]{(-\tfrac{1}{2}(b-11) - \tfrac{1}{2}\sqrt{[(b-11^2) - 16^2]})},$$

all three of which will be real for $(b-11)^2 \leqslant 16^2$.

(51) In line with note (50), read '27 vel −5' (27 or −5).

(52) 'quam'.

(53) But not thereby, of course, necessarily any the less admissible: indeed, where both positive and negative roots exist, it is convention which dictates the choice of variable so that positive roots correspond always to possible solutions. Newton himself in the sequel largely ignores the present distinction, particularly in geometrical contexts where 'positive' and 'negative' senses to directions of lines are arbitrarily assigned. (In the tenth of the following worked problems, for instance, both 'true' and 'false' positions of *C*—to the right or left of *E* in the line *DE*, corresponding to positive and negative values of $EC = z$ respectively—are equally allowed.)

The unknown quantity is, however, said to be a root of the equation, and each equation possesses as many roots as the unknown quantity's dimensions in it. Thus, seeing that $x^3 - x^2 - 14x + 24 = 0$ ascends to three dimensions, it will have three roots. They are 2, 3 and -4. For any one of these when written in x's place in the equation will make all its terms jointly equal to nothing. The reason for this is that an equation is formed from the continued product of its roots together. Set $x = 2$, $x = 3$ and $x = -4$, or (as is exactly the same) $x - 2 = 0$, $x - 3 = 0$ and $x + 4 = 0$, and from the continued multiplication of these there will come to be $x^3 - x^2 - 14x + 24 = 0$. But two or four of these roots, or more still if the equation is of more dimensions, may on occasion lose their possibility, and then they are usually said to be imaginary or impossible. Thus the equation $x^3 - 3x^2 - 9x + b = 0$ will, if b be not greater than $3^{(50)}$ nor less than -1,[50] have three real roots. But two of these will, if b become 3 or -1,[51] prove to be equal; subsequently, should b surpass these limits, they will lose their possibility. For a pair of possible roots come to be equal before they lose their possibility. Each problem has as many solutions as the equation to which it is brought possesses real roots. Of these the positives may be considered as 'true' solutions, the negatives as 'false' ones,[53] and all those solutions are investigated simultaneously by bringing the problem down to an equation and extracting the roots of this.

Occasionally an equation's root is forthcoming by adding or taking away certain quantities on its either side and then extracting the root of each. In this manner we are at liberty to find the roots of every quadratic. If, for instance, ...[54]

---

(54) The manuscript breaks off with the two suspension hyphens shown. We should understand something like 'Ut si habeatur $xx - px + q = 0$, adde utrinqʒ $\frac{1}{4}pp - q$ et exurget

$$xx - px + \tfrac{1}{4}pp = \tfrac{1}{4}pp - q$$

adeoqʒ extracta utrobiqʒ radice $x - \frac{1}{2}p = \pm\sqrt{\frac{1}{4}pp - q}$ seu $x = \frac{1}{2}p \pm \sqrt{\frac{1}{4}pp - q}$' (If, for instance, there be had $x^2 - px + q = 0$, add $\frac{1}{4}p^2 - q$ to each side and there will ensue $x^2 - px + \frac{1}{4}p^2 = \frac{1}{4}p^2 - q$ and therefore, with the square root extracted on each side,

$$x - \tfrac{1}{2}p = \pm\sqrt{[\tfrac{1}{4}p^2 - q]} \quad \text{or} \quad x = \tfrac{1}{2}p \pm \sqrt{[\tfrac{1}{4}p^2 - q]}).$$

Since this paragraph both here seems out of place and is comprehended in 'Reg. 7' of the following section on the 'Reductio Æquationum', it may well be that Newton would subsequently have omitted it had he ever completed the present revise of his *Arithmetica*.

Corresponding to the grading of equations (after reduction to 'simplest' form) by the dimension of the unknown quantity in their greatest terms 'nullo respectu ad quantitates cognitas habito, nec ad intermedios terminos' (see **1**, **2**, §1: Newton's page 30), a stray contemporary sheet (ULC. Add. 3964.8: 22ʳ/22ᵛ, reproduced in appendix below) contains two drafts expounding the equivalent grading of problems according to the degree of the 'simplest' (surd-free) equations to which their conditions can be reduced. If, as we think not unlikely, their substance was intended to be inserted in the present revise of the *Arithmetica*, here would be its natural place.

[INVENTIO ÆQUATIONUM.][55]

[56]Æquationum inventio quæ est Artis hujus pars longe difficillima et a nemine docetur,[57] consistit in excogitatione scripturæ Algebraicæ qua conditiones problematum designari possint haud secus quam conceptus nostri characteribus græcis vel latinis. Est enim Algebra nihil aliud quàm sermo mathematicus designandis quantitatum relationibus accommodatus. In hoc sermone quantitates locum habent verborum et æquationes sententiarum. Designentur quantitates tam notæ quàm ignotæ symbolis quibusvis tanquam vocabulis Algebraicis, et sententiæ quibus problemata enunciantur,[58] ex sermone latino vel alio quovis populari in hunc sermonem translatæ fient æquationes desideratæ. Et quamvis sermo vulgaris quo problemata proponuntur aliquando minus aptus fuerit qui scribatur Algebraicè, tamen commodis mutationibus adhibitis et ad sensum potius quam verborum sonos attendendo, versio reddetur facilis. Sic enim quælibet apud gentes loquendi formæ propria habent idiomata, quæ ubi obvenerint, translatio ex unis in alias non verbo tenus instituenda est sed ex sensu determinanda.[59] In quæstionibus tamen non Geometricis si conditiones singulæ distinctis sententijs vel sententiarum clausulis enuncientur atcp partes, tota et proportionalia per æqualitates inter partes et tota et inter rationes factave mediorum et extremorum [multiplicatione][60] semper designentur, non video quid moram injiciat Tyronibus. Tentent igitur vires suas primū in quæstionibus non Geometricis, ad normam exemplorum sequentium.

Prob. 1.[61]

Inveniendi sunt tres numeri continuè proportionales quorum summa sit 20 et quadratorum summa 140.

---

(55) This previously demoted Newtonian title (see note (46)) is here reinserted by us to accentuate the contrast of the following text with the complementary 'Reductio [et Solutio!] Æquationum' which follows. It will be evident that its opening paragraphs revise the parent *Arithmetica*'s section 'Quomodo Quæstio aliqua [Arithmetica] ad æquationem redigatur' (**1**, 2, §1: Newton's pages 39–41, with worked examples on following pages 41–52), while its latter portion dealing with 'Problemata Geometrica' is a much refashioned summary of the succeeding section 'Quomodo Quæstiones Geometricæ ad æquationem redigantur' (*ibid.*: Newton's pages 52–65), ignoring its closing remarks (*ibid.*: Newton's pages 65–7) on elementary analytical geometry and the algebraic representation of a given geometrical locus by a corresponding defining equation in some standard (perpendicular) Cartesian coordinate system.

(56) Newton first began: 'Æquationum inventionem...conabor sic exponere. Ea consistit in excogitatione expressionum Algebraicarum quibus...' (The finding of equations...I shall attempt to expound in the following manner. It consists in devising algebraic expressions by which...).

(57) This phrase echoes Newton's earlier observation to John Collins on 27 September 1670 regarding Kinckhuysen's *Algebra ofte Stelkonst* that 'having composed somthing pretty largely

[The finding of equations][55]

[56]The finding of equations—the part of this art which is by far the most difficult and yet is explained by no one[57]—consists in devising an algebraic script by which the conditions of problems may be designated analogously to the way in which we denote our concepts in Greek or roman characters. In this language quantities fill the rôle of words and equations that of sentences. Let quantities both known and unknown be designated by any symbols you wish as their algebraic words, and the sentences in which problems are enunciated[58] will, when translated from Latin or any other popular tongue into this language, then become the equations desired. And though the vernacular in which problems are propounded prove to be not at all suited to being written in algebraic form, yet by introducing appropriate changes and paying heed to the sense of words rather than to their spoken form, the change-over will be rendered an easy one. As a parallel, to be sure, the various national forms of speech have their own peculiar idioms and, when these are met with, translation from one to another is not to be instituted by a mere verbal transliteration, but must be determined from the sense.[59] In the case of non-geometrical questions, however, if individual conditions are enunciated in distinct sentences or clauses of sentences, with parts, wholes and proportionals always designated by equalities between parts and wholes and between ratios or by effecting the multiplication of middles and extremes, I see nothing to block the path for beginners. Let them therefore first test their strength on non-geometrical questions, on the pattern of the following examples.

Problem 1[61]

It is required to find three numbers in continued proportion, whose sum shall be 20 and the sum of their squares 140.

---

about reducing problems to an æquation when I came to consider his examples...I found most of them solved not by any generall Analytical method but by particular & contingent inventions, wch though many times more concise then a generall method would allow, yet in my judgment are lesse propper to instruct a learner' (*Correspondence of Isaac Newton*, **1**, 1959: 43–4; compare II: 423, note (110)).

(58) This amplifies the colourless cancelled phrase 'conditiones problematum' (the conditions of problems).

(59) All but a repetition word for word of an equivalent passage in the parent *Arithmetica* (**1**, 2, §1: Newton's page 41). We there (*ibid.*: note (128)) gave factual confirmation of Newton's deep (and indeed lasting) interest in language structure and the various connotative 'moodes' of speech.

(60) The omission of this necessary word, made by Humphrey Newton in his initial transcript of the present section, was overlooked by Newton when he subsequently added several minor corrections to this passage.

(61) A repetition, in style and content, of the worked example on (Newton's) pages 39–40 of the parent *Arithmetica* (**1**, 2, §1).

Ut hoc fiat ponantur *x y* et *z* nomina numerorum trium quæsitorum et Quæstio e Latinis literis in Algebraicas sic vertetur.

Quæstio Latinè enunciata.	*Eadem Algebraicè.*
Quæruntur tres numeri his conditionibus,	$x? \, y? \, z?$
Ut sint continue proportionales	$x . y :: y . z$, sive $xz = yy$.
Ut omnium summa sit 20	$x + y + z = 20$.
Et ut quadratorum summa sit 140.	$xx + yy + zz = 140$.

Sic deducitur quæstio ad tres æquationes $xz = yy$, $x + y + z = 20$, et

$$xx + yy + zz = 140,$$

quarum ope quantitates tres ignotæ *x*, *y* et *z* per regulas infra[62] tradendas invenientur. Tot enim semper inveniri debent æquationes quot sunt quantitates ignotæ in ipsis.

Prob. 2.[63]

Secanda est magnitudo sectione divina.

Enunciatio explicatior Latinè.	*Algebraicè.*
Datæ magnitudinis - - -	a
inveniendæ sunt partes - -	$x + y = a?$
quæ sint in geometrica progressione	$x . y :: y . a$, id est
cum tota. - - - -	$\dfrac{x}{y} = \dfrac{y}{a}$ seu $ax = yy$.

Duæ sunt hic ignotæ quantitates *x* et *y* et perinde duæ æquationes $x + y = a$ et $ax = yy$ inventæ quarum ope quantitates illæ per regulas post tradendas determinentur.

Prob. 3.[64]

Latinè.	*Algebraicè.*
Vendit quis alicui centum modios tritici et quinqʒ dolia vini libris quinquaginta,	$100x + 5y = 50l$.
alteri centum ulnas[65] serici et viginti quinqʒ modios tritici libris triginta,	$100z + 25x = 30l$.
tertio decem dolia vini et quadringentas ulnas serici libris centum et sexaginta:	$10y + 400z = 160l$.
Requiritur quanti Venditor modium tritici, dolium vini & ulnam serici æstimavit.	$x? \, y? \, z?$

(62) In the following section on the 'Reductio Æquationum' where, along with those of the eleven following problems, the reduction and solution of the present 'Prob. 1' is given.

(63) This classical problem, given geometrical construction in Euclid, *Elements*, II, 11 (compare III: 407, note (16)) and again—an interpolation?—VI, 30, is new as it stands in Newton's mathematical papers. The term 'divine section' (for one in extreme and mean

To do this, set x, y and z as the names of the three numbers sought, and the question will be transliterated from verbal into algebraic form in this manner:

The question verbally enunciated	*The same algebraically*
Three numbers are sought,	x? y? z?
subject to these conditions:	
that they be in continued proportion;	$x:y = y:z$, that is, $xz = y^2$;
that their total sum shall be 20;	$x+y+z = 20$;
and that the sum of their squares be 140.	$x^2+y^2+z^2 = 140$.

In this way the question is brought to the three equations $xz = y^2$, $x+y+z = 20$ and $x^2+y^2+z^2 = 140$, by whose help the three unknown quantities x, y and z will be ascertained by means of the rules subsequently[62] to be delivered. For there ought always to be found as many equations as they contain unknown quantities.

Problem 2[63]

It is required to cut a magnitude in 'divine' (golden) section.

A more detailed verbal enunciation	*Algebraically*
Of a given magnitude	a
parts have to be found	$x+y = a$?
which are to be in geometrical	$x:y = y:a$, that is,
progression with the whole.	$x/y = y/a$ or $ax = y^2$.

There are here two unknown quantities x and y, and correspondingly two equations $x+y = a$ and $ax = y^2$ are found by whose aid those quantities shall be determined by the rules to be delivered subsequently.

Problem 3[64]

Verbally	*Algebraically*
A person sells someone a hundred measures of wheat and fifty barrels of wine for fifty pounds,	$100x+5y = 50l$,
another a hundred lengths[65] of silk and twenty-five measures of wheat for thirty pounds,	$100z+25x = 30l$,
a third ten barrels of wine and four hundred lengths of silk for a hundred and sixty pounds:	$10y+400z = 160l$.
how much the seller reckoned a measure of wheat, a barrel of wine and a length of silk to be worth is sought.	x? y? z?

proportion) was popularized by Luca Pacioli in his treatise *De Diuina Proportione* (Venice, 1509) and subsequently taken up by Ramus and Kepler; see D. E. Smith, *History of Mathematics*, **2** (Boston, 1925): 291, note 3.

 (64) Again new as it stands, but evidently adapted from the 'Exemplum' given of arithmetical 'Prob 9' in the parent *Arithmetica* (**1**, 2, §1: Newton's pages 46–7).

 (65) An ell or arm's length, now conventionally taken as $1\frac{1}{4}$ yards.

Prob 4.[66]

Latinè.	*Algebraicè.*
Distribuenti nummos inter mendicantes	x numerus denariorum.
	y numerus mendicantiū.
desunt tres solidi quo minus det singulis solida singula.	$12^{denar} \times y = x + 36^{denar}.$
Dat igitur singulis novem et supersunt sex denarij:	$9^d \times y = x - 6^d.$
Requiritur numerus mendicantium.	y?

Aliquando plures latent Propositiones in una sententiæ clausula, quæ omnes eruendæ sunt et distinctè enunciandæ, ut fit in sequenti Problemate.

Prob 5.[67]

Mercator quidem nummos ejus triente quotannis adauget demptis 100^{lib} quas annuatim impendet in familiam & post tres annos fit duplo ditior. Quæruntur nummi.

Enunciatio distinctior.	*Versio Algebraica.*
Mercator habet nummos quosdam	x?
Ex quibus anno primo expendit 100^{lib}	$x - 100 = p.$
Et reliquum adauget triente	$p + \frac{1}{3}p = \frac{4}{3}p = q.$
Annoꝗ secundo expendit 100^{lib}	$q - 100 = r.$
Et reliquum adauget triente	$\frac{4}{3}r = s.$
Et sic anno tertio expendit 100^{lib}	$s - 100 = t.$
Et reliquo trientem lucratus est	$\frac{4}{3}t = v.$
Fitꝗ duplo ditior quam sub initio.	$v = 2x.$

Prob. 7.[68]

Emendus est reditus annuus ducentarum librarum in annos septem proximè sequentes ad rationem usuræ annuæ librarum sex per centum: Requiritur justum totius pretium in parata pecunia.

Pendet autem hujus resolutio ab hac analogia continua quod sit, ut 106^{lib} post annum unum solvendæ ad 100^{lib} in parata pecunia, ita reditus quilibet annuus ad pretium suum anno uno anticipatum et pretium illud ad pretium anno altero anticipatum & sic in infinitum. Quo intellecto, declaratio explicatior ita fiet.

---

(66) Slightly adapted (with a suitable inflation in the value of the hand-out and the number of beggars) from 'Prob. 4' on page 42 of Newton's parent *Arithmetica*.

(67) This repeats, both in style and content, Newton's second worked example on pages 40–1 of the parent *Arithmetica*, except that here no continuous reduction (and consequent solution) of the resulting algebraic equations is made.

(68) Except that the period of payment is now stretched to seven years, this is effectively a particular case ($a = 200$, $x = z = 100/106$) of Newton's concluding arithmetical 'Prob 17'

Problem 4[66]

Verbally	Algebraically
A man sharing a sum of money among beggars	x number of pence,
	y number of beggars:
lacks three shillings to be able to give each a shilling,	$12y = x + 36$ (pence),
so he gives each ninepence and has sixpence left:	$9y = x - 6$ (pence).
The number of beggars is required.	y?

Sometimes several propositions are concealed in one clause of a sentence, and all these have to be hunted out and enunciated in distinct fashion, as happens in the following problem:

Problem 5[67]

A certain merchant each year increases his capital by a third, less £100 which he spends annually on his household, and after three years becomes twice as rich. What is his capital?

More distinct enunciation	Algebraic version
A merchant has a certain capital:	x?
of this the first year he spends £100,	$x - 100 = p$,
and increases the rest by a third;	$p + \frac{1}{3}p = \frac{4}{3}p = q$;
and the second year he spends £100,	$q - 100 = r$
and increases the rest by a third;	$\frac{4}{3}r = s$;
and likewise the third year he spends £100,	$s - 100 = t$,
making a profit of one-third on the rest;	$\frac{4}{3}t = v$;
and comes to be twice as rich as at the start.	$v = 2x$.

Problem 7[68]

There has to be bought an annuity returning two hundred pounds in each of the next seven following years at an annual interest rate of six pounds per hundred. What is the fair price of the whole for ready cash?

Now the solution of this depends on the following continued proportion: as £106 to be paid after one year are to £100 in ready cash, so is any annual return to its price when anticipated by one year, and that price to the price when anticipated by a second year, and so on indefinitely. Once this is understood, the more detailed statement will be thus:

---

on page 52 in the parent *Arithmetica* (itself subsequently to be added to the present revise in unchanged form; see note (132) below). Again, no preliminary reduction is made of the algebraic equivalents. The number '7' set by Newton on this problem (here reproduced in its correct manuscript sequence) suggests that he initially—before cancelling it—intended to interchange it with the following 'Prob. 6'.

latinè.	*algebraicè.*
In ratione usuræ 6 per cent	$106 . 100 :: 1 . z.$
emendus est parata pecunia	$y.$
reditus annuus librarum ducentarū	$200^{\text{lib}} = a.$
solvendus septies.	$y = p + q + r + s + t + v + w$ pretijs
id est	redituum septem.
post annum unum	$1 . z :: a^{\text{lib}}.p,$ seu $za = p.$
duos	$p \quad .q, \qquad zp = q.$
tres	$q \quad .r, \qquad zq = r.$
quatuor	$r \quad .s, \qquad zr = s.$
quincȝ	$s \quad .t, \qquad zs = t.$
sex	$t \quad .v, \qquad zt = v.$
septem.	$v \quad .w, \qquad zv = w.$
Quæritur pretium.	$y\,?$

Eodem recidit si dato pretio quæratur ratio usuræ $\dfrac{1}{z}$ vel reditus annu[u]s a.[69]

Siquando conditiones rerum naturalium ingrediuntur statum quæstionis, reducendæ sunt hæ ad leges quæ possunt algebraicè scribi, ut fit in sequentibus.

Prob 6.[70]

Lapide in puteum cadente, requiritur altitudo putei ex dato tempore inter dimissionem lapidis et auditum sonum fundi percussi.

Ut resolutio hic procedat cognoscendum est quantum spatium quovis tempore grave cadendo & sonus eundo conficiant: ut quod dato tempore p, grave percurrat spatium q et sonus spatium r et quod alio quovis tempore spatium sono confectum sit tempori proportionale et spatium gravi cadente confectum sit in duplicata illa ratione seu proportionale quadrato temporis. His positis Problema sic enunciabitur.

Latine.	*Algebraicè.*
Requiritur altitudo putei	$x\,?$
ex dato tempore toto	$t = y + z.$
cujus pars lapide cadente	$yy . pp :: x . q,$ seu $qyy = ppx.$
pars sono fundi percussi ascendente consumitur.[71]	$z . p :: x . r, \qquad px = rz.$

---

(69) It follows immediately that a, y and z are connected by the reduced equation

$$y = (z + z^2 + z^3 + z^4 + z^5 + z^6 + z^7)a \; [= z\,(1 - z^7)\,a/(1 - z)],$$

whence to find the inverse rate of gain z (given the annual return a and purchase price y) will require extracting the (unique) positive root of a seventh-degree equation—one which has no exact algebraic solution and for the extraction of whose root in approximate terms (say, by the Gregorian iteration of $z = (az^8 + y)/(a + y)$, $0 < z < 1$; see IV: 660–1, note (12)) Newton has

Verbally	*Algebraically*
At an interest rate of 6 per cent	$1/z = 106/100$.
there has to be bought for ready money	y.
an annuity returning £200	$a = $ £200.
to be paid seven times;	$y = p+q+r+s+t+v+w$, the
that is,	price of seven repayments, where
after year one,	$1:z = a:p$, or $za = p$ (pounds),
– – two,	$= p:q$, $zp = q$,
– – three,	$= q:r$, $zq = r$,
– – four,	$= r:s$, $zr = s$,
– – five,	$= s:t$, $zs = t$,
– – six,	$= t:v$, $zt = v$,
– – seven.	$= v:w$, $zv = w$.
What is its price?	y?

It comes to the same if, for a given price, the rate of interest $1/z$ or annual repayment a be sought.[69]

Should ever the conditions of natural phenomena enter the circumstances of a question, these must be reduced to laws which can be written algebraically, as happens in the following.

Problem 6[70]

Where a stone falls down a well, it is required to find the depth of the well from the given time between the stone's release and hearing the sound made by its striking the well's bottom.

That the resolution may here proceed, there must be known the exact distances covered in any time by a body as it falls and by a sound as it travels: say, that in a given time p the body traverses the distance q and the sound the distance r, while in any other time the distance covered by a sound is proportional to the time and the distance covered by a falling body is in that ratio doubled, that is, proportional to the square of the time. With these suppositions the problem will be thus enunciated:

Verbally	*Algebraically*
The depth of a well is required,	x?
given the total time spent[71]	$t = y+z$, where
partly in the stone falling,	$y^2:p^2 = x:q$, or $qy^2 = px^2$,
partly in the rising of the sound of the bottom's	$z:p = x:r$, $px = rz$.
being struck.	

made no provision in the extant manuscripts of his *Arithmetica*. This difficulty was, we may readily guess, his reason for subsequently cancelling the problem.

(70) This is Newton's 'geometrical'[!] Problem 45 on page 116 of the parent *Arithmetica*, with the (unreduced) algebraic equivalents of its conditions made explicit.

(71) 'conficitur' (completed) was first written.

Prob 7.[72]

Auri, argenti, æris et misturæ ex his tribus dantur pretia, pondera in aere et pondera in aqua, requiritur quantitas auri in mistura.

Sciendum est hic quod differentia gravitatum corporis in aere et aqua sit gravitas aquæ æqualis corpori & inde datis corporum magnitudinibus problema sic enunciabitur.

Latine.	*Algebraicè.*
Auri, argenti, æris et misturæ ex his omnibus dantur magnitudines	$a, b, c, d = x + y + z.$
pretia	$e, f, g, h = \dfrac{ex}{a} + \dfrac{fy}{b} + \dfrac{gz}{c}.$
gravitates	$k, l, m, n = \dfrac{kx}{a} + \dfrac{ly}{b} + \dfrac{mz}{c}.$
Requiritur quantitas auri in mistura.	x?

Problemata Geometrica præcedentibus difficiliora sunt eò, quod in eorum statu includuntur non solum magnitudines sed etiam Figurarum species & formæ quæ nequeunt algebraicè scribi. Pro formis igitur substituendæ sunt æquipollentes magnitudines et magnitudinum relationes quibus formæ illæ determinantur. Et ut casus præcipuos enumerem,[73]

[1.] Si anguli magnitudo in Quæstione est, conditio illa enunciatur algebraicè per proportionem laterum trianguli alicujus, cujus iste est angulus.

[2.] Si Figura datur specie,[74] enunciatur id algebraicè per datas rationes laterum & diagonalium aliarumv́e linearū a quibus species illa determinatur.

[3.] Si duæ Figuræ sunt similes[,] enunciatur similitudo[75] per proportionalitatem laterum & aliarum correspondentiū linearum.

[4.] Si triangulum aliquod dicitur esse rectangulum, effertur hoc algebraicè per æqualitatem inter quadratum hypotenusæ & quadrata laterum.

Si perpendiculum in triangulo aliquo demitti dicitur[,] sæpe designatur id algebraicè per Prop 12 et 13 lib. 2 Elem.[76]

---

(72) A generalization of 'Prob 10' on page 47 of the parent *Arithmetica* (see **1**, 2, §1: note (147)), now adding a third substance (copper) to the admixture and supposing known the relative price of both the admixture and its components.

(73) Compare (Newton's) pages 56–8 of the parent *Arithmetica* (**1**, 2, §1). Geometrical relationships, rather than their equivalent algebraic structures, now become pre-eminent in the following enumeration.

(74) That is, 'in kind' and hence with the ratios of its sides and other rectilinear elements given.

(75) Newton has deleted 'expressio fit' (its expression is effected).

(76) A cancelled first conclusion reads in elaboration 'ponendo differentiam quadratorum laterum æqualem differentiæ quadratorum segmentorum basis, vel quod perinde est æqualem

Problem 7.[72]

Given the prices of gold, silver and copper and of an admixture of these three, together with their weights in air and in water, what is the quantity of gold in the admixture?

You must here be aware that the difference in weight of a body in air and in water is the weight of water equal to the body (in mass); in consequence, given the sizes of the bodies, the problem will be enunciated in this manner:

Verbally	*Algebraically*
Of gold, silver, copper and an admixture	
of all these three there are given the	
sizes,	a, b, c and $d = x+y+z,$
prices,	e, f, g and $h = (e/a)\,x + (f/b)\,y + (g/c)\,z,$
weights.	k, l, m and $n = (k/a)\,x + (l/b)\,y + (m/c)\,z.$
What is the quantity of gold in the admixture?	x?

Geometrical problems are more difficult than the preceding ones for the reason that in their circumstances are included not only magnitudes but also species and forms of figures which cannot be algebraically written. In place of these forms, therefore, we must substitute equivalent magnitudes and the relationships between these magnitudes by which those forms are determined. To enumerate the outstanding cases:[73]

[1.] If the size of an angle forms part of the question, that condition is stated algebraically through a proportion between the sides of some triangle of which it is an angle.

[2.] If a figure is given in species,[74] that is stated algebraically by means of given ratios between its sides and diagonals or other lines by which the species is determined.

[3.] If two figures are similar, the similarity is stated[75] by a proportionality between their sides and other corresponding lines.

[4.] If some triangle is said to be right-angled, this is brought out by an equality between the square of the hypotenuse and the squares of the sides.

If a perpendicular is said to be let fall in some triangle, it is often designated algebraically by means of *Elements*, Book II, Propositions 12 and 13.[76]

---

rectangulo sub basi et dupla distantia perpendiculi a medio basis' (by setting the difference of the squares of the sides equal to the difference of the squares of the base-segments, or, what is exactly the same, equal to the rectangle contained by the base and twice the distance of the perpendicular from the mid-point of the base). It is difficult to know why Newton has deleted this present paragraph—virtually a repeat of one on pages 57–8 of the parent text (see note (73))—for it is several times invoked in 'Prob 12' below.

[5.] Si puncta sunt in circumferentia circuli, exprimitur id algebraicè per æquales distantias a centro, vel per æquales angulos ad circumferentiam, vel etiam per æqualia rectangula linearum[77] per puncta illa ductarum et inde usqᵹ ad communem concursum sumptarum, vel deniqᵹ per aliam quamvis definitivam proprietatem circuli. Atqᵹ idem de Coni sectionibus cæterisqᵹ figuris curvilineis intelligendum est.

Et ut hæ et similes expressiones commode procedant, sæpe complendæ sunt figuræ ducendo et producendo lineas vel per puncta quædam transeuntes, vel positione aut inclinatione datas vel alteris aut parallelas aut perpendiculares aut alia assignata lege inclinatas, ut conficiantur triangula specie vel magnitudine data aut alteris similia, aut rectangula aliamve notam quamvis proprietatem habentia. Et tum demum notando linearum summas, differentias, proportiones, aliasqᵹ proprietates quæ ex statu Quæstionis et forma Schematis emanant, colligentur æquationes desideratæ, ut in exemplis sequentibus.

Prob 8.[78]

Mensurata locorum A, B distantia, et angulis CAB, CBA observatione cognitis; inveniendum est punctum D in quod perpendiculum de loco ignoto C ad lineam AB demissum incidet, et perpendiculi illius longitudo exhibenda est.

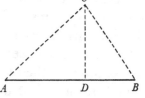

Hic dantur triangulorum *ADC BDC* anguli omnes atqᵹ adeo proportiones laterum. Assumantur ergo proportiones illæ, puta *AD* ad *DC* ut *d* ad *e* et *BD* ad *DC* ut *d* ad *f* et resolutio ita se habebit.

Latine.	*Algebraicè.*
Data linea *AB* quæruntur partes *AD, BD* et perpendiculum *CD* ea lege ut triangula *ADC BDC* dentur specie.	$a = x + y$? z ? $x . z :: d . e$ seu $ex = dz$. $y . z :: d . f$ seu $fy = dz$.

Prob 9.[79]

Invenire triangulum ABC cujus datur area, ambitus et angulus A.

Datum angulum *A* subtendo perpendiculo *CD* ut constituatur datum specie triangulum *ADC*. Sic datur ratio *AD* ad *DC*: sit ista *d* ad *e*. Scribatur item *aa* pro area et *b* pro ambitu trianguli et conditiones problematis sic enunciabuntur.

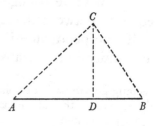

---

(77) 'segmentorum' (segments) was first written.

[5.] If points are in the circumference of a circle, that is expressed algebraically by equal distances from the centre, or by equal angles at the circumference, or again by equal products of lines[77] drawn through those points and
taken thence as far as their common meet, or finally by any other defining
property of the circle. And the same is to be understood for conic sections and
other curvilinear figures.

And so that these and similar representations may go ahead without obstacle,
figures must often be completed by drawing and extending lines either passing
through certain points, or given in position or slope, or either parallel or
perpendicular to others or inclined to them under some assigned condition, in
order to make up triangles which are given in species or size, or are similar to
others, or are right-angled, or possess some other known property. And then at
last, on noting the sums and differences of the lines, their proportions and
other properties which derive from the status of the question and the
diagram's form, will the desired equations be gathered, as in the following
examples.

Problem 8[78]

*Where the distance between the positions A and B is measured and the angles $C\widehat{A}B$,
$C\widehat{B}A$ have been ascertained by observation, it is required to find the point D onto which the
perpendicular dropped from the unknown position C to the line AB shall fall, and also to
display the length of that perpendicular.*

Here all the angles in the triangles *ADC*, *BDC* are given and hence the ratios
of the sides. Assume those ratios, therefore,—say, *AD* to *DC* as *d* to *e*, and *BD* to
DC as *d* to *f*—and the resolution will take on the following form:

Verbally	*Algebraically*
Given the line *AB* we require its parts *AD*, *BD* and the perpendicular *CD*,	$a = x+y$? z? where
subject to the triangles *ADC*, *BDC* being given in species.	$x:z = d:e$ or $ex = dz$, $y:z = d:f$ or $fy = dz$.

Problem 9[79]

To find a triangle ABC whose area, perimeter and angle $\hat{A}$ is given.

I subtend the given angle $\hat{A}$ by the perpendicular *CD* so as to constitute the
triangle *ADC* given in species. Thus the ratio of *AD* to *DC* is given: let it be as
d to *e*. Likewise write a^2 for the area and *b* for the triangle's perimeter, and the
conditions of the problem will be enunciated thus:

---

(78) The opening geometrical 'Prob. 1' on (Newton's) pages 67–8 of the parent *Arithmetica*
(**1**, 2, §1). Once again the corresponding algebraic equivalents are given in unreduced form,
their resolution being delayed to the next section.

(79) Essentially geometrical 'Prob. 4' on page 69 of the parent text (**1**, 2, §1).

Latinè.	*Algebraicè.*
Quæruntur *AC, BC, AB, CD*	$v?\ x?\ y?\ z?$
et ipsius *AB* partes *AD, BD*	$y = s + t.$
sic ut anguli ad *D* recti sint	$ss + zz = vv.\ \ tt + zz = xx.$
et detur angulus *A*	$s.v::d.e,$ seu $es = dv.$
et ambitus *ACBA*	$v + x + y = b.$
et area $\frac{1}{2}AB \times CD.$	$\frac{1}{2}yz = aa.$

Prob 10.[80]

Problema aliquod eò deducitur ut a datis punctis A, B ad rectam DE positione datam ducendæ sint rectæ duæ AC BC quarum differentia sit dato N æqualis. Quæritur punctum C.

Ut Problematis hujus conditiones deriventur in triangula demitto a datis punctis *A, B* ad rectam *DE* perpendicula *AD, BE*, dein sic pergo.

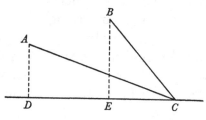

Latinè.	*Algebraicè.*
Dantur *AD, BE, DE, N*:	$a.\ b.\ c.\ n.$
Quæruntur *AC, BC, DC, EC*;	$v?\ x?\ y?\ z?$
ita ut *N* differentia sit ipsarum *AC, BC,*	$n = v - x.$
et *DE* differentia ipsarum *DC EC,*	$c = y - z.$
et triangula *ADC BEC* sint rectangula.	$a^2 + y^2 = v^2.\ \ b^2 + z^2 = x^2.$

Idem aliter.

Juncta *AB* occurrat ipsi *DE* in *D* et ad ipsam demittatur perpendiculum *CG*.[81]

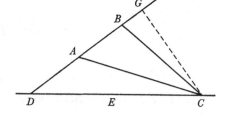

---

(80) 'Prob 11' on pages 72–3 of the parent *Arithmetica*.

(81) It is clear that Newton abandoned this variant way of solution only because it cannot straightforwardly be transposed into an algebraic equivalent in terms of the constants and unknowns previously assigned, for it is certainly viable. Thus, if we now put $DA = \alpha$, $DB = \beta$ and $DC = Y$, $DG = Z$, then, on retaining $AC = v$ and $BC = x$ and setting $k = \cos \hat{D}$, we readily deduce $Y^2 - Z^2 = x^2 - (Z-\alpha)^2 = v^2 - (Z-\beta)^2$ and $Z = kY$ with (as before) $v = x + n$; and, when any three of v, x, Y and Z are eliminated between these equations, there ensues one involving the remaining unknown quantity alone. By eliminating v, x and Z, for instance, we

Verbally	*Algebraically*
There are sought *AC, BC, AB, CD*	$v?\ x?\ y?\ z?$
and the parts *AD, BD* of *AB*	$y = s+t$, where
such that the angles at *D* are right,	$s^2+z^2 = v^2,\ t^2+z^2 = x^2,$
angle *Â* is given,	$s:v = d:e$ or $es = dv$,
and also the perimeter *ACBA*	$v+x+y = b$ and
and the area $\frac{1}{2}AB \times CD$.	$\frac{1}{2}yz = a^2.$

Problem 10[80]

Some problem is reduced to that of drawing from the given points A, B to the straight line DE given in position two straight lines AC, BC whose difference shall be equal to the given magnitude N. The point C is required.

To transfer the conditions of this problem to triangles I let fall perpendiculars *AD, BE* from the given points *A, B* to the straight line *DE*, and then proceed in this way:

Verbally	*Algebraically*
There are given *AD, BE, DE, N*:	$a,\ b,\ c,\ n:$
and *AC, BC, DC, EC* are sought	$v?\ x?\ y?\ z?$ where
such that *N* be the difference of *AC* and *BC*	$n = v-x,$
and *DE* the difference of *DC* and *EC*,	$c = y-z,$
and the triangles *ADC, BEC* be right.	$a^2+y^2 = v^2,\ b^2+z^2 = x^2.$

The same another way

Join *AB* and let it meet *DE* in *D* and then let fall to it the perpendicular *CG*.[81]

---

obtain successively $x^2-(Z-\alpha)^2 = (x+n)^2-(Z-\beta)^2$ or $x = ((\alpha-\beta)\,Z-\frac{1}{2}(\alpha^2-\beta^2+n^2))/n$, and thence $(1-k^2)\,Y^2 = (k(\alpha-\beta)\,Y-\frac{1}{2}(\alpha^2-\beta^2+n^2))^2/n^2-(kY-\alpha)^2$, so that

$$(1-k^2(\alpha-\beta)^2/n^2)\,Y^2+k((\alpha-\beta)\,(\alpha^2-\beta^2+n^2)-2\alpha)\,Y+\alpha^2-\tfrac{1}{4}(\alpha^2-\beta^2+n^2)^2/n^2 = 0.$$

On substituting in this $\alpha = a\lambda,\ \beta = b\lambda,\ k = \sqrt{[1-1/\lambda^2]}$ and

$$Y = z+b\sqrt{[\lambda^2-1]} = z+bc/(b-a), \quad \text{where} \quad \lambda = \sqrt{[1+c^2/(b-a)^2]},$$

there results the quadratic

$$(1-c^2/n^2)\,z^2+c(1-(a^2-b^2+c^2)/n^2)\,z+b^2-\tfrac{1}{4}(1-(a^2-b^2+c^2)/n^2)^2n^2 = 0$$

obtained, directly from the preceding approach, in the next section; compare note (117) below.

Prob. 11.[82]

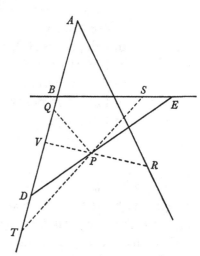

Problema aliquod eò deducitur ut in recta quadam positione data DE inveniendum sit punctum P a quo ad alias positione datas tres rectas AQ, AR, BS ducantur in datis angulis aliæ totidem rectæ PQ, PR, PS quæ sint continuè proportionales: complenda est resolutio Problematis.

Productis aliquibus lineis compleo triangula positione vel specie data quæ resolutioni suffecerint, dein sic pergo.

Latinè.	*Algebraicè.*
Quæruntur PQ, PR, PS, DV, AV, DT, BT, VR, VP, PT, ST.	$z, y, x, v, t, s, r, q, p, n, m?$
e quibus sunt PQ, PR, PS continuè proportionales	$zx = yy.$
et PR, PV, PS, PT partes ipsarum RV, ST.	$p = y + q. \; m = x + n.$
Ad earum determinationem dantur	
AD summa ipsarum AV, DV	$a = t + v.$
et BD differentia ipsarum BT, DT.	$b = r - s.$
et rationes PQ, DV, DT, VP, PT ad	$z . v . s . p . n :: c . d . e . f . g,$ seu
invicem puta ut c, d, e, f, g.	$\dfrac{z}{c} = \dfrac{v}{d} = \dfrac{s}{e} = \dfrac{p}{f} = \dfrac{h}{g}.$
et ratio AV ad VR puta ut h ad d	$dt = hq.$
et ratio BT ad ST puta ut i ad d.	$dr = im.$[83]

(82) This new problem effectively determines the intersection of the 3-line locus (P)—a hyperbola as shown—defined by $PQ \times PS = PR^2$ (see IV: 219–20) with the given straight line DE. (It will be clear that the locus is tangent to the base lines $AB(PQ = 0)$ and $BC(PS = 0)$ at the points, A and C respectively, in which they are met by $AC \; (PR = 0)$.) Having earlier omitted in his present revise the section in the parent *Arithmetica* (1, 2, §1: Newton's pages 65–7; compare note (55) above) dealing with the representation of a geometrical curve by its algebraic defining equation in some given system of Cartesian coordinates, Newton cannot here, unfortunately, make use of the 'natural' reduction of the problem to determining the eliminant of the Cartesian equations of the locus (P) and line DE (see next note), and in default has to have recourse

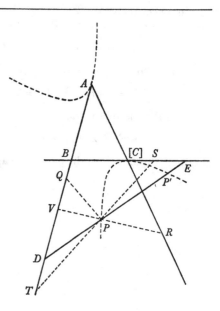

Problem 11[82]

Some problem is reduced to that of finding in a certain straight line DE given in position a point P from which there are to be drawn at given angles, to three other straight lines AQ, AR, BS given in position, an equal number of further straight lines PQ, PR, PS so that they are in continued proportion: the resolution of the problem is to be completed.

With the extension of several lines I complete the triangles given in position or species which should prove adequate for the resolution, and then I proceed this way:

Verbally	*Algebraically*
There are required *PQ, PR, PS, DV, AV, DT, BT, VR, VP, PT, ST*;	$z, y, x, v, t, s, r, q, p, n, m$?
of these *PQ, PR, PS* are in continued proportion,	$zx = y^2$,
while *PR, PV* and *PS, PT* are the parts of *RV, ST*.	$p = y+q, m = x+n$, where
In order to determine these there are given:	
AD, the sum of *AV* and *DV*;	$t+v = a$,
BD, the difference of *BT* and *DT*;	$r-s = b$,
the ratios of *PQ, DV, DT, VP, PT* to one	$z:v:s:p:n = c:d:e:f:g$, or
another—say, as *c, d, e, f, g*;	$z/c = v/d = s/e = p/f = n/g$,
the ratio of *AV* to *VR*—say, as *h* to *d*;	$dt = hq$,
and the ratio of *BT* to *ST*—say, as *i* to *d*.	$dr = im$.[83]

---

to a somewhat clumsy *ad hoc* algebraic reduction which is (though perhaps not obviously seen to be so at first glance) its equivalent. See also 1, 2, §1: note (235).

(83) The constancy of these ratios results from the evident fact that, as *P* moves freely in the line *DE*, the figure *PQVDT* and the triangles *AVR* and *BTS* have their angles given, and so in Newtonian terms (see note (74)) are given in species. It would have been more in keeping with Descartes' discussion of the 4-line locus in his *Geometrie* [= *Discours de la Methode* (Leyden, 1637): 297–413]: 304–12/325–34 (compare D. T. Whiteside, 'Patterns of Mathematical Thought in the later Seventeenth Century' (*Archive for History of Exact Sciences*, 1, 1961: 179–388): 290–5) to have reduced the problem to the intersections of the 3-line locus (*P*) and the line *DE*, defined in some appropriate system of coordinates—an approach not here available to Newton, however (see note (82)). Thus, on setting $PQ:QV = c:\lambda$, whence it is readily shown that λ is the greater root—in the figure illustrated—of

$$\lambda^2 + (d+e+(f^2-g^2)/(d+e))\lambda + f^2 - c^2 = 0,$$

we may determine point *P* by the oblique Cartesian coordinates $AQ = v$, $QP = z$, whence $AV = v+(\lambda/c)z$, $VP = (f/c)z$, $TP = (g/c)z$, $BT = v+[(\lambda+d+e)/c]z+b-a$ and

$$AD = v+[(\lambda+d)/c]z \quad \text{with} \quad VR = (d/h)AV, \quad TS = (d/i)BT$$

and $PR = VR-VP$, $PS = TS-TP$; accordingly the defining equations of the locus (*P*):

$$PQ \times PS = PR^2$$

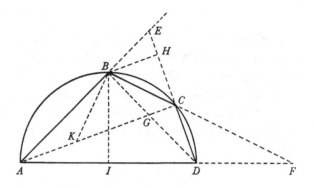

Prob 12.[84]

Determinandus est semicirculus ADCB qui capiat tres datas longitudines AB, BC, CD.

[*Latinè.*]	[*Algebraicè.*]
Dantur *AB, BC, CD*.	*a. b. c.*
Junge *AD, BD*.	*x? y?*
et erit triangulum *ABD* rectangulum	$xx = aa + yy$.
Productæ lineæ [*CD*] sit pars exterior *CH*	*z?*
terminata ad *BH* perpendiculum[85] trianguli *BCD*.	$yy = bb + cc + 2cz$.
Et ob æquales angulos *BAD BCH* similia erunt triangula *ABD, BCH*.	$x \cdot a :: b \cdot z$, seu $xz = ab$.

Hoc modo complendo schematum constructiones & singulas problematum conditiones enumerando eascq scribendo Algebraicè, prodeunt semper æquationes quæ resolutioni sufficiunt. Quinetiam plures æquationes licebit colligere et tum demum habito delectu superfluas rejicere. Coeant *AD BC* in *F*; *AB DC* in *E*; *AC BD* in *G*. Ad *AD* demitte perpendiculum [*B*]*I* et fiat angulus *ABK* æqualis angulo *DBC*. Jam

Quæruntur *AF, DF, BF, CF, AE, BE*,	*v, t, s, r, q, p*,
DE, CE, AC, AG, GC, AK, KC, BG, GD,	*o, n, m, l, k, i, h, g, f*,
AI, DI, FI, BI, e quibus	*e, d, α, β* [?]
BC, CF; *AD, DF*; *AI, IF*; *AI, ID* partes sunt ipsarū *BF, AF, AD*	$b + r = s. \ x + t = v = e + \alpha.$ $e + d = x.$
et *EC, CD*; *EB, BA* partes ipsarum *ED, EA*	$n + c = o. \ p + a = q.$
et *AK, KC*; *AG, GC*; *BG, CD* partes ipsarū *AC, BD*.	$i + h = m = l + k.$ $g + f = y.$
Sunt autem similia triangula *ABF, CDF*, ut et *AGB, DGC*;	$\dfrac{a}{c} = \dfrac{v}{r} = \dfrac{s}{t} = \dfrac{l}{f} = \dfrac{g}{k}.$

and the line *DE* (fixed by *AD* = *a*) are, on putting $k = (d^2 + fh)/c$ and $l = (de - gi)/c$, respectively $z(d(v + [(\lambda + d)/c]z - a) + lz + bd)/i = (d(v + [(\lambda + d)/c] z) - kz)^2/h^2$ and

$$v + [(\lambda + d)/c] z = a.$$

Problem 12[84]

The semicircle ADCB which shall take the three given lengths AB, BC, CD is to be determined.

Verbally	*Algebraically*
AB, BC, CD are given	a, b, c.
Join AD, BD	x? y?
and then the triangle ABD will be right.	$x^2 = a^2 + y^2$.
Of line DC produced let CH be the external part	z?
terminated at BH, the perpendicular[85] of triangle BCD.	$y^2 = b^2 + c^2 + 2cz$,
Then, because of the equal angles $\widehat{BAD}$, $\widehat{BCH}$, the triangles ABD, BCH will be similar.	$x:a = b:z$, or $xz = ab$.

In this way, by completing the constructions in the diagrams and enumerating the individual conditions in the problems and writing them algebraically, there always result equations which are sufficient for their resolution. To be sure, we will be free to gather several equations and then, having made our choice of them, subsequently to reject the superfluous ones. Let AD, BC meet in F; AB, DC in E; and AC, BD in G. To AD let fall the perpendicular BI and make angle $\widehat{ABK}$ equal to angle $\widehat{DBC}$. Now

Let there be sought AF, DF, BF, CF, AE, BE,	v, t, s, r, q. p,
DE, CE, AC, AG, GC, AK, KC, BG, GD,	o, n, m, l, k, i, h, g, f,
AI, DI, FI, BI. Of these	e, d, α, β? where
BC, CF; AD, DF; AI, IF; AI, ID are parts of	$b + r = s$, $x + t = v = e + \alpha$,
BF, AF, AD;	$e + d = x$,
and EC, CD; EB, BA parts of ED, EA;	$n + c = o$, $p + a = q$,
and AK, KC; AG, GC; BG, CD parts of AC, BD.	$i + h = m = l + k$, $g + f = y$;
Further, the triangles ABF, CDF and also	$a/c = v/r = s/t = l/f = g/k$,
AGB, DGC are similar;	

When v is eliminated, there comes at once to be $z(lz + bd)/i = (ad - kz)^2/h^2$, that is,

$$(h^2 l - k^2 i)\, z^2 + 2(aik + \tfrac{1}{2}bh^2)\, dz - a^2 d^2 i = 0,$$

as Newton—a trivial omission apart—computes in his final reduction of the problem in the next section (see note (118)).

(84) A modified presentation of the Schootenian problem elaborated on (Newton's) pages 55/59–65 of the parent *Arithmetica* (see **1**, 2, §1: note (179)). The first mode of the present exposition corresponds to the earlier pages 55 and 59, while the latter approaches codify those on pages 59 – 65. Except that here the algebraic equivalents are unreduced (their common solution is once more reserved to the next section) there are no novelties in Newton's revised development of the problem.

(85) That is, the altitude from B to the base CD.

et BDF, ACF et EAC, EDB;

et EAD, ECB et AGD, BGC;

et BKC, ABD, ABI, DBI

　& ABK, DBC.

Et rectangula sunt triangula ACD, ABG, GCD,

　AIB, BID, BIF,

　EBD, ECA.

Et in triangulis BAF, BDF, EAD, EBC demittuntur perpendicula BI, AC, DB, BH.

$$\frac{m}{y}=\frac{v}{s}=\frac{r}{t}=\frac{q}{o}=\frac{n}{p}.$$

$$\frac{x}{b}=\frac{q}{n}=\frac{o}{p}=\frac{l}{g}=\frac{f}{k}.$$

$$\frac{b}{h}=\frac{y}{x}=\frac{\beta}{a}=\frac{d}{y};\; \frac{x}{a}=\frac{a}{e}=\frac{y}{\beta};$$

$$\frac{a}{y}=\frac{e}{\beta}=\frac{\beta}{d}=\frac{i}{c}.$$

$$mm+cc=xx.$$
$$aa+gg=ll.$$
$$kk+cc=ff.$$
$$ee+\beta\beta=aa.$$
$$\beta\beta+dd=yy.$$
$$\beta\beta+\alpha\alpha=ss.$$
$$pp+yy=oo.$$
$$mm+nn=qq.$$
$$aa+vv-2ev=ss$$
$$=yy+tt+2dt.$$
$$qq=xx+oo-2co.$$
$$xx=qq+oo-2qp.$$
$$pp=bb+nn-2nz.$$

Et eodem modo ducere liceret plures lineas in schemate et colligere plures æquationes.

　Tentent igitur Tyrones imprimis schemata construere et inde æquationes quotcunq colligere, tum seligant æquationes quæ resolutioni suffecerint idq hac methodo. Seligant imprimis æquationem quamvis quæ quantitates notas quam plurimas, ignotas quam paucissimas continet; dein huic adjungant alias gradatim quæ numerum notarum quantitatum quam maximè, ignotarum quam minimè augeant. Et sic pergant usq dum selectæ numero æquent quantitates suas ignotas et problema inclusis omnibus conditionibus determinatum reddant. Sic plures excogitare licebit modos solvendi problema unumquodq: quales sunt in hoc exemplo modi sequentes.[86]

　Modus 1 initio positus. $xz=ab$. $yy=bb+cc+2cz$. $xx=aa+yy$.

　Modus 2. $\frac{a}{c}=\frac{v}{r}$. $b+r=s$. $\frac{a}{c}=\frac{s}{t}$. $x+t=v$. $\frac{x}{a}=\frac{a}{e}$. $ss=aa+vv-2ev$.

　Modus 3. $\frac{a}{c}=\frac{v}{r}$. $b+r=s$. $\frac{a}{c}=\frac{s}{t}$. $x+t=v$. $\frac{m}{y}=\frac{v}{s}$. $mm+cc=xx=aa+yy$.

　Modus 4. $\frac{a}{c}=\frac{l}{f}$. $\frac{x}{b}=\frac{l}{g}$. $\frac{a}{c}=\frac{g}{k}$. $l+k=m$. $g+f=y$. $mm+cc=xx=aa+yy$.

　Modus 5. $\frac{a}{y}=\frac{i}{c}$. $\frac{b}{h}=\frac{y}{x}$. $i+h=m$. $mm+cc=xx=aa+yy$.

also *BDF*, *ACF* and *EAC*, *EDB*;

also *EAD*, *ECB* and *AGD*, *BGC*;

also *BKC*, *ABD*, *ABI*, *DBI*

　and *ABK*, *DBC*

And the triangles *ACD*, *ABG*, *GCD* are right-angled;

　and *AIB*, *BID*, *BIF*;

　and *EBD*, *ECA*.

And in the triangles *BAF*, *BDF*, *EAD*, *EBC*

the perpendiculars *BI*, *AC*, *DB*, *BH* are let fall.

$$m/y = v/s = r/t = q/o = n/p,$$
$$x/b = q/n = o/p = l/g = f/k,$$
$$b/h = y/x = \beta/\alpha = d/y,$$
$$x/a = a/e = y/\beta,$$
$$a/y = e/\beta = \beta/d = i/c;$$
$$m^2+c^2 = x^2,\ a^2+g^2 = l^2,$$
$$k^2+c^2 = f^2,$$
$$e^2+\beta^2 = a^2,\ \beta^2+d^2 = y^2,$$
$$\beta^2+\alpha^2 = s^2,$$
$$p^2+y^2 = o^2,\ m^2+n^2 = q^2;$$
$$a^2+v^2-2ev = s^2$$
$$= y^2+t^2+2dt,$$
$$q^2 = x^2+o^2-2co,$$
$$x^2 = q^2+o^2-2qp,$$
$$p^2 = b^2+n^2-2nz.$$

And in the same way it would be permissible to draw further lines in the diagram and gather yet more equations.

In the first instance, therefore, beginners should strive to construct diagrams and from these gather any number of equations, and then they should select the equations adequate to the resolution: and that by this method. Let them select first of all any equation containing as many known quantities and as few unknown ones as possible; then let them add others to this, one by one, which shall increase the number of known quantities as much and of unknowns as little as possible: let them proceed in this way until those selected are equal in number to the unknown quantities contained in them and so, including all its conditions, render the problem determinate. It will thus be permissible to contrive several modes of solving each single problem; in the present example, for instance, there are the following:[86]

Mode 1 (that set out initially). $xz = ab$, $y^2 = b^2+c^2+2cz$, $x^2 = a^2+y^2$.

Mode 2. $a/c = v/r$, $b+r = s$, $a/c = s/t$, $x+t = v$, $x/a = a/e$, $s^2 = a^2+v^2-2ev$.

Mode 3. $a/c = v/r$, $b+r = s$, $a/c = s/t$, $x+t = v$, $m/y = v/s$, $m^2+c^2 = x^2 = a^2+y^2$.

Mode 4. $a/c = l/f$, $x/b = l/g$, $a/c = g/k$, $l+k = m$, $g+f = y$,
$$m^2+c^2 = x^2 = a^2+y^2.$$

Mode 5. $a/y = i/c$, $b/h = y/x$, $i+h = m$, $m^2+c^2 = x^2 = a^2+y^2$.

---

(86) Here only the quantities a, b and c are known: to determine the others as many independent equations are needed as there are unknowns (z, y, x, v, t, s, ..., d, α, β) present in each mode.

Modus 6. $\frac{a}{y}=\frac{i}{c}\cdot\frac{b}{h}=\frac{y}{x}=\frac{d}{y}\cdot i+h=m.\ mm+cc=xx.\ \frac{x}{a}=\frac{a}{e}\cdot e+d=x.$

Modus 7. $n+c=o.\ p+a=q.\ \frac{x}{b}=\frac{q}{n}=\frac{o}{p}\cdot qq=xx+oo-2co.$

&c.

Et tum demum licebit modos inter se comparare et eligere simplicissimum. Qua ratione Tyro non solum facultatem solvendi problemata[87] sed etiam vim inveniendi solutiones optimas paulatim acquiret.

Reductio Æquationum.[88]

Inventis æquationibus superest earum reductio ad formas desideratas. Ea fit partim transformando plures æquationes in unam, partim concinnando æquationem resultantem. Reductio posterior debet prius intelligi et continetur his regulis.

Reg. 1. Delendæ semper sunt quantitates quæ se mutuo per additionem vel sub[d]uctionem destruunt, et conjungendæ quæ possunt coalescere. Veluti si habeatur

$$5b-3a+[2]x=5a+3x$$

subductis utrinꝗ $2x$ et $5a$ restabit $5b-8a=x$. Atꝗ ita $\frac{2ab+bx}{a}-b=a+b$

conjungendo $\frac{2ab}{a}-b$, seu $2b-b$, fit $\frac{bx}{a}+b=a+b$ et hæc delendo utrinꝗ b evadit

$\frac{bx}{a}=a.$

Reg. 2. Delendi sunt etiam per multiplicationem et divisionem communes omnium terminorum divisores et factores. Sic habito $15bb=24ab+3bx$ divido terminos omnes per factorem[89] b et fit $15b=24a+3x_{[,]}$ deinde per factorem[89] 3 et prodit $5b=8a+x$. Et habito $\frac{b^3}{ac}-\frac{bbx}{cc}=\frac{xx}{c}$ multiplico [terminos] omnes per c et prodit $\frac{b^3}{a}-\frac{bbx}{c}=xx.$ Et habito $\frac{z^3}{2ab-bb}=\frac{az^2-a^3}{2az-bz}$ multiplico per $2a-b$ et fit $\frac{z^3}{b}=\frac{az^2-a^3}{z}.$

[90]Excipe factores ex cognitis e[t] incognitis quantitatibus conflatos. Si æqu[atio] in tales resolvitur, retinendus est eorū aliquis rejectis cæteris. Ut si

---

(87) A needlessly restrictive 'Geometrica' (geometrical) is here deleted.

(88) This much revised, altered and augmented section is built on corresponding sections in the parent *Arithmetica*. In particular the following *Regulæ* 1, 2 (first paragraph) and 4–7 repeat corresponding Rules 1–5 and 7 in the earlier 'De concinnanda Æquatione solitaria' (**1**, 2, §1: Newton's pages 31–3), and the succeeding *Regulæ* 8–11, condense the substance

Mode 6. $a/y = i/c$, $b/h = y/x = d/y$, $i+h = m$, $m^2+c^2 = x^2$, $x/a = a/e$, $e+d = x$.

Mode 7. $n+c = o$, $p+a = q$, $x/b = q/n = o/p$, $q^2 = x^2+o^2-2co$.

And so on.

It will then subsequently be permissible to compare the modes one with another and choose the simplest. By this procedure the beginner will gradually acquire not only a facility in solving[87] problems but also the power of finding the best solutions.

The reduction of equations[88]

Once equations have been found, it remains to reduce them to desired form. That is done partly by transforming several equations into one, partly by refining the resulting equation. The latter reduction ought the sooner to be understood and is contained in these rules:

Rule 1. Quantities which mutually destroy one another by addition or subtraction must always be deleted, and those which can coalesce must always be combined. For instance, if there be had $5b-3a+2x = 5a+3x$, after $2x$ and $5a$ are taken away on each side there will remain $5b-8a = x$. Similarly $(2ab+bx)/a-b = a+b$ comes, on combining $2ab/a-b$, that is, $2b-b$, to be $bx/a+b = a+b$, and on deleting b on each side this becomes $bx/a = a$.

Rule 2. Common divisors and factors of all terms must also be deleted by multiplication and division. Thus, where $15b^2 = 24ab+3bx$ is had, I divide all the terms by the[89] factor b and there comes $15b = 24a+3x$, then by the[89] factor 3 and there results $5b = 8a+x$. And where $b^3/ac-b^2x/c^2 = x^2/c$ is had, I multiply all the terms by c and there results $b^3/a-b^2x/c = x^2$. And where

$$z^3/(2ab-b^2) = (az^2-a^3)/(2az-bz),$$

I multiply by $2a-b$ and there comes to be $z^3/b = (az^2-a^3)/z$.

[90]Do not include factors composed of known and unknown quantities. If an equation is resolved into such factors, some one of these must be retained and

---

of the subsequent sections on 'Exterminatio quantitatis incognitæ per æqualitatem valorum ejus/substituendo pro ea valorem suum/quæ plurium in utracq æquatione dimensionum existit' (*ibid.*: Newton's pages 34–7): *Regulæ* 2 (second paragraph), 3, 12 and 13 are newly contrived (partially on the inserted leaf f. 93 d, here silently incorporated into Humphrey Newton's transcript along with certain other minor corrections made by him in that initial copy; see note (2) above). The application of these rules in sequel to reducing (and indeed explicitly solving) the sets of algebraic equations which 'translate' the twelve problems propounded in the previous section is, of course, novel.

(89) 'datum' (given) is cancelled.

(90) This additional paragraph (on f. 93 d; see note (88)) was initially set out as a separate 'Reg   . *Loco æquationis inventæ usurpandus est aliquis ipsius divisor si modo divisores habeat*' (*Rule* [2′ ?]. *In place of the equation found, some divisor of it is to be employed if it should prove to have divisors*).

æquatio $x^3 - 7x + 6 = 0$ resolvitur in factores $x - 2$ et $xx + 2x - 3$[91] retinendus est is qui Problematis conditionibus satisfaciet et rejiciendus est alter. Inventio autem divisorum hujus generis in superioribus ostensa est.

Reg. 3. Æquatio secundum dimensiones quantitatis quæsitæ ordinanda est et ea de causa termini[92] *quivis de alterutra parte æquationis in alteram prout commodum fuerit per Additionem et Subductionem transferendi.* Ut si habetur $aa - 3ay = ab - bb + by$ et quæritur y, aufero utrobiᴕ terminos $-3ay$ et $ab - bb$ seu quod perinde est transfero eos ad contrarias partes æquationis cum signis suis $+$ et $-$ mutatis eò, ut ex una parte consistant termini omnes multiplicati per y et ex altera reliqui omnes: et prodit $aa - ab + bb = 3ay + by$, seu $\begin{matrix} aa \\ -ab \\ +bb \end{matrix} = \begin{matrix} 3a \\ +b \end{matrix} y$. Haud secus æquatio

$abx + a^3 - aax = abb - 2abx - x^3$ per debitam transpositionem et ordinationem evadit $x^3 = \dfrac{aa}{-3ab} x \dfrac{-a^3}{+aab}$, vel $x^3 \dfrac{-aa}{+3ab} x \dfrac{+a^3}{-aab} = 0$.

Reg. 4. Si in Denominatore fractionis alicujus irreducibilis reperiatur litera ad cujus dimensiones æquatio ordinanda est, debent omnes æquationis termini per istum denominatorem aut per ejus factorem aliquem cui inest illa litera multiplicari. Ut si æquatio $\dfrac{ax}{a-x} + b = x$ secundum dimensiones x ordinanda sit, multiplicentur omnes ejus termini per $a - x$ denominatorem fractionis $\dfrac{ax}{a-x}$ siquidem x inibi reperiatur et prodibit $ax + ab - bx = ax - xx$, seu $ab - bx = -xx$, et facta utriusᴕ partis translatione $xx = bx - ab$. Atᴕ ita si habeatur $\dfrac{a^3 - abb}{2cy - cc} = y - c$ terminiᴕ juxta dimensiones y ordinandi sint, multiplicentur per denominatorem $2cy - cc$ vel saltem per factorem $2y - c$ quo y tollatur e denominatore, et exurget $\dfrac{a^3 - abb}{c} = 2yy - 3cy + cc$ et ordinando $\dfrac{a^3 - abb}{c} - cc + 3cy = 2yy$. Ad eundem modum $\dfrac{aa}{x} - a = x$ multiplicando per x evadit $aa - ax = xx$, et $\dfrac{aabb}{cxx} = \dfrac{xx}{a+b-x}$ multiplicando primo per xx[5] dein per $a + b - x$ evadit $\dfrac{a^3bb + aab^3 - aabbx}{c} = x^4$.

Reg. 5. Sicui surdæ quantitati irreducibili inest litera ad cujus dimensiones æquatio ordinari debet, transferendi sunt cæteri omnes termini ad æquationis partes contrarias cum signis mutatis et pars utraᴕ in se semel ducenda si radix quadratica sit vel bis si ea sit cubica &c. Sic ad ordinandum juxta dimensiones x æquationem $\sqrt{aa - ax} + a = x$,

---

(91) This in turn factorizes as $(x - 1)(x + 3)$. No doubt Newton would have framed a more carefully chosen example if he had realized this. In sequel he first continued: 'rejiciendus est

the rest rejected. If, for instance, the equation $x^3 - 7x + 6 = 0$ is resolved into the factors $x - 2$ and $x^2 + 2x - 3$,[91] the one satisfying the conditions of the problem must be retained and the other is to be rejected. The finding of divisors of this kind has, however, been exhibited in a preceding section.

Rule 3. An equation must be ordered according to the dimensions of the quantity sought, and on this account any[92] *terms are, as it proves convenient, to be transferred by addition and subtraction from either side of the equation to the other.* For instance, if there is had $a^2 - 3ay = ab - b^2 + by$ and y is sought, I take away the terms $-3ay$ and $ab - b^2$ from either side—or, what is exactly the same, I transfer them to the opposite side of the equation with their signs $+$ and $-$ changed—in order that on one side there shall stand all terms multiplied by y and on the other all the rest: there results $a^2 - ab + b^2 = 3ay + by$, or $a^2 - ab + b^2 = (3a + b)\, y$. No differently, the equation $abx + a^3 - a^2x = ab^2 - 2abx - x^3$ comes, by appropriate transposition and ordering, to be $x^3 = (a^2 - 3ab)\, x - (a^3 - a^2b)$ or

$$x^3 - (a^2 - 3ab)\, x + (a^3 - a^2b) = 0.$$

Rule 4. If in the denominator of some irreducible fraction there be found the letter according to whose dimensions the equation is to be ordered, all terms in the equation ought to be multiplied by that denominator, or some factor of it in which that letter is present. For instance, if the equation $ax/(a - x) + b = x$ has to be ordered in sequence of the dimensions of x, let all its terms be multiplied by $a - x$, the denominator of the fraction $ax/(a - x)$, seeing that x is found therein, and there will result

$$ax + ab - bx = ax - x^2 \quad \text{or} \quad ab - bx = -x^2,$$

and on making transference of each side $x^2 = bx - ab$. And thus, if there be had $(a^3 - ab^2)/(2cy - c^2) = y - c$ and the terms have to be ordered according to the dimensions of y, let them be multiplied by the denominator $2cy - c^2$, or its factor $2y - c$ at least, so as to remove y from the denominator, and there will arise $(a^3 - ab^2)/c = 2y^2 - 3cy + c^2$, and on ordering it $(a^3 - ab^2)/c - c^2 + 3cy = 2y^2$. In much the same way $a^2/x - a = x$ comes, on multiplying by x, to be $a^2 - ax = x^2$, and $a^2b^2/cx^2 = x^2/(a + b - x)$, on multiplying first by x^2 and then by $a + b - x$, comes to be $(a^3b^2 + a^2b^3 - a^2b^2x)/c = x^4$.

Rule 5. If in any irreducible surd quantity there is present the letter according to whose dimensions the equation ought to be ordered, all the rest of the terms must be transferred to the opposite side of the equation with their signs changed and then each side multiplied once into itself if the root be square, twice if it be cube, and so on. Thus to order according to

---

horum alter et alter ponendus nihilo æqualis et loco æquationis usurpandus. Uter sit rejiciendus ex natura Problematis innotescet' (of these one is to be rejected and the other set equal to nothing and employed in the equation's place. Which shall need to be rejected will come to be known from the nature of the problem).

(92) 'cognati' (kindred) is cancelled.

transfero a ad alteras partes fitcg $\sqrt{aa-ax}=x-a$, et quadratis partibus, $aa-ax=xx-2ax+aa$, seu $0=xx-ax$ hoc est $x=a$. Sic etiam

$$\sqrt{3}:\overline{aax+2axx-x^3}-a+x=0,$$

transponendo $-a+x$ evadit $\sqrt{3}:\overline{aax+2axx-x^3}=a-x$, et partibus cubicè multiplicatis $aax+2axx-x^3=a^3-3aax+3axx-x^3$, seu $xx=4ax-aa$. Et sic $y=\sqrt{ay+yy-a\sqrt{ay-yy}}$ quadratis partibus evadit $yy=ay+yy-a\sqrt{ay-yy}$ et terminis debitè transpositis $ay=a\sqrt{ay-yy}$ seu $y=\sqrt{ay-yy}$, et partibus iterum quadratis $yy=ay-yy$, et transposito $-yy$ oritur $2yy=ay$ sive $2y=a$. Siquando plures sunt radicales quantitates finge singulas totidem literis æquales et ex æquationibus resultantibus literas illas per methodum sequentiū Regularum[93] extermina. Quemadmodum si sit $\sqrt{ay}-\sqrt{aa-ay}=2a+\sqrt[3]{ayy}$, scribendo t pro $\sqrt{ay}$, v pro $\sqrt{aa-ay}$, & x pro $\sqrt[3]{ayy}$ habebuntur æquationes $t-v=2a+x$, $tt=ay$, $vv=aa-ay$, et $x^3=ayy$, ex quibus tollendo gradatim t v et x resultabit tandem æquatio libera ab omni asymmetria.

Reg. 6. Terminis secundum dimensiones literæ alicujus ope præcedentium regularum dispositis, si maxima ejusdem literæ dimensio per cognitam quamlibet quantitatem multiplicatur vel dividitur debet tota æquatio per eandem dividi vel multiplicari: Qua ratione quantitas incognita unius dimensionis[94] semper obtineri potest. Sic $2y=a$ dividendo partem utramcg per 2 dat $y=\tfrac{1}{2}a$ et $\dfrac{bx}{a}=a$ dividendo per $\dfrac{b}{a}$ dat $x=\dfrac{aa}{b}$ et $aa-ab+bb=3ay+by$ dividendo per $3a+b$ dat $y=\dfrac{aa-ab+bb}{3a+b}$. Item

$$ayy=2xxy+axx$$

dividendo per a producit $yy=\dfrac{2xx}{a}y+xx$. Et $\begin{matrix}2ac\\-cc\end{matrix}x^3\begin{matrix}+a^3\\+aac\end{matrix}xx\begin{matrix}-2a^3c\\+aacc\end{matrix}x-a^3cc=0$

dividendo per $2ac-cc$ evadit $x^3+\dfrac{a^3\atop aac}xx{-2a^3c\atop+aacc}x-a^3cc}{2ac-cc}=0$. sive

$$x^3+\dfrac{a^3+aac}{2ac-cc}xx-aax-\dfrac{a^3c}{2a-c}=0.$$

Reg. 7. Æquatio si fieri possit in duas partes distribuenda est ut radix ex utracg possit extrahi. Id in æquationibus secundæ formæ[95] sic fit. Terminus altissimus affirmativè sumptus una cum termino medio æquatur termino infimo. Addatur

---

(93) Newton first wrote 'sequentis capitis' (following chapter). He evidently made the decision to add the substance of the parent sections on the 'Exterminatio quantitatis incognitæ' (see note (88)) as *Regulæ* complementary to the present Rules 1–7 only at a fairly late stage in composing the present revise.

(94) A curious phrase. Newton clearly means that the unknown quantity will be the root of an equation, the coefficient of whose highest term is unity.

the dimensions of x the equation $\sqrt{[a^2-ax]}+a = x$, I transfer a to the other side and there comes $\sqrt{[a^2-ax]} = x-a$; then, when the sides are squared,

$$a^2-ax = x^2-2ax+a^2 \quad \text{or} \quad 0 = x^2-ax, \quad \text{that is,} \quad x = a.$$

So also $\sqrt[3]{[a^2x+2ax^2-x^3]}-a+x = 0$ comes, on transposing $-a+x$, to be $\sqrt[3]{[a^2x+2ax^2-x^3]} = a-x$ and, when the sides are cubed,

$$a^2x+2ax^2-x^3 = a^3-3a^2x+3ax^2-x^3, \quad \text{that is,} \quad x^2 = 4ax-a^2.$$

And thus $y = \sqrt{[ay+y^2-a\sqrt{(ay-y^2)}]}$, when its sides are squared, comes to be $y^2 = ay+y^2-a\sqrt{(ay-y^2)}$ and, with its terms appropriately transposed,

$$ay = a\sqrt{(ay-y^2)} \quad \text{or} \quad y = \sqrt{(ay-y^2)};$$

then, with its sides again squared, $y^2 = ay-y^2$, and with $-y^2$ transposed there ensues $2y^2 = ay$ or $2y = a$. Whenever there are several radical quantities, conceive that these are singly equal to corresponding letters and then eliminate those letters from the resulting equations by the method in the following rules.[93] For instance, if there be $\sqrt{[ay]}-\sqrt{[a^2-ay]} = 2a+\sqrt[3]{[a^2y]}$, on writing t in place of $\sqrt{[ay]}$, v in place of $\sqrt{[a^2-ay]}$ and x in place of $\sqrt[3]{[ay^2]}$ there will be obtained the equations $t-v = 2a+x$, $t^2 = ay$, $v^2 = a^2-ay$ and $x^3 = ay^2$, and on removing t, v and x stage by stage from these there will at length result an equation free of all irrationality.

Rule 6. If, after the terms have, with the help of the preceding rules, been arranged according to the dimensions of some letter, the greatest dimension of this same letter is multiplied or divided by any known quantity, the whole equation ought to be divided or multiplied by it. By this procedure an unknown quantity having unit dimension[94] can always be obtained. Thus $2y = a$ on dividing each side by 2 yields $y = \frac{1}{2}a$, $bx/a = a$ on dividing by b/a yields $x = a^2/b$, and $a^2-ab+b^2 = (3a+b)y$ on dividing by $3a+b$ gives $y = (a^2-ab+b^2)/(3a+b)$. Likewise $ay^2 = 2x^2y+ax^2$ on dividing by a produces $y^2 = (2x^2/a)y+x^2$, while

$$(2ac-c^2)x^3 + (a^3+a^2c)x^2 - (2a^3c-a^2c^2)x - a^3c^2 = 0$$

on dividing by $2ac-c^2$ comes to be

$$x^3 + ((a^3+a^2c)x^2 - (2a^3c-a^2c^2)x - a^3c^2)/(2ac-c^2) = 0,$$

that is, $x^3 + ((a^3+a^2c)/(2ac-c^2))x^2 - a^2x - a^3c/(2a-c) = 0.$

Rule 7. If it can be done, an equation is to be apportioned into two parts such that the root of each can be extracted. That is achieved in equations of the second form[95] in this way. The highest term, taken positively, together with the middle term is equal

---

(95) Involving up to a quadratic power of the unknown, as in the second line of the tabulation on page 560 above.

utrinꝗ quadratum semissis termini medij divisum per terminum altissimum et radicem quadraticam utrobiꝗ licebit extrahere. Sic æquatio $xx=4x-1$ sub-ducendo utrinꝗ $4x$ fit $xx-4x=-1$, dein addendo utrinꝗ quadratum ex $2x$ applicatum ad xx hoc est addendo utrinꝗ 4 fit $xx-4x+4=3$ et extracta utrobiꝗ radice $x-2=\pm\sqrt{3}$, seu $x=2\pm\sqrt{3}$ hoc est $x=2+\sqrt{3}$ vel $x=2-\sqrt{3}$. [96]Nam radix omnis quadraticæ est duplex; una radix affirmativa est et altera negativa. Sic etiam æquatio $yy=\dfrac{2xxy}{a}+xx$, transponendo medium terminum

fit $yy-\dfrac{2xxy}{a}=xx$ et addendo utrobiꝗ quadratum ipsius $\dfrac{xxy}{a}$[97] divisum per yy

hoc est addendo utrobiꝗ $\dfrac{x^4}{aa}$ fit $yy-\dfrac{2xxy}{a}+\dfrac{x^4}{aa}=xx+\dfrac{x^4}{aa}$ et extrahendo utrobiꝗ

radicem evadit $y-\dfrac{xx}{a}=\pm\sqrt{xx+\dfrac{x^4}{aa}}$ seu $y=\dfrac{xx}{a}\pm\sqrt{xx+\dfrac{x^4}{aa}}$. Quinetiam æquatio

quartæ formæ[98] cujus termini impares desunt reducitur per eandem regulam. Proponatur æquatio $y^4+6abyy=2a^4$. Addatur utrobiꝗ quadratum ex $3abyy$ divisum per y^4 et fiet $y^4+6abyy+9aabb=9aabb+2a^4$ seu $yy=-3ab\pm\sqrt{9aabb+2a^4}$ et extracta denuo radice $y=[\pm]\sqrt{-3ab\pm\sqrt{9aabb+2a^4}}$. Casus difficiliores alibi tractabo.[99]

His regulis concinnantur æquationes seorsum: sequentibus verò conferuntur inter se et quantitates incognitæ exterminantur.

Reg. 8. Habitis duabus vel pluribus æquationibus totidem incognitas quantitates involventibus, seligitur tum quantitas incognita quæ facilius exterminatur, tum etiam æquatio quæ quantitatem illam modo simplicissimo involvit, et valor illius quantitatis quæritur in hac æquatione et substituitur in cæteris. Deinde ex æquationibus resultantibus exterminatur alia quævis ignota quantitas per eandem methodum. Et sic pergitur donec omnes ignotæ exterminentur una dempta quæ in unica æquatione ultimò resultante manebit. Nam numerus æquationum qualibet operatione perinde ac numerus ignotarum quantitatum diminuitur. Ut si fuerint $x+y=a$ et $ax=by$, posterior per Reg 6 dat $y=\dfrac{ax}{b}$ et scribendo $\dfrac{ax}{b}$ pro y, prior dat $x+\dfrac{ax}{b}=a$ seu $\dfrac{bx+ax}{b}=a$. Unde per Reg 6 est $x=\dfrac{ab}{b+a}$. Rursus si fuerint $xx+xy=yy$ et $2xy+yy=ax$, posterior æquatio per Reg 3 et 6 dat $x=\dfrac{yy}{a-2y}$. Quo substituto prior dat

$$\frac{y^4}{aa-4ay+4yy}+\frac{y^3}{a-2y}=yy.$$

---

(96) A cancelled first sequel reads: 'Ubi notes quod radicis signum pono ambiguum eò, quod' (Here you should note that I set the sign of the root as ambiguous because...).

(97) Humphrey Newton's transcript here reads '$\dfrac{2xxy}{a}$' in error.

to the lowest term: add to each side the square of half the middle term divided by the highest term and we will then be free to extract the square root of each side. Thus the equation $x^2 = 4x - 1$ comes, on subtracting $4x$ from each side, to be $x^2 - 4x = -1$, and then on adding to each side the square of $2x$ divided by x^2, that is, adding 4 to each, there comes $x^2 - 4x + 4 = 3$ and, when the root of each side is extracted, $x - 2 = \pm\sqrt{3}$ or $x = 2 \pm \sqrt{3}$, that is, $x = 2 + \sqrt{3}$ or $x = 2 - \sqrt{3}$. [96]For the root of every quadratic is twofold: one root is positive and the other negative. So, too, the equation $y^2 = (2x^2/a)\,y + x^2$ on transposing the middle term becomes $y^2 - (2x^2/a)\,y = x^2$ and by adding to each side the square of $(x^2/a)\,y$ divided by y^2, that is, adding x^4/a^2 to each, there comes

$$y^2 - 2(x^2/a)\,y + x^4/a^2 = x^2 + x^4/a^2,$$

and on extracting the root of each side there proves to be

$$y - x^2/a = \pm\sqrt{[x^2 + x^4/a^2]} \quad \text{or} \quad y = x^2/a \pm \sqrt{[x^2 + x^4/a^2]}.$$

To be sure, an equation of the fourth form[98] in which odd terms are lacking is reduced by the same rule. Let the equation $y^4 + 6aby^2 = 2a^4$ be proposed. Add to each side the square of $3aby^2$ divided by y^4 and there will come

$$y^4 + 6aby^2 + 9a^2b^2 = 2a^4 + 9a^2b^2 \quad \text{or} \quad y^2 = -3ab \pm \sqrt{(2a^4 + 9a^2b^2)},$$

and when the root is extracted afresh $y = [\pm]\sqrt{[-3ab \pm \sqrt{(2a^4 + 9a^2b^2)}]}$. More difficult cases I shall treat elsewhere.[99]

By these rules equations are refined individually: by the following ones, in contrast, they are compared one with another and their unknown quantities eliminated.

Rule 8. Where two or more equations involving an equal number of unknown quantities are had, there is selected both an unknown quantity rather easily eliminable and also the equation involving that quantity in the simplest way, then the value of that quantity is ascertained in this equation and substituted in the rest. Next, from the resulting equations is eliminated any other unknown quantity by the same method. And the same procedure is carried out till all unknowns are eliminated but for the one which will remain in the unique equation finally resulting. For the number of equations is decreased by any of these operations equally as the number of unknown quantities is diminished. If, for example, there were $x + y = a$ and $ax = by$, the latter by Rule 6 gives $y = ax/b$, while on writing ax/b in y's place the former yields $x + ax/b = a$ or $(b + a)\,x/b = a$; whence by Rule 6 there is $x = ab/(b + a)$. Again, if there were $x^2 + xy = y^2$ and $2xy + y^2 = ax$, the latter equation by Rules 3 and 6 gives $x = y^2/(a - 2y)$, and when this is substituted the former yields $y^4/(a^2 - 4ay + 4y^2) + y^3/(a - 2y) = y^2$,

---

(98) One involving the unknown up to its fourth power; compare note (95).

(99) In a projected later portion of the present revised *Arithmetica* which (compare note (3)) was never written. This more general 'Æquationum Reductio per divisores surdos' is elaborated on (Newton's) pages 187–203 in the parent text; see **1, 2,** §1: note (538).

hoc est[100] per Reg 2, $\dfrac{yy}{aa-4ay+4yy}+\dfrac{y}{a-2y}=1$, et per Reg 4

$$yy+ay-2yy=aa-4ay+4yy$$

id est per Reg. 1, $5yy-5ay+aa=0$ et per Reg 6 et 7, $y=\frac{1}{2}a\pm\sqrt{\dfrac{9aa}{20}}$. Adhæc si fuerint $xx+az=yy$, $2az+yy=ax$ et $az=xy$, per tertiam æquationem prodit $z=\dfrac{xy}{a}$ et scribendo $\dfrac{xy}{a}$ pro z æquatio[nes] prima et secuna evadunt $xx+xy=yy$ & $2xy+yy=ax$. ex quibus quantitas x exterminatur ut supra.

Reg 9. Siquando quantitas exterminanda ascendit in singulis æquationibus ad duas vel plures dimensiones, seligitur æquatio ubi quantitas illa minus ascendit et in ea valor potestatis altissimæ quæritur et substituitur in alia quavis: tum in æquatione resultante quæritur valor potestatis altissimæ et substituitur in æquatione prius selecta: postea in æquatione novissime resultante quæritur valor potestatis altissimæ et substituitur in æquatione prius resultante[,] et sic pergitur usqg dum resultet æquatio in qua quantitas exterminanda sit unius tantum dimensionis atqg adeo cujus beneficio quantitas illa per Reg 1 exterminari possit. Ut si obvenerint æquationes $x^3+2y^3=axy$ et $xx+xy=yy$. et exterminanda sit x, posterior dabit $xx=yy-xy$: quo substituto prior evadet primùm $xyy-xxy+2y^3=axy$, dein $xyy-y^3+[x]yy+2y^3=axy$ hoc est[100] per Reg 1 et 2, $2xy+yy=ax$. Quo habito cætera peraguntur ut in Regula octava. Rursus si fuerint $x^3+2y^3=axy$ et $x^3+xxy+xyy+y^3=axy$[,] per priorem fit $x^3=axy-2y^3$ eoqg substituto posterior evadit $axy-2y^3+xxy+xyy+y^3=axy$, hoc est per Reg 1 $xxy+xyy-y^3=0$ et per Reg 2 et 3 $xx+xy=yy$. Quo invento cætera peraguntur ut in exemplo priore.

Reg. 10. Siquando æquationum aliqua caret terminis imparibus dispone in alia quavis terminos pares ad unam partem et impares ad alteram, et partibus quadratis resultabit æquatio tertia[101] sine terminis imparibus cujus beneficio reductio solet breviter evadere. Ut si habeantur æquationes $x^4-4xxyy\genfrac{}{}{0pt}{}{+y^4}{+a^4}=0$ & $xx+xy-yy=0$ dispone terminos posterior[i]s in hunc modum $xx-yy=-xy$ et partibus quadratis orietur $x^4-2xxyy+y^4=xxyy$, et per Reg 2 et 3 $x^4=3xxyy-y^4$. Jam in æquatione prima pro x^4 scribe $3xxyy-y^4$ et deletis delendis orietur $-xxyy+a^4=0$ seu $xxyy=a^4$ et radice per Reg 7 utrobiqg extractâ $xy=aa$,[102] et per Reg 6 $y=\dfrac{aa}{x}$ [102]. Scribatur

---

(100) Assuming (to satisfy 'Reg 2' fully) that $y \neq 0$.

(101) There is some danger hereby of introducing unwanted duplicate roots in the final eliminant.

(102) These equations should be '$xy=\pm aa$' and, in consequence, '$y=\pm\dfrac{aa}{x}$' respectively, hence yielding the correct eliminant $x^4\pm a^2x^2=a^4$ or $(x^4+a^2x^2-a^4)(x^4-a^2x^2-a^4)=0$, and therefore $x=\pm\sqrt{[\frac{1}{2}(\pm1\pm\sqrt{5})]}\,a$.

that is, by Rule 2,[100] $y^2/(a^2-4ay+4y^2)+y/(a-2y)=1$, and so, by Rule 4, $y^2+ay-2y^2=a^2-4ay+4y^2$, that is, by Rule 1, $5y^2-5ay+a^2=0$ and so, by Rules 6 and 7, $y=\frac{1}{2}a\pm\sqrt{[\frac{9}{20}a^2]}$. Further, if there were $x^2+az=y^2$, $2az+y^2=ax$ and $az=xy$, by the third equation there results $z=xy/a$ and then, on writing xy/a in z's place, the first and second equations come to be $x^2+xy=y^2$ and $2xy+y^2=ax$; from these the quantity x is eliminated as above.

    Rule 9. Whenever the quantity to be eliminated ascends in individual equations to two or more dimensions, the equation in which that quantity ascends least is selected, and the value of the highest power in it is ascertained and substituted in any other; then in the resulting equation the value of the highest power is ascertained and substituted in the equation before selected; subsequently, in the equation most recently resulting the value of the highest power is ascertained and substituted in the equation previously resulting: and so the procedure is continued until there results an equation in which the quantity to be eliminated is of but one dimension, and hence by its aid that quantity can, by Rule 1, be eliminated. Should there, for instance, occur the equations $x^3+2y^3=axy$ and $x^2+xy=y^2$, and x need to be eliminated, the latter will yield $x^2=y^2-xy$; and, when this is substituted, the former will come first to be $xy^2-x^2y+2y^3=axy$ and then

$$xy^2-y^3+xy^2+2y^3=axy,$$

that is, by Rules 1 and 2, $2xy+y^2=ax$. Once this is obtained, the rest is accomplished as in the eighth Rule. Again, if there were

$$x^3+2y^3=axy \quad \text{and} \quad x^3+x^2y+xy^2+y^3=axy,$$

by the former there comes $x^3=axy-2y^3$ and then, when this is substituted, the latter comes to be $axy-2y^3+x^2y+xy^2+y^3=axy$, that is,[100] by Rule 1, $x^2y+xy^2-y^3=0$ and so, by Rules 2 and 3, $x^2+xy=y^2$. Once this is found, the rest is accomplished as in the former example.

    Rule 10. Whenever some one of the equations lacks odd terms, arrange in any other even terms on one side and odd ones on the other, and when the sides are squared there will result a third equation[101] *without odd terms: by the aid of this the reduction usually proves to be shorter.* For instance, if the equations

$$x^4-4a^2y^2+y^4+a^4=0 \quad \text{and} \quad x^2+xy-y^2=0$$

be had, arrange the latter's terms in this fashion $x^2-y^2=-xy$ and then, after the sides are squared, there will ensue $x^4-2x^2y^2+y^4=x^2y^2$ and so, by Rules 2 and 3, $x^4=3x^2y^2-y^4$. Now in the first equation in place of x^4 write $3x^2y^2-y^4$ and, on deleting what needs to be, there will arise

$$-x^2y^2+a^4=0 \quad \text{or} \quad x^2y^2=a^4;$$

then, after the root is, by Rule 7, extracted on each side, $xy=a^2$ [102] and so, by Rule 6, $y=a^2/x$.[102] Let a^2/x be written in y's place, therefore, and the previous

igitur $\dfrac{aa}{x}$ pro y, et æquatio superior $xx + xy - yy = 0$ evadet $xx + aa - \dfrac{a^4}{xx} = 0$, quæ per Reg 4 fit $x^4 + aaxx = a^4$, et per Reg 7 $x = \sqrt{-\tfrac{1}{2}aa + \sqrt{\tfrac{5}{4}a^4}}$.[103]

Reg 11. Exterminatur aliquando quantitas incognita addendo vel subducendo æqualia æqualibus aut multiplicando vel dividendo æqualia per æqualia aut etiam ponendo æqualia inter se quæ æquantur alicui tertio. Sic habitis æquationibus $2x = y + 5$ et $x = y + 2$ demantur æqualia ex æqualibus et prodibit $x = 3_{[,]}$ et habitis $ay = bx$ et $\dfrac{xx}{y} = b + x$ multiplicentur æqualia per æqualia et orietur $axx = bbx + bxx$ vel quæratur in utraqꝫ valor ipsius y et fiet $\dfrac{bx}{a}\,(=y) = \dfrac{xx}{b+x}$. Habitis autem $yy = \dfrac{2xxy}{a} + xx$ & $yy = 2xy + \dfrac{x^4}{aa}$ ut y exterminetur, extrahe in utraqꝫ radicem y sicut in Reg 7 ostensum est, et prodibunt $y = \dfrac{xx}{a} + \sqrt{\dfrac{x^4}{aa} + xx}$, et $y = x + \sqrt{\dfrac{x^4}{aa} + xx}$. Pone jam æqualia inter se quæ sunt ipsi y æquales et fiet $\dfrac{xx}{a} + \sqrt{\dfrac{x^4}{aa} + xx} = x + \sqrt{\dfrac{x^4}{aa} + xx}$ et per Reg 1 $\dfrac{xx}{a} = x$ vel $xx = ax$ et $x = a$.[104] Hujusmodi Reductiones necessariæ non sunt sed laborem tamen aliquando sublevant. Quo spectat etiam Regula sequens.

Reg. 12. Pro datis quantitatibus complexis assumere et scribere licet simplices quascunꝫ. Sic æquatio $xx + \dfrac{2ab - 2bb}{a+b}\,x = \dfrac{abb - b^3}{a+b}$ scribendo c pro $\dfrac{ab - bb}{a+b}$ fit $xx + 2cx = bc$ et radice per Reg 7 extracta $x = -c + \sqrt{bc + cc}$.[105]

Reg. 13. Postquam æquationes ad unam reducuntur debet ea per superiores Regulas concinnari$_{[,]}$ deinde per Regulam secundam ad dimensiones pauciores[106] si fieri possit reduci: ac deniꝫ si relatio aliqua inter radices æquationis inveniri possit, eruentur radices illæ fingendo aliam earundem dimensionū æquationem cujus radices habeant relationem illā inter se.[107] Ut si[108]

Hæ sunt præcipuæ Regulæ quibus æquationes ad formam quamvis desideratam reducimus. Ut Lector usum earum in solutione Problematum jam clarius percipiat, et exercitatione multiplici mandet omnia memoriæ atqꝫ

---

(103) In line with the previous note read '...fit $x^4 \pm aaxx = a^4$ et ... $x = \sqrt{\mp \tfrac{1}{2}aa + \sqrt{\tfrac{5}{4}a^4}}$' (where it is understood that the radicals are ambiguous in sign).

(104) Or, of course, $x(=y) = 0$. See **1**, 2, §1: note (114) and compare II: 407, note (77).

(105) Again understand that the radical is ambiguous in sign. This newly contrived *Regula* is used below in the (erroneous) conclusion of 'Prob. 11'.

(106) Newton first wrote 'ad formam [simplicissimam]' (to [simplest] form) and then more explicitly 'ad dimensiones quam paucissimas' (to as few dimensions as can be).

equation $x^2 + xy - y^2 = 0$ will come to be $x^2 + a^2 - a^4/x^2 = 0$; by Rule 4 this becomes $x^4 + a^2 x^2 = a^4$ and, by Rule 7, $x = \sqrt{[-\frac{1}{2}a^2 + \sqrt{(\frac{5}{4}a^4)}]}$.[103]

Rule 11. On occasion an unknown quantity is eliminated by adding or subtracting equals to or from equals, or again by multiplying or dividing equals, by equals or even by setting equal to one another magnitudes equal to some third one. So, having the equations $2x = y + 5$ and $x = y + 2$, take equals from equals and there will result $x = 3$; and, having $ay = bx$ and $x^2/y = b + x$, multiply equals by equals and there will ensue $ax^2 = b^2 x + bx^2$, or seek the value of y in each and there will come $bx/a(=y) = x^2/(b+x)$. Having, however, $y^2 = 2(x^2/a)\,y + x^2$ and $y^2 = 2xy + x^4/a^2$, to eliminate y extract the root y in each, as shown in Rule 7, and there will result $y = x^2/a + \sqrt{[x^4/a^2 + x^2]}$ and $y = x + \sqrt{[x^4/a^2 + x^2]}$. Now set what are equal to y equal to one another and there will come to be

$$x^2/a + \sqrt{[x^4/a^2 + x^2]} = x + \sqrt{[x^4/a^2 + x^2]},$$

and so, by Rule 1, $x^2/a = x$ or $x^2 = ax$ and $x = a$.[104] Reductions of this sort are not imperative but on occasion, however, they lighten the drudgery. The following rule is also pertinent to this.

Rule 12. In place of given complicated quantities we are at liberty to assume and write any simple ones whatsoever. Thus the equation

$$x^2 + ((2ab - 2b^2)/(a+b))\,x = (ab^2 - b^3)/(a+b)$$

comes, on writing c in place of $(ab - b^2)/(a+b)$, to be $x^2 + 2cx = bc$ and so, with the root extracted by Rule 7, $x = -c + \sqrt{[bc + c^2]}$.[105]

Rule 13. After the equations are reduced to a single one, it ought to be refined by means of the preceding rules; subsequently, if this is feasible, reduced by Rules 2 and 3 to fewer dimensions;[106] *and then finally, if some relationship between the roots of the equation can be found, those roots will be derived by conceiving another equation of the same dimensions whose roots shall have that relationship to each other.*[107] For instance, if [108]

These are the principal rules by which we may reduce equations to any desired form. So that the reader may now the more clearly perceive their use in the solution of problems, and by multiple exercises entrust all to memory and

---

(107) A final phrase ' *& quæ ad easdem dimensiones ascendat cum æquatione priore*' (*and which shall mount to the same dimensions as those of the first equation*) is deleted, doubtless because it is unnecessarily restrictive.

(108) Newton breaks off with the example undefined. Twenty years before, in his May 1665 paper 'Concerning Equations when the ratio of their rootes is considered', he had outlined—essentially by determining an invariant transformation of its roots—a pertinent reduction of a quartic equation (deduced from the Apollonian verging problem 'recited by D: Cartes pag 83 [of his *Geometria*, ₂1659]' and here, by coincidence, resolved in Newton's following 'Prob. [21]') which, suitably amended would have filled the rôle admirably; see I: 509–11, especially notes (9) and (13).

habitum acquirat resolvendi Problemata, visum est reductiones æquationum subjungere ad quas Problemata duodecim in capite præcedente[109] deduxeram.

Prob. 1. Dantur æquationes $xz=yy$. $x+y+z=20$. $xx+yy+zz=140$. Prima per Reg 6 dat $z=\dfrac{yy}{x}$ [>] quo valore per Reg 8 scripto in secunda et tertia pro z prodeunt $x+y+\dfrac{yy}{x}=20$ et $xx+yy+\dfrac{y^4}{xx}=140$, id est per Reg 4 & 3,

$$xx \begin{matrix}+y\\-20\end{matrix} x+yy=0 \quad \text{et} \quad x^4 \begin{matrix}+yy\\-140\end{matrix} xx+y^4=0.$$

In posteriore desunt termini impares et propterea per Reg 10 dispono terminos prioris in hunc modum $xx+yy=20x-xy$ et partibus quadratis prodit

$$x^4+2xxyy+y^4=400x^2-40xxy+xxyy.$$

Id est (per Reg 1 et 3) $x^4=\begin{matrix}-yy\\-40y\\+400\end{matrix} x^2-y^4$. Hunc ipsius x^4 valorem per Reg 9 substituo in æquatione superiore $x^4 \begin{matrix}+yy\\-140\end{matrix} xx+y^4=0$, et per Reg 1 deletis delendis prodit $\begin{matrix}40y\\-260\end{matrix} xx=0$ seu per Reg 2 et 3, $40y=260$ id est per Reg 6 $y=6\frac{1}{2}$.

Invento y datur simul x per æquationem superiorem $xx \begin{matrix}+y\\-20\end{matrix} x+yy=0$ quippe quæ per Reg 8 evadit $xx-13\frac{1}{2}x+42\frac{1}{4}=0$ et per Reg 7 $x=6\frac{3}{4}\pm\sqrt{3\frac{5}{16}}$.

Prob 2. Dantur æquationes $x+y=a$ et $ax=yy$. Posterior per Reg 6 dat $x=\dfrac{yy}{a}$ et inde per Reg 8 prior dat $\dfrac{yy}{a}+y=a$ seu per Reg 6 $yy+ay=aa$ et per Reg 7 $y=-\frac{1}{2}a+\sqrt{\frac{5}{4}aa}$. Unde simul habetur $x=a-y$.[110]

Prob. 3. Dantur æquationes $100x+5y=50l$. $100z+25x=30l$.

$$10y+400z=160l.$$

id est per Reg 2 æquationes $20x+y=10l$. $20z+5x=6l$. et $y+40z=16l$. Prima per Reg 3 dat $y=10l-20x$. Quo per Reg 8 scripto in æquatione tertia fit $10l-20x+40z=16l$ seu per Reg 3 et 6 $z=\dfrac{3l+10x}{20}$: eoꝗ per Reg 8 scripto in æquatione $20z+5x=6l$ prodit $3l+10x+5x=6l$ seu per Reg 3 et 6, $x=\frac{1}{5}l$. Et inde simul dantur $y=10l-20x$ et $z=\dfrac{3l+10x}{20}$[111] ut supra.

Prob 4. Extant æquationes $12y=x+36$ et $9y=x-6$. Posterior per Reg 3 dat $x=9y+6$ et inde prior per Reg 8 exterminando x fit $12y=9y+6+36$ seu $3y=42$ et $y=14$.

acquire a disposition for resolving problems, I have thought fit to append the reductions of the equations to which I brought the twelve problems in the previous section.[109]

Problem 1. There are given the equations

$$xz = y^2, \quad x+y+z = 20, \quad x^2+y^2+z^2 = 140.$$

The first, by Rule 6, gives $z = y^2/x$, and when this value is, by Rule 8, written in the second and third in z's place there result

$$x+y+y^2/x = 20 \quad \text{and} \quad x^2+y^2+y^4/x^2 = 140,$$

that is, by Rules 4 and 3, $x^2+(y-20)x+y^2 = 0$ and $x^4+(y^2-140)x^2+y^4 = 0$. In the latter odd terms are lacking and accordingly, by Rule 10, I arrange the former's terms in this manner $x^2+y^2 = (20-y)x$, and when the sides are squared there results $x^4+2x^2y^2+y^4 = (400-40y+y^2)x^2$; that is, by Rules 1 and 3, $x^4 = (-y^2-40y+400)x^2-y^4$. This value of x^4 I substitute, in accord with Rule 9, in the preceding equation $x^4+(y^2-140)x^2+y^4 = 0$ and, on deleting what needs to be by Rule 1, there results $(40y-260)x^2 = 0$ and so, by Rules 2 and 3, $40y = 260$, that is, by Rule 6, $y = 6\frac{1}{2}$. Once y is found, x is given at once by the preceding equation $x^2+(y-20)x+y^2 = 0$, seeing that by Rule 8 it proves to be $x^2-13\frac{1}{2}x+42\frac{1}{4} = 0$ and so, by Rule 7, $x = 6\frac{3}{4} \pm \sqrt{3\frac{5}{16}}$.

Problem 2. There are given the equations $x+y = a$ and $ax = y^2$. The latter, by Rule 6, gives $x = y^2/a$ and thence, by Rule 8, the former gives $y^2/a+y = a$, that is, by Rule 6, $y^2+ay = a^2$ and so, by Rule 7, $y = -\frac{1}{2}a + \sqrt{[\frac{5}{4}a^2]}$. Whence there is at once obtained $x = a-y$.[110]

Problem 3. There are given the equations $100x+5y = 50l$, $100z+25x = 30l$ and $10y+400z = 160l$, that is, by Rule 2, the equations $20x+y = 10l$, $20z+5x = 6l$ and $y+40z = 16l$. The first by Rule 3 yields $y = 10l-20x$. And when this, by Rule 8, is written in the third equation there comes

$$10l-20x+40z = 16l,$$

that is, by Rules 3 and 6, $z = \frac{1}{20}(3l+10x)$: and with this, by Rule 8, written in the equation $20z+5x = 6l$ there results $3l+10x+5x = 6l$ and so, by Rules 3 and 6, $x = \frac{1}{5}l$. From this there are at once given $y = 10l-20x$ and

$$z = \tfrac{1}{20}(3l+10x)^{[111]}$$

as above.

Problem 4. There exist the equations $12y = x+36$ and $9y = x-6$. The latter, by Rule 3, gives $x = 9y+6$ and thence the first, on eliminating x by Rule 8, becomes $12y = 9y+6+36$ or $3y = 42$ and so $y = 14$.

---

(109) That expounding his approach to the 'Inventio Æquationum'; see pages 564–84 above.

(110) Namely '$\frac{3}{2}a - \sqrt{\frac{5}{4}aa}$'.

(111) Whence '$y = 6l$' and '$z = \frac{1}{4}l$'.

Prob 5. Extant æquationes $x-100=p$. $\frac{4p}{3}=q$. $q-100=r$. $\frac{4}{3}r=s$. $s-100=t$.

$\frac{4}{3}t=v$. $v=2x$. Ergo per Reg 8, $\frac{4x-400}{3}=q$. $\frac{4x-400}{3}-100\left(=\frac{4x-700}{3}\right)=r$.

$\frac{16x-2800}{9}=s$. $\frac{16x-2800}{9}-100\left(=\frac{16x-3700}{9}\right)=t$. $\frac{64x-14800}{27}=v=2x$. Ergo

per Reg 2, $64x-14800=54x$ et per Reg 3, $10x=14800$ et $x=1480$.

Prob 6. Inventæ sunt æquationes $t=y+z$. $qyy=ppx$. $px=[r]z$. Ergo $z=\frac{px}{r}$ et

$t(=y+z)=y+\frac{px}{r}$ seu $y=t-\frac{px}{r}$. Scribatur $t-\frac{px}{r}$ pro y in æquatione $qyy=ppx$ et

fiet $qtt-\frac{2qtp}{r}x+\frac{qpp}{rr}xx=ppx$ et per reductionem $\frac{qpp}{rr}xx-\frac{2qtp}{r}x+qtt=0$ seu

$$xx\frac{-2qtr-prr}{pq}x+\frac{ttrr}{pp}=0.$$ Per Reg 12 scribe a pro $\frac{tr}{p}$ et b pro $\frac{\overset{-pp}{[r]r}}{q}$ et fiet

$xx\frac{-2a}{-b}x+aa=0$. Unde per Reg 7 est $x=a+\frac{1}{2}b\pm\sqrt{ab+\frac{1}{4}bb}$.

Prob. 7. Extant æquationes $d=x+y+z$. $h=\frac{ex}{a}+\frac{fy}{b}+\frac{gz}{c}$. $n=\frac{kx}{a}+\frac{ly}{b}+\frac{mz}{c}$. Per

primam est $z=d-x-y_{[,]}$ quo in secunda ac tertia substituto fit

$$h=\frac{ex}{a}+\frac{fy}{b}+\frac{gd-gx-gy}{c}$$

et $n=\frac{kx}{a}+\frac{ly}{b}+\frac{[m]d-mx-my}{c}$. Per Reg 12 scribe p pro $h-\frac{gd}{c}$, q pro $\frac{e}{a}-\left[\frac{g}{c}\right]$. r

pro $\frac{f}{b}-\frac{g}{c}$. s pro $n-\frac{md}{c}$. t pro $\frac{k}{a}-\frac{m}{c}$. et $\frac{v}{a}$ pro $\frac{l}{b}-\frac{m}{c}$ et æquationes illæ evadent

$p=\frac{qx+ry}{a}$ et $s=\frac{tx+vy}{a}$ seu $ap=qx+ry$ et $as=tx+vy$. Per priorem est $\frac{ap-qx}{r}=y$.

Ergo per posteriorem $as=tx+\frac{apv-qvx}{r}$. Unde $ars-apv=rtx-qvx$ et per Reg 6,

$\frac{ars-apv}{rt-qv}=x$.[112]

Prob. 8. Habentur æquationes $a=x+y$, $ex=dz$. et $fy=dz$. Ergo per Reg 11

$ex=fy$ et $\frac{ex}{f}=y$. $a=x+\frac{ex}{f}$. et $\frac{af}{f+e}=x$.

---

(112) This solution is readily cast into more familiar modern notation, since

$$rs-pv=-\begin{vmatrix} p & r \\ s & v \end{vmatrix}=-a\begin{vmatrix} d & 1 & 1 \\ h & f/b & g/c \\ n & l/b & m/c \end{vmatrix} \quad \text{and} \quad rt-qv=-\begin{vmatrix} q & r \\ t & v \end{vmatrix}=-a^2\begin{vmatrix} 1 & 1 & 1 \\ e/a & f/b & g/c \\ k/a & l/b & m/c \end{vmatrix}.$$

Problem 5. There exist the equations $x - 100 = p$, $\frac{4}{3}p = q$, $q - 100 = r$, $\frac{4}{3}r = s$, $s - 100 = t$, $\frac{4}{3}t = v$ and $v = 2x$. Therefore, by Rule 8, $\frac{1}{3}(4x - 400) = q$,

$$\tfrac{1}{3}(4x - 400) - 100 \ (\text{or } \tfrac{1}{3}(4x - 700)) = r, \quad \tfrac{1}{9}(16x - 2800) = s,$$

$\frac{1}{9}(16x - 2800) - 100$ (or $\frac{1}{9}(16x - 3700)) = t$, $\frac{1}{27}(64x - 14\,800) = v = 2x$. Therefore, by Rule 2, $64x - 14\,800 = 54x$ and so, by Rule 3, $10x = 14\,800$ and $x = 1480$.

Problem 6. There were found the equations $t = y + z$, $qy^2 = p^2x$, $px = rz$. Therefore $z = px/r$ and so t (or $y + z$) $= y + px/r$ or $y = t - px/r$. Write $t - px/r$ in y's place in the equation $qy^2 = p^2x$ and there will come to be

$$qt^2 - (2qtp/r)\,x + (qp^2/r^2)\,x^2 = p^2x,$$

and by reduction $(qp^2/r^2)\,x^2 - (2qtp/r + p^2)\,x + qt^2 = 0$ or

$$x^2 - ((2qtr + pr^2)/pq)\,x + t^2r^2/p^2 = 0.$$

By Rule 12 write a in place of tr/p and b in that of r^2/q and there will come $x^2 - (2a + b)\,x + a^2 = 0$. Whence by Rule 7 it is $x = a = \frac{1}{2}b \pm \sqrt{[ab + \frac{1}{4}b^2]}$.

Problem 7. There exist the equations $d = x + y + z$,

$$h = (e/a)\,x + (f/b)\,y + (g/c)\,z \quad \text{and} \quad n = (k/a)\,x + (l/b)\,y + (m/c)\,z.$$

By the first there is $z = d - x - y$, and when this is substituted in the second and third there comes $h = (e/a)\,x + (f/b)\,y + (g/c)\,(d - x - y)$ and

$$n = (k/a)\,x + (l/b)\,y + (m/c)\,(d - x - y).$$

By Rule 12 write p for $h - gd/c$, q/a for $e/a - g/c$, r/a for $f/b - g/c$, s for $n - md/c$, t/a for $k/a - m/c$ and v/a for $l/b - m/c$, and then those equations will prove to be $p = (qx + ry)/a$ and $s = (tx + vy)/a$, or $ap = qx + ry$ and $as = tx + vy$. By the former there is $(ap - qx)/r = y$ and therefore by the latter $as = tx + v(ap - qx)/r$. Whence $a(rs - pv) = (rt - qv)\,x$ and so, by Rule 6, $x = a(rs - pv)/(rt - qv)$.[112]

Problem 8. There are had the equations $a = x + y$, $ex = dz$ and $fy = dz$. Therefore, by Rule 11, $ex = fy$ and so $y = (e/f)\,x$, $a = x + (e/f)\,x$ and $x = af/(f + e)$.

Similarly, since
$$-a \begin{vmatrix} 1 & d & 1 \\ e/a & h & g/c \\ k/a & n & m/c \end{vmatrix} = - \begin{vmatrix} p & q \\ s & t \end{vmatrix} = qs - pt$$

and, where $\alpha = h - (e/a)\,d$, $\beta = n - (k/a)\,d$, $\gamma/a = e/a - f/b$ and $\delta/a = k/a - l/b$,

also
$$-a \begin{vmatrix} 1 & 1 & d \\ e/a & f/b & h \\ k/a & l/b & n \end{vmatrix} = - \begin{vmatrix} \alpha & \gamma \\ \beta & \delta \end{vmatrix} = \beta\gamma - \alpha\delta,$$

it follows that $y = a(pt - qs)/(rt - qv)$ and $z = a(\alpha\delta - \beta\gamma)/(rt - qv)$. We have not reproduced a cancelled first version of this paragraph, differing only in the substitution of y and z for the present x and y and in the concomitant replacement of e, f, g and k, l, m by f, g, e and l, m, k respectively in the definitions of p, q, r, s, t, v.

Prob. 9.[113] Dantur æquationes $y = s + t$, $ss + zz = vv$, $tt + zz = xx$, $es = d[v]$, $v + x + y = b$ et $\frac{1}{2}yz = aa$. Per quartam est $s = \frac{dv}{e}$, et inde per primam et secundam

$y = \frac{dv}{e} + t$ & $\frac{ddvv}{ee} + zz = vv$. seu $eezz = \frac{+ee}{-dd}vv$. Pone $ee - dd = ff$ et erit $eezz = ffvv$. et

$ez = fv$[114] et $z = \frac{fv}{e}$. Ergo per tertiam et sextam $tt + \frac{ffvv}{ee} = xx$ et $\frac{fvy}{2e} = aa$. Per

octavam est $y - \frac{dv}{e} = t$ et inde per decimam quartam $yy - \frac{2dvy}{e} + \frac{ddvv}{ee} + \frac{ffvv}{ee} = xx$.

Ubi si pro ff restituas $ee - dd$ fiet $yy - \frac{2dvy}{e} + vv = xx$. Per decimam quintam

est $v = \frac{2eaa}{fy}$ et inde per quintam et decimam octavam $\frac{2eaa}{fy} + x + y = b$, et

$yy - \frac{4daa}{f} + \frac{4eea^4}{ffyy} = xx$. seu $yy \frac{+x}{-b} y + \frac{2eaa}{f} = 0$ et $y^4 - \frac{4daa}{f}y^2 + \frac{4eea^4}{ff} = 0$. Ubi cùm

$$[-xx]$$

termini impares desint dispono terminos æquationis prioris in hunc modum

$yy + \frac{2eaa}{f} = by - xy$ et partibus quadratis fit

$$y^4 + \frac{4ea^2}{f}yy + \frac{4eea^4}{ff} = bbyy - 2bxyy + xxyy$$

seu $y^4 + \frac{4eaa}{f}yy + \frac{4eea^4}{ff} = 0$. qua subducta de æquatione vigesima tertia[115]

$$-xx$$
$$+2bx$$
$$-bb$$

restat $-\frac{4daa}{f}yy \left[-\frac{4eaa}{f}yy \right] - 2bxyy + bbyy = 0$[116] seu $2bx = bb - \frac{4daa}{f} \left[-\frac{4eaa}{f} \right]$

et $x = \frac{1}{2}b - \frac{2daa}{bf} \left[-\frac{2eaa}{bf} \right]$. Inventa x dantur cæteræ ignotæ quantitates per æquationes superiores.

Prob 10. Extant æquationes $n = v - x$, $c = y - z$, $aa + yy = vv$, $bb + [zz] = xx$. Per primam et secundam est $v = n + x$ et $y = c + z$. Ergo per tertiam

$$aa + cc + 2cz + zz = nn + 2nx + xx.$$

Aufer quartam et manebit $aa + cc - bb + 2cz = nn + 2nx$. Pro $aa + cc - bb - nn$

---

(113) To make Newton's somewhat clumsy numbering of the equations (as presented in sequence) in the following paragraph consistent with his intention, it is necessary to omit the simple redefinition '$ee - dd = ff$' from the reckoning. The 'eighth' equation will be '$y = \frac{dv}{e} + t$',

Problem 9.[113] There are given the equations $y = s+t$, $s^2+z^2 = v^2$, $t^2+z^2 = x^2$, $es = dv$, $v+x+y = b$ and $\frac{1}{2}yz = a^2$. By the fourth there is $s = (d/e)\,v$ and thence, by the first and second, $y = (d/e)\,v+t$ and $(d^2/e^2)\,v^2+z^2 = v^2$, that is,

$$e^2z^2 = (e^2-d^2)\,v^2.$$

Put $e^2-d^2 = f^2$ and there will be $e^2z^2 = f^2v^2$, and so $ez = fv$[114] and $z = (f/e)\,v$. Therefore by the third and sixth equation $t^2+(f^2/e^2)\,v^2 = x^2$ and $\frac{1}{2}(f/e)\,vy = a^2$. By the eighth there is $y-(d/e)\,v = t$ and thence, by the fourteenth,

$$y^2-2(d/e)\,vy+(d^2/e^2)\,v^2+(f^2/e^2)\,v^2 = x^2.$$

If in this you now restore e^2-d^2 in place of f^2 there will come to be

$$y^2-2(d/e)\,vy+v^2 = x^2.$$

By the fifteenth equation there is $v = 2(e/f)\,a^2/y$ and thence, by the fifth and eighteenth,

$$2(e/f)\,a^2/y+x+y = b \quad \text{and} \quad y^2-4(d/f)\,a^2+4(e^2/f^2)\,a^4/y^2 = x^2,$$

that is, $y^2+(x-b)\,y+2(e/f)\,a^2 = 0$ and $y^4-(4(d/f)\,a^2+x^2)\,y^2+4(e^2/f^2)\,a^4 = 0$. Since odd terms are lacking in the latter, I arrange the terms of the former equation in this manner $y^2+2(e/f)\,a^2 = (b-x)\,y$ and, when its sides are squared, there comes $y^4+4(e/f)\,a^2y^2+4(e^2/f^2)\,a^4 = (b^2-2bx+x^2)\,y^2$, that is,

$$y^4+(4(e/f)\,a^2-x^2+2bx-b^2)\,y^2+4(e^2/f^2)\,a^4 = 0.$$

And when this is subtracted from the twenty-third equation[115] there remains $-(4(d/f)\,a^2[+4(e/f)\,a^2]+2bx-b^2)\,y^2 = 0$,[116] that is, $2bx = b^2-4(d[+e])\,a^2/f$ and so $x = \frac{1}{2}b-2(d[+e])\,a^2/bf$. Once x is found, the remaining unknown quantities are yielded by the preceding equations.

Problem 10. There exist the equations $n = v-x$, $c = y-z$, $a^2+y^2 = v^2$ and $b^2+z^2 = x^2$. By the first and second there is $v = n+x$ and $y = c+z$. Therefore by the third equation $a^2+c^2+2cz+z^2 = n^2+2nx+x^2$. Take away the fourth and there will remain $a^2+c^2-b^2+2cz = n^2+2nx$. In place of $a^2+c^2-b^2-n^2$ assume

---

for instance,—being eighth in order—but the 'fourteenth' is meant to be '$tt+\dfrac{ffvv}{ee}=xx$' and the 'eighteenth' must be taken as '$yy-\dfrac{2dvy}{e}+vv=xx$'.

(114) Newton assumes the ratio e/f to be positive, so that the second root $ez = -fv$ is here inadmissible.

(115) The previous quartic in y, that is.

(116) Newton's equation lacks the term '$-\dfrac{4eaa}{f}yy$' here inserted, an omission which vitiates the manuscript sequel (here duly emended).

assume et scribe $2bf$ et fiet $2bf + 2cz = 2nx$ seu $\dfrac{bf+cz}{n} = x$. Unde per quartam est

$bb + zz = \dfrac{bbff + 2bfcz + cczz}{nn}$.[117]

Prob. 11. Dantur æquationes $zx = yy$. $p = y + q$. $m = x + n$. $a = t + v$. $b = r - s$.

$\dfrac{z}{c} = \dfrac{v}{d} = \dfrac{s}{e} = \dfrac{p}{f} = \dfrac{n}{g}$. $dt = hq$. $dr = im$. Per posteriores sunt $v = \dfrac{dz}{c}$. $s = \dfrac{ez}{c}$. $p = \dfrac{fz}{c}$. $n = \dfrac{gz}{c}$.

$q = \dfrac{dt}{h}$ et $m = \dfrac{dr}{i}$. Et inde per priores $\dfrac{fz}{c} = y + \dfrac{dt}{h}$. $\dfrac{dr}{i} = x + \dfrac{gz}{c}$. $a = t + \dfrac{dz}{c}$. $b = r - \dfrac{ez}{c}$.

Per primam et ultimas duas sunt $x = \dfrac{yy}{z}$, $t = a - \dfrac{dz}{c}$, $r = b + \dfrac{ez}{c}$ et inde hæ duæ

$\dfrac{fz}{c} = y + \dfrac{dt}{h}$ et $\dfrac{dr}{i} = x + \dfrac{gz}{c}$ evadunt $\dfrac{fz}{c} = y + \dfrac{da}{h} - \dfrac{ddz}{ch}$ & $\dfrac{db}{i} + \dfrac{dez}{ic} = \dfrac{yy}{z} + \dfrac{gz}{c}$. Pro $\dfrac{fh+dd}{c}$

et $\dfrac{de-gi}{c}$ assume et scribe k et l et æquationes novissimæ evadent $\dfrac{kz}{h} = y + \dfrac{da}{h}$ et

$\dfrac{lz}{i} = \dfrac{yy}{z} - \dfrac{db}{i}$. Per priorem est $y = \dfrac{kz-da}{h}$ et inde per posteriorem

$$\frac{lz}{i} = \frac{kkzz - 2kdaz + ddaa}{hhz} - \frac{db}{i},$$

seu $\dfrac{hhl}{-kki} zz + 2kdaiz = ddaai$.[118] Pro $\dfrac{kdai}{hhl-kki}$ scribe α et æquatio novissima

evadet $zz + 2\alpha z = \dfrac{ada}{k}$ et extracta radice $z = -\alpha + \sqrt{\alpha\alpha + \dfrac{ada}{k}}$.

Prob. 12. Dantur æquationes $xx = aa + yy$. $yy = bb + cc + 2cz$ et $xz = ab$. Per

tertiam est $z = \dfrac{ab}{x}$ et inde per secundam $yy = bb + cc + \dfrac{2cab}{x}$ et per primam

$xx = aa + bb + cc + \dfrac{2cab}{x}$, id est $x^3 = \begin{array}{c} aa \\ +bb \\ +cc \end{array} x + 2abc$.[119] Eandem æquationem ex alijs

sex supra inventis æquationum modis tentet Lector[120] exercitationis gratia derivare.

Completæ jam sunt duodecim problematum resolutiones et responsa[121] inventa quæ ita se habent.

1. Tres numeri continuè proportionales sunt $6\frac{3}{4} + \sqrt{3\frac{5}{16}}$, $6\frac{1}{2}$, et $6\frac{3}{4} - \sqrt{3\frac{5}{16}}$.

2. Lineæ secandæ pars major est $-\frac{1}{2}a + \sqrt{\frac{5}{4}aa}$.

3. Modius tritici valet $\frac{1}{5}^{\text{lib}}$, seu quatuor solidis.

---

(117) That is, $(n^2 - c^2) z^2 - 2bcfz + b^2(n^2 - f^2) = 0$ (compare note (81)), whence (see note (124) below) $z = b(cf \pm n\sqrt{[c^2 + f^2 - n^2]})/(n^2 - c^2)$.

(118) A term '$+bdhhz$' is omitted from the left-hand side of this equation and the accuracy of Newton's further argument is spoilt. In sequel we need to take

$$(kdai + \tfrac{1}{2}bdh^2)/(h^2l - k^2i) = \alpha \quad \text{and} \quad d^2a^2i/(h^2l - k^2i)\alpha = \beta$$

and write $2bf$ and there will come to be $2bf + 2cz = 2nx$ or $x = (bf + cz)/n$. Hence by the fourth equation there is $b^2 + z^2 = (b^2f^2 + 2bcfz + c^2z^2)/n^2$.[117]

Problem 11. There are given the equations $zx = y^2$, $p = y + q$, $m = x + n$, $a = t + v$, $b = r - s$, $z/c = v/d = s/e = p/f = n/g$, $dt = hq$ and $dr = im$. By the latter ones there is then $v = (d/c)\, z$, $s = (e/c)\, z$, $p = (f/c)\, z$, $n = (g/c)\, z$, $q = (d/h)\, t$ and $m = (d/i)\, r$; and thence by the former ones

$$(f/c)\, z = y + (d/h)\, t, \quad (d/i)\, r = x + (g/c)\, z, \quad a = t + (d/c)\, z \quad \text{and} \quad b = r - (e/c)\, z.$$

By the first equation and the two last $x = y^2/z$, $t = a - (d/c)\, z$ and $r = b + (e/c)\, z$, and thereby these two, $(f/c)\, z = y + (d/h)\, t$ and $(d/i)\, r = x + (g/c)\, z$, come to be $(f/c)\, z = y + da/h - (d^2/ch)\, z$ and $db/i + (de/ic)\, z = y^2/z + (g/c)\, z$. In place of $(fh + d^2)/c$ and $(de - gi)/c$ assume and write k and l, and then the most recent equations will prove to be $(k/h)\, z = y + da/h$ and $(l/i)\, z = y^2/z - db/i$. By the former there is $y = (kz - da)/h$ and thence by the latter

$$(l/i)\, z = (k^2z^2 - 2kdaz + d^2a^2)/h^2z - db/i,$$

that is, $(h^2l - k^2i)\, z^2 + 2kdaiz = d^2a^2i$.[118] For $kdai/(h^2l - k^2i)$ write α and the most recent equation will come to be $z^2 + 2\alpha z = ad\alpha/k$ and, when the root is extracted, $z = -\alpha + \sqrt{[\alpha^2 + ad\alpha/k]}$.

Problem 12. There are given the equations $x^2 = a^2 + y^2$, $y^2 = b^2 + c^2 + 2cz$ and $xz = ab$. By the third there is $z = ab/x$ and thence by the second

$$y^2 = b^2 + c^2 + 2cab/x,$$

and so by the first $x^2 = a^2 + b^2 + c^2 + 2cab/x$, that is, $x^3 = (a^2 + b^2 + c^2)\, x + 2abc$.[119] Let the reader[120] attempt to derive the same equation from the other six above-found modes of equations as an exercise.

The resolutions of the twelve problems are now complete and their answers[121] found; they are had as follows:

1. The three numbers in continued proportion are $6\frac{3}{4} + \sqrt{3\frac{5}{16}}$, $6\frac{1}{2}$ and $6\frac{3}{4} - \sqrt{3\frac{5}{16}}$.

2. The greater portion of the line to be cut is $-\frac{1}{2}a + \sqrt{[\frac{5}{4}a^2]}$.

3. A measure of wheat is worth $\mathcal{L}\frac{1}{5}$ or four shillings.

---

to achieve the solution $z = -\alpha + \sqrt{[\alpha(\alpha + \beta)]}$ given below (see note (125)).

(119) Newton doubtless intended to elaborate the exact algebraic solution of the general reduced cubic—and, indeed, general quartic—in a projected future section, never drafted, on the lines of the 'reductiones quædam cubicarum [et biquadraticarum] æquationum vulgò notæ' set out by him in the parent *Arithmetica* (**1**, **2**, §1: Newton's pages 203–7).

(120) Observe that here (and twice more in the sequel) Newton addresses the private student, and no longer—if ever before, in fact (compare **1**, **2**, §1: note (580))—a collective academic audience gathered to hear him lecture.

(121) 'simul' (simultaneously) is cancelled.

4. Mendici sunt quatuordecim.

5. Mercator sub initio habebat 1480^{lib}.

6. Altitudo putei est $a + \frac{1}{2}b - \sqrt{ab + \frac{1}{4}bb}$ positis $\dfrac{tr}{p} = a$ et $\dfrac{[r]r}{q} = b$.

7. Quantitas auri in mistura est $\dfrac{asr - apv}{rt - qv}$.$^{(122)}$

8. Longitudo segmenti *AD* est $\dfrac{af}{e + f}$.

9. Trianguli latus *BC* dato angulo obversum est $\frac{1}{2}b - \dfrac{2daa}{bf} \left[- \dfrac{2eaa}{bf} \right]$.$^{(123)}$

10. Longitudo *EC* est $\dfrac{bcf \pm b\sqrt{[nn]ff + ccnn - n^4}}{nn - cc}$.$^{(124)}$

11. Longitudo *PQ* est $-\alpha + \sqrt{\alpha\alpha[+]\alpha\beta}$.$^{(125)}$

12. Circuli quæsiti diameter est ea ipsius *x* longitudo qua

$$x^3 \quad \text{et} \quad \sqrt{aa + bb + cc} \times x + 2abc$$

evadunt æqualia.

METHODUS.$^{(126)}$

Postquam Lector Problemata in æquationes derivare et æquationes illas reducere didicit, consulendum erit succinctis et artificiosis operationibus quibus computus et levior reddatur et elegantior. Eò spectat hæc Regula. Quantitates aliquæ a quibus cæteræ omnes nullo inter notas et ignotas habito discrimine facillime et per computum maximè simplex derivari possint, ipso operis initio assumendæ sunt et nomina ijsdem imponenda. Tum ex assumptis colligendæ sunt cæteræ quibus aliàs imponenda forent nomina, et enunciando deinceps relationes assumptarum et collectarum ad invicem colligentur æquationes quæ jam pauciores erunt atcʒ adeo minori labore reduci possunt quàm superius ubi plures ignotæ quantitates assumebantur. Itacʒ quoties obveniunt totum et partes non impono nomina his omnibus sed vel totum designo per summam partium vel partem aliquam per excessum totius supra partes reliquas. Quoties obveniunt quantitates proportionales designo extremum alterutrum per factum sub medijs divisum per extremum alterum. Quoties datum est rectangulum sub duabus ignotis exprimo alterutram ignotarum per datum illud

---

(122) Where *p*, *q*, *r*, *s*, *t* and *v* are understood to have their previously assigned meaning.

(123) The term '$-\dfrac{2eaa}{bf}$' is our insertion (compare note (116)).

(124) Evidently through a computational slip on Newton's part (since, unlike *b*, the factor '$\sqrt{nn}$' is not taken outside the radical) the transcribed manuscript reads

$$\text{'}\ldots \pm b\sqrt{ccff + ccnn - n^4}\text{'}$$

in the numerator. On the corrected form see note (117).

4. There are fourteen beggars.

5. The merchant had £1480 at the start.

6. The depth of the well is $a + \frac{1}{2}b - \sqrt{[ab + \frac{1}{4}b^2]}$, supposing $tr/p = a$ and $r^2/q = b$.

7. The quantity of gold in the admixture is $a(sr - pv)/(rt - qv)$.[122]

8. The length of the segment AD is $af/(e + f)$.

9. The triangle's side BC facing the given angle is $\frac{1}{2}b - 2(d[+e])\,a^2/bf$.[123]

10. The length of EC is $(bcf \pm b\sqrt{[(f^2 + c^2)\,n^2 - n^4]})/(n^2 - c^2)$.[124]

11. The length of PQ is $-\alpha + \sqrt{[\alpha^2 + \alpha\beta]}$.[125]

12. The diameter of the required circle is the length x for which x^3 and $(a^2 + b^2 + c^2)\,x + 2abc$ prove to be equal.

METHOD[126]

After the reader has learnt how to distil problems into equations and to reduce those equations, attention will need to be paid to concise, skilful procedures by which the computation is rendered both lighter (smoother) and more elegant. The following rule bears on this. A few quantities from which all the rest might, without distinction between known and unknown ones, most easily and by the most straightforward computation be derivable are to be assumed at the very outset of the work and names set upon them. Then from these assumed quantities others are to be gathered on which we would otherwise have to impose names, and by subsequently expressing the relationships of those assumed and those deduced to each other there will be gathered equations which will now be fewer and reducible with less effort than previously, when more unknown quantities were assumed. Accordingly, each time there occurs a whole and its parts, I do not set names on all these but either designate the whole by the sum of its parts or denote some part by the excess of the whole over the parts remaining. Each time proportional quantities occur, I designate one or other extreme by the product of the middles divided by the other extreme. Each time there is given the product of two unknowns, I express one or other of the unknowns by

---

(125) Following on his previously vitiated solution (see note (118)), it will be evident that Newton here intends us to set $\alpha = kdai/(h^2l - k^2i)$ and $\beta = ad/k$, rather than to insert their correct values. Since Newton invariably takes an undirected line-length to be positive in sign, the manuscript reading ' $-\alpha + \sqrt{\alpha\alpha - \alpha\beta}$ ' must be a slip (on Humphrey Newton's part?).

(126) In this much reworked and expanded section (which concludes the extant manuscript) Newton amplifies his earlier guiding rule in the parent *Arithmetica* (**1**, 2, §1: page 61) for 'translating' a given problem most simply, elegantly and succinctly into an effective algebraic equivalent. The examples to which he applies this refined 'method' are all—we need scarcely say it—narrowly based on corresponding geometrical 'questions' in that primary text (see notes (128), (129), (130), 132) and (134)).

rectangulum divisum per alteram. Quoties occurrit [triangulum][127] rectangulum pono summam quadratorum laterum pro quadrato hypotenusæ vel
excessum quadrati hypotenusæ supra quadratum lateris alterutrius pro
quadrato lateris alterius, et sic in cæteris. Hujus igitur methodi vestigijs insistendo,

In primo præcedentium Problematum ubi tres quæruntur proportionales
numeri x, y, et z pro tertio non scribo z sed $\frac{yy}{x}$ pergendo in hunc modum.[128]

Quæruntur tres numeri continuè proportionales	$\left[x?\ y?\ \frac{yy}{x}? \right]$
quorum summa sit 20	$x+y+\dfrac{yy}{x}=20.$
et summa quadratorum 140.	$xx+yy+\dfrac{y^4}{xx}=140.$

In quinto problemate colligo quantitates alias ex alijs ut sequitur.[129]

Mercator habet nummos quosdam ex quibus anno primo impendit 100^{lib}	$x.$ $x-100.$
in familiā et reliquum adauget triente	$x-100+\dfrac{x-100}{3}$ seu $\dfrac{4x-400}{3}.$
annoꝗ secundo impendit 100^{lib}	$\dfrac{4x-400}{3}-100$ seu $\dfrac{4x-700}{3}.$
et reliquum adauget triente	$\dfrac{4x-700}{3}+\dfrac{4x-700}{9}$ seu $\dfrac{16x-2800}{9}.$
et sic anno tertio impendit 100^{lib}	$\dfrac{16x-2800}{9}-100$ seu $\dfrac{16x-3700}{9}.$
et reliquo trientem lucratus est	$\dfrac{16x-3700}{9}+\dfrac{16x-3700}{27}$ seu $\dfrac{64x-14800}{27}.$
fitꝗ duplo ditior quam sub initio.	$\dfrac{64x-14800}{27}=2x.$

In decimo si AD, BE, DE, N et EC respectivè nominentur a, b, c, n et x,
resolutio sic procedit.[130] Erit $DC(=DE+EC)=a+x.$

$BC^q(=BE^q+EC^q)=bb+xx.$ $AC^q(=AD^q+DC^q)=2aa+2ax+xx.$

& $N(=AC-BC)=\sqrt{[2]aa+2ax+xx}-\sqrt{bb+xx}=n.$

Sic primum Problema deducitur ad duas æquationes, quint[u]m vero ac
decimum ad unas, et Reductio levior jam superest. Tentet Lector exercitationis

---

(127) This necessary word is omitted (by Humphrey Newton?) in the transcribed
manuscript.

(128) This is copied from Newton's page 40 in the parent *Arithmetica* (1, 2, §1).

(129) An exact copy of the equivalent reduction on Newton's pages 40–1 of the parent text.

that given product divided by the second. Each time a right-angled triangle occurs, I put the sum of the squares of the sides for the square of the hypotenuse or set the excess of the square of the hypotenuse over the side of one or other side for the square of the second side. And the like in other cases. By treading in the steps of this method, therefore,

— in the first of the preceding problems, where three numbers x, y and z in proportion are sought, for the third I no longer write z but y^2/x, proceeding in this manner:[128]

There are required three numbers in continued proportion,	$x, y, y^2/x$?
the sum of which shall be 20	$x+y+y^2/x = 20,$
and the sum of their squares 140.	$x^2+y^2+y^4/x^2 = 140.$

— in the fifth problem I gather certain quantities from others as follows:[129]

A merchant has a certain sum of money,	$x.$
of which the first year he spends £100	$x-100,$
on his family, increasing the rest by a third;	$x-100+\frac{1}{3}(x-100)$ or $\frac{1}{3}(4x-400);$
and the second year he spends £100 increasing the remainder by a third;	$\frac{1}{3}(4x-400)-100$ or $\frac{1}{3}(4x-700),$ $\frac{1}{3}(4x-700)+\frac{1}{9}(4x-700)$ or $\frac{1}{9}(16x-2800);$
likewise the third year he spends £100,	$\frac{1}{9}(16x-2800)-100$ or $\frac{1}{9}(16x-3700),$
gaining from the remainder a third,	$\frac{1}{9}(16x-3700)+\frac{1}{27}(16x-3700)$ or $\frac{1}{27}(64x-14800),$
and becomes twice as rich as at the start.	$\frac{1}{27}(64x-14800) = 2x.$

— in the tenth, if AD, BE, DE, N and EC be named respectively a, b, c, n and x, the solution proceeds this way:[130] there will be

$$DC(\text{or } DE+EC) = a+x, \quad BC^2(\text{or } BE^2+EC^2) = b^2+x^2,$$

$AC^2(\text{or } AD^2+DC^2) = 2a^2+2ax+x^2$, and so

$$n(\text{or } N = AC-BC) = \sqrt{[2a^2+2ax+x^2]}-\sqrt{[b^2+x^2]}.$$

Thus the first problem is brought down to two equations, the fifth and tenth to single ones, and a less weighty reduction is now left. Let the reader as an

---

(130) Except for an interchange in the assigned denominations of the given and unknown line-lengths, this is essentially the reduction used in geometrical 'Prob. 11' of the parent *Arithmetica* (**1**, 2, §1: Newton's page 72). In immediate sequel Newton commits the blunder of setting DC to be $a+x$, and the remaining calculation is faulted. Correctly, on making $DC = c+x$ there ensues $\sqrt{[a^2+c^2+2cx+x^2]}-\sqrt{[b^2+x^2]} = n$, whence $cx+bf = n\sqrt{[b^2+x^2]}$ and so $(n^2-c^2)x^2-2bcfx+b^2(n^2-f^2) = 0$ on taking, as before, $2bf = a^2-b^2+c^2-n^2$ (compare note (117)).

gratia cætera novem ad æquationes quàm paucissimas prima operatione similiter deducere.

Difficultas hujus Methodi consistit in electione quantitatum ex quibus cæteræ computo simplicissimo derivari possint. Ad hoc faciendum debebit Artifex problema sic mente evolvere quasi id antea solutum et confectum esset. Nullum inter quantitates notas et ignotas cogitet discrimen, sed omnes consideret tanquam si darentur nec solvendum esset problema sed post completam solutionem examinandum. Hoc modo comparando quantitates inter se et earum nexus et mutuas relationes quarum beneficio aliquæ ex alijs derivari possint sedulo evolvendo, incidet tandem in methodum aliquā resolutionis, aut forte, peritus si fuerit, in plures methodos quarum optimam eligere debebit.[131]

... ...[132]

Siquando duæ ex radicibus æquationis prodituræ cognitam aliquam habeant relationem ad invicem, quæro alias quantitates quæ eodem modo se habeant ad utramcp, quales sunt earum summa vel differentia vel rectangulum vel quadratorum summa vel differentia. Sic enim æquatio prodibit simplicior.[133] Ejus rei exemplum est problema sequens.

Prob. [*21.*][134]

Angulum rectum BCF longitudine data EF subtendere quæ ad datum punctum A ab anguli lateribus æquidistans convergat.

Evolvendo casus radicum sentio quod longitudo *EF* modis quatuor inter rectas *CE CF* duci potest qui omnes eodem calculo determinantur.[135] Locari

---

(131) The remainder of the manuscript page is left blank—for an intended future insertion by Newton?—and then, without introduction or heading of any kind, begins 'Prob. 1' of twenty problems copied word for word from the parent *Arithmetica* (see next note). In line with that text (compare **1**, 2, §1: Newton's page 67) we would have expected some such transitional phrase as 'Cujus rei exempla habentur in sequentibus problematibus quæ in pleniorem illustrationem hujus Methodi congeruntur'.

(132) In the portion (pages 69–91) of the manuscript here omitted Humphrey Newton has transcribed the text of twenty problems from the parent *Arithmetica* without introducing any variation of any kind and leaving (unfilled) blanks at appropriate places for Newton himself—as was his invariable habit—to trace in their corresponding figures. Of these, Problems 1–11 repeat previous 'arithmetical' Problems 1–3/5–12 (**1**, 2, §1: Newton's pages 42–9; the earlier Problem 4 was also transcribed in sequence and then, perhaps because of its total lack of difficulty, later cancelled); Problems 12 and 13 resite preceding 'geometrical' Problems 58 and 59 (**1**, 2, §1: pages 134–40) in logical appendix to that—now Problem 11—on which they logically depend (see *ibid.*: note (442)); Problem 14 reproduces 'arithmetical' Problem 17 from the parent text (*ibid.*: page 52); and, finally, Problems 15–20 repeat the earlier group of 'geometrical' Problems 5–10 (*ibid.*: pages 69–72). A final 'Prob. [21]', reproduced in sequel, contains several novelties, if broadly based on the earlier 'geometrical' Problem 13 (*ibid.*: pages 74–6).

exercise attempt similarly to bring the other nine down to as few equations as possible at the first (stage in) operation.

The difficulty in this method consists in the choice of the quantities from which the remaining ones are to be derivable by the simplest computation. To do this the skilled algebraist will need to roll a problem over in his mind just as though it had previously been solved and completed. He should be mindful of no distinction between known and unknown quantities, but should consider all as if they were given and as though the problem were not to be solved but to be examined after the completion of its solution. By comparing quantities one with another in this way and diligently turning over both their interconnexions and mutual relationships, with the help of which some may be derivable from others, he will at length chance upon some method of solution—or perhaps, if he be expert, upon several methods, the best of which he will need to choose.[131]

$$\ldots \qquad \ldots \qquad \ldots \qquad \ldots^{(132)}$$

Whenever two of the roots of the equation about to ensue have some known relationship to one another, I seek other quantities which are involved in the same way with each—their sum or difference, for instance, or their product, or the sum or difference of their squares. For in this way the equation will turn out to be simpler.[133] An example in point is the following problem.

Problem 21[134]

To subtend a right angle $B\widehat{C}F$ by a given length EF, so that it shall be directed through the given point A equidistant from the sides of the angle.

By mulling over the cases of roots I perceive that the length *EF* can be drawn in four ways between the straight lines *CE*, *CF*, and that all these are determined

---

(133) A first revised statement of Newton's simplifying 'rule of mates' (see **1**, 2, §1: note (245)), which he forthwith proceeds, as before, to exemplify in the case of the simple Apollonian verging problem first noted by him twenty years before in Descartes' *Geometrie* (see 1: 509, note (9)). A more elaborate discussion of the possibilities of applying this 'Regula fratrum' is found in the stray contemporary manuscript reproduced in [2] below.

(134) Compare note (132). Newton's essentially congruent draft of this reworking of 'geometrical' Problem 13 in the parent *Arithmetica* (**1**, 2, §1: Newton's pages 74–6), and of the preceding paragraph which introduces it, is contained on ff. 93b/93c of a loose insert at this point in the manuscript (compare note (2) above). There the problem was first enunciated: '*A dati quadrati ABCD angulo quovis A rectam AF ducere cujus pars EF longitudine data oppositis quadrati lateribus productis BC DC interjaceat*' (From any corner *A* of the given square *ABCD* to draw a straight line *AF* whose portion *EF*, given in length, shall lie between the extended opposing sides *BC*, *DC* of the square).

(135) Compare 1: 509–10. This observation, first made by Schooten in his *Commentarii in Librum III*, Q of Descartes' *Geometrie* (*Geometria*, ₁1649: 266–70 = ₂1659: 310–15), had been noted by Newton twenty years before (see 1: 509–10) but is lacking in the parent 'Prob. 13' of his *Arithmetica* (**1**, 2, §1: pages 74–5).

enim potest hæc longitudo vel in angulo *BCF* ad *EF*, vel in angulo opposito *DCe* ad *fe*, vel deniҩ in angulo *BCD* ad $^2F^2E$ et $^2f^2e$. Adeo ut si verbi gratia ad determinandum positionem *EF* quæratur longitudo *BE* obvenerit æquatio

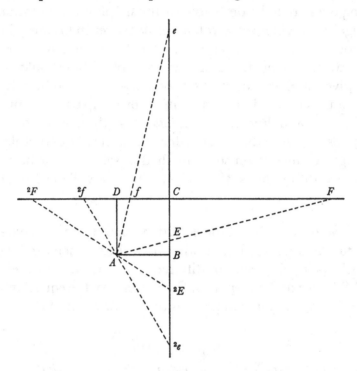

dimensionum quatuor cum radicibus totidem *BE*, *Be*, B^2E, B^2e. Animadverto præterea quod *AB* sit medium proportionale inter binas radices *BE Be* ut et inter binas B^2E B^2e.[136] Et hac de causa radices hasce prætermitto et quæro binarum puta *BE* et *Be* semisummam et semidifferentiam. Sint hæ *z* et *x*, sitҩ *AB* = *a* et *EF* = *c* et erit *BE* = *z* − *x* et *Be* = *z* + *x* et harum rectangulum *zz* − *xx* = *aa*. Unde *zz* = *aa* + *xx*. Erit etiam *CE* = *a* − *z* + *x*, et *CF* (= *Ce*) = *z* + *x* − *a*,[137] et harum quadrata simul sumpta 2*aa* + 2*zz* − 4*az* + 2*xx* = *cc*. Ubi si pro 2*xx* scribatur 2*zz* − 2*aa* fiet 4*zz* − 4*az* = *cc*. Inde datur *z*,[138] & addendo utrinҩ 4*az* fit

$$cc + 4az = 4zz = 4aa + 4xx \quad \text{et} \quad \tfrac{1}{4}cc + az - aa = xx.$$

Inventis autem *x* et *z*, habentur radices *BE* = *z* − *x* & *Be* = *z* + *x*.[139]

---

(136) Since the pairs of triangles *Abe*, *EBA* and *AB* ^2e, ^2EBA are similar.

(137) In his preliminary draft (see note (134))—here initially copied and then subsequently altered to its correct final form—Newton first wrote '*a* + *z* + *x*' by a slip, thereafter deducing that '& harum quadrata simul sumpta 2*aa* + 2*zz* + 4*ax* + 2*xx* = *cc*. Ubi si pro 2*zz* scribatur 2*aa* + 2*xx* fiet 4*aa* + 4*ax* + 4*xx* = *cc*. Per hanc æquationem inveniri potest *x*[,] deinde per hanc

by the same computation.[135] To be sure, this length can be located either in the angle $\widehat{BCF}$ at *EF*, or in the opposite angle $\widehat{DCe}$ at *fe*, or finally in the angle $\widehat{BCD}$ at $^2F^2E$ or $^2f^2e$. So much so that if, for example, to determine the position *EF* the length of *BE* be sought, there will come out an equation of four dimensions having an equal number of roots *BE*, *Be*, B^2E and B^2e. I notice, furthermore, that *AB* is the mean proportional between the pair of roots *BE*, *Be* and also between the pair B^2E, B^2e.[136] And for this reason I pass these roots by and seek, say, the half-sum and half-difference of the pair *BE*, *Be*. Let these be *z* and *x*, and let $AB = a$ and $EF = c$; then will there be $BE = z-x$ and $Be = z+x$, and the product of these $z^2 - x^2 = a^2$. Hence $z^2 = a^2 + x^2$. There will also be $CE = a - z + x$ and $CF(= Ce) = z + x - a$,[137] and the squares of these taken together $2a^2 + 2z^2 - 4az + 2x^2 = c^2$. Here, if in place of $2x^2$ there be written $2z^2 - 2a^2$, there will come $4z^2 - 4az = c^2$: from this *z* is given,[138] and by adding $4az$ to each side there comes to be $c^2 + 4az = 4z^2 = 4a^2 + 4x^2$ and so

$$x^2 = az - a^2 + \tfrac{1}{4}c^2.$$

Once *x* and *z* are found, however, the roots $BE = z - x$ and $Be = z + x$ are had.[139]

$\tfrac{1}{4}cc - ax = aa + xx = zz$ habebitur *z* ac deniꝗ ponend[o] $z - x = BE$ et $z + x = Be$ habebuntur quantitates quæsitæ' (and the squares of these taken together $2a^2 + 2z^2 + 4ax + 2x^2 = c^2$. Here if in place of $2z^2$ there be written $2a^2 + 2x^2$, there will come $4a^2 + 4ax + 4x^2 = c^2$. By means of this equation *x* can be found, and then by means of this, $\tfrac{1}{4}c^2 - ax = a^2 + x^2 = z^2$, there will be had *z*, while finally by setting $z - x = BE$ and $z + x = Be$ the quantities required will be obtained).

(138) Namely, $z = \tfrac{1}{2}(a \pm \sqrt{[a^2 + c^2]})$.

(139) It follows that $x^2 = \tfrac{1}{2}a(-a \pm \sqrt{[a^2 + c^2]}) + \tfrac{1}{4}c^2$, whence *BE*, *Be* have the respective values $\tfrac{1}{2}a \pm \tfrac{1}{2}\sqrt{[a^2 + c^2]} \mp \sqrt{(-\tfrac{1}{2}a^2 + \tfrac{1}{4}c^2 \pm \tfrac{1}{2}a\sqrt{[a^2 + c^2]})}$ found by Descartes in his *Geometrie*, ₁1637: 387–8 (= *Geometria*, ₂1659: 82–4), and the corresponding lengths of *AE*, *Ae* are

$$-\tfrac{1}{2}c \pm \sqrt{(a^2 + \tfrac{1}{4}c^2 \pm a\sqrt{[a^2 + c^2]})}$$

(compare II: 436, note (132) and page 75 of the parent *Arithmetica*). In his preliminary draft (see note (134)) Newton began a new paragraph in sequel 'At si' (But if...) and then broke off. What little is known—or can be conjectured—of the intended content of the unwritten sections of the present Book 1 and of the succeeding Book 2 of the present revised *Arithmetica* is set out in the introduction (pages 536–7 above); see also note (3). We here append the text of a stray contemporary manuscript dealing more fully with the 'rule of mates' instanced in the preceding problem, and which might well have been composed as its replacement or elaboration.

[2]⁽¹⁴⁰⁾

Regula Fratrum.⁽¹⁴¹⁾

In resolutione Problematum id maxime agendū est ut conclusiones evadant quam simplicissimæ. Sic enim Reductiones æquationum effugientur quibus operationes perplexæ red[d]untur et inelegantes et quæ nonnunquam molestissimæ⁽¹⁴²⁾ sunt et in æquationibus dimensionum plurium⁽¹⁴³⁾ vix aut ne vix quidem per regulas hactenus inventas institui⁽¹⁴⁴⁾ possunt. Has autem effugere soleo per sequentem Regulam.⁽¹⁴⁵⁾

Siquando duæ sint quantitates gemellæ id est quæ eodem modo se habeant ad conditiones Problematis quæq cognitam aliquam habeant relationem ad invicem, his non impono nomina sed earum loco usurpo quantitates quæ eodem modo se habent ad utramque; quales sunt earum summa vel differentia vel rectangulum vel quadratorum summa vel differentia.⁽¹⁴⁶⁾ Animum quoq maxime attendo ad rectas et circulos quæ per puncta duo transeunt similibus conditionibus prædita, nec non ad rectas quæ ab ejusmodi punctis hinc inde æqualiter distant.⁽¹⁴⁷⁾

Exempl 1.⁽¹⁴⁸⁾ *Si trianguli alicujus darentur tria quælibet ex his*[,] *angulo ad verticem, area, basi, perpendiculo in basem, perimetro, differentia angulorum ad basem, differentia vel rectangulum segmentorum basis, summa vel differentia vel rectangulum laterum, ½ differentia segmentorū anguli ad verticem, differentia inter basem et summam laterum, et inde quærerentur cætera*: pro fratribus habenda essent et latera duo, et segmenta duo basis, et

---

(140) Add. 3963.5: 40^r–41^r, a much corrected and reordered autograph draft entered on three sides of a folded folio sheet.

(141) Newton first wrote in the provisional heading 'Reductiones fugiendæ & Resolutionis elegantia' (Reductions to avoid and elegance in resolution), then tentatively replaced it by 'Resolutio simplex et elegans' (A simple and elegant resolution) before deciding on this final title (of a section to follow on [1], replacing its final 'Prob. [21]'?).

(142) 'difficillimæ' (most difficult) was first written.

(143) Originally 'plusquam quatuor' (more than four). In fact, no equation higher than a quartic is treated of—even implicitly so—in Newton's following examples and the cancelled phrase could have no significance in their context. The standard solutions of the general cubic and quartic equations are sketched in his parent *Arithmetica* (**1**, 2, §1: Newton's pages 203–7). Though he seems to have had some hope that a Tschirnhaus reduction might reduce certain classes of higher equations to ones of lower dimension (compare *ibid.*: note (591)), he doubtless suspected—he could scarcely have proved—that no algebraic solution of the general quintic (and of equations of still higher dimension) was possible.

(144) This replaces 'compleri' (completed).

(145) As the title affirms, understand 'fratrum' (of brothers). Indeed Newton first continued in immediate sequel with a repeated subhead '*Regula fratrum*', following with the definition: 'Fratres voco puncta vel lineas quæ eodem modo se habent ad conditiones problematis' (I call 'brothers' points or lines which are involved in the same manner in the conditions of the problem). He then decided—not quite logically?—to make the brothers twin!

(146) An original further alternative 'vel expositio rectæ aut circuli qui per earum correspondentes terminos transeat' (or ones specifying a straight line or circle which shall pass

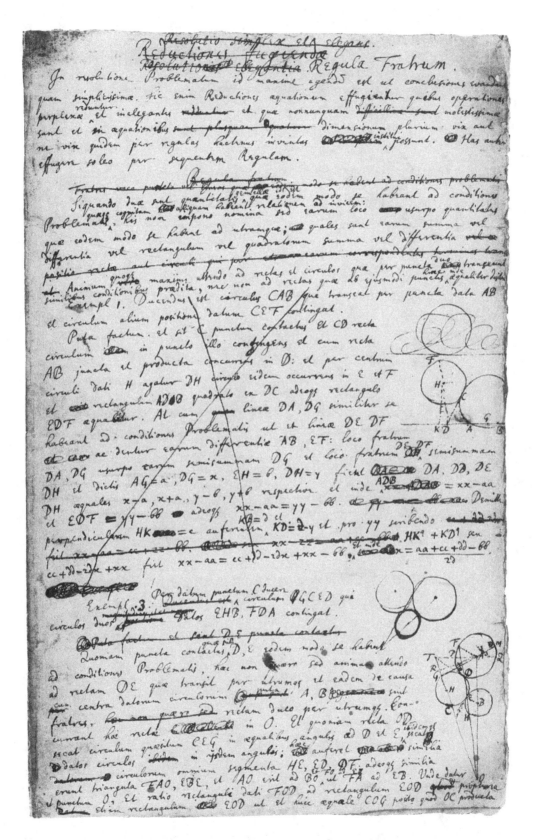

Plate IV. Opening page of the 'Regula Fratrum' (**2**, [2]).

[2][(140)]

THE RULE OF BROTHERS[(141)]

In the resolution of problems the action most needed is to contrive that the conclusions turn out as simple as possible. For in this way reductions of equations will be evaded in which the operations are rendered intricate and graceless, ones which not infrequently are most troublesome[(142)] and which in equations of several[(143)] dimensions can scarcely ever—or not even that—be achieved[(144)] by the rules so far found. These I usually escape, however, by means of the following rule.[(145)]

Whenever there are two twin quantities—ones, that is, which are involved in the same way in the conditions of the problem and have some known relationship to each other—, I do not set names on these, but in their place I employ quantities which are related in the same way to each: their sum or difference, for instance, or the sum or difference of their squares.[(146)] I also pay very particular attention to straight lines and circles passing through two points possessed of similar conditions, and in addition to straight lines equally distant from points of this sort on their either side.[(147)]

EXAMPLE 1.[(148)] *If in some triangle there were to be given any three of these: the angle at the vertex, the area, base, the perpendicular onto the base, the perimeter, the difference of the angles at the base, the difference or product of the base segments, the sum or difference or product of the (inclined) sides, the half-difference of the segments of the angle at the vertex, or the difference between the base and sum of the sides, and from them the rest were required,* you would need to reckon as brothers both the two sides, and the two base

---

through their corresponding end-points) was subsequently replaced by the following sentence (a late addition in the manuscript, squashed in between here and the next paragraph).

(147) As in his earlier enunciation of this rule for identifying 'twin' quantities in a problem, whose coupling will, by appropriate choice of a variable related 'in the same way' to each, greatly simplify and shorten its reduction, Newton again lays himself open to Lagrange's charge that 'un tel choix de l'inconnue est assés peu naturel, &...ce n'est...qu'après coup qu'on peut le faire' (1, 2, §1: note (245)). The answer Newton would surely return is that an experienced mathematician very often does have a foreknowledge of the type of solution required by a particular problem presented to him, but without knowing its exact quantity, and that he will employ such simplifying procedures as that now codified in the 'Regula fratrum' to lessen—in advance—the subsequent labour of computing the intricacies of the exact solution. Lagrange's impatience—that of a man scornful of all but the most general and powerful of mathematical techniques—is understandable, but at the same time it has blinded him a little to the validity (as distinct from the wide utility) of Newton's rule. In any event, the following worked examples (here reordered from their manuscript sequence to accord with Newton's final numbering of them) should go some way towards providing 'l'évidence que l'on est en droit d'exiger dans ces sortes de matières' which Lagrange went on to insist was required.

(148) A late addition on the manuscript's final page (f. 41ʳ). Newton's evident hurry will explain its baldness and offhand presentation.

anguli duo ad basem, et segmenta duo anguli ad verticem[,] ideoꝗ horum nullum directè quærerem sed si binorum daretur ½ summa quære[re]m ½ differentiam, si daretur ½ differentia quærerem ½ summa, si neutra daretur quærerem utramꝗ si modò calculus sic facile proceder[e]t, vel quærerem tertium aliquod quod eodem modo se haberet ad utrumꝗ. Calculum in tot casibus tentet qui volet.[149]

EXEMPL. 2.[150] *Ab angulo A quadrati dati ABCD rectam AF ducere cujus pars longitudine data EF angulum externum oppositorum laterum BCF subtendet.*

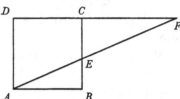

Hic longitudines *AE*, *AF* eodem modo se habent ad conditiones Problematis ideoꝗ pro gemellis habendæ sunt. Habent autem datam differentiam *EF* ideoꝗ Problema resolvi posset quærendo earum summam. Sic etiam *BE* et *DF* gemellæ sunt et habent datum medium proportionalem *AB*.[151] Sed et *CE*, *CF* quoꝗ gemellæ sunt et datam habent summam quadratorum,[152] suntꝗ ad invicem ut *AD* ad *DC*+*CF*.[153] Sint harum semisumma *z* & semidifferentia *x* sitꝗ *AD*=*a* et *EF*=*b* et erit *z*+*x*=*CF*, *z*−*x*=*CE*. $2zz+2xx=CF^q+CE^q=EF^q=bb$ et *a*+*z*+*x*=*DF* adeoꝗ *a* ad *a*+*z*+*x* ut *z*−*x* ad *z*+*x* et multiplicatis extremis et medijs *az*+*ax*=*az*−*ax*+*zz*−*xx* seu *xx*+2*ax*=*zz*. Unde æquatio superior tollendo 2*zz* fit 4*xx*+4*ax*=*bb*. Unde ¼*bb*=*ax*+*xx* et ¼*bb*+*ax*(=*xx*+2*ax*)=*zz*. Per penultimam æquationem datur *x* ac deinde per ultimam datur *z*.[154]

EXEMPL 1. *Ducendus est circulus CAB qui transeat per puncta data A B et circulum alium positione datum CEF contingat.*[155]

Puta factum et sit *C* punctum contactus et *CD* recta circulum in puncto illo contingens et cum recta *AB* juncta et producta concurrens in *D*: et per centrum circuli dati *H* agatur *DH* circulo eidem occurrens in *E* et *F* et rectangulum *ADB*

---

(149) A typical Newtonian refusal to be bothered with niceties. Geometrical instances of such coupled 'fratres' will be found in the constructions outlined, for example, on IV: 246–54.

(150) Originally '3' (when this example was set, on f. 40ᵛ, after the two following problems on Apollonian circle tangencies). The enunciation of this classical verging problem, one which has already appeared several times in Newton's mathematical papers (and, in particular, on Newton's pages 74–5 of the parent *Arithmetica*) is very close to that in Newton's draft of 'Prob. [21]' in [1]; see note (134) above.

(151) At once, since the triangles *BEA* and *DAF* are similar, *BE*:*AB* = *AD* (or *AB*):*DF* and therefore *BE*×*DF* = *AB*². From a more sophisticated viewpoint the constancy of the product *BE*×*DF* is an immediate corollary of the points *E*, *F* being in perspective through *A* and hence in 1, 1 correspondence (see IV: 243, note (36)). For a solution to the present verging problem using this approach compare I: 511, note (13). Newton first continued in sequel: 'Et idem dicendum est de [*CE* et *CF*]' (And the same[!] is to be said of *CE* and *CF*).

(152) Namely, the square of the corresponding hypotenuse *EF*.

(153) A cancelled first continuation reads: 'Quinetiam ducendo quadrati diagonalem *AC* et ad hanc a punctis *E*, *F* demittendo perpendiculares exhiberi possent gemellæ plures' (And

segments, and the two angles at the base, and the two segments of the angle at the vertex; on that account I would look for none of these directly but, if the half-sum of a pair were given, I would seek their half-difference and, if their half-difference were given, I would seek their half-sum, and if neither were given I would seek both, provided the computation were readily to proceed in this way, or I would seek some third magnitude which would be related in the same manner to both. Let him who may wish to do so try the computation in the several corresponding cases.[149]

EXAMPLE 2.[150] *From the corner A of the given square ABCD to draw a straight line AF such that its portion EF given in length shall subtend the external angle* $\widehat{BCF}$ *of the opposite sides.*

Here the lengths AE, AF are involved in the same way in the conditions of the problem, and accordingly they are to be reckoned as twin. They have, however, the given difference EF and consequently the problem might be resolved by seeking their sum. So also BE and DF are twin, having the given mean proportional AB.[151] Yet again, CE and CF are also twin, having the sum of their squares[152] given, while they are to one another as AD to $DC+CF$.[153] Let the half-sum of these be z and their half-difference x, and let $AD = a$ and $EF = b$; there will then be $CF = z+x$, $CE = z-x$ and so

$$2z^2+2x^2 = (CF^2+CE^2 = EF^2\,\text{or})\,b^2,$$

while $DF = a+z+x$ and hence a is to $a+z+x$ as $z-x$ to $z+x$, and so, on multiplying extremes and middles, $az+ax = az-ax+z^2-x^2$ or $x^2+2ax = z^2$. Whence, on eliminating $2z^2$, the previous equation becomes $4x^2+4ax = b^2$, and therefrom $x^2+ax = \frac{1}{4}b^2$, and so $z^2 = (x^2+2ax =)\,ax+\frac{1}{4}b^2$. By the last equation but one x is given and subsequently z is given by the last.[154]

EXAMPLE 1. *A circle CAB is to be drawn which shall pass through the given points A, B and touch another circle CEF given in position.*[155]

Suppose it done. Let C be the point of contact and CD the straight line touching the circle at that point and meeting the extension of the line joining A and B in D; then through the centre H of the given circle draw DH intersecting this same circle in E and F, and the product $DA \times DB$ will be equal to the square

---

to be sure, by drawing the square's diagonal AC and letting fall perpendiculars to this from the points E, F further twins might be displayed)—such as the perpendiculars themselves and the distances of their feet from A.

(154) Much as before (compare note (139)), it follows that $x = -\frac{1}{2}a \pm \frac{1}{2}\sqrt{[a^2+b^2]}$ and so $z = \pm\sqrt{(-\frac{1}{2}a^2+\frac{1}{4}b^2\pm\frac{1}{2}a\sqrt{[a^2+b^2]})}$.

(155) Originally Newton's opening example, this is a variant algebraic reduction—widely different in detail, if not essential geometrical structure, from that of 'Prob 39' on pages 101–2 of the parent *Arithmetica*—more akin to the synthetic revise of the latter set out as 'Prob 3' of his contemporary 'Quæstionum solutio Geometrica' (IV: 256; compare 1, 2, §1: note (327) above).

quadrato ex DC adeoqꝫ rectangulo EDF æquabitur. At cum lineæ DA, $D[B]$ similiter se habeant ad conditiones Problematis ut et lineæ DE DF ac dentur earum differentiæ AB, EF: loco fratrum DA, $D[B]$ usurpo earum semisummam DG et loco fratrum DE, DF semisummam DH et dictis $AG=a$, $DG=x$, $EH=b$, $DH=y$ fient DA, DB, DE, $D[F]$ æquales $x-a$, $x+a$, $y-b$, $y+b$ respective et inde $ADB=xx-aa$ et $EDF=yy-bb$ adeoqꝫ $xx-aa=yy-bb$.[156]

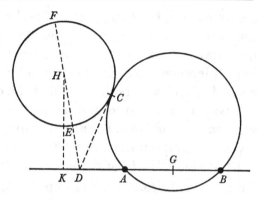

Demitte perpendiculum $HK=c$ auferentem $K[G]=d$ et $KD=d-[x]$[157] et pro yy scribendo HK^q+KD^q seu $cc+dd-2dx+xx$ fiet $xx-aa=cc+dd-2dx+xx-bb$. Et inde $x=\dfrac{aa+cc+dd-bb}{2d}$.[158]

EXEMPL. 3.[159] *Per datum punctum C ducere circulum $GCED$ qui circulos duos datos*[160] *EHB, FDA contingat.*

[161]Quoniam puncta contactus quæ sint D, E eodem modo se habent ad conditiones Problematis, hæc non quæro sed animum attendo ad rectam DE quæ transit per utrumqꝫ et eadem de causa cum centra datorum circulorum puta A, B sint fratres,[162] rectam duco per utrumqꝫ. Concurrant hæ rectæ in O.[163] Et quoniam recta OD secat circulum quæsitum CEG in æqualibus angulis ad D et E, ibidemqꝫ secat datos circulos in ijsdem angulis: hæc auferet similia circulorum omnium segmenta HE, ED, DF, adeoqꝫ similia erunt triangula FAO, $EB[O]$, et AO erit ad BO et FO ad EO ut FA ad EB. Unde datur punctum

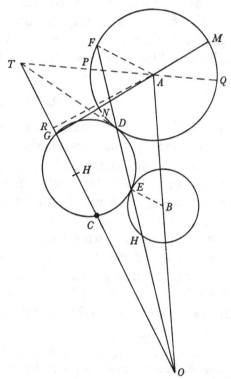

---

(156) '*et* $yy-xx=bb-aa$' (and so $y^2-x^2 = b^2-a^2$) is deleted.

(157) By a slip Newton first wrote '$KB=d$' and '$KD=d-y$'. A brief abortive sequel, immediately cancelled, is not reproduced.

(158) This will fix the point D itself uniquely, but two tangents DC, DC' (not shown) can be drawn from D to the given circle and hence two circles (through A, B and one of C, C') can be constructed to touch it. The present 'Exempl. 1' was cancelled when, in reordering these

of DC and hence to the product $DE \times DF$. Since, however, the lines DA, DB and also the lines DE, DF are similarly involved in the conditions of the problem, and because their differences AB, EF are given, in place of the brothers DA, DB I employ their half-sum DG and in place of the brothers DE, DF their half-sum DH; and then, on calling $AG = a$, $DG = x$, $EH = b$ and $DH = y$, DA, DB, DE, DF will become equal to $x-a$, $x+a$, $y-b$, $y+b$ respectively, and thereby $DA \times DB = x^2 - a^2$ and $DE \times DF = y^2 - b^2$, and consequently

$$x^2 - a^2 = y^2 - b^2.^{(156)}$$

Let fall the perpendicular $HK = c$ cutting off $KG = d$ and so $KD = d-x$,[157] and on writing $HK^2 + KD^2$, that is, $c^2 + d^2 - 2dx + x^2$, in place of y^2 there will come to be $x^2 - a^2 = c^2 + d^2 - 2dx + x^2 - b^2$, and thence

$$x = (a^2 + c^2 + d^2 - b^2)/2d.^{(158)}$$

EXAMPLE 3.[159] *Through the given point C to draw a circle GCED which shall touch two circles EHB, FDA given.*[160]

[161]Seeing that the points of contact—let these be D and E—are related in the same manner to the conditions of the problem, I do not seek these but turn my attention to the straight line DE passing through both; and for the same reason, since the centres of the given circles—A and B, say—are brothers,[162] I draw a straight line through both. Let these straight lines meet in O.[163] And seeing that the straight line OD intersects the required circle CEG in equal angles at D and E, and also at those points intersects the given circles in the same angles, it will cut off similar segments HE, ED, DF in all the circles; hence the triangles FAO, EBO will be similar, and so AO will be to BO and FO to EO as FA to EB. Hence

---

illustrative problems (see note (147)), Newton absorbed its text into the following more general problem.

(159) Originally '2' (when initially set as a companion to the—then uncancelled—'Exempl 1') and thereafter '1' (when this predecessor was absorbed into it), the final number was allotted when (compare note (147)) the manuscript was later reordered in its sequence. Newton first began the following enunciation: 'Ducendus est circulus $GCED$...' (A circle $GCED$ must be drawn...).

(160) 'magnitudine et positione' (in magnitude and position) is deleted. The divergence of the following construction from that of the equivalent (synthetically restyled) 'Prob 41' on pages 105–6 of the parent *Arithmetica* will be evident; compare our restoration of its preliminary algebraic analysis (1, 2, §1: note (341)).

(161) Newton first began with a copy of his opening to the preceding (cancelled) 'Exempl 1': 'Puta factum et sint D, E puncta contactus' (Suppose it done. Let D, E be the points of contact...).

(162) This replaces 'germana' (full brothers). The psychologically orientated biographer of Newton will make something of his reluctance to admit brothers german—as distinct from maternal brethren?—into his argument. In sequel Newton has deleted 'hæc non quæro sed' (these I do not seek out, but).

(163) Namely (as the figure is drawn) the external centre of similitude of the two given circles.

O, et ratio rectanguli dati *FOD* ad rectangulum *EOD* & propterea etiam rectangulum *EOD* ut et huic æquale *COG* posito quod *OC* producta incidat in circulum quæsitum ad *G*. Applicetur hoc rectangulū ad ipsius latus datum *OC* et dabitur latus alterum *OG*.[164] Et sic Problema eò reducitur ut per data duo puncta *C*, *G* circulum *DGCE* ducamus qui tangat alterutrum datorum circulorum puta circulum *D*[*F*]*A*. Nam si tangat alterutrum tanget etiam alterum.

Si circuli *DGCE* et *D*[*F*]*A* se mutuò secarent, animum adverterē ad rectam quæ per eorum intersectiones transiret et eadem de causa animum jam attendo ad rectam quæ tangit utrumcȝ in puncto contactus. Nam recta per intersectiones transiens ubi intersectiones illæ coeunt[165] fit tangens circulorum. Duco igitur circulorum tangentem communem *DT* occurrentē productæ *OC* in *T*, et acta *TA* quæ circulo dato occurrat in punctis *P* et *Q*, rectangula *GTC* *PTQ* quadrato ex *TD* atcȝ adeo sibi ipsis æqualia erunt. Jam cum rectæ *TG*, *TC* ut et rectæ *TP*, *TQ* gemellæ sint, datascȝ habeant differentias *GC*, *PQ*, nomina non istis sed istarum semisummis impono. Sint igitur fratrum *TG*, *TC* semisumma *TH* = *x*, semidifferentia *GH* = *a* et fratrum *TP*, *TQ* semisumma *TA* = *y*, semidifferentia *AP* = *b*, et ipsæ *TG*, *TC*, *TP*, *TQ* erunt *x*−*a*, *x*+*a*, *y*−*b*, *y*+*b* respectivè et æqualia ipsarum rectangula *GTC* & *PTQ* erunt *xx*−*aa* et *yy*−*bb*. Ad *OT* demitte perpendicularem *AR* = *c* auferentem *RH* = *d*, et pro *yy* scribendo *AR^q*+*RT^q* seu *cc*+*xx*−2*dx*+*dd* fiet *xx*−*aa* = *cc*+*xx*−2*dx*+*dd*−*bb* seu $x = \dfrac{aa + cc + dd - bb}{2d}$. [166] Unde facile colligitur constructio talis. Age *GA* circulo dato occurrentem in *M* et *N* et erit *GT* ad *GN* ut *GM* ad 2*RH*.[167]

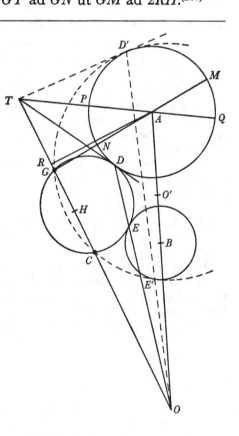

(164) And hence also the point *G*, the correlate of *C* in the inversion (of power √[*AO* × *BO*]) which transforms the given circles, of centres *A* and *B*, into each other and in which the point-pairs *D*, *E* and *F*, *H* correspond.

(165) 'et fiunt punctum contactus' (and become a point of contact) is cancelled.

(166) This analysis absorbs that of the preceding (cancelled) 'Exempl 1'; compare note (158).

(167) For we readily deduce that

$$2d(x-a) = (d-a)^2 + c^2 - b^2,$$

whence

$$2RH \times GT = (GR^2 + AR^2 \text{ or}) \ GA^2 - AP^2,$$

that is, *GT*/(*GA*−*AP*) = (*GA*+*AP*)/2*RH*. It will be clear that the second tangent *TD'* from *T* to the circle of centre *A* will define a second circle (through *D'*, *G* and *C*) touching the given circles in *D'* and *E'* respectively, where *E'* is the mate of

the point O is given, and so is the ratio of the given product $OF \times OD$ to the product $OE \times OD$, and accordingly also the product $OE \times OD$ and its equal $OC \times OG$, supposing that OC produced intersects the required circle at G. Divide this product by its given 'side' (member) OC and there will be given its other 'side' OG.[164] And in this way the problem is reduced to drawing through the two given points C and G a circle $DGCE$ which shall touch one or other of the given circles, say the circle DFA. For, should it touch one, it will also touch the other.

Were the circles $DGCE$ and DFA mutually to intersect, I would turn my attention to the straight line which would pass through their intersections, and on the same ground I now pay heed to the straight line which touches both at a contact point: for the straight line passing through their intersections becomes, when those intersections coincide,[165] a tangent to the circles. I therefore draw the circle's common tangent DT, meeting OC produced in T, and, when TA is drawn so as to meet the given circle in the points P and Q, the products $TG \times TC$ and $TP \times TQ$ will be equal to the square of TD and hence to each other. Now since the straight lines TG, TC and also the lines TP, TQ are twin, having given differences GC and PQ, I set names not upon these but upon their half-sums. Let, therefore, the half-sum TH of the brothers TG, TC be $= x$, their half-difference $GH = a$, and the half-sum TA of the brothers TP, TQ be $= y$, their half-difference $AP = b$, and then TG, TC, TP, TQ themselves will be $x-a$, $x+a$, $y-b$, $y+b$ respectively and their equal products $TG \times TC$ and $TP \times TQ$ will be x^2-a^2 and y^2-b^2. To OT let fall the perpendicular $AR = c$ cutting off $RH = d$, and on writing $AR^2 + RT^2$, that is, $c^2 + x^2 - 2dx + d^2$, in place of y^2 there will come to be $x^2 - a^2 = c^2 + x^2 - 2dx + d^2 - b^2$, or $x = (a^2 + c^2 + d^2 - b^2)/2d$.[166] It is hence easy to collect the following construction: draw GA meeting the given circle in M and N, and then GT will be to GN as GM to $2RH$.[167]

D' in the inversion of pole O and power $\sqrt{[AO \times BO]}$. Since a second pair of solutions derive analogously from the internal centre O' of similitude of the two given circles (by constructing the two circles which touch the circle of centre A and also pass through G while C' is in GO' such that $AO'/O'B = GO'/O'C'$), the original problem will have four solutions altogether.

Though we have not distinguished these in our translation, the mid-point H of the chord CG of the circle requiring to be constructed should not be confused with the identically marked end-point of the segment HE in the circle of centre B.

In resolutione Problematis hujus potui quoqȝ animum attendisse ad rectam quæ in medio consistit inter puncta gemella G, C et ad rectam GC normalis est. In ea enim locatur centrum circuli quæsiti. Et sic Problema eo recidit, ut in recta illa inveniatur punctum a quo si rectæ ad puncta data A et G ducantur, hæ datam habeant differentiam AF.[168] Nam punctum illud erit centrum circuli quæsiti. Pro gemellis vero nunc habenda sunt puncta A et G.[169]

---

(168) Namely the radius of the (given) circle of centre A. The present reduction is that used in 'Prob 39' of the parent *Arithmetica* (1, 2, §1: Newton's pages 101–2).

(169) A cancelled concluding phrase reads 'ut et rectæ duæ ab ijsdem vel ad centrum quæsitum ductæ vel ad rectā in qua centrum illud jacet perpendiculariter demissæ, nec non segmenta rectæ hujus inter centrum illud et perpendicula jacentia' (as also the two straight lines let fall from these either to the required centre or perpendicular to the straight line in which that centre lies, and again the segments of this line lying between that centre and [the feet of] these perpendiculars).

In the resolution of this problem I could also have given my attention to the straight line standing midway between the twin points G and C at right angles to the line GC: for in it is located the centre of the required circle. And thus the problem reduces to finding a point in that line such that, if straight lines be drawn from it to the given points A and G, these shall have the given difference AF:[168] for that point will be the centre of the required circle. The points A and G are now, of course, to be considered as twin.[169]

APPENDIX. THE GRADING OF
GEOMETRICAL PROBLEMS[1]

[1684?]

Extracts from a stray autograph[2] in the University Library, Cambridge

[1]

[3]Ubi verò Problema aliquod sic ad hujusmodi[4] quantitatum æquationem devenit[5] & æquatio illa ad formam simplicissimā & commodissimam reducitur, potest gradus problematis exinde dignosci. Nam si nullū est in æqu. latus surdum, index maximæ dimensionis quæsitæ quantitatis est etiam index gradus problematis. Sic in æquatione superiore[6] numerus 3 est index utriusᴄᴣ. At si aliquod ad hujus conclusionem deducitur quæ a quantitatū æqualitate dicitur æquatio[7]

(1) In marked contrast to his critical remarks in the parent *Arithmetica* regarding the validity of classifying problems in this way (see **1**, **2**, §1: Newton's pages 211–14), Newton here introduces the naïve notion that all problems reducible to algebraic equations are to be graded—by implication, effectively so?—according to the degree of those equations. No mention is here made of the inconsistencies which result from taking the relative ease and 'simplicity' of their construction into consideration, and the mood is far removed from any iconoclastic assertion that 'Ad simplicitatem constructionis expressiones Algebraicæ nil conferunt' (*ibid.*: Newton's page 213). Leaning perhaps over-heavily on the internal evidence of its text, we conjecture (compare **2**, [1]: note (54)) that its substance was intended by Newton to be inserted at the end of the introductory section on the 'Æquatio et ejus radices' in his revised *Arithmetica*.

(2) Add. 3964.8: 22ʳ/22ᵛ. The preceding date is conjectured on the basis (see note (1)) that the present roughly written and extensively corrected draft texts are narrowly connected with the allied portion of the 'Arithmeticæ Universalis Liber Primus', assigned above (**2**, [1]: note (1)) to the spring of 1684, but it is more generally supported by an examination of Newton's handwriting. The original use of the sheet whose mathematical content is here reproduced was the more mundane (but, to Newton, equally important) one of penning a sternly phrased letter to an unidentified correspondent whose uncommunicative brother had fallen behind in payment of money to Newton: ''Tis now a month since yoʳ last, & I have not yet heard from yoʳ Brother. I question not but you have seen him by this time & therefore... I beg yʳ favour of a line or two from you to let me know how things stand now & when I may expect to have some money & yᵉ security you told me of. The least troublesome way for me will be to take bond, but yᵉ bondsmen I desire to know, & in what readiness things are.' We readily surmise that the intended recipient was a Lincolnshire acquaintance—perhaps Newton's one-time schoolmate Arthur Storer, whose brother Oliver (together with their father) held the tenancy of Newton's Woolsthorpe farm, allowing its condition to fall badly into decay; compare Newton's letter of 11 January 1687/8 to an 'arbitrator' of their subsequent dispute (*Correspondence of Isaac Newton*, **2**, 1960: 502–4).

(3) Newton first began 'Et simili computo'.

(4) Understand the general 'formulæ' in the preceding revised *Arithmetica* (page 560 above), in which the equation is listed in the decreasing dimensional order of its terms with the

[2]

Ad hujusmodi conclusiones quæ[8] æquationes dicuntur, solent hodierni Geometræ deducere omnia Problemata.[9] Et postquam æquatio ad formam simplicissimam reducitur maxima quæsitæ quantitatis dimensio gradum Problematis indicare solet:[10] Ut hic dimensio triplex ipsius x^3 gradum tertium. Verum tamen[11] si in æquatio[ne] reperiatur latus aliquod surdum quadraticum debet gradus ille duplicari, si cubicum triplicari[,] si biquadraticū vel duo quadratica quadruplicari & sic in reliquis.[12] Sic Problema quarti est gradus quod ad æquationem $xx = ax - \overline{a^3b}|^{\frac{1}{2}}$ deducitur. sexti quod ad æquationem $xx = x, \; \overline{a^3 - b^3}^{\frac{1}{3}} + ab$. octavi quod ad æquationem $xx = x, \; \overline{ab}|^{\frac{1}{2}} - \overline{a^4 + b^4}|^{\frac{1}{2}}$.[13] Sed hisce una cum exceptionibus quibusdam explicandis non immoror.

Verū[14] ubi latera aliqua ex quibus in se ductis æquatio generatur[15] possunt

---

coefficient of the highest term unity. Newton has in sequel deleted the amplification 'conclusionem ubi aliquæ quantitates sunt æquales alijs seu'.

(5) 'deducitur' was first written.

(6) This otherwise unidentified cubic may well have been one or other of those,

$$x^3 - x^2 - 14x + 24 = 0 \quad \text{and} \quad x^3 - 3x^2 - 9x + b = 0,$$

instanced in the related section on the 'Æquatio et ejus radices' in the revised *Arithmetica* (page 562 above).

(7) The draft breaks off in mid-sentence and is followed immediately by its revision.

(8) Newton began to write 'a qua[n]titatum [æqualitate?]'.

(9) In sequel is cancelled 'Dein quomodo quæsitæ quantitates ex his eliciantur docent. Verum hoc Genus' and then 'Et æquationum dimensionibus æstima[nt gradus Problematum æquales?]'. Compare 1, 2, §1: note (620).

(10) Newton first wrote more vaguely '...potest gradus problematis ex ejus dimensionibus colligi'.

(11) A first continuation here reads 'debet index maximæ illius dimensionis multiplicari per denominatorem indicis illius la[teris] si latus quadraticum est'.

(12) In a first exemplification Newton wrote in sequel: 'Ut si æquatio sit $xx = ax - \overline{a^3b}|^{\frac{1}{2}}$ propter surdum latus quadrat[ic]um $\overline{a^{[3]}b}|^{\frac{1}{2}}$ duplico dimensiones duas quæsitæ quantitatis, dicoq̃ problema quod ad hanc æquationē deducitur quarti gradus esse. Nam pone $\overline{a^3b}|^{\frac{1}{2}} = ey$ seu $a^3b = eeyy$ et æquatio illa resolvitur in duas duarum dimensionū $xx = ax - ey$ et $\dfrac{a^3b}{ee} = yy$.

Posterior dabit duplex y. Potest enim latus quadraticum pro arbitrio affirmativè vel negativè sumi et sic $\overline{a^{[3]}b}|^{\frac{1}{2}}$ hic affirmativ[us vel negativus est].'

(13) A further aborted illustration reads at this point:

'At æquatio $\quad x = \overline{a^3 + ab^5\frac{1}{2}}|^{\frac{1}{3}} + \overline{a^3 - ab^5\frac{1}{2}}|^{\frac{1}{3}} \quad$ cujus rei rationes satis obviæ sunt.'

The gap should be filled by 'noni' since, on setting $ab^5 = y^6$ and cubing, there results $x^3 = 2a^3 + 3x\sqrt[3]{[(a^3 + y^3)(a^3 - y^3)]}$, whence

$$(x^3 - 2a^3)^3 - 27x^3(a^6 - y^6) = 0.$$

(14) 'tamen' was added in preliminary drafts which ineffectually continued first with 'quoties æquationes' and then 'ubi æquatio si ex lateribus simpliciorum graduū'.

(15) 'ita inter se conjungi solent ut' is cancelled.

hoc modo seorsim in terminis Analyticis exprimi, debet æquatio in eadem resolvi. Et latus illud solum retineri in quod responsum desideratum continetur atcʒ gradus problematis ex illius gradu æstimari. Ut si in aliquo Problemate incidissem in æquationem $x^3 {-ab \atop -aa} x + aab = 0$ resolverem hanc in latera

$$xx + ax - ab = 0 \quad \text{et} \quad x - a = 0.$$

quorum prius secundi est gradus, posterius primi. Et perinde ut responsum desideratū in hoc vel illo latere continetur problema primi vel secundi gradus esse ducerem. Et quomodo alia æquationū latera inveniantur docent[16] Algebræ scriptores.

[17]Verum antequam gradus Problematis ex æquatione dignosci potest, reduci debet æquatio ad formam simplicissimā et quomodo tales reductiones perficiantur docent Algebræ scriptores. Id tantum noto quod ubi æquatio reducitur ad in[18]

---

(16) This replaces the less cautious 'fuse docuerunt', an assertion which is (compare [1]: note (30)) decidedly untrue in the case of writers on algebra before Newton himself.

(17) Newton first began this unfinished (final?) paragraph with 'Habita simplicissima æquatione ad quam Problema reduci potest si ea secundi est [gradus, est etiam problema secundi?]', and then continued with an equally abortive 'Reductionem æqu[ationum docent Algebræ scriptores?]'.

(18) The manuscript breaks off in mid-word. We may perhaps restore the intended conclusion to Newton's sentence as '...ad in[dicatam formam, maxima dimensio quæsitæ quantitatis est index gradus problematis]'.

INDEX OF NAMES

Adam, C., 266n

Adams, J. C., 535

Al-Nasawi, Ali ibn Ahmad (Almochtasso Abdulhasan), 464n

Anderson, Alexander, 63n

Anne (Queen of England), 9n

Apollonius of Perga, 15, 214n, 268n, 314–15, 393n, 464n, 497n, 513n

Archimedes, 30, 50n, 146n, 298n, 427n, 428–9, 456n, 460n, 462n, 464–5, 470–1, 475n

Aubrey, John, 532n

Aylesbury, Thomas, 26n

Bachmakova, I. G., 387n, 408n

Baillet, Adrien, 26n

Baily, Francis, xiii, xiv, 32n

Balam, Richard, 57n

Barrow, Isaac, 4, 5n, 7, 32n, 63n, 166n, 193n, 464n

Beaudeux, Noël, 19, 403n

Beaugrand, Jean de, 465n

Bernard, John Peter, 18n

Bernoulli, Jakob, 20, 283n, 520n, 522n

Bernoulli, Johann, 13n, 15n, 16–17, 20, 23–4n, 245n, 380n, 469n, 520n, 522n

Bernoulli, Niklaus I, 16, 24–5n, 380–1n

Borelli, Giovanni Alfonso, 464n

Boscovich, Roger Joseph, 302–3n

Boulliau, Ismaël, 515n

Boyer, C. B., 246n

Cajori, F., 58n, 60n, 62–3n, 269n, 560n

Campbell, Colin, 517n

Campbell, George, 18

Castiglione, Giovanni, 18, 303n, 403n

Clavius, Christopher, 193n

Clerselier, Claude, 266n

Collins, John, 7, 22n, 96–7n, 112n, 184n, 216–17n, 279n, 411n, 474n, 564n

Colson, John, 18

Conduitt, John, xii, xiii, 9n, 14–15n

Copernicus, Nicolaus, 298–9, 509n, 514–15, 516n

Costabel, P., 418n

Cotes, Roger, 6n, 7–8, 15, 32n, 54n

Crompton, James, 524n

Cunn, Samuel, 16–17n, 22n

Décoré, G. A., 18

Dedekind, R., 387n

Delambre, J. B. J., 19, 278n

Descartes, René, 6n, 26, 27n, 30, 31n, 58n, 62n, 84n, 87n, 107n, 117n, 148n, 159n, 162n, 166n, 181n, 183n, 199n, 215n, 231n, 245n, 266–8n, 270n, 274n, 278n, 295n, 306n, 346n, 352–3n, 355–6n, 393n, 409n, 411n, 414–15, 418–20n, 424–5n, 427n, 429n, 459n, 470n, 475n, 488–9, 495–6n, 498n, 524n, 535, 549n, 554n, 579n, 595n, 609n, 611n

Diocles, 465–6n

Dugas, R., 148–9n

Duillier, Nicolas Fatio de, 9n

Dulaurens, François, 112n

Edleston, Joseph, 6–9n, 12n, 32n

Elliott, R. W. V., 134n

Eneström, G., 27n

Eratosthenes, 469n

'Espinasse, M., 521n

Euclid of Alexandria, 11, 26n, 63n, 102n, 159n, 163n, 166n, 193n, 244–5, 429n, 495–6n, 498n, 509n, 566n

Euler, Leonhard, 18, 403n, 419n, 520n

Eutocius, 456n, 475n

Fardelli, Michelangelo, 25n

Fermat, Pierre de, 465n

Ferro, Scipione del, 412n

Flamsteed, John, xiii, xiv, 5, 6, 8n, 19, 21, 32n, 33n, 54n, 524–5n, 528n

Folkes, Martin, 18n

Fontenelle, Bernard le Bovier de, xii

Foster, Samuel, 464n

Gauss, C. F., 340n

Gerhardt, C. I., 13n, 16n, 24n, 27–30n, 167n, 381n

Girard, Albert, 58n, 361n, 559n

Glorioso, Camillo, 96–7n

's Gravesande, Willem Jakob, 15n, 18

Greenstreet, W. J., 14n

Gregory, David, xii, 9–11, 15, 20, 427n, 520n, 522n, 536n

Gregory, James, 4, 413n, 516–17n, 570n

Grisio, Salvatore, 97n

Gua de Malves, Jean Paul de, 27n
Gutschoven, Gerard van, 181n

Haestrecht, Godefroy de, 266–8n
Hall, A. R., xiii, 149n
Hall (Line), Francis, 532n
Hall, M. B., xiii
Haller, Albrecht von, 17n
Halley, Edmond, xii, xiv, 11n, 12, 13–14n, 15, 16–17n, 18, 21n, 22, 23n, 31, 422n, 438n, 516–17n, 524–5n
Hare, St John, xii
Harriot, Thomas, 26n, 27, 51n, 63n, 158n, 359n, 438n, 516n
Harris, John, 15
Heath, T. L., 253n, 423n, 427n, 456n, 462n, 464–5n, 469-70n, 475n
Heiberg, J. L., 456n, 465n, 475n
Heraclitus, 393n, 513n
Herivel, J. W., 149n, 525n
Hermann, Jakob, 16, 23–5n, 27–9n
Hero(n) of Alexandria, 50n
Hiero of Syracuse, 146
Hilbert, David, 387n
Hiscock, W. G., xii, 9–11n
Hofmann, J. E., 162n, 520n
Hooke, Robert, 6n, 148n, 211n, 299n, 520–1n
Horne, Thomas, 6n, 7
Horrocks (Horrox), Jeremiah, 211n, 299n
Horsley, Samuel, 18, 448n
Huygens, Christiaan, 20, 30n, 148–9n, 211n, 283n, 286n, 302n, 520–2n, 550n

Jones, William, 7n

Kästner, Abraham Gotthelf, 27n
Kepler, Johannes, 210n, 298–9n, 514–15n, 567n
Kersey, John, 21n, 96n
Kinckhuysen, Gerard, 3, 55n, 84n, 87n, 113n, 118n, 120–1n, 123n, 129n, 134–5n, 144n, 150–1n, 153–4n, 159n, 184–7n, 189n, 195n, 203–4n, 207n, 209n, 340n, 352–3n, 355–8n, 393n, 409–10n, 413n, 416n, 459n, 564n
Kronecker, L., 387n

Lacroix, S. F., 246n
Lagny, Thomas Fantet de, 31n
Lagrange, J. L., 19n, 204–5n, 613n
La Hire, Philippe de, 474n, 517n
Laurens, François du, *see* Dulaurens, François
Lecchi, G. A., 18, 403n

Leibniz, Gottfried Wilhelm v on, 7n, 13n 16–17, 20, 23, 24–9n, 30, 158n, 166n 380–1n, 520n, 522n
Le Paige, C., 216n
Lower, William, 243n
Lucas, Henry, 8–9n

Machin, John, 14–15n
Maclaurin, Colin, 18, 307n, 361n, 387n, 402n
Maghetti, Benedetto, 97n
Magliabecci, Antonio, 30n
Maguire, James, 17n
McKenzie, D. F., 9n, 12n, 55n
Meibomius, Marcus, 409n
Menæchmus, 475n
Mercator (Kauffman), Nicolaus, 3n, 135n, 352n, 515n
Mersenne, Marin, 427n, 457n, 475n
Moivre, Abraham de, 15, 18
Molther, Johannes, 457n
Murdoch, Patrick, 19n, 361n
Musgrave, James, 14n

Newton, Humphrey, xi, xii, 55n, 535, 538n, 547–8n, 557n, 559–60n, 565n, 585n, 608n
Nicomedes, 456–7n, 469n
Nobis, H. M., 17n
Norwood, Richard, 63n

Oldenburg, Henry, 7n, 22n, 216–17n, 517n
Oughtred, William, 51n, 57–8n, 62–3n, 96n, 158n, 190n, 193n, 287n, 291n

Pacioli, Luca, 567n
Pannekoek, A., 516n
Pappus, 168n, 193n, 227n, 253n, 309n, 422–3, 424n, 426–7, 456–7n, 460n, 462–3, 464n, 468–9, 470n, 496n, 513n
Pardies, Ignace Gaston, 279n, 282n, 520–1n
Pellet, Thomas, 539n
Prestet, Jean, 27n
Proclus, 465n
Ptolemy, Claudius, 167n, 174n, 177n, 516n

Ramée (Ramus), Pierre de la, 567n
Raphson, Joseph, 12n, 16–17, 22n, 93n
Ravier, É., 30n
Rawlyns, Richard, 57n
Recorde, Robert, 63n
Rigaud, S. P., xiii, 97n, 217n, 517n
Roberval, Gilles Personne de, 26n, 465n
Ruffner, J. A., 21n, 525n

Schooten, Frans van, 10, 26, 62n, 144n, 146n, 162n, 173n, 195n, 214n, 222n, 266n, 268–9n, 418–19n, 427n, 471n, 481n, 581n, 609n
Schooten, Pieter van, 166n
Seidenberg, A., 457n
Sharp, Abraham, 8n
Simpson, Thomas, 303n
Simson, Robert, 250n
Sluse, René-François de, 148n, 215–17n, 243n, 474n
Smalbrook(e), Richard (?), 9n
Smith, D. E., 87n, 567n
Smith (Newton, *née* Aiscough), Hannah, 6
Stampioen de Jonge, Jan, 266n
Stanhope, Philip (2nd Earl), 361n
Stevin, Simon, 62n, 279n, 521n
Storer, Arthur, 622n
Storer, Oliver, 622n
Sylvester, J. J., 349n

Tannery, P., 266n
Taylor, C., 249n
Thâbit ibn Qurra, 464n
Thomas, I. B., 423n, 465n
Tooke, Benjamin, 14n, 23
Tooke, Samuel, 14n
Toomer, G. J., 465n
Towneley, Richard, 216–17n
Tropfke, J., 50n, 193n, 226n, 246n, 340n
Truesdell, C. A., 287n, 520n, 522n
Tschirnhaus, Walter Ehrenfried von, 413n, 612n

Turnbull, H. W., 413n, 517n
Turnor, Edmund, xii

Viète, François, 31n, 57–8n, 63n, 92–3n, 157–8n, 262n, 411n, 422n, 459–60n
Villamil, R. de, 14n
Vitruvius Pollio, Marcus, 146n

Waessenaer, Jacob van, 266n, 418n
Waller, Richard, 521n
Wallis, John, xii, 7, 13n, 21n, 26n, 30n, 148n, 158n, 211n, 215–16n, 299n, 303n, 409n, 532n
Waring, Edward, 18, 387n, 419n
Warner, Walter, 26n, 63n, 359n
Weland, Woldegk, 51n
Whiston, William, xii, 3, 8–9, 11–13, 14n, 15, 17–18, 20–1, 22n, 23–4, 27n, 29n, 31n, 45n, 54–508 (*passim*), 518n, 521n, 535, 536n, 539n
Whiteside, D. T., 305n, 515n, 579n
Wilder, Theaker, 17n
Wilkins, John, 532n
Wing, Vincent, 515n
Wollenschläger, K., 15n
Wren, Christopher, 30n, 148–9n, 210–11n, 216n, 299n, 300n, 520n, 524–5n, 530n

Yuschkevich, A. P., 19

Zanotti, Eustachio, 302n